THEORETICAL AND MATHEMATICAL PHYSICS
Problems and Solutions

THEORETICAL AND MATHEMATICAL PHYSICS
Problems and Solutions

Willi-Hans Steeb
University of Johannesburg
South Africa

World Scientific

NEW JERSEY · LONDON · SINGAPORE · BEIJING · SHANGHAI · HONG KONG · TAIPEI · CHENNAI

Published by

World Scientific Publishing Co. Pte. Ltd.

5 Toh Tuck Link, Singapore 596224

USA office: 27 Warren Street, Suite 401-402, Hackensack, NJ 07601

UK office: 57 Shelton Street, Covent Garden, London WC2H 9HE

Library of Congress Cataloging-in-Publication Data
Names: Steeb, W.-H., author.
Title: Problems and solutions in theoretical and mathematical physics /
    Willi-Hans Steeb (University of Johannesburg, South Africa).
Description: Fourth (extended and revised) edition. | New Jersey : World Scientific, 2018. |
    Includes bibliographical references and index.
Identifiers: LCCN 2018032340| ISBN 9789813275379 (hardcover : alk. paper) |
    ISBN 9813275375 (hardcover : alk. paper) | ISBN 9789813275966 (pbk. : alk. paper) |
    ISBN 9813275960 (pbk. : alk. paper)
Subjects: LCSH: Mathematical physics--Problems, exercises, etc.
Classification: LCC QC20.82 .S72 2018 | DDC 530.15--dc23
LC record available at https://lccn.loc.gov/2018032340

British Library Cataloguing-in-Publication Data
A catalogue record for this book is available from the British Library.

For any available supplementary material, please visit
https://www.worldscientific.com/worldscibooks/10.1142/11132#t=suppl

Printed in Singapore

Preface

The purpose of this book is to supply a collection of problems together with their detailed solution which will prove to be valuable to students as well as to research workers in the fields of mathematics, physics, engineering and other sciences. Each chapter provides a short introduction. The topics range in difficulty from elementary to advanced. Almost all problems are solved in detail and most of the problems are self-contained. All relevant definitions are given. Stimulating supplementary problems are also provided in each chapter. Students can learn important principles and strategies required for problem solving. Teachers will also find this text useful as a supplement, since important concepts and techniques are developed in the problems. The material was tested in my lectures given around the world.

Both introductory problems for undergraduate and advanced undergraduate students are provided. More advanced problems together with their detailed solutions are collected, to meet the needs of graduate students and researchers. Problems included cover most of the new fields in theoretical and mathematical physics such as tensor product, Lax representation, Bäcklund transformation, soliton equations, Hilbert space theory, uncertainty relation, entanglement, spin systems, Bose system. Fermi systems differential forms, Lie algebra valued differential forms, Hirota technique, Painlevé test, the Bethe ansatz, the Yang-Baxter relation, wavelets, gauge theory, differential geometry, string theory, chaos, fractals, complexity, ergodic theory, etc.

In the reference section some other books are listed having useful problems for students in theoretical physics and mathematical physics.

Any useful suggestions and comments are welcome.

Email addresses of the author:

steebwilli@gmail.com
steeb_wh@yahoo.com

Home page of the author:

http://issc.uj.ac.za

Contents

Notation

$\emptyset$	empty set
$A \subset B$	subset A of set B
$A \cap B$	the intersection of the sets A and B
$A \cup B$	the union of the sets A and B
$\mathbb{N}$	natural numbers
$\mathbb{Z}$	integers
$\mathbb{Q}$	rational numbers
$\mathbb{R}$	real numbers
$\mathbb{R}^+$	nonnegative real numbers
$\mathbb{C}$	complex numbers
$\mathbb{R}^n$	n-dimensional Euclidian space
$\mathbb{C}^n$	n-dimensional complex linear space
$\mathbb{H}$	Hilbert space
i	$:= \sqrt{-1}$
$\Re(z)$	real part of the complex number z
$\Im(z)$	imaginary part of the complex number z
$\mathbf{x} \in \mathbb{R}^n$	element $\mathbf{x}$ of $\mathbb{R}^n$ (column vector)
$\mathbf{0}$	zero vector (column vector)
$f \circ g$	composition of two mappings $(f \circ g)(x) = f(g(x))$
u	dependent variable
t	independent variable (time variable)
x	independent variable (space variable)
$\mathbf{x}^T = (x_1, x_2, \ldots, x_n)$	vector of independent variables, ^{T} means transpose
$\mathbf{u}^T = (u_1, u_2, \ldots, u_n)$	vector of dependent variables, ^{T} means transpose
$\| \cdot \|$	norm
$\mathbf{x} \cdot \mathbf{y} \equiv \mathbf{x}^T \mathbf{y}$	scalar product (inner product) in vector space $\mathbb{R}^n$
$\langle, \rangle$	scalar product in Hilbert space
$\mathbf{x} \times \mathbf{y}$	vector product in vector space $\mathbb{R}^3$
det	determinant of a square matrix
tr	trace of a square matrix
0_n	$n \times n$ zero matrix
I_n	$n \times n$ unit matrix (identity matrix)
I	identity operator

$[\,,\,]$	commutator
$[\,,\,]_+$	anticommutator
δ_{jk}	Kronecker delta with $\delta_{jk} = 1$ for $j = k$ and $\delta_{jk} = 0$ for $j \neq k$
$\epsilon_{jk\ell}$	total antisymmetric tensor $\epsilon_{123} = 1$
$\mathrm{sgn}(x)$	the sign of x, 1 if $x > 0$, -1 if $x < 0$, 0 if $x = 0$
λ	eigenvalue
ϵ	real parameter
$\otimes$	Kronecker product, tensor product
$\oplus$	direct sum
$\wedge$	Grassmann product (exterior product, wedge product)
$\star$	Hodge duality operator
L	Lagrange function
H	Hamilton function
$\mathcal{L}$	Lagrange density
$\hat{H}$	Hamilton operator
$\hat{N}$	Number operator
$\hat{P}$	Momentum operator
$L_2(\Omega)$	Hilbert space of square integrable functions
$\langle\,,\,\rangle$	scalar product in Hilbert space
δ	delta function
$\mathbf{k}$	wave vector
$\mathbf{v}$	velocity
ω	frequency
$\mathbf{E}$	electric field strength (electric intensity)
$\mathbf{B}$	magnetic flux density (magnetic induction)
$\mathbf{S}$	Poynting vector $sec^{-3}.kg$
$\mathbf{A}$	Vector potential, magnetic
$\hbar$	reduced Planck constant
G	gravitational constant
ϵ_0	permittivity, electric
ϵ_r	permittivity, relative
μ_0	permeability, magnetic
$c_0 = 1/\sqrt{\epsilon_0\mu_0}$	speed of light in vacuum, $m.s^{-1}$
D	diffusion coefficient, $m^2.s^{-1}$
T	Temperature
k_B	Boltzmann constant
β	$1/(k_B T)$
S_1, S_2, S_3	Spin matrices for spin $s = 1/2$, $s = 1, \ldots$
$\sigma_1, \sigma_2, \sigma_3$	Pauli spin matrices
$b^\dagger, b$	Bose creation and annihilation operators
$c^\dagger, c$	Fermi creation and annihilation operators

Chapter 1

Sums, Products and Discrete Fourier Transform

Arithmetic series, geometric series and harmonic series play a central role in problems in theoretical and mathematical physics.

An *arithmetic series* is the sum of a sequence $\{ s_k \}$, $k = 0, 1, \ldots$ in which each term is calculated from the previous one by adding (or subtracting) a constant c. This means $s_k = s_{k-1} + c = s_{k-2} + 2c = \cdots = s_0 + c \cdot k$, where $k = 1, 2, \ldots$. In particular one has ($n \geq 1$)

$$s_0 + (s_0 + d) + (s_0 + 2d) + \cdots + (s_0 + (n-1)d) = \frac{1}{2}n(2s_0 + (n-1)d).$$

A *geometric series* is a series with a constant ratio between successive terms. For example

$$\sum_{n=0}^{N-1} e^{in\alpha}, \quad N \in \mathbb{N}, \ \alpha \in \mathbb{R}.$$

The *harmonic series* is the infinite series $\sum_{k=1}^{\infty} \frac{1}{k}$ which diverges to infinity. The *Riemann zeta function* is given by

$$\zeta(s) = \sum_{k=1}^{\infty} \frac{1}{k^s}, \quad \Re(s) \leq 1.$$

1

Let N be a positive integer, $N \geq 2$. The discrete one-dimensional Fourier transform for a given sequence $x(n)$, $(n = 0, 1, \ldots, N-1)$ is defined by

$$\hat{x}(k) := \frac{1}{N} \sum_{n=0}^{N-1} x(n) e^{-i2\pi kn/N}$$

where $k = 0, 1, \ldots, N-1$. The discrete Fourier transform is linear.

The *Bernoulli numbers* B_0, B_1, B_2, ... are defined by the power series expansion

$$\frac{x}{e^x - 1} = \sum_{j=0}^{\infty} \frac{B_j}{j!} x^j \equiv B_0 + \frac{B_1}{1!} x + \frac{B_2}{2!} x^2 + \cdots.$$

One finds $B_0 = 1$, $B_1 = -1/2$, $B_2 = 1/6$, $B_3 = 0$, $B_4 = -1/30$ and $B_j = 0$ if $j \geq 3$ and j odd. The Bernoulli numbers are utilized in the *Euler summation formula*

$$\sum_{j=1}^{n} f(j) = \int_1^n f(t)dt + \frac{1}{2}(f(n)+f(1)) + \sum_{k=1}^{n} \frac{B_{2k}}{(2k)!}(f^{(2k-1)}(n) - f^{(2k-1)}(1)) + R_m(n)$$

where

$$|R_m(n)| \leq \frac{4}{(2\pi)^{2m}} \int_1^m |f^{(2m)}(t)|dt.$$

The *Poisson summation formula* is given by

$$\sum_{k=-\infty}^{\infty} f(k) = \sum_{n=-\infty}^{\infty} \left(\int_{\mathbb{R}} e^{2\pi inx} f(x)dx \right).$$

Let $z \in \mathbb{C}$. The *Taylor expansions* for $\exp(z)$, $\cosh(z)$ and $\sinh(z)$ are given by

$$\exp(z) = \sum_{k=0}^{\infty} \frac{z^k}{k!} = \lim_{k\to\infty} \left(1 + \frac{z}{k}\right)^k$$

$$\cosh(z) = \sum_{k=0}^{\infty} \frac{z^{2k}}{(2k)!} = 1 + \frac{z^2}{2!} + \frac{z^4}{4!} + \cdots$$

$$\sinh(z) = \sum_{k=0}^{\infty} \frac{z^{2k+1}}{(2k+1)!} = z + \frac{z^3}{3!} + \frac{z^5}{5!} + \cdots.$$

Another sum appearing often in theoretical physics is

$$1 + 2\sum_{j=1}^{n} \cos(j\alpha) = \frac{\sin((n+1/2)\alpha)}{\sin(\alpha/2)}.$$

Problem 1. We consider *geometric series*.
(i) Let $n \geq 0$. Calculate the finite sum

$$S(n, x) = \sum_{k=0}^{n} x^k$$

where $x \neq 0$.
(ii) Let $N \in \mathbb{N}$ and $\alpha \in \mathbb{R}$. Calculate the finite sum

$$S(\alpha) = \sum_{n=0}^{N-1} e^{in\alpha}$$

where $\alpha \in \mathbb{R}$ and $N \in \mathbb{N}$.
(iii) Let $z = \exp(i\alpha)$, where $\alpha \in \mathbb{R}$ and n be a positive integer. Calculate the finite sum

$$\sum_{k=-n}^{n} z^k.$$

(iv) Let a, b be nonnegative integers, $b > a$ and $x \in \mathbb{R}$. Calculate the finite sum

$$\sum_{j=a}^{b} e^{jx}.$$

(v) Calculate the infinite sum

$$S = \sum_{n=1}^{\infty} \frac{1}{3^n}.$$

Solution 1. (i) Let $x = 1$, then obviously $S(n, 1) = \sum_{k=0}^{n} 1 = n + 1$. Now let $x \neq 1$. Since

$$xS(n, x) = \sum_{k=0}^{n} x^{k+1} = x + x^2 + \cdots + x^n + x^{n+1}$$

we obtain from $xS(n, x) - S(n, x)$ that

$$S(n, x) = \frac{1 - x^{n+1}}{1 - x}.$$

(ii) Let $\alpha = 2m\pi$ where $m \in \mathbb{Z}$. Since $e^{2inm\pi} = 1$, $n \in \mathbb{N}_0$ we find $S(2m\pi) = N$. Now let $\alpha \neq 2m\pi$, where $m \in \mathbb{Z}$. Owing to $e^{in\alpha} \equiv (e^{i\alpha})^n$ the sum (1) over n is a geometric series. Consequently,

$$S(\alpha) = \sum_{n=0}^{N-1} e^{in\alpha} = \frac{1 - e^{iN\alpha}}{1 - e^{i\alpha}} = \frac{\sin(N\alpha/2)}{\sin(\alpha/2)} e^{i(N-1)\alpha/2}.$$

The sums

$$\sum_{n=1}^{N-1} n e^{in\alpha}, \qquad \sum_{n=1}^{N-1} n^2 e^{in\alpha}$$

can now easily be calculated since

$$\frac{dS}{d\alpha} = i \sum_{n=1}^{N-1} n e^{in\alpha}, \qquad \frac{d^2 S}{d\alpha^2} = -\sum_{n=1}^{N-1} n^2 e^{in\alpha}.$$

(iii) For $\alpha \neq 0$ we have

$$\sum_{k=-n}^{n} z^k = z^{-n} \sum_{k=0}^{2n} z^k = z^{-n} \frac{z^{2n+1} - 1}{z - 1} = \frac{z^{n+1} - z^{-n}}{z - 1}$$

$$= \frac{z^{n+1/2} - z^{-(n+1/2)}}{z^{1/2} - z^{-1/2}} = \frac{\sin((n+1/2)\alpha)}{\sin(\alpha/2)}.$$

For $\alpha = 0$ the sum is $2n + 1$.

(iv) For $x \neq 0$ we have

$$\sum_{j=a}^{b} e^{jx} = e^{ax}(1 + e^x + e^{2x} + \cdots + e^{(b-a)x}) = e^{ax} \frac{1 - e^{(b-a+1)x}}{1 - e^x}$$

$$= \frac{e^{ax}}{1 - e^x} + \frac{e^{bx}}{1 - e^{-x}}.$$

For $x = 0$ the sum is $b - a + 1$.

(v) We can use the ratio test or the nth root test to prove that the series converges absolutely. We have

$$S = \sum_{n=1}^{\infty} \frac{1}{3^n} \equiv \frac{1}{3} + \frac{1}{9} + \frac{1}{27} + \cdots \equiv \frac{1}{3}\left(1 + \frac{1}{3} + \frac{1}{9} + \cdots\right) \equiv \frac{1}{3}(1 + S).$$

Consequently, $S = \frac{1}{2}$.

Problem 2. Let $n, m \in \mathbb{N}$ with $n > m$. Show that

$$\sum_{k=m}^{n} u_k(v_{k+1} - v_k) \equiv u_k v_k \big|_m^{n+1} - \sum_{k=m}^{n} v_{k+1}(u_{k+1} - u_k) \tag{1}$$

where $u_k v_k \big|_m^{n+1} = u_{n+1} v_{n+1} - u_m v_m$. Equation (1) is called the formula for *summation by parts*.

Solution 2. We start from the identity

$$u_{k+1} v_{k+1} - u_k v_k \equiv u_k(v_{k+1} - v_k) + v_{k+1}(u_{k+1} - u_k).$$

Therefore

$$\sum_{k=m}^{n}(u_{k+1}v_{k+1} - u_k v_k) \equiv \sum_{k=m}^{n} u_k(v_{k+1} - v_k) + \sum_{k=m}^{n} v_{k+1}(u_{k+1} - u_k).$$

Obviously, the left-hand side is given by

$$\sum_{k=m}^{n}(u_{k+1}v_{k+1} - u_k v_k) \equiv u_{n+1}v_{n+1} - u_m v_m.$$

Problem 3.   Let n be a positive integer and $f(j) = j(j-1)(j-2)$ with $j = 1, 2, \ldots, n$. Let $a_j := f(j+1) - f(j)$. By calculating $\sum_{j=1}^{n} a_j$ show that

$$\sum_{j=1}^{n} j^2 = \frac{n(n+1)(2n+1)}{6}.$$

Solution 3.   From $a_j := f(j+1) - f(j)$ we obtain

$$\sum_{j=1}^{n} a_j = f(n+1) - f(1) = f(n+1) = (n+1)n(n-1).$$

Now $a_j = f(j+1) - f(j) = (j+1)j(j-1) - j(j-1)(j-2) = 3j^2 - 3j$. It follows that

$$\sum_{j=1}^{n} a_j = 3\sum_{j=1}^{n} j^2 - 3\sum_{j=1}^{n} j = (n+1)n(n-1).$$

Since $\sum_{j=1}^{n} j = n(n+1)/2$ we obtain

$$\sum_{j=1}^{n} j^2 = \frac{(n+1)n(n-1)}{3} + \frac{n(n+1)}{2} = \frac{(n+1)n(2n+1)}{6}.$$

Problem 4.   Let $n > 2$ and $A, h > 0$. Find V_n given by

$$V_n = \frac{Ah}{n^3} \sum_{j=1}^{n-1} j^2.$$

Then find $\lim_{n \to \infty} V_n$. This refers to the volume of a *pyramid*, where h is the height and A is the area of the base.

Solution 4.   With

$$\sum_{j=1}^{m} j^2 = \frac{1}{6} m(m+1)(2m+1)$$

we obtain

$$V_n = \frac{1}{3} Ah \left(1 - \frac{1}{n} \right) \left(1 - \frac{1}{2n} \right).$$

Hence $V_\infty = Ah/3$.

Problem 5.   Let $x \in \mathbb{R}$ and $|x| < 1$. Then we know that

$$\sum_{k=0}^{\infty} x^k = \frac{1}{1-x}.$$

(i) Use this result to calculate $\sum_{k=1}^{\infty} \tanh^{2k}(r)$, $r \in \mathbb{R}$.
(ii) Use the result from (i) to calculate $\sum_{k=1}^{\infty} k \tanh^{2k}(r)$.

Solution 5.   (i) Setting $x = \tanh^2(r)$ we have

$$\sum_{k=1}^{\infty} \tanh^{2k}(r) = \sum_{k=0}^{\infty} \tanh^{2k}(r) - 1 = \sum_{k=0}^{\infty} x^k - 1$$

$$= \frac{1}{1-x} - 1 = \frac{x}{1-x} = \frac{\tanh^2(r)}{1 - \tanh^2(r)}$$

$$= \sinh^2(r).$$

(ii) Since

$$\frac{d}{dr} \sinh^2(r) = 2\sinh(r)\cosh(r)$$

differentiation of $\tanh^{2k}(r)$ with respect to r provides

$$\frac{d}{dr} \sum_{k=1}^{\infty} \tanh^{2k}(r) = \frac{2}{\cosh^2(r)} \sum_{k=1}^{\infty} k \tanh^{2k-1}(r)$$

$$= \frac{2}{\cosh^2(r)\tanh(r)} \sum_{k=1}^{\infty} k \tanh^{2k}(r)$$

$$= 2\sinh(r)\cosh(r).$$

Thus using $\tanh(r) \equiv \sinh(r)/\cosh(r)$ we finally arrive at

$$\sum_{k=1}^{\infty} k \tanh^{2k}(r) = \sinh^2(r)\cosh^2(r).$$

Problem 6. Let $n \in \mathbb{N}$. Evaluate the finite sum

$$\sum_{k=1}^{n} \frac{k^2}{2^k}. \tag{1}$$

Solution 6. If we define

$$S(x) := \sum_{k=1}^{n} k^2 x^k$$

then the sum (1) is given by $S(1/2)$. We know that

$$\sum_{k=0}^{n} x^k = \frac{1 - x^{n+1}}{1 - x}, \qquad x \neq 1.$$

Differentiating each side with respect to x yields

$$\sum_{k=1}^{n} kx^{k-1} = \frac{1 - (n+1)x^n + nx^{n+1}}{(1-x)^2}.$$

Multiplying each side of this equation by x, differentiating a second time, and multiplying the result by x yields

$$S(x) = \sum_{k=1}^{n} k^2 x^k = \frac{x(1+x) - x^{n+1}(nx - n - 1)^2 - x^{n+2}}{(1-x)^3}.$$

From this equation it follows that

$$S\left(\frac{1}{2}\right) = \sum_{k=1}^{n} \frac{k^2}{2^k} = 6 - \frac{1}{2^{n-2}}\left(\frac{1}{2}n - n - 1\right)^2 - \frac{1}{2^{n-1}} = 6 - \left(\frac{n^2 + 4n + 6}{2^n}\right).$$

Problem 7. Calculate the finite sum

$$\sum_{k=-N}^{N} \frac{1}{1 + e^{\lambda \epsilon(k)}} \tag{1}$$

where $N \in \mathbb{N}$, $\epsilon(-k) = -\epsilon(k)$ and λ is a real parameter. This sum plays a role in solid state physics.

Solution 7. We define

$$S(N, \lambda) := \sum_{k=-N}^{N} \frac{1}{1 + e^{\lambda \epsilon(k)}}.$$

Taking the derivative of S with respect to λ we obtain

$$\frac{dS}{d\lambda} = -\sum_{k=-N}^{N} \frac{\epsilon(k)e^{\lambda\epsilon(k)}}{(1+e^{\lambda\epsilon(k)})^2} \equiv -\sum_{k=-N}^{N} \frac{\epsilon(k)}{2+2\cosh(\lambda\epsilon(k))}.$$

Since $\epsilon(-k) = -\epsilon(k)$ and $\cosh(x)$ is an even function we find $dS/d\lambda = 0$. From $S(N,\lambda)$ we obtain the initial condition for the differential equation

$$S(N, \lambda = 0) = \sum_{k=-N}^{N} \frac{1}{1+1} = \frac{1}{2}(2N+1).$$

Consequently, the solution of the linear differential equation with the initial condition is given by $S(N,\lambda) = N + 1/2$. This means the sum (1) is independent of λ.

Problem 8.   Consider

$$z = \sum_{n \in I} a_n e^{-ix_n} \tag{1}$$

where I is a finite set and a_n, $x_n \in \mathbb{R}$. Calculate $z\bar{z}$, where $\bar{z}$ denotes the complex conjugate of z.

Solution 8.   From (1) we obtain $\bar{z} = \sum_{m \in I} a_m e^{ix_m}$. Therefore

$$z\bar{z} = \sum_{n \in I} \sum_{m \in I} a_n a_m e^{-ix_n} e^{ix_m} = \sum_{n \in I} \sum_{m \in I} a_n a_m e^{i(x_m - x_n)}.$$

With $e^{i\alpha} = \cos(\alpha) + i\sin(\alpha)$ it follows that

$$z\bar{z} = \sum_{n \in I} \sum_{m \in I} a_n a_m (\cos(x_m - x_n) + i\sin(x_m - x_n)).$$

Since $\sin(-\alpha) \equiv -\sin(\alpha)$ we have $\sin(x_m - x_n) \equiv -\sin(x_n - x_m)$. Therefore

$$\sum_{n \in I} \sum_{m \in I} a_n a_m \sin(x_m - x_n) = 0.$$

Consequently,

$$z\bar{z} = \sum_{n \in I} \sum_{m \in I} a_n a_m \cos(x_m - x_n).$$

Problem 9.   (i) Show that

$$(\cos(\theta) + i\sin(\theta))^n \equiv \cos(n\theta) + i\sin(n\theta) \tag{1}$$

where $\theta \in \mathbb{R}$ and $n \in \mathbb{N}$. Identity (1) is sometimes called *De Moivre's theorem*.

(ii) Show that $\cos(2\theta) \equiv \cos^2(\theta) - \sin^2(\theta)$, $\sin(2\theta) \equiv 2\sin(\theta)\cos(\theta)$ using De Moivre's Formula, where $\theta \in \mathbb{R}$.

Solution 9. (i) We apply the *principle of mathematical induction*. The identity (1) is true for $n = 1$. Assume that the identity is true for k, i.e.

$$(\cos(\theta) + i\sin(\theta))^k \equiv \cos(k\theta) + i\sin(k\theta).$$

Multiplying both sides by $\cos(\theta) + i\sin(\theta)$ yields

$$(\cos(\theta) + i\sin(\theta))^{k+1} = (\cos(k\theta) + i\sin(k\theta))(\cos(\theta) + i\sin(\theta))$$

or

$$(\cos(\theta) + i\sin(\theta))^{k+1} = \cos((k+1)\theta) + i\sin((k+1)\theta).$$

Consequently, if the result is true for $n = k$ then it is also true for $n = k+1$. Since the result is true for $n = 1$, it has to be true for $n = 1+1 = 2$, $n = 2+1$ etc. and so has to be true for all $n \in \mathbb{N}$. We have $(e^{i\theta})^n \equiv e^{in\theta}$, where $\theta \in \mathbb{R}$ and $n \in \mathbb{N}$.

(ii) Let n be a positive integer. We set $z = re^{i\theta}$. Then

$$z^n = (re^{i\theta})^n = r^n e^{in\theta} = r^n(\cos(n\theta) + i\sin(n\theta)).$$

If $|z| = r = 1$, then $e^{in\theta} \equiv \cos(n\theta) + i\sin(n\theta) \equiv (\cos(\theta) + i\sin(\theta))^n$. Now let $n = 2$. Then it follows that

$$\cos(2\theta) + i\sin(2\theta) = e^{2i\theta} = (e^{i\theta})^2 = (\cos(\theta) + i\sin(\theta))^2$$
$$= \cos^2(\theta) - \sin^2(\theta) + 2i\sin(\theta)\cos(\theta).$$

The real and imaginary parts provide the trigonometric relations given above. We can find trigonometric identities for arbitrary n. For example for $n = 3$ we find the identities

$$\cos(3\theta) \equiv 4\cos^3(\theta) - 3\cos(\theta), \quad \sin(3\theta) \equiv 3\sin(\theta) - 4\sin^3(\theta)$$

by considering the real and imaginary parts.

Problem 10. Let a be a positive constant and

$$S := \left\{ \frac{a}{2}(1,1,1), \ \frac{a}{2}(1,1,-1), \ \frac{a}{2}(1,-1,1), \ \frac{a}{2}(-1,1,1), \right.$$
$$\left. \frac{a}{2}(1,-1,-1), \ \frac{a}{2}(-1,1,-1), \ \frac{a}{2}(-1,-1,1), \ \frac{a}{2}(-1,-1,-1) \right\}.$$

Calculate the finite sum

$$\sum_{\boldsymbol{\Delta} \in S} e^{i(\mathbf{k}\cdot\boldsymbol{\Delta})}$$

where $\mathbf{k} \cdot \mathbf{\Delta} := k_1\Delta_1 + k_2\Delta_2 + k_3\Delta_3$.

Solution 10.

$$\sum_{\mathbf{\Delta} \in S} e^{i(\mathbf{k} \cdot \mathbf{\Delta})} \equiv \sum_{\mathbf{\Delta} \in S} e^{i(k_1\Delta_1 + k_2\Delta_2 + k_3\Delta_3)}$$

$$\equiv e^{i(k_1a/2 + k_2a/2 + k_3a/2)} + e^{-i(k_1a/2 + k_2a/2 + k_3a/2)}$$
$$+ e^{i(k_1a/2 + k_2a/2 - k_3a/2)} + e^{-i(k_1a/2 + k_2a/2 - k_3a/2)}$$
$$+ e^{i(k_1a/2 - k_2a/2 + k_3a/2)} + e^{-i(k_1a/2 - k_2a/2 + k_3a/2)}$$
$$+ e^{i(-k_1a/2 + k_2a/2 + k_3a/2)} + e^{-i(-k_1a/2 + k_2a/2 + k_3a/2)}.$$

Thus

$$\sum_{\mathbf{\Delta} \in S} e^{i\mathbf{k} \cdot \mathbf{\Delta}} \equiv 2[\cos(k_1a/2 + k_2a/2 + k_3a/2) + \cos(k_1a/2 + k_2a/2 - k_3a/2)$$

$$+ \cos(k_1a/2 - k_2a/2 + k_3a/2) + \cos(-k_1a/2 + k_2a/2 + k_3a/2)].$$

For $b, c, d \in \mathbb{R}$ we have the identity

$$\cos(b+c+d) + \cos(b+c-d) + \cos(b-c+d) + \cos(-b+c+d) \equiv 4\cos(b)\cos(c)\cos(d).$$

Thus we obtain

$$\sum_{\mathbf{\Delta} \in S} e^{i(\mathbf{k} \cdot \mathbf{\Delta})} \equiv 8\cos\left(\frac{k_1a}{2}\right)\cos\left(\frac{k_2a}{2}\right)\cos\left(\frac{k_3a}{2}\right).$$

Problem 11.   Sum the finite series $S := a_0 + a_1 + \cdots + a_T$, where $a_0 = 2$ and $a_1 = 5$ for

$$a_{t+2} = 5a_{t+1} - 6a_t \qquad (1)$$

with $t = 0, 1, 2, \ldots$. Apply the technique of *generating functions*. Let

$$F(x) := a_0 + a_1x + a_2x^2 + \cdots + a_tx^t + \cdots \qquad (2)$$

be the ansatz for the generating function.

Solution 11.   The first few terms of the a_t-sequence are

$$2, \ 5, \ 13, \ 35, \ 97, \ 275, \ 793, \ldots .$$

A general formula for the nth term is not apparent. We obtain from (2)

$$-5xF(x) = -5a_0x - 5a_1x^2 - \cdots - 5a_{t+1}x^{t+2} - \cdots$$
$$6x^2F(x) = 6a_0x^2 + 6a_1x^3 + \cdots + 6a_tx^{t+2} + \cdots .$$

Adding (2) and these equations and using (2), we obtain

$$(1 - 5x + 6x^2)F(x) = a_0 + (a_1 - 5a_0)x$$

so that

$$F(x) = \frac{2 - 5x}{(1 - 2x)(1 - 3x)} \equiv \frac{1}{1 - 2x} + \frac{1}{1 - 3x}.$$

Making use of the geometric series provides

$$F(x) = \sum_{t=0}^{\infty} (2x)^t + \sum_{t=0}^{\infty} (3x)^t \equiv \sum_{t=0}^{\infty} (2^t + 3^t)x^t.$$

Thus $a_t = 2^t + 3^t$ for $t = 0, 1, 2, \ldots$. Consequently, the sum is given by

$$S = a_0 + a_1 + \cdots + a_T = \sum_{t=0}^{T} (2^t + 3^t) = \sum_{t=0}^{T} 2^t + \sum_{t=0}^{T} 3^t.$$

It follows that

$$S = \frac{2^{T+1} - 1}{2 - 1} + \frac{3^{T+1} - 1}{3 - 1} = 2^{T+1} - 1 + \frac{3^{T+1} - 1}{2} = \frac{2^{T+2} + 3^{T+1} - 3}{2}.$$

The problem can also be solved by solving the linear difference equation with constant coefficients (1) using the ansatz $a_t \propto r^t$ and the initial conditions $a_0 = 2$ and $a_1 = 5$.

Problem 12. (i) Let

$$A_j := \frac{1}{x - a_j}, \qquad B_{kj} := \frac{1}{a_k - a_j} \tag{1}$$

where $k \neq j$. Show that

$$A_n B_{ns} + A_s B_{sn} = A_n A_s. \tag{2}$$

(ii) Show that the identity

$$\sum_{\substack{j,k=1 \\ j \neq k}}^{n} \frac{1}{x - a_j} \frac{1}{x - a_k} \equiv 2 \sum_{\substack{j,k=1 \\ j \neq k}}^{n} \frac{1}{x - a_k} \frac{1}{a_k - a_j}$$

holds. Using the identity (2) we can write this equation as

$$\sum_{\substack{j,k=1 \\ j \neq k}}^{n} A_j A_k \equiv 2 \sum_{\substack{j,k=1 \\ j \neq k}}^{n} A_k B_{kj}.$$

Solution 12.   (i) By straightforward calculation we find

$$A_n B_{ns} + A_s B_{sn} = \frac{1}{a_n - a_s} \left(\frac{1}{x - a_n} - \frac{1}{x - a_s} \right)$$

$$= \frac{1}{a_n - a_s} \left(\frac{x - a_s - x + a_n}{(x - a_n)(x - a_s)} \right)$$

$$= \frac{1}{(x - a_n)(x - a_s)} = A_n A_s.$$

(ii) We find that

$$\sum_{\substack{j,k=1 \\ j \neq k}}^{n} A_j A_k \equiv \sum_{\substack{j,k=1 \\ j \neq k}}^{n-1} A_j A_k + \sum_{s=1}^{n-1} A_n A_s + \sum_{t=1}^{n-1} A_t A_n.$$

Using (2) we obtain

$$\sum_{\substack{j,k=1 \\ j \neq k}}^{n} A_j A_k \equiv 2 \sum_{\substack{j,k=1 \\ j \neq k}}^{n-1} A_k B_{kj} + 2A_n \sum_{s=1}^{n-1} A_s$$

and

$$A_n \sum_{s=1}^{n-1} A_s \equiv \sum_{s=1}^{n-1} (A_n B_{ns} + A_s B_{sn}) \equiv \sum_{s=1}^{n-1} A_n B_{ns} + \sum_{t=1}^{n-1} A_t B_{tn}.$$

Therefore identity (2) follows.

Problem 13.   Show that every real number $r > 0$ can be represented as the *Cantor series*

$$r = \sum_{\nu=1}^{\infty} \frac{c_\nu}{\nu!} = c_1 + \frac{c_2}{2!} + \frac{c_3}{3!} + \cdots \tag{1}$$

where $0 \leq c_\nu \leq \nu - 1$, $\nu > 1$, $c_\nu \in \mathbb{N} \cup \{0\}$.

Solution 13.   We write the real number r in the form

$$r = [r] + \frac{\rho_2}{2} = c_1 + \frac{\rho_2}{2} \tag{2}$$

where $[r]$ denotes the largest integer $\leq r$. Thus $[r] = c_1$ and $\rho_2 < 2$. We set

$$\rho_2 = [\rho_2] + \frac{\rho_3}{3} = c_2 + \frac{\rho_3}{3}, \qquad \rho_{n-1} = [\rho_{n-1}] + \frac{\rho_n}{n} = c_{n-1} + \frac{\rho_n}{n} \tag{3}$$

and

$$\rho_n = [\rho_n] + \frac{\rho_{n+1}}{n+1} = c_n + \frac{\rho_{n+1}}{n+1}. \tag{4}$$

Thus we have $c_m \leq \rho_m < m$ for $m = 2, 3, 4, \ldots, n$. Inserting (3) and (4) into (2) yields

$$r = c_1 + \frac{c_2}{2} + \frac{c_3}{2 \cdot 3} + \cdots + \frac{c_n}{n!} + \frac{\rho_{n+1}}{(n+1)!}.$$

Using $c_m \leq \rho_m < m$ we find

$$0 \leq r - \left(c_1 + \frac{c_2}{2} + \frac{c_3}{3!} + \cdots + \frac{c_n}{n!} \right) < \frac{1}{n!}.$$

Thus (1) follows.

Problem 14. Calculate the infinite series

$$\sum_{k=1}^{\infty} \frac{1}{2k^2 - k}.$$

Hint: Use partial fraction decomposition and the series

$$\ln(1 + x) = \sum_{k=1}^{\infty} (-1)^{k+1} \frac{x^k}{k}, \qquad -1 < x \leq 1.$$

Solution 14. The *partial fraction decomposition* provides

$$\frac{1}{2k^2 - k} \equiv \frac{2}{2k - 1} - \frac{1}{k} \equiv \frac{2}{2k - 1} - \frac{2}{2k}.$$

Therefore

$$\sum_{k=1}^{\infty} \frac{1}{2k^2 - k} = 2 \sum_{k=1}^{\infty} \left(\frac{1}{2k - 1} - \frac{1}{2k} \right) = 2 \left(1 - \frac{1}{2} + \frac{1}{3} - \frac{1}{4} + \cdots \right).$$

We have an *alternating series* which is convergent. Using the expansion for $\ln(1 + x)$ we find

$$\sum_{k=1}^{\infty} \frac{1}{2k^2 - k} = 2 \ln(2).$$

Problem 15. A series $\sum_{j=0}^{\infty} a_j$, of real or complex numbers, is said to have a $(C, 1)$ sum (a *Cesáro sum*), s, if $t_n \to s$ as $n \to \infty$, where

$$t_n := \frac{s_0 + s_1 + \cdots + s_n}{n + 1}$$

and the s_k are the partial sums $s_k := a_0 + a_1 + \cdots + a_k$. Consider the alternating series $\sum_{j=0}^{\infty}(-1)^j = 1 - 1 + 1 - 1 + \cdots$. The series is not convergent. Calculate the Cesáro sum.

Solution 15. We have

$$t_{2n} = \frac{(n+1)}{2n+1}, \qquad t_{2n+1} = \frac{(n+1)}{2n+2}.$$

Thus $t_{2n} \to 1/2$, $t_{2n+1} \to 1/2$ as $n \to \infty$.

Problem 16. Let $\phi \in \mathbb{R}$. Calculate the infinite series

$$f(\phi) = \sum_{j=1}^{\infty} \sin\left(\frac{2\phi}{3^j}\right) \sin\left(\frac{\phi}{3^j}\right).$$

Use the identity $\sin(\phi_1)\sin(\phi_2) \equiv \frac{1}{2}\cos(\phi_1 - \phi_2) - \frac{1}{2}\cos(\phi_1 + \phi_2)$.

Solution 16. Using the identity we have for the kth partial sum

$$S_k = \sum_{j=1}^{k} \sin\left(\frac{2\phi}{3^j}\right) \sin\left(\frac{\phi}{3^j}\right) = \sum_{j=1}^{k} \left(\frac{1}{2}\cos\left(\frac{\phi}{3^j}\right) - \frac{1}{2}\cos\left(\frac{\phi}{3^{j-1}}\right)\right)$$

$$= \frac{1}{2}\cos\left(\frac{\phi}{3^k}\right) - \frac{1}{2}\cos(\phi).$$

Thus

$$\sum_{j=1}^{\infty} \sin\left(\frac{2\phi}{3^j}\right) \sin\left(\frac{\phi}{3^j}\right) = \lim_{k \to \infty}\left(\frac{1}{2}\cos\left(\frac{\phi}{3^k}\right) - \frac{1}{2}\cos(\phi)\right) = \frac{1}{2}(1 - \cos(\phi)).$$

Problem 17. Let $z = e^{i\pi/n}$ with $n \geq 2$. Calculate the finite sum

$$S_n(z) = 2\sum_{k=1}^{n} kz^{2k-1}.$$

Solution 17. We define

$$R_n(z) := \sum_{k=1}^{n} z^{2k}$$

and $z \neq 1$. Since $R_n(z) - z^2 R_n(z) = z^2 - z^{2n+2}$ we obtain

$$R_n(z) = \frac{z^2 - z^{2n+2}}{1 - z^2}.$$

Differentiation of $R_n(z)$ with respect to z yields

$$\frac{dR_n(z)}{dz} = \frac{2z - (2n+2)z^{2n+1} + 2nz^{2n+3}}{(1-z^2)^2} = 2\sum_{k=1}^{n} kz^{2k-1} = S_n(z).$$

Since $z^{2n} = 1$ we arrive at

$$S_n(z) = -\frac{2nz}{1-z^2} = -\frac{2n}{1/z - z} = \frac{2n}{e^{i\pi/n} - e^{-i\pi/n}} = \frac{n}{i\sin(\pi/n)}.$$

Problem 18.   Let f_k $(k = 1, 2, \ldots, n)$ be differentiable functions. Assume that $f_k \neq 0$ for $k = 1, 2, \ldots, n$. Let

$$S(x) = \prod_{k=1}^{n} f_k(x) \equiv f_1(x)f_2(x)\cdots f_n(x). \tag{1}$$

Calculate the derivative of S with respect to x.

Solution 18.   We apply the *product rule* for differentiation

$$\frac{d}{dx}(h(x)g(x)) = \frac{dh}{dx}g + h\frac{dg}{dx}$$

to S, i.e.

$$\frac{dS}{dx} = \frac{d}{dx}\prod_{k=1}^{n} f_k(x).$$

Thus

$$\frac{dS}{dx} \equiv \frac{df_1}{dx}f_2\cdots f_n + f_1\frac{df_2}{dx}f_3\cdots f_n + \cdots + f_1 f_2 \cdots f_{n-1}\frac{df_n}{dx}$$

or

$$\frac{dS}{dx} \equiv \frac{1}{f_1}\frac{df_1}{dx}S(x) + \frac{1}{f_2}\frac{df_2}{dx}S(x) + \cdots + \frac{1}{f_n}\frac{df_n}{dx}S(x).$$

Consequently

$$\frac{dS}{dx} = S(x)\sum_{j=1}^{n}\frac{1}{f_j}\frac{df_j}{dx} \equiv S(x)\sum_{j=1}^{n}\frac{d}{dx}(\ln(f_j(x))).$$

Problem 19.   Let n be a positive integer with $n \geq 2$ and $\lambda := \exp(2\pi i/n)$. We have

$$\prod_{j=1}^{n-1}(x - \lambda^j) = \frac{x^n - 1}{x - 1}. \tag{1}$$

Show that

$$\sum_{j=1}^{n-1} \frac{1}{x - \lambda^j} = \frac{d}{dx} \ln\left(\frac{x^n - 1}{x - 1}\right). \tag{2}$$

Solution 19.   Taking the logarithm of (1) we obtain

$$\sum_{j=1}^{n-1} \ln(x - \lambda^j) = \ln\left(\frac{x^n - 1}{x - 1}\right). \tag{3}$$

Taking the derivative of (3) with respect to x yields (2).

Problem 20.   (i) Calculate the finite sum

$$\sum_{k=0}^{N-1} e^{i2\pi k(n-m)/N} \tag{1}$$

where $n, m \in \mathbb{N} \cup \{0\}$, $N \in \mathbb{N}$ and $0 \leq n, m \leq N - 1$.
(ii) Find the discrete inverse Fourier transform.

Solution 20.   (i) If $n = m$, then $\sum_{k=0}^{N-1} 1 = N$. If $n \neq m$, then $n - m = q$, where $0 < q \leq N - 1$. Then the sum (1) is a geometric series. Therefore

$$\sum_{k=0}^{N-1} e^{i2\pi kq/N} = 0.$$

Consequently,

$$\sum_{k=0}^{N-1} e^{i2\pi k(n-m)/N} = N\delta_{nm}$$

where δ_{nm} denotes the *Kronecker delta*.
(ii) From the definition of the discrete Fourier transform it follows that

$$\hat{x}(k)e^{i2\pi km/N} = \frac{1}{N} \sum_{n=0}^{N-1} x(n)e^{i2\pi k(m-n)/N}.$$

Summation over k of both sides gives

$$\sum_{k=0}^{N-1} \hat{x}(k)e^{i2\pi km/N} = \frac{1}{N} \sum_{k=0}^{N-1}\sum_{n=0}^{N-1} x(n)e^{i2\pi k(m-n)/N}$$

$$= \frac{1}{N} \sum_{n=0}^{N-1} x(n) \sum_{k=0}^{N-1} e^{i2\pi k(m-n)/N}.$$

Using the result from (i) we obtain

$$\sum_{k=0}^{N-1} \hat{x}(k)e^{i2\pi km/N} = \frac{1}{N}\sum_{n=0}^{N-1} x(n)N\delta_{mn} = x(m).$$

Therefore

$$x(n) = \sum_{k=0}^{N-1} \hat{x}(k)e^{i2\pi kn/N}.$$

Problem 21. Consider the sequence

$$x(n) = \cos(2\pi n/N)$$

where $N = 8$ and $n = 0, 1, \ldots, N-1$. Find $\hat{x}(k)$ ($k = 0, 1, \ldots, N-1$), i.e. find the discrete Fourier transform.

Solution 21. We have

$$\hat{x}(k) = \frac{1}{8}\sum_{n=0}^{7} \cos(2\pi n/8)e^{-i2\pi kn/8}.$$

Using the identity $\cos(2\pi n/8) \equiv (e^{i2\pi n/8} + e^{-i2\pi n/8})/2$ we have

$$\hat{x}(k) = \frac{1}{16}\sum_{n=0}^{7} (e^{i2\pi n(1-k)/8} + e^{-i2\pi n(1+k)/8}).$$

Consequently,

$$\hat{x}(k) = \begin{cases} 1/2 & \text{for} \quad k = 1 \\ 1/2 & \text{for} \quad k = 7 \\ 0 & \text{otherwise} \end{cases}.$$

Problem 22. Let $N \geq 1$ and $\mathbf{x} = (x(0), x(1), \ldots, x(N-1))$. Consider the sequence of the discrete Fourier transform

$$\hat{\mathbf{x}} = (\hat{x}(0), \hat{x}(1), \ldots, \hat{x}(N-1)).$$

We define the scalar products

$$(\mathbf{x}, \mathbf{x}) := \sum_{n=0}^{N-1} \bar{x}(n)x(n), \qquad (\hat{\mathbf{x}}, \hat{\mathbf{x}}) := \sum_{k=0}^{N-1} \bar{\hat{x}}(k)\hat{x}(k) \qquad (1)$$

where $\bar{x}(n)$ denotes the complex conjugate of $x(n)$. Show that

$$(\mathbf{x}, \mathbf{x}) = N(\hat{\mathbf{x}}, \hat{\mathbf{x}}).$$

Solution 22.   From (1) we have

$$(\mathbf{x}, \mathbf{x}) = \sum_{n=0}^{N-1} \bar{x}(n)x(n) = \sum_{k=0}^{N-1}\sum_{l=0}^{N-1} \bar{\hat{x}}(k)\hat{x}(l) \sum_{n=0}^{N-1} e^{i2\pi n(l-k)/N}.$$

Using the identity $\sum_{n=0}^{N-1} e^{i2\pi n(l-k)/N} = N\delta_{lk}$ we find

$$(\mathbf{x}, \mathbf{x}) = N \sum_{k=0}^{N-1} \bar{\hat{x}}(k)\hat{x}(k).$$

Thus $(\mathbf{x}, \mathbf{x}) = N(\hat{\mathbf{x}}, \hat{\mathbf{x}})$.

Problem 23.   Let $x(m)$, $y(m)$ be two sequences with $m = 0, 1, \ldots, N-1$. Show that the discrete Fourier transform of a *circular convolution*

$$(x \odot y)(n) = \sum_{m=0}^{N-1} x(m)y(n-m) = \sum_{m=0}^{N-1} x(n-m)y(m)$$

is the product of the two discrete Fourier transforms of $x(n)$ and $y(m)$. For a circular convolution we have $x(m+N) = x(m)$, $y(m+N) = y(m)$.

Solution 23.   Let $u(n) = (x \odot y)(n)$. Then

$$\hat{u}(k) = \frac{1}{N} \sum_{n=0}^{N-1} u(n)e^{-i2\pi kn/N} = \frac{1}{N} \sum_{n=0}^{N-1}\sum_{m=0}^{N-1} x(m)y(n-m)e^{-i2\pi kn/N}.$$

Using $n - m = p \pmod{N}$ we obtain

$$\hat{u}(k) = \frac{1}{N} \sum_{m=0}^{N-1} \left(\sum_{p+m=0 \bmod N}^{N-1} x(m)y(p)e^{-i2\pi k(p+m)/N} \right)$$

$$= \left(\frac{1}{N} \sum_{m=0}^{N-1} x(m)e^{-i2\pi km/N} \right) \sum_{p+m=0 \bmod N}^{N-1} y(p)e^{-i2\pi kp/N}$$

$$= \hat{x}(k) \sum_{p+m=0 \bmod N}^{N-1} y(p)e^{-i2\pi kp/N}.$$

Next we have to calculate the finite sum

$$\sum_{p+m=0 \bmod N}^{N-1} y(p)e^{-i2\pi kp/N}.$$

Note that we have mod N summation. This means we sum over all terms $p + m = j \bmod N$. For example if $N = 4$ we sum over all combinations (p, m) with $p < N$, $m < N$, i.e.

$$(0,0), \quad (1,3), \quad (2,2), \quad (3,1), \quad (0,1), \quad (1,0), \quad (2,3), \quad (3,2),$$
$$(0,2), \quad (2,0), \quad (1,1), \quad (3,3), \quad (0,3), \quad (3,0), \quad (1,2), \quad (2,1).$$

Thus

$$\sum_{p+m=0 \bmod N}^{N-1} y(p)e^{-i2\pi kp/N} = N \sum_{p=0}^{N-1} y(p)e^{-i2\pi kp/N} = N^2 \hat{y}(k).$$

Thus the result follows.

Problem 24. The two-dimensional *discrete Fourier transform* of a two-dimensional array is computed by first performing a one-dimensional Fourier transform of the rows of the array and then a one-dimensional Fourier transform of the columns of the result (or vice versa). The absolute value of the Fourier coefficients does not change under a translation of the two-dimensional pattern. The Fourier transform also has the property that it preserves angles, that is, similar patterns in the original domain are also similar in the Fourier domain.

Definition. Let $x(n_1, n_2)$ denote an array of real values, where n_1, n_2 are integers such that $0 \le n_1 \le N_1 - 1$ and $0 \le n_2 \le N_2 - 1$. The two-dimensional discrete Fourier transform $\hat{x}(k_1, k_2)$ of $x(n_1, n_2)$ is defined by

$$\hat{x}(k_1, k_2) := \frac{1}{N_1} \frac{1}{N_2} \sum_{n_1=0}^{N_1-1} \sum_{n_2=0}^{N_2-1} x(n_1, n_2) \exp\left(-\frac{2\pi}{N_1} i n_1 k_1 - \frac{2\pi}{N_2} i n_2 k_2 \right)$$

where $0 \le k_1 \le N_1 - 1$ and $0 \le k_2 \le N_2 - 1$.

Consider two arrays $x(n_1, n_2)$ and $y(n_1, n_2)$ of real values. Assume that

$$y(n_1, n_2) = x(n_1 + d_1, n_2 + d_2)$$

where d_1 and d_2 are two given integers. The addition $n_1 + d_1$ is performed modulo N_1 and $n_2 + d_2$ is performed modulo N_2 (torus). Show that

$$\|\hat{y}(k_1, k_2)\| = \|\hat{x}(k_1, k_2)\|.$$

Solution 24. We have

$$\hat{y}(k_1, k_2) = \frac{1}{N_1} \frac{1}{N_2} \sum_{n_1=0}^{N_1-1} \sum_{n_2=0}^{N_2-1} x(n_1 + d_1, n_2 + d_2) \exp\left(-\frac{2\pi}{N_1} i n_1 k_1 - \frac{2\pi}{N_2} i n_2 k_2 \right).$$

With the change of indices $n_1' = (n_1 + d_1) \bmod N_1$, $n_2' = (n_2 + d_2) \bmod N_2$ we obtain

$$\hat{y}(k_1, k_2) = \frac{1}{N_1} \frac{1}{N_2} \sum_{n_1'=0}^{N_1-1} \sum_{n_2'=0}^{N_2-1} x(n_1', n_2') \exp\left(2\pi i \left(-\sum_{j=1}^{2} \frac{n_j' k_j}{N_j} + \sum_{j=1}^{2} \frac{d_j k_j}{N_j}\right)\right).$$

This can be written as

$$\hat{y}(k_1, k_2) = \exp\left(\frac{2\pi}{N_1} i d_1 k_1\right) \exp\left(\frac{2\pi}{N_2} i d_2 k_2\right) \hat{x}(k_1, k_2).$$

This expression tells us that the Fourier coefficients of the array $y(n_1, n_2)$ are the same as the coefficients of the array $x(n_1, n_2)$ except for a phase factor

$$\exp\left(-\frac{2\pi}{N} i d_1 k_1\right) \exp\left(-\frac{2\pi}{N} i d_2 k_2\right).$$

Taking the absolute value results in $\|\hat{y}(k_1, k_2)\| = \|\hat{x}(k_1, k_2)\|$. Thus the absolute values of the Fourier coefficients for both patterns are identical.

Problem 25. Let $N \geq 1$. The discrete one-dimensional *cosine transform* of the given sequence $x(n)$, $n = 0, 1, \ldots, N - 1$ is defined by

$$C(0) := \frac{1}{\sqrt{N}} \sum_{n=0}^{N-1} x(n),$$

$$C(k) := \sqrt{\frac{2}{N}} \sum_{n=0}^{N-1} x(n) \cos\left(\frac{(2n + 1)k\pi}{2N}\right), \qquad k = 1, \ldots, N - 1.$$

Find the inverse discrete one-dimensional cosine transform.

Solution 25. The inverse one-dimensional cosine transform is given by

$$x(n) = \frac{1}{\sqrt{N}} C(0) + \sqrt{\frac{2}{N}} \sum_{k=1}^{N-1} C(k) \cos\left(\frac{(2n + 1)k\pi}{2N}\right).$$

To find the inverse cosine transform we apply the identity

$$\cos(\alpha)\cos(\beta) \equiv \frac{1}{2}\cos(\alpha + \beta) + \frac{1}{2}\cos(\alpha - \beta)$$

with the special case $\cos^2(\alpha) \equiv (\cos(2\alpha) + 1)/2$. Furthermore we use that

$$\sum_{n=0}^{N-1} \cos\left(\frac{(2n + 1)k\pi}{2N}\right) = 0, \qquad k = 1, \ldots, N - 1$$

and

$$\frac{1}{\sqrt{N}} \sum_{n=0}^{N-1} C(0) = \sqrt{N} C(0).$$

Problem 26.   Consider the coupled system of linear difference equations

$$W_{r+1}(k,l,1) = W_r(k-1,l,1) + e^{-i\pi/4} W_r(k,l-1,2) + e^{i\pi/4} W_r(k,l+1,4)$$
$$W_{r+1}(k,l,2) = e^{i\pi/4} W_r(k-1,l,1) + W_r(k,l-1,2) + e^{-i\pi/4} W_r(k+1,l,3)$$
$$W_{r+1}(k,l,3) = e^{i\pi/4} W_r(k,l-1,2) + W_r(k+1,l,3) + e^{-i\pi/4} W_r(k,l+1,4)$$
$$W_{r+1}(k,l,4) = e^{-i\pi/4} W_r(k-1,l,1) + e^{i\pi/4} W_r(k+1,l,3) + W_r(k,l+1,4)$$

where $k,l \in \{0,1,\ldots,L-1\}$ and $L \equiv 0$ (periodic boundary conditions). This system of equations can be written in the matrix form

$$W_{r+1}(k,l,\nu) = \sum_{k',l',\nu'} M(kl\nu, k'l'\nu') W_r(k',l',\nu')$$

where $\nu, \nu' = 1, 2, 3, 4$. Diagonalize M with respect to k and l by applying the discrete Fourier transform.

Solution 26.   The *discrete Fourier transform* in two dimensions is given by

$$\hat{W}_r(p,q,\nu) = \sum_{k=0}^{L-1} \sum_{l=0}^{L-1} e^{-2\pi i(pk+ql)/L} W_r(k,l,\nu).$$

The inverse discrete Fourier transformation follows as

$$W_r(k,l,\nu) = \frac{1}{L^2} \sum_{p=0}^{L-1} \sum_{q=0}^{L-1} e^{2\pi i(pk+ql)/L} \hat{W}_r(p,q,\nu).$$

Taking Fourier components on both sides of the system of equations we find that each equation contains only $\hat{W}_r(p,q,\nu)$ with the same p,q. Consequently, the matrix M is diagonal with respect to p and q. For given p,q we find the 4×4 matrix

$$M(pq\nu|pq\nu') = \begin{pmatrix} \beta^{-p} & \alpha^{-1}\beta^{-q} & 0 & \alpha\beta^q \\ \alpha\beta^{-p} & \beta^{-q} & \alpha^{-1}\beta^p & 0 \\ 0 & \alpha\beta^{-q} & \beta^p & \alpha^{-1}\beta^q \\ \alpha^{-1}\beta^{-p} & 0 & \alpha\beta^p & \beta^q \end{pmatrix}$$

where $\alpha := e^{i\pi/4}$ and $\beta := e^{2\pi i/L}$.

Problem 27.   Let ϵ be a real parameter and $a_n, b_n \in \mathbb{R}$. Assume that

$$\exp\left(\sum_{n=1}^{\infty} \frac{(i\epsilon)^n b_n}{n!}\right) = \sum_{n=0}^{\infty} \frac{(i\epsilon)^n a_n}{n!} \tag{1}$$

where $a_0 = 1$. Find the relationship between the coefficients a_n and b_n.

Solution 27. An arbitrary term of the exponential function on the left hand side of (1) is given by

$$\frac{1}{k!} \left(\sum_{n=1}^{\infty} \frac{(i\epsilon)^n b_n}{n!} \right)^k = \frac{1}{k!} \left(\sum_{n_1=1}^{\infty} \frac{(i\epsilon)^{n_1} b_{n_1}}{n_1!} \right) \cdots \left(\sum_{n_k=1}^{\infty} \frac{(i\epsilon)^{n_k} b_{n_k}}{n_k!} \right)$$

$$= \frac{1}{k!} \sum_{n_1=1}^{\infty} \sum_{n_2=1}^{\infty} \cdots \sum_{n_k=1}^{\infty} \frac{(i\epsilon)^{n_1+n_2+\cdots+n_k} b_{n_1} b_{n_2} \cdots b_{n_k}}{n_1! n_2! \ldots n_k!}.$$

Therefore

$$\exp \left(\sum_{n=1}^{\infty} \frac{(i\epsilon)^n b_n}{n!} \right) \equiv 1 + \sum_{n=1}^{\infty} \frac{(i\epsilon)^n b_n}{n!} + \frac{1}{2!} \sum_{n_1=1}^{\infty} \sum_{n_2=1}^{\infty} \frac{(i\epsilon)^{n_1+n_2} b_{n_1} b_{n_2}}{n_1! n_2!} + \cdots$$

$$+ \frac{1}{k!} \sum_{n_1=1}^{\infty} \sum_{n_2=1}^{\infty} \cdots \sum_{n_k=1}^{\infty} \frac{(i\epsilon)^{n_1+n_2+\cdots+n_k} b_{n_1} b_{n_2} \cdots b_{n_k}}{n_1! n_2! \cdots n_k!} + \cdots$$

$$= \sum_{n=0}^{\infty} \frac{(i\epsilon)^n a_n}{n!}.$$

Equating terms of the same power in $i\epsilon$ we obtain for the first three terms

$$(i\epsilon)^1 : \qquad a_1 = b_1$$
$$(i\epsilon)^2 : \qquad a_2 = b_2 + b_1^2$$
$$(i\epsilon)^3 : \qquad a_3 = b_3 + 3b_2 b_1 + b_1^3.$$

It follows that $b_1 = a_1$, $b_2 = a_2 - a_1^2$, $b_3 = a_3 - 3a_2 a_1 + 2a_1^3$. Let X be a random variable with probability density function $f_X(x)$. The n-th moment of X is defined as

$$\langle X^n \rangle := \int x^n f_X(x) dx$$

where the integral is over the entire range of X and $n = 0, 1, 2, \ldots$. Therefore $\langle X^0 \rangle = 1$. The *characteristic function* $\phi_X(k)$ is defined as

$$\phi_X(k) := \sum_{n=0}^{\infty} \frac{(ik)^n \langle X^n \rangle}{n!}.$$

The *cumulant expansion* is given by

$$\phi_X(k) = \exp \left(\sum_{n=1}^{\infty} \frac{(ik)^n C_n(X)}{n!} \right).$$

Thus the equations give the connection between $C_n(X)$ and $\langle X^n \rangle$. The cumulant expansion also plays an important rôle in high temperature expansion in statistical physics.

Problem 28. Let
$$\hat{H} = \hat{H}_0 + \hat{H}_1$$
be a Hamilton operator describing a many-body Fermi or Bose system. The *grand thermodynamic potential* is given by

$$\Omega(\beta) := -\frac{1}{\beta} \ln \operatorname{tr}(\exp(-\beta(\hat{H} - \mu \hat{N}))) = -\frac{1}{\beta} \ln(Z)$$

where μ is the chemical potential, $\hat{N}$ is the number operator and Z is the grand thermodynamic partition function. Assume that the grand thermodynamic potential for the unperturbed Hamilton operator $\hat{H}_0$

$$\Omega_0(\beta) := -\frac{1}{\beta} \ln(\operatorname{tr}(\exp(-\beta(\hat{H}_0 - \mu \hat{N}))))$$

can be calculated.
(i) Let $\exp(-\beta(\hat{H} - \mu \hat{N})) \equiv (\exp(-\beta(\hat{H}_0 - \mu \hat{N})))S(\beta)$. Find $S(\beta)$.
(ii) Calculate
$$Z(\beta) = \operatorname{tr}(\exp(-\beta(\hat{H}_0 - \mu \hat{N}))S(\beta)). \tag{1}$$
(iii) From (i) we find

$$S(\beta) = 1 + \sum_{n=1}^{\infty} (-1)^n \int_0^\beta d\tau_1 \int_0^{\tau_1} d\tau_2 \cdots \int_0^{\tau_{n-1}} d\tau_n \hat{H}_1(\tau_1)\hat{H}_1(\tau_2)\ldots\hat{H}_1(\tau_n) \tag{2}$$

where
$$\hat{H}_1(\beta) := \exp(\beta(\hat{H}_0 - \mu \hat{N}))\hat{H}_1 \exp(-\beta(\hat{H}_0 - \mu \hat{N}))$$

(the so-called *interaction picture*). It can be shown that $S(\beta)$ can be written as

$$S(\beta) = \sum_{n=0}^{\infty} \frac{(-1)^n}{n!} \int_0^\beta d\tau_1 \int_0^\beta d\tau_2 \cdots \int_0^\beta d\tau_n T_\tau[\hat{H}_1(\tau_1)\hat{H}_1(\tau_2)\ldots\hat{H}_1(\tau_n)]$$

where T_τ is the *time-ordering operator*. We have

$$T_\tau[A(\tau_1)B(\tau_2)] := \begin{cases} A(\tau_1)B(\tau_2), \text{ if } \tau_2 < \tau_1 \\ B(\tau_2)A(\tau_1), \text{ if } \tau_1 < \tau_2 \end{cases}.$$

We set

$$\langle T_\tau \exp(-\int_0^\beta \hat{H}_1(\tau)d\tau)\rangle_0 \equiv \exp(\langle T_\tau \exp(-\int_0^\beta \hat{H}_1(\tau)d\tau) - 1\rangle_c). \tag{3}$$

Express $\langle\ldots\rangle_c$ in terms of $\langle\ldots\rangle_0$. This is the so-called *cumulant expansion*. The expansions given by (1) and (2) cannot be used for calculating Ω. The cumulant expansion must be used.

Solution 28. (i) We set

$$\exp(-\beta(\hat{H} - \mu\hat{N})) = (\exp(-\beta(\hat{H}_0 - \mu\hat{N})))S(\beta) =: \phi(\beta).$$

Taking the derivative of this equation with respect to β gives

$$\frac{\partial\phi}{\partial\beta} = -(\hat{H} - \mu\hat{N})\phi.$$

Therefore

$$\frac{\partial S}{\partial\beta} = -\hat{H}_1(\beta)S.$$

Obviously we have the "initial condition" $S(\beta = 0) \equiv S(0) = 1$. The integration of this equation with the initial condition gives

$$S(\beta) = 1 - \int_0^\beta \hat{H}_1(\tau)S(\tau)d\tau.$$

By iterating this equation we arrive at

$$S(\beta) = 1 - \int_0^\beta \hat{H}_1(\tau)d\tau + \int_0^\beta d\tau_1 \int_0^{\tau_1} d\tau_2 \hat{H}_1(\tau_1)\hat{H}_1(\tau_2) + \cdots$$

$$+ (-1)^n \int_0^\beta d\tau_1 \int_0^{\tau_1} d\tau_2 \cdots \int_0^{\tau_{n-1}} d\tau_n \hat{H}_1(\tau_1)\hat{H}_1(\tau_2)\cdots\hat{H}_1(\tau_n) + \cdots.$$

(ii) For the grand thermodynamical partition function Z we obtain

$$Z(\beta) = \text{tr}(\exp(-\beta(\hat{H}_0 - \mu\hat{N}))S(\beta))$$

$$\equiv \frac{[\text{tr}(\exp(-\beta(\hat{H}_0 - \mu\hat{N})))][\text{tr}(\exp(-\beta(\hat{H}_0 - \mu\hat{N}))S(\beta))]}{\text{tr}(\exp(-\beta(\hat{H}_0 - \mu\hat{N})))}.$$

Therefore we find

$$Z(\beta) = e^{-\beta\Omega_0}\left(1 - \int_0^\beta \langle\hat{H}_1(\tau)\rangle_0 d\tau + \int_0^\beta d\tau_1 \int_0^{\tau_1} d\tau_2 \langle\hat{H}_1(\tau_1)\hat{H}_1(\tau_2)\rangle_0 + \cdots\right)$$

where

$$\langle\cdots\rangle_0 := \frac{\text{tr}(\cdots e^{-\beta(\hat{H}_0 - \mu\hat{N})})}{\text{tr}(e^{-\beta(\hat{H}_0 - \mu\hat{N})})}.$$

Taking the logarithm of both sides of this equation we find

$$\Omega = \Omega_0 - \frac{1}{\beta}\ln\left(1 - \int_0^\beta \langle\hat{H}_1(\tau)\rangle_0 d\tau + \int_0^\beta d\tau_1 \int_0^{\tau_1} d\tau_2 \langle\hat{H}_1(\tau_1)\hat{H}(\tau_2)\rangle_0 - \cdots\right).$$

(iii) Taking the logarithm of both sides of (3) with $\hat{H}_1 \to \epsilon\hat{H}_1$ it follows that we can set

$$\phi(\epsilon) = \sum_{n=1}^{\infty} \frac{(-1)^n \epsilon^n}{n!} \int_0^\beta d\tau_1 \cdots \int_0^\beta d\tau_n \langle T_\tau \hat{H}_1(\tau_1) \cdots \hat{H}_1(\tau_n) \rangle_c$$

and

$$\phi(\epsilon) = \ln \langle T_\tau \exp(-\epsilon \int_0^\beta \hat{H}_1(\tau) d\tau) \rangle_0$$

where ϵ is a real parameter. This equation can be written as

$$\phi(\epsilon) = \lim_{\alpha \to 0} \ln \langle T_\tau \exp((\epsilon + \alpha)(- \int_0^\beta \hat{H}_1(\tau) d\tau)) \rangle_0.$$

Since $\exp(\epsilon \partial/\partial\alpha) f(\alpha) \equiv f(\alpha + \epsilon)$ we find

$$\phi(\epsilon) = \lim_{\alpha \to 0} \sum_{n=1}^{\infty} \frac{\epsilon^n}{n!} \left(\frac{\partial}{\partial\alpha}\right)^n \ln \langle T_\tau \exp(\alpha(- \int_0^\beta \hat{H}_1(\tau) d\tau)) \rangle_0.$$

Comparing powers of ϵ^n we can express $\langle \cdots \rangle_c$ in terms of $\langle \cdots \rangle_0$. For the first two terms we find

$$\int_0^\beta d\tau \langle \hat{H}_1(\tau) \rangle_c = \int_0^\beta d\tau \langle \hat{H}_1(\tau) \rangle_0$$

$$\int_0^\beta d\tau_1 \int_0^\beta d\tau_2 \langle T_\tau \hat{H}_1(\tau_1) \hat{H}_1(\tau_2) \rangle_c = \int_0^\beta d\tau_1 \int_0^\beta d\tau_2 \langle T_\tau \hat{H}_1(\tau_1) \hat{H}_1(\tau_2) \rangle_0$$
$$- \left(\int_0^\beta d\tau \langle \hat{H}_1(\tau) \rangle_0\right)^2.$$

Problem 29.   Let $c_j^\dagger$, c_j ($j = 1, 2$) be Fermi creation and annihilation operators, respectively. Note that $(c_j^\dagger)^2 = 0$, $(c_j)^2 = 0$, where 0 is the zero operator and $[c_j^\dagger, c_k]_+ = \delta_{jk} I$, where I is the identity operator. Calculate the sum

$$S = \sum_{j_1, j_2, j_3, j_4 = 1}^{2} a_{j_1 j_2} a_{j_3 j_4} c_{j_1}^\dagger c_{j_2} c_{j_3}^\dagger c_{j_4}.$$

Solution 29.   Owing to the properties of the Fermi operators given above only the following ten combinations of $(j_1 j_2 j_3 j_4)$ contribute to the sum

$$(1111), (1112), (1122), (1221), (1222),$$

$$(2111), (2112), (2211), (2221), (2222).$$

Furthermore, utilizing $c_j^\dagger c_j c_j^\dagger c_k = c_j^\dagger c_k$ we arrive at

$$S = (a_{11}^2 + a_{12}a_{21})c_1^\dagger c_1 + (a_{22}^2 + a_{12}a_{21})c_2^\dagger c_2$$
$$+ 2\det(A)c_1^\dagger c_1 c_2^\dagger c_2 + a_{12}\mathrm{tr}(A)c_1^\dagger c_2 + a_{21}\mathrm{tr}(A)c_2^\dagger c_1$$

where $\mathrm{tr}(A)$ denotes the trace of A and $\det(A)$ the determinant of A.

Problem 30. Find the sum

$$\sum_{j,k,\ell=1}^{3} \varepsilon_{jk\ell} x_j dx_k \wedge dx_\ell$$

with $\varepsilon_{123} = \varepsilon_{312} = \varepsilon_{231} = 1$, $\varepsilon_{213} = \varepsilon_{321} = \varepsilon_{132} = -1$ and 0 otherwise. Here $\wedge$ is the *exterior product*, i.e. $dx_j \wedge dx_k = -dx_k \wedge dx_j$. This means $dx_j \wedge dx_j = 0$.

Solution 30. We have

$$\sum_{j,k,\ell=1}^{3} \varepsilon_{jk\ell} x_j dx_k \wedge dx_\ell = 2(x_3 dx_1 \wedge dx_2 + x_2 dx_3 \wedge dx_1 + x_1 dx_2 \wedge dx_3).$$

Supplementary Problems

Problem 1. Let N be a positive integer. Consider the set

$$S := \left\{ \frac{2\pi k}{2N} \; : \; k = -N, -N+1, \ldots, 0, \ldots, N-1 \right\}.$$

Show that $\sum_{k \in S} \cos(k) = 0$ utilizing $\cos(-x) = \cos(x)$ and $\cos(x + \pi) = -\cos(x)$.

Problem 2. Find the sum of the series $\sum_{j=0}^{\infty} \frac{1}{(4j)!}$. Start with the expansions for $\cos(x)$ and $\cosh(x)$

$$\cos(x) = \sum_{j=0}^{\infty} \frac{(-1)^j x^{2j}}{(2j)!}, \qquad \cosh(x) = \sum_{j=0}^{\infty} \frac{x^{2j}}{(2j)!}$$

and $\cos(x) + \cosh(x)$. Then set $x = 1$.

Problem 3. Let $a > 0$ with dimension length. Show that

$$\frac{a}{2\pi} \sum_{m=-\infty}^{+\infty} e^{ikam} = \sum_{\ell=-\infty}^{+\infty} \delta\left(k - \frac{2\pi}{a}\ell\right)$$

with k dimension 1/length and δ denotes the *Dirac delta function*.

Problem 4. Show that

$$\sum_{j=1}^{\infty} \frac{\cos(jx)}{j^2} = \frac{\pi^2}{6} - \frac{\pi x}{2} + \frac{x^2}{4}.$$

Problem 5. Let $N \geq 2$. Show that the N Fourier basis functions are given by

$$f_k(x) = \frac{1}{\sqrt{N}} \exp(i(2\pi k/N)x)$$

where k varies from $-(N-1)/2$ to $(N-1)/2$ in unit steps are orthonormal over the interval $-N/2, N/2$ of length N. Show that

$$\sum_{k=-(N-1)/2}^{k=(N-1)/2} f_k^*(x)f_k(y) = \frac{1}{N}\frac{\sin(\pi(x-y))}{\sin(\pi(x-y)/N)}.$$

Problem 6. Show that

$$\sum_{k=1}^{\infty} \frac{k}{3 \cdot 5 \cdot 7 \cdots (2k+1)} = \frac{1}{3} + \frac{2}{3 \cdot 5} + \frac{3}{3 \cdot 5 \cdot 7} + \cdots = \frac{1}{2}. \qquad (1)$$

Hint: Consider the partial sums $(n \geq 1)$

$$S_n = \sum_{k=1}^{n} \frac{k}{3 \cdot 5 \cdot 7 \cdots (2k+1)}$$

and the difference $1/2 - S_n$. The formula $S_n = \frac{1}{2}(1 - 1/(3 \cdot 5 \cdots (2n+1)))$ can be proved by induction.

Problem 7. (i) Let $n \in \mathbb{N}$ and $x \in \mathbb{R}$ with $|x| < 1$. Show that $(1+x)^n$ can be expressed using the exponential function and a series expansion for $\ln(1+x)$, i.e.

$$(1+x)^n = \exp(n\ln(1+x)) = \exp\left(n\sum_{k=1}^{\infty} \frac{(-1)^{k-1}x^k}{k}\right).$$

(ii) Let $n \geq 1$ and $x \in \mathbb{R}$. Show that

$$\prod_{r=1}^{n-1} (x^2 - 2x \cos(\pi r/n) + 1) = \frac{x^{2n} - 1}{x^2 - 1}.$$

(iii) Show that

$$\prod_{j=1}^{\infty} \frac{1 - x^{n+1}}{1 - x^n} = \sum_{k=0}^{\infty} x^{k(k+1)/2}.$$

Problem 8.   Let k_B be the Boltzmann constant and $\beta = 1/(k_B T)$.
(i) Show that

$$Z(\beta) = \sum_{n=1}^{\infty} \exp(-\beta \hbar \omega n) = \frac{1}{1 - \exp(-\beta \hbar \omega)}.$$

(ii) Consider the *partition function*

$$Z(\beta) = 2 \prod_{n=1}^{\infty} \left(1 + \left(\frac{\beta \hbar \omega}{\pi(2n - 1)} \right)^2 \right).$$

Show that by utilizing the identity

$$\cosh(x/2) \equiv \prod_{n=1}^{\infty} \left(1 + \frac{x^2}{\pi^2(2n - 1)^2} \right)$$

one has $Z(\beta) = 2 \cosh(\beta \hbar \omega/2)$.

Problem 9.   Let ω be a fixed frequency and $\beta = 1/(k_B T)$. The normalized eigenvectors (*Bell states*) of the Hamilton operator

$$\hat{H} = \hbar \omega \begin{pmatrix} 0 & 0 & 0 & 1 \\ 0 & 0 & 1 & 0 \\ 0 & 1 & 0 & 0 \\ 1 & 0 & 0 & 0 \end{pmatrix} \equiv \hbar \omega \sigma_1 \otimes \sigma_1$$

are

$$|\psi_1\rangle = \frac{1}{\sqrt{2}} (1 \quad 0 \quad 0 \quad 1)^T, \quad |\psi_2\rangle = \frac{1}{\sqrt{2}} (1 \quad 0 \quad 0 \quad -1)^T,$$

$$|\psi_3\rangle = \frac{1}{\sqrt{2}} (0 \quad 1 \quad 1 \quad 0)^T, \quad |\psi_4\rangle = \frac{1}{\sqrt{2}} (0 \quad 1 \quad -1 \quad 0)^T$$

with the corresponding eigenvalues $E_1 = \hbar\omega$, $E_2 = -\hbar\omega$, $E_3 = \hbar\omega$, $E_4 = -\hbar\omega$. Show that $K(\beta) = \sum_{j=1}^{4} |\psi_j\rangle\langle\psi_j| e^{-\beta E_j}$ is given by

$$K(\beta) = \begin{pmatrix} \cosh(\beta\hbar\omega) & 0 & 0 & -\sinh(\beta\hbar\omega) \\ 0 & \cosh(\beta\hbar\omega) & -\sinh(\beta\hbar\omega) & 0 \\ 0 & -\sinh(\beta\hbar\omega) & \cosh(\beta\hbar\omega) & 0 \\ -\sinh(\beta\hbar\omega) & 0 & 0 & \cosh(\beta\hbar\omega) \end{pmatrix}.$$

Show that $K(\beta = 0) = I_4$, $K(\beta_1)K(\beta_2) = K(\beta_1 + \beta_2)$ and (*Bloch equation*)

$$\hat{H}K(\beta) = -\frac{\partial K(\beta)}{\partial\beta}.$$

Chapter 2

Transformations, Functions and Maps

Let X, Y be two non-empty sets. Then a map of X into Y is a function f with X as domain and Y as codomain, i.e. f maps X into Y. We write $f : X \to Y$. With $x \in X$ and $f(x) = y \in Y$ y is called the value of the function f corresponding to the argument x. If f is a map of X into Y we say that f is surjective (onto) when $f(X) = Y$; injective (one-one) when for all $x, x' \in X$, $f(x) = f(x')$ implies $x = x'$; bijective (one-one, onto) when it is both injective and surjective.

A bijection of a finite set S onto itself is called a *permutation* of S.

The composition of maps is denoted by $\circ$. Let $f_1 : S_1 \to S_2$, $f_2 : S_2 \to S_3$ be maps of S_1 into S_2 and S_2 into S_3, respectively. The composition of the maps f_1 and f_2 is given by

$$(f_2 \circ f_1)(x) = f_2(f_1(x)).$$

Let S be a nonempty set and $f : S \to S$. Then x^* is called a *fixed point* if $f(x^*) = x^*$. Let $x^*, y^* \in S$ and $x^* \neq y^*$. If $f(x^*) = y^*$ and $f(y^*) = f(f(x^*)) = x^*$ we have a periodic orbit of period 2.

The superposition principle applies to linear difference equations. If two solutions of the linear difference equation are given, then the sum of the two solutions is also a solution of the linear difference equation. A solution can also be multiplied with a constant to be a solution again.

Recursive sequences (also called difference equations) are as follows: Let $f : \mathbb{R}^k \to \mathbb{R}$ $(k \geq 1)$ and $a_0, a_1, \ldots, a_{k-1} \in \mathbb{R}$. One defines the recursive sequence $(x_n)_{n \in \mathbb{N}_0}$ by $x_j = a_j$ for all $j = \{0, 1, \ldots, k-1\}$ and

$$x_{j+k} = f(x_j, x_{j+1}, \ldots, x_{j+k-1})$$

for all $j \geq 0$.

The *Galilean transformation* is given by $(k = 1, 2, 3)$

$$x_k \mapsto x_k' = x_k + v_k t, \quad t \mapsto t' = t$$

with the velocity $\mathbf{v} = (v_1, v_2, v_3)$. Let c be the speed of light in vacuum. The *Lorentz transformation* parallel to the first spatial axis is given by $(\beta = v/c, v < c)$

$$x_1' = \frac{x - vt}{\sqrt{1 - \beta^2}}, \quad x_2' = x_2, \quad x_3' = x_3,$$

$$t' = \frac{t - \beta x_1/c}{\sqrt{1 - \beta^2}}$$

with

$$(x_1')^2 + (x_2')^2 + (x_3')^2 - c^2 (t')^2 = x_1^2 + x_2^2 + x_3^2 - c^2 t^2.$$

Polar coordinates are given by

$$x_1(r, \theta) = r \cos(\theta), \quad x_2(r, \theta) = r \sin(\theta)$$

with $r \geq 0$. *Spherical coordinates* are given by

$$x_1(r, \phi, \theta) = r \cos(\phi) \sin(\theta),$$
$$x_2(r, \phi, \theta) = r \sin(\phi) \sin(\theta),$$
$$x_3(r, \phi, \theta) = r \cos(\theta)$$

where $0 \leq \phi < 2\pi$, $0 < \theta < \pi$, $r \geq 0$. Another useful coordinate system in $\mathbb{R}^3$ is *parabolic coordinates*

$$x_1(\xi, \eta, \phi) = \xi \eta \cos(\phi),$$
$$x_2(\xi, \eta, \phi) = \xi \eta \sin(\phi),$$
$$x_3(\xi, \eta, \phi) = \frac{1}{2}(\xi^2 - \eta^2)$$

where $\xi \geq 0$, $\eta \geq 0$.

Problem 1.   In population dynamics one considers the logistic growth $u_{t+1} = au_t(1 - u_t/P)$ $(t = 0, 1, \ldots)$ where a denotes the exponential growth rate and P is the carrying capacity (maximal population size). For $u_t = P$ the population growth stops. Setting $x_t = u_t/P$ and $a = 4$ we obtain $x_{t+1} = 4x_t(1 - x_t)$. Consider $f : [0, 1] \to [0, 1]$, $f(x) = 4x(1 - x)$ or equivalently

$$x_{t+1} = 4x_t(1 - x_t) \tag{1}$$

where $t = 0, 1, 2, \ldots$ and $x_0 \in [0, 1]$. This nonlinear difference equation (map) is the so-called *logistic equation* or *logistic map*.
(i) Show that $x_t \in [0, 1]$ for $t = 0, 1, 2, \ldots$
(ii) Find the *fixed points*, i.e. solve $f(x^*) = x^*$.
(iii) Assume that f is differentiable (which is the case for the logistic map). Then

$$y_{t+1} = \frac{df}{dx}(x_t)y_t$$

is called the (difference) *variational equation* or *linearized equation*. Give the variational equation for (1).
(iv) Study the stability of the fixed points.
(v) Show that the general solution is given by

$$x_t = \frac{1}{2} - \frac{1}{2}\cos(2^t \cos^{-1}(1 - 2x_0)) \tag{2}$$

where x_0 is the initial value.
(vi) Find the periodic orbits.

Solution 1.   (i) Let $x \in [0, 1]$. Then $x(1 - x) \leq 1/4$. Consequently, $4x(1 - x) \leq 1$. If $x \neq 1/2$, then $x(1 - x) < 1/4$. In other words, the function $g(x) = x(1 - x)$ has one maximum at $x = 1/2$ with $g(1/2) = 1/4$.
(ii) The fixed points x^* are solutions of the algebraic equation

$$4x^*(1 - x^*) = x^*.$$

We obtain as fixed points $x_1^* = 0$, $x_2^* = 3/4$. The fixed points are time-independent solutions.
(iii) Since $f(x) = 4x(1 - x)$ we have

$$\frac{df}{dx} = 4 - 8x$$

and we obtain the variational equation $y_{t+1} = (4 - 8x_t)y_t$ $(t = 0, 1, 2, \ldots)$.
(iv) Inserting the fixed point $x_1^* = 0$ into the variational equation yields the linear difference equation $y_{t+1} = 4y_t$. The solution is given by $y_t = 4^t y_0$. Therefore the fixed point $x_1^* = 0$ is unstable, since $y_t \to \infty$ as $t \to \infty$. Inserting the fixed point $x_2^* = 3/4$ into the variational equation yields

$$y_{t+1} = -2y_t.$$

Therefore this fixed point is also unstable, since $|y_t| \to \infty$ as $t \to \infty$.
(v) Let $\alpha = \cos^{-1}(1 - 2x_0)$. Then

$$x_t = \frac{1}{2} - \frac{1}{2}\cos(2^t\alpha).$$

It follows that

$$x_{t+1} = \frac{1}{2} - \frac{1}{2}\cos(2^{t+1}\alpha) = \frac{1}{2} - \frac{1}{2}(2\cos^2(2^t\alpha) - 1) = 1 - \cos^2(2^t\alpha).$$

The left hand side of (1) is given by

$$4x_t(1 - x_t) = 4x_t - 4x_t^2 = 1 - \cos^2(2^t\alpha).$$

This proves that (2) is the general solution of (1), where x_0 is the initial value.

(vi) The periodic orbits are given by the initial values

$$x_0 = \frac{1}{2} - \frac{1}{2}\cos\left(\frac{r\pi}{2^s}\right)$$

where r and s are positive integers. We have

$$\arccos(1 - 2x_0) = \arccos\left(\cos\left(\frac{r\pi}{2^s}\right)\right) = \frac{r\pi}{2^s}.$$

It follows that

$$x_t = \frac{1}{2} - \frac{1}{2}\cos\left(\frac{2^t r\pi}{2^s}\right).$$

Problem 2. Consider the *logistic map* $f : [0, 1] \to [0, 1]$, $f(x) = 4x(1-x)$. Show that the solution of the algebraic equation $f(f(x^*)) = x^*$ gives the fixed points of f and a *periodic orbit* of f.

Solution 2. The fixed points of f are given by $x_1^* = 0$, $x_2^* = 3/4$. Since $f(f(x)) = 16x(1 - x)(1 - 4x(1 - x))$ we find that the solutions are

$$x_1^* = 0, \qquad x_2^* = \frac{3}{4}, \qquad x_3^* = \frac{5 - \sqrt{5}}{8}, \qquad x_4^* = \frac{5 + \sqrt{5}}{8}.$$

Thus x_1^* and x_2^* are the fixed points of f. Let

$$x_0 = \frac{5 - \sqrt{5}}{8}.$$

Then $x_1 = (5 + \sqrt{5})/8$, $x_2 = (5 - \sqrt{5})/8 = x_0$. Thus we have a periodic orbit with period 2. $f(f(x))$ is the second iterate of f. When we consider higher iterates we find other periodic orbits of f.

Problem 3.   Consider the logistic map $f : [0, 1] \rightarrow [0, 1]$, $f(x) = 4x(1-x)$.
(i) The *Ljapunov exponent* λ is given by

$$\lambda(x_0) := \lim_{T \to \infty} \frac{1}{T} \sum_{t=0}^{T-1} \ln \left| \frac{df(x)}{dx} \right|_{x=x_t}$$

where x_0 is the initial value, i.e. the Ljapunov exponent depends on the initial value. The Ljapunov exponent measures the exponential rate at which the derivative grows. Find the maximal Ljapunov exponent.
(ii) The *time average* is defined by

$$\langle x_t \rangle := \lim_{T \to \infty} \frac{1}{T} \sum_{t=0}^{T} x_t.$$

For almost all initial values we find $\langle x_t \rangle = 1/2$. The *autocorrelation function* is defined by

$$C_{xx}(\tau) := \lim_{T \to \infty} \frac{1}{T} \sum_{t=0}^{T} (x_t - \langle x_t \rangle)(x_{t+\tau} - \langle x_t \rangle)$$

where $\tau = 0, 1, 2, \ldots$. Find the autocorrelation function.

Solution 3.   (i) Since $df/dx = 4 - 8x$ we find that the Ljapunov exponent is given by $\lambda(x_0) = \ln(2)$ for almost all initial values, i.e. the periodic orbits are excluded.
(ii) Using the exact solution

$$x_t = \frac{1}{2} - \frac{1}{2} \cos(2^t \cos^{-1}(1 - 2x_0))$$

we find that the autocorrelation function is given by

$$C_{xx}(\tau) = \begin{cases} \frac{1}{8} & \text{for} \quad \tau = 0 \\ 0 & \text{otherwise} \end{cases}$$

Problem 4.   Consider $f : [0, 1] \mapsto [0, 1]$, $f(x) = 4x(1 - x)$ and let $\phi : [0, 1] \mapsto [0, 1]$ be defined by

$$\phi(x) = \frac{2}{\pi} \arcsin(\sqrt{x}).$$

Calculate $\widetilde{f} = \phi \circ f \circ \phi^{-1}$, where $\circ$ denotes the *composition of functions*.

Solution 4. The inverse ϕ^{-1} of ϕ is defined on $[0,1]$ and we find

$$\phi^{-1}(x) = \sin^2\left(\frac{\pi x}{2}\right) \equiv \frac{1 - \cos(\pi x)}{2}$$

since $\arcsin \circ \sin(y) = y$, $\sin \circ \arcsin(x) = x$, where $y \in [-\pi/2, \pi/2]$. To find $\widetilde{f}$ we set $y(x) = \sin^2(\pi x/2)$, $v(y) = 4y(1 - y)$, $w(v) = 2\arcsin(\sqrt{v})/\pi$. Inserting $y(x)$ into $v(y)$ gives

$$v(x) = 4\sin^2\left(\frac{\pi x}{2}\right)\left(1 - \sin^2\left(\frac{\pi x}{2}\right)\right) = \left(2\sin\left(\frac{\pi x}{2}\right)\cos\left(\frac{\pi x}{2}\right)\right)^2.$$

Now

$$\sqrt{v(x)} = 2\sin\left(\frac{\pi x}{2}\right)\cos\left(\frac{\pi x}{2}\right) = \sin(\pi x).$$

For $x \in [0, 1/2]$ we have $\pi x \in [0, \pi/2]$ and therefore we find for this range

$$w(x) = \frac{2}{\pi}\arcsin(\sin(\pi x)) = 2x.$$

For $x \in [1/2, 1]$ we have $\pi x \in [\pi/2, \pi]$ and therefore we find for this range

$$w(x) = \frac{2}{\pi}\arcsin(\sin(\pi x)) = 2(1 - x).$$

Consequently,

$$\widetilde{f}(x) = \begin{cases} 2x & \text{if } x \in [0, 1/2] \\ 2(1 - x) & \text{if } x \in [1/2, 1]. \end{cases}$$

The mapping $\widetilde{f}$ is called the *tent map*.

Problem 5. (i) Let $f : [-1, 1] \mapsto [-1, 1]$, $f(x) := 1 - 2x^2$. Let $-1 \le a \le b \le 1$ and

$$\mu([a, b]) := \frac{1}{\pi}\int_a^b \frac{dx}{\sqrt{1 - x^2}}. \tag{1}$$

Calculate $\mu([-1, 1])$.
(ii) Show that $\mu(f^{-1}([a, b])) = \mu([a, b])$, where $f^{-1}([a, b])$ denotes the set S which is mapped under f to $[a, b]$, i.e. $f(S) = [a, b]$. The quantity μ is called the *invariant measure* of the map f.

Solution 5. (i) Since

$$\int_a^b \frac{dx}{\sqrt{1 - x^2}} = \arcsin(b) - \arcsin(a)$$

and $\arcsin(1) = \pi/2$, $\arcsin(-1) = -\pi/2$ we obtain $\mu([-1, 1]) = 1$.

(ii) The inverse function f^{-1} is not globally defined. We set $f_1 : [-1, 0] \mapsto [-1, 1]$ with $f_1(x) = 1 - 2x^2$. Then $f_1^{-1} : [-1, 1] \mapsto [-1, 0]$,

$$f_1^{-1}(x) = -\sqrt{(1 - x)/2}$$

with $f_1^{-1}(-1) = -1$ and $f_1^{-1}(1) = 0$. Analogously, we set $f_2 : [0, 1] \mapsto [-1, 1]$ with $f_2(x) = 1 - 2x^2$. Then $f_2^{-1} : [-1, 1] \mapsto [0, 1]$,

$$f_2^{-1}(x) = \sqrt{(1 - x)/2}$$

with $f_2^{-1}(-1) = 1$ and $f_2^{-1}(1) = 0$. It follows that

$$f_1^{-1}(a) = -\sqrt{(1 - a)/2}, \qquad f_1^{-1}(b) = -\sqrt{(1 - b)/2},$$
$$f_2^{-1}(a) = \sqrt{(1 - a)/2}, \qquad f_2^{-1}(b) = \sqrt{(1 - b)/2}.$$

Thus f maps the intervals

$$[-\sqrt{(1 - a)/2}, -\sqrt{(1 - b)/2}], \qquad [\sqrt{(1 - b)/2}, \sqrt{(1 - a)/2}]$$

into the interval $[a, b]$. We notice that $\pi\mu([a, b]) = \arcsin(b) - \arcsin(a)$. Furthermore $\arcsin(-x) \equiv -\arcsin(x)$. Now

$$\int_{-\sqrt{(1-a)/2}}^{-\sqrt{(1-b)/2}} \frac{dx}{\sqrt{1 - x^2}} + \int_{\sqrt{(1-b)/2}}^{\sqrt{(1-a)/2}} \frac{dx}{\sqrt{1 - x^2}} = \arcsin(b) - \arcsin(a).$$

The condition can also be written as

$$\frac{1}{\pi} \frac{1}{\sqrt{1 - x^2}} = \frac{d}{dx} \int_{f^{-1}([-1,x])} \frac{ds}{\pi\sqrt{1 - s^2}}.$$

Problem 6.   Let $g : [0, 1] \mapsto [0, 1]$ be defined by (*tent map*)

$$g(x) := \begin{cases} 2x & \text{for } 0 \leq x \leq 1/2 \\ 2(1 - x) & \text{for } 1/2 < x \leq 1 \end{cases}.$$

Let $0 \leq a \leq b \leq 1$ and

$$\nu([a, b]) := \int_a^b dx.$$

(i) Show that $\nu(g^{-1}([a, b])) = \nu([a, b])$, where $g^{-1}([a, b])$ is the set S which is mapped under g to $[a, b]$, i.e. $g(S) = [a, b]$.
(ii) Find the Ljapunov exponent of the tent map.

Solution 6.   (i) Obviously

$$\nu([a, b]) = \int_a^b dx = b - a.$$

The set $g^{-1}([a, b])$ is given by

$$g^{-1}([a, b]) = [a/2, b/2] \cup [1 - b/2, 1 - a/2]$$

where $0 \le a \le b \le 1$. Therefore

$$\nu(g^{-1}([a, b])) = \int_{a/2}^{b/2} dx + \int_{1-b/2}^{1-a/2} dx = b - a = \nu([a, b]).$$

(ii) Since g is not differentiable at $x = 1/2$ we define the *Ljapunov exponent* as

$$\lambda(x_0) := \lim_{T \to \infty} \lim_{\epsilon \to 0} \frac{1}{T} \ln \left| \frac{g^{(T)}(x_0 + \epsilon) - g^{(T)}(x_0)}{\epsilon} \right|$$

where x_0 is the initial value and $g^{(T)}$ is the T-th iterate of the function g. Inserting the tent map into this expression we obtain $\lambda(x_0) = \ln(2)$ for almost all initial values, i.e. the initial conditions $x_0 \in \mathbb{Q} \cap [0, 1]$ are excluded, where $\mathbb{Q}$ denotes the rational numbers. Thus the Ljapunov exponent is equal to $\ln(2)$ if the initial value $x_0 \in [0, 1]$ is an irrational number. If $x_0 \in \mathbb{Q} \cap [0, 1]$, then we obtain a periodic orbit. For example, if $x_0 = \frac{1}{7}$, then we find $x_1 = \frac{2}{7}$, $x_2 = \frac{4}{7}$, $x_3 = \frac{6}{7}$, $x_4 = \frac{2}{7}$, ...

Problem 7. Find all analytic functions $f : \mathbb{R} \to \mathbb{R}$ satisfying

$$f(2x + 1) = x^2.$$

Solution 7. We set $y = 2x + 1$. The $x = \frac{1}{2}(y - 1)$. Hence

$$x^2 = \frac{1}{4}(y^2 - 2y + 1).$$

It follows that $f(x) = \frac{1}{4}(x^2 - 2x + 1)$.

Problem 8. Let F be a closed set on a complete metric space X. A *contracting mapping* is a mapping $f : F \to F$ such that

$$d(f(x), f(y)) \le k d(x, y), \qquad 0 \le k < 1, \quad d \text{ distance in } X.$$

One also says that f is *Lipschitzian* of order $k < 1$.
(i) Prove the following theorem: A contracting mapping f has strictly one fixed point, i.e. there is one and only one point x^* such that $x^* = f(x^*)$.
(ii) Apply the theorem to the linear equation $A\mathbf{x} = \mathbf{b}$, where A is an $n \times n$ matrix over the real numbers and $\mathbf{x} = (x_1, x_2, \ldots, x_n)^T$.

Solution 8.   (i) The proof is by iteration. Let $x_0 \in F$. Then

$$f(x_0) \in F, \ldots, f^{(n)}(x_0) = f(f^{(n-1)}(x_0)) \in F$$

and

$$d(f^{(n)}(x_0), f^{(n-1)}(x_0)) \leq kd(f^{(n-1)}(x_0), f^{(n-2)}(x_0))$$
$$\leq \cdots \leq k^{n-1}d(f(x_0), x_0).$$

Since $k < 1$ the sequence $f^{(n)}$ is a *Cauchy sequence* and tends to a limit x^* when n tends to infinity

$$x^* = \lim_{n \to \infty} f^{(n)}(x_0) = \lim_{n \to \infty} f(f^{(n-1)}(x_0)) = f(x^*).$$

The uniqueness of x^* results from the defining property of contracting mappings: Assume that there is another point y^* such that $y^* = f(y^*)$, then on the one hand

$$d(f(y^*), f(x^*)) = d(y^*, x^*)$$

but on the other hand $d(f(y^*), f(x^*)) \leq kd(y^*, x^*)$, $k < 1$. Therefore $d(y^*, x^*) = 0$ and $y^* = x^*$.

(ii) We set $c_{jk} := -a_{jk} + \delta_{jk}$, where a_{jk} are the matrix elements of A and δ_{jk} is the Kronecker delta. Then the linear equation $A\mathbf{x} = \mathbf{b}$ takes the form $\mathbf{x} = C\mathbf{x} + \mathbf{b}$. If

$$\sum_{k=1}^{n} |c_{jk}| < 1, \qquad j = 1, 2, \ldots, n$$

then the theorem can be applied.

Problem 9.   Let M be a manifold or simply $\mathbb{R}^n$. Assume the map $f : M \to M$ is continuous. μ is called a *Borel probability measure* if μ assigns to every Borel subset $E \subset M$ a certain probability or measure that we denote by $\mu(E)$. A property is said to hold a.e. (almost everywhere) if it is enjoyed by every $x \in M$ except possibly on a set E with $\mu(E) = 0$. Given a probabilistic dynamical system $f : (M, \mu) \to (M, \mu)$. Such a system is said to be *ergodic* if

$$f^{-1}(E) = E \;\Rightarrow\; \mu(E) = 0 \quad \text{or} \quad \mu(E) = 1.$$

That is, an ergodic system cannot be decomposed into two nontrivial subsystems that do not interact with each other.

Birkhoff Ergodic Theorem. For any integrable function $\phi : (M, \mu) \to \mathbb{R}$, the *time average*

$$\frac{1}{n} \sum_{i=0}^{n-1} \phi(f^{(i)}(x)) \tag{1}$$

converges for a.e. x. Denote this limit by $\phi^*(x)$. If the system is ergodic, then $\phi^*(x)$ equals a.e. the *space average*

$$\int_M \phi(x)d\mu(x).\tag{2}$$

Apply the theorem to the logistic map $f : [0, 1] \to [0, 1]$, $f(x) = 4x(1 - x)$, where $\phi(x) = \ln|df/dx|$.

Solution 9.   It can be shown that

$$d\mu(x) = \frac{1}{\pi\sqrt{x(1-x)}}dx$$

is an ergodic *invariant measure* and that $\ln|df/dx|$ is integrable. Substituting $\phi(x) = \ln|df/dx|$ in the ergodic theorem, we have for a.e. x

$$\frac{1}{n}\ln\left|\frac{f^{(n)'}(x)}{dx}\right| = \frac{1}{n}\ln\prod_{i=0}^{n-1}\left|f'(f^{(i)}(x))\right| = \frac{1}{n}\sum_{i=0}^{n-1}\phi(f^{(i)}(x)).$$

For $n \to \infty$ we obtain

$$\int_0^1 \ln|4 - 8x|\frac{1}{\pi\sqrt{x(1-x)}}dx = \ln(2)$$

where we used that $df/dx = 4 - 8x$. This says that for a.e. x, $|(f^{(n)})'(x)|$ is in the order of 2^n for large enough n. The number $\ln(2)$ is called the *Ljapunov exponent*.

Problem 10.   Given a compact set X embedded in $\mathbb{R}^n$. The *capacity* of X can be calculated by counting the number of grid boxes of side $\epsilon = B^{-\ell}$ that cover X. $B > 1$ and $\ell > 0$ are integers. Let $N(\ell)$ be the number of such boxes. The capacity dimension is assumed to satisfy

$$\lim_{\ell \to \infty}\frac{\ln N(\ell)}{\ell \cdot \ln(B)} = D_c.\tag{1}$$

The definition of the capacity dimension is

$$D_c := \overline{\lim_{\epsilon \to 0}}\frac{\ln(\bar{N}(\epsilon))}{\ln(1/\epsilon)}\tag{2}$$

where $\bar{N}(\epsilon)$ is the minimum number of boxes of side ϵ of any type that covers the set X. If the limit in the definition exists the limit in (1) will also exist. Discuss how D_c could be calculated numerically.

Solution 10. Denote by G_ℓ the set of all grid boxes of side $B^{-\ell}$. Numerical calculations of the capacity are based on the heuristic idea that the set X has finite D_c-dimensional measure V. The quantity V satisfies

$$V \approx (\bar{N}(B^{-\ell})) \cdot (B^{-\ell})^{D_c}$$

for sufficiently large ℓ. Since $N(B^{-\ell}) \leq N(\ell) \leq N(B^{-\ell}) \cdot c$ for $c > 1$, a constant independent of X or ℓ, implies that

$$V \approx O(1) \cdot N(\ell) \cdot [B^{-\ell}]^{D_c}$$

where $O(1)$ holds as $\ell \to \infty$. Thus if V is constant and $O(1)$ is also assumed to be a constant then

$$\ln(N(\ell)) \approx \ln(\text{const} \cdot V) + \ell \cdot \ln B \cdot D_c$$

that is, for large ℓ a plot of $\ln N$ versus $\ell \cdot \ln B$ is approximately a straight line with slope D_c. The replacement of $O(1)$ by a constant introduces a possible error in the computation that results from the use of boxes to cover X.

Problem 11. Consider a function $f(j)$, where j is integer-valued. Find multiplication rules of the form

$$(f * g)(j) = \sum_{k,\ell} c_{jk\ell} f(k) g(\ell)$$

and difference operators Δ of the form

$$\Delta f(j) = \sum_k d_{jk} f(k)$$

where the $c_{jk\ell}$ and the d_{jk} are real constants such that the following properties hold:
(i) $*$ is commutative and associative
(ii) if f is constant, then $f * g = fg$ (normal multiplication)
(iii) if f is constant, then $\Delta f = 0$
(iv) the *Leibniz rule* $\Delta(f * g) = f * (\Delta g) + (\Delta f) * g$ holds
(v) the function f is periodic, i.e. there is a fixed positive N such that $f(j + N) = f(j)$.

Solution 11. A solution which satisfies these conditions is

$$(f * g)(j) = \langle f \rangle g(j) + \langle g \rangle f(j) - \langle f \rangle \langle g \rangle$$

where $\langle f \rangle$ is the average $\langle f \rangle := \frac{1}{N} \sum_{j=1}^N f(j)$. The *difference operator* Δ is the *forward difference* $\Delta f(j) := f(j + 1) - f(j)$.

Problem 12. Consider the nonlinear difference equation

$$x_{t+1} = 2x_t(1 - x_t) \tag{1}$$

where $t = 0, 1, 2, \ldots$ and $x_0 \in [0, 1]$. Hence $x_t \in [0, 1]$.
(i) Show that the exact solution of the initial-value problem is given by

$$x_t = \frac{1}{2} - \frac{1}{2}(1 - 2x_0)^{2^t}. \tag{2}$$

(ii) Find the fixed points of the equation. Study the stability of the fixed points.

Solution 12. (i) If $x_0 \in [0, 1]$, then $x_t \in [0, 1]$ for $t = 1, 2, \ldots$. From (2) it follows that

$$x_{t+1} = \frac{1}{2} - \frac{1}{2}(1 - 2x_0)^{2^{t+1}} \equiv \frac{1}{2} - \frac{1}{2}(1 - 2x_0)^{2 \cdot 2^t}.$$

On the other hand we have

$$2x_t(1 - x_t) = 2\left[\frac{1}{2} - \frac{1}{2}(1 - 2x_0)^{2^t}\right]\left[1 - \frac{1}{2} + \frac{1}{2}(1 - 2x_0)^{2^t}\right]$$
$$= \frac{1}{2}\left[1 - (1 - 2x_0)^{2 \cdot 2^t}\right].$$

This proves that (2) is a solution of the initial value problem of (1).
(ii) The fixed points are determined by the equation $x^* = 2x^*(1 - x^*)$. It follows that $x_1^* = 0$ and $x_2^* = 1/2$ are the fixed points. Let $0 < x_0 < 1$. Then from solution (2) we find that $x_t \to 1/2$ as $t \to \infty$. This means that the fixed point x_2^* is stable and the fixed point x_1^* is unstable.

Problem 13. Let

$$x_{t+1} = \frac{ax_t + b}{cx_t + d}, \qquad t = 0, 1, \ldots \tag{1}$$

where $c \neq 0$ and

$$D := \det\begin{pmatrix} a & b \\ c & d \end{pmatrix} = ad - bc \neq 0.$$

(i) Let

$$x_t = y_t - \frac{d}{c}. \tag{2}$$

Find the difference equation for y_t.
(ii) Let $y_t = w_{t+1}/w_t$. Find the difference equation for w_t.

Solution 13.   (i) Inserting the transformation (2) into (1) yields

$$y_{t+1} = \frac{a+d}{c} - \frac{D}{c^2 y_t} \quad \Rightarrow \quad y_{t+1} y_t - \frac{a+d}{c} y_t + \frac{D}{c^2} = 0.$$

(ii) The $y_t = w_{t+1}/w_t$ reduces this nonlinear difference equation to the linear second order difference equation

$$w_{t+2} - \frac{a+d}{c} w_{t+1} + \frac{D}{c^2} w_t = 0.$$

Problem 14.   (i) Let r_t satisfy the nonlinear difference equation

$$r_{t+1}(1 + a r_t) = 1 \tag{1}$$

with $t = 0, 1, 2, \ldots$, $r_0 = 0$ and $a \neq 0$. Show that r_k can be expressed as the *continued fraction*

$$r_t = \cfrac{1}{1 + \cfrac{a}{1 + \cfrac{a}{1 + \cfrac{\cdot}{\cdot + \cfrac{\cdot}{\cdot + \cfrac{a}{1}}}}}} \tag{2}$$

which terminates at the k-th stage. Show that (1) can be linearized by

$$r_t = \frac{n_t}{n_{t+1}}. \tag{3}$$

(ii) Find the solution of the nonlinear difference equation

$$r_{t+1}(b + r_t) = 1 \tag{4}$$

with $r_0 = 0$ and $b \neq 0$.

Solution 14.   (i) From (1) it follows that $r_1 = 1$ and $r_t = 1/(1 + a r_{t-1})$, where $t \geq 1$. Since

$$r_{t-1} = \frac{1}{1 + a r_{t-2}}$$

etc., we obtain (2). By making the substitution (3) the nonlinear difference equation (1) reduces to the linear difference equation with constant coefficients $n_{t+1} - n_t - a n_{t-1} = 0$, $t = 1, 2, \ldots$, where $n_0 = 0$ and $n_1 = 1$.

(ii) From (4) we obtain

$$r_t = \frac{1}{b + r_{t-1}}, \qquad r_{t-1} = \frac{1}{b + r_{t-2}}.$$

Hence the solution can be written as a continued fraction. The nonlinear difference equation (4) can be linearized using ansatz (3). We find

$$n_{t+2} - bn_{t+1} - n_t = 0$$

where $n_0 = 0$ and $n_1 = 1$. The infinite continued fraction

$$\cfrac{1}{b + \cfrac{1}{b + \cfrac{1}{b + \cdots}}}$$

converges to $\frac{1}{2}(\sqrt{b^2 + 4} - b)$ when $b > 0$, and to $-\frac{1}{2}(\sqrt{b^2 + 4} + b)$ when $b < 0$. In particular, we have

$$\sqrt{2} = 1 + \cfrac{1}{2 + \cfrac{1}{2 + \cdots}}$$

Problem 15. Find the solution of the nonlinear difference equation

$$y_{t+1} = 2y_t^2 - 1 \qquad (1)$$

where $t = 0, 1, 2, \ldots$.

Solution 15. Making the substitution $y_t = \cos(u_t)$ transforms (1) into

$$\cos(u_{t+1}) = \cos(2u_t)$$

where we have used the identity $\cos^2(\alpha) \equiv \frac{1}{2}(1 + \cos(2\alpha))$. Hence the solution is either $u_{t+1} = 2u_t + 2m\pi$ or $u_{t+1} = -2u_t + 2n\pi$, where m and n are arbitrary integers. From the second alternative, we obtain the solution of (1)

$$y_t = \cos\left(2^t\theta + (-1)^t \frac{2n\pi}{3}\right)$$

where θ is an arbitrary constant and n an arbitrary integer. The solution corresponding to the first alternative is contained in this one.

Problem 16. Show that the solution of the initial-value problem of the nonlinear difference equation

$$x_{t+1} = \frac{4x_t(1 - x_t)(1 - k^2 x_t)}{(1 - k^2 x_t^2)^2}, \qquad t = 0, 1, \ldots \qquad (1)$$

$(x_0 \in [0,1], 0 \le k^2 \le 1)$ is given by

$$x_t = \text{sn}^2(2^t\text{sn}^{-1}(\sqrt{x_0}, k), k), \quad t = 0, 1, \dots \tag{2}$$

where sn denotes a *Jacobi elliptic function*. Consequently, we need the *addition theorems* for Jacobi elliptic functions. In the following the modulus k is omitted. The addition property of the Jacobi elliptic function sn is given by

$$\text{sn}(u \pm v, k) \equiv \frac{\text{sn}(u,k)\text{cn}(v,k)\text{dn}(v,k) \pm \text{cn}(u,k)\text{sn}(v,k)\text{dn}(u,k)}{1 - k^2\text{sn}^2(u,k)\text{sn}^2(v,k)}.$$

From this equation we obtain as special case

$$\text{sn}(2u, k) \equiv \frac{2\text{sn}(u,k)\text{cn}(u,k)\text{dn}(u,k)}{1 - k^2\text{sn}^4(u,k)}.$$

The addition property of the Jacobi elliptic function cn is given by

$$\text{cn}(u \pm v, k) \equiv \frac{\text{cn}(u,k)\text{cn}(v,k) \mp \text{sn}(u,k)\text{sn}(v,k)\text{dn}(u,k)\text{dn}(v,k)}{1 - k^2\text{sn}^2(u,k)\text{sn}^2(v,k)}.$$

Thus we obtain as special case

$$\text{cn}(2u, k) \equiv \frac{\text{cn}^2(u,k) - \text{sn}^2(u,k)\text{dn}^2(u,k)}{1 - k^2\text{sn}^4(u,k)}.$$

The addition property of the Jacobi elliptic function dn is given by

$$\text{dn}(u \pm v, k) \equiv \frac{\text{dn}(u,k)\text{dn}(v,k) \mp k^2\text{sn}(u,k)\text{sn}(v,k)\text{cn}(u,k)\text{cn}(v,k)}{1 - k^2\text{sn}^2(u,k)\text{sn}^2(v,k)}.$$

From this identity we obtain as special case

$$\text{dn}(2u, k) \equiv \frac{\text{dn}^2(u,k) - k^2\text{sn}^2(u,k)\text{cn}^2(u,k)}{1 - k^2\text{sn}^4(u,k)}.$$

Solution 16.   Let $\alpha := 2^t\text{sn}^{-1}(\sqrt{x_0}, k)$. Then $x_{t+1} = \text{sn}^2(2\alpha, k)$. Applying the addition theorems given above we find that (2) is the solution to (1).

Problem 17.   Consider a first-order difference equation

$$x_{t+1} = f(x_t), \quad t = 0, 1, \dots \tag{1}$$

and a second order difference equation

$$x_{t+2} = g(x_t, x_{t+1}), \quad t = 0, 1, \dots \tag{2}$$

If

$$g(x, f(x)) = f(f(x)) \qquad (3)$$

then (1) is called an *invariant* of (2).

Show that $f(x) = 2x^2 - 1$ is an invariant of $g(x, y) = y - 2x^2 + 2y^2$.

Solution 17. The left-hand side of (3) is given by

$$g(x, f(x)) = 2x^2 - 1 - 2x^2 + 2(2x^2 - 1)^2 = -1 + 2(2x^2 - 1)^2.$$

The right-hand side of (3) is given by $f(f(x)) = 2(2x^2 - 1)^2 - 1$. Thus we see that f is an invariant of g.

Problem 18. The *Fibonacci trace map* is given by

$$x_{t+3} = 2x_{t+2}x_{t+1} - x_t. \qquad (1)$$

Show that the Fibonacci trace map admits the invariant

$$I(x_t, x_{t+1}, x_{t+2}) = x_t^2 + x_{t+1}^2 + x_{t+2}^2 - 2x_t x_{t+1} x_{t+2} - 1. \qquad (2)$$

Solution 18. From (1) we obtain by shifting $t \to t - 1$

$$x_{t+2} = 2x_{t+1}x_t - x_{t-1}.$$

Inserting this equation into (2) yields

$$x_t^2 + x_{t+1}^2 + (2x_{t+1}x_t - x_{t-1})^2 - 2x_t x_{t+1}(2x_{t+1}x_t - x_{t-1}) - 1 =$$

$$x_t^2 + x_{t+1}^2 - 4x_{t-1}x_t x_{t+1} + x_{t-1}^2 + 2x_{t-1}x_t x_{t+1} - 1.$$

Thus we find the expression $x_{t-1}^2 + x_t^2 + x_{t+1}^2 - 2x_{t-1}x_t x_{t+1} - 1$. Shifting $t \to t+1$ we obtain $x_t^2 + x_{t+1}^2 + x_{t+2}^2 - 2x_t x_{t+1} x_{t+2} - 1$ which is the invariant (2).

Problem 19. Find the exact solution of the initial value problem for the system of difference equations

$$x_{1t+1} = (2x_{1t} - 2x_{2t} - 1)(2x_{1t} + 2x_{2t} - 1), \quad x_{2t+1} = 4x_{2t}(2x_{1t} - 1)$$

where $t = 0, 1, 2, \ldots$.

Solution 19. Using the addition theorems for sine, cosine, cosh and sinh we obtain

$$x_{1t} = \cos^2(u_t) \cosh^2(v_t) - \sin^2(u_t) \sinh^2(v_t)$$
$$x_{2t} = -2\sin(u_t)\cos(u_t)\sinh(v_t)\cosh(v_t)$$

where $u_t = 2^t u_0$ and $v_t = 2^t v_0$. The values u_0 and v_0 are determined by

$$x_{10} = \cos^2(u_0)\cosh^2(v_0) - \sin^2(u_0)\sinh^2(v_0)$$
$$x_{20} = -2\sin(u_0)\cos(u_0)\sinh(v_0)\cosh(v_0).$$

Given x_{10} and x_{20} this is a system of transcendental equations.

Problem 20.   Let $k = 0, 1, \ldots$. Consider the integral

$$I_k(\alpha) = \int_0^\pi \frac{\cos(k\theta) - \cos(k\alpha)}{\cos(\theta) - \cos(\alpha)} d\theta.$$

Hence $I_0(\alpha) = 0$, $I_1(\alpha) = \pi$. Find a second order difference equation of the form

$$c_1(\alpha)I_{k+1} + c_2(\alpha)I_k + c_3(\alpha)I_{k-1}, \quad k = 1, 2, \ldots$$

and solve it with the initial conditions $I_0(\alpha) = 0$, $I_1(\alpha) = \pi$.

Solution 20.   For the sum $c_1 I_{k+1} + c_2 I_k + c_3 I_{k-1}$ we find the integral

$$c_1 I_{k+1} + c_2 I_k + c_3 I_{k-1} =$$

$$\int_0^\pi \frac{((c_1 + c_3)\cos(\theta) + c_2)\cos(k\theta) + ((c_3 - c_1)\sin(\theta))\sin(k\theta)}{\cos(\theta) - \cos(\alpha)} d\theta.$$

If we choose c_1, c_2, c_3 as $c_1 = c_3$, $c_2 = -2c_1\cos(\alpha)$ the integral is given by

$$2c_1 \int_0^\pi \cos(k\theta)d\theta = 0 \quad \text{for} \quad k = 1, 2, \ldots$$

Hence the difference equation is

$$I_{k+1} - 2\cos(\alpha)I_k + I_{k-1} = 0, \quad k = 1, 2, \ldots .$$

This is a linear second order difference equation and together with the initial conditions $I_0(\alpha) = 0$, $I_1(\alpha) = \pi$ the solution is

$$I_k(\alpha) = \frac{\pi \sin(k\alpha)}{\sin(\alpha)}.$$

Problem 21.   (i) Find the solution of the linear difference equation with constant coefficients

$$x_{t+2} - x_{t+1} - x_t = 0 \tag{1}$$

where $t = 0, 1, 2 \ldots$ and $x_0 = 0$, $x_1 = 1$. The numbers x_t are called *Fibonacci numbers*.

(ii) Find

$$g = \lim_{t \to \infty} \frac{x_{t+1}}{x_t}. \tag{2}$$

(iii) Solve the linear recurrence relation

$$x_{n+1} = 1 + \sum_{j=0}^{n-1} x_j, \qquad x_0 = 1.$$

We have $x_1 = 1$, since the empty sum is zero.

Solution 21. (i) Since (1) is a linear difference equation with constant coefficients we can solve (1) with the ansatz $x_t = ar^t$, where a is a constant. Inserting the ansatz into (1) yields the quadratic equation $r^2 - r - 1 = 0$. The solution to this quadratic equation is given by

$$r_{1,2} = \frac{1}{2} \pm \sqrt{\frac{5}{4}} = \frac{1}{2}\left(1 \pm \sqrt{5}\right).$$

Consequently, the solution to (1) is given by $x_t = a_1 r_1^t + a_2 r_2^t$, where a_1 and a_2 are the two constants of "integration". Imposing the initial condition $x_0 = 0$, $x_1 = 1$ yields

$$x_t = \frac{1}{\sqrt{5}}\left[\left(\frac{1}{2}(1 + \sqrt{5})\right)^t - \left(\frac{1}{2}(1 - \sqrt{5})\right)^t\right].$$

(ii) From (2) we find

$$g = \lim_{t \to \infty} \frac{x_{t+1}}{x_t} = \lim_{t \to \infty} \frac{x_{t+2}}{x_{t+1}} = \lim_{t \to \infty} \frac{x_{t+1} + x_t}{x_{t+1}} = 1 + \lim_{t \to \infty} \frac{x_t}{x_{t+1}}$$

$$= 1 + \lim_{t \to \infty} \frac{1}{x_{t+1}/x_t}.$$

Thus g is the solution of the equation $g = 1 + 1/g$ with $g > 1$.
(iii) For $n \geq 1$, we have

$$x_{n+1} = 1 + \sum_{j=0}^{n-1} x_j = \left(1 + \sum_{j=0}^{n-2} x_j\right) + x_{n-1} = x_n + x_{n-1}$$

and so by induction x_n is the Fibonacci number for n.

Problem 22. A coin is tossed n times. What is the probability that two heads will turn up in succession somewhere in the sequence of throws?

Solution 22. Let P_n denote the probability that two consecutive heads do not appear in n throws. Obviously $P_1 = 1$, $P_2 = 3/4$. If $n > 2$, there are two cases. If the first throw is tails, then two consecutive heads will not appear in the remaining $n-1$ tosses with probability P_{n-1} (by our choice of notation). If the first throw is heads, the second toss must be tails to avoid two consecutive heads, and then two consecutive heads will not appear in the remaining $n-2$ throws with probability P_{n-2}. Thus, we obtain the linear difference equation

$$P_n = \frac{1}{2}P_{n-1} + \frac{1}{4}P_{n-2}, \qquad n > 2.$$

This difference equation can be transformed to a more familiar form by multiplying each side by 2^n

$$2^n P_n = 2^{n-1}P_{n-1} + 2^{n-2}P_{n-2}.$$

Defining $S_n := 2^n P_n$ we obtain $S_n = S_{n-1} + S_{n-2}$. This is the linear difference equation for the *Fibonacci sequence*. Note that $S_n = F_{n+2}$. Thus, the probability is given by $Q_n = 1 - P_n = 1 - F_{n+2}/2^n$, $n \geq 1$.

Problem 23.   (i) Let A be an $n \times n$ matrix with constant entries. Then

$$\mathbf{x}_{t+1} = A\mathbf{x}_t \tag{1}$$

is a system of n linear difference equations with constant coefficients, where $t = 0, 1, 2, \ldots$. Show that the solution of the initial-value problem is given by

$$\mathbf{x}_t = A^t\mathbf{x}_0 \tag{2}$$

where $\mathbf{x}_0$ is the initial column vector in $\mathbb{R}^n$.
(ii) Consider the 4×4 matrix

$$A = \begin{pmatrix} 1 & 0 & 0 & 0 \\ 1 & 1 & 0 & 0 \\ 0 & 1 & 1 & 0 \\ -1 & -1 & 0 & 1 \end{pmatrix} \equiv I_4 + \begin{pmatrix} 0 & 0 & 0 & 0 \\ 1 & 0 & 0 & 0 \\ 0 & 1 & 0 & 0 \\ -1 & -1 & 0 & 0 \end{pmatrix}.$$

Find the solution of the difference equation (1).

Solution 23.   (i) From (2) it follows that

$$\mathbf{x}_{t+1} = A^{t+1}\mathbf{x}_0 = A^t A\mathbf{x}_0 = AA^t\mathbf{x}_0 = A\mathbf{x}_t.$$

(ii) We set $A = I_4 + B$, where I_4 is the 4×4 unit matrix. Therefore

$$B = \begin{pmatrix} 0 & 0 & 0 & 0 \\ 1 & 0 & 0 & 0 \\ 0 & 1 & 0 & 0 \\ -1 & -1 & 0 & 0 \end{pmatrix}.$$

With $[I_4, B] = 0_4$ we obtain

$$A^t = I_4 + tB + \frac{t(t-1)}{2} B^2 + \cdots + B^t.$$

Since

$$B^2 = \begin{pmatrix} 0 & 0 & 0 & 0 \\ 0 & 0 & 0 & 0 \\ 1 & 0 & 0 & 0 \\ -1 & 0 & 0 & 0 \end{pmatrix}$$

and B^3 is the 4×4 zero matrix we obtain

$$A^t = I_4 + tB + \frac{t(t-1)}{2} B^2.$$

Therefore the solution of the initial value problem is given by

$$\mathbf{x}_t = \left(I_4 + tB + \frac{t(t-1)}{2} B^2 \right) \mathbf{x}_0.$$

Problem 24. (i) Show that the two-dimensional map

$$\mathbf{f} : \quad x_{k+1} = \lambda x_k + \cos(\theta_k), \quad \theta_{k+1} = 2\theta_k \bmod (2\pi), \quad k = 0, 1, \ldots \quad (1)$$

when $2 > \lambda > 1$, has two attractors, at $x = \pm\infty$. Show that the map has no finite attractor.

(ii) Find the *boundary of basins of attractions*.

Solution 24. (i) There are no finite attractors because the eigenvalues of the *Jacobian matrix* of the two-dimensional map (1) are 2 and λ, where $\lambda > 1$. Thus

$$\mathbf{f}^{(n)}(x_0, \theta_0) = (x_n, \theta_n \bmod (2\pi)) \quad (2)$$

and x_n either tends to $+\infty$ or $-\infty$ as $n \to \infty$, except for the (unstable) boundary set $x = g(\theta)$, for which x_n remains finite.

(ii) To find this boundary set, we note that $\theta_k = 2^k \theta_0 \bmod (2\pi)$. The map $\mathbf{f}$ is noninvertible, but we can select any x_n and find one orbit that ends at (x_n, θ_n), by using the above θ_k and taking

$$x_{k-1} = \lambda^{-1} x_k - \lambda^{-1} \cos(2^{k-1}\theta_0).$$

For the given (x_n, θ_0) we find that this orbit started at

$$x_0 = \lambda^{-n} x_n - \sum_{l=0}^{n-1} \lambda^{-l-1} \cos(2^l \theta_0).$$

The boundary between the two basins are those (x_0, θ_0) such that x_n is finite as $n \to \infty$. Thus the x and θ are related by

$$x = -\sum_{l=0}^{\infty} \lambda^{-l-1} \cos(2^l \theta) \equiv g(\theta).$$

Since $\lambda > 1$, this sum converges absolutely and uniformly. On the other hand

$$\frac{dg(\theta)}{d\theta} = \frac{1}{2} \sum_{l=0}^{\infty} (2/\lambda)^{l+1} \sin(2^l \theta)$$

and the sum diverges, since $\lambda < 2$. Hence $g(\theta)$ is nondifferentiable. The curve has a fractal dimension $d_c = 2 - (\ln(\lambda))(\ln(2))^{-1}$.

Problem 25.   The *Anosov map* is defined as follows: $\Omega = [0, 1)^2$,

$$\phi(x, y) = (x + y, x + 2y) \Leftrightarrow \begin{pmatrix} x \\ y \end{pmatrix} \mapsto \begin{pmatrix} 1 & 1 \\ 1 & 2 \end{pmatrix} \begin{pmatrix} x \\ y \end{pmatrix} \quad \text{mod} \, 1.$$

(i) Show that the map preserves Lebesgue measure.
(ii) Show that ϕ is invertible. Show that the entire sequence can be recovered from one term.
(iii) Show that ϕ is *mixing* utilizing the Hilbert space $L_2([0, 1] \times [0, 1])$.

Solution 25.   (i) Since the *Jacobian determinant* of the matrix of the map is equal to one, i.e.

$$\det \begin{pmatrix} 1 & 1 \\ 1 & 2 \end{pmatrix} = 1$$

the map ϕ preserves the Lebesgue measure.
(ii) The inverse of the matrix which belongs to the Lie group $SL(2, \mathbb{R})$

$$\begin{pmatrix} 1 & 1 \\ 1 & 2 \end{pmatrix} \quad \text{is given by} \quad \begin{pmatrix} 2 & -1 \\ -1 & 1 \end{pmatrix}.$$

Since ϕ is invertible, the entire sequence can be recovered from one term. Thus ϕ maps Ω $1 - 1$ onto itself.
(iii) We observe that the n-th iterate is given by

$$\phi^{(n)}(x, y) = (a_{2n-2}x + a_{2n-1}y, a_{2n-1}x + a_{2n}y)$$

where the a_n are the *Fibonacci numbers* given by $a_0 = a_1 = 1$ and $a_{n+1} = a_n + a_{n-1}$ for $n \geq 1$. To check this notice that

$$a_{2n-2} + 2a_{2n-1} + a_{2n} = a_{2n} + a_{2n+1}.$$

Consider the Hilbert space $L_2([0,1) \times [0,1))$. Let f and g be two elements of this Hilbert space defined by

$$f(x,y) := \exp(2\pi i(px + qy)) \equiv \exp(2\pi ipx)\exp(2\pi iqy)$$
$$g(x,y) := \exp(2\pi i(rx + sy)) \equiv \exp(2\pi irx)\exp(2\pi isy)$$

where $p, q, r, s \in \mathbb{Z}$. Since $(k \in \mathbb{Z})$

$$\int_0^1 \exp(2\pi ikx)dx = 0$$

unless $k = 0$. It follows that

$$\int_0^1 \int_0^1 f(x,y)g(\phi^{(n)}(x,y))dxdy = 0$$

unless $ra_{2n-2} + sa_{2n-1} + p = 0$, $ra_{2n-1} + sa_{2n} + q = 0$. Now the difference equation $b_{n+1} - b_n - b_{n-1} = 0$ has a two-parameter family of solutions given by

$$b_n = C_1 \left(\frac{1+\sqrt{5}}{2}\right)^n + C_2 \left(\frac{1-\sqrt{5}}{2}\right)^n$$

so the *Fibonacci numbers* are given by

$$a_n = \frac{1}{2}\left(\frac{1+\sqrt{5}}{2}\right)^n + \frac{1}{2}\left(\frac{1-\sqrt{5}}{2}\right)^n \qquad \text{for } n \geq 0.$$

From the last formula we see that

$$\lim_{n\to\infty} (a_{2n}/a_{2n-1}) = \lim_{n\to\infty} (a_{2n-1}/a_{2n-2}) = \frac{1+\sqrt{5}}{2}.$$

The two equations for a_n cannot hold for infinitely many n unless $p = q = r = s = 0$. The last result implies that if

$$f(x,y) = \sum_{j=1}^k a_j \exp(2\pi i(p_jx + q_jy)), \quad g(x,y) = \sum_{j=1}^l b_j \exp(2\pi i(r_jx + s_jy)).$$

Then as $n \to \infty$,

$$\int_0^1 \int_0^1 f(x,y)g(\phi^{(n)}(x,y))dxdy \to 0.$$

Since the f and g for which the last equation holds are dense in

$$L_0^2([0,1]^2) := \left\{ f : \int_0^1 \int_0^1 f(x,y)dxdy = 0, \quad \int_0^1 \int_0^1 f^2(x,y)dxdy < \infty \right\}$$

it follows that the map ϕ is mixing.

Problem 26. Let $\mathbb{C}$ be the complex plane. Let $c \in \mathbb{C}$. The *Mandelbrot set M* is defined as follows

$$M := \{ c \in \mathbb{C} : c, c^2 + c, (c^2 + c)^2 + c, \ldots \not\to \infty \}. \tag{1}$$

(i) Show that to find the Mandelbrot set one has to study the recursion relation

$$z_{t+1} = z_t^2 + c \tag{2}$$

where $t = 0, 1, 2, \ldots$ and $z_0 = 0$.
(ii) Write the recursion relation in terms of real and imaginary parts. For a given $c \in \mathbb{C}$ (or $(c_1, c_2) \in \mathbb{R}^2$) we can now study whether or not c belongs to M.
(iii) Show that $(c_1, c_2) = (0, 0)$ belongs to M.

Solution 26. (i) This is obvious since

$$z_1 = c, \quad z_2 = c^2 + c, \quad z_3 = (c^2 + c)^2 + c$$

etc.
(ii) Using $z = x + iy$ and $c = c_1 + ic_2$ with $x, y, c_1, c_2 \in \mathbb{R}$ we can write (2) as

$$x_{t+1} = x_t^2 - y_t^2 + c_1, \qquad y_{t+1} = 2x_t y_t + c_2$$

(iii) With the initial value $(x_0, y_0) = (0, 0)$ and $(c_1, c_2) = (0, 0)$ we obtain $(x_1, y_1) = (0, 0)$. Therefore $z_t = 0$ for all t. The Mandelbrot set lies within $|c| < 2$. For $|c| > 2$ the sequence diverges.

Problem 27. Let $a \in \mathbb{R}$. Consider the transformation

$$\tilde{t}(t, x) = \frac{1}{a} e^{ax} \sinh(at), \quad \tilde{x}(t, x) = \frac{1}{a} (e^{ax} \cosh(at) - 1)$$

with $\lim_{a \to 0} \tilde{t} = t$, $\lim_{a \to 0} \tilde{x} = x$. Find the inverse of the transformation.

Solution 27. The inverse transformation is given by

$$t(\tilde{t}, \tilde{x}) = \frac{1}{a} \operatorname{arctanh} \left(\frac{a\tilde{t}}{a\tilde{x} + 1} \right), \quad x(\tilde{t}, \tilde{x}) = \frac{1}{2a} \ln \left((a\tilde{x} + 1)^2 - (a\tilde{t})^2 \right).$$

Problem 28. (i) Consider the potential

$$U(q) = \frac{1}{4} q^4 + \frac{\mu_0}{3} q^3 + \frac{\mu_1}{2} q^2 + \mu_2 q \tag{1}$$

where μ_0, μ_1 and μ_2 are real parameters. Show that the term $\mu_0 q^3/3$ can be eliminated with the help of the transformation

$$\tilde{q}(q) = q + \delta. \qquad (2)$$

(ii) Let

$$U(q, \mu_1, \mu_2) = \frac{1}{4}q^4 + \frac{\mu_1}{2}q^2 + \mu_2 q.$$

Find the parameter region where the potential $U(q) = q^4/4$ changes qualitatively with respect to μ_1 and μ_2.

(iii) Show that the *van der Waals equation*

$$(V - b)\left(P + \frac{a}{V^2}\right) = RT$$

can be expressed as $q^3 + \mu_1 q + \mu_2 = 0$, where a, b and R are positive constants. Here P, V and T are the pressure, volume and temperature, respectively.

Solution 28. (i) From (2) we obtain

$$q^3 = \tilde{q}^3 - 3\delta\tilde{q}^2 + \cdots, \qquad q^4 = \tilde{q}^4 - 4\delta\tilde{q}^3 + \cdots .$$

Therefore the condition to eliminate the cubic term in U is $\mu_0 = 3\delta$.

(ii) The condition

$$\frac{\partial U}{\partial q} = 0$$

gives the *cubic equation* $q^3 + \mu_1 q + \mu_2 = 0$. This cubic equation defines a surface in the (q, μ_1, μ_2) space. The solution of the cubic equation is given by

$$q_1 = y_1 + y_2,$$

$$q_2 = -\frac{y_1 + y_2}{2} + \frac{y_1 - y_2}{2} i\sqrt{3}, \qquad q_3 = -\frac{y_1 + y_2}{2} - \frac{y_1 - y_2}{2} i\sqrt{3}$$

where

$$y_1 := \sqrt[3]{-\frac{\mu_2}{2} + \sqrt{D}}, \qquad y_2 := \sqrt[3]{-\frac{\mu_2}{2} - \sqrt{D}}.$$

Here

$$D := \left(\frac{\mu_2}{2}\right)^2 + \left(\frac{\mu_1}{3}\right)^3$$

is the *discriminant*. Since μ_1 and μ_2 are real we find (a) one root is real and the two others are complex conjugate if $D > 0$, (b) all roots are real and at least two are equal if $D = 0$, (c) all roots are real and unequal if $D < 0$. We now have to investigate whether the real solutions lead to maxima, minima or points of inflexion for the potential (3). If $D > 0$, the potential (3) has one minimum at $q = 0$. There is no qualitative change of the potential $U(q) = q^4/4$. If $D < 0$, the potential (3) has two minima symmetric to $q = 0$. The point $q = 0$ is a local maximum of the potential U. In this case we have a qualitative change of the potential (3). Consequently, $D = 0$ is a bifurcation line in the μ_1-μ_2 plane. This problem is the so-called *cusp catastrophe* in catastrophe theory.

(iii) The van der Waals equation can be written near the critical point as $(V - V_c)^3 = 0$ or $V^3 - 3V_cV^2 + 3V_c^2V - V_c^3 = 0$, where V_c is the critical pressure. This equation is compared with the van der Waals equation with $T = T_c$ and $P = P_c$, i.e.

$$(V - b)\left(P_c + \frac{a}{V^2}\right) = RT_c.$$

From this equation it follows that

$$V^3 - \left(b + \frac{RT_c}{P_c}\right)V^2 + \frac{a}{P_c}V - \frac{ab}{P_c} = 0.$$

Comparing the coefficients we obtain

$$3V_c = b + \frac{RT_c}{P_c}, \qquad 3V_c^2 = \frac{a}{P_c}, \qquad V_c^3 = \frac{ab}{P_c}.$$

The solution of this system of equations is given by

$$RT_c = \frac{8a}{27b}, \qquad P_c = \frac{a}{27b^2}, \qquad V_c = 3b.$$

We now introduce the normalized quantities

$$\bar{P} := \frac{P}{P_c}, \qquad \bar{T} := \frac{T}{T_c}, \qquad \bar{V} := \frac{V}{V_c}.$$

We find

$$\left(\bar{P} + \frac{3}{\bar{V}^2}\right)\left(\bar{V} - \frac{1}{3}\right) = \frac{8}{3}\bar{T}.$$

Introducing the density $\bar{X} := \frac{1}{\bar{V}}$ leads to

$$(\bar{P} + 3\bar{X}^2)\left(\frac{1}{\bar{X}} - \frac{1}{3}\right) = \frac{8}{3}\bar{T}.$$

With the transformation $p := \bar{P} - 1$, $x := \bar{X} - 1$, $t := \bar{T} - 1$ we obtain $x^3 + \frac{1}{3}(8t + p)x + \frac{1}{3}(8t - 2p) = 0$. Therefore $x^3 + \mu_1 x + \mu_2 = 0$, where

$$\mu_1 := \frac{1}{3}(8t + p), \qquad \mu_2 := \frac{1}{3}(8t - 2p).$$

Problem 29.   For a sphere of radius r and mass density ρ the mass that must be concentrated at its centre is ($\lambda \geq 0$)

$$M(\lambda) = \frac{4\pi r\rho}{\lambda^2}(\cosh(\lambda r) - \sinh(\lambda r)/(\lambda r)).$$

Find $\lim_{\lambda \to 0} M(\lambda)$.

Solution 29.   Since

$$\cosh(\lambda r) = 1 + \frac{\lambda^2 r^2}{2!} + \frac{\lambda^4 r^4}{4!} + \cdots, \qquad \sinh(\lambda r) = \frac{\lambda r}{1!} + \frac{\lambda^3 r^3}{3!} + \cdots$$

we have

$$\cosh(\lambda r) - \frac{\sinh(\lambda r)}{\lambda r} = \frac{1}{3}\lambda^2 r^2 + \cdots.$$

Thus $M(\lambda = 0) = \frac{4}{3}\pi r^2 \rho$.

Problem 30.   Find the maximum area of all triangles that can be inscribed in an *ellipse* given by $x^2/a^2 + y^2/b^2 = 1$ with semiaxes a and b. Describe the triangles that have maximum area.

Solution 30.   We represent the *ellipse* using the parametric representation

$$x(\tau) = a\cos(\tau), \qquad y(\tau) = b\sin(\tau)$$

where $\tau \in [0, 2\pi)$. A triple of points on the ellipse is given by

$$(a\cos(\tau_j), b\sin(\tau_j)), \quad j = 1, 2, 3.$$

Thus the area A of an inscribed triangle is given by

$$A = \frac{1}{2}\det\begin{pmatrix} 1 & a\cos(\tau_1) & b\sin(\tau_1) \\ 1 & a\cos(\tau_2) & b\sin(\tau_2) \\ 1 & a\cos(\tau_3) & b\sin(\tau_3) \end{pmatrix} = \frac{ab}{2}\det\begin{pmatrix} 1 & \cos(\tau_1) & \sin(\tau_1) \\ 1 & \cos(\tau_2) & \sin(\tau_2) \\ 1 & \cos(\tau_3) & \sin(\tau_3) \end{pmatrix}.$$

This is ab times the area of a triangle inscribed in the unit circle. Hence, the area is maximal if $\tau_2 = \tau_1 + 2\pi/3$ and $\tau_3 = \tau_2 + 2\pi/3$. The area A is given by

$$A = \frac{3\sqrt{3}}{4}ab$$

where we used the fact that $\sin(2\pi/3) = \sqrt{3}/2$ and $\cos(2\pi/3) = -1/2$. Thus the area is maximal when the corresponding triangle inscribed in the unit circle is regular.

Problem 31.   Show that the linear one-dimensional diffusion equation

$$\frac{\partial u}{\partial t} = D \frac{\partial^2 u}{\partial x^2}$$

is invariant under the *Galilean transformation*

$$t'(x,t) = t, \quad x'(x,t) = x + vt, \quad u'(x'(x,t), t'(x,t)) = u(x,t)E(x,t)$$

where we set

$$E(x,t) := \exp\left(-\frac{1}{2D}\left(\frac{1}{2}v^2 t + vx\right)\right).$$

Here v is a constant velocity and D the diffusion constant with dimension $meter^2 . sec^{-1}$. Note that

$$\frac{\partial E}{\partial x} = -\frac{v}{2D}E, \quad \frac{\partial^2 E}{\partial x^2} = \frac{v^2}{4D^2}E, \quad \frac{\partial E}{\partial t} = -\frac{v^2}{4D}E$$

and

$$\left(D\frac{\partial^2 E}{\partial x^2} + v\frac{\partial E}{\partial x} - \frac{\partial E}{\partial t}\right)u = 0, \qquad 2D\frac{\partial E}{\partial x} + vE = 0.$$

Solution 31.   First we note that

$$\frac{\partial x'}{\partial x} = 1, \quad \frac{\partial x'}{\partial t} = v, \quad \frac{\partial t'}{\partial x} = 0, \quad \frac{\partial t'}{\partial t} = 1.$$

Applying the *chain rule* and *product rule* we have

$$\frac{\partial u'}{\partial t} = \frac{\partial u'}{\partial x'}\frac{\partial x'}{\partial t} + \frac{\partial u'}{\partial t'}\frac{\partial t'}{\partial t} = \frac{\partial E}{\partial t}u + E\frac{\partial u}{\partial t} \;\Rightarrow\; v\frac{\partial u'}{\partial x'} + \frac{\partial u'}{\partial t'} = \frac{\partial E}{\partial t}u + E\frac{\partial u}{\partial t}.$$

Furthermore

$$\frac{\partial u'}{\partial x} = \frac{\partial u'}{\partial x'}\frac{\partial x'}{\partial x} + \frac{\partial u'}{\partial t'}\frac{\partial t'}{\partial x} = \frac{\partial E}{\partial x}u + E\frac{\partial u}{\partial x} \;\Rightarrow\; \frac{\partial u'}{\partial x'} = \frac{\partial E}{\partial x}u + E\frac{\partial u}{\partial x}.$$

Taking the derivative with respect to x of the last equation we find

$$\frac{\partial^2 u'}{\partial x'^2} = \frac{\partial^2 E}{\partial x^2}u + 2\frac{\partial E}{\partial x}\frac{\partial u}{\partial x} + E\frac{\partial^2 u}{\partial x^2}.$$

It follows that

$$v\frac{\partial u'}{\partial x'} + \frac{\partial u'}{\partial t'} = \frac{\partial E}{\partial t}u + E\frac{\partial u}{\partial t} \;\Rightarrow\; \frac{\partial u'}{\partial t'} = D\frac{\partial^2 u'}{\partial x'^2}.$$

Problem 32.   (i) Consider the *Lorentz transformation*

$$x' = \frac{1}{\sqrt{1-\beta^2}}(x - vt), \qquad t' = \frac{1}{\sqrt{1-\beta^2}}\left(t - \frac{v}{c^2}x\right)$$

where $\beta = v/c$. Assume that $k'_x x' - \omega' t' = k_x x - \omega t$, where ω is the frequency and $\mathbf{k} = (k_x, k_y, k_z)$ is the wave vector. Find the transformation between k'_x, ω' and k_x, ω.

(ii) Consider the special case $\omega = ck_x$ for the transformation.

(iii) Simplify the transformation for $v/c \ll 1$.

Solution 32. (i) Inserting the Lorentz transformation into $k'_x x' - \omega' t'$ yields

$$k'_x x' - \omega' t' = k'_x \frac{1}{\sqrt{1-\beta^2}}(x - vt) - \omega' \frac{1}{\sqrt{1-\beta^2}}\left(t - \frac{v}{c^2}x\right)$$

$$= \left(\frac{k'_x}{\sqrt{1-\beta^2}} + \frac{\omega' v}{\sqrt{1-\beta^2 c^2}}\right)x - \left(\frac{vk'_x}{\sqrt{1-\beta^2}} + \frac{\omega'}{\sqrt{1-\beta^2}}\right)t$$

$$= k_x x - \omega t.$$

Comparing with respect to x and t provides the transformation

$$k_x = \frac{1}{\sqrt{1-\beta^2}}\left(k'_x + \frac{v}{c^2}\omega'\right) \qquad \omega = \frac{1}{\sqrt{1-\beta^2}}(vk'_x + \omega').$$

The inverse transformation in matrix notation is

$$\begin{pmatrix} k'_x \\ \omega' \end{pmatrix} = \frac{1}{\sqrt{1-\beta^2}}\begin{pmatrix} 1 & -v/c^2 \\ -v & 1 \end{pmatrix}\begin{pmatrix} k_x \\ \omega \end{pmatrix}.$$

(ii) With the assumption $\omega = ck_x$ the transformation reduces to

$$\omega = \frac{\omega'}{\sqrt{1-\beta^2}}\left(1 + \frac{v}{c}\right).$$

(iii) Since $v \ll c$ and $\beta^2 = v^2/c^2$ we have $\sqrt{1-\beta^2} \approx 1$. Therefore

$$\omega = \omega'\left(1 + \frac{v}{c}\right).$$

Problem 33. *Rindler coordinates* are coordinates appropriate for an observer undergoing constant proper acceleration a. For an unprimed inertial frame set of coordinates one assigns the acceleration frame observer primed coordinates (the Rindler coordinates)

$$ct = x_0 = \left(\frac{c^2}{a} + x'_1\right)\sinh\left(\frac{at'}{c}\right) = \left(\frac{c^2}{a} + x'_1\right)\sinh\left(\frac{ax'_0}{c^2}\right)$$

$$x_1 = \left(\frac{c^2}{a} + x'_1\right)\cosh\left(\frac{at'}{c}\right) - \frac{c^2}{a} \equiv \left(\frac{c^2}{a} + x'_1\right)\cosh\left(\frac{ax'_0}{c^2}\right) - \frac{c^2}{a}$$

with $x_0' = ct'$ and a has the dimension of an acceleration $(meter \cdot sec^{-2})$. Express the metric tensor field (Minkowski) $g = dx_0 \otimes dx_0 - dx_1 \otimes dx_1$ in Rindler coordinates.

Solution 33.   We obtain

$$\tilde{g} = \left(1 + \frac{ax_1'}{c^2}\right) dx_0' \otimes dx_0' - dx_1' \otimes dx_1'.$$

Problem 34.   (i) Show that $g = dx_1 \otimes dx_1 + dx_2 \otimes dx_2$ is invariant under

$$\begin{pmatrix} x_1' \\ x_2' \end{pmatrix} = \begin{pmatrix} \cos(\alpha) & -\sin(\alpha) \\ \sin(\alpha) & \cos(\alpha) \end{pmatrix} \begin{pmatrix} x_1 \\ x_2 \end{pmatrix}.$$

(ii) Show that $g = dx_0 \otimes dx_0 - dx_1 \otimes dx_1$ is invariant under

$$\begin{pmatrix} x_0' \\ x_1' \end{pmatrix} = \begin{pmatrix} \cosh(\alpha) & \sinh(\alpha) \\ \sinh(\alpha) & \cosh(\alpha) \end{pmatrix} \begin{pmatrix} x_0 \\ x_1 \end{pmatrix}.$$

Solution 34.   (i) With

$$dx_1' = \cos(\alpha)dx_1 - \sin(\alpha)dx_2, \quad dx_2' = \sin(\alpha)dx_1 + \cos(\alpha)dx_2$$

and $\sin^2(\alpha) + \cos^2(\alpha) = 1$ we obtain $g = dx_1' \otimes dx_1' + dx_2' \otimes dx_2'$.
(ii) With

$$dx_1' = \cosh(\alpha)dx_1 + \sinh(\alpha)dx_2, \quad dx_2' = \sinh(\alpha)dx_1 + \cosh(\alpha)dx_2$$

and $\cosh^2(\alpha) - \sinh^2(\alpha) = 1$ we obtain $g = dx_0' \otimes dx_0' - dx_1' \otimes dx_1'$.

Problem 35.   Let

$$\mathbb{S}^2 := \{\, (x_1, x_2, x_3) \in \mathbb{R}^3 \;:\; x_1^2 + x_2^2 + x_3^2 = 1 \,\}.$$

An element $\eta \in \mathbb{S}^2$ can be written as

$$\eta = (\cos(\phi)\sin(\theta), \sin(\phi)\sin(\theta), \cos(\theta))$$

where $\phi \in [0, 2\pi)$ and $\theta \in [0, \pi]$. The *stereographic projection* is a map

$$\Pi : \mathbb{S}^2 \setminus \{\, (0, 0, -1) \,\} \to \mathbb{R}^2$$

given by

$$x_1(\theta, \phi) = \frac{2\sin(\theta)\cos(\phi)}{1 + \cos(\theta)}, \qquad x_2(\theta, \phi) = \frac{2\sin(\theta)\sin(\phi)}{1 + \cos(\theta)}.$$

(i) Let $\theta = 0$ and ϕ be arbitrary. Find x_1, x_2. Give a geometric interpretation.

(ii) Find the inverse of the map, i.e. find $\Pi^{-1} : \mathbb{R}^2 \to \mathbb{S}^2 \setminus \{ (0, 0, -1) \}$.

Solution 35. (i) Since $\sin(0) = 0$ we find $x_1 = x_2 = 0$, i.e. the point $(0, 0, 1)$ is mapped to the origin $(0, 0)$.

(ii) Using division we find $\phi(x_1, x_2) = \arctan(x_2/x_1)$. Since

$$x_1^2 + x_2^2 = \frac{4 \sin^2(\theta)}{(1 + \cos(\theta))^2}, \qquad \tan(\theta/2) = \frac{\sin(\theta)}{1 + \cos(\theta)}$$

we obtain $\theta(x_1, x_2) = 2 \arctan(\sqrt{x_1^2 + x_2^2}/2)$.

Problem 36. Let m_A, m_B, $\mathbf{R}_A$, $\mathbf{R}_B$ be the masses and centre-of mass coordinates of mass A and B, respectively. We set $m = m_A + m_B$. Find the inverse of the transformation

$$\mathbf{r}(\mathbf{R}_A, \mathbf{R}_B) = \mathbf{R}_A - \mathbf{R}_B, \qquad \mathbf{R}(\mathbf{R}_A, \mathbf{R}_B) = \frac{1}{m}(m_A \mathbf{R}_A + m_B \mathbf{R}_B).$$

Solution 36. Solving the system of linear equations we obtain

$$\mathbf{R}_A = \mathbf{R} + \frac{m_B}{m} \mathbf{r}, \qquad \mathbf{R}_B = \mathbf{R} - \frac{m_B}{m} \mathbf{r}.$$

Problem 37. Let

$$\mathbb{S}^n := \{ (x_1, x_2, \ldots, x_{n+1}) \in \mathbb{R}^{n+1} : x_1^2 + x_2^2 + \cdots + x_{n+1}^2 = 1 \}.$$

(i) Show that $\mathbb{S}^3$ can be considered as a subset of $\mathbb{C}^2$ ($\mathbb{C}^2 \cong \mathbb{R}^4$)

$$\mathbb{S}^3 = \{ (z_1, z_2) \in \mathbb{C}^2 : |z_1|^2 + |z_2|^2 = 1 \}.$$

(ii) The *Hopf map* $\pi : \mathbb{S}^3 \to \mathbb{S}^2$ is defined by

$$\pi(z_1, z_2) := (\bar{z}_1 z_2 + \bar{z}_2 z_1, -i\bar{z}_1 z_2 + i\bar{z}_2 z_1, |z_1|^2 - |z_2|^2).$$

Find the parametrization of $\mathbb{S}^3$, i.e. find $z_1(\theta, \phi)$, $z_2(\theta, \phi)$ and thus show that indeed π maps $\mathbb{S}^3$ onto $\mathbb{S}^2$.

(iii) Show that $\pi(z_1, z_2) = \pi(z_1', z_2')$ if and only if $z_j' = e^{i\alpha} z_j$ ($j = 1, 2$) and $\alpha \in \mathbb{R}$.

Solution 37. (i) Let $z_1 = x_1 + iy_1$ and $z_2 = x_2 + iy_2$, where $x_1, x_2, y_1, y_2 \in \mathbb{R}$. Then from $|z_1|^2 + |z_2|^2 = 1$ it follows that $x_1^2 + y_1^2 + x_2^2 + y_2^2 = 1$.

(ii) Since $|z_1|^2 + |z_2|^2 = 1$ we have the parametrization

$$z_1(\theta, \phi) = \cos(\theta/2)e^{i\phi_1}, \qquad z_2(\theta, \phi) = \sin(\theta/2)e^{i\phi_2}$$

where $0 \leq \theta \leq \pi$ and $\phi_1, \phi_2 \in \mathbb{R}$. Thus

$$\pi(\cos(\theta/2)e^{i\phi_1}, \sin(\theta/2)e^{i\phi_2}) = (\sin(\theta)\cos(\phi_2-\phi_1), \sin(\theta)\sin(\phi_2-\phi_1), \cos(\theta)).$$

(iii) From $\pi(z_1, z_2) = \pi(z_1', z_2')$ we obtain the three equations

$$\overline{z}_1 z_2 + \overline{z}_2 z_1 = \overline{z}_1' z_2' + \overline{z}_2' z_1'$$
$$-\overline{z}_1 z_2 + \overline{z}_2 z_1 = -\overline{z}_1' z_2' + \overline{z}_2' z_1'$$
$$|z_1|^2 - |z_2|^2 = |z_1'|^2 - |z_2'|^2.$$

Inserting $|z_1|^2 + |z_2|^2 = |z_1'|^2 + |z_2'|^2 = 1$ into the last equation yields $|z_1| = |z_1'|$ and $|z_2| = |z_2'|$. Adding the first two equations provides $z_1 \overline{z}_2 = z_1' \overline{z}_2'$. Thus we obtain the solution $z_1' = e^{i\alpha}z_1$ and $z_2' = e^{i\alpha}z_2$, where $\alpha \in \mathbb{R}$.

Problem 38. The n-dimensional complex projective space $\mathbb{CP}^n$ is the set of all complex lines on $\mathbb{C}^{n+1}$ passing through the origin. Let f be the map that takes nonzero vectors in $\mathbb{C}^2$ to vectors in $\mathbb{R}^3$ by

$$f(z_1, z_2) = \left(\frac{z_1 \overline{z}_2 + \overline{z}_1 z_2}{z_1 \overline{z}_1 + \overline{z}_2 z_2}, \frac{z_1 \overline{z}_2 - \overline{z}_1 z_2}{i(z_1 \overline{z}_1 + \overline{z}_2 z_2)}, \frac{z_1 \overline{z}_1 - \overline{z}_2 z_2}{z_1 \overline{z}_1 + \overline{z}_2 z_2} \right).$$

The map f defines a bijection between $\mathbb{CP}^1$ and the unit sphere in $\mathbb{R}^3$. Consider the normalized vectors in $\mathbb{C}^2$

$$\begin{pmatrix} 1 \\ 0 \end{pmatrix}, \qquad \begin{pmatrix} 0 \\ 1 \end{pmatrix}, \qquad \frac{1}{\sqrt{2}}\begin{pmatrix} 1 \\ 1 \end{pmatrix}, \qquad \frac{1}{\sqrt{2}}\begin{pmatrix} 1 \\ -1 \end{pmatrix}, \qquad \frac{1}{\sqrt{2}}\begin{pmatrix} i \\ -i \end{pmatrix}.$$

Apply f to these vectors in $\mathbb{C}^2$.

Solution 38. Since all vectors in $\mathbb{C}^2$ are normalized we have $z_1 \overline{z}_1 + z_2 \overline{z}_2 = 1$. We find

$$\begin{pmatrix} 0 \\ 0 \\ 1 \end{pmatrix}, \qquad \begin{pmatrix} 0 \\ 0 \\ -1 \end{pmatrix}, \qquad \begin{pmatrix} 1 \\ 0 \\ 0 \end{pmatrix}, \qquad \begin{pmatrix} -1 \\ 0 \\ 0 \end{pmatrix}, \qquad \begin{pmatrix} -1 \\ 0 \\ 0 \end{pmatrix}.$$

Problem 39. Suppose that the functions g and h are continuously differentiable and $dh/dx \neq 0$ on an open interval I that contains a (except possibly at a). Suppose that

$$\lim_{x \to a} g(x) = 0 \quad \text{and} \quad \lim_{x \to a} h(x) = 0$$

or that

$$\lim_{x \to a} g(x) = \pm\infty, \qquad \lim_{x \to a} h(x) = \pm\infty.$$

Thus with g/h we have an indeterminate form of type $0/0$ or ∞/∞. Then

$$\lim_{x \to a} \frac{g(x)}{h(x)} = \lim_{x \to a} \frac{dg(x)/dx}{dh(x)/dx}$$

if the limit on the right hand side exists (or is ∞ or $-\infty$). This is called *L'Hospital's rule*.

(i) Consider $f(x) = \sin(x)/x$. Determine $f(0)$ using L'Hospital's rule.

(ii) Let $f : \mathbb{R}^+ \to \mathbb{R}$, $f(x) = x\ln(x)$ for $x > 0$. Determine $f(0)$ using L'Hospital's rule.

(iii) Let

$$f(\theta) = \frac{\cos(\pi \cos(\theta)/2)}{\sin(\theta)} \qquad \text{for} \quad \theta \neq n\pi \tag{1}$$

where $n \in \mathbb{Z}$. Determine $f(n\pi)$ using L'Hospital's rule.

Solution 39. (i) We have $d(\sin(x))/dx = \cos(x)$, $dx/dx = 1$ and $\cos(0) = 1$. Thus $f(0) = 1$.

(ii) Since

$$\lim_{x \to +0} x\ln(x) \equiv \lim_{x \to +0} \frac{\ln(x)}{1/x}$$

we have $g(x) = \ln(x)$ and $h(x) = 1/x$. It follows that

$$\frac{dg}{dx} = \frac{1}{x}, \qquad \frac{dh}{dx} = -\frac{1}{x^2}.$$

Therefore

$$\lim_{x \to +0} x\ln(x) = \lim_{x \to +0} \frac{\frac{1}{x}}{\frac{-1}{x^2}} = \lim_{x \to +0} \frac{-x^2}{x} = 0.$$

(iii) Let $g(\theta) = \cos(\pi \cos(\theta)/2)$, $h(\theta) = \sin(\theta)$. Then

$$\frac{dg}{d\theta} = \sin(\pi \cos(\theta)/2)\frac{\pi}{2}\sin(\theta), \qquad \frac{dh}{d\theta} = \cos(\theta).$$

Since

$$\left.\frac{dg}{d\theta}\right|_{\theta=n\pi} = 0, \qquad \left.\frac{dh}{d\theta}\right|_{\theta=n\pi} \neq 0$$

we obtain $\lim_{\theta \to n\pi} f(\theta) = 0$.

Problem 40. (i) Calculate $\lim_{\beta \to \infty} \tanh(\beta x)$, where $x \in \mathbb{R}$.

(ii) Calculate

$$\lim_{\beta \to \infty} \frac{1}{\beta} \ln(4\cosh(\beta x)\cosh(\beta y))$$

where $x, y \in \mathbb{R}$.

(iii) Find
$$\lim_{\beta \to \infty} \frac{\sinh(\beta x)}{\cosh(\beta x) + \cosh(\beta y)}$$
where $\beta > 0$.

(iv) Calculate
$$\lim_{x \to y} \frac{\sin(x) - \sin(y)}{x - y}.$$

Solution 40. (i) Let $x = 0$, then $\tanh(0) = 0$. If $x > 0$, then
$$\lim_{\beta \to \infty} \tanh(\beta x) = 1.$$

If $x < 0$, then $\lim_{\beta \to \infty} \tanh(\beta x) = -1$. Thus
$$\lim_{\beta \to \infty} \tanh(\beta x) = \begin{cases} 1 & x > 0 \\ 0 & x = 0 \\ -1 & x < 0. \end{cases}$$

One writes $\lim_{\beta \to \infty} \tanh(\beta x) = \operatorname{sgn}(x)$.

(ii) Applying L'Hospital's rule we have
$$\lim_{\beta \to \infty} \frac{1}{\beta} \ln[4 \cosh(\beta x) \cosh(\beta y)] = \lim_{\beta \to \infty} [x \tanh(\beta x) + y \tanh(\beta y)]$$
$$= x \operatorname{sgn}(x) + y \operatorname{sgn}(y) = |x| + |y|.$$

(iii) Applying the addition theorem we find
$$\lim_{\beta \to \infty} \frac{\sinh(\beta x)}{\cosh(\beta x) + \cosh(\beta y)} = \lim_{\beta \to \infty} \frac{\sinh\left[\beta\left(\left(\frac{x}{2} + \frac{y}{2}\right) + \left(\frac{x}{2} - \frac{y}{2}\right)\right)\right]}{\cosh(\beta x) + \cosh(\beta y)}$$
$$= \frac{1}{2}\operatorname{sgn}(x + y) + \frac{1}{2}\operatorname{sgn}(x - y)$$

where we have used the identity
$$\frac{\sinh\left[\beta\left(\left(\frac{x+y}{2}\right) + \left(\frac{x-y}{2}\right)\right)\right]}{\cosh(\beta x) + \cosh(\beta y)} =$$
$$\frac{\sinh\left(\beta\frac{x+y}{2}\right)\cosh\left(\beta\frac{x-y}{2}\right) + \cosh\left(\beta\frac{x+y}{2}\right)\sinh\left(\beta\frac{x-y}{2}\right)}{2\cosh\left(\beta\frac{x+y}{2}\right)\cosh\left(\beta\frac{x-y}{2}\right)}.$$

(iv) We set $x = y + \epsilon$. Then
$$\lim_{\epsilon \to 0} \frac{\sin(y + \epsilon) - \sin(y)}{\epsilon} = \lim_{\epsilon \to 0} \frac{\sin(y)\cos(\epsilon) + \cos(y)\sin(\epsilon) - \sin(y)}{\epsilon}.$$

Applying L'Hospital's rule yields

$$\lim_{\epsilon \to 0} \frac{-\sin(y)\sin(\epsilon) + \cos(y)\cos(\epsilon)}{1} = \cos(y).$$

Problem 41. Let $a > 0$. Consider *elliptical coordinates* η, ξ defined by

$$x_1(\eta, \xi) = a\cos(\xi)\cosh(\eta), \quad x_2(\eta, \xi) = a\sin(\xi)\sinh(\eta)$$

where $0 \le \eta < \infty$ and $0 \le \xi < 2\pi$. Find $x_1^2 + x_2^2$.

Solution 41. We have

$$\begin{aligned}
x_1^2 + x_2^2 &= a^2(\cos^2(\xi)\cosh^2(\eta) + \sin^2(\xi)\sinh^2(\eta)) \\
&= a^2((1 - \sin^2(\xi))(1 + \sinh^2(\eta)) + \sin^2(\xi)\sinh^2(\eta)) \\
&= a^2(1 - \sin^2(\xi) + \sinh^2(\eta)).
\end{aligned}$$

Problem 42. An electrical network has the form shown in the following figure, where R_1 and R_2 are given constant resistances. V is the voltage supplied by a source.

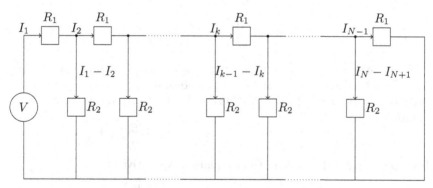

(i) Write down *Kirchhoff's law* for the first loop, the k-th loop and the $(N + 1)$-th loop.
(ii) Give the general solution for the difference equation of the k-th loop.
(iii) Determine the constants in the general solution of the difference equation with the boundary conditions (first and $(N + 1)$-th loop).

Solution 42. (i) For the first loop which includes the applied voltage Kirchhoff's law provides us with

$$I_1 R_1 + (I_1 - I_2)R_2 - V = 0. \tag{1}$$

For the k-th loop we find

$$I_k R_1 + (I_k - I_{k+1})R_2 - (I_{k-1} - I_k)R_2 = 0. \tag{2}$$

Kirchhoff's law for the last loop $((N+1)$-th loop) gives

$$I_{N+1}R_1 - (I_N - I_{N+1})R_2 = 0. \tag{3}$$

Equation (2) can be written as the difference equation

$$I_{k+1} - \left(2 + \frac{R_1}{R_2}\right)I_k + I_{k-1} = 0 \tag{4}$$

where $k = 1, 2, \ldots, N$.

(ii) Equation (4) is a linear difference equation with constant coefficients. It can therefore be solved with the ansatz

$$I_k = ar^k. \tag{5}$$

The first and last loop provide the boundary conditions. Inserting the ansatz (5) into (4) leads to the algebraic equation

$$r^2 - \left(2 + \frac{R_1}{R_2}\right)r + 1 = 0.$$

The solution of this quadratic equation is given by

$$r_{1,2} = 1 + \frac{R_1}{2R_2} \pm \sqrt{\frac{R_1}{R_2} + \frac{R_1^2}{4R_2^2}}.$$

Consequently, the general solution to the difference equation (4) is given by $I_k = ar_1^k + br_2^k$, where a and b are two constants.

(iii) We impose the boundary conditions (1) and (3). From (1) and (3) we obtain

$$I_2 = \frac{I_1(R_1 + R_2) - V}{R_2}, \qquad I_N = \frac{(R_1 + R_2)I_{N+1}}{R_2}.$$

If we set $k = 2$ and $k = N$ in these equations we arrive at

$$I_2 = ar_1^2 + br_2^2, \qquad I_N = ar_1^N + br_2^N.$$

It follows that

$$a = \frac{I_2 r_2^N - I_N r_2^2}{r_1^2 r_2^N - r_2^2 r_1^N}, \qquad b = \frac{I_N r_1^2 - I_2 r_1^N}{r_1^2 r_2^N - r_2^2 r_1^N}.$$

Therefore

$$I_k = \left(\frac{I_2 r_2^N - I_N r_2^2}{r_1^2 r_2^N - r_2^2 r_1^N}\right)r_1^k + \left(\frac{I_N r_1^2 - I_2 r_1^N}{r_1^2 r_2^N - r_2^2 r_1^N}\right)r_2^k.$$

where r_1, r_2 are given above.

Problem 43. The junction between p- and n-type semiconductors has properties which make it the basis of many electronic devices. For a p-n junction *diode* we have the equation

$$d = \left(\frac{2\epsilon_0\epsilon_r(n_A + n_D)(V_D - U)}{en_An_D} \right)^{1/2}$$

where d has the dimension of a length (thickness of the depletion layer) and

$e = 1.6021 \cdot 10^{-19}\, A\cdot s$ elementary charge

$\epsilon_0 = 8.8542 \cdot 10^{-12}\, m^{-3} \cdot sec^4 \cdot kg^{-1} \cdot A^2$ permittivity of the vacuum

$\epsilon_r = 16$ for Germanium

$n_A = n_D = 10^{22}\, m^{-3}$ donor and acceptor concentrations

$V_D = 0.358\, V$ diffusion voltage

$a = 10^{-6}\, m^2$ area of the junction

Write a C++ program that finds d in dependence of the voltage U with $-10V \leq U \leq 0V$ and step size $0.5\, V$ and the capacity

$$C(d) = \epsilon_0\epsilon_r a/d.$$

All the units are in the MKSA system. Thus the result for d will be in meters and the result for the capacity C will be in Farad $(= s^4 A^2/m^2 kg)$.

Solution 43.

```
// diode.cpp
#include <iostream>
#include <cmath>      // for sqrt
using namespace std;

int main(void)
{
  const double e = 1.6021E-19;
  const double eps0 = 8.8542E-12;
  const double epsr = 16;
  const double n = 1.0E22;
  const double vd = 0.358;
  const double a = 1.0E-6;
  double v, d, c;
  cout << "voltage " << "     " << "capacity " << endl;
  v = -10.0;
  while(v <= 0.0)
  {
    d = sqrt(2.0*eps0*epsr*(n+n)*(vd-v)/(e*n*n)); c = eps0*epsr*a/d;
    cout << v << "      " << c << " F" << endl;
```

```
v = v + 0.5;
    }
}
```

Supplementary Problems

Problem 1.   Show that the term

$$T(\mu, \nu) = \sinh^2(\mu)\cos^2(\nu) + \cosh^2(\mu)\sin^2(\nu)$$

can be simplified to

$$T(\mu, \nu) = \sinh^2(\mu) + \sin^2(\nu)$$

utilizing $\cos^2(\nu) \equiv 1 - \sin^2(\nu)$ and $\cosh^2(\mu) = 1 + \sinh^2(\mu)$.

Problem 2.   Let $c > 0$ and $f : \mathbb{R} \to \mathbb{R}$ be a polynomial. Show that

$$\exp\left(c\frac{d^2}{dx^2}\right)f(x) = \frac{1}{\sqrt{4\pi c}}\int_{-\infty}^{+\infty}\exp\left(-\frac{(x-y)^2}{4c}\right)f(y)dy.$$

So if $f(x) = x^2$ we obtain $x^2 + 2c$.

Problem 3.   Let $\omega > 0$. Find

$$\lim_{(\omega t)\to 0}\frac{\sin^2(\omega t)}{(\omega t)^2}.$$

Problem 4.   Let $c > 0$ and $0 \le v_1 < c,\ 0 \le v_2 < c$. We define the composition

$$v_1 \star v_2 := \frac{v_1 + v_2}{1 + v_1 v_2/c^2}.$$

Is the composition associative?

Problem 5.   Let $f : \mathbb{R}^2 \to \mathbb{R}$, $g : \mathbb{R}^2 \to \mathbb{R}$ be analytic functions. We define the *star product*

$$f(x_1, x_2) \star g(x_1, x_2) :=$$

$$\lim_{x_1'\to x_1, x_2'\to x_2}\exp\left(\frac{\partial}{\partial x_1}\frac{\partial}{\partial x_2'} - \frac{\partial}{\partial x_1'}\frac{\partial}{\partial x_2}\right)f(x_1, x_2)g(x_1', x_2').$$

Let $f(x_1, x_2) = \sin(x_1 + x_2)$, $g(x_1, x_2) = \sin(x_1 - x_2)$. Find the star product.

Problem 6. The *Kustaanheimo-Stiefel transformation* is defined by the map from $\mathbb{R}^4$ (coordinates u_1, u_2, u_3, u_4) to $\mathbb{R}^3$ (coordinates x_1, x_2, x_3)

$$x_1(u_1, u_2, u_3, u_4) = 2(u_1 u_3 - u_2 u_4)$$
$$x_2(u_1, u_2, u_3, u_4) = 2(u_1 u_4 + u_2 u_3)$$
$$x_3(u_1, u_2, u_3, u_4) = u_1^2 + u_2^2 - u_3^2 - u_4^2$$

together with the constraint $u_2 du_1 - u_1 du_2 - u_4 du_3 + u_3 du_4 = 0$.
(i) Show that $r^2 = x_1^2 + x_2^2 + x_3^2 = u_1^2 + u_2^2 + u_3^2 + u_4^2$.
(ii) Show that

$$\Delta_3 = \frac{1}{4r}\Delta_4 - \frac{1}{4r^2}V^2$$

where

$$\Delta_3 = \frac{\partial^2}{\partial x_1^2} + \frac{\partial^2}{\partial x_2^2} + \frac{\partial^2}{\partial x_3^2}, \quad \Delta_4 = \frac{\partial^2}{\partial u_1^2} + \frac{\partial^2}{\partial u_2^2} + \frac{\partial^2}{\partial u_3^2} + \frac{\partial^2}{\partial u_4^2}$$

and V is the vector field

$$V = u_2 \frac{\partial}{\partial u_1} - u_1 \frac{\partial}{\partial u_2} - u_4 \frac{\partial}{\partial u_3} + u_3 \frac{\partial}{\partial u_4}.$$

(iii) Consider the differential one form $\alpha = u_2 du_1 - u_1 du_2 - u_4 du_3 + u_3 du_4$. Find $d\alpha$. Find $L_V \alpha$, where $L_V(.)$ denotes the Lie derivative.
(iv) Let $g(x_1(u_1, u_2, u_3, u_4), x_2(u_1, u_2, u_3, u_4), x_3(u_1, u_2, u_3, u_4))$ be a smooth function. Show that $L_V g = 0$.

Problem 7. Let $x_0 = ct$. Consider the one-dimensional *sine-Gordon equation*

$$\frac{\partial^2 u}{\partial x_0^2} - \frac{\partial^2 u}{\partial x_1^2} = \sin(u).$$

Express the equation in *light-cone coordinates*

$$\xi = \frac{1}{2}(x_0 + x_1), \quad \eta = \frac{1}{2}(x_1 - x_0), \quad \tilde{u}(\xi(x_0, x_1), \eta(x_0, x_1)) = u(x_0, x_1).$$

Problem 8. Let $a > 0$. Show that $f(x) = a^x$ is a solution of

$$f(x + y) = f(x)f(y).$$

Problem 9. Consider the analytic function $f : \mathbb{R} \to [-\pi, \pi]$

$$f(x) = \arctan(x).$$

Show that

$$f(x) + f(y) = f\left(\frac{x+y}{1-xy}\right).$$

What happens at $xy = 1$?

Problem 10. (i) Consider the operators

$$\hat{Q} = q + i\hbar\frac{\partial}{\partial q}, \quad \hat{P} = -i\hbar\frac{\partial}{\partial q}.$$

Find the commutator $[\hat{Q}, \hat{P}]$.
(ii) Consider the operators

$$\hat{Q} = \frac{1}{2}q + i\hbar\frac{\partial}{\partial q}, \quad \hat{P} = -2i\hbar\frac{\partial}{\partial q}.$$

Find the commutator $[\hat{Q}, \hat{P}]$.
(iii) Consider the operators

$$\hat{Q} = i\hbar\frac{\partial}{\partial p} + \frac{1}{2}q, \quad \hat{P} = -i\hbar\frac{\partial}{\partial q} + \frac{1}{2}p.$$

Find the commutator $[\hat{Q}, \hat{P}]$.

Problem 11. Let $\alpha \in \mathbb{R}$ and

$$T = x_1 x_2 + \frac{\partial^2}{\partial x_1 \partial x_2}.$$

Show that

$$\exp(i\alpha T)x_1 \exp(-i\alpha T) = x_1 \cosh(\alpha) + i\sinh(\alpha)\frac{\partial}{\partial x_2}$$

$$\exp(i\alpha T)x_2 \exp(-i\alpha T) = x_2 \cosh(\alpha) + i\sinh(\alpha)\frac{\partial}{\partial x_1}.$$

Problem 12. Let $\mathbf{v}(\mathbf{x})$ be a velocity field. Show that

$$\mathbf{v} \cdot (\mathbf{v} \cdot \nabla)\mathbf{v} \equiv \mathbf{v} \cdot \nabla\left(\frac{v^2}{2}\right)$$

where $\cdot$ denotes the scalar product and

$$\mathbf{v}(\mathbf{x}) \cdot \nabla := v_1(\mathbf{x})\frac{\partial}{\partial x_1} + v_2(\mathbf{x})\frac{\partial}{\partial x_2} + v_3(\mathbf{x})\frac{\partial}{\partial x_3}.$$

Problem 13. Let V be a smooth vector field in $\mathbb{R}^n$ and $f_1, f_2 : \mathbb{R}^n \to \mathbb{R}$ be smooth functions.
(i) Show that $e^{\tau V}(f_1 + f_2) = e^{\tau V} f_1 + e^{\tau V} f_2$.
(ii) Show that $e^{\tau V}(f_1 f_2) = (e^{\tau V} f_1)(e^{\tau V} f_2)$.

Problem 14. The *translation operator* in phase space is given by

$$\hat{T}(X, P) := \exp(i(P\hat{x} - X\hat{p})/\hbar).$$

Show that

$$\hat{T}(X, P)\hat{T}(X', P') = \exp(i(X'P - XP')/(2\hbar))\hat{T}(X + X', P + P')$$

and $\hat{T}^\dagger(X, P) = \hat{T}(-X, -P)$.

Problem 15. Consider a four dimensional *torus* $\mathbb{S}^3 \otimes \mathbb{S}^1$ defined by

$$\left(\sqrt{x_1^2 + x_2^2 + x_3^2 + u^2} - a\right)^2 + v^2 = 1$$

where $a > 1$ is the constant radius of $\mathbb{S}^3$. Show that the torus can be parametrized as

$$x_1(\psi, \rho, \phi_1, \phi_2) = (a + \cos(\psi))\rho\cos(\phi_1)$$
$$x_2(\psi, \rho, \phi_1, \phi_2) = (a + \cos(\psi))\rho\sin(\phi_1)$$
$$x_3(\psi, \rho, \phi_1, \phi_2) = (a + \cos(\psi))\sqrt{1 - \rho^2}\cos(\phi_2)$$
$$u(\psi, \rho, \phi_1, \phi_2) = (a + \cos(\psi))\sqrt{1 - \rho^2}\sin(\phi_2)$$
$$v(\psi, \rho, \phi_1, \phi_2) = \sin(\psi)$$

where $\psi[0, 2\pi)$, $\phi_1 \in [0, 2\pi)$, $\phi_2 \in [0, 2\pi)$, $\rho \in [0, 1]$.

Problem 16. Let M be the mass of the nucleus and m the electron mass. Let $\mathbf{R}_0$ be the position vector of the nucleus and $\mathbf{R}_1$, $\mathbf{R}_2$ be the position vectors of the two electrons. Consider the three particle nonrelativistic Schrödinger equation

$$\hat{H}u(\mathbf{R}_0, \mathbf{R}_1, \mathbf{R}_2) = Eu(\mathbf{R}_0, \mathbf{R}_1, \mathbf{R}_2)$$

where

$$\hat{H} = -\frac{\hbar^2}{2M}\nabla_{R_0}^2 - \frac{\hbar^2}{2m}\nabla_{R_1}^2 - \frac{\hbar^2}{2m}\nabla_{R_2}^2$$
$$-\frac{Ze^2}{4\pi\epsilon_0(|\mathbf{R}_0 - \mathbf{R}_1|)} - \frac{Ze^2}{4\pi\epsilon_0(|\mathbf{R}_0 - \mathbf{R}_2|)} + \frac{e^2}{4\pi\epsilon_0(|\mathbf{R}_1 - \mathbf{R}_2|)}.$$

Let $\mu = mM/(m + M)$ be the *reduced mass* and $a_\mu = (m/\mu)a_0$ be the reduced *Bohr radius* with $a_0 = (4\pi\epsilon_0\hbar^2)/(me^2)$. Express the equation in *Jacobi coordinates*

$$\mathbf{r} := (\mathbf{R}_1 - \mathbf{R}_0)/a_\mu,$$

$$\mathbf{x} := \Lambda(\mathbf{R}_2 - \mathbf{R}_0 - y\mathbf{R}_1 - \mathbf{R}_0))/a_\mu, \quad \mathbf{X} := \Lambda(\mathbf{R}_0 + y(\mathbf{R}_1 + \mathbf{R}_2 - \mathbf{R}_0))/a_\mu$$

where $y = \mu/M$, $\Lambda = 1/(1 - y^2)$. Show that the derivatives transform as

$$\nabla_{\mathbf{R}_0} = -a_\mu^{-1}(\nabla_{\mathbf{r}} + \Lambda(1 - y)(\nabla_{\mathbf{x}} - \nabla_{\mathbf{X}}))$$
$$\nabla_{\mathbf{R}_1} = a_\mu^{-1}(\nabla_{\mathbf{r}} - \Lambda y\nabla_{\mathbf{x}} + \Lambda y\nabla_{\mathbf{X}})$$
$$\nabla_{\mathbf{R}_2} = a_\mu^{-1}(\Lambda\nabla_{\mathbf{x}} + \Lambda y\nabla_{\mathbf{X}}).$$

Problem 17. Consider the one-dimensional map $(t = 0, 1, 2, \ldots)$

$$x_{t+1} = \frac{x_t + 2}{x_t + 1}, \quad x_0 \geq 0.$$

Find the fixed points of the map. Let $x_0 = 0$. Does $\lim_{t\to\infty} x_t$ tend to a fixed point?

Problem 18. Consider the map $\mathbf{f} : \mathbb{R}^2 \to \mathbb{R}^2$

$$f_1(x_1, x_2) = x_1 + x_2, \quad f_2(x_1, x_2) = x_1 x_2$$

or written as difference equation $x_{1,\tau+1} = x_{1,\tau} + x_{2,\tau}$, $x_{2,\tau+1} = x_{1,\tau}x_{2,\tau}$. Find the fixed points of the map. Then solve the difference equation for $x_{1,0} = 1/4$, $x_{2,0} = 1/5$.

Problem 19. Show that the solution of the initial value problem of the system of difference equations

$$\begin{pmatrix} x_{1,\tau+1} \\ x_{2,\tau+1} \end{pmatrix} = \begin{pmatrix} 1 & -1 \\ 1 & 1 \end{pmatrix} \begin{pmatrix} x_{1,\tau} \\ x_{2,\tau} \end{pmatrix}$$

is given by

$$\begin{pmatrix} x_{1,\tau} \\ x_{2,\tau} \end{pmatrix} = (\sqrt{2})^\tau \begin{pmatrix} \cos(\pi\tau/4)x_{1,0} - \sin(\pi\tau/4)x_{2,0} \\ \sin(\pi\tau/4)x_{1,0} + \cos(\pi\tau/4)x_{2,0} \end{pmatrix}.$$

Problem 20. A fixed charge Q is located on the z-axis with coordinates $\mathbf{r}_a = (0, 0, d/2)$, where d is the interfocal distance of the *prolate spheroidal* *coordinates*

$$x(\eta, \xi, \phi) = \frac{1}{2}d((1 - \eta^2)(\xi^2 - 1))^{1/2}\cos(\phi)$$

$$y(\eta, \xi, \phi) = \frac{1}{2}d((1 - \eta^2)(\xi^2 - 1))^{1/2}\sin(\phi)$$

$$z(\eta, \xi, \phi) = \frac{1}{2}d\eta\xi$$

where $-1 \leq \eta \leq +1$, $1 \leq \xi \leq \infty$, $0 \leq \phi \leq 2\pi$. Show that the *Coulomb potential* $V(\mathbf{r}) = Q/(|\mathbf{r} - \mathbf{r}_a|)$ expressed in prolate spheroidal coordinates is given by

$$V = \frac{2Q}{d}\frac{\xi + \eta}{\xi^2 - \eta^2}.$$

Problem 21.   Consider the one-dimensional Schrödinger equation (eigenvalue problem)

$$-\frac{\hbar^2}{2m}\frac{d^2u(x)}{dx^2} + V(x)u(x) = Eu(x).$$

For numerical studies the eigenvalue equation is replaced by the difference equation

$$-\frac{\hbar^2}{2m}\left(\frac{u_{n+1} + u_{n-1} - 2u_n}{\delta^2}\right) + V_n u_n = Eu_n$$

where $\delta := x_{n+1} - x_n$ (step size), $u_n := u(x_n)$, $V_n = V(x_n)$. Imposing the boundary conditions the set of linear equations can be solved as an eigenvalue problem of a symmetric matrix over the real numbers. Show that the error in the representation relative to $(2mE/\hbar^2)u_n$ is

$$\frac{1}{12}\delta^2 u_n^{(4)}.$$

The Numerov-Cooley approximation uses the second order difference equation

$$-\frac{\hbar^2}{2m}\left(\frac{u_{n+1} + u_{n-1} - 2u_n}{\delta^2}\right) =$$

$$\frac{5}{6}(E - V_n)u_n + \frac{1}{12}(E - V_{n+1})u_{n+1} + \frac{1}{12}(E - V_{n-1})u_{n-1}.$$

Show that relative error using this approximation is $29\delta^4 u_n^{(6)}/300$. Note that this approximation gives an asymmetric matrix equation.

Problem 22.   Let $\tau = 0, 1, \ldots$. Find solutions of the system of recursion relations

$$R(\tau + 1) = 1/(1/R_a(\tau + 1) + 1/R_b(\tau + 1))$$
$$R_a(\tau + 1) = 1 + 1/(1/R(\tau) + 1/(1 + R(\tau)))$$
$$R_b(\tau + 1) = 1 + (1 + R(\tau))/2$$

where $R(0) = 1$.

Problem 23.   Let $p_1, p_2, p_3 \geq 0$ and $p_1 + p_2 + p_3 = 1$. Solve the linear *Leslie model*

$$
\begin{pmatrix} x_{1,\tau+1} \\ x_{2,\tau+1} \\ x_{3,\tau+1} \end{pmatrix} = \begin{pmatrix} p_1 & p_2 & p_3 \\ 1 & 0 & 0 \\ 0 & 1 & 0 \end{pmatrix} \begin{pmatrix} x_{1,\tau} \\ x_{2,\tau} \\ x_{3,\tau} \end{pmatrix}, \quad \tau = 0, 1, 2, \ldots
$$

given the initial values $x_{1,0}, x_{2,0}, x_{3,0}$.

Chapter 3

Algebraic and Transcendental Equations

A complex polynomial is a complex function of the form

$$p(z) = c_n z^n + c_{n-1} z^{n-1} + \cdots + c_0$$

where $c_0, c_1, \ldots, c_n$ are complex numbers and n is a natural number. A root, or zero, of this polynomial is a complex number w such that $p(w) = 0$. The fundamental theorem of algebra states that any complex polynomial must have a complex root.

Let $n \geq 1$. The solutions of the equation

$$z^n = 1$$

are given by $z_k = e^{2\pi i k/n}$ $(k = 0, 1, \ldots, n-1)$. They form an abelian group under multiplication. We have $1 + z + z^2 + \cdots + z^{n-1} = 0$.

Theorem of Vieta. Let $x_1, x_2, \ldots, x_n$ be the n roots of

$$x^n + a_1 x^{n-1} + \cdots + a_{n-1} x + a_n = 0.$$

Then

$$\sum_{j=1}^n x_j = -a_1, \qquad \sum_{j,k=1; j<k}^n x_j x_k = a_2,$$

$$\sum_{j,k,\ell=1; j<k<\ell}^n x_j x_k x_\ell = -a_3, \quad \cdots \quad , \prod_{j=1}^n x_j = (-1)^n a_n.$$

The *quadratic equation* $(a, b \in \mathbb{R})$ $z^2 + az + b = 0$ has the solution

$$z_\pm = \frac{1}{2}(-a \pm \sqrt{a^2 - 4b})$$

with $z_+ + z_- = -a$, $z_+ z_- = b$.

A *cubic equation* can be cast into the form

$$z^3 + a_1 z^2 + a_2 z + a_0 = 0.$$

Let

$$Q := \frac{1}{9}(3a_2 - a_1^2), \quad R := \frac{1}{54}(9a_1 a_2 - 27a_3 - 2a_1^3)$$

$$S := (R + \sqrt{Q^3 + R^2})^{1/3}, \quad T := (R - \sqrt{Q^3 + R^2})^{1/3}.$$

The solutions are given by

$$z_1 = S + T - \frac{1}{3}a_1$$

$$z_2 = -\frac{1}{2}(S+T) - \frac{1}{3}a_1 + \frac{1}{2}i\sqrt{3}(S-T), \quad z_3 = -\frac{1}{2}(S+T) - \frac{1}{3}a_1 - \frac{1}{2}i\sqrt{3}(S-T).$$

A *quartic equation* $x^4 + a_1 x^3 + a_2 x^2 + a_3 x + a_4 = 0$ can be cast into the form

$$y^4 + py^2 + qy + r = 0$$

where p, q, r depend on a_1, a_2, a_3, a_4. The *cubic resolvent* is

$$z^3 + 2pz^2 + (p^2 - 4r)z - q^2 = 0.$$

If z_1, z_2, z_3 are the roots of the cubic resolvent, then

$$y_1 = \frac{1}{2}(\sqrt{z_1} + \sqrt{z_2} + \sqrt{z_3}), \quad y_2 = \frac{1}{2}(\sqrt{z_1} - \sqrt{z_2} - \sqrt{z_3}),$$

$$y_3 = \frac{1}{2}(-\sqrt{z_1} + \sqrt{z_2} - \sqrt{z_3}), \quad y_4 = \frac{1}{2}(-\sqrt{z_1} - \sqrt{z_2} + \sqrt{z_3})$$

are the solutions of $y^4 + py^2 + qy + r = 0$.

In most cases numerical methods must be applied to find the roots of a function f. One of the main methods is *Newton's method*. For the one-dimensional case it is given by the difference equation

$$x_{\tau+1} = x_\tau - \frac{f(x_\tau)}{f'(x_\tau)}, \quad \tau = 0, 1, \dots$$

where $f : \mathbb{R} \to \mathbb{R}$ is a differentiable function for which we want to find the root $f(x) = 0$.

Problem 1. Let $n \in \mathbb{N}$. Solve $z^n = 1$.

Solution 1. Since every complex number z can be written as $z = re^{i\phi}$ we have

$$z^n = r^n e^{in\phi}.$$

From $z^n = 1$ we obtain $r = 1$ and $e^{in\phi} \equiv \cos(n\phi) + i\sin(n\phi) = 1$. Therefore

$$z_k = \cos\left(\frac{2\pi k}{n}\right) + i\sin\left(\frac{2\pi k}{n}\right)$$

are the roots of 1, where $k = 0, 1, \ldots, n-1$. They from an abelian group under multiplication and one has $1 + z + z^2 + \cdots + z^{n-1} = 0$.

Problem 2. Let d be a positive distance, v, V be velocities $v \neq V$ and T_1, T_2 time-intervals. Assume that

$$\frac{d}{V+v} = T_1, \qquad \frac{d}{V-v} = T_2.$$

Find d/V.

Solution 2. From

$$v = \frac{d - VT_1}{T_1}, \qquad v = \frac{-d + VT_2}{T_2}$$

we can eliminate the velocity v and find

$$\frac{d}{V} = \frac{2T_1 T_2}{T_1 + T_2}.$$

Problem 3. Consider the function

$$f(x_1, x_2, \alpha) = x_1 \sin(\alpha) + x_2 \cos(\alpha) - 1.$$

Solve the system of equations $f(x_1, x_2, \alpha) = 0$, $\partial f(x_1, x_2, \alpha)/\partial\alpha = 0$.

Solution 3. Since $\partial f/\partial\alpha = x_1 \cos(\alpha) - x_2 \sin(\alpha) = 0$ we obtain the solution $x_1 = \sin(\alpha)$, $x_2 = \cos(\alpha)$, i.e. $x_1^2 + x_2^2 = 1$. Study the case $f(x_1, x_2, \alpha) = x_1 \sinh(\alpha) - x_2 \cosh(\alpha) + 1$.

Problem 4. The gravitational acceleration g at the surface of the earth is given by $9.81\, meter \cdot sec^{-2}$. Let $v = 333\, meter \cdot sec^{-1}$ be the speed of sound in air. Solve the quadratic equation

$$\frac{g}{2}t^2 = (T - t)v$$

with respect to t, where t denotes a time and $T = 4\,sec$.

Solution 4.   We have

$$t^2 + \frac{2v}{g}t - \frac{2T}{g}v = 0 \quad \Rightarrow \quad \left(t + \frac{v}{g}\right)^2 = \frac{2T}{g} + \frac{v^2}{g^2}$$

and therefore

$$t = -\frac{v}{g} + \sqrt{\frac{v}{g}\left(2T + \frac{v}{g}\right)}.$$

Obviously in the context of the problem only the positive square root makes sense. Inserting the values we obtain $t \approx 3.8\,sec$. What does the problem describe?

Problem 5.   Consider the cubic equation $x^3 + ax + b = 0$ and $b \neq 0$. Note that the term with x^2 can be removed by a linear transformation. Solve the cubic equation with the ansatz (nonlinear transformation)

$$x = \frac{\alpha}{y} + \beta y, \quad \alpha, \beta \neq 0.$$

Solution 5.   Inserting the ansatz into the cubic equation and setting $a = -3\alpha\beta$ provides

$$\beta^3 y^6 + b y^3 + \alpha^3 = 0.$$

Setting $z = y^3$ we can reduce the equation to a quadratic equation

$$\beta^3 z^2 + bz + \alpha^3 = 0.$$

Problem 6.   Find all solutions of the quartic equation

$$x^4 + x^3 + x^2 + x + 1 = 0. \tag{1}$$

Solution 6.   Method 1. Equation (1) can be solved by dividing by x^2, substituting $y = x + 1/x$, and then applying the quadratic formula. Thus, we have

$$x^2 + \frac{1}{x^2} + x + \frac{1}{x} + 1 = 0$$

$$\left(x^2 + 2 + \frac{1}{x^2}\right) + \left(x + \frac{1}{x}\right) + (1 - 2) = 0$$

$$\left(x + \frac{1}{x}\right)^2 + \left(x + \frac{1}{x}\right) - 1 = 0$$

$$y^2 + y - 1 = 0.$$

The roots of this equation are $y_1 = (-1 + \sqrt{5})/2$, $y_2 = (-1 - \sqrt{5})/2$. It remains to determine x by solving the two equations

$$x + \frac{1}{x} = y_1, \quad \text{and} \quad x + \frac{1}{x} = y_2$$

which are equivalent to $x^2 - y_1 x + 1 = 0$, $x^2 - y_2 x + 1 = 0$. The four roots found by solving these are

$$x_1 = \frac{-1 + \sqrt{5}}{4} + i\frac{\sqrt{10 + 2\sqrt{5}}}{4}, \quad x_2 = \frac{-1 + \sqrt{5}}{4} - i\frac{\sqrt{10 + 2\sqrt{5}}}{4},$$

$$x_3 = \frac{-1 - \sqrt{5}}{4} + i\frac{\sqrt{10 - 2\sqrt{5}}}{4}, \quad x_4 = \frac{-1 - \sqrt{5}}{4} - i\frac{\sqrt{10 - 2\sqrt{5}}}{4}.$$

Method 2. Another approach to this problem is to multiply each side of the original equation by $x - 1$. Since

$$(x - 1)(x^4 + x^3 + x^2 + x + 1) = x^5 - 1$$

an equivalent problem is to find all x (other than $x = 1$) which satisfy $x^5 = 1$. These are the five fifth roots of unity, given by

$$x_1 = \cos(2\pi/5) + i\sin(2\pi/5), \quad x_2 = \cos(4\pi/5) + i\sin(4\pi/5),$$

$$x_3 = \cos(6\pi/5) + i\sin(6\pi/5), \quad x_4 = \cos(8\pi/5) + i\sin(8\pi/5)$$

and $x_5 = 1$.

Problem 7. Archimedes' principle states that any body completely or partially submerged in a fluid (with mass density ρ_0) at rest is acted upon by an upward force (also called buoyant force and measured in Newton ($meter \cdot sec^{-2} \cdot kg$)) the magnitude of which is equal to the weight of the fluid displaced by the body. The acceleration g due to gravity at the surface of the earth is given by $g = 9.80665 \, meter \cdot sec^{-2}$. Consider the systems of equations

$$mg = (V_0 + Ah_0)\rho_0 g, \quad mg = (V_0 + Ah_1)\rho_1 g, \quad mg = (V_0 + Ah)\rho g$$

where ρ_0 is the density of water and ρ_1, ρ_0, h_0, h_1 are given. Find h.

Solution 7. We can first eliminate g, then V_0 and A. This provides the solution

$$h = h_0 + (h_1 - h_0)\frac{\rho_1(\rho - \rho_0)}{\rho(\rho_1 - \rho_0)}.$$

Problem 8.   Let c be the speed of light, m_0 the rest mass of the electron and h the Planck constant. Consider the three equations (Compton effect)

$$\frac{hc}{\lambda_0} = \frac{hc}{\lambda} + (m - m_0)c^2, \quad \frac{h}{\lambda_0} = \frac{h}{\lambda}\cos(\phi) + mv\cos(\theta), \quad 0 = \frac{h}{\lambda} + mv\sin(\theta).$$

Find $\lambda - \lambda_0$ as function of ϕ.

Solution 8.   From the second and third equation we obtain

$$\frac{h^2}{\lambda_0^2} + \frac{h^2}{\lambda^2} - \frac{2h^2}{\lambda\lambda_0}\cos(\phi) = mv^2.$$

The first equation can be written as

$$\frac{h^2}{\lambda_0^2} + \frac{h^2}{\lambda^2} - \frac{2h^2}{\lambda\lambda_0} = (m - m_0)^2c^2.$$

Applying the identity $1 - \cos(\phi) \equiv 2\sin^2(\phi/2)$ it follows that

$$\lambda - \lambda_0 = \frac{2h}{m_0 c}\sin^2(\phi/2)$$

where $\lambda_c = h/(m_0 c)$ is called the *Compton wavelength*.

Problem 9.   Consider the fractional linear transformation

$$x = \frac{at + b}{ct + d}.$$

Find a, b, c, d such that $x = 1 \to t = 0$, $x = 2 \to t = 1$, $x = 3 \to t = \infty$.

Solution 9.   We obtain the three equations

$$1 = \frac{b}{d}, \quad 2 = \frac{a + b}{c + d}, \quad 3 = \frac{a}{c}$$

with the four unknowns a, b, c, d. It follows that $a = 3d$, $b = d$, $c = d$. Hence

$$x = \frac{3t + 1}{t + 1}.$$

Problem 10.   Given a set of N real numbers $x_1, x_2, \ldots, x_N$. It is often useful to express the sum of the j powers

$$s_j = x_1^j + x_2^j + \cdots + x_N^j, \quad j = 0, 1, 2, \ldots$$

in terms of the *elementary symmetric functions*

$$\sigma_1 = \sum_{i=1}^{N} x_i, \quad \sigma_2 = \sum_{i<j}^{N} x_i x_j, \quad \sigma_3 = \sum_{i<j<k}^{N} x_i x_j x_k, \quad \dots, \quad \sigma_N = x_1 x_2 \cdots x_N.$$

Consider the special case with three numbers x_1, x_2, x_3. Then the elementary symmetric functions are given by

$$\sigma_1 = x_1 + x_2 + x_3, \quad \sigma_2 = x_1 x_2 + x_1 x_3 + x_2 x_3, \quad \sigma_3 = x_1 x_2 x_3.$$

We know that the elementary symmetric functions are the coefficients (up to sign) of the polynomial with the roots x_1, x_2, x_3. In other words the values of x_1, x_2, x_3 each satisfy the polynomial equation

$$x^3 - \sigma_1 x^2 + \sigma_2 x - \sigma_3 = 0. \tag{1}$$

Find a recursion relation for

$$s_j := x_1^j + x_2^j + x_3^j, \qquad j = 0, 1, 2, \dots$$

and give the initial values. Calculate s_3 and s_4.

Solution 10. We can multiply equation (1) by any power of x to provide the equation

$$x^j - \sigma_1 x^{j-1} + \sigma_2 x^{j-2} - \sigma_3 x^{j-3} = 0, \quad j = 3, 4, \dots.$$

Thus we find the recursion relation

$$s_j - \sigma_1 s_{j-1} + \sigma_2 s_{j-2} - \sigma_3 s_{j-3} = 0, \quad j = 3, 4, \dots \tag{2}$$

with the initial conditions $s_0 = 3$, $s_1 = \sigma_1$, $s_2 = \sigma_1^2 - 2\sigma_2$. Inserting these initial conditions into the recursion relation (2) yields

$$s_3 = \sigma_1^3 - 3\sigma_1\sigma_2 + 3\sigma_3$$

and

$$s_4 = \sigma_1^4 - 4\sigma_1^2\sigma_2 + 2\sigma_2^2 + 4\sigma_1\sigma_3.$$

Since σ_1 is of degree 1, σ_2 is of degree 2, and σ_3 is of degree 3, each of these expressions is homogeneous.

Problem 11. Let L be a given positive real number. Solve the system of two coupled nonlinear equations

$$1 = x^2 L + 2x^2 y + L x^2 y^2, \qquad 0 = -x^2 + 2x^2 y + L x^2 y^2.$$

Solution 11.   Subtracting the second equation from the first equation we obtain $1 = x^2(1 + L)$ or $x^2 = 1/(1 + L)$. Inserting $x^2 = 1/(1 + L)$ into the second equation we obtain

$$x = \frac{\pm 1}{\sqrt{L + 1}}, \qquad y = \frac{-1 \pm \sqrt{L + 1}}{L}.$$

Problem 12.   Let $t_1 = 120 \, sec$, $t_2 = 180 \, sec$, $t_3 = 240 \, sec$. ω_1, ω_2, ω_3 are frequencies. Solve the system of four equations

$$t_1(\omega_1 + \omega_2) = 1, \quad t_2(\omega_1 + \omega_3) = 1, \quad t_3(\omega_2 + \omega_3) = 1, \quad t(\omega_1 + \omega_2 + \omega_3) = 1$$

with respect to the time t.

Solution 12.   Addition of the first three equations provides

$$2(\omega_1 + \omega_2 + \omega_3) = 1/t_1 + 1/t_2 + 1/t_3 \;\Rightarrow\; \omega_1 + \omega_2 + \omega_3 = \frac{1}{2}\left(\frac{1}{t_1} + \frac{1}{t_2} + \frac{1}{t_3} \right).$$

It follows that

$$\frac{t}{2}\left(\frac{1}{t_1} + \frac{1}{t_2} + \frac{1}{t_3} \right) = 1 \;\Rightarrow\; t = \frac{2t_1 t_2 t_3}{t_1 t_2 + t_2 t_3 + t_3 t_1} = (80/13) \, sec.$$

Problem 13.   Find the solution of the equation

$$\cos(x) + \cos(3x) + \cdots + \cos((2n - 1)x) = 0 \tag{1}$$

where $n \in \mathbb{N}$.

Solution 13.   Obviously, (1) is invariant under $x \to x + \pi$. Now (1) can be written in the form

$$\Re(e^{ix} + e^{3ix} + \cdots + e^{(2n-1)ix}) = 0$$

where $\Re$ denotes the real part of a complex number. The sum on the left-hand side is a geometric series. Therefore we obtain for the real part

$$\frac{\sin(2nx)}{2\sin(x)} = 0.$$

Consequently,

$$x = \frac{\pi}{2n}, \quad \frac{2\pi}{2n}, \quad \frac{3\pi}{2n}, \quad \cdots \quad , \quad \frac{(2n - 1)\pi}{2n} \quad \text{modulo } \pi.$$

The solution of this problem also gives the extrema of the function

$$f(x) = \sin(x) + \frac{1}{3}\sin(3x) + \cdots + \frac{1}{2n-1}\sin((2n-1)x).$$

The function f is a *Fourier approximation*.

Problem 14. (i) Solve the equation $\sin((2n+1)\phi) = 0$, where $n \in \mathbb{Z}$ and $\phi \in \mathbb{R}$.
(ii) Solve the equation $\sin(z) = 0$, where $z \in \mathbb{C}$.

Solution 14. (i) From $\sin(x) = 0$ with $x \in \mathbb{R}$ we obtain $x = k\pi$ with $k \in \mathbb{Z}$. Therefore $(2n+1)\phi = k\pi$ or $\phi = \frac{k\pi}{2n+1}$.
(ii) Since

$$\sin(z) \equiv \sin(x+iy) \equiv \sin(x)\cos(iy) + \cos(x)\sin(iy)$$
$$\equiv \sin(x)\cosh(y) + i\cos(x)\sinh(y)$$

we have $\sin(x)\cosh(y) + i\cos(x)\sinh(y) = 0$. It follows that

$$\sin(x)\cosh(y) = 0, \qquad \cos(x)\sinh(y) = 0.$$

Since $\cosh(y) \neq 0$ for $y \in \mathbb{R}$, the first equation can only be satisfied if $x = k\pi$, where $k \in \mathbb{Z}$. Since $\cos(k\pi) \neq 0$ if $k \in \mathbb{Z}$, the second equation can only be satisfied if $y = 0$. Consequently, the solution to $\sin(z) = 0$ is $z = x + iy$ with $x = k\pi$ and $y = 0$.

Problem 15. Let $0 < \theta < 1$ and

$$\sum_{k=1}^{\infty}\sum_{j=0}^{k-1} \theta^{j+k+2} = 1.$$

Find θ.

Solution 15. We have

$$\sum_{k=1}^{\infty}\sum_{j=0}^{k-1}\theta^{j+k+2} = \sum_{k=1}^{\infty}\theta^{k+2}\sum_{j=0}^{k-1}\theta^j = \sum_{k=1}^{\infty}\theta^{k+2}\left(\frac{1-\theta^k}{1-\theta}\right)$$

$$= \frac{\theta^2}{1-\theta}\left(\sum_{k=1}^{\infty}\theta^k - \sum_{k=1}^{\infty}(\theta^2)^k\right)$$

$$= \frac{\theta^2}{1-\theta}\left(\left(\frac{1}{1-\theta}-1\right) - \left(\frac{1}{1-\theta^2}-1\right)\right)$$

$$= \frac{\theta^2}{1-\theta}\left(\frac{\theta}{1-\theta}-\frac{\theta^2}{1-\theta^2}\right) = \frac{\theta^3}{(1-\theta)(1-\theta^2)}$$

$$= 1.$$

Thus we obtain the quadratic equation $\theta^2 + \theta - 1 = 0$ with the solution (golden mean number) $\theta = (-1 + \sqrt{5})/2$.

Problem 16.   A curve $f(x, y) = 0$ admits a *rational parametrization* if there exist rational functions α and β such

$$f(\alpha(\tau), \beta(\tau)) = 0$$

i.e.

$$f(x, y) = 0 \Leftrightarrow \begin{cases} x = \alpha(\tau) \\ y = \beta(\tau) \end{cases}$$

Such a curve $f(x, y) = 0$ is called *unicursal*. The curve $x^n + y^n - 1 = 0$ is not unicursal for $n \geq 3$, otherwise *Fermat-Wiles* would be wrong.
(i) Find the rational parametrization for the unit circle $x^2 + y^2 - 1 = 0$.
(ii) Find the rational parametrization for the *nodal cubic* $y^3 + x^3 - xy = 0$.

Solution 16.   (i) The circle admits the rational parametrization

$$\alpha(\tau) = \frac{1 - \tau^2}{1 + \tau^2}, \qquad \beta(\tau) = \frac{2\tau}{1 + \tau^2}.$$

With $\tau = \tan(\theta/2)$ we obtain $\alpha(\theta) = \cos(\theta)$ and $\beta(\theta) = \sin(\theta)$.
(ii) The nodal cubic admits the rational parametrization

$$\alpha(\tau) = \frac{\tau}{1 + \tau^3}, \qquad \beta(\tau) = \frac{\tau^2}{1 + \tau^3}.$$

Problem 17.   A *diophantine equation* is a polynomial equation with integral coefficients as unknowns. Find the integral solutions of

$$x^2 + y^2 = z^2. \tag{1}$$

Solution 17.   Obviously $x = y = z = 0$ is a solution. With $z \neq 0$ we obtain

$$\left(\frac{x}{z}\right)^2 + \left(\frac{y}{z}\right)^2 = 1, \qquad x/z, \ y/z \in \mathbb{Q}.$$

Thus

$$\left(\frac{x}{z}\right)^2 + \left(\frac{y}{z}\right)^2 = 1 \Leftrightarrow \begin{cases} \dfrac{x}{z} = \dfrac{1 - t^2}{1 + t^2} \\ \dfrac{y}{z} = \dfrac{2t}{1 + t^2} \end{cases}$$

and

$$((x/z), (x/y)) \in \mathbb{Q} \times \mathbb{Q} \Leftrightarrow t = y/(1 + x) \in \mathbb{Q}.$$

Let $t = a/b$ $(a, b \in \mathbb{Z})$ and $b \neq 0$. Then the integral solutions of $x^2 + y^2 = z^2$ are

$$x = a^2 - b^2, \qquad y = 2ab, \qquad z = a^2 + b^2$$

for all $a, b \in \mathbb{Z}$.

Problem 18. Consider the system of polynomial equations

$$\begin{aligned}
f_1(\mathbf{x}) &= x_1 - x_2 - x_3 = 0 \\
f_2(\mathbf{x}) &= x_1 + x_2 - x_3^2 = 0 \\
f_3(\mathbf{x}) &= x_1^2 + x_2^2 - 1 = 0.
\end{aligned}$$

Use a *Gaussian elimination-like method* to find the solutions of this system of polynomial equations.

Solution 18. We choose x_1 in the first polynomial as the first term for eliminating terms in the two other polynomials

$$\begin{aligned}
g_1(\mathbf{x}) &= f_1(\mathbf{x}) = x_1 - x_2 - x_3 \\
g_2(\mathbf{x}) &= f_2(\mathbf{x}) - f_1(\mathbf{x}) = 2x_2 - x_3^2 + x_3 \\
g_3(\mathbf{x}) &= f_3(\mathbf{x}) - (x_1 + x_2 + x_3)f_1(\mathbf{x}) = 2x_2^2 + 2x_2x_3 + x_3^2 - 1.
\end{aligned}$$

We choose the variable x_2 in g_2 as the most important variable. Then we multiply g_2 by another polynomial and subtract it from $2g_2$ in order to eliminate the terms containing x_2. We do the same for the third polynomial

$$\begin{aligned}
h_1(\mathbf{x}) &= 2g_1(\mathbf{x}) + g_2(\mathbf{x}) = 2x_1 - x_3^2 - x_3 \\
h_2(\mathbf{x}) &= g_2(\mathbf{x}) = 2x_2 - x_3^2 + x_3 \\
h_3(\mathbf{x}) &= 2g_3(\mathbf{x}) - (2x_2 + x_3^2 + x_3)g_2(\mathbf{x}) = x_3^4 + x_3^2 - 2.
\end{aligned}$$

The new set of equations is in upper triangular-form. The last polynomial is only in x_3, the second one is only in x_2 and x_3, and the first one is a polynomial in x_1 and x_3.

Problem 19. Let $\otimes$ be the Kronecker product.
(i) Can one find (normalized) vectors $\mathbf{x}$ and $\mathbf{y}$ in $\mathbb{R}^2$ and $\mathbb{R}^3$, respectively such that

$$\frac{1}{2}\begin{pmatrix} 1 \\ 1 \\ 0 \\ 0 \\ -1 \\ -1 \end{pmatrix} = \begin{pmatrix} x_1 \\ x_2 \end{pmatrix} \otimes \begin{pmatrix} y_1 \\ y_2 \\ y_3 \end{pmatrix} \equiv \begin{pmatrix} x_1y_1 \\ x_1y_2 \\ x_1y_3 \\ x_2y_1 \\ x_2y_2 \\ x_2y_3 \end{pmatrix} \; ?$$

(ii) Can one find (normalized) vectors $\mathbf{x}$ and $\mathbf{y}$ in $\mathbb{R}^2$ and $\mathbb{R}^3$, respectively such that

$$\frac{1}{2}\begin{pmatrix} 1 \\ 1 \\ 0 \\ 0 \\ -1 \\ -1 \end{pmatrix} = \begin{pmatrix} y_1 \\ y_2 \\ y_3 \end{pmatrix} \otimes \begin{pmatrix} x_1 \\ x_2 \end{pmatrix} \equiv \begin{pmatrix} y_1 x_1 \\ y_1 x_2 \\ y_2 x_1 \\ y_2 x_2 \\ y_3 x_1 \\ y_3 x_2 \end{pmatrix} ?$$

Solution 19.   (i) We obtain the six equations

$$1/2 = x_1 y_1, \ 1/2 = x_1 y_2, \ 0 = x_1 y_3, \ 0 = x_2 y_1, \ -1/2 = x_2 y_2, \ -1/2 = x_2 y_3.$$

From $0 = x_1 y_3$ we conclude that either $x_1 = 0$ or $y_3 = 0$ (or both equal to 0). This contradicts that $1/2 = x_1 y_1$, $-1/2 = x_2 y_3$. So the system of equations admits no solution.

(ii) From $x_1 y_2 = 0$, $x_2 y_2 = 0$ and the four other equations we obtain $y_2 = 0$. Then $x_1 y_1 = 1/2$, $x_2 y_1 = 1/2$, $x_1 y_3 = -1/2$, $x_2 y_3 = -1/2$ admits the normalized vectors

$$\mathbf{x} = \frac{1}{\sqrt{2}}\begin{pmatrix} 1 \\ 1 \end{pmatrix}, \quad \mathbf{y} = \frac{1}{\sqrt{2}}\begin{pmatrix} 1 \\ 0 \\ -1 \end{pmatrix}.$$

Problem 20.   Let c be the speed of light, h be the Planck constant and k_B be the Boltzmann constant. Planck's law describes the spectral radiance of electromagnetic tadiation at all frequencies from a black body at temperature T $(^\circ K)$

$$I(\nu, T) = \frac{2h\nu}{c^2} \cdot \frac{1}{\exp(h\nu/(k_B T)) - 1}.$$

To find the maximum of $I(\nu, T)$ for a fixed T we have to solve $dI/d\nu = 0$. With the dimensionless quantity $\alpha = h\nu/(k_B T)$ we obtain

$$3(e^\alpha - 1) = \alpha e^\alpha \quad \Rightarrow \quad f(\alpha) = 3e^\alpha - 3 - \alpha e^\alpha = 0.$$

Apply Newton's method to solve $f(\alpha) = 0$.

Solution 20.   With

$$\frac{df}{d\alpha} = 2e^\alpha - \alpha e^\alpha.$$

Hence

$$\alpha_{T+1} = \alpha_T - \frac{3e^{\alpha_T} - \alpha_T e^{\alpha_T} - 3}{e^{\alpha_T}(2 - \alpha_T)}, \quad T = 0, 1, \ldots$$

Selecting $\alpha_0 = 3.0$ as initial value we obtain after some iterations that $\alpha \approx 2.82$, i.e. $h\nu = 2.82\, k_B T$.

Problem 21. Given the 3×3 matrix

$$A = \begin{pmatrix} 1 & 0 & 1 \\ 0 & 1 & 0 \\ 1 & 0 & 1 \end{pmatrix}.$$

Find the eigenvalues of A by solving the system of equations

$$\lambda_1 + \lambda_2 + \lambda_3 = \mathrm{tr}(A)$$
$$\lambda_1^2 + \lambda_2^2 + \lambda_3^2 = \mathrm{tr}(A^2)$$
$$\lambda_3^2 + \lambda_2^3 + \lambda_3^3 = \mathrm{tr}(A^3).$$

Solution 21. Applying the Maxima program

```
load("nchrpl");
A: matrix([1,0,1],[0,1,0],[1,0,1]);
A2: A . A;
A3: A . A . A;
trA: mattrace(A);
trA2: mattrace(A2);
trA3: mattrace(A3);
solve([l1+l2+l3=trA,l1*l1+l2*l2+l3*l3=trA2,
       l1*l1*l1+l2*l2*l2+l3*l3*l3=trA3],[l1,l2,l3]);
```

We obtain the solution $\lambda_1 = 0$, $\lambda_2 = 1$, $\lambda_3 = 2$.

Supplementary Problems

Problem 1. Let (x_1, x_2, x_3) be a given position in $\mathbb{E}^3$. Solve the system of equations

$$-\sin(\theta_1)(\ell_2 \cos(\theta_2) + \ell_3 \cos(\theta_3)) = x_1$$
$$\cos(\theta_1)(\ell_2 \cos(\theta_2) + \ell_3 \cos(\theta_3)) = x_2$$
$$\ell_2 \sin(\theta_2) + \ell \sin(\theta_3)) = x_3$$

where ℓ_2, ℓ_3 denote some length.
(i) Apply a *rational parametrization*

$$\cos(\theta_j) = \frac{1 - t_j^2}{1 + t_j^2}, \qquad \sin(\theta_j) = \frac{2t_j}{1 + t_j^2}, \qquad j = 1, 2, 3$$

(ii) Set $s_j = \cos(\theta_j)$, $c_j = \cos(\theta_j)$ and then add the equations

$$s_1^2 + c_1^2 = 1, \quad s_2^2 + c_2^2 = 1, \quad s_3^2 + c_3^2 = 1$$

to the system.

Problem 2.   Given $h_1, h_2, h_3, h_4 \in \mathbb{R}$. Solve the system of equations

$$h_1 = \lambda_1 + \mu_1 + \lambda_1\mu_1, \quad h_2 = \lambda_1 + \mu_2 + \lambda_1\mu_2,$$

$$h_3 = \lambda_2 + \mu_1 + \lambda_2\mu_1, \quad h_4 = \lambda_2 + \mu_2 + \lambda_2\mu_2.$$

Problem 3.   Consider the analytic function $f : \mathbb{R} \to \mathbb{R}$, $f(x) = \sin(x)$. Show that the only solution of the fixed point equation $f(x) = x$ is given by $x = 0$.

Problem 4.   Consider the analytic function $f : \mathbb{R} \to \mathbb{R}$

$$f(x) = 1 + x + \cos(x) \cosh(x).$$

Show that the fixed point equation $f(x) = x$ has infinitely many solutions.

Problem 5.   Find all 2×2 matrices X, Y such that

$$[X, Y] = \sigma_3 \equiv \begin{pmatrix} 1 & 0 \\ 0 & -1 \end{pmatrix}.$$

Let $\alpha \in \mathbb{R}$. Show that

$$X = \begin{pmatrix} 0 & \cosh(\alpha) \\ \sinh(\alpha) & 0 \end{pmatrix}, \quad Y = \begin{pmatrix} 0 & \sinh(\alpha) \\ \cosh(\alpha) & 0 \end{pmatrix}$$

is a solution.

Problem 6.   Let $\mathbf{x}$, $\mathbf{y}$ be vectors in $\mathbb{R}^3$. Find all solutions of $\mathbf{x} \times \mathbf{y} = \mathbf{x} + \mathbf{y}$.

Chapter 4

Vector and Matrix Calculus

Let $\mathbb{F}$ be a field, for example the set of real numbers $\mathbb{R}$ or the set of complex numbers $\mathbb{C}$. Let $m, n \geq 1$ be two integers. An array A of numbers in $\mathbb{F}$

$$\begin{pmatrix} a_{11} & a_{12} & a_{13} & \cdots & a_{1n} \\ a_{21} & a_{22} & a_{23} & \cdots & a_{2n} \\ \vdots & \vdots & \vdots & \ddots & \vdots \\ a_{m1} & a_{m2} & a_{m3} & \cdots & a_{mn} \end{pmatrix} = (a_{ij})$$

is called an $m \times n$ *matrix* with entry a_{ij} in the ith row and jth column. A *row vector* is a $1 \times n$ matrix. A *column vector* is an $n \times 1$ matrix. If $a_{ij} = 0$ for all i, j, then A is called a zero matrix. I_n denotes the $n \times n$ identity matrix with the diagonal elements equal to 1 and 0 otherwise.

Let $A = (a_{ij})$ and $B = (b_{ij})$ be two $m \times n$ matrices. We define $A + B$ to be the $m \times n$ matrix whose entry in the i-th row and j-th column is $a_{ij} + b_{ij}$. Matrix multiplication is only defined between two matrices if the number of columns of the first matrix is the same as the number of rows of the second matrix. If A is an $m \times n$ matrix and B is an $n \times p$ matrix, then the matrix product AB is an $m \times p$ matrix defined by

$$(AB)_{ij} = \sum_{r=1}^{n} a_{ir} b_{rj}$$

for each pair i and j, where $(AB)_{ij}$ denotes the (i, j)-th entry in AB. Matrix addition is commutative. Matrix multiplication is associative.

Let $A = (a_{ij})$ and $B = (b_{ij})$ be two $m \times n$ matrices with entries in some field. Then their *Hadamard product* is the entrywise product of A and B, that is the $m \times n$ matrix $A \bullet B$ whose (i,j)-th entry is $a_{ij} b_{ij}$.

An $n \times n$ matrix A over $\mathbb{C}$ is called *normal* if $AA^* = A^*A$. An $n \times n$ matrix H over $\mathbb{C}$ is called *hermitian* if $H^* = H$. An $n \times n$ matrix over $\mathbb{C}$ is called a *projection matrix* if $\Pi^2 = \Pi$ and $\Pi^* = \Pi$. An $n \times n$ matrix over $\mathbb{C}$ is called *unitary* if $U^* = U^{-1}$.

Let A, B be $n \times n$ matrices. The commutator is defined as

$$[A, B] = AB - BA.$$

The anticommutator is defined as

$$[A, B]_+ = AB + BA.$$

The trace of an $n \times n$ matrix A is defined as

$$\text{tr}(A) = \sum_{j=1}^{n} a_{jj} = \sum_{j=1}^{n} \lambda_j$$

where λ_j are the eigenvalues of A counting multiplicities. The determinant of an $n \times n$ matrix A is defined as

$$\det(A) = \sum_{\sigma \in S_n} \left(\text{sgn}(\sigma) \prod_{j=1}^{n} a_{j,\sigma_j} \right) = \prod_{j=1}^{n} \lambda_j.$$

Note that $\det(\exp(A)) \equiv \exp(\text{tr}(A))$. The exponential of a square matrix can be defined as $\exp(A) := \sum_{j=0}^{\infty} A^j / (j!)$.

Let $\mathbf{u}$ and $\mathbf{v}$ be vectors in $\mathbb{R}^3$. Then the *vectors product* is defined as

$$\mathbf{u} \times \mathbf{v} := \begin{pmatrix} u_2 v_3 - u_3 v_2 \\ u_3 v_1 - u_1 v_3 \\ u_1 v_2 - u_2 v_1 \end{pmatrix}.$$

For any three vectors $\mathbf{v}_1$, $\mathbf{v}_2$, $\mathbf{v}_3$ in $\mathbb{R}^3$ we have (*Jacobi identity*)

$$(\mathbf{v}_1 \times \mathbf{v}_2) \times \mathbf{v}_3 + (\mathbf{v}_3 \times \mathbf{v}_1) \times \mathbf{v}_2 + (\mathbf{v}_2 \times \mathbf{v}_3) \times \mathbf{v}_1 = \mathbf{0}.$$

In the *singular value decomposition* an $m \times n$ matrix A can be written as

$$A = U \Sigma V^T$$

where U is an $m \times m$ orthogonal matrix, V is an $n \times n$ orthogonal matrix, Σ is an $m \times n$ diagonal matrix with nonnegative entries and T denotes transpose.

Problem 1. Consider the normalized vectors in $\mathbb{R}^2$ ($\theta_1, \theta_2 \in [0, 2\pi)$)

$$\begin{pmatrix} \cos(\theta_1) \\ \sin(\theta_1) \end{pmatrix}, \quad \begin{pmatrix} \cos(\theta_2) \\ \sin(\theta_2) \end{pmatrix}.$$

Find the condition on θ_1 and θ_2 such that the vector in $\mathbb{R}^2$

$$\begin{pmatrix} \cos(\theta_1) \\ \sin(\theta_1) \end{pmatrix} + \begin{pmatrix} \cos(\theta_2) \\ \sin(\theta_2) \end{pmatrix}$$

is normalized.

Solution 1. From the condition that the vector

$$\begin{pmatrix} \cos(\theta_1) + \cos(\theta_2) \\ \sin(\theta_1) + \sin(\theta_2) \end{pmatrix}$$

is normalized it follows that

$$(\sin(\theta_1) + \sin(\theta_2))^2 + (\cos(\theta_1) + \cos(\theta_2))^2 = 1.$$

Thus we have $\sin(\theta_1)\sin(\theta_2) + \cos(\theta_1)\cos(\theta_2) = -1/2$. It follows that

$$\cos(\theta_1 - \theta_2) = -\frac{1}{2}.$$

Therefore, $\theta_1 - \theta_2 = 2\pi/3$ or $\theta_1 - \theta_2 = 4\pi/3$.

Problem 2. Can one find five vectors $\mathbf{v}_j$ ($j = 1, 2, 3, 4, 5$) in $\mathbb{R}^3$ such that the angle between any two of these vectors is obtuse? An *obtuse angle* θ is one that is larger then $\pi/2$ but less than π. We can set $\mathbf{v}_5 = (0\ 0\ -1)^T$ without loss of generality.

Solution 2. We set $\mathbf{v}_j = (v_{j1}\ v_{j2}\ v_{j3})^T$. Suppose that all angles are obtuse. Then we have for the scalar product

$$\mathbf{v}_j^T \mathbf{v}_k = |\mathbf{v}_j||\mathbf{v}_k|\cos(\theta_{jk})$$

where θ_{jk} is the (obtuse) angle between $\mathbf{v}_j$ and $\mathbf{v}_k$. Taking $k = 5$ we have $v_{j3} > 0$ for $j = 1, 2, 3, 4$. Consider now the projection $\mathbf{v}_j \to \mathbf{w}_j = (v_{j1}\ v_{j2}\ 0)^T$. Since the $\mathbf{w}_j$ span at most a two-dimensional vector space some pairs of $\mathbf{w}_1$, $\mathbf{w}_2$, $\mathbf{w}_3$, $\mathbf{w}_4$ form a non-obtuse angle. We assume that $\mathbf{w}_1$, $\mathbf{w}_2$ is such a pair, so that $0 \le \mathbf{w}_1^T \mathbf{w}_2 = w_{11}w_{21} + w_{12}w_{22}$. It follows that

$$\mathbf{v}_1^T \mathbf{v}_2 = v_{11}v_{21} + v_{12}v_{22} + v_{13}v_{23} > 0.$$

and the angle θ_{12} between $\mathbf{v}_1$ and $\mathbf{v}_2$ is acute, i.e. $\theta_{12} \in (0, \pi/2)$. Thus we have a contradiction. Thus not all the angles formed by pairs can be obtuse.

Problem 3.   Let $\mathbf{k}$ be a normalized vector in $\mathbb{R}^3$. Let $\mathbf{x}$ be a vector in $\mathbb{R}^3$. Is

$$\mathbf{x} \equiv (\mathbf{k} \cdot \mathbf{x})\mathbf{k} + \mathbf{k} \times (\mathbf{x} \times \mathbf{k}) ?$$

Here $\mathbf{k} \cdot \mathbf{x} = k_1 x_1 + k_2 x_2 + k_3 x_3$ is the *scalar product* and $\times$ denotes the *vector product*.

Solution 3.   We have

$$\mathbf{k} \times (\mathbf{x} \times \mathbf{k}) = \begin{pmatrix} x_1(k_2^2 + k_3^2) - k_1(k_2 x_2 + k_3 x_3) \\ x_2(k_1^2 + k_3^2) - k_2(k_1 x_1 + k_3 x_3) \\ x_3(k_1^2 + k_2^2) - k_3(k_1 x_1 + k_2 x_2) \end{pmatrix}$$

and

$$(\mathbf{k} \cdot \mathbf{x})\mathbf{k} = \begin{pmatrix} k_1(k_1 x_1 + k_2 x_2 + k_3 x_3) \\ k_2(k_1 x_1 + k_2 x_2 + k_3 x_3) \\ k_3(k_1 x_1 + k_2 x_2 + k_3 x_3) \end{pmatrix}.$$

Since $\|\mathbf{k}\| = \sqrt{k_1^2 + k_2^2 + k_3^2} = 1$ we find that the identity is true.

Problem 4.   Consider the three vectors in $\mathbb{E}^3$

$$\mathbf{a}_1 = \frac{a}{2}\begin{pmatrix} 1 & 1 & -1 \end{pmatrix}, \quad \mathbf{a}_2 = \frac{a}{2}\begin{pmatrix} -1 & 1 & 1 \end{pmatrix}, \quad \mathbf{a}_3 = \frac{a}{2}\begin{pmatrix} 1 & -1 & 1 \end{pmatrix} \quad (1)$$

where a is a positive constant. Find the vectors $\mathbf{b}_j$ $(j = 1, 2, 3)$ such that

$$\mathbf{b}_j \cdot \mathbf{a}_k = 2\pi \delta_{jk} \tag{2}$$

where (scalar product of $\mathbf{b}_j$ and $\mathbf{a}_k$)

$$\mathbf{b}_j \cdot \mathbf{a}_k := \sum_{n=1}^{3} b_{jn} a_{kn}, \qquad \delta_{jk} := \begin{cases} 1 & j = k \\ 0 & \text{otherwise} \end{cases}.$$

The vectors $\{\mathbf{a}_j : j = 1, 2, 3\}$ are a basis of the *body-centered cubic lattice*. The vectors $\{\mathbf{b}_j : j = 1, 2, 3\}$ are called the *reciprocal basis*.

Solution 4.   In matrix notation (2) can be written as $BA = 2\pi I_3$, where B is the matrix whose rows are the vectors $\mathbf{b}_j$, A is the matrix whose columns are the vectors $\mathbf{a}_j$ and I_3 is the 3×3 unit matrix. Therefore

$$A = \frac{a}{2}\begin{pmatrix} 1 & -1 & 1 \\ 1 & 1 & -1 \\ -1 & 1 & 1 \end{pmatrix}.$$

We find $\det(A) \neq 0$. Consequently $B = 2\pi A^{-1}$. Using *Gauss elimination* we find

$$B = \frac{2\pi}{a}\begin{pmatrix} 1 & 1 & 0 \\ 0 & 1 & 1 \\ 1 & 0 & 1 \end{pmatrix}.$$

Finally we arrive at

$$\mathbf{b}_1 = \frac{2\pi}{a}\begin{pmatrix}1 & 1 & 0\end{pmatrix}, \quad \mathbf{b}_2 = \frac{2\pi}{a}\begin{pmatrix}0 & 1 & 1\end{pmatrix}, \quad \mathbf{b}_3 = \frac{2\pi}{a}\begin{pmatrix}1 & 0 & 1\end{pmatrix}.$$

The vectors

$$\frac{a}{2}\begin{pmatrix}1 & 1 & 0\end{pmatrix}, \quad \frac{a}{2}\begin{pmatrix}0 & 1 & 1\end{pmatrix}, \quad \frac{a}{2}\begin{pmatrix}1 & 0 & 1\end{pmatrix}$$

are a basis of the *face-centered cubic lattice*.

Problem 5. Let A be a linear transformation of a vector space V into itself. Suppose $\mathbf{x} \in V$ is such that $A^n\mathbf{x} = 0$, $A^{n-1}\mathbf{x} \neq 0$ for some positive integer n. Show that the vectors $\mathbf{x}, A\mathbf{x}, \ldots, A^{n-1}\mathbf{x}$ are linearly independent.

Solution 5. Suppose that there are scalars $c_0, c_1, \ldots, c_{n-1}$ such that

$$c_0\mathbf{x} + c_1 A\mathbf{x} + \cdots + c_k A^k\mathbf{x} + \cdots + c_{n-1}A^{n-1}\mathbf{x} = 0.$$

Applying A^{n-1} to both sides, we obtain

$$c_0 A^{n-1}\mathbf{x} + c_1 A^n\mathbf{x} + \cdots + c_k A^{n-1+k}\mathbf{x} + \cdots + c_{n-1}A^{n-1+n-1}\mathbf{x} = 0$$

where we used $A0 = 0$. Thus $c_0 A^{n-1}\mathbf{x} = 0$ and therefore $c_0 = 0$. By the *induction principle* (multiplying by A^{n-k-1}) we find that all $c_k = 0$. Thus the set is linearly independent.

Problem 6. Find all 2×2 matrices A with

$$\det(A) = a_{11}a_{22} - a_{12}a_{21} = 1 \tag{1}$$

and

$$A\frac{1}{\sqrt{2}}\begin{pmatrix}1\\1\end{pmatrix} = \frac{1}{\sqrt{2}}\begin{pmatrix}1\\1\end{pmatrix}. \tag{2}$$

Solution 6. From (2) we find the system of linear equations

$$a_{11} + a_{12} = 1, \qquad a_{21} + a_{22} = 1.$$

From (1) we have $a_{11}a_{22} - a_{12}a_{21} = 1$. Eliminating a_{22} we obtain

$$a_{11} + a_{12} = 1, \qquad a_{11}(1 - a_{21}) - a_{12}a_{21} = 1.$$

Eliminating a_{12} we obtain $a_{11} - a_{21} = 1$ with $a_{12} = 1 - a_{11}$ and $a_{22} = 2 - a_{11}$. Thus

$$A = \begin{pmatrix} a_{11} & 1 - a_{11} \\ a_{11} - 1 & 2 - a_{11} \end{pmatrix}$$

with a_{11} arbitrary. These matrices form a *group* under matrix multiplication. For $a_{11} = 1$ we obtain the unit matrix.

Problem 7. Let

$$\mathbf{x} = (\,x_1 \quad x_2 \quad x_3\,)^T, \qquad \mathbf{y} = (\,y_1 \quad y_2 \quad y_3\,)^T$$

be two normalized column vectors in $\mathbb{R}^3$. Assume that $\mathbf{x}^T\mathbf{y} = 0$, i.e. the vectors are orthogonal. Is the vector $\mathbf{x} \times \mathbf{y}$ a unit vector again?

Solution 7. We have

$$\mathbf{x} \times \mathbf{y} = \begin{pmatrix} x_2 y_3 - x_3 y_2 \\ x_3 y_1 - x_1 y_3 \\ x_1 y_2 - x_2 y_1 \end{pmatrix} \tag{1}$$

$$x_1^2 + x_2^2 + x_3^2 = 1, \qquad y_1^2 + y_2^2 + y_3^2 = 1 \tag{2}$$

and

$$\mathbf{x}^T\mathbf{y} = x_1 y_1 + x_2 y_2 + x_3 y_3 = 0. \tag{3}$$

Now the square of the norm of the vector $\mathbf{x} \times \mathbf{y}$ is given by

$$\begin{aligned} \|\mathbf{x} \times \mathbf{y}\|^2 = {}& x_2^2 y_3^2 + x_3^2 y_2^2 + x_3^2 y_1^2 + x_1^2 y_3^2 + x_1^2 y_2^2 + x_2^2 y_1^2 \\ & -2(x_2 x_3 y_2 y_3 + x_1 x_3 y_1 y_3 + x_1 x_2 y_1 y_2). \end{aligned}$$

From (2) it follows that $x_3^2 = 1 - x_1^2 - x_2^2$, $y_3^2 = 1 - y_1^2 - y_2^2$. From (3) it follows that $x_3 y_3 = -x_1 y_1 - x_2 y_2$. By rearranging and squaring this equation we obtain

$$2 x_1 y_1 x_2 y_2 = x_3^2 y_3^2 - x_1^2 y_1^2 - x_2^2 y_2^2 = 1 - y_1^2 - y_2^2 - x_1^2 - x_2^2 + x_1^2 y_2^2 + x_2^2 y_1^2.$$

Inserting these equations into $\|\mathbf{x} \times \mathbf{y}\|^2$ gives $\|\mathbf{x} \times \mathbf{y}\|^2 = 1$. Thus the vector $\mathbf{x} \times \mathbf{y}$ is normalized.

Problem 8. (i) Find four vectors a_1, a_2, a_3, a_4 in $\mathbb{R}^3$ such that

$$\mathbf{a}_j^T \mathbf{a}_k = \frac{4}{3}\delta_{jk} - \frac{1}{3} = \begin{cases} 1 & \text{for } j = k \\ -1/3 & \text{for } j \neq k \end{cases}.$$

(ii) Calculate the sums

$$\sum_{j=1}^{4} \mathbf{a}_j, \qquad \frac{3}{4}\sum_{j=1}^{4} \mathbf{a}_j \mathbf{a}_j^T.$$

Solution 8. (i) Geometrically speaking, such a quartet of vectors consists of the vectors pointing from the center of a cube to nonadjacent corners. We also may picture these four vectors as the normal vectors for the faces of the *tetrahedron* that is defined by the other four corners of the cube. Owing to the conditions the four vectors are normalized. Thus

$$
\mathbf{a}_1 = \frac{1}{\sqrt{3}} \begin{pmatrix} 1 \\ 1 \\ 1 \end{pmatrix}, \quad
\mathbf{a}_2 = \frac{1}{\sqrt{3}} \begin{pmatrix} 1 \\ -1 \\ -1 \end{pmatrix}, \quad
\mathbf{a}_3 = \frac{1}{\sqrt{3}} \begin{pmatrix} -1 \\ 1 \\ -1 \end{pmatrix}, \quad
\mathbf{a}_4 = \frac{1}{\sqrt{3}} \begin{pmatrix} -1 \\ -1 \\ 1 \end{pmatrix}.
$$

(ii) We have

$$
\sum_{j=1}^{4} \mathbf{a}_j = \mathbf{0}, \qquad \frac{3}{4} \sum_{j=1}^{4} \mathbf{a}_j \mathbf{a}_j^T = I_3.
$$

Problem 9. Consider the normalized vector $\mathbf{v}_0 = \begin{pmatrix} 1 & 0 & 0 \end{pmatrix}^T$ in $\mathbb{R}^3$. Find three normalized vectors $\mathbf{v}_1$, $\mathbf{v}_2$, $\mathbf{v}_3$ such that

$$
\sum_{j=0}^{3} \mathbf{v}_j = \mathbf{0}, \qquad \mathbf{v}_j^T \mathbf{v}_k = -\frac{1}{3} \quad (j \neq k).
$$

Solution 9. Owing to the form of $\mathbf{v}_0$ we have due to the second condition that $v_{1,1} = v_{2,1} = v_{3,1} = -1/3$. We obtain

$$
\mathbf{v}_1 = \begin{pmatrix} -1/3 \\ 2\sqrt{2}/3 \\ 0 \end{pmatrix}, \qquad
\mathbf{v}_2 = \begin{pmatrix} -1/3 \\ -\sqrt{2}/3 \\ \sqrt{6}/3 \end{pmatrix}, \qquad
\mathbf{v}_3 = \begin{pmatrix} -1/3 \\ -\sqrt{2}/3 \\ -\sqrt{6}/3 \end{pmatrix}.
$$

Problem 10. Let x^μ be the standard coordinates in $\mathbb{R}^4$ with $\mu = 0, 1, 2, 3$. We define new coordinates x^{PQ} $(P, Q = 1, 2)$ by

$$
\begin{pmatrix} x^{11} & x^{12} \\ x^{21} & x^{22} \end{pmatrix} := \begin{pmatrix} x^0 - ix^3 & -ix^1 - x^2 \\ -ix^1 + x^2 & x^0 + ix^3 \end{pmatrix}.
$$

Let $Z^\alpha = (\omega^P, \pi_Q)$ be four complex coordinates on *twistor space* $\mathbb{C}^4$, i.e.

$$
Z^0 = \omega^1, \quad Z^1 = \omega^2, \quad Z^2 = \pi_1, \quad Z^3 = \pi_2.
$$

The basic equation expressing the relationship between Z^α and x^μ is

$$
\omega^P = \sum_{Q=1}^{2} x^{PQ} \pi_Q. \tag{1}
$$

The reality structure on Z^α is given by the *antilinear map*

$$Z^\alpha \mapsto Z^{*\alpha} = (\overline{Z^1}, -\overline{Z^0}, \overline{Z^3}, -\overline{Z^2}),$$

$$\omega^P \mapsto \omega^{*P} = (\overline{\omega^2}, -\overline{\omega^1}), \quad \pi_P \mapsto \pi_P^* = (\overline{\pi_2}, -\overline{\pi_1}).$$

Show that equation (1) is preserved under this map, in the sense that $\omega^P = \sum_{Q=1}^2 x^{PQ}\pi_Q$ if and only if

$$\omega^{*P} = \sum_{Q=1}^2 x^{PQ}\pi_Q^*.$$

Solution 10.   We have $\overline{x^{11}} = x^{22}$ and $\overline{x^{12}} = -x^{21}$. From $\omega^P = \sum_{Q=1}^2 x^{PQ}\pi_Q$ we have

$$\omega^1 = x^{11}\pi_1 + x^{12}\pi_2, \qquad \omega^2 = x^{21}\pi_1 + x^{22}\pi_2$$

and from $\omega^{*P} = \sum_{Q=1}^2 x^{PQ}\pi_Q^*$ we have

$$\omega^{*1} = x^{11}\pi_1^* + x^{12}\pi_2^*, \qquad \omega^{*2} = x^{21}\pi_1^* + x^{22}\pi_2^*.$$

Applying the antilinear map we have

$$\omega^{*1} = x^{11}\pi_1^* + x^{12}\pi_2^*$$
$$\Leftrightarrow \overline{\omega^2} = x^{11}\overline{\pi_2} - x^{12}\overline{\pi_1}$$
$$\Leftrightarrow \omega^2 = \overline{x^{11}}\pi_2 - \overline{x^{12}}\pi_1$$
$$\Leftrightarrow \omega^2 = x^{22}\pi_2 + x^{21}\pi_1.$$

Analogously we show that from $\omega^{*2} = x^{21}\pi_1^* + x^{22}\pi_2^*$ follows $\omega^1 = x^{12}\pi_2 + x^{11}\pi_1$ and that $\omega^P = \sum_{Q=1}^2 x^{PQ}\pi_Q$ follows from $\omega^{*P} = \sum_{Q=1}^2 x^{PQ}\pi_Q^*$.

Problem 11.   (i) Let R be a *nonsingular* $n \times n$ matrix (i.e. R^{-1} exists). Let A and B be two arbitrary $n \times n$ matrices. Assume that $R^{-1}AR$ and $R^{-1}BR$ are diagonal matrices. Show that the commutator of A and B is given by $[A, B] = 0_n$.
(ii) Let X be an arbitrary $n \times n$ matrix. Let U be a nonsingular $n \times n$ matrix. Assume that $UXU^{-1} = X$. Show that $[X, U] = 0_n$.

Solution 11.   (i) Since $R^{-1}AR$ and $R^{-1}BR$ are diagonal matrices it follows that $[R^{-1}AR, R^{-1}BR] = 0_n$. This means that $R^{-1}AR$ and $R^{-1}BR$ commute. Therefore

$$R^{-1}ARR^{-1}BR - R^{-1}BRR^{-1}AR = 0_n.$$

It follows that $R^{-1}ABR - R^{-1}BAR = R^{-1}[A, B]R = 0_n$. Finally

$$RR^{-1}[A, B]RR^{-1} = [A, B] = 0_n.$$

(ii) From $UXU^{-1} = X$ we obtain $UXU^{-1} - X = 0_n$. Multiplying from the right with U, we find $UX - XU \equiv [U, X] = 0_n$.

Problem 12.   Let A and B be two $n \times n$ matrices. Assume that B is nonsingular. This means that $\det(B) \neq 0$. Therefore the inverse of B exists. Show that $[A, B^{-1}] = -B^{-1}[A, B]B^{-1}$.

Solution 12.   We have $[A, B^{-1}] \equiv AB^{-1} - B^{-1}A$. Since $BB^{-1} = I_n$ and $B^{-1}B = I_n$, where I_n is the $n \times n$ unit matrix, we can write

$$[A, B^{-1}] = AB^{-1} - B^{-1}A = B^{-1}BAB^{-1} - B^{-1}ABB^{-1}.$$

Thus $[A, B^{-1}] = -B^{-1}(AB - BA)B^{-1} = -B^{-1}[A, B]B^{-1}$.

Problem 13.   Let $a, b, c \in \mathbb{R}$ and let

$$A(a, b, c) = \begin{pmatrix} a & b & 0 & 0 & \cdots & 0 & 0 & 0 \\ c & a & b & 0 & \cdots & 0 & 0 & 0 \\ 0 & c & a & b & \cdots & 0 & 0 & 0 \\ & & \ddots & & \ddots & & & \\ \vdots & & & \ddots & & \ddots & & \\ & & & & \ddots & & \ddots & \\ 0 & 0 & 0 & 0 & \cdots & c & a & b \\ 0 & 0 & 0 & 0 & \cdots & 0 & c & a \end{pmatrix}$$

be an $n \times n$ tridiagonal matrix. In other words

$$A_{jk} = \begin{cases} b & \text{if} & j = k + 1 \\ c & \text{if} & j = k - 1 \\ a & \text{if} & j = k \\ 0 & \text{otherwise} \end{cases}$$

with $j, k = 1, 2, \ldots, n$. Calculate $\det(A(a, b, c))$.

Solution 13.   Let $D_n(a, b, c) := \det(A(a, b, c))$. By expanding in minors on the first row, we find the linear second-order difference equation with constant coefficients

$$D_{n+2}(a, b, c) = aD_{n+1}(a, b, c) - bcD_n(a, b, c) \tag{1}$$

where $n = 1, 2, \ldots$. Obviously the initial values of this difference equation are given by

$$D_1(a, b, c) = a, \qquad D_2(a, b, c) = a^2 - bc.$$

Since the linear difference equation (1) has constant coefficients we can solve it with the ansatz

$$D_n(a, b, c) = kr^n$$

where k is a constant, or we may employ the recurrence relations for the *Chebyshev polynomials* to obtain

$$D_n(a, b, c) = \begin{cases} (bc)^{n/2} U_n \left(\frac{a}{2\sqrt{bc}} \right) & \text{if } bc \neq 0 \\ a^n & \text{if } bc = 0 \end{cases}$$

where U_n is the n-th degree Chebyshev polynomial of the second kind given by

$$U_n(s) := \frac{(s + \sqrt{s^2 - 1})^{n+1} - (s - \sqrt{s^2 - 1})^{n+1}}{2\sqrt{s^2 - 1}}.$$

We notice that

$$a^n = \lim_{bc \to 0} (bc)^{n/2} U_n \left(\frac{a}{2\sqrt{bc}} \right).$$

The solution can also be given as

$$D_n(a, b, c) = a^n \prod_{j=1}^{n} \left(1 - \frac{2\sqrt{bc}}{a} \cos \left(\frac{j\pi}{n+1} \right) \right).$$

The ansatz $D_n(a, b, c) = kr^n$ leads to $r^2 = ar - bc$ with the solution

$$r_{1,2} = \frac{a}{2} \pm \sqrt{\frac{a^2}{4} - bc}.$$

Problem 14.   Let C be an $n \times n$ matrix.
(i) Let A, B be two $n \times n$ matrices. Assume that $\text{tr}(A) = 0$, $\text{tr}(B) = 0$. Can we conclude that $\text{tr}(AB) = 0$?
(ii) Let A and B be two $n \times n$ matrices. Prove that $\text{tr}(AB) \neq (\text{tr}A)(\text{tr}B)$ in general.

Solution 14.   (i) The answer is no. Let

$$A = B = \begin{pmatrix} 0 & 1 \\ 1 & 0 \end{pmatrix} \Rightarrow AB = \begin{pmatrix} 1 & 0 \\ 0 & 1 \end{pmatrix}.$$

Then $\text{tr}(A) = \text{tr}(B) = 0$ and $\text{tr}(AB) = 2$.

(ii) As a counterexample we can consider the matrices given in (i).

Problem 15.   Let A be an $n \times n$ *hermitian matrix*, i.e.

$$A = A^* \equiv \bar{A}^T \tag{1}$$

where $^-$ denotes the complex conjugate and ^{T} the transpose.
(i) Show that $A + iI_n$ is invertible, where I_n denotes the $n \times n$ unit matrix.
(ii) Show that $U := (A - iI_n)(A + iI_n)^{-1}$ is a unitary matrix.

Solution 15.   (i) Since A is hermitian we can find a unitary matrix V such that VAV^* is a diagonal matrix, where $V^* = V^{-1}$. The diagonal elements are real (they are the eigenvalues of A). Thus

$$V(A + iI_n)V^* = VAV^* + iVI_nV^* = VAV^* + iI_n$$
$$= \text{diag}(\lambda_1 + i, \lambda_2 + i, \ldots, \lambda_n + i).$$

The inverse of this diagonal matrix exists and therefore the inverse of $A+iI_n$ exists since $\det(XYZ) \equiv \det(X)\det(Y)\det(Z)$ for $n \times n$ matrices X, Y, Z.
(ii) Recall that U is a *unitary matrix* if $U^*U = I_n$, where $U^* \equiv \bar{U}^T$. Note that $A - iI_n$ and $A + iI_n$ commute with each other, i.e.

$$[A - iI_n, A + iI_n] = 0_n.$$

Now

$$U^* = [(A+iI_n)^{-1}]^*(A-iI_n)^* = [(A+iI_n)^*]^{-1}(A+iI_n) = (A-iI_n)^{-1}(A+iI_n).$$

Consequently,

$$U^*U = (A - iI_n)^{-1}(A + iI_n)(A - iI_n)(A + iI_n)^{-1}$$
$$= (A - iI_n)^{-1}(A - iI_n)(A + iI_n)(A + iI_n)^{-1} = I_n.$$

Thus U is a unitary matrix.

Problem 16.   Let X, Y and Z be arbitrary $n \times n$ matrices and $[X, Y] := XY - YX$ be the commutator. Then

$$[X, Y + Z] = [X, Y] + [X, Z], \quad [X, Y] = -[Y, X].$$

Calculate $[X, [Y, Z]] + [Z, [X, Y]] + [Y, [Z, X]]$.

Solution 16.   Since

$$[X, [Y, Z]] = [X, YZ - ZY] = XYZ - YZX - XZY + ZYX$$

$$[Z, [X, Y]] = [Z, XY - YX] = ZXY - XYZ - ZYX + YXZ$$

$$[Y, [Z, X]] = [Y, ZX - XZ] = YZX - ZXY - YXZ + XZY$$

we obtain, by adding these equations

$$[X, [Y, Z]] + [Z, [X, Y]] + [Y, [Z, X]] = 0_n.$$

This equation is called the *Jacobi identity*. The $n \times n$ matrices over the real or complex numbers form a *Lie algebra* under the commutator. The $n \times n$ matrices over the real or complex numbers form an *associative algebra* with unit element (the unit matrix) under matrix multiplication.

Problem 17.   Let A, B and H be three $n \times n$ matrices. Assume that

$$[A, H] = 0_n, \qquad [B, H] = 0_n.$$

Calculate $[[A, B], H]$.

Solution 17.   For arbitrary $n \times n$ matrices X, Y and Z we have (Jacobi identity)

$$[X, [Y, Z]] + [Z, [X, Y]] + [Y, [Z, X]] = 0_n.$$

Now let $X \equiv A$, $Y \equiv B$ and $Z \equiv H$. Since $[A, H] = 0_n$ and $[B, H] = 0_n$ it follows that $[H, [A, B]] = 0_n$. Since $[H, [A, B]] = -[[A, B], H]$ we obtain $[[A, B], H] = 0_n$. If we assume $[H, A] = A$, $[B, H] = B$ we also find $[[A, B], H] = 0_n$.

Problem 18.   Let A and B be two $n \times n$ matrices. Assume that A^{-1} exists. Find the expansion of $(A - \epsilon B)^{-1}$ as a power series in ϵ, where ϵ is a real parameter. Set

$$(A - \epsilon B)^{-1} := \sum_{n=0}^{\infty} \epsilon^n L_n. \tag{1}$$

Solution 18.   Multiplying (1) on the left by $A - \epsilon B$ we obtain

$$I_n = \sum_{n=0}^{\infty} \epsilon^n (A - \epsilon B) L_n = A L_0 + \sum_{n=1}^{\infty} \epsilon^n (A L_n - B L_{n-1})$$

where I_n is the $n \times n$ unit matrix. By equating coefficients of powers of ϵ we find that $L_0 = A^{-1}$ and $L_n = A^{-1} B L_{n-1}$, where $n = 1, 2, \dots$. Consequently,

$$(A - \epsilon B)^{-1} = A^{-1} + \epsilon A^{-1} B A^{-1} + \epsilon^2 A^{-1} B A^{-1} B A^{-1} + \cdots. \tag{2}$$

If $A = I_n$, then expansion (2) takes the form

$$(I_n - \epsilon B)^{-1} = I_n + \epsilon B + \epsilon^2 B^2 + \cdots$$

Discuss the radius of convergence of this series.

Problem 19. (i) Let A and B be two hermitian matrices. Is the commutator $[A, B]$ again a hermitian matrix?
(ii) Let A and B be two skew-hermitian matrices. Is the commutator $[A, B]$ again a skew-hermitian matrix?

Solution 19. (i) Since A and B are hermitian matrices we have $A^* = A$, $B^* = B$. Now

$$([A, B])^* = (AB - BA)^* = (AB)^* - (BA)^* = B^*A^* - A^*B^*.$$

Thus $([A, B])^* = BA - AB = [B, A]$. Consequently, in general the commutator of two hermitian matrices is not a hermitian matrix, since $[A, B] \neq [B, A]$ in general. Note that $i[A, B]$ is a hermitian matrix.
(ii) Since A and B are two *skew-hermitian matrices* we have $A^* = -A$, $B^* = -B$. Now $([A, B])^* = (AB - BA)^* = (AB)^* - (BA)^*$. Thus

$$([A, B])^* = B^*A^* - A^*B^* = BA - AB = -[A, B].$$

Consequently, the commutator of two skew-hermitian matrices is again a skew-hermitian matrix. Notice that $\text{tr}([X, Y]) = 0$ for two arbitrary $n \times n$ matrices X and Y over $\mathbb{R}$ or $\mathbb{C}$.

Problem 20. Let A, B and C be three arbitrary $n \times n$ matrices. Then $\text{tr}(AB) = \text{tr}(BA)$ and therefore $\text{tr}([A, B]) = 0$. Furthermore one has (*cyclic invariance*)

$$\text{tr}(ABC) = \text{tr}(CAB) = \text{tr}(BCA).$$

Find 2×2 matrices A, B and C such that $\text{tr}(ABC) \neq \text{tr}(ACB)$.

Solution 20. Let

$$A = \begin{pmatrix} 1 & 0 \\ 0 & 0 \end{pmatrix}, \qquad B = \begin{pmatrix} 0 & 1 \\ 0 & 0 \end{pmatrix}, \qquad C = \begin{pmatrix} 0 & 0 \\ 1 & 0 \end{pmatrix}.$$

Then

$$ABC = \begin{pmatrix} 1 & 0 \\ 0 & 0 \end{pmatrix}, \qquad ACB = \begin{pmatrix} 0 & 0 \\ 0 & 0 \end{pmatrix}.$$

Consequently $\text{tr}(ABC) = 1$, $\text{tr}(ACB) = 0$.

Problem 21. (i) Let $A(\epsilon)$ be an $n \times n$ matrix. The entries depend smoothly on a real parameter ϵ. Assume that $dA/d\epsilon$ and $A^{-1}(\epsilon)$ exist for all ϵ. Show that

$$\frac{dA^{-1}}{d\epsilon} = -A^{-1}(\epsilon)\frac{dA}{d\epsilon}A^{-1}(\epsilon). \tag{1}$$

(ii) Let $B(\epsilon)$ be an $n \times n$ matrix. The entries depend smoothly on a real parameter ϵ. Show that in general

$$B(\epsilon)\frac{dB}{d\epsilon} \neq \frac{dB}{d\epsilon}B(\epsilon). \tag{1}$$

Solution 21. (i) We start from the identity $A(\epsilon)A^{-1}(\epsilon) = I_n$, where I_n is the $n \times n$ unit matrix. Taking the derivative of this identity with respect to ϵ yields

$$\frac{dA}{d\epsilon}A^{-1}(\epsilon) + A(\epsilon)\frac{dA^{-1}}{d\epsilon} = 0_n$$

since $dI_n/d\epsilon = 0_n$. Therefore

$$\frac{dA}{d\epsilon}A^{-1}(\epsilon) = -A(\epsilon)\frac{dA^{-1}}{d\epsilon}.$$

Multiplying both sides from the left by $A^{-1}(\epsilon)$ we find (1).
(ii) Let

$$B(\epsilon) := \begin{pmatrix} \epsilon & \epsilon^2 \\ \epsilon^3 & 1 \end{pmatrix} \Rightarrow \frac{dB}{d\epsilon} = \begin{pmatrix} 1 & 2\epsilon \\ 3\epsilon^2 & 0 \end{pmatrix}.$$

Thus

$$B\frac{dB}{d\epsilon} = \begin{pmatrix} \epsilon + 3\epsilon^4 & 2\epsilon^2 \\ 3\epsilon^2 + \epsilon^3 & 2\epsilon^4 \end{pmatrix}, \qquad \frac{dB}{d\epsilon}B = \begin{pmatrix} \epsilon + 2\epsilon^4 & 2\epsilon + \epsilon^2 \\ 3\epsilon^3 & 3\epsilon^4 \end{pmatrix}.$$

Thus (1) holds in general. In particular we have

$$\frac{d}{d\epsilon}B^2 = B\frac{dB}{d\epsilon} + \frac{dB}{d\epsilon}B \neq 2B\frac{dB}{d\epsilon}$$

in general.

Problem 22. Let A be a square finite-dimensional matrix over $\mathbb{R}$ such that

$$AA^T = I_n. \tag{1}$$

Show that

$$A^T A = I_n. \tag{2}$$

Does (2) also hold for infinite dimensional matrices?

Solution 22.   (i) Since $\det(A) = \det(A^T)$ and $\det(I_n) = 1$ we obtain from (1) that $(\det(A))^2 = 1$. Therefore the inverse of A exists and we have $A^T = A^{-1}$ with $A^{-1}A = AA^{-1} = I_n$.
(ii) The answer is no. Let

$$A = \begin{pmatrix} 0 & 1 & 0 & 0 & 0 & \dots \\ 0 & 0 & 1 & 0 & 0 & \dots \\ 0 & 0 & 0 & 1 & 0 & \dots \\ \vdots & \vdots & \vdots & \vdots & \ddots & \dots \end{pmatrix}.$$

Then the transpose matrix A^T of A is given by

$$A^T = \begin{pmatrix} 0 & 0 & 0 & 0 & 0 & \dots \\ 1 & 0 & 0 & 0 & 0 & \dots \\ 0 & 1 & 0 & 0 & 0 & \dots \\ 0 & 0 & 1 & 0 & 0 & \dots \\ \vdots & \vdots & \vdots & \vdots & \ddots & \dots \end{pmatrix}.$$

It follows that $AA^T = \text{diag}(1, 1, 1, \dots) \equiv I$, $A^T A = \text{diag}(0, 1, 1, \dots)$, where I is the infinite-dimensional unit matrix. Hence $A^T A \neq AA^T$.

Problem 23.   Consider the boundary value problem (*Dirichlet boundary conditions*)

$$\frac{d^2u}{dx^2} + u = 0, \qquad u(0) = u(1) = 1$$

for the interval $[0, 1]$. The exact solution is given by

$$u(x) = \cos(x) + \frac{1 - \cos(1)}{\sin(1)} \sin(x).$$

Find a matrix equation for the discretization

$$\frac{d^2u}{dx^2} \longrightarrow \frac{u_{i-1} - 2u_i + u_{i+1}}{h^2}, \qquad u \longrightarrow u_i$$

where $h = 0.1$ and $u_i := u(i \cdot h)$.

Solution 23.   Since the interval is $[0, 1]$ and $h = 0.1$ we have $u_0, u_1, \dots, u_{10}$, where u_0 and u_{10} are given by the boundary conditions. We have

$$\begin{aligned} i = 1: & \quad u_0 - 2u_1 + u_2 + h^2 u_1 = 0 \\ i = 2: & \quad u_1 - 2u_2 + u_3 + h^2 u_2 = 0 \\ i = 3: & \quad u_2 - 2u_3 + u_4 + h^2 u_3 = 0 \\ & \quad \vdots \\ i = 9: & \quad u_8 - 2u_9 + u_{10} + h^2 u_9 = 0. \end{aligned}$$

The first equation can be brought into the form $-2u_1 + u_2 + h^2 u_1 = -u_0$. The last equation can be brought into the form $u_8 - 2u_9 + h^2 u_9 = -u_{10}$. The boundary conditions $u(0) = 1$, $u(1) = 1$ provides $u_0 = 1$, $u_{10} = 1$. Thus we get the matrix form of the discretization with a 9×9 matrix

$$
\begin{pmatrix}
\alpha & 1 & 0 & 0 & 0 & 0 & 0 & 0 & 0 \\
1 & \alpha & 1 & 0 & 0 & 0 & 0 & 0 & 0 \\
0 & 1 & \alpha & 1 & 0 & 0 & 0 & 0 & 0 \\
0 & 0 & 1 & \alpha & 1 & 0 & 0 & 0 & 0 \\
0 & 0 & 0 & 1 & \alpha & 1 & 0 & 0 & 0 \\
0 & 0 & 0 & 0 & 1 & \alpha & 1 & 0 & 0 \\
0 & 0 & 0 & 0 & 0 & 1 & \alpha & 1 & 0 \\
0 & 0 & 0 & 0 & 0 & 0 & 1 & \alpha & 1 \\
0 & 0 & 0 & 0 & 0 & 0 & 0 & 1 & \alpha
\end{pmatrix}
\begin{pmatrix}
u_1 \\ u_2 \\ u_3 \\ u_4 \\ u_5 \\ u_6 \\ u_7 \\ u_8 \\ u_9
\end{pmatrix}
=
\begin{pmatrix}
-1 \\ 0 \\ 0 \\ 0 \\ 0 \\ 0 \\ 0 \\ 0 \\ -1
\end{pmatrix}
$$

with the abbreviation $\alpha := -2 + h^2$. The matrix on the left-hand side is a tridiagonal matrix. The inverse of the matrix exists. The solution algorithm for the tridiagonal equation is called the tridiagonal solution (a variant of Gaussian elimination).

Problem 24. Let A and L be two $n \times n$ matrices. Calculate $e^L A e^{-L}$. Consider $A(\epsilon) := e^{\epsilon L} A e^{-\epsilon L}$, where ϵ is a real parameter and apply *parameter differentiation*.

Solution 24. Taking the derivative of $A(\epsilon)$ with respect to ϵ yields

$$
\frac{dA}{d\epsilon} = L e^{\epsilon L} A e^{-\epsilon L} - e^{\epsilon L} A e^{-\epsilon L} L = [L, A(\epsilon)].
$$

The second derivative gives

$$
\frac{d^2 A}{d\epsilon^2} = \left[L, \frac{dA}{d\epsilon} \right] = [L, [L, A(\epsilon)]]
$$

and so on. Consequently, we can write the matrix $e^L A e^{-L} = A(1)$ as a *Taylor series expansion* about the origin, i.e.

$$
A(1) = A(0) + \frac{1}{1!} \frac{dA(0)}{d\epsilon} + \frac{1}{2!} \frac{d^2 A(0)}{d\epsilon^2} + \cdots
$$

where $A(0) \equiv A$ and

$$
\frac{dA(0)}{d\epsilon} \equiv \left. \frac{dA(\epsilon)}{d\epsilon} \right|_{\epsilon=0}.
$$

Thus we find

$$
e^L A e^{-L} \equiv A + [L, A] + \frac{1}{2!} [L, [L, A]] + \frac{1}{3!} [L, [L, [L, A]]] + \cdots .
$$

Problem 25. Let A and B be two $n \times n$ matrices. Assume that

$$[[A, B], A] = 0_n, \qquad [[A, B], B] = 0_n. \tag{1}$$

Show that

$$\exp(A) \exp(B) \equiv \exp(A + B + \frac{1}{2}[A, B]). \tag{2}$$

Start with

$$T(\epsilon) := e^{\epsilon A} e^{\epsilon B} \tag{3}$$

where ϵ is a real parameter. Apply *parameter differentiation*.

Solution 25. Differentiating (3) with respect to ϵ gives

$$\frac{dT}{d\epsilon} = A e^{\epsilon A} e^{\epsilon B} + e^{\epsilon A} B e^{\epsilon B} = (A + e^{\epsilon A} B e^{-\epsilon A}) T(\epsilon). \tag{4}$$

Using (1), we find $e^{\epsilon A} B e^{-\epsilon A} = B - [B, A]\epsilon$. Therefore

$$\frac{dT}{d\epsilon} = (A + B + [A, B]\epsilon) T(\epsilon).$$

Consequently, T is the solution to this matrix differential equation with the initial condition $T(\epsilon = 0) = I_n$, where I_n is the $n \times n$ unit matrix. The initial condition follows from (3). Since the matrices $A + B$ and $[A, B]$ commute, the matrix differential equation (4) can be integrated as if the matrices were merely numbers, to give the solution

$$T(\epsilon) = e^{\epsilon(A+B)} e^{\frac{1}{2}\epsilon^2[A,B]}.$$

The identity (2) follows by setting $\epsilon = 1$.

Problem 26. Let $\boldsymbol{\sigma} := (\sigma_1, \sigma_2, \sigma_3)$, where

$$\sigma_1 := \begin{pmatrix} 0 & 1 \\ 1 & 0 \end{pmatrix}, \qquad \sigma_2 := \begin{pmatrix} 0 & -i \\ i & 0 \end{pmatrix}, \qquad \sigma_3 := \begin{pmatrix} 1 & 0 \\ 0 & -1 \end{pmatrix}$$

are the *Pauli spin matrices*. Let $\mathbf{a} := (a_1, a_2, a_3)$ where $a_j \in \mathbb{R}$ and

$$\mathbf{a} \cdot \boldsymbol{\sigma} := a_1 \sigma_1 + a_2 \sigma_2 + a_3 \sigma_3.$$

(i) Calculate $(\mathbf{a} \cdot \boldsymbol{\sigma})^2$, $(\mathbf{a} \cdot \boldsymbol{\sigma})^3$ and $(\mathbf{a} \cdot \boldsymbol{\sigma})^4$.
(ii) Calculate $\exp(\mathbf{a} \cdot \boldsymbol{\sigma})$.

Solution 26. (i) First we notice that $\sigma_1^2 = I_2$, $\sigma_2^2 = I_2$, $\sigma_3^2 = I_2$, where I_2 is the 2×2 unit matrix. Furthermore the anticommutator vanishes

$$\sigma_1 \sigma_2 + \sigma_2 \sigma_1 = 0_2, \qquad \sigma_1 \sigma_3 + \sigma_3 \sigma_1 = 0_2, \qquad \sigma_2 \sigma_3 + \sigma_3 \sigma_2 = 0_2$$

where 0_2 is the 2×2 zero matrix. Thus

$$(\mathbf{a} \cdot \boldsymbol{\sigma})^2 = (a_1\sigma_1 + a_2\sigma_2 + a_3\sigma_3)(a_1\sigma_1 + a_2\sigma_2 + a_3\sigma_3) = (a_1^2 + a_2^2 + a_3^2)I_2.$$

In the following we set $a^2 \equiv \mathbf{a}^2 := a_1^2 + a_2^2 + a_3^2$. It follows that

$$(\mathbf{a} \cdot \boldsymbol{\sigma})^3 = (\mathbf{a} \cdot \boldsymbol{\sigma})^2(\mathbf{a} \cdot \boldsymbol{\sigma}) = (a^2 I_2)(\mathbf{a} \cdot \boldsymbol{\sigma}) = a^2(\mathbf{a} \cdot \boldsymbol{\sigma})$$
$$(\mathbf{a} \cdot \boldsymbol{\sigma})^4 = (\mathbf{a} \cdot \boldsymbol{\sigma})^2(\mathbf{a} \cdot \boldsymbol{\sigma})^2 = (a^2 I_2)(a^2 I) = a^4 I_2.$$

(ii) From the definition

$$\exp(\mathbf{a} \cdot \boldsymbol{\sigma}) := \sum_{k=0}^{\infty} \frac{(\mathbf{a} \cdot \boldsymbol{\sigma})^k}{k!}$$

and the result from (i) we have

$$\exp(\mathbf{a} \cdot \boldsymbol{\sigma}) = \left(1 + \frac{a^2}{2!} + \frac{a^4}{4!} + \cdots\right) I_2 + \left(1 + \frac{a^2}{3!} + \frac{a^4}{5!} + \cdots\right) \mathbf{a} \cdot \boldsymbol{\sigma}.$$

Thus

$$\exp(\mathbf{a} \cdot \boldsymbol{\sigma}) = (\cosh(a))I_2 + \frac{1}{a}(\sinh(a))(\mathbf{a} \cdot \boldsymbol{\sigma})$$

for $a \neq 0$. If $\mathbf{a} = (0, 0, 0)$, we have

$$\exp(\mathbf{a} \cdot \boldsymbol{\sigma}) = \begin{pmatrix} 1 & 0 \\ 0 & 1 \end{pmatrix} = I_2.$$

Problem 27. Let $f : \mathbb{R} \to \mathbb{R}$ be an analytic function. Let $\theta \in \mathbb{R}$, $\mathbf{n}$ be a normalized vector in $\mathbb{R}^3$ and $\sigma_1, \sigma_2, \sigma_3$ be the Pauli spin matrices. We define $\mathbf{n} \cdot \boldsymbol{\sigma} := n_1\sigma_1 + n_2\sigma_2 + n_3\sigma_3$. Then

$$f(\theta\mathbf{n} \cdot \boldsymbol{\sigma}) \equiv \frac{1}{2}(f(\theta) + f(-\theta))I_2 + \frac{1}{2}(f(\theta) - f(-\theta))(\mathbf{n} \cdot \boldsymbol{\sigma}).$$

Apply this identity to $f(x) = \sin(x)$.

Solution 27. Since $f(-x) = \sin(-x) = -\sin(x)$ we have $f(x) + f(-x) = 0$ and $f(x) - f(-x) = 2\sin(x)$. Thus we obtain $f(\theta\mathbf{n} \cdot \boldsymbol{\sigma}) = \sin(\theta)\mathbf{n} \cdot \boldsymbol{\sigma}$.

Problem 28. Consider the symmetric 2×2 matrix

$$A = \begin{pmatrix} -1 & 2 \\ 2 & -1 \end{pmatrix}. \tag{1}$$

Calculate $\exp(\epsilon A)$ where $\epsilon \in \mathbb{R}$.

Solution 28. The matrix A is symmetric over $\mathbb{R}$. Therefore, there exists an orthogonal matrix U such that UAU^{-1} is a diagonal matrix. The diagonal elements of UAU^{-1} are the eigenvalues of A. The matrix U^* we find from the normalized eigenvectors of A. Since A is symmetric the eigenvalues are real. We set $D = UAU^{-1}$ with

$$D = \begin{pmatrix} d_{11} & 0 \\ 0 & d_{22} \end{pmatrix} \Rightarrow \exp(D) = \begin{pmatrix} e^{d_{11}} & 0 \\ 0 & e^{d_{22}} \end{pmatrix}.$$

From $D = UAU^{-1}$ it follows that

$$\exp(\epsilon D) = \exp(\epsilon U A U^{-1}) = U \exp(\epsilon A)U^{-1}.$$

Therefore $\exp(\epsilon A) = U^{-1}\exp(\epsilon D)U$. The matrix U is constructed by means of the eigenvalues and normalized eigenvectors of A. The eigenvalues of A are given by $\lambda_1 = 1$, $\lambda_2 = -3$. The corresponding normalized eigenvectors are

$$\mathbf{v}_1 = \frac{1}{\sqrt{2}}\begin{pmatrix} 1 \\ 1 \end{pmatrix}, \qquad \mathbf{v}_2 = \frac{1}{\sqrt{2}}\begin{pmatrix} 1 \\ -1 \end{pmatrix}.$$

Consequently, the matrix U^* is given by

$$U^* = \frac{1}{\sqrt{2}}\begin{pmatrix} 1 & 1 \\ 1 & -1 \end{pmatrix}.$$

It follows that $U^* = U = U^{-1}$. Finally we arrive at

$$\exp(\epsilon A) = U^* e^{\epsilon D} U = \frac{1}{2}\begin{pmatrix} e^\epsilon + e^{-3\epsilon} & e^\epsilon - e^{-3\epsilon} \\ e^\epsilon - e^{-3\epsilon} & e^\epsilon + e^{-3\epsilon} \end{pmatrix}.$$

The solution of the initial value problem of the autonomous system of linear ordinary differential equations

$$\frac{du_1}{d\epsilon} = -u_1 + 2u_2, \qquad \frac{du_2}{d\epsilon} = 2u_1 - u_2$$

is given by

$$\begin{pmatrix} u_1(\epsilon) \\ u_2(\epsilon) \end{pmatrix} = e^{\epsilon A}\begin{pmatrix} u_1(\epsilon = 0) \\ u_2(\epsilon = 0) \end{pmatrix}.$$

Problem 29. Let $\epsilon \in \mathbb{R}$. Calculate $f(\epsilon) = e^{-\epsilon\sigma_2}\sigma_3 e^{\epsilon\sigma_2}$. Hint. Differentiate the matrix-valued function f with respect to ϵ and solve the initial value problem of the resulting ordinary differential equation.

Solution 29.   Since $[\sigma_2, \sigma_3] = 2i\sigma_1$, $[\sigma_2, \sigma_1] = -2i\sigma_3$ we obtain

$$\frac{df}{d\epsilon} = -e^{-\epsilon\sigma_2}\sigma_2\sigma_3 e^{\epsilon\sigma_2} + e^{-\epsilon\sigma_2}\sigma_3\sigma_2 e^{\epsilon\sigma_2} = -e^{-\epsilon\sigma_2}[\sigma_2, \sigma_3]e^{\epsilon\sigma_2}$$

$$= -2ie^{-\epsilon\sigma_2}\sigma_1 e^{\epsilon\sigma_2}$$

and

$$\frac{d^2 f}{d\epsilon^2} = 2i(e^{-\epsilon\sigma_2}\sigma_2\sigma_1 e^{\epsilon\sigma_2} - e^{-\epsilon\sigma_2}\sigma_1\sigma_2 e^{\epsilon\sigma_2}) = -2ie^{-\epsilon\sigma_2}2i\sigma_3 e^{\epsilon\sigma_2} = 4f(\epsilon).$$

Thus we have to solve the linear differential equation with constant coefficients $d^2 f/d\epsilon^2 = 4f$. The initial conditions are $f(0) = \sigma_3$ and $df(0)/d\epsilon = -2i\sigma_1$. We obtain the solution

$$f(\epsilon) = \frac{1}{2}(\sigma_3 - i\sigma_1)e^{2\epsilon} + \frac{1}{2}(\sigma_3 + i\sigma_1)e^{-2\epsilon}.$$

Problem 30.   Let A, B be $n \times n$ matrices over $\mathbb{C}$ and $\alpha \in \mathbb{C}$. The *Baker-Campbell-Hausdorff formula* states that

$$e^{\alpha A}Be^{-\alpha A} = B + \alpha[A, B] + \frac{\alpha^2}{2!}[A, [A, B]] + \cdots = \sum_{j=0}^{\infty}\frac{\alpha^j}{j!}\{A^j, B\} = \widetilde{B}(\alpha)$$

where $[A, B] := AB - BA$ and $\{A^j, B\} = [A, \{A^{j-1}, B\}]$ is the repeated commutator.
(i) Extend the formula to $e^{\alpha A}B^k e^{-\alpha A}$, where $k \geq 1$.
(ii) Extend the formula to $e^{\alpha A}e^B e^{-\alpha A}$.

Solution 30.   (i) Using that $e^{-\alpha A}e^{\alpha A} = I_n$ we have

$$e^{\alpha A}B^k e^{-\alpha A} = e^{\alpha A}Be^{-\alpha A}e^{\alpha A}B^{k-1}e^{-\alpha A} = \widetilde{B}(\alpha)e^{\alpha A}B^{k-1}e^{-\alpha A} = (\widetilde{B}(\alpha))^k.$$

(ii) Using the result from (i) we obtain $e^{\alpha A}e^B e^{-\alpha A} = \exp(\widetilde{B}(\alpha))$.

Problem 31.   Let $a > 0$ with dimension length. Find the determinant of the matrix

$$M(a) = \frac{a}{2}\begin{pmatrix} 1 & 0 & 1 \\ 1 & 1 & 0 \\ 0 & 1 & 1 \end{pmatrix}$$

which plays a role for the fcc-lattice. Give an interpretation.

Solution 31.   We find $\det(M(a)) = a^3/4$ which is the volume spanned by the vectors

$$\begin{pmatrix} a/2 \\ a/2 \\ 0 \end{pmatrix}, \quad \begin{pmatrix} 0 \\ a/2 \\ a/2 \end{pmatrix}, \quad \begin{pmatrix} a/2 \\ 0 \\ a/2 \end{pmatrix}.$$

Problem 32. Consider the Hilbert space of 2×2 matrices over $\mathbb{C}$ with the scalar product

$$\langle X, Y \rangle := \text{tr}(XY^*).$$

Let $\alpha, \beta \in \mathbb{R}$ and

$$A(\alpha) = \frac{1}{\sqrt{2}} \begin{pmatrix} \cos(\alpha) & \sin(\alpha) \\ \sin(\alpha) & -\cos(\alpha) \end{pmatrix}, \quad B(\beta) = \frac{1}{\sqrt{2}} \begin{pmatrix} \cos(\beta) & \sin(\beta) \\ \sin(\beta) & -\cos(\beta) \end{pmatrix}.$$

Find the condition on α, β such that $\langle A(\alpha), B(\beta) \rangle = \frac{1}{2}$. The matrices $A(\alpha)$, $B(\beta)$ contain $\frac{1}{\sqrt{2}}\sigma_3$ with $\alpha = 0$, $\beta = 0$ and $\frac{1}{\sqrt{2}}\sigma_1$ with $\alpha = \pi/2$, $\beta = \pi/2$.

Solution 32. Note that $A(\alpha) = A^*(\alpha)$, $B(\beta) = B^*(\beta)$. Then

$$\langle A(\alpha), B(\beta) \rangle = \text{tr}(A(\alpha)B(\beta)) = \cos(\alpha)\cos(\beta) + \sin(\alpha)\sin(\beta) = \cos(\alpha - \beta).$$

Hence $\cos(\alpha - \beta) = \frac{1}{2}$ with the solutions $\alpha - \beta = \pi/3, 5\pi/3 \pmod{2\pi}$. Study the case

$$\langle A(\alpha) \otimes A(\alpha), B(\beta) \otimes B(\beta) \rangle = \frac{1}{4}.$$

Extend the matrix to

$$A(\alpha, \phi) = \frac{1}{\sqrt{2}} \begin{pmatrix} \cos(\alpha) & e^{i\phi}\sin(\alpha) \\ e^{-i\phi}\sin(\alpha) & -\cos(\alpha) \end{pmatrix}.$$

Problem 33. (i) Let $\alpha \in \mathbb{R}$ and

$$A(\alpha) = \begin{pmatrix} 0 & \alpha \\ \alpha & 0 \end{pmatrix}.$$

Find $\sin(A)$ utilizing the *Cayley-Hamilton theorem*.
(ii) Consider the matrix

$$A = \begin{pmatrix} \lambda_1 & c \\ 0 & \lambda_2 \end{pmatrix}$$

where $\lambda_1, \lambda_2, c \in \mathbb{R}$. Calculate $\sin(A)$.

Solution 33. (i) We have

$$\sin(A) = b_1 A + b_0 I_2 = \begin{pmatrix} b_0 & b_1\alpha \\ b_1\alpha & b_0 \end{pmatrix}.$$

The eigenvalues of $A(\alpha)$ are $\lambda = \pm\alpha$. Then

$$\sin(\lambda_+) = \sin(\alpha) = b_1\lambda_+ + b_0, \quad \sin(\lambda_-) = -\sin(\alpha) = b_1\lambda_- + b_0.$$

Thus we have to solve the two equations

$$\sin(\alpha) = b_1\alpha + b_0, \qquad -\sin(\alpha) = -b_1\alpha + b_0$$

and find $b_0 = 0$, $b_1 = \frac{\sin(\alpha)}{\alpha}$. Finally

$$\sin(A) = \begin{pmatrix} 0 & b_1\alpha \\ b_1\alpha & 0 \end{pmatrix} + \begin{pmatrix} b_0 & 0 \\ 0 & b_0 \end{pmatrix} = \begin{pmatrix} 0 & \sin(\alpha) \\ \sin(\alpha) & 0 \end{pmatrix}.$$

(ii) For the case $\lambda_1 = \lambda_2 = \lambda$ we obviously find

$$\sin\begin{pmatrix} \lambda & c \\ 0 & \lambda \end{pmatrix} = \begin{pmatrix} \sin(\lambda) & c\cos(\lambda) \\ 0 & \sin(\lambda) \end{pmatrix}.$$

Consider now the case $\lambda_1 \neq \lambda_2$. First we note that

$$\begin{pmatrix} \lambda_1 & c \\ 0 & \lambda_2 \end{pmatrix}^n = \begin{pmatrix} \lambda_1^n & s \\ 0 & \lambda_2^n \end{pmatrix}$$

where $s = (\lambda_1^{n-1} + \lambda_1^{n-2}\lambda_2 + \lambda_1^{n-3}\lambda_2^2 + \cdots + \lambda_2^{n-1})c$. Thus

$$\frac{s(\lambda_1 - \lambda_2)}{\lambda_1 - \lambda_2} = \frac{1}{\lambda_1 - \lambda_2}(\lambda_1^n - \lambda_2^n)c.$$

Thus

$$\sin(A) = \begin{pmatrix} \sin(\lambda_1) & c(\sin(\lambda_1) - \sin(\lambda_2))/(\lambda_1 - \lambda_2) \\ 0 & \sin(\lambda_2) \end{pmatrix}.$$

Problem 34.   Consider the vectors in $\mathbb{R}^2$

$$\mathbf{v}(\alpha) = \begin{pmatrix} \cos(\alpha) \\ \sin(\alpha) \end{pmatrix}, \quad \mathbf{u}(\beta) = \begin{pmatrix} \cos(\beta) \\ \sin(\beta) \end{pmatrix}.$$

Find α, β such that $|\mathbf{v}^*(\alpha)\mathbf{u}(\beta)|^2 = \frac{1}{2}$.

Solution 34.   We have $\mathbf{v}(\alpha)\mathbf{u}(\beta) = \cos(\alpha - \beta)$. Hence $\cos^2(\alpha - \beta) = \frac{1}{2}$. A solution is $\alpha - \beta = \pi/4$.

Problem 35.   Let A be an $n \times n$ matrix over $\mathbb{C}$. Does the matrix equation

$$A + A^* + AA^* = 0_n$$

only admit the trivial solution $A = 0_n$?

Solution 35.   No. Consider the *polar form* $A = HU$, where H is hermitian and U is unitary. Assume that H is nonzero, then there exist a nonzero real eigenvalue λ of H with corresponding eigenvector $\mathbf{v}$. Thus

$$\mathbf{v}^*(HU + U^*H + H^2)\mathbf{v} = \lambda(\mathbf{v}^*U\mathbf{v} + \mathbf{v}^*U^*\mathbf{v}) + \lambda^2\mathbf{v}^*\mathbf{v} = 0.$$

By assumption $\lambda \neq 0$ so that

$$\lambda = -\frac{(v^*Uv + v^*U^*v)}{v^*v} \neq 0.$$

For example, let $n = 2$, $U = \sigma_1$ (Pauli spin matrix) and $v = \frac{1}{\sqrt{2}}(1 \quad 1)^T$. Then

$$H = \begin{pmatrix} -1 & -1 \\ -1 & -1 \end{pmatrix} \Rightarrow A = HU = H.$$

Thus

$$\begin{pmatrix} -1 & -1 \\ -1 & -1 \end{pmatrix} + \begin{pmatrix} -1 & -1 \\ -1 & -1 \end{pmatrix} + \begin{pmatrix} 2 & 2 \\ 2 & 2 \end{pmatrix} = \begin{pmatrix} 0 & 0 \\ 0 & 0 \end{pmatrix}.$$

Problem 36. Consider the orthogonal group $O(n, \mathbb{R}) \subset \mathbb{R}^{n \times n}$ with the linear product $(v, w \in O(n, \mathbb{R}))$

$$\langle v, w \rangle := \mathrm{tr}(v^T w).$$

The orthogonal projection $\Pi(g) : \mathbb{R}^{n \times n} \to T_g O(n)$ is given by

$$\Pi(g)v := \frac{1}{2}(v - gv^T g).$$

Let $n = 2$ and

$$g = \begin{pmatrix} 0 & 1 \\ 1 & 0 \end{pmatrix}, \quad v = \begin{pmatrix} a & b \\ c & d \end{pmatrix}.$$

Find the orthogonal projection.

Solution 36. We have

$$\Pi(g)v = \frac{1}{2}\left(\begin{pmatrix} a & b \\ c & d \end{pmatrix} - \begin{pmatrix} 0 & 1 \\ 1 & 0 \end{pmatrix} \begin{pmatrix} a & c \\ b & d \end{pmatrix} \begin{pmatrix} 0 & 1 \\ 1 & 0 \end{pmatrix} \right) = \frac{1}{2} \begin{pmatrix} a - d & 0 \\ 0 & a - d \end{pmatrix}.$$

Consider the case that (Hadamard matrix)

$$g = \frac{1}{\sqrt{2}} \begin{pmatrix} 1 & 1 \\ 1 & -1 \end{pmatrix}.$$

Problem 37. Let I_2 be the 2×2 identity matrix. Find all solutions of

$$X^2 = I_2.$$

We call X the *square root* of the 2×2 identity matrix.

Solution 37.   We find the trivial solutions $X = I_2$ and $X = -I_2$. Furthermore with $b \neq 0$

$$X = \begin{pmatrix} a & b \\ (1-a^2)/b & -a \end{pmatrix}.$$

This case includes the Pauli spin matrices σ_1, σ_2, σ_3. Finally

$$X = \begin{pmatrix} 1 & 0 \\ c & -1 \end{pmatrix}, \quad X = \begin{pmatrix} -1 & 0 \\ c & 1 \end{pmatrix}, \quad X = \begin{pmatrix} 1 & d \\ 0 & -1 \end{pmatrix}, \quad X = \begin{pmatrix} -1 & d \\ 0 & 1 \end{pmatrix}$$

are solutions. These matrices are nonnormal if $c \neq 0$ and $d \neq 0$.

Supplementary Problems

Problem 1.   Let $x \in \mathbb{R}$. Do the vector

$$\mathbf{v}_1 = \begin{pmatrix} 1 \\ x \\ x^2 \\ x^3 \end{pmatrix}, \quad \mathbf{v}_2 = \begin{pmatrix} 0 \\ 1 \\ 2x \\ 3x^2 \end{pmatrix}, \quad \mathbf{v}_3 = \begin{pmatrix} 0 \\ 0 \\ 2 \\ 6x \end{pmatrix}, \quad \mathbf{v}_4 = \begin{pmatrix} 0 \\ 0 \\ 0 \\ 6 \end{pmatrix}$$

form an orthonormal basis in $\mathbb{R}^6$? Find the determinant of the 4×4 matrix

$$\begin{pmatrix} \mathbf{v}_1^T\mathbf{v}_1 & \mathbf{v}_1^T\mathbf{v}_2 & \mathbf{v}_1^T\mathbf{v}_3 & \mathbf{v}_1^T\mathbf{v}_4 \\ \mathbf{v}_2^T\mathbf{v}_1 & \mathbf{v}_2^T\mathbf{v}_2 & \mathbf{v}_2^T\mathbf{v}_3 & \mathbf{v}_2^T\mathbf{v}_4 \\ \mathbf{v}_3^T\mathbf{v}_1 & \mathbf{v}_3^T\mathbf{v}_2 & \mathbf{v}_3^T\mathbf{v}_3 & \mathbf{v}_3^T\mathbf{v}_4 \\ \mathbf{v}_4^T\mathbf{v}_1 & \mathbf{v}_4^T\mathbf{v}_2 & \mathbf{v}_4^T\mathbf{v}_3 & \mathbf{v}_4^T\mathbf{v}_4 \end{pmatrix}.$$

Problem 2.   Find the set of all four (column) vectors $\mathbf{u}_1$, $\mathbf{u}_2$, $\mathbf{v}_1$, $\mathbf{v}_2$ in $\mathbb{R}^2$ such that the following conditions are satisfied

$$\mathbf{v}_1^T\mathbf{u}_2 = 0, \quad \mathbf{v}_2^T\mathbf{u}_1 = 0, \quad \mathbf{v}_1^T\mathbf{u}_1 = 1, \quad \mathbf{v}_2^T\mathbf{u}_2 = 1.$$

One obtains the following four conditions

$$v_{11}u_{21} + v_{12}u_{22} = 0, \qquad v_{21}u_{11} + v_{22}u_{12} = 0,$$

$$v_{11}u_{11} + v_{12}u_{12} = 1, \qquad v_{21}u_{21} + v_{22}u_{22} = 1.$$

We have four equations with eight unknowns. Find solutions of these equations.

Problem 3.   The *character table* of C_3 ($\omega = \exp(-i2\pi/3)$) is given by

A	1	1	1
E	1	ω	$\bar{\omega}$
E'	1	$\bar{\omega}$	ω

Is the 3×3 matrix

$$U = \frac{1}{\sqrt{3}} \begin{pmatrix} 1 & 1 & 1 \\ 1 & \omega & \bar{\omega} \\ 1 & \bar{\omega} & \omega \end{pmatrix}$$

unitary?

Problem 4. Consider a *tetrahedron* described by the three nonzero vectors $\mathbf{v}_1$, $\mathbf{v}_2$, $\mathbf{v}_3$. We also assume the vectors are linearly independent. Show that the normals to the faces defined by two of these three vectors, normalized to the area of the face, are given by

$$\mathbf{n}_1 = \frac{1}{2}\mathbf{v}_2 \times \mathbf{v}_3, \quad \mathbf{n}_2 = \frac{1}{2}\mathbf{v}_3 \times \mathbf{v}_1, \quad \mathbf{n}_3 = \frac{1}{2}\mathbf{v}_1 \times \mathbf{v}_2.$$

Show that

$$\mathbf{v}_1 = \frac{2}{3V}\mathbf{n}_2 \times \mathbf{n}_3, \quad \mathbf{v}_2 = \frac{2}{3V}\mathbf{n}_3 \times \mathbf{n}_1, \quad \mathbf{v}_3 = \frac{2}{3V}\mathbf{n}_1 \times \mathbf{n}_2$$

where the volume V is given by $V = \frac{1}{3!}(\mathbf{v}_1 \times \mathbf{v}_2) \cdot \mathbf{v}_3$.

Problem 5. Let σ_1, σ_2, σ_3 be the Pauli spin matrices. Consider the sixteen 4×4 matrices

$$I_{16}, \quad \Gamma_1 = \begin{pmatrix} \sigma_1 & 0_2 \\ 0_2 & \sigma_1 \end{pmatrix}, \quad \Gamma_2 = \begin{pmatrix} \sigma_2 & 0_2 \\ 0_2 & \sigma_2 \end{pmatrix},$$

$$\Gamma_3 = \begin{pmatrix} 0_2 & \sigma_3 \\ \sigma_3 & 0_2 \end{pmatrix}, \quad \Gamma_4 = \begin{pmatrix} 0_2 & -i\sigma_3 \\ i\sigma_3 & 0_2 \end{pmatrix}, \quad \Gamma_5 = \begin{pmatrix} -\sigma_3 & 0_2 \\ 0_2 & \sigma_3 \end{pmatrix},$$

$$\Gamma_{[\mu,\nu]} = \frac{1}{2}i(\Gamma_\mu\Gamma_\nu - \Gamma_\nu\Gamma_\mu), \quad \mu, \nu = 1, 2, 3, 4, 5.$$

Show that

$$\Gamma_\nu\Gamma_\mu + \Gamma_\mu\Gamma_\nu = 2\delta_{\nu\mu}I_{16}$$

$$[\Gamma_\nu, \Gamma_{[\lambda,\mu]}] = 2i\delta_{\nu\lambda}\Gamma_\mu - 2i\delta_{\nu\mu}\Gamma_\lambda$$

$$[\Gamma_{[\nu,\mu]}, \Gamma_{[\lambda,\sigma]}] = 2i\delta_{\mu\lambda}\Gamma_{[\nu,\sigma]} - 2i\delta_{\nu\lambda}\Gamma_{[\mu,\sigma]} + 2i\delta_{\nu\sigma}\Gamma_{[\nu,\lambda]} - 2i\delta_{\mu\sigma}\Gamma_{[\nu,\lambda]}.$$

Do the 16 matrices form an orthonormal basis in the Hilbert space of the 4×4 matrices?

Problem 6.   Show that the vector $\mathbf{v}$ in $\mathbb{R}^4$ is normalized

$$\mathbf{v} = \begin{pmatrix} \cos(\theta_1) \\ \sin(\theta_1)\cos(\theta_2) \\ \sin(\theta_1)\sin(\theta_2)\cos(\theta_3) \\ \sin(\theta_1)\sin(\theta_2)\sin(\theta_3) \end{pmatrix}.$$

Problem 7.   Show that the equation of a *hyperplane* passing through the points $\mathbf{x}_1, \mathbf{x}_2, \ldots, \mathbf{x}_n$ in $\mathbb{R}^n$ can be given in the form

$$\det \begin{pmatrix} 1 & 1 & 1 & \cdots & 1 \\ \mathbf{x} & \mathbf{x}_1 & \mathbf{x}_2 & \cdots & \mathbf{x}_n \end{pmatrix} = 0.$$

Apply it to $n = 4$ with (Bell basis)

$$\mathbf{x}_1 = \frac{1}{\sqrt{2}}\begin{pmatrix} 1 \\ 0 \\ 0 \\ 1 \end{pmatrix}, \quad \mathbf{x}_2 = \frac{1}{\sqrt{2}}\begin{pmatrix} 1 \\ 0 \\ 0 \\ -1 \end{pmatrix},$$

$$\mathbf{x}_3 = \frac{1}{\sqrt{2}}\begin{pmatrix} 0 \\ 1 \\ 1 \\ 0 \end{pmatrix}, \quad \mathbf{x}_4 = \frac{1}{\sqrt{2}}\begin{pmatrix} 0 \\ 1 \\ -1 \\ 0 \end{pmatrix}.$$

Problem 8.   Find all 2×2 matrices A and B which commute with $AB - BA$.

Problem 9.   (i) Let U be an $n \times n$ unitary and hermitian matrix. Show that

$$\Pi = \frac{1}{2}(I_n - U)$$

is a projection matrix. Show that $\Pi \otimes \Pi$ is a projection matrix.
(ii) Let $\sigma_1, \sigma_2, \sigma_3$ be the Pauli spin matrices. Show that the matrices

$$\frac{1}{2}(I_2 + \sigma_j), \quad \frac{1}{2}(I_2 - \sigma_j), \quad j = 1, 2, 3$$

are projection matrices. Show that the matrices

$$\frac{1}{2}(I_4 + \sigma_j \otimes \sigma_k), \quad \frac{1}{2}(I_4 - \sigma_j \otimes \sigma_k), \quad j, k = 1, 2, 3$$

are projection matrices.

(iii) Show that the matrices

$$\frac{1}{2}(I_8 + \sigma_j \otimes \sigma_k \otimes \sigma_\ell), \quad \frac{1}{2}(I_8 - \sigma_j \otimes \sigma_k \otimes \sigma_\ell), \quad j, k, \ell = 1, 2, 3$$

are projection matrices.

(iv) Let U be a unitary and hermitian $n \times n$ matrix. Show that

$$\Pi_+ = \frac{1}{2}(I_n + U), \quad \Pi_- = \frac{1}{2}(I_n - U)$$

are *projection matrices*. Show that $\Pi_+ \Pi_- = 0_n$.

Problem 10. Consider the matrices

$$A = \begin{pmatrix} 0 & 1 & 0 \\ 1 & 0 & 1 \\ 0 & 1 & 0 \end{pmatrix}, \quad B = \begin{pmatrix} 1 & 0 & 1 \\ 0 & 1 & 0 \\ 1 & 0 & 1 \end{pmatrix}.$$

Write down all six 3×3 permutation matrices P_j ($j = 0, 1, \ldots, 5$) with

$$P_0 = \begin{pmatrix} 1 & 0 & 0 \\ 0 & 1 & 0 \\ 0 & 0 & 1 \end{pmatrix}, \quad P_5 = \begin{pmatrix} 0 & 0 & 1 \\ 0 & 1 & 0 \\ 1 & 0 & 0 \end{pmatrix}.$$

(i) Find the permutation matrices in this set such that $P_j A P_j^T = A$.

(ii) Find the permutation matrices in this set such that $P_j B P_j^T = B$.

Problem 11. Consider the vectors in $\mathbb{R}^3$

$$\mathbf{v}_1 = \begin{pmatrix} 0 \\ \sin(\theta) \\ \cos(\theta) \end{pmatrix}, \quad \mathbf{v}_2 = \begin{pmatrix} 0 \\ -\cos(\theta) \\ \sin(\theta) \end{pmatrix}.$$

Find the vector product $\mathbf{v}_1 \times \mathbf{v}_2$. Discuss.

Problem 12. Consider the 3×3 matrix

$$\frac{1}{\sqrt{2}} \begin{pmatrix} -1/\sqrt{3} & -1/\sqrt{3} & -1/\sqrt{3} \\ -1/\sqrt{2} & 1/\sqrt{2} & 0 \\ -1/\sqrt{6} & -1/\sqrt{6} & \sqrt{2}/\sqrt{3} \end{pmatrix}$$

which plays a role in particle physics. Show that the columns are pairwise orthonormal.

Problem 13. Let $n \geq 2$ and $\omega = e^{2\pi i/n}$. Consider the $n \times n$ matrices

$$D = \text{diag}(1 \ \omega \ \cdots \ \omega^{n-1}), \quad \Gamma = \begin{pmatrix} 0 & 0 & \cdots & 1 \\ 1 & 0 & \cdots & 0 \\ \vdots & \ddots & \vdots & 0 \\ 0 & \cdots & 1 & 0 \end{pmatrix}.$$

So Γ is a permutation matrix with $\Gamma^n = I_n$. Furthermore $D^n = I_n$. Find the commutator $[D, \Gamma]$.

Problem 14. Is the matrix

$$
M(\alpha, \beta) = \begin{pmatrix} \cos(\alpha)\cos(\beta) & -\sin(\beta) & -\sin(\alpha)\cos(\beta) \\ \cos(\alpha)\sin(\beta) & \cos(\beta) & -\sin(\alpha)\sin(\beta) \\ \sin(\beta) & 0 & \cos(\alpha) \end{pmatrix}
$$

invertible?

Problem 15. Let A be a positive definite real $n \times n$ matrix. Show that

$$
\ln(\det(A)) \le \operatorname{tr}(A) - n.
$$

For which case we have equality?

Problem 16. Show that

$$
\det \begin{pmatrix} x & y & z & 1 \\ x_1 & y_1 & z_1 & 1 \\ x_2 & y_2 & z_2 & 1 \\ x_3 & y_3 & z_3 & 1 \end{pmatrix} = 0
$$

defines a *plane* in $\mathbb{R}^3$ passing through the points (x_1, y_1, z_1), (x_2, y_2, z_2), (x_3, y_3, z_3).

Problem 17. Show that

$$
S(\zeta, \phi) = \begin{pmatrix} \cos(\zeta)\cos(\phi) & -\sin(\phi) & -\sin(\zeta)\cos(\phi) \\ \cos(\zeta)\sin(\phi) & \cos(\phi) & -\sin(\zeta)\sin(\phi) \\ \sin(\zeta) & 0 & \cos(\zeta) \end{pmatrix}
$$

is an orthogonal matrix, i.e. $S(\zeta, \phi)S(\zeta, \phi)^T = I_3$.

Problem 18. Consider the vectors in $\mathbb{C}^2$

$$
\mathbf{v}_1 = \begin{pmatrix} \cos(\alpha/2) \\ ie^{i\phi/2}\sin(\alpha/2) \end{pmatrix}, \quad \mathbf{v}_2 = \begin{pmatrix} \cos(\alpha/2) \\ -ie^{-i\phi/2}\sin(\alpha/2) \end{pmatrix}.
$$

Show that $\mathbf{v}_1^T \mathbf{v}_2 = 1$, $\mathbf{v}_1^T \mathbf{v}_1 = e^{i\phi}\cos(\alpha)$, $\mathbf{v}_2^T \mathbf{v}_2 = e^{-i\phi}\cos(\alpha)$.

Problem 19. Let $\mathbf{v}_0, \mathbf{v}_1, \mathbf{v}_2, \mathbf{v}_3$ be an orthonormal basis in the Hilbert space $\mathbb{C}^4$. Show that the vectors

$$
\mathbf{u}_0 = \frac{1}{2}(\mathbf{v}_0 + \mathbf{v}_1 + \mathbf{v}_2 + \mathbf{v}_3), \quad \mathbf{u}_1 = \frac{1}{2}(\mathbf{v}_0 - \mathbf{v}_1 + \mathbf{v}_2 - \mathbf{v}_3),
$$

$$\mathbf{u}_2 = \frac{1}{2}(\mathbf{v}_0 + \mathbf{v}_1 - \mathbf{v}_2 - \mathbf{v}_3), \quad \mathbf{u}_3 = \frac{1}{2}(\mathbf{v}_0 - \mathbf{v}_1 - \mathbf{v}_2 + \mathbf{v}_3)$$

also form an orthonormal basis in $\mathbb{C}^4$.

Problem 20. Find the conditions on $\alpha, \phi, \beta, \psi$ such that

$$|\mathbf{v}^*(\alpha, \phi)\mathbf{u}(\beta, \psi)| = \frac{1}{2}$$

for the vectors in $\mathbb{C}^2$

$$\mathbf{v}(\alpha, \phi) = \begin{pmatrix} e^{i\phi}\cos(\alpha) \\ \sin(\alpha) \end{pmatrix}, \quad \mathbf{u}(\beta, \psi) = \begin{pmatrix} e^{i\psi}\cos(\beta) \\ \sin(\beta) \end{pmatrix}.$$

Problem 21. Consider the Hilbert space $\mathbb{C}^n$. Let $k \leq n$ and $\{\mathbf{v}_1, \mathbf{v}_2, \ldots, \mathbf{v}_k\}$ be a set of k vectors in $\mathbb{C}^n$. Then the *Gram matrix* is the $k \times k$ matrix given by

$$G = (g_{j\ell}), \quad g_{j\ell} = \langle \mathbf{v}_j, \mathbf{v}_\ell \rangle$$

where $\langle , \rangle$ denotes the scalar product in $\mathbb{C}^n$. Consider $n = 4$ and the four vectors

$$\begin{pmatrix} 1 \\ z \\ z^2 \\ z^3 \end{pmatrix}, \begin{pmatrix} 0 \\ 1 \\ 2z \\ 3z^2 \end{pmatrix}, \begin{pmatrix} 0 \\ 0 \\ 2 \\ 6z \end{pmatrix}, \begin{pmatrix} 0 \\ 0 \\ 0 \\ 6 \end{pmatrix}.$$

Find the Gram matrix and its trace and determinant.

Problem 22. Let

$$\mathbf{v}(\theta) = \begin{pmatrix} \cos(\theta) \\ \sin(\theta) \end{pmatrix}.$$

Find all 2×2 matrices A, B such that $\mathbf{v}^* AB\mathbf{v} = (\mathbf{v}^* A\mathbf{v})(\mathbf{v}^* B\mathbf{v})$.

Problem 23. Let A, B be 2×2 hermitian matrices and

$$|\psi\rangle = \begin{pmatrix} \cos(\theta) \\ \sin(\theta) \end{pmatrix}.$$

Find the minima of the function

$$f(\theta) = \|AB - A\langle\psi|B|\psi\rangle - \langle\psi|A|\psi\rangle B + \langle\psi|A|\psi\rangle\langle\psi|B|\psi\rangle I_2\|.$$

Problem 24. Let $\alpha \in \mathbb{R}$. Show that the vectors

$$\mathbf{v}_1 = \frac{1}{\sqrt{1 + 2\sinh^2(\alpha)}}\begin{pmatrix} \cosh(\alpha) \\ \sinh(\alpha) \end{pmatrix}, \quad \mathbf{v}_2 = \frac{1}{\sqrt{1 + 2\sinh^2(\alpha)}}\begin{pmatrix} \sinh(\alpha) \\ -\cosh(\alpha) \end{pmatrix}$$

form an orthonormal basis in $\mathbb{C}^2$. Find $\mathbf{v}_1\mathbf{v}_1^T$, $\mathbf{v}_2\mathbf{v}_2^T$, $\mathbf{v}_1\mathbf{v}_2^T$.

Problem 25. Consider the 2×2 *elementary matrices*

$$E_{11} = \begin{pmatrix} 1 & 0 \\ 0 & 0 \end{pmatrix}, \quad E_{22} = \begin{pmatrix} 0 & 0 \\ 0 & 1 \end{pmatrix}, \quad E_{12} = \begin{pmatrix} 0 & 1 \\ 0 & 0 \end{pmatrix}, \quad E_{21} = \begin{pmatrix} 0 & 0 \\ 1 & 0 \end{pmatrix}$$

with the commutator $[E_{12}, E_{21}] = E_{11} - E_{22}$. Let $\theta \in \mathbb{R}$. Show that

$$\exp\left(-\theta(E_{12} + E_{21})\right) =$$

$$\exp(-\tanh(\theta)E_{12})\exp(\ln(\cosh(\theta))(E_{22} - E_{11}))\exp(-\tanh(\theta)E_{21}).$$

Problem 26. (i) Let A, B be $n \times n$ matrices. Show that using exponential theory such that

$$\exp(\tau(A + B)) = \exp(\tau A/2)\exp(\tau B)\exp(\tau A/2) + O(\tau^3).$$

(ii) Let A, B be $n \times n$ matrices over $\mathbb{C}$ and $C = A + B$. Show that

$$e^{-\tau C} = \lim_{n \to \infty} \left(e^{-\tau A/n}e^{-\tau B/n}\right)^n, \quad \tau \geq 0.$$

Problem 27. Let

$$\omega = \exp(2\pi i/3) \equiv \cos(2\pi/3) + i\sin(2\pi i/3) \equiv -\frac{1}{2} + \frac{1}{2}i\sqrt{3}$$

with $\omega^3 = 1$. Consider the Hadamard matrices (which are all unitary)

$$H_0 = \begin{pmatrix} 1 & 0 & 0 \\ 0 & 1 & 0 \\ 0 & 0 & 1 \end{pmatrix}, \quad H_1 = \frac{1}{\sqrt{3}}\begin{pmatrix} 1 & 1 & 1 \\ 1 & \omega & \omega^2 \\ 1 & \omega^2 & \omega \end{pmatrix},$$

$$H_2 = \frac{1}{\sqrt{3}}\begin{pmatrix} 1 & 1 & 1 \\ \omega^2 & \omega & 1 \\ 1 & \omega & \omega^2 \end{pmatrix}, \quad H_3 = \frac{1}{\sqrt{3}}\begin{pmatrix} 1 & 1 & 1 \\ \omega & \omega^2 & 1 \\ 1 & \omega^2 & \omega \end{pmatrix}.$$

The columns of each matrix form an orthonormal basis in $\mathbb{C}^3$. Show that one has four mutually unbiased bases.

Problem 28. Consider the two matrices

$$A = \begin{pmatrix} 0 & a_{12} \\ a_{21} & 0 \end{pmatrix}, \quad B = \begin{pmatrix} 0 & b_{12} \\ b_{21} & 0 \end{pmatrix}.$$

Find the condition such that $[A, B] = 0_2$.

Chapter 5

Matrices and Eigenvalue Problems

Let A be an $n \times n$ matrix over $\mathbb{C}$. The *eigenvalue equation* is given by

$$A\mathbf{x} = \lambda\mathbf{x}$$

with $\mathbf{x} \neq \mathbf{0}$. Here λ is called the eigenvalue and $\mathbf{x}$ is called the corresponding eigenvector. The eigenvalues are determined by

$$\det(A - \lambda I_n) = 0.$$

Thus for any eigenvector $\mathbf{x}$ of A, $A\mathbf{x}$ is a scalar multiple of $\mathbf{x}$. Hence it follows that

$$\lambda = \frac{\mathbf{x}^* A \mathbf{x}}{\mathbf{x}^* \mathbf{x}}.$$

The eigenvalues of a hermitian matrix $A = A^*$ are real. Let A be an arbitrary $n \times n$ hermitian matrix. Assume that all eigenvalues of A are different. Then the eigenvectors are pairwise orthogonal.

The eigenvalues of a skew-hermitian matrix $A^* = -A$ are given by $\mu i, -\mu i$ with $\mu \in \mathbb{R}$.

The eigenvalues of a unitary matrix have the form $e^{i\phi}$ with $\phi \in \mathbb{R}$.

If the matrix is an upper (or lower) triangular matrix, then the eigenvalues are given by the elements on the diagonal.

If the matrix is a projection matrix, then the eigenvalues can only be 0 or 1.

If A is an invertible $n \times n$ matrix and λ is an eigenvalue of A, then $1/\lambda$ is an eigenvalue of A^{-1}.

The eigenvalues of an $n \times n$ density matrix ρ (which is a positive semi-definite matrix with $\text{tr}(\rho) = 1$) are given by $\lambda_j \in [0,1]$ and $\sum_{j=1}^{n} \lambda_j = 1$.

Let A be an $m \times n$ matrix over $\mathbb{C}$. Then AA^* and A^*A have the same nonzero eigenvalues.

Let $\lambda_1, \ldots, \lambda_n$ be the eigenvalues of the $n \times n$ matrix A. Then

$$\text{tr}(A) = \lambda_1 + \cdots + \lambda_n$$
$$\text{tr}(A^2) = \lambda_1^2 + \cdots + \lambda_n^2$$
$$\vdots$$
$$\text{tr}(A^n) = \lambda_1^n + \cdots + \lambda_n^n.$$

We have
$$\det(A) = \lambda_1 \lambda_2 \cdots \lambda_n.$$

Given a normal $n \times n$ matrix A, i.e.

$$AA^* = A^*A.$$

Let $\lambda_1, \lambda_2, \ldots, \lambda_n$ be the eigenvalues of A and $\mathbf{v}_1, \mathbf{v}_2, \ldots, \mathbf{v}_n$ be the pairwise orthonormal normalized eigenvectors of A. The A can be expressed as (*spectral theorem*)

$$A = \lambda_1 \mathbf{v}_1 \mathbf{v}_1^* + \lambda_2 \mathbf{v}_2 \mathbf{v}_2^* + \cdots + \lambda_n \mathbf{v}_n \mathbf{v}_n^*.$$

If λ is an eigenvalue of an $n \times n$ matrix A, then e^λ is an eigenvalue of e^A. In general, let $f : \mathbb{C} \to \mathbb{C}$ be an analytic function, then $f(\lambda)$ is an eigenvalue of $f(A)$.

Let A be an $n \times n$ matrix over $\mathbb{C}$,

$$p(\lambda) = \det(A - \lambda I_n)$$

and $p(\lambda) = 0$ the *characteristic equation*. The *Cayley-Hamilton theorem* states that substituting the matrix A into the characteristic polynomial results in the $n \times n$ zero matrix, i.e.

$$p(A) = 0_n.$$

Problem 1. (i) Show that the eigenvalues of a hermitian matrix are real. (ii) Show that the eigenvalues of a skew-hermitian matrix are purely imaginary or zero. (iii) Show that the eigenvalues λ_j of a unitary matrix satisfy $|\lambda_j| = 1$.

Solution 1. (i) Since A is hermitian we have

$$A^* = A \tag{1}$$

where $A^* \equiv \bar{A}^T$. Now we have the identity

$$(A\mathbf{x})^* \mathbf{x} \equiv \mathbf{x}^* A^* \mathbf{x} \equiv \mathbf{x}^*(A^* \mathbf{x}). \tag{2}$$

Inserting (1) and the eigenvalue equation into (2) gives $(\lambda \mathbf{x})^* \mathbf{x} = \mathbf{x}^*(\lambda \mathbf{x})$. Consequently $\bar{\lambda}(\mathbf{x}^*\mathbf{x}) = \lambda(\mathbf{x}^*\mathbf{x})$. Since $\mathbf{x}^*\mathbf{x} \neq 0$ we have $\bar{\lambda} = \lambda$. Therefore λ must be real.
(ii) Since A is skew-hermitian we have $A^* = -A$. Using this property we have

$$(A\mathbf{x})^* \mathbf{x} \equiv \mathbf{x}^* A^* \mathbf{x} \equiv \mathbf{x}^*(A^* \mathbf{x}) \equiv -\mathbf{x}^*(A\mathbf{x}).$$

Inserting the eigenvalue equation yields $\bar{\lambda}(\mathbf{x}^*\mathbf{x}) = -\lambda(\mathbf{x}^*\mathbf{x})$. Since $\mathbf{x}^*\mathbf{x} \neq 0$ we have $\bar{\lambda} = -\lambda$. Thus the eigenvalues are purely imaginary or zero.
(iii) Since U is a unitary matrix we have $U^* = U^{-1}$, where U^{-1} is the inverse of U. Let $U\mathbf{x} = \lambda \mathbf{x}$ be the eigenvalue equation. It follows that $(U\mathbf{x})^* = (\lambda \mathbf{x})^*$ or $\mathbf{x}^* U^* = \mathbf{x}^* \bar{\lambda}$. Therefore

$$\mathbf{x}^* U^* U \mathbf{x} = \bar{\lambda}\lambda \mathbf{x}^* \mathbf{x}.$$

Since $U^*U = I_n$ we have $\mathbf{x}^*\mathbf{x} = \bar{\lambda}\lambda \mathbf{x}^*\mathbf{x}$. Since $\mathbf{x}^*\mathbf{x} \neq 0$ we have $\bar{\lambda}\lambda = 1$. Thus λ can be written as $\lambda = e^{i\alpha}$, $\alpha \in \mathbb{R}$.

Problem 2. Consider the hermitian and unitary matrix

$$A = \begin{pmatrix} 0 & i \\ -i & 0 \end{pmatrix} = -\sigma_2.$$

(i) Calculate the eigenvalues and normalized eigenvectors of the matrix A.
(ii) Are the eigenvectors orthogonal? Do the eigenvectors form a basis in $\mathbb{C}^2$?
(iii) Find a unitary matrix U such that

$$U^* A U = \begin{pmatrix} \lambda_1 & 0 \\ 0 & \lambda_2 \end{pmatrix}.$$

Obviously, λ_1 and λ_2 are the eigenvalues of A.

Solution 2.   (i) The *eigenvalues* are determined by the equation

$$\det(A - \lambda I_2) = 0$$

where I_2 is the 2×2 unit matrix. This equation leads to the quadratic equation $\lambda^2 - 1 = 0$. Thus we obtain the eigenvalues $\lambda_1 = 1$, $\lambda_2 = -1$. The matrix A is hermitian and thus the eigenvalues must be real. The eigenvectors are determined by solving the linear equation

$$A \begin{pmatrix} x_1 \\ x_2 \end{pmatrix} = \lambda \begin{pmatrix} x_1 \\ x_2 \end{pmatrix}.$$

Therefore the normalized eigenvector which belongs to λ_1 is given by

$$\mathbf{v}_1 = \frac{1}{\sqrt{2}} \begin{pmatrix} 1 \\ -i \end{pmatrix}.$$

The normalized eigenvector which belongs to λ_2 is given by

$$\mathbf{v}_2 = \frac{1}{\sqrt{2}} \begin{pmatrix} 1 \\ i \end{pmatrix}.$$

(ii) The two eigenvectors are orthogonal, i.e.

$$\frac{1}{\sqrt{2}} (1 \quad i) \frac{1}{\sqrt{2}} \begin{pmatrix} 1 \\ i \end{pmatrix} = 0.$$

Thus they form an orthonormal basis in the vector space $\mathbb{C}^2$.
(iii) The normalized eigenvectors given above lead to the unitary matrix

$$U = \frac{1}{\sqrt{2}} \begin{pmatrix} 1 & 1 \\ -i & i \end{pmatrix} \Rightarrow U^* = \frac{1}{\sqrt{2}} \begin{pmatrix} 1 & i \\ 1 & -i \end{pmatrix} \Rightarrow U^*AU = \begin{pmatrix} 1 & 0 \\ 0 & -1 \end{pmatrix}$$

where $U^* = U^{-1}$.

Problem 3.   Calculate the eigenvalues λ_j of the matrices

$$A = \begin{pmatrix} 0 & 0 & 1 \\ 0 & 1 & 0 \\ 1 & 0 & 0 \end{pmatrix}, \quad B = \begin{pmatrix} 0 & 0 & 0 & 1 \\ 0 & 0 & 1 & 0 \\ 0 & 1 & 0 & 0 \\ 1 & 0 & 0 & 0 \end{pmatrix}.$$

Use the property that the matrices are symmetric over $\mathbb{R}$ and orthogonal. Moreover use the fact that the trace of an $n \times n$ matrix is equal to the sum of the eigenvalues of the matrix counting multiplicities.

Solution 3. Since the matrices A and B are symmetric over $\mathbb{R}$ it follows that the eigenvalues λ_j are real. From the property that the matrices are orthogonal it follows that the eigenvalues are ± 1. From

$$\text{tr}(A) = \lambda_1 + \lambda_2 + \lambda_3 = 1$$

we find that the eigenvalues of the matrix A are given by $\{\,1, 1, -1\,\}$. Since

$$\text{tr}(B) = \lambda_1 + \lambda_2 + \lambda_3 + \lambda_4 = 0$$

we find that the eigenvalues of B are given by $\{\,1, 1, -1, -1\,\}$. Extend to n-dimensions.

Problem 4. (i) Let A be the symmetric matrix

$$A = \begin{pmatrix} 5 & -2 & -4 \\ -2 & 2 & 2 \\ -4 & 2 & 5 \end{pmatrix}$$

over $\mathbb{R}$. Calculate the eigenvalues and eigenvectors of A. Are the eigenvectors orthogonal to each other? If not, try to find orthogonal eigenvectors using the *Gram-Schmidt algorithm*.
(ii) Let B be an arbitrary symmetric $n \times n$ matrix over $\mathbb{R}$. Show that the eigenvectors which belong to different eigenvalues are orthogonal.

Solution 4. (i) Since the matrix A is symmetric over $\mathbb{R}$ we find that the eigenvalues are real. The eigenvalues are determined by $\det(A - \lambda I_3) = 0$. From this equation we obtain the characteristic polynomial

$$-\lambda^3 + 12\lambda^2 - 21\lambda + 10 = 0.$$

The eigenvalues are $\lambda_1 = 1$, $\lambda_2 = 1$, $\lambda_3 = 10$ with the corresponding eigenvectors

$$\mathbf{v}_1 = \begin{pmatrix} -1 \\ -2 \\ 0 \end{pmatrix}, \quad \mathbf{v}_2 = \begin{pmatrix} -1 \\ 0 \\ -1 \end{pmatrix}, \quad \mathbf{v}_3 = \begin{pmatrix} 2 \\ -1 \\ -2 \end{pmatrix}.$$

We find $\langle \mathbf{v}_1, \mathbf{v}_3 \rangle = 0$, $\langle \mathbf{v}_2, \mathbf{v}_3 \rangle = 0$, $\langle \mathbf{v}_1, \mathbf{v}_2 \rangle = 1$, where $\langle , \rangle$ denotes the scalar product, i.e. $\langle \mathbf{v}_j, \mathbf{v}_k \rangle := \mathbf{v}_j{}^T \mathbf{v}_k$ and T denotes the transpose. To apply the Gram-Schmidt algorithm we choose $\mathbf{v}'_1 = \mathbf{v}_1$, $\mathbf{v}'_2 = \mathbf{v}_2 + \alpha \mathbf{v}_1$ such that

$$\alpha = -\frac{(\mathbf{v}_1, \mathbf{v}_2)}{(\mathbf{v}_1, \mathbf{v}_1)} = -\frac{1}{5}.$$

Consequently,

$$\mathbf{v}'_2 = \begin{pmatrix} -4/5 \\ 2/5 \\ -1 \end{pmatrix}.$$

The vectors $\mathbf{v}_1$, $\mathbf{v}'_2$, $\mathbf{v}_3$ are orthogonal.

(ii) From the eigenvalue equations $B\mathbf{u}_j = \lambda_j \mathbf{u}_j$, $B\mathbf{u}_k = \lambda_k \mathbf{u}_k$ we obtain

$$\mathbf{u}_k^T B \mathbf{u}_j = \lambda_j \mathbf{u}_k^T \mathbf{u}_j, \qquad \mathbf{u}_j^T B \mathbf{u}_k = \lambda_k \mathbf{u}_j^T \mathbf{u}_k.$$

Subtracting the two equations yields $0 = (\lambda_j - \lambda_k)\mathbf{u}_k^T \mathbf{u}_j$ since $\mathbf{u}_k^T \mathbf{u}_j = \mathbf{u}_j^T \mathbf{u}_k$ and $\mathbf{u}_k^T B\mathbf{u}_j = \mathbf{u}_j^T B\mathbf{u}_k$. It follows that $\mathbf{u}_k^T \mathbf{u}_j \equiv \langle \mathbf{u}_k, \mathbf{u}_j \rangle = 0$ since $\lambda_j \neq \lambda_k$ by assumption.

Problem 5.   Let

$$A = \begin{pmatrix} 0 & c & -b \\ -c & 0 & a \\ b & -a & 0 \end{pmatrix}$$

where $a, b, c \in \mathbb{R}$ and $a, b, c \neq 0$. Calculate the eigenvalues and eigenvectors of A.

Solution 5.   (i) The matrix A is skew symmetric over $\mathbb{R}$. Therefore the eigenvalues must be purely imaginary or zero. Since $\det(A) = 0$ and

$$\det(A) = \lambda_1 \lambda_2 \lambda_3$$

where λ_1, λ_2, λ_3 denote the eigenvalues we conclude that at least one of the eigenvalues must be zero. Thus the eigenvalues are 0, ki and $-ki$ where $k \in \mathbb{R}$. Eigenvalues of matrices over the field $\mathbb{R}$ occur in complex conjugate pairs. In the present case it also follows from the fact that

$$\text{tr}(A) = \lambda_1 + \lambda_2 + \lambda_3 = 0.$$

To find k we have to solve $\det(A - \lambda I_3) = 0$. We obtain

$$\lambda^3 + \lambda(a^2 + b^2 + c^2) = 0.$$

Therefore $\lambda_1 = 0$, $\lambda_2 = i\sqrt{a^2 + b^2 + c^2}$, $\lambda_3 = -i\sqrt{a^2 + b^2 + c^2}$. The eigenvector of the eigenvalue $\lambda_1 = 0$ is determined by

$$A \begin{pmatrix} x_1 \\ x_2 \\ x_3 \end{pmatrix} = 0 \begin{pmatrix} x_1 \\ x_2 \\ x_3 \end{pmatrix} \equiv \begin{pmatrix} 0 \\ 0 \\ 0 \end{pmatrix}.$$

Consequently $cx_2 - bx_3 = 0$, $-cx_1 + ax_3 = 0$, $bx_1 - ax_2 = 0$. The solution is given by

$$\begin{pmatrix} x_1 \\ x_2 \\ x_3 \end{pmatrix} = \begin{pmatrix} a \\ b \\ c \end{pmatrix}.$$

The eigenvector of the eigenvalue $\lambda_2 = ik$ is determined by

$$cx_2 - bx_3 = ikx_1, \qquad -cx_1 + ax_3 = ikx_2, \qquad bx_1 - ax_2 = ikx_3$$

where $k := \sqrt{a^2 + b^2 + c^2}$. We find

$$\begin{pmatrix} x_1 \\ x_2 \\ x_3 \end{pmatrix} = \begin{pmatrix} ac - ibk \\ bc + iak \\ c^2 - k^2 \end{pmatrix}.$$

The eigenvector of the eigenvalue $\lambda_2 = -ik$ is determined by

$$cx_2 - bx_3 = -ikx_1, \quad -cx_1 + ax_3 = -ikx_2, \quad bx_1 - ax_2 = -ikx_3.$$

Solving this system of equations yields

$$\begin{pmatrix} x_1 \\ x_2 \\ x_3 \end{pmatrix} = \begin{pmatrix} ac + ibk \\ bc - iak \\ c^2 - k^2 \end{pmatrix}.$$

Problem 6. Let A be an $n \times n$ matrix. Show that

$$\det(\exp(A)) \equiv \exp(\operatorname{tr}(A)). \tag{1}$$

Solution 6. Any $n \times n$ matrix can be brought into a triangular form by a *similarity transformation*. This means there is an invertible $n \times n$ matrix R such that

$$R^{-1}AR = T \tag{2}$$

where T is a triangular matrix with diagonal elements t_{jj} which are the eigenvalues of A. We set $t_{jj} = \lambda_j$. From (2) it follows that $A = RTR^{-1}$ and therefore $\exp(A) = \exp(RTR^{-1}) = R(\exp(T))R^{-1}$. Since T is triangular, the diagonal elements of the k-th power of T are λ_j^k where k is a positive integer. Consequently, the diagonal elements of $\exp(T)$ are $\exp(\lambda_j)$. Since the determinant of a triangular matrix is equal to the product of its diagonal elements we find

$$\det(\exp(T)) = \exp(\lambda_1 + \lambda_2 + \cdots + \lambda_n) = \exp(\operatorname{tr}(T)).$$

Since $\operatorname{tr}(T) = \operatorname{tr}(R^{-1}AR) = \operatorname{tr}(A)$ and

$$\det(\exp(T)) = \det(R(\exp(T))R^{-1}) = \det(\exp(RTR^{-1})) = \det(\exp(A))$$

we obtain (1). The identity $\exp(RTR^{-1}) \equiv R(\exp(T))R^{-1}$ can be seen from

$$\exp(RTR^{-1}) = \sum_{k=0}^{\infty} \frac{(RTR^{-1})^k}{k!}$$

and $RR^{-1} = R^{-1}R = I_n$.

Problem 7.   Let A be a 4×4 symmetric matrix. Assume that the eigenvalues are given by 0, 1, 2, and 3 with the corresponding normalized eigenvectors (*Bell basis*)

$$\frac{1}{\sqrt{2}}\begin{pmatrix} 1 \\ 0 \\ 0 \\ 1 \end{pmatrix}, \quad \frac{1}{\sqrt{2}}\begin{pmatrix} 1 \\ 0 \\ 0 \\ -1 \end{pmatrix}, \quad \frac{1}{\sqrt{2}}\begin{pmatrix} 0 \\ 1 \\ 1 \\ 0 \end{pmatrix}, \quad \frac{1}{\sqrt{2}}\begin{pmatrix} 0 \\ 1 \\ -1 \\ 0 \end{pmatrix}. \tag{1}$$

Find the matrix A. Apply two different methods.

Solution 7.   Since A is a symmetric matrix over $\mathbb{R}$ there exists an orthogonal matrix O such that $D = OAO^T$, where D is the diagonal matrix $D = \text{diag}(0, 1, 2, 3)$. The orthogonal matrix O^T is given by the normalized eigenvectors of A, i.e.

$$O^T = \frac{1}{\sqrt{2}}\begin{pmatrix} 1 & 1 & 0 & 0 \\ 0 & 0 & 1 & 1 \\ 0 & 0 & 1 & -1 \\ 1 & -1 & 0 & 0 \end{pmatrix} \Rightarrow O = \frac{1}{\sqrt{2}}\begin{pmatrix} 1 & 0 & 0 & 1 \\ 1 & 0 & 0 & -1 \\ 0 & 1 & 1 & 0 \\ 0 & 1 & -1 & 0 \end{pmatrix}.$$

Since $O^T = O^{-1}$ we find from (2) that $A = O^T D O$. We obtain

$$A = \frac{1}{2}\begin{pmatrix} 1 & 0 & 0 & -1 \\ 0 & 5 & -1 & 0 \\ 0 & -1 & 5 & 0 \\ -1 & 0 & 0 & 1 \end{pmatrix}.$$

Applying the *spectral theorem* we have

$$A = \lambda_1 \mathbf{v}_1 \mathbf{v}_1^* + \lambda_2 \mathbf{v}_2 \mathbf{v}_2^* + \lambda_3 \mathbf{v}_3 \mathbf{v}_3^* + \lambda_4 \mathbf{v}_4 \mathbf{v}_4^* = \mathbf{v}_2 \mathbf{v}_2^* + 2\mathbf{v}_3 \mathbf{v}_3^* + 3\mathbf{v}_4 \mathbf{v}_4^*$$

where $\mathbf{v}_1$, $\mathbf{v}_2$, $\mathbf{v}_3$, $\mathbf{v}_4$ are the pairwise orthonormal normalized eigenvectors of A.

Problem 8.   Let $\hat{B}$ and $\hat{C}$ be two linear bounded operators with a discrete spectrum (for example two finite-dimensional $n \times n$ matrices). Assume that

$$[\hat{B}, \hat{C}]_+ = 0 \tag{1}$$

where the *anticommutator* is defined as $[\hat{B}, \hat{C}]_+ := \hat{B}\hat{C} + \hat{C}\hat{B}$. Let $\mathbf{u}$ be an eigenvector of both $\hat{B}$ and $\hat{C}$. What can be said about the corresponding eigenvalues?

Solution 8. From the eigenvalue equations $\hat{B}\mathbf{u} = B\mathbf{u}$, $\hat{C}\mathbf{u} = C\mathbf{u}$, where B and C are the eigenvalues of $\hat{B}$ and $\hat{C}$, respectively, we obtain using (1)

$$[\hat{B}, \hat{C}]_+\mathbf{u} = (\hat{B}\hat{C} + \hat{C}\hat{B})\mathbf{u} = \hat{B}(\hat{C}\mathbf{u}) + \hat{C}(\hat{B}\mathbf{u}).$$

Thus

$$[\hat{B}, \hat{C}]_+\mathbf{u} = \hat{B}C\mathbf{u} + \hat{C}B\mathbf{u} = BC\mathbf{u} + CB\mathbf{u}.$$

Finally $[\hat{B}, \hat{C}]_+\mathbf{u} = 2BC\mathbf{u} = \mathbf{0}$. Consequently, since $\mathbf{u} \neq \mathbf{0}$ we have the solutions $(B = 0, C \neq 0)$, $(B \neq 0, C = 0)$, $(B = 0, C = 0)$.

Problem 9. Let A be an $n \times n$ matrix with eigenvalue λ.
(i) Show that λ^2 is an eigenvalue of A^2.
(ii) Show that e^λ is an eigenvalue of e^A.
(iii) Show that $\sin(\lambda)$ is an eigenvalue of $\sin(A)$.
(iv) Assume that A^{-1} exists. Show that $1/\lambda$ is an eigenvalue of A^{-1}.

Solution 9. (i) From the eigenvalue equation $A\mathbf{x} = \lambda\mathbf{x}$ we obtain

$$A^2\mathbf{x} = A(A\mathbf{x}) = A(\lambda\mathbf{x}) = \lambda A\mathbf{x} = \lambda^2\mathbf{x}.$$

Obviously, $A^3\mathbf{x} = \lambda^3\mathbf{x}$ etc.
(ii) Using the expansion

$$e^A := \sum_{k=0}^{\infty} \frac{A^k}{k!}$$

and the result from (i) we find that e^A has the eigenvalue e^λ.
(iii) Using the expansion

$$\sin(A) := \sum_{k=0}^{\infty} \frac{(-1)^k A^{2k+1}}{(2k+1)!}$$

and the result from (i) we find that $\sin(\lambda)$ is an eigenvalue of $\sin(A)$.
(iv) From the eigenvalue equation (1) we find $A^{-1}(A\mathbf{x}) = A^{-1}(\lambda\mathbf{x})$. Thus $\mathbf{x} = \lambda A^{-1}\mathbf{x}$. Since $\mathbf{x} \neq \mathbf{0}$ we find that $1/\lambda$ is an eigenvalue of A^{-1}.

Problem 10. A norm of an $n \times n$ matrix A over the real numbers can be defined as

$$\|A\| := \sup_{\|\mathbf{x}\|=1} \|A\mathbf{x}\| \tag{1}$$

where $\|\ \|$ on the right-hand side denotes the Euclidean norm. Let

$$A = \begin{pmatrix} 1 & 1 \\ 2 & 2 \end{pmatrix}.$$

(i) Find $\|A\|$.
(ii) Find the eigenvalues of $A^T A$ and compare with the result of (i).

Solution 10.   (i) We apply the *Lagrange multiplier method*. Let

$$\mathbf{x} := \begin{pmatrix} x_1 \\ x_2 \end{pmatrix} \Rightarrow \mathbf{x}^T := (x_1 \quad x_2).$$

Using matrix multiplication we obtain

$$\|A\mathbf{x}\|^2 = (A\mathbf{x})^T A\mathbf{x} = \mathbf{x}^T A^T A\mathbf{x} = 5x_1^2 + 10x_1x_2 + 5x_2^2.$$

The constraint is $\|\mathbf{x}\| = 1 \Leftrightarrow x_1^2 + x_2^2 = 1$. Thus we have

$$f(x_1, x_2) = 5x_1^2 + 10x_1x_2 + 5x_2^2 + \lambda(x_1^2 + x_2^2 - 1)$$

where λ is the Lagrange multiplier. From the equations

$$\frac{\partial f}{\partial x_1} = 10x_1 + 10x_2 + 2\lambda x_1 = 0, \qquad \frac{\partial f}{\partial x_2} = 10x_1 + 10x_2 + 2\lambda x_2 = 0$$

we find $x_1^2 = x_2^2 = \frac{1}{2}$. The square of the norm is given by

$$\|A\|^2 = 5 \cdot \frac{1}{2} + 10 \cdot \frac{1}{2} + 5 \cdot \frac{1}{2} = 10.$$

(ii) We have

$$A^T A = \begin{pmatrix} 5 & 5 \\ 5 & 5 \end{pmatrix}.$$

The rank of the matrix $A^T A$ is 1 and $\operatorname{tr}(A^T A) = 10$. Thus we find that the eigenvalues of $A^T A$ are given by 0 and 10. Thus the square of the norm is the largest eigenvalue of $A^T A$. Is this true in general?

Problem 11.   Let A be an $n \times n$ matrix over $\mathbb{R}$. Assume that A^{-1} exists. To compute A^{-1} we can use the iteration

$$X_{t+1} = 2X_t - X_t A X_t, \qquad t = 0, 1, 2, \ldots$$

with $X_0 = \alpha A^T$. The scalar α is chosen such that $0 < \alpha < 2/\sigma_1$ with σ_1 to be the largest *singular value* of A. The singular values of a square matrix B are the positive roots of the eigenvalues of the hermitian matrix B^*B (or $B^T B$ if the matrix B is real). Apply the iteration to the 2×2 matrix

$$A = \begin{pmatrix} 1 & 1 \\ 1 & 0 \end{pmatrix}.$$

Solution 11. The largest eigenvalue of $A^T A$ is $\lambda = (3 + \sqrt{5})/2$. Thus we choose $\alpha = 1/2$ since

$$\frac{1}{2} < \frac{2\sqrt{2}}{\sqrt{3 + \sqrt{5}}}.$$

We obtain

$$X_1 = 2 \cdot \frac{1}{2} \begin{pmatrix} 1 & 1 \\ 1 & 0 \end{pmatrix} - \frac{1}{4} \begin{pmatrix} 1 & 1 \\ 1 & 0 \end{pmatrix} \begin{pmatrix} 1 & 1 \\ 1 & 0 \end{pmatrix} \begin{pmatrix} 1 & 1 \\ 1 & 0 \end{pmatrix} = \begin{pmatrix} 1/4 & 1/2 \\ 1/2 & -1/4 \end{pmatrix}.$$

For X_2 we find

$$X_2 = \begin{pmatrix} 3/16 & 11/16 \\ 11/16 & -1/2 \end{pmatrix}.$$

The series converges to the inverse of A

$$A^{-1} = \begin{pmatrix} 0 & 1 \\ 1 & -1 \end{pmatrix}.$$

Problem 12. (i) Let A, B be 2×2 matrices over $\mathbb{R}$ and vectors $\mathbf{x}$, $\mathbf{y}$ in $\mathbb{R}^2$ such that

$$A\mathbf{x} = \mathbf{y}, \qquad B\mathbf{y} = \mathbf{x}$$

and $\mathbf{x}^T \mathbf{y} = 0$, $\mathbf{x}^T \mathbf{x} = 1$, $\mathbf{y}^T \mathbf{y} = 1$. Show that both AB and BA have the eigenvalue $+1$.
(ii) Find all 2×2 matrices A, B which satisfy the conditions given in (i). Use

$$\mathbf{x} = \begin{pmatrix} \cos(\alpha) \\ \sin(\alpha) \end{pmatrix}, \qquad \mathbf{y} = \begin{pmatrix} -\sin(\alpha) \\ \cos(\alpha) \end{pmatrix}.$$

Solution 12. (i) We have

$$B(A\mathbf{x}) = B\mathbf{y} = \mathbf{x} = (BA)\mathbf{x}, \quad A(B\mathbf{y}) = A\mathbf{x} = \mathbf{y} = (AB)\mathbf{y}.$$

Thus the matrices AB and BA have the eigenvalue $+1$.
(ii) From

$$A\mathbf{x} = \begin{pmatrix} a_{11} & a_{12} \\ a_{21} & a_{22} \end{pmatrix} \begin{pmatrix} \cos(\alpha) \\ \sin(\alpha) \end{pmatrix} = \begin{pmatrix} -\sin(\alpha) \\ \cos(\alpha) \end{pmatrix} = \mathbf{y}$$

we obtain two equations for the unknown a_{11}, a_{12}, a_{21}, a_{22}

$$a_{11} \cos(\alpha) + (a_{12} + 1)\sin(\alpha) = 0, \quad (a_{21} - 1)\cos(\alpha) + a_{22} \sin(\alpha) = 0.$$

We eliminate a_{12} and a_{21}. The cases $\cos(\alpha) = 0$, $\sin(\alpha) \neq 0$ and $\sin(\alpha) = 0$, $\cos(\alpha) \neq 0$ are obvious. Assume now $\sin(\alpha) \neq 0$ and $\cos(\alpha) \neq 0$. Then we obtain the matrix

$$A = \begin{pmatrix} a_{11} & -1 - a_{11}\cos(\alpha)/\sin(\alpha) \\ 1 - a_{22}\sin(\alpha)/\cos(\alpha) & a_{22} \end{pmatrix}.$$

Similarly for the matrix B.

Problem 13.   Find all 2×2 matrices over $\mathbb{R}$ such that the eigenvectors do not span $\mathbb{R}^2$.

Solution 13.   Let

$$A = \begin{pmatrix} a & b \\ c & d \end{pmatrix}$$

be such a matrix. Thus A has at least one eigenvalue. If the eigenvectors of A do not span $\mathbb{R}^2$ then A has exactly one eigenvalue λ. Thus $\text{tr}(A) = a+d = 2\lambda$ and $\det(A) = ad - bc = \lambda^2$ from which follows $(a+d)^2 = 4(ad - bc)$ which can be written as $(a-d)^2 = -bc$. Let $\mathbf{x} := \begin{pmatrix} x_1 & x_2 \end{pmatrix}^T$ be the corresponding eigenvector, i.e.

$$ax_1 + bx_2 = \lambda x_1 \quad \Rightarrow \quad ax_1 = \lambda x_1 - bx_2$$
$$cx_1 + dx_2 = \lambda x_2 \quad \Rightarrow \quad dx_2 = \lambda x_2 - cx_1.$$

Let $\mathbf{x}_\perp := \begin{pmatrix} -x_2 & x_1 \end{pmatrix}^T$, i.e. $\mathbf{x}^T\mathbf{x}_\perp = 0$. If the eigenvectors of A do not span $\mathbb{R}^2$ then $\mathbf{x}^T A\mathbf{x}_\perp \neq 0$. Substituting the equations above yields

$$\mathbf{x}^T A\mathbf{x}_\perp = bx_1^2 + (dx_2)x_1 - (ax_1)x_2 + cx_2^2 = (b-c)(x_1^2 + x_2^2) \neq 0.$$

Since $\mathbf{x} \neq \mathbf{0}$ we have $b - c \neq 0$. Thus we have the conditions

$$(a-d)^2 = -bc, \qquad b \neq c$$

which could also be written as $(\text{tr}(A))^2 = 4\det(A)$, $A \neq A^T$.

Problem 14.   (i) Let

$$\mathbf{x} = \begin{pmatrix} x_1 \\ x_2 \\ x_3 \end{pmatrix}$$

where $x_j \in \mathbb{R}$. Show that at least one the eigenvalues of the 3×3 matrix

$$A := \begin{pmatrix} x_1 \\ x_2 \\ x_3 \end{pmatrix} \begin{pmatrix} x_1 & x_2 & x_3 \end{pmatrix}$$

is equal to zero. Let $\mathbf{x}$ be a nonzero column vector in $\mathbb{R}^n$ and $n \geq 2$. Consider the $n \times n$ matrix $\mathbf{xx}^T$. Find one nonzero eigenvalue and the corresponding eigenvector of this matrix.

(ii) Show that if A is an $n \times m$ matrix and if B is an $m \times n$ matrix, then $\lambda \neq 0$ is an eigenvalue of the $n \times n$ matrix AB if and only if λ is an eigenvalue of the $m \times m$ matrix BA. Show that if $m = n$ then the conclusion is true even for $\lambda = 0$.

Solution 14. (i) We have

$$A = \begin{pmatrix} x_1^2 & x_1 x_2 & x_1 x_3 \\ x_1 x_2 & x_2^2 & x_2 x_3 \\ x_1 x_3 & x_2 x_3 & x_3^2 \end{pmatrix}.$$

Straightforward calculation yields $\det(A) = 0$. Since $\det(A) = \lambda_1 \lambda_2 \lambda_3$ we can conclude that at least one eigenvalue is 0. What are the other two eigenvalues? Since matrix multiplication is associative we have

$$(\mathbf{xx}^T)\mathbf{x} = \mathbf{x}(\mathbf{x}^T\mathbf{x}) = (\mathbf{x}^T\mathbf{x})\mathbf{x}.$$

Thus $\mathbf{x}$ is an eigenvector and $\mathbf{x}^T\mathbf{x}$ is the nonzero eigenvalue.

(ii) If $\lambda \neq 0$ is an eigenvalue of AB and if $\mathbf{v}$ is an eigenvector of AB, then $AB\mathbf{v} = \lambda\mathbf{v} \neq \mathbf{0}$ since $\lambda \neq 0$ and $\mathbf{v} \neq \mathbf{0}$. Thus $B\mathbf{v} \neq \mathbf{0}$. Therefore $BAB\mathbf{v} = \lambda B\mathbf{v}$ and λ is an eigenvalue of BA. If A and B are square matrices and $\lambda = 0$ is an eigenvalue of AB with eigenvector $\mathbf{v}$, then we have $\det(AB) = \det(BA) = 0$. Thus $\lambda = 0$ is an eigenvalue of BA.

Problem 15. We know that a hermitian matrix has only real eigenvalues. Can we conclude that a matrix with only real eigenvalues is hermitian?

Solution 15. We cannot conclude that a matrix with only real eigenvalues is hermitian. For example the non-hermitian matrix

$$A = \begin{pmatrix} 0 & 1 & 0 \\ 1 & 0 & 1 \\ 0 & 2 & 0 \end{pmatrix}$$

admits the real eigenvalues $\sqrt{3}$, $-\sqrt{3}$, 0. The matrix is nonnormal.

Problem 16. Let $a, b \in \mathbb{R}$. Find, by inspection, two eigenvectors and the corresponding eigenvalues of the 4×4 matrix

$$\begin{pmatrix} a & 0 & 0 & b \\ 0 & a & 0 & b \\ 0 & 0 & a & b \\ b & b & b & 0 \end{pmatrix}.$$

Solution 16.   Obviously $(1 \quad -1 \quad 0 \quad 0)^T$, $(1 \quad 0 \quad -1 \quad 0)^T$ are eigenvectors with the corresponding eigenvalues a and a. Note that the scalar product of the two vectors is nonzero.

Problem 17.   Consider the following 3×3 matrix A and vector $\mathbf{v}$ in $\mathbb{R}^3$

$$A = \begin{pmatrix} 0 & 1 & 0 \\ 1 & 0 & 1 \\ 0 & 1 & 0 \end{pmatrix}, \qquad \mathbf{v} = \begin{pmatrix} \sin(\alpha) \\ \sin(2\alpha) \\ \sin(3\alpha) \end{pmatrix}$$

where $\alpha \in \mathbb{R}$ and $\alpha \neq n\pi$ with $n \in \mathbb{Z}$. Show that using this vector we can find the eigenvalues and eigenvectors of A.

Solution 17.   We find

$$A\mathbf{v} = \begin{pmatrix} \sin(2\alpha) \\ \sin(\alpha) + \sin(3\alpha) \\ \sin(2\alpha) \end{pmatrix}.$$

Under the assumption that $\mathbf{v}$ is an eigenvector of A for specific α's we find the condition

$$\begin{pmatrix} \sin(2\alpha) \\ \sin(\alpha) + \sin(3\alpha) \\ \sin(2\alpha) \end{pmatrix} = \lambda \begin{pmatrix} \sin(\alpha) \\ \sin(2\alpha) \\ \sin(3\alpha) \end{pmatrix}.$$

Thus we have to solve the three equations

$$\sin(2\alpha) = \lambda \sin(\alpha), \quad \sin(\alpha) + \sin(3\alpha) = \lambda \sin(2\alpha), \quad \sin(2\alpha) = \lambda \sin(3\alpha).$$

We have to study the case $\lambda = 0$ and $\lambda \neq 0$. For $\lambda = 0$ we have to solve the two equations $\sin(2\alpha) = 0$, $\cos(2\alpha)\sin(\alpha) = 0$ with the solution $\alpha = \pi/2$. Thus the eigenvalue $\lambda = 0$ has the normalized eigenvector

$$\frac{1}{\sqrt{2}} \begin{pmatrix} 1 \\ 0 \\ -1 \end{pmatrix}.$$

For $\lambda \neq 0$ we obtain the equation $2\sin(\alpha) = \lambda^2 \sin(\alpha)$ after elimination of $\sin(2\alpha)$ and $\sin(3\alpha)$ using the first and third equation. Since $\alpha \neq n\pi$ we have that $\sin(\alpha) \neq 0$. Thus $\lambda^2 = 2$ with the eigenvalues $\lambda = -\sqrt{2}$ and $\lambda = \sqrt{2}$ and the corresponding normalized eigenvectors

$$\frac{1}{2} \begin{pmatrix} 1 \\ -\sqrt{2} \\ 1 \end{pmatrix}, \qquad \frac{1}{2} \begin{pmatrix} 1 \\ \sqrt{2} \\ 1 \end{pmatrix}.$$

Can this technique be extended to the 4×4 matrix and vector $\mathbf{v} \in \mathbb{R}^4$

$$A = \begin{pmatrix} 0 & 1 & 0 & 0 \\ 1 & 0 & 1 & 0 \\ 0 & 1 & 0 & 1 \\ 0 & 0 & 1 & 0 \end{pmatrix}, \qquad \mathbf{v} = \begin{pmatrix} \sin(\alpha) \\ \sin(2\alpha) \\ \sin(3\alpha) \\ \sin(4\alpha) \end{pmatrix}$$

and even higher dimensions?

Problem 18. Let B be a 2×2 matrix with eigenvalues λ_1 and λ_2. Find the eigenvalues of the 4×4 matrix

$$X = \begin{pmatrix} 0 & 0 & 1 & 0 \\ 0 & 0 & 0 & 1 \\ b_{11} & b_{12} & 0 & 0 \\ b_{21} & b_{22} & 0 & 0 \end{pmatrix}.$$

Let $\mathbf{v}$ be an eigenvector of B with eigenvalue λ. What can be said about an eigenvector of the 4×4 matrix X given by eigenvector $\mathbf{v}$ and eigenvalue of B?

Solution 18. The characteristic equation $\det(X - rI_4) = 0$ is given by

$$r^4 - r^2(b_{11} + b_{22}) + (b_{11}b_{22} - b_{12}b_{21}) = 0$$

or $r^4 - r^2 \mathrm{tr}(B) + \det(B) = 0$. Thus since $\mathrm{tr}(B) = \lambda_1 + \lambda_2$ and $\det(B) = \lambda_1\lambda_2$ we have

$$r^4 - r^2(\lambda_1 + \lambda_2) + \lambda_1\lambda_2 = 0.$$

This quartic equation can be reduced to a quadratic equation. We find the eigenvalues $r_1 = \sqrt{\lambda_1}$, $r_2 = -\sqrt{\lambda_1}$, $r_3 = \sqrt{\lambda_2}$, $r_4 = -\sqrt{\lambda_2}$. Let $\mathbf{v}$ be an eigenvector of B with eigenvalue λ, i.e. $B\mathbf{v} = \lambda\mathbf{v}$. Then we have

$$\begin{pmatrix} 0 & 0 & 1 & 0 \\ 0 & 0 & 0 & 1 \\ b_{11} & b_{12} & 0 & 0 \\ b_{21} & b_{22} & 0 & 0 \end{pmatrix} \begin{pmatrix} v_1 \\ v_2 \\ \sqrt{\lambda}v_1 \\ \sqrt{\lambda}v_2 \end{pmatrix} = \sqrt{\lambda} \begin{pmatrix} v_1 \\ v_2 \\ \sqrt{\lambda}v_1 \\ \sqrt{\lambda}v_2 \end{pmatrix}.$$

Thus

$$\begin{pmatrix} v_1 \\ v_2 \\ \sqrt{\lambda}v_1 \\ \sqrt{\lambda}v_2 \end{pmatrix}$$

is an eigenvector of X.

Problem 19.   Consider the infinite-dimensional symmetric matrix

$$
A = \begin{pmatrix}
0 & 1 & 0 & 0 & \cdots \\
1 & 0 & 1 & 0 & \cdots \\
0 & 1 & 0 & 1 & \cdots \\
& & \ddots & & \ddots \\
& & & \ddots & \\
& & & & \ddots
\end{pmatrix}.
$$

In other words

$$
A_{jk} = \begin{cases}
1 & \text{if} \quad j = k+1 \\
1 & \text{if} \quad j = k-1 \\
0 & \text{otherwise}
\end{cases}
$$

with $j, k = 1, 2, \ldots$. Find the spectrum of this infinite-dimensional matrix.

Solution 19.   Let A_n be the $n \times n$ truncated matrix of A. Then the eigenvalue problem for A_n is given by $A_n \mathbf{x} = \lambda \mathbf{x}$ with

$$
A_n = \begin{pmatrix}
0 & 1 & 0 & 0 & \cdots & 0 & 0 & 0 \\
1 & 0 & 1 & 0 & \cdots & 0 & 0 & 0 \\
0 & 1 & 0 & 1 & \cdots & 0 & 0 & 0 \\
& \ddots & & \ddots & & & \ddots & \\
\vdots & & \ddots & & \ddots & & & \\
& & & \ddots & & \ddots & & \\
0 & 0 & 0 & 0 & \cdots & 1 & 0 & 1 \\
0 & 0 & 0 & 0 & \cdots & 0 & 1 & 0
\end{pmatrix}.
$$

First we calculate the eigenvalues of A_n. Then we study A_n as $n \to \infty$. The eigenvalue problem leads to $D_n(\lambda) = 0$ where

$$
D_n(\lambda) \equiv \det \begin{pmatrix}
-\lambda & 1 & 0 & 0 & \cdots & 0 & 0 & 0 \\
1 & -\lambda & 1 & 0 & \cdots & 0 & 0 & 0 \\
0 & 1 & -\lambda & 1 & \cdots & 0 & 0 & 0 \\
& \vdots & & & & & & \\
0 & 0 & 0 & 0 & \cdots & 1 & -\lambda & 1 \\
0 & 0 & 0 & 0 & \cdots & 0 & 1 & -\lambda
\end{pmatrix}.
$$

We try to find a difference equation for $D_n(\lambda)$, where $n = 1, 2, \ldots$. We obtain

$$D_n(\lambda) = -\lambda \det \begin{pmatrix} -\lambda & 1 & 0 & 0 & \cdots & 0 & 0 & 0 \\ 1 & -\lambda & 1 & 0 & \cdots & 0 & 0 & 0 \\ 0 & 1 & -\lambda & 1 & \cdots & 0 & 0 & 0 \\ 0 & 0 & 1 & -\lambda & \cdots & 0 & 0 & 0 \\ & & \vdots & & & & & \\ 0 & 0 & 0 & 0 & \cdots & 1 & -\lambda & 1 \\ 0 & 0 & 0 & 0 & \cdots & 0 & 1 & -\lambda \end{pmatrix}$$

$$- \det \begin{pmatrix} 1 & 1 & 0 & 0 & \cdots & 0 & 0 & 0 \\ 0 & -\lambda & 1 & 0 & \cdots & 0 & 0 & 0 \\ 0 & 1 & -\lambda & 1 & \cdots & 0 & 0 & 0 \\ & \vdots & & & & & & \\ 0 & 0 & 0 & 0 & \cdots & 1 & -\lambda & 1 \\ 0 & 0 & 0 & 0 & \cdots & 0 & 1 & -\lambda \end{pmatrix}.$$

The first determinant on the right-hand side is equal to $D_{n-1}(\lambda)$. For the second determinant we find (expansion of the first row)

$$\det \begin{pmatrix} 1 & 1 & 0 & 0 & \cdots & 0 & 0 & 0 \\ 0 & -\lambda & 1 & 0 & \cdots & 0 & 0 & 0 \\ 0 & 1 & -\lambda & 1 & \cdots & 0 & 0 & 0 \\ & \vdots & & & & & & \\ 0 & 0 & 0 & 0 & \cdots & 1 & -\lambda & 1 \\ 0 & 0 & 0 & 0 & \cdots & 0 & 1 & -\lambda \end{pmatrix} = D_{n-2}(\lambda).$$

Consequently, we obtain a second order linear difference equation

$$D_n(\lambda) = -\lambda D_{n-1}(\lambda) - D_{n-2}(\lambda)$$

with the "initial conditions" $D_1(\lambda) = -\lambda$, $D_2(\lambda) = \lambda^2 - 1$. To solve this linear difference equation we make the ansatz

$$D_n(\lambda) = e^{in\theta}$$

where $n = 1, 2, \ldots$. Inserting the this ansatz into the difference equation yields

$$e^{in\theta} = -\lambda e^{i(n-1)\theta} - e^{i(n-2)\theta}.$$

It follows that $e^{i\theta} = -\lambda - e^{-i\theta}$. Consequently $\lambda = -2\cos(\theta)$. The general solution to the linear difference equation is given by

$$D_n(\lambda) = C_1 \cos(n\theta) + C_2 \sin(n\theta)$$

where C_1 and C_2 are constants and $\lambda = -2\cos(\theta)$. Imposing the initial conditions it follows that

$$D_n(\lambda) = \frac{\sin((n+1)\theta)}{\sin(\theta)}.$$

Since $D_n(\lambda) = 0$ we have to solve the equation

$$\frac{\sin((n+1)\theta)}{\sin(\theta)} = 0.$$

The solutions to this equation are given by $\theta = k\pi/(n+1)$ with $k = 1, 2, \ldots, n$. Since $\lambda = -2\cos(\theta)$, we find the eigenvalues

$$\lambda_k = -2\cos\left(\frac{k\pi}{n+1}\right)$$

with $k = 1, 2, \ldots, n$. Consequently, $|\lambda_k| \leq 2$. If $n \to \infty$, then there are infinitely many λ_k with $|\lambda_k| \leq 2$ and $\lambda_k - \lambda_{k+1} \to 0$. Therefore

$$\text{spectrum}(A) = [-2, 2]$$

i.e. we have a *continuous spectrum*.
Another approach to find the spectrum is as follows. Let $A = B + B^T$, where

$$B = \begin{pmatrix} 0 & 0 & 0 & 0 & \cdots \\ 1 & 0 & 0 & 0 & \cdots \\ 0 & 1 & 0 & 0 & \cdots \\ 0 & 0 & 1 & 0 & \cdots \\ & & \vdots & & \end{pmatrix}, \qquad B^T = \begin{pmatrix} 0 & 1 & 0 & 0 & \cdots \\ 0 & 0 & 1 & 0 & \cdots \\ 0 & 0 & 0 & 1 & \cdots \\ & & \vdots & & \end{pmatrix}.$$

Then $B^T B = I$, where I is the infinite unit matrix. Notice that $BB^T \neq I$. We use the following notation

$$Cf = \lambda f \quad \text{means} \quad \|Cf_n - \lambda f_n\| \to 0.$$

Now $Bf = \lambda f \Rightarrow B^T Bf = B^T \lambda f$. Therefore

$$If = \lambda B^T f \quad \Rightarrow \quad f = \lambda B^T f \quad \Rightarrow \quad B^T f = \frac{1}{\lambda} f.$$

From $Bf = \lambda f$ it also follows that

$$\|Bf\|^2 = (Bf, Bf) = (f, B^T Bf) = (f, f) = \|f\|^2.$$

On the other hand $(Bf, Bf) = \bar{\lambda}\lambda(f, f) = |\lambda|^2 \|f\|^2$. Since $\|f\| > 0$ we find that $|\lambda|^2 = 1$. Therefore

$$Af = (B + B^T)f = \left(\lambda + \frac{1}{\lambda}\right)f = (\lambda + \bar{\lambda})f = 2(\cos(\phi))f.$$

This means

$$\bigwedge_{\phi \in \mathbb{R}} \|(A - 2(\cos\phi)I)f_n\| \to 0$$

or

$$A \begin{pmatrix} \sin(\phi) \\ \sin(2\phi) \\ \sin(3\phi) \\ \vdots \end{pmatrix} = 2\cos(\phi) \begin{pmatrix} \sin(\phi) \\ \sin(2\phi) \\ \sin(3\phi) \\ \vdots \end{pmatrix}.$$

The vector on the left-hand side is not an element of the Hilbert space $\ell_2(\mathbb{N})$. For the first two rows we have the identities

$$\sin(2\phi) \equiv 2\sin(\phi)\cos(\phi), \qquad \sin(\phi) + \sin(3\phi) \equiv 2\cos(\phi)\sin(2\phi).$$

Here $\ell_2(\mathbb{N})$ is the *Hilbert space* of all infinite vectors (sequences)

$$\mathbf{u} = (u_1, u_2, \ldots)^T$$

of complex numbers u_j such that $\sum_{j=1}^{\infty} |u_j|^2 < \infty$.

Problem 20. Let σ_1, σ_2, σ_3 be the Pauli spin matrices. Find the eigenvalues of the 8×8 matrix $(\sigma_1 \otimes \sigma_1 + \sigma_2 \otimes \sigma_2 + \sigma_3 \otimes \sigma_3) \otimes \sigma_1$.

Solution 20. The 8×8 matrix can be written as a *direct sum*

$$\begin{pmatrix} 0 & 1 \\ 1 & 0 \end{pmatrix} \oplus \begin{pmatrix} 0 & -1 & 0 & 2 \\ -1 & 0 & 2 & 0 \\ 0 & 2 & 0 & -1 \\ 2 & 0 & -1 & 0 \end{pmatrix} \oplus \begin{pmatrix} 0 & 1 \\ 1 & 0 \end{pmatrix}.$$

Thus the eigenvalues can be calculated from the two 2×2 matrices and the 4×4 matrix. We find the eigenvalues -1 (3 times), $+1$ (3 times), $+3$, -3.

Problem 21. Let $\phi \in \mathbb{R}$. Consider the 4×4 matrix

$$A(\phi) = \begin{pmatrix} 0 & 1 & 0 & 0 \\ 0 & 0 & 1 & 0 \\ 0 & 0 & 0 & 1 \\ e^{i\phi} & 0 & 0 & 0 \end{pmatrix}.$$

(i) Is the matrix unitary?
(ii) Find the eigenvalues and normalized eigenvectors of $A(\phi)$.
(iii) Can the matrix be written as the Kronecker product of two 2×2 matrices?

Solution 21. (i) Yes. We have $A(\phi)A^*(\phi) = I_4$.

(ii) The eigenvalues are $ie^{i\phi/4}$, $-e^{i\phi/4}$, $-ie^{i\phi/4}$, $e^{i\phi/4}$ with the corresponding normalized eigenvectors

$$\frac{1}{2}\begin{pmatrix} 1 \\ ie^{i\phi/4} \\ -e^{i\phi/2} \\ -ie^{3i\phi/4} \end{pmatrix}, \quad \frac{1}{2}\begin{pmatrix} 1 \\ -e^{i\phi/4} \\ e^{i\phi/2} \\ -e^{3i\phi/4} \end{pmatrix}, \quad \frac{1}{2}\begin{pmatrix} 1 \\ -ie^{i\phi/4} \\ -e^{i\phi/2} \\ ie^{3i\phi/4} \end{pmatrix}, \quad \frac{1}{2}\begin{pmatrix} 1 \\ e^{i\phi/4} \\ e^{i\phi/2} \\ e^{3i\phi/4} \end{pmatrix}.$$

(iii) No.

Problem 22. Consider the two matrices

$$S_1 = \frac{1}{\sqrt{2}}\begin{pmatrix} 0 & 1 & 0 \\ 1 & 0 & 1 \\ 0 & 1 & 0 \end{pmatrix}, \quad T_1 = \begin{pmatrix} 0 & 0 & 0 \\ 0 & 0 & -i \\ 0 & i & 0 \end{pmatrix}.$$

Both are hermitian, play a role for spin-1 and admit the eigenvalues $+1$, 0, -1. Find a unitary matrix U such that $US_1U^* = T_1$ applying the spectral theorem.

Solution 22. Calculating the normalized eigenvectors of S_1 and T_1 we have the spectral decomposition of S_1 as

$$S_1 = 1\begin{pmatrix} 1/2 \\ 1/\sqrt{2} \\ 1/2 \end{pmatrix}(1/2 \quad 1/\sqrt{2} \quad 1/2) + 0\begin{pmatrix} 1/\sqrt{2} \\ 0 \\ -1/\sqrt{2} \end{pmatrix}(1/\sqrt{2} \quad 0 \quad -1/\sqrt{2})$$

$$-1\begin{pmatrix} 1/2 \\ -1/\sqrt{2} \\ 1/2 \end{pmatrix}(1/2 \quad -1/\sqrt{2} \quad 1/2)$$

and of T_1 as

$$T_1 = 1\begin{pmatrix} 0 \\ 1/\sqrt{2} \\ i/\sqrt{2} \end{pmatrix}(0 \quad 1/\sqrt{2} \quad -i/\sqrt{2}) + 0\begin{pmatrix} 1 \\ 0 \\ 0 \end{pmatrix}(1 \quad 0 \quad 0)$$

$$-1\begin{pmatrix} 0 \\ 1/\sqrt{2} \\ -i/\sqrt{2} \end{pmatrix}(0 \quad 1/\sqrt{2} \quad i/\sqrt{2}).$$

It follows that the unitary matrix U is given by

$$U = \begin{pmatrix} 0 \\ 1/\sqrt{2} \\ i/\sqrt{2} \end{pmatrix}(1/2 \quad 1/\sqrt{2} \quad 1/2) + \begin{pmatrix} 1 \\ 0 \\ 0 \end{pmatrix}(1/\sqrt{2} \quad 0 \quad -1/\sqrt{2})$$

$$+ \begin{pmatrix} 0 \\ 1/\sqrt{2} \\ -i/\sqrt{2} \end{pmatrix}(1/2 \quad -1/\sqrt{2} \quad 1/2)$$

$$= \begin{pmatrix} 1/\sqrt{2} & 0 & -1/\sqrt{2} \\ 1/\sqrt{2} & 0 & 1/\sqrt{2} \\ 0 & i & 0 \end{pmatrix}.$$

Hence

$$U^* = \begin{pmatrix} 1/\sqrt{2} & 1/\sqrt{2} & 0 \\ 0 & 0 & -i \\ -1/\sqrt{2} & 1/\sqrt{2} & 0 \end{pmatrix}.$$

Problem 23. Let $n \geq 1$ and $m \geq 1$. Consider the $T = (t_{j_1 \dots j_m})$ order-m *tensor* of size $(n \times \cdots \times n)$ (m-times), $(j_1, \dots, j_m = 1, \dots, n)$. One defines the operator on $\mathbf{v} \in \mathbb{C}^n$ written as

$$(T\mathbf{v}^{m-1})_k := \sum_{j_2=1}^{n} \cdots \sum_{j_m=1}^{n} t_{kj_2 \dots j_m} v_{j_2} \cdots v_{j_m}, \quad k = 1, \dots, n.$$

The $(E-)$ eigenvector of T are the fixed points (up to scaling) of this operator

$$T\mathbf{v}^{m-1} = \lambda \mathbf{v} \quad \text{where} \quad \mathbf{v} \neq \mathbf{0}.$$

Let $m = 3$, $n = 2$ with $t_{122} = 1$, $t_{211} = 1$ and all other entries are 0. Solve the eigenvalue problem.

Solution 23. We obtain

$$(T\mathbf{v}^2)_1 = a_{122}v_2^2 = v_2^2 = \lambda v_1, \quad (T\mathbf{v}^2)_2 = a_{211}v_1^2 = v_1^2 = \lambda v_2.$$

Now $\lambda = 0$ is not a solution, since $v_1^2 = v_2^2 = 0$ implies $v_1 = v_2 = 0$. If $\lambda \neq 0$, then $v_1 \neq 0$ and $v_2 \neq 0$. We also have $v_1^3 = v_2^3$.

Problem 24. Consider the symmetric 3×3 matrix

$$A = \begin{pmatrix} 0 & 3 & 4 \\ 3 & 1 & 5 \\ 4 & 5 & 2 \end{pmatrix}.$$

Find $\operatorname{tr}(A)$, $\operatorname{tr}(A^2)$, $\operatorname{tr}(A^3)$. Then solve the system of equations

$$\operatorname{tr}(A) = \lambda_1 + \lambda_2 + \lambda_3, \quad \operatorname{tr}(A^2) = \lambda_1^2 + \lambda_2^2 + \lambda_3^2, \quad \operatorname{tr}(A^3) = \lambda_1^3 + \lambda_2^3 + \lambda_3^3$$

to find the eigenvalues of A. Apply the Buchberger algorithm to simplify the system. Use Maxima.

Solution 24. We have (we denote the eigenvalues by x_1, x_2, x_3)

$$x_1 + x_2 + x_3 = 3, \quad x_1^2 + x_2^2 + x_3^2 = 105, \quad x_1^3 + x_2^3 + x_3^3 = 717.$$

Applying the Maxima program

```
/* grobner.mac */
load(grobner);
poly_reduction(
[x1+x2+x3-3,x1*x1+x2*x2+x3*x3-105,x1*x1*x1+x2*x2*x2+x3*x3*x3-717],
[x1,x2,x3]);
poly_reduced_grobner(
[x1+x2+x3-3,x1*x1+x2*x2+x3*x3-105,x1*x1*x1+x2*x2*x2+x3*x3*x3-717],
[x1,x2,x3]);
```

Both `poly_reduction()` and `poly_reduced_grobner()` provides the same result, namely

$$x_1 + x_2 + x_3 - 3 = 0$$
$$x_2^3 + x_2 x_3 - 3x_2 + x_3^2 - 3x_3 - 48 = 0$$
$$x_3^3 - 3x_3^2 - 48x_3 - 86 = 0.$$

Obviously the third equation is the characteristic equation and provides the three eigenvalues. The solutions are

$$x_1 = -2.526848, \quad x_2 = -3.691884, \quad x_3 = 9.218732.$$

Supplementary Problems

Problem 1. Let A be an $n \times n$ matrix over $\mathbb{C}$ and $f : \mathbb{C} \to \mathbb{C}$ be analytic in a region D containing the spectrum of A. Then the matrix $f(A)$ can be defined as the *Cauchy integral formula*

$$f(A) = \frac{1}{2\pi i} \int_{\partial D} (zI_n - A)^{-1} f(z) dz.$$

Let

$$A = \begin{pmatrix} 0 & 0 & 1 \\ 0 & 1 & 0 \\ 1 & 0 & 0 \end{pmatrix}$$

with the spectrum $+1$ (twice) and -1. Find $\exp(A)$ applying the Cauchy integral formula.

Problem 2. Let $\alpha > 0$ and A be an $n \times n$ matrix over $\mathbb{C}$. Show that

$$e^{-\alpha A} = \frac{i}{2\pi} \int_C e^{-\alpha \lambda} (A - \lambda I_n)^{-1} d\lambda$$

where C is the contour in the complex λ plane which encloses all eigenvalues of the matrix A. Let

$$A = \begin{pmatrix} 0 & 1 \\ 0 & 0 \end{pmatrix}.$$

Calculate the right-hand side.

Problem 3. Let A be an $m \times n$ matrix over $\mathbb{C}$. Show that AA^* and A^*A have the same nonzero eigenvalues.

Problem 4. Let A and B be real symmetric matrices. Hence A and B have only real eigenvalues. Can we conclude that AB has only real eigenvalues?

Problem 5. Let

$$\mathbf{w}_1 = \begin{pmatrix} 1 & 0 \end{pmatrix}^T, \quad \mathbf{w}_2 = \begin{pmatrix} 0 & 1 \end{pmatrix}^T$$

and

$$\mathbf{u} = \sum_{j_1,j_2,j_3=1}^{2} t_{j_1 j_2 j_3} \mathbf{w}_{j_1} \otimes \mathbf{w}_{j_2} \otimes \mathbf{w}_{j_3} = \mathbf{w}_1 \otimes \mathbf{w}_2 \otimes \mathbf{w}_2 + \mathbf{w}_2 \otimes \mathbf{w}_1 \otimes \mathbf{w}_1$$

$$= \begin{pmatrix} 0 & 0 & 0 & 1 & 1 & 0 & 0 & 0 \end{pmatrix}^T$$

i.e. $t_{122} = 1$, $t_{211} = 1$ and all other coefficients are equal to 0. Can the vector $\mathbf{u} \in \mathbb{C}^8$ be written as the Kronecker product of a vector in $\mathbb{C}^2$ and a vector in $\mathbb{C}^4$? Consider both cases $\mathbb{C}^2 \otimes \mathbb{C}^4$ and $\mathbb{C}^4 \otimes \mathbb{C}^2$.

Problem 6. Consider the 4×4 matrix

$$A = \begin{pmatrix} -1/2 & 1/2 & 1/2 & 1/2 \\ 1/2 & -1/2 & 1/2 & 1/2 \\ 1/2 & 1/2 & -1/2 & 1/2 \\ 1/2 & 1/2 & 1/2 & -1/2 \end{pmatrix}.$$

Find the eigenvalues of A without calculating the eigenvalues. Utilize the information from A^2, $\text{tr}(A)$ and that the matrix A is symmetric over $\mathbb{R}$. Then find the eigenvalues of $A \otimes A$ and $A \otimes I_4 + I_4 \otimes A$.

Problem 7. Let $\omega = e^{2\pi i/3}$ and

$$D = \begin{pmatrix} 1 & 0 & 0 \\ 0 & \omega & 0 \\ 0 & 0 & \omega^2 \end{pmatrix}, \quad \Gamma = \begin{pmatrix} 0 & 1 & 0 \\ 0 & 0 & 1 \\ 1 & 0 & 0 \end{pmatrix}.$$

Find the eigenvalues and normalized eigenvectors of

$$\hat{H} = \hbar\omega_1(D \otimes I_3 + I_3 \otimes D) + \hbar\omega_2(\Gamma \otimes \Gamma^T + \Gamma^T \otimes \Gamma).$$

Problem 8.   Find the eigenvalues and eigenvectors of

$$
A = \begin{pmatrix}
a_{11} & 0 & 0 & a_{14} \\
0 & a_{22} & a_{23} & 0 \\
0 & a_{32} & a_{33} & 0 \\
a_{41} & 0 & 0 & a_{44}
\end{pmatrix}
$$

by calculating the eigenvalues and eigenvectors of

$$
\begin{pmatrix} a_{11} & a_{14} \\ a_{41} & a_{44} \end{pmatrix}, \quad
\begin{pmatrix} a_{22} & a_{23} \\ a_{32} & a_{33} \end{pmatrix}.
$$

Problem 9.   Let S_1, S_2, S_3 be the spin-$\frac{1}{2}$ matrices

$$
S_1 = \frac{1}{2} \begin{pmatrix} 0 & 1 \\ 1 & 0 \end{pmatrix}, \quad
S_2 = \frac{1}{2} \begin{pmatrix} 0 & -i \\ i & 0 \end{pmatrix}, \quad
S_3 = \frac{1}{2} \begin{pmatrix} 1 & 0 \\ 0 & -1 \end{pmatrix}.
$$

Solve the eigenvalue problem for the Hamilton operator

$$
\hat{H} = \hbar\omega_1 (S_1 \otimes S_2 \otimes S_3 + S_3 \otimes S_1 \otimes S_2 + S_2 \otimes S_3 \otimes S_1)
$$
$$
+ \hbar\omega_2 (S_3 \otimes I_2 \otimes I_2 + I_2 \otimes S_3 \otimes I_2 + I_2 \otimes I_2 \otimes S_3).
$$

Problem 10.   Consider the symmetric binary matrices

$$
A = \begin{pmatrix} 0 & 1 & 0 \\ 1 & 0 & 1 \\ 0 & 1 & 0 \end{pmatrix}, \quad
B = \begin{pmatrix} 1 & 0 & 1 \\ 0 & 1 & 0 \\ 1 & 0 & 1 \end{pmatrix}.
$$

Find the eigenvalues and eigenvectors of A and B. Find the eigenvalues and eigenvectors of the anti-commutator $[A, B]_+$. Discuss.

Problem 11.   Let H be a hermitian 3×3 matrix with eigenvalues λ_0, λ_1, λ_2 and $\lambda_0 \leq \lambda_1 \leq \lambda_2$. The corresponding normalized eigenvectors are v_0, v_1, v_2. What are the conditions on H such that

$$
v_0 = \frac{1}{\sqrt{2}} (v_1 + v_2)?
$$

Problem 12.   Show that the 3×3 matrix

$$
A(\theta, \phi) = \begin{pmatrix}
\cos(\theta) & \sin(\theta)e^{-i\phi}/\sqrt{2} & 0 \\
\sin(\theta)e^{i\phi}/\sqrt{2} & 0 & \sin(\theta)e^{-i\phi}/\sqrt{2} \\
0 & \sin(\theta)e^{i\phi}/\sqrt{2} & -\cos(\theta)
\end{pmatrix}
$$

is hermitian, $\mathrm{tr}(A(\theta,\phi)) = 0$, and the eigenvalues are given by $+1$, 0, -1 with the corresponding normalized eigenvectors

$$\mathbf{v}_1 = \begin{pmatrix} (1+\cos(\theta))e^{-i\phi}/2 \\ \sin(\theta)/\sqrt{2} \\ (1-\cos(\theta))e^{i\phi}/2 \end{pmatrix}, \quad \mathbf{v}_0 = \begin{pmatrix} -\sin(\theta)e^{i\phi}/\sqrt{2} \\ \cos(\theta) \\ \sin(\theta)e^{i\phi}/\sqrt{2} \end{pmatrix},$$

$$\mathbf{v}_{-1} = \begin{pmatrix} (1+\cos(\theta))e^{-i\phi}/2 \\ -\sin(\theta)/\sqrt{2} \\ (1+\cos(\theta))e^{i\phi}/2 \end{pmatrix}.$$

Problem 13. Show that the vectors form an orthonormal basis in $\mathbb{C}^3$

$$\mathbf{v}_1 = \frac{1}{\sqrt{3}}\begin{pmatrix} 1 \\ 1 \\ 1 \end{pmatrix}, \quad \mathbf{v}_2 = \frac{1}{\sqrt{2}}\begin{pmatrix} -1 \\ 0 \\ 1 \end{pmatrix}, \quad \mathbf{v}_3 = \frac{1}{\sqrt{6}}\begin{pmatrix} 1 \\ -2 \\ 1 \end{pmatrix}.$$

Assume that the vectors are eigenvectors of a 3×3 hermitian matrix H with (real) eigenvalues λ_1, λ_2, λ_3. Reconstruct the matrix H.

Problem 14. Let U be an $n \times n$ unitary matrix and A, B be $n \times n$ hermitian matrices with the same eigenvalues. Assume that $UAU^{-1} = B$. Show that

$$\sum_{j=1}^{n} \lambda_j (U\mathbf{a}_j\mathbf{a}_j^* - \mathbf{b}_j\mathbf{b}_j^* U) = 0_n$$

where $\mathbf{a}_j$, $\mathbf{b}_j$ are the normalized eigenvectors of A and B, respectively.

Problem 15. (i) Let $x \geq 0$. Find the eigenvalues and eigenvectors of the 3×3 matrix

$$A(x) = \begin{pmatrix} 1 & \sqrt{x} & \sqrt{x} \\ \sqrt{x} & 0 & x \\ \sqrt{x} & x & 0 \end{pmatrix}.$$

(ii) Find the eigenvalues and eigenvectors of the 4×4 matrix

$$A(x_1, x_2, x_3) = \begin{pmatrix} 0 & 1 & 0 & 0 \\ 0 & 0 & 1 & 0 \\ 0 & 0 & 0 & 1 \\ x_1 & x_2 & x_3 & 0 \end{pmatrix}.$$

Chapter 6

Complex Analysis

A complex number z can be written as

$$z = x + iy, \qquad x, y \in \mathbb{R}$$

where x is the real part $\Re(z)$, y is the imaginary part $\Im(z)$ and $i = \sqrt{-1}$. The number $\bar{z} = x - iy$ is known as the *complex conjugate* of z. A complex number z also admits the polar coordinate representation

$$z = |z|e^{i\phi} = |z|(\cos(\phi) + i\sin(\phi)).$$

The length, $r = |z|$, of the segment $(0,0) - (x,y)$ is known as the absolute value or modulus of z and is denoted by $|z|$. We have $|z| = \sqrt{z\bar{z}}$. The angle which the segment Oz makes with the Ox axis is known as the *argument* (amplitude or angle) of z and is denoted by $\arg(z)$. We have

$$\tan(\arg(z)) = \tan(\phi) = \frac{y}{x}$$

and $\arg z$ is defined as a real number modulo 2π provided $z \neq 0$. Some authors take $0 \leq \arg(z) < 2\pi$ while others opt for the range $-\pi < \arg(z) \leq \pi$. Let

$$z_1 = \rho_1 e^{i\phi_1} \equiv \rho_1(\cos(\phi_1) + i\sin(\phi_1)), \quad z_2 = \rho_2 e^{i\phi_2} \equiv \rho_2(\cos(\phi_2) + i\sin(\phi_2))$$

with $\rho_1, \rho_2 \geq 0$. Then $z_1 z_2 = \rho_1 \rho_2(\cos(\phi_1 + \phi_2) + i\sin(\phi_1 + \phi_2))$. Let $z_2 \neq 0$. Then

$$\frac{z_1}{z_2} = \frac{\rho_1}{\rho_2}(\cos(\phi_1 - \phi_2) + i\sin(\phi_1 - \phi_2)).$$

Both Cauchy's integral formula and the residue theorem play a central role in solving problems in complex analysis.

Let $f : \mathbb{C} \to \mathbb{C}$ be analytic inside and on a circle having its centre at $z = a$. Then for all points z in the circle we have a *Taylor series expansion* of f given by

$$f(z) = f(a) + \sum_{n=1}^{\infty} \frac{f^{(n)}(a)}{n!} (z - a)^n$$

where $f^{(n)}$ denotes the n-th derivative.

A necessary condition that the complex function $f(z) = u(x, y) + iv(x, y)$ be analytic in a region R is that u, v satisfy the *Cauchy-Riemann equations*

$$\frac{\partial u}{\partial x} = \frac{\partial v}{\partial y}, \quad \frac{\partial u}{\partial y} = -\frac{\partial v}{\partial x}.$$

Cauchy's integral formula. Let Γ be a simple closed positively oriented contour. If f is analytic in some simply connected domain D containing Γ and z_0 is any point inside Γ, then

$$f(z_0) = \frac{1}{2\pi i} \int_{\Gamma} \frac{f(z)}{z - z_0} dz.$$

Residue Theorem. If Γ is a simple closed positively oriented contour and f is analytic inside and on Γ, except at the points $z_1, z_2, \ldots, z_n$ inside Γ, then

$$\int_{\Gamma} f(z) dz = 2\pi i \sum_{j=1}^{n} \text{Res}(z_j).$$

If f has a pole of order m at z_0, then

$$\text{Res}(f; z_0) = \lim_{z \to z_0} \frac{1}{(m-1)!} \frac{d^{m-1}}{dz^{m-1}} ((z - z_0)^m f(z)).$$

Suppose that the function f is holomorphic on an open annulus $A(a; r, R)$. Then f has a *Laurent series expansion*

$$f(z) = \sum_{k=-\infty}^{\infty} c_k (z - a)^k$$

which converges absolutely in the annulus and uniformly on compact sub-annuli. The coefficients c_k of the Laurent expansion are determined by uniquely as

$$c_k = \frac{1}{2\pi i} \int_{\gamma} \frac{f(\zeta)}{\zeta - a} d\zeta, \qquad k = 0, \pm 1, \pm 2, \pm 3, \ldots$$

where γ is any positively oriented Jordan curve in the annulus which wraps around the point a.

Problem 1.   (i) Let $i := \sqrt{-1}$. Calculate i^i.
(ii) Let $z = x + iy$ with $x, y \in \mathbb{R}$. Calculate $|e^{iz}|$.

Solution 1.   (i) Let z_1 and z_2 be complex numbers. Assume that $z_1 \neq 0$.
One defines

$$z_1^{z_2} := e^{z_2 \ln(z_1)}.$$

Now

$$\ln(z) = \ln(r) + i(\theta + 2k\pi), \qquad k \in \mathbb{Z}$$

where $z = x + iy$, $x, y \in \mathbb{R}$ and $r := \sqrt{x^2 + y^2}$. Therefore $\ln(z)$ is an
infinitely many valued function. The *principal branch* of $\ln(z)$ is defined as
$\ln(r) + i\theta$, where $0 \leq \theta < 2\pi$. Consequently, for the principal branch we
have

$$i^i = e^{i \ln(i)} = e^{ii\pi/2} = e^{-\pi/2}.$$

(ii) Since $z = x + iy$ with $x, y \in \mathbb{R}$ the complex conjugate is given by
$\bar{z} = x - iy$. Therefore

$$|e^{iz}| := \sqrt{e^{iz}e^{-i\bar{z}}} = \sqrt{e^{i(x+iy)}e^{-i(x-iy)}} = \sqrt{e^{-2y}} = e^{-y}.$$

Problem 2.   Let n be an integer. Then $e^{1+2n\pi i} = e$. If we write

$$(e^{1+2n\pi i})^{1+2n\pi i} = e^{1+2n\pi i} = e$$

and

$$(e^{1+2n\pi i})^{1+2n\pi i} = e^{1+4n\pi i-4n^2\pi^2} = ee^{-4n^2\pi^2}$$

it follows that $e^{-4n^2\pi^2} = 1$. Solve this paradox.

Solution 2.   Define the *imaginary remainder* $\Im r(z)$ and the *imaginary
quotient* $\Im q(z)$ by

$$\Im(z) = \Im(r(z)) + 2\pi\Im(q(z))$$

where $\Im(r(z)) \in (-\pi, \pi]$ and $\Im(q(z)) \in \mathbb{Z}$. We have $(e^u)^v = e^{uv}e^{-v2\pi i\Im q(u)}$
since

$$(e^u)^v = e^{\ln(e^u)^v} = e^{(\Re(u)+i\Im r(u))v} = e^{(u-i2\pi\Im q(u))v} = e^{uv}e^{-vi2\pi\Im q(u)}.$$

Thus the expression $(e^{1+2n\pi i})^{1+2n\pi i} = e^{1+4n\pi i-4n^2\pi^2}$ should be replaced by

$$(e^{1+2n\pi i})^{1+2n\pi i} = e^{1+4n\pi i-4n^2\pi^2}e^{-(1+2n\pi i)2\pi i\Im q(1+2n\pi i)}$$

$$= e^{1-4n^2\pi^2}e^{-(1+2n\pi i)2\pi in} = e^{1-4n^2\pi^2}e^{-2n\pi i+4n^2\pi^2} = e.$$

In problem 1 we considered the domain $[0, 2\pi)$ for the principal branch. In
this problem we consider the domain $(-\pi, \pi]$ for the principal branch.

Problem 3. Let $z = x + iy$, where $x, y \in \mathbb{R}$. Find the real and imaginary part of

$$\cos\left(\frac{z}{2}\right) \overline{\sin\left(\frac{z}{2}\right)}.$$

Solution 3. Since

$$\cos\left(\frac{z}{2}\right) \equiv \frac{e^{i(x+iy)/2} + e^{-i(x+iy)/2}}{2}$$

and

$$\overline{\sin\left(\frac{z}{2}\right)} \equiv \sin\left(\frac{\bar{z}}{2}\right) \equiv \frac{e^{i(x-iy)/2} - e^{-i(x-iy)/2}}{2i}$$

we obtain

$$\cos\left(\frac{z}{2}\right) \sin\left(\frac{\bar{z}}{2}\right) \equiv \frac{e^{ix} - e^{-ix} + e^{y} - e^{-y}}{4i}$$

or

$$\cos\left(\frac{z}{2}\right) \sin\left(\frac{\bar{z}}{2}\right) \equiv \frac{1}{2}\sin(x) + \frac{1}{2i}\sinh(y) \equiv \frac{1}{2}\sin(x) - \frac{i}{2}\sinh(y).$$

Hence, the real part is $\frac{1}{2}\sin(x)$ and the imaginary part is $-\frac{1}{2}\sinh(y)$.

Problem 4. Set $z = re^{i\phi}$. Calculate $(z \neq 1)$ the imaginary part

$$\Im\left(\frac{1+z}{1-z}\right).$$

Solution 4. We have

$$\Im\left(\frac{1+z}{1-z}\right) = \Im\left(\frac{1+z}{1-z}\frac{1-\bar{z}}{1-\bar{z}}\right) = \Im\left(\frac{1+z-\bar{z}-z\bar{z}}{1-z-\bar{z}+z\bar{z}}\right)$$

$$= \Im\left(\frac{1+r(e^{i\phi}-e^{-i\phi})-r^2}{1-r(e^{i\phi}+e^{-i\phi})+r^2}\right) = \Im\left(\frac{1+2ir\sin(\phi)-r^2}{1-2r\cos(\phi)+r^2}\right)$$

$$= \frac{2r\sin(\phi)}{1-2r\cos(\phi)+r^2}.$$

Problem 5. Let μ, ν be two complex numbers. Assume that

$$|\mu|^2 - |\nu|^2 = 1.$$

Find a parametrization.

Solution 5. A parametrization is $(r \in \mathbb{R})$ $\mu = \cosh(r)$, $\nu = e^{i\phi} \sinh(r)$ since $\cosh^2(r) - \sinh^2(r) = 1$ and $e^{i\phi}e^{-i\phi} = 1$.

Problem 6. Let $\alpha \in \mathbb{R}$ and $z_1, z_2 \in \mathbb{C}$. Calculate

$$R = z_1^* z_2 e^{(e^{i\alpha}-1)z_1 z_1^*} e^{(e^{-i\alpha}-1)z_2 z_2^*} + z_1 z_2^* e^{(e^{-i\alpha}-1)z_1 z_1^*} e^{(e^{i\alpha}-1)z_2 z_2^*}$$

using $z_j = r_j e^{i\phi_j}$ and $e^{i\phi_j} = \cos(\phi_j) + i\sin(\phi_j)$. The expression plays a role in quantum optics. Obviously R is real. Consider then the special case $\phi_1 = \phi_2$.

Solution 6. Using $\cos(-x) = \cos(x)$ and $\sin(-x) = -\sin(x)$ we have

$$R = r_1 r_2 (e^{i(\phi_2 - \phi_1)} e^{(e^{i\alpha}-1)r_1^2} e^{(e^{-i\alpha}-1)r_2^2} + e^{i(\phi_1 - \phi_2)} e^{(e^{-i\alpha}-1)r_1^2} e^{(e^{i\alpha}-1)r_2^2})$$

$$= r_1 r_2 \cos(\phi_2 - \phi_1)(e^{(e^{i\alpha}-1)r_1^2 + (e^{-i\alpha}-1)r_2^2} + e^{(e^{-i\alpha}-1)r_1^2 + (e^{i\alpha}-1)r_2^2})$$

$$+ i r_1 r_2 \sin(\phi_2 - \phi_1)(e^{(e^{i\alpha}-1)r_1^2 + (e^{-i\alpha}-1)r_2^2} - e^{(e^{-i\alpha}-1)r_1^2 + (e^{i\alpha}-1)r_2^2})$$

$$= 2 r_1 r_2 \cos(\phi_2 - \phi_1) e^{(\cos(\alpha)-1)(r_1^2 + r_2^2)} \cos((r_1^2 - r_2^2)\sin(\alpha))$$

$$- 2 r_1 r_2 \sin(\phi_2 - \phi_1) e^{(\cos(\alpha)-1)(r_1^2 + r_2^2)} \sin((r_1^2 - r_2^2)\sin(\alpha)) .$$

For the special case $\phi_1 = \phi_2$ we obtain

$$R = 2 r_1 r_2 e^{(\cos(\alpha)-1)(r_1^2 + r_2^2)} \cos((r_1^2 - r_2^2)\sin(\alpha)) .$$

Problem 7. Show that for all $x, y \in \mathbb{R}$

$$|e^{ix} - e^{iy}| = 2 \left| \sin\left(\frac{x-y}{2}\right) \right| .$$

Solution 7. We have

$$|e^{ix} - e^{iy}| = \left| e^{i(x+y)/2} \right| \cdot \left| e^{i(x-y)/2} - e^{-i(x-y)/2} \right|$$

$$= |e^{i(x-y)/2} - e^{-i(x-y)/2}| = 2 \left| i \sin\left(\frac{x-y}{2}\right) \right|$$

$$= 2 \left| \sin\left(\frac{x-y}{2}\right) \right| .$$

Problem 8. Let $x, y \in \mathbb{R}$ and

$$A = \begin{pmatrix} x & -y \\ y & x \end{pmatrix} .$$

Calculate $\exp(A)$. Use the fact that there is a field *isomorphism* between the complex numbers $z = x + iy$ and the 2×2 matrices given by A.

Solution 8. We calculate e^z and then use the isomorphism

$$e^z = e^{x+iy} = e^x e^{iy} = e^x(\cos(y) + i\sin(y)) = e^x \cos(y) + ie^x \sin(y).$$

Thus the real part of e^z is given by $e^x \cos(y)$ and the imaginary part of e^z is given by $e^x \sin(y)$. Owing to the isomorphism we have

$$e^A = \begin{pmatrix} e^x \cos(y) & -e^x \sin(y) \\ e^x \sin(y) & e^x \cos(y) \end{pmatrix}.$$

Problem 9. Given the equation $e^x - e^{-x} = -2iy$. Show that

$$e^x = -iy \pm \sqrt{1 - y^2}.$$

Solution 9. Multiplying the equation by e^x yields $(e^x)^2 + 2iye^x - 1 = 0$. Solving this quadratic equation for e^x provides the result.

Problem 10. Let $z = x + iy$, $\bar{z} = x - iy$ where $x, y \in \mathbb{R}$.
(i) Show that

$$\frac{\partial}{\partial z} = \frac{1}{2}\left(\frac{\partial}{\partial x} - i\frac{\partial}{\partial y}\right), \qquad \frac{\partial}{\partial \bar{z}} = \frac{1}{2}\left(\frac{\partial}{\partial x} + i\frac{\partial}{\partial y}\right). \tag{1}$$

(ii) Find $dz \wedge d\bar{z}$, where $\wedge$ denotes the *exterior product*. The exterior product satisfies $dx \wedge dx = 0$, $dy \wedge dy = 0$, $dx \wedge dy = -dy \wedge dx$.

Solution 10. (i) Let f be a smooth function of x and y ($x, y \in \mathbb{R}$). Then

$$df = \frac{\partial f}{\partial x}dx + \frac{\partial f}{\partial y}dy. \tag{2}$$

Since $dz = d(x + iy) = dx + idy$, $d\bar{z} = d(x - iy) = dx - idy$ we obtain

$$dx = \frac{1}{2}(dz + d\bar{z}), \qquad dy = \frac{1}{2i}(dz - d\bar{z}). \tag{3}$$

Inserting (3) into (2) yields

$$df = \frac{\partial f}{\partial x}\frac{1}{2}(dz + d\bar{z}) + \frac{\partial f}{\partial y}\frac{1}{2i}(dz - d\bar{z})$$

$$= \frac{1}{2}\left(\frac{\partial f}{\partial x} - i\frac{\partial f}{\partial y}\right)dz + \frac{1}{2}\left(\frac{\partial f}{\partial x} + i\frac{\partial f}{\partial y}\right)d\bar{z}.$$

Since

$$df = \frac{\partial f}{\partial z} dz + \frac{\partial f}{\partial \bar{z}} d\bar{z}$$

we find (1). Note that

$$\frac{\partial}{\partial x} = \frac{\partial}{\partial z} + \frac{\partial}{\partial \bar{z}}, \qquad \frac{\partial}{\partial y} = i \left(\frac{\partial}{\partial z} - \frac{\partial}{\partial \bar{z}} \right).$$

(ii) We obtain $dz \wedge d\bar{z} = d(x + iy) \wedge d(x - iy) = -2i dx \wedge dy$.

Problem 11.   Discuss the mapping of the periodic strip

$$-\pi < \Re(z) \leq +\pi$$

by the analytic function $w(z) = \sin(z)$. Find the images of the straight lines $\Re(z) = \text{const}$ and of the straight segments $\Im(z) = \text{const}$.

Solution 11.   Since $z = x + iy$ and $w(z(x,y)) = u(x,y) + iv(x,y)$, where u and v are real-valued functions and

$$\sin(z) = \sin(x + iy) = \frac{e^y + e^{-y}}{2} \sin(x) + i \frac{e^y - e^{-y}}{2} \cos(x)$$

we obtain

$$u(x,y) = \frac{e^y + e^{-y}}{2} \sin(x), \qquad v(x,y) = \frac{e^y - e^{-y}}{2} \cos(x).$$

Let y_0 be a fixed positive number. If $y = y_0$ and x increases from $-\pi$ to π, then the function w describes an ellipse with foci at $\pm i$, since

$$\frac{u^2}{C^2} = \sin^2(x), \qquad \frac{v^2}{D^2} = \cos^2(x)$$

and $\sin^2(x) + \cos^2(x) = 1$ with $C = (e^y + e^{-y})/2$, $D = (e^y - e^{-y})/2$. The ellipse is described once in the clockwise direction starting at its lowest point. For $y = -y_0$ we obtain the same ellipse described in the opposite direction starting at its highest point. Through each point of the plane (except those along the real segment $[-1, 1]$) passes exactly one such ellipse. If $y_0 = 0$ we obtain

$$u(x, y_0 = 0) = \sin(x), \qquad v(x, y_0 = 0) = 0.$$

Hence we describe the segment $[-1, 1]$ twice (from 0 to -1, to 0, to 1, to 0). To obtain the complete image of the periodic strip, we take two copies of the w plane, cut one along the positive imaginary axis, the other along the negative imaginary axis and both along the real segment $[-1, 1]$. If we join the banks of the horizontal slits crosswise, we find the two-sheeted Riemann surface which is the one-to-one image of the periodic strips. The lines $\Re(z) = \text{const}$ are taken into hyperbolas orthogonal to the ellipses.

Problem 12. Calculate

$$\int_0^\infty \frac{x\sin(ax)}{x^2 + \lambda^2}\,dx$$

where $a, \lambda^2 \in \mathbb{R}$ and $\lambda^2 > 0$.

Solution 12. We solve the problem by considering the integration in the complex plane. First we notice that

$$\int_0^\infty \frac{x\sin(ax)}{x^2 + \lambda^2}\,dx = \frac{1}{2}\int_{-\infty}^{+\infty} \frac{x\sin(ax)}{x^2 + \lambda^2}\,dx = \frac{1}{2}\Im\int_{-\infty}^{+\infty} \frac{xe^{iax}}{x^2 + \lambda^2}\,dx$$

where $\Im$ denotes the imaginary part. Now we consider

$$\int_C \frac{ze^{iaz}}{z^2 + \lambda^2}\,dz$$

where C is the contour consisting of the line along the x-axis from $-R$ to $+R$ and the semicircle Γ above the x-axis having this line as diameter. Thus we have

$$\int_C \frac{ze^{iaz}}{z^2 + \lambda^2}\,dz = \int_{-R}^{+R} \frac{xe^{iax}}{x^2 + \lambda^2}\,dx + \int_\Gamma \frac{Re^{i\phi}e^{iaRe^{i\phi}}}{R^2e^{2i\phi} + \lambda^2}Re^{i\phi}i\,d\phi \qquad (1)$$

since $z = Re^{i\phi}$, $dz = Re^{i\phi}i\,d\phi$. Now $z^2 + \lambda^2 \equiv (z + i\lambda)(z - i\lambda)$, where only the pole at $z = i\lambda$ lies inside C. The left-hand side is calculated with the help of the *residue theorem*. The residue a_1 of a complex function f at $z = a$, where $z = a$ is a pole of order k, is calculated as follows

$$a_{-1} := \lim_{z \to a} \frac{1}{(k-1)!} \frac{d^{k-1}}{dz^{k-1}}((z-a)^k f(z)).$$

If $k = 1$ (simple pole) we have $a_{-1} = \lim_{z \to a}(z - a)f(z)$. The *residue* is given by

$$\mathrm{Res}_{z=i\lambda}\frac{ze^{iaz}}{z^2 + \lambda^2} := \lim_{z \to i\lambda} \frac{z(z - i\lambda)e^{iaz}}{(z + i\lambda)(z - i\lambda)} = \frac{i\lambda e^{iai\lambda}}{2i\lambda} = \frac{1}{2}e^{-a\lambda}.$$

The *residue theorem* states that

$$\oint_C f(z)\,dz = 2\pi(a_{-1} + b_{-1} + c_{-1} + \cdots)$$

where $a_{-1}, b_{-1}, c_{-1}, \ldots$ are the residues inside C. Consequently

$$\int_C \frac{ze^{iaz}}{z^2 + \lambda^2} dz = \pi i e^{-a\lambda}.$$

If $R \to \infty$, the second integral on the right hand side of (1) vanishes. Therefore

$$\int_{-\infty}^{+\infty} \frac{xe^{iax}}{x^2 + \lambda^2} dx = \pi i e^{-a\lambda}, \qquad \Im\left(\int_{-\infty}^{+\infty} \frac{xe^{iax}}{x^2 + \lambda^2} dx\right) = \pi e^{-a\lambda}.$$

Finally

$$\int_0^\infty \frac{x \sin(ax)}{x^2 + \lambda^2} dx = \frac{\pi}{2} e^{-a\lambda}.$$

Problem 13. (i) Calculate the Fourier transform $\hat{f}(\omega)$ of $f(t) = e^{-|t|}$.
(ii) Calculate the inverse Fourier transform of $\hat{f}(\omega)$.

Solution 13. (i) From the definition of the *Fourier transform*

$$\hat{f}(\omega) := \int_{-\infty}^{+\infty} f(t) e^{i\omega t} dt$$

we obtain

$$\hat{f}(\omega) = \int_{-\infty}^0 e^t e^{i\omega t} dt + \int_0^\infty e^{-t} e^{i\omega t} dt \equiv \int_{-\infty}^0 e^{(i\omega+1)t} dt + \int_0^\infty e^{(i\omega-1)t} dt.$$

Integration yields the analytic function $\hat{f}(\omega) = 2/(1 + \omega^2)$.
(ii) The inverse Fourier transform is given by

$$f(t) = \frac{1}{2\pi} \int_{-\infty}^{+\infty} \hat{f}(\omega) e^{-i\omega t} d\omega = \frac{1}{\pi} \int_{-\infty}^{+\infty} \frac{1}{1 + \omega^2} e^{-i\omega t} d\omega.$$

To calculate this integral we extend ω in the complex domain and consider

$$\frac{1}{\pi} \oint_C \frac{e^{-izt}}{1 + z^2} dz.$$

Since the integrand has the two poles $z = \pm i$ and

$$\frac{1}{1 + z^2} \equiv -\frac{1}{2i} \frac{1}{z + i} + \frac{1}{2i} \frac{1}{z - i}$$

we consider two paths C_1 and C_2. The path C_1 is the contour consisting of the line along the x-axis from $-R$ to R and the semicircle Γ above the

x-axis having this line as diameter. The path C_2 is the contour consisting of the line along the x-axis from $-R$ to R and the semicircle Γ below the x-axis having this line as diameter. For the path C_1 we have

$$\frac{1}{\pi} \oint_{C_1} \frac{e^{-izt}}{1+z^2} dz \equiv -\frac{1}{2i\pi} \oint_{C_1} \frac{e^{-izt}}{z+i} dz + \frac{1}{2i\pi} \oint_{C_1} \frac{e^{-izt}}{z-i} dz.$$

The first integral on the right-hand side is equal to zero, since the pole $z = -i$ is not inside C_1. Consequently,

$$\frac{1}{\pi} \oint_{C_1} \frac{e^{-izt}}{1+z^2} dz = \frac{1}{2i\pi} \oint_{C_1} \frac{e^{-izt}}{z-i} dz.$$

Thus $f(t) = e^t$ for $t \in (-\infty, 0]$. Analogously, for the path C_2 we have

$$\frac{1}{\pi} \oint_{C_2} \frac{e^{-izt}}{1+z^2} dz = -\frac{1}{2i\pi} \oint_{C_2} \frac{e^{-izt}}{z+i} dz + \frac{1}{2i\pi} \oint_{C_2} \frac{e^{-izt}}{z-i} dz.$$

The second integral on the right-hand side is equal to zero, since the pole $z = i$ is not inside C_2. Consequently,

$$\frac{1}{\pi} \oint_{C_2} \frac{e^{-izt}}{1+z^2} dz = -\frac{1}{2i\pi} \oint_{C_2} \frac{e^{-izt}}{z+i} dz.$$

Thus $f(t) = e^{-t}$ for $t \in [0, \infty)$. Consequently, $f(t) = e^{-|t|}$.

Problem 14. (i) Expand the function $\sqrt{(z-1)(z-2)}$ for $|z| > 2$ into its *Laurent series*.
(ii) Expand the function $\ln(1/(1-z))$ for $|z| > 1$ into its Laurent series.

Solution 14. (i) For $|z| > 2$ we have

$$\sqrt{(z-1)(z-2)} = \pm z \left(1 - \frac{1}{z}\right)^{1/2} \left(1 - \frac{2}{z}\right)^{1/2}.$$

Applying the binomial expansions for the square root, we can write for $|z| > 2$

$$\pm z \left(1 - \binom{1/2}{1}\frac{1}{z} + \binom{1/2}{2}\frac{1}{z^2} - \cdots\right) \left(1 - \binom{1/2}{1}\frac{2}{z} + \binom{1/2}{2}\frac{2^2}{z^2} - \cdots\right)$$

$$= \pm \left(c_0 z - c_1 + \frac{c_2}{z} - \frac{c_3}{z^2} + \cdots\right)$$

where

$$c_n = \binom{1/2}{n} + 2\binom{1/2}{n-1}\binom{1/2}{1} + 2^2\binom{1/2}{n-2}\binom{1/2}{2} + \cdots + 2^n \binom{1/2}{n}.$$

(ii) The function does not have a Laurent expansion for $|z| > 1$, since the function is not single-valued there. If we put $z' = 1/z$, then we find

$$\ln\left(\frac{-z'}{1-z'}\right) = \ln(-z') + \ln\left(\frac{1}{1-z'}\right) = \ln\left(-\frac{1}{z}\right) + \frac{1}{z} + \frac{1}{2z^2} + \frac{1}{3z^3} + \cdots.$$

Problem 15.   The *unilateral z-transform* is defined by

$$Z(z) := \sum_{n=0}^{\infty} x(n)z^{-n}$$

i.e. the input starts at $n = 0$. The inverse z-transform is given by

$$x(n) = \frac{1}{2\pi i} \oint Z(z)z^{n-1}dz$$

where $\oint$ denotes integration around a circular path in the complex plane centered at the origin. Apply the z-*transform* to find the solution of the linear third order difference equation $x_{n+3} - 3x_{n+2} + 3x_{n+1} - x_n = 0$ with the initial conditions $x_0 = 0$, $x_1 = 1$, $x_2 = 4$.

Solution 15.   We have

$$Z(x_{n+\ell})(z) = z^\ell Z(x_n)(z) - \sum_{\nu=0}^{\ell-1} v^2 z^{\ell-\nu}, \quad \ell = 1, 2, 3.$$

Thus

$$z^3 Z(x_n)(z) - z^2 - 4z - 3(z^2 Z(x_n)(z) - z) + 3z Z(x_n)(z) - Z(x_n)(z) = 0.$$

Therefore

$$Z(x_n)(z) = \frac{z(z+1)}{(z-1)^3}, \quad |z| > 1.$$

Since

$$\frac{z(z+1)}{(z-1)^3} \equiv \frac{1}{z-1} + \frac{3}{(z-1)^2} + \frac{2}{(z-1)^3}$$

we obtain

$$x_n = \binom{n-1}{0} + 3\binom{n-1}{1} + 2\binom{n-1}{2}.$$

For $n > 3$ we have $x_n = 1 + 3(n-1) + (n-1)(n-2) = n^2$.

Problem 16.   Assume that f is a meromorphic function in the finite plane $\mathbb{C}$ and has a finite number of poles $a_1, a_2, \ldots, a_m$ none of which is an integer. Assume, moreover, that

$$\lim_{z \to \infty} z f(z) = 0.$$

Then

$$\lim_{N \to \infty} \sum_{n=-N}^{n=N} f(n)$$

exists and

$$\lim_{N \to \infty} \sum_{n=-N}^{n=N} f(n) = -\sum_{k=1}^{m} \text{Res}(a_k; \pi f(z) \cot(\pi z))$$

where Res denotes the residue and $\cot(\pi z) \equiv \cos(\pi z)/\sin(\pi z)$. Calculate

$$\sum_{n=-\infty}^{\infty} \frac{1}{n^2 + n + 1}.$$

Solution 16. The function $f(z) = 1/(z^2 + z + 1)$ is meromorphic in the finite plane $\mathbb{C}$ and has two poles, because

$$z^2 + z + 1 \equiv (z - a_1)(z - a_2)$$

with $a_1 = -(1 + i\sqrt{3})/2$, $a_2 = \bar{a}_1$. The residue of $\pi \cot(\pi z)/(z^2 + z + 1)$ at $z = a_1$ is given by

$$\lim_{z \to a_1} \frac{(z - a_1)\pi \cot \pi z}{(z - a_1)(z - \bar{a}_1)} = \frac{\pi \cot(\pi a_1)}{a_1 - \bar{a}_1} = -\frac{i}{\sqrt{3}}\pi \cot\left(\frac{\pi(1 + i\sqrt{3})}{2}\right).$$

Similarly, the residue of $\pi \cot(\pi z)/(z^2 + z + 1)$ at $z = \bar{a}_1$ is given by

$$\lim_{z \to \bar{a}_1} \frac{(z - \bar{a}_1)\pi \cot(\pi z)}{(z - a_1)(z - \bar{a}_1)} = \frac{\pi \cot(\pi \bar{a}_1)}{\bar{a}_1 - a_1} = \frac{i}{\sqrt{3}}\pi \cot\left(\frac{\pi(1 - i\sqrt{3})}{2}\right).$$

Therefore

$$\sum_{n=-\infty}^{\infty} \frac{1}{n^2 + n + 1} = -\frac{2\pi i}{\sqrt{3}} \tan\left(\frac{i\pi\sqrt{3}}{2}\right) = \frac{2\pi}{\sqrt{3}} \tanh\left(\frac{\pi\sqrt{3}}{2}\right)$$

where we have used the identity $\cot(\pi/2 + \alpha) \equiv -\tan(\alpha)$.

Problem 17. Let

$$\{w, z\} := \frac{w'''}{w'} - \frac{3}{2}\left(\frac{w''}{w'}\right)^2$$

be the *Schwarzian derivative* of w, where $w' \equiv dw/dz$. Let y_1 and y_2 be two linearly independent solutions of the equation

$$y'' + Q(z)y = 0 \tag{1}$$

which are defined and holomorphic in some simply connected domain D in the complex plane.

(i) Show that $w(z) = y_1(z)/y_2(z)$ satisfies the third order differential equation

$$\{w, z\} = 2Q(z) \tag{2}$$

at all points of D where $y_2(z) \neq 0$.

(ii) Show that if $w(z)$ is a solution of (2) and is holomorphic in some neighbourhood of a point $z_0 \in D$, then one can find two linearly independent solutions, $u(z)$ and $v(z)$ of (2) defined in D so that

$$w(z) = \frac{u(z)}{v(z)}.$$

If $v(z_0) = 1$ the solutions u and v are uniquely defined.

Solution 17. (i) The expression $y_1 y_2' - y_2 y_1'$ is called the *Wronskian*. We may assume that the Wronskian of y_1 and y_2 is identically one. Then

$$w'(z) = (y_2(z))^{-2} \quad \Rightarrow \quad \frac{w''(z)}{w'(z)} = -2\frac{y_2'(z)}{y_2(z)}$$

and

$$\left(\frac{w''(z)}{w'(z)}\right)' = -2\frac{y_2''(z)}{y_2(z)} + 2\left(\frac{y_2'(z)}{y_2(z)}\right)^2 = 2Q(z) + \frac{1}{2}\left(\frac{w''(z)}{w'(z)}\right)^2$$

from which the first assertion follows.

(ii) Suppose that a solution of equation (2) is given by its initial values

$$w(z_0), \qquad w'(z_0), \qquad w''(z_0)$$

at a point z_0 in D. We may assume that $w'(z_0) \neq 0$ since otherwise $Q(z)$ could not be holomorphic at $z = z_0$. We can now choose two linearly independent solutions, $u(z)$ and $v(z)$, of (2) so that at $z = z_0$ the quotient

$$w_1(z) = \frac{u(z)}{v(z)}$$

has the initial values $w(z_0)$, $w'(z_0)$, $w''(z_0)$ the same as $w(z)$. If $v(z_0) = 1$, then the solutions $u(z)$ and $v(z)$ are uniquely determined, and we must have $w(z) \equiv w_1(z)$.

Problem 18. A function which is analytic everywhere in the finite complex plane (i.e. everywhere except at ∞) is called an *entire function*. For example, the functions $\sin(z)$, e^z and z^4 are entire functions. An entire function can be represented by a Taylor series which has an infinite radius of convergence. Conversely, if a power series has an infinite radius of convergence, it represents an entire function.

Solve the first order linear differential equation

$$\left(\frac{d}{dz} - \frac{\epsilon + \kappa^2}{z + \kappa} + \kappa \right) w(z) = 0 \tag{1}$$

where κ and ϵ are real constants. Impose the condition that w is an entire function. The differential equation (1) is the *Bargmann representation* of the *displaced harmonic oscillator*.

Solution 18. Method 1. Using the power series ansatz

$$w(z) = \sum_{n=0}^{\infty} b_n (z + \kappa)^{n+s}$$

yields $s = \epsilon + \kappa^2$ and the recurrence relation for the expansion coefficients b_n's is given by

$$b_n = \frac{(-\kappa)}{n} b_{n-1}.$$

This recurrence relation can easily be solved

$$b_n = \frac{(-\kappa)^n}{\Gamma(n+1)} b_0$$

where Γ denotes the gamma function, i.e. $\Gamma(n+1) = n!$ with $n = 0, 1, 2, \ldots$. The series $\sum_n b_n z^n$ therefore converges for all z and

$$w(z) = b_0 (z + \kappa)^{\epsilon + \kappa^2} \sum_{n=0}^{\infty} \frac{(-\kappa)^n}{\Gamma(n+1)} (z + \kappa)^n = b_0 (z + \kappa)^{\epsilon + \kappa^2} e^{-\kappa(z+\kappa)}.$$

Imposing the condition that w is an entire function leads to $\epsilon + \kappa^2 = n$, where $n = 0, 1, 2, \ldots$. In physics this is called a *quantization condition*. Method 2. Equation (1) can be directly integrated since

$$\frac{dw}{dz} = \left(\frac{\epsilon + \kappa^2}{z + \kappa} - \kappa \right) w(z).$$

It follows that

$$\int \frac{dw}{w} = \int \left(\frac{\epsilon + \kappa^2}{z + \kappa} - \kappa \right) dz \implies w(z) = C(z + \kappa)^{\epsilon + \kappa^2} e^{-\kappa z}.$$

The requirement that w be entire implies that there be no branch-point singularities at $z = -\kappa$ and therefore $\epsilon + \kappa^2 = n$.

Problem 19.   Assume $Ox_1x_2x_3$ is the system of rectangular coordinates whose axes Ox_1, Ox_2 coincide with the real and imaginary axes Ox, Oy of the complex plane $\mathbb{C}$. Assume, moreover, that the ray emanating from the north pole $N(0,0,1)$ of the unit sphere $\mathbb{S}^2$

$$x_1^2 + x_2^2 + x_3^2 = 1 \tag{1}$$

and intersecting $\mathbb{S}^2$ at $A(x_1, x_2, x_3)$ intersects $\mathbb{C}$ at the point z. The point $z = x + iy$ is then called the *stereographic projection* of $A(x_1, x_2, x_3)$ whereas A is called the spherical image of z. The stereographic projection is given by

$$x_1 = \frac{z + \bar{z}}{1 + |z|^2}, \quad x_2 = \frac{z - \bar{z}}{i(1 + |z|^2)}, \quad x_3 = \frac{|z|^2 - 1}{1 + |z|^2}, \quad z = \frac{x_1 + ix_2}{1 - x_3}.$$

(i) Find the spherical images of $e^{i\alpha}$ where $\alpha \in \mathbb{R}$.
(ii) Let θ, ϕ be the geographical latitude and longitude of A respectively. Show that the stereographic projection of A has the representation

$$z = e^{i\phi} \tan\left(\frac{1}{4}\pi + \frac{1}{2}\theta\right). \tag{2}$$

(iii) The distance $\sigma(z_1, z_2)$ between two points on $\mathbb{S}^2$ whose stereographic projections are z_1, z_2 is called the *spherical distance* or *chordal distance* between z_1 and z_2. Show that

$$\sigma(z_1, z_2) = \frac{2|z_1 - z_2|}{\sqrt{(1 + |z_1|^2)(1 + |z_2|^2)}} \tag{3}$$

is the Euclidean distance of the images of z_1 and z_2.

Solution 19.   (i) Since $z = e^{i\alpha}$ we have $\bar{z} = e^{-i\alpha}$ and $z\bar{z} = 1$. Therefore

$$z + \bar{z} = 2\cos(\alpha) \qquad z - \bar{z} = 2i\sin(\alpha).$$

It follows that $x_1 = \cos(\alpha)$, $x_2 = \sin(\alpha)$, $x_3 = 0$.
(ii) Since *spherical coordinates* are given by

$$x_1(\theta, \phi) = \sin(\theta)\cos(\phi), \quad x_2(\theta, \phi) = \sin(\theta)\sin(\phi), \quad x_3(\theta, \phi) = \cos(\theta)$$

we obtain

$$z = \frac{x_1 + ix_2}{1 - x_3} = \frac{\sin(\theta)\cos(\phi) + i\sin(\theta)\sin(\phi)}{1 - \cos(\theta)} = \frac{e^{i\phi}\sin(\theta)}{1 - \cos(\theta)}.$$

Obviously, (2) follows.

(iii) Let **x** and **y** be two points on the unit sphere (1). Then

$$x_1^2 + x_2^2 + x_3^2 = 1, \qquad y_1^2 + y_2^2 + y_3^2 = 1. \tag{4}$$

It follows that

$$\|\mathbf{x} - \mathbf{y}\| = \sqrt{(x_1 - y_1)^2 + (x_2 - y_2)^2 + (x_3 - y_3)^2}$$
$$= \sqrt{2 - 2x_1 y_1 - 2x_2 y_2 - 2x_3 y_3}$$

where we have used (4). Under the stereographic projection we have

$$z_1 \to (x_1, x_2, x_3), \qquad z_2 \to (y_1, y_2, y_3).$$

Now

$$1 + |z_1|^2 = \frac{2}{1 - x_3} \qquad 1 + |z_2|^2 = \frac{2}{1 - y_3} \tag{5}$$

and with $z_1 = (x_1 + ix_2)/(1 - x_3)$, $z_2 = (y_1 + iy_2)/(1 - y_3)$ we obtain

$$|z_1 - z_2| = \frac{\sqrt{2(1 - x_3)^2(1 - y_3^2) - 2(1 - x_3)(1 - y_3)(x_1 y_1 + x_2 y_2)}}{(1 - x_3)(1 - y_3)}.$$

Inserting this equation and (5) into (3) leads to $\sigma(z_1, z_2) = \|\mathbf{x} - \mathbf{y}\|$. The stereographic projection can be extended to arbitrary dimensions. Let $\mathbb{S}^n$ denote the unit n-sphere in $\mathbb{R}^{n+1}$ and let $\mathbf{q} = (0, \ldots, 0, 1)$ denote the "north pole" of $\mathbb{S}^n$. Let

$$\mathbf{f} : \mathbb{R}^n \to \mathbb{S}^n$$

be the map which sends each $\mathbf{p} \in \mathbb{R}^n$ into the point different from $\mathbf{q}$ where the line through $(\mathbf{p}, 0) \in \mathbb{R}^{n+1}$ and $\mathbf{q}$ cuts $\mathbb{S}^n$. Since

$$\boldsymbol{\alpha}(t) = t(\mathbf{p}, 0) + (1 - t)\mathbf{q} = (t\mathbf{p}, 1 - t)$$

is a parametrization of the line through $(\mathbf{p}, 0)$ and $\mathbf{q}$, and since $\|\boldsymbol{\alpha}(t)\| = 1$ if and only if $t = 0$ or $t = 2/(\|\mathbf{p}\|^2 + 1)$ the map $\mathbf{f}$ is given by

$$\mathbf{f}(x_1, x_2, \ldots, x_n) = \frac{(2x_1, 2x_2, \ldots, 2x_n, x_1^2 + x_2^2 + \cdots + x_n^2 - 1)}{x_1^2 + x_2^2 + \cdots + x_n^2 + 1}.$$

The mapping $\mathbf{f}$ is a parametrized surface which maps $\mathbb{R}^n$ one-to-one onto $\mathbb{S}^n \setminus \{\mathbf{q}\}$. The chart $\mathbf{f}^{-1}$ is called the stereographic projection from $\mathbb{S}^n \setminus \{\mathbf{q}\}$ onto the equatorial hyperplane.

Problem 20. A *Möbius transformation* is a function of the form

$$w(z) = \frac{az + b}{cz + d}$$

where a, b, c, d are complex numbers and $ad - bc \neq 0$. Find the Möbius transformation that carries the point 1 to i, the point 0 to 1 and the point i to ∞.

Solution 20.   Since 0 goes to 1, we must have $b/d = 1$. Thus

$$w(z) = \frac{az + b}{cz + b}.$$

Since i goes to ∞, $z = i$ must make the denominator zero. Hence we have $ci + b = 0$, $b = -ic$, and therefore

$$w(z) = \frac{az - ic}{cz - ic}.$$

Finally, since 1 goes to i

$$i = \frac{a - ic}{c - ic}.$$

Thus $a = c(1 + 2i)$. The coefficient c is at our disposal, with $c \neq 0$. Let $c = 1$. Then we arrive at

$$w(z) = \frac{(1 + 2i)z - i}{z - i}.$$

A Möbius transformation is a one-to-one map of the extended z plane to the extended w plane, with $w = 0$ when $az + b = 0$ and $w = \infty$ when $cz + d = 0$. The transformation is a composition of magnifications, rotations, translations, and a reciprocal transformation $w(z) = 1/z$.

Problem 21.   Give the domain in the complex plane where the infinite series

$$\sum_{n=1}^{\infty} \frac{1}{n^2} \exp\left(\frac{nz}{z-2}\right) \tag{1}$$

converges.

Solution 21.   Let

$$f(z) = \exp\left(\frac{z}{z-2}\right).$$

The series (1) can then be cast into the form

$$\sum_{n=1}^{\infty} \frac{1}{n^2}(f(z))^n.$$

Thus using the theory of power series, it converges if and only if $|f(z)| \leq 1$. This inequality holds when $\Re(z/(z - 2)) \leq 0$. Thus we have to find the region which is sent into the closed left half-plane by the linear fractional map $h(z) = z/(z-2)$. The inverse of this map is the map $g(z) = 2z/(z-1)$. Since $g(0) = 0$ and $g(\infty) = 2$, the image of the imaginary axis under g is a circle passing through the points 0 and 2. We can conclude that $|f(z)| \leq 1$ if and only if $|z - 1| \leq 1$ and $z \neq 2$, which is the region of convergence of the series (1).

Problem 22. Let

$$f(z) = \frac{P(z)}{Q(z)}$$

where P and Q are polynomials with real coefficients of degree m and n, respectively with $n \geq m + 2$. If the polynomial Q has simple zeros at the points $t_1, t_2, \ldots, t_r$ on the x-axis, then the *Cauchy principal value* (PV) of the integral is

$$PV \int_{-\infty}^{\infty} \frac{P(x)}{Q(x)} dx = 2\pi i \sum_{j=1}^{k} \text{Res} \left[\frac{P}{Q}, z_j \right] + \pi i \sum_{j=1}^{r} \text{Res} \left[\frac{P}{Q}, t_j \right]$$

where $z_1, z_2, \ldots, z_k$ are the poles of the function f that lie in the upper half-plane. Here Res denotes the residue. Evaluate

$$PV \int_{-\infty}^{\infty} \frac{x}{x^3 - 8} dx.$$

Solution 22. Solving $z^3 = 8$ yields the three roots

$$z_1 = 2, \quad z_2 = -1 - i\sqrt{3}, \quad z_3 = -1 + i\sqrt{3}.$$

Thus z_1 lies on the real axis and z_3 lies in the upper half plane and $k = r = 1$. Next we compute the residues at $z_1 = 2$ and $z_3 = -1 + i\sqrt{3}$. We find

$$\text{Res}[f, 2] = \frac{1}{6}, \quad \text{Res}[f, -1 + i\sqrt{3}] = -\frac{i}{3(i + \sqrt{3})} = -\frac{1}{12} - \frac{i}{4\sqrt{3}}.$$

Thus

$$PV \int_{-\infty}^{\infty} \frac{x}{x^3 - 8} dx = 2\pi i \left(-\frac{1}{12} - \frac{i}{4\sqrt{3}} \right) + \frac{\pi i}{6} = \frac{\pi}{2\sqrt{3}}.$$

Supplementary Problems

Problem 1. The *Hilbert transform* of a function $f \in L_2(\mathbb{R})$ is defined as

$$H(f)(y) = \frac{1}{\pi} PV \int_{-\infty}^{\infty} \frac{f(x)}{x - y} dx$$

where *PV* stands for *Cauchy principal value*. Show that the Hilbert transform of $f(x) = 1/(1 + x^4)$ is given by

$$H(f)(y) = -\frac{1}{\sqrt{2}} \frac{y(1 + y^2)}{1 + y^4}.$$

Problem 2.   The *Fourier transform* is given by

$$\hat{f}(k) = \int_{\mathbb{R}} f(x)e^{ikx}dx, \quad f(x) = \frac{1}{2\pi} \hat{f}(k)e^{-ikx}dk.$$

(i) Show that for the *Gaussian distribution* we have

$$\frac{1}{\sqrt{2\pi\sigma^2}} \exp\left(-\frac{(x - E)^2}{2\sigma^2}\right) \Leftrightarrow \exp\left(ikE - \frac{1}{2}\sigma^2 k^2\right).$$

(ii) Show that for *Poisson distribution* we have

$$\sum_{n=0}^{\infty} \frac{1}{n!}\lambda^n \exp(-\lambda)\delta(x - n) \Leftrightarrow \exp(\lambda(e^{ik} - 1)).$$

(iii) Show that for *Lorentzian distribution* we have

$$\frac{\Gamma}{\pi} \frac{1}{(x - m)^2 + \Gamma^2} \Leftrightarrow \exp(imk - |k\Lambda|).$$

(iv) Show for the product of two Gaussian distributions we have

$$\exp\left(ik(E_1 + E_2) - \frac{1}{2}(\sigma_1^2 + \sigma_2^2)k^2\right).$$

Chapter 7

Integration

Let $f : [a, b] \mapsto \mathbb{R}$ and let g be a function such that $dg/dx = f(x)$ for all $x \in [a, b]$. Then (*Newton's integral*)

$$\int_a^b f(x)dx = g(b) - g(a).$$

The fundamental theorem of calculus states that if a function f is Riemann integrable on the interval $I = [a, b]$ and g is a differentiable function on (a, b) such that $dg/dx = f(x)$ for all $x \in (a, b)$ and $\lim_{x \to a^+} g(x)$, $\lim_{x \to b^-} g(x)$ exist, then

$$\int_a^b f(x)dx = \lim_{x \to b^-} g(x) - \lim_{x \to a^+} g(x).$$

If f and g are differentiable functions on the interval $[a, b]$ and if the function df/dx and dg/dx are Riemann integrable on $[a, b]$ then (*integration by parts*)

$$\int_a^b f(x)\frac{dg}{dx}dx = f(b)g(b) - f(a)g(a) - \int_a^b \frac{df}{dx}g(x)dx.$$

Mean value theorem. Let $f : [a, b] \to [a, b]$ be a continuous function and differentiable on (a, b), then there is a point $x^* \in (a, b)$ such that $f(b) - f(a) = (b - a)df(x = x^*)/dx$.

Every function which is Riemann integrable is Lebesgue integrable and the values of the two integrals are equal. A standard example of a function on an interval $[a, b] \subset \mathbb{R}$ which is Lebesgue integrable but not Riemann integrable is the function $\chi_{\mathbb{Q}}$, where χ is the indicator function.

Problem 1.   The *time average* of a continuous function f is

$$\langle f \rangle := \lim_{T \to \infty} \frac{1}{2T} \int_{-T}^{T} f(t)dt.$$

Find the time-average of the functions

$$f_1(t) = \cos(\omega t)\sin(\omega t), \quad f_2(t) = \cos^2(\omega t), \quad f_3(t) = \sin^2(\omega t).$$

Solution 1.   We find

$$\langle f_1 \rangle = \lim_{T \to \infty} \frac{1}{2T} \int_{-T}^{T} \cos(\omega t)\sin(\omega t)dt = 0$$

$$\langle f_2 \rangle = \lim_{T \to \infty} \frac{1}{2T} \int_{-T}^{T} \cos^2(\omega t)dt = \lim_{T \to \infty} \frac{1}{2T} \int_{-T}^{T} \left(\frac{1}{2} + \frac{1}{2}\cos(2\omega t) \right) dt = \frac{1}{2}$$

$$\langle f_3 \rangle = \lim_{T \to \infty} \frac{1}{2T} \int_{-T}^{T} \sin^2(\omega t)dt = \lim_{T \to \infty} \frac{1}{2T} \int_{-T}^{T} \left(\frac{1}{2} - \frac{1}{2}\cos(2\omega t) \right) dt = \frac{1}{2}.$$

Problem 2.   Consider $i_1(t) = I^2 \sin^2(\omega t)$, $i_2(t) = I^2 \sin^2(\omega t + \phi)$, where ω is a fixed frequency. Calculate

$$\langle i_1(t)i_2(t) \rangle := \lim_{T \to \infty} \frac{1}{T} \int_0^T i_1(t)i_2(t)dt$$

and $\langle i_1(t)i_2(t) \rangle - \langle i_1(t) \rangle \langle i_2(t) \rangle$.

Solution 2.   Since

$$\sin^2(\omega t + \phi) \equiv (\sin(\omega t)\cos(\phi) + \sin(\phi)\cos(\omega t))^2$$

we find

$$\langle i_1(t)i_2(t) \rangle = I^4 \lim_{T \to \infty} \frac{1}{T} \int_0^T \sin^2(\omega t)(\sin(\omega t)\cos(\phi) + \sin(\phi)\cos(\omega t))^2 dt$$

$$= \frac{I^4}{4} + \frac{I^4}{8}\cos(2\phi).$$

We obtain

$$\langle i_1 i_2 \rangle - \langle i_1 \rangle \langle i_2 \rangle = \frac{I^4}{8}\cos(2\phi).$$

Problem 3.   Let $T > 0$ and $\omega = 2\pi/T$. Let $m, n \in \mathbb{N}$ and $\alpha, \beta \in \mathbb{R}$. Calculate

$$I(\alpha, \beta) = \frac{1}{T} \int_0^T c_m \sin(m\omega t + \phi_m - \alpha)c_n \sin(n\omega t + \phi_n - \beta)dt.$$

Solution 3. Owing to the identity

$$\sin(m\omega t + \phi_m - \alpha)\sin(n\omega t + \phi_n - \beta) \equiv$$

$$\frac{1}{2}\cos((m-n)\omega t + \phi_m - \phi_n - \alpha + \beta) - \frac{1}{2}\cos((m+n)\omega t + \phi_m + \phi_n - \alpha - \beta)$$

we obtain

$$I(\alpha, \beta) = \begin{cases} \frac{1}{2}c_m^2 \cos(\alpha - \beta) & \text{for } m = n \\ 0 & \text{for } m \neq n \end{cases}$$

Problem 4. (i) Calculate

$$I = \int_0^\infty e^{-x^2} dx.$$

To evaluate the integral transform the single integral into a *double integral*.
(ii) Let $n \in \mathbb{N}$. Calculate the integral

$$I = \int_0^{\pi/2} \sin^{2n}(\theta)\cos^{2n+1}(\theta)d\theta$$

by considering the *double integral*

$$\int\int_D (r\sin(\theta))^{2n}(r\cos(\theta))^{2n+1}e^{-r^2} r\,dr\,d\theta$$

where D is the first quadrant.

Solution 4. (i) We have

$$I^2 = \int_0^\infty \left(\int_0^\infty e^{-x^2} dx\right) e^{-y^2} dy = \int_0^\infty \int_0^\infty e^{-x^2} e^{-y^2} dx\,dy$$

$$= \int_0^\infty \int_0^\infty e^{-(x^2+y^2)} dx\,dy.$$

Introducing *polar coordinates* $x(r, \phi) = r\cos(\phi)$, $y(r, \phi) = r\sin(\phi)$ with $0 \leq \phi < \pi/2$, $0 \leq r < \infty$ we obtain

$$I^2 = \int_0^{\pi/2} \int_0^\infty e^{-r^2} r\,dr\,d\phi = \int_0^{\pi/2} \left. -\frac{1}{2}e^{-r^2}\right]_0^\infty d\phi = \frac{1}{2}\int_0^{\pi/2} d\phi = \frac{1}{4}\pi.$$

It follows that $I = \sqrt{\pi}/2$.
(ii) We can write

$$I \int_0^\infty r^{4n+2}e^{-r^2} dr = \int\int_D (r\sin(\theta))^{2n}(r\cos(\theta))^{2n+1}e^{-r^2} r\,dr\,d\theta$$

$$= \left(\int_0^\infty y^{2n}e^{-y^2} dy\right)\left(\int_0^\infty x^{2n+1}e^{-x^2} dx\right).$$

Changing all integration variables to s, we have

$$I = \frac{\int_0^\infty s^{2n} e^{-s^2} ds \int_0^\infty s^{2n+1} e^{-s^2} ds}{\int_0^\infty s^{4n+2} e^{-s^2} ds} = \frac{I_{2n} I_{2n+1}}{I_{4n+2}}$$

where

$$I_k = \int_0^\infty s^k e^{-s^2} ds.$$

Now we have $I_1 = 1/2$ and integration by parts provides

$$I_k = \frac{k-1}{2} I_{k-2}, \quad k \geq 3.$$

It follows that

$$I = \frac{I_{2n} \cdot n \cdot (n-1) \cdots 1 \cdot I_1}{\frac{4n+1}{2} \cdot \frac{4n-1}{2} \cdots \frac{2n+1}{2} \cdot I_{2n}} = \frac{n! 2^n}{(4n+1)(4n-1) \cdots (2n+1)}.$$

Problem 5.   Given that

$$\int_0^\infty \frac{\sin(x)}{x} dx = \frac{1}{2}\pi.$$

Find

$$I = \int_0^\infty \frac{\sin^2(x)}{x^2} dx.$$

Let $\epsilon \geq 0$ be a real parameter. Calculate the more general integral

$$I(\epsilon) = \int_0^\infty \frac{\sin^2(\epsilon x)}{x^2} dx, \quad \epsilon \geq 0$$

by using a technique called *parameter differentiation.*

Solution 5.   Differentiating each side of the previous equation with respect to ϵ (we can interchange differentiation and integration), we obtain

$$\frac{dI(\epsilon)}{d\epsilon} = \int_0^\infty \frac{2x \sin(\epsilon x) \cos(\epsilon x)}{x^2} dx = \int_0^\infty \frac{\sin(2\epsilon x)}{x} dx.$$

We set $y = 2\epsilon x$. Therefore we find $dy = 2\epsilon dx$ and

$$\frac{dI(\epsilon)}{d\epsilon} = \int_0^\infty \frac{\sin y}{y} dy = \frac{1}{2}\pi.$$

Integrating each side gives $I(\epsilon) = \frac{1}{2}\pi\epsilon + C$, where C is the constant of integration. Since $I(0) = 0$, we obtain $C = 0$. Thus $I(\epsilon) = \pi\epsilon/2$, $\epsilon \geq 0$. Setting $\epsilon = 1$ yields $I(1) = I = \pi/2$.

Problem 6. (i) Let $\alpha_j \in \mathbb{C}$. Calculate

$$I = \int_0^\infty d\tau_2 \int_0^{\tau_2} d\tau_1 \exp(\alpha_1 \tau_1 + \alpha_2 \tau_2) \tag{1}$$

where $\Re(\alpha_2) < 0$ and $\Re(\alpha_1 + \alpha_2) < 0$. Here $\Re$ denotes the real part.
(ii) Calculate

$$I = \int_0^\infty d\tau_n \int_0^{\tau_n} d\tau_{n-1} \cdots \int_0^{\tau_2} d\tau_1 \exp\left(\sum_{m=1}^n \alpha_m \tau_m \right) \tag{2}$$

where $\Re(\sum_{m=k}^n \alpha_m) < 0$ for all $k \le n$.

Solution 6. (i) Equation (1) can be written in the form

$$\int_0^\infty d\tau_2 \int_0^{\tau_2} d\tau_1 \exp(\alpha_1\tau_1 + \alpha_2\tau_2) = \int_0^\infty d\tau_2 \int_0^\infty d\tau_1 \Theta(\tau_2 - \tau_1) \exp(\alpha_1\tau_1 + \alpha_2\tau_2)$$

where

$$\Theta(\tau_2 - \tau_1) = \begin{cases} 1 \text{ for } & \tau_2 > \tau_1 \\ 0 \text{ otherwise.} \end{cases}$$

Θ is called the *step function*. By the substitution $x_1 = \tau_1$, $x_2 = \tau_2 - \tau_1$ we obtain

$$I = \left(\int_0^\infty dx_2 e^{\alpha_2 x_2} \right) \left(\int_0^\infty dx_1 e^{(\alpha_1 + \alpha_2)x_1} \right) = \frac{1}{\alpha_2(\alpha_1 + \alpha_2)}.$$

(ii) By the substitution $x_{n-1} = \tau_{n-1}$, $x_n = \tau_n - \tau_{n-1}$ we write (2) in the form

$$I = \left(\int_0^\infty dx_n \exp(\alpha_n x_n) \right) \left(\int_0^\infty dx_{n-1} \int_0^{x_{n-1}} d\tau_{n-2} \cdots \int_0^{\tau_2} d\tau_1 \right.$$

$$\left. \times \exp\left(\sum_{m=1}^{n-2} \alpha_m \tau_m \right) \exp((\alpha_n + \alpha_{n-1})x_{n-1}) \right).$$

Repeating this process we can write the integral as

$$I = \left(\int_0^\infty dx_n \exp(\alpha_n x_n) \right) \left(\int_0^\infty dx_{n-1} \exp(\alpha_n + \alpha_{n-1})x_{n-1} \right) \cdots$$

$$\times \left(\int_0^\infty dx_1 \exp\left(\left(\sum_{m=1}^n \alpha_m \right) x_1 \right) \right).$$

We obtain

$$I = (-1)^n (\alpha_n)^{-1} (\alpha_n + \alpha_{n-1})^{-1} \cdots \left(\sum_{m=1}^n \alpha_m \right)^{-1}.$$

since for $s < 0$ we have

$$\int_0^\infty e^{sx}dx = \frac{1}{s}e^{sx}\Big|_0^\infty = -\frac{1}{s}.$$

Problem 7. Let A be an $n \times n$ hermitian matrix. Prove the identity

$$e^{A^2} \equiv \int_{-\infty}^{+\infty} \exp(-\pi x^2 I_n - 2Ax\sqrt{\pi})dx \tag{1}$$

where I_n is the $n \times n$ unit matrix.

Solution 7. Since A is an $n \times n$ hermitian matrix there is an $n \times n$ unitary matrix U such that the matrix U^*AU is diagonal, where $U^* = U^{-1}$. We write

$$\tilde{A} = U^*AU = \mathrm{diag}(\lambda_1, \lambda_2, \ldots, \lambda_n). \tag{2}$$

Obviously, $\lambda_1, \ldots, \lambda_n$ are the eigenvalues of the matrix A. Since A is hermitian the eigenvalues are real. From (1) we obtain

$$U^* e^{A^2} U \equiv U^* \left(\int_{-\infty}^{+\infty} dx \exp(-\pi x^2 I - 2Ax\sqrt{\pi}) \right) U. \tag{3}$$

Since $U^* e^{A^2} U \equiv e^{U^* A^2 U} \equiv e^{U^* AUU^* AU} \equiv e^{\tilde{A}^2}$ and

$$U^* \left(\int_{-\infty}^{+\infty} \exp(-\pi x^2 I_n - 2Ax\sqrt{\pi}) \right) U = \int_{-\infty}^{+\infty} \exp(-\pi x^2 I_n - 2\tilde{A}x\sqrt{\pi})dx$$

we obtain from identity (3) that

$$e^{\tilde{A}^2} \equiv \int_{-\infty}^{+\infty} \exp(-\pi x^2 I - 2\tilde{A}x\sqrt{\pi})dx.$$

From (2) we obtain

$$\tilde{A}^2 = \mathrm{diag}(\lambda_1^2, \lambda_2^2, \ldots, \lambda_n^2), \quad e^{\tilde{A}^2} = \mathrm{diag}(e^{\lambda_1^2}, e^{\lambda_2^2}, \ldots, e^{\lambda_n^2}).$$

Thus

$$\int_{-\infty}^{+\infty} \exp(-\pi x^2 I - 2\tilde{A}x\sqrt{\pi})dx$$

$$= \int_{-\infty}^{+\infty} dx \, \mathrm{diag}\left(e^{-\pi x^2 - 2\lambda_1 x\sqrt{\pi}}, \ldots, e^{-\pi x^2 - 2\lambda_n x\sqrt{\pi}} \right)$$

$$= \mathrm{diag}\left(\int_{-\infty}^{+\infty} e^{-\pi x^2 - 2\lambda_1 x\sqrt{\pi}}dx, \ldots, \int_{-\infty}^{+\infty} e^{-\pi x^2 - 2\lambda_n x\sqrt{\pi}}dx \right).$$

Since

$$\int_{-\infty}^{+\infty} e^{-\pi x^2 - 2\lambda_i x \sqrt{\pi}} dx = \exp(\lambda_i^2)$$

we have

$$\int_{-\infty}^{+\infty} \exp(-\pi x^2 I_n - 2\tilde{A}x\sqrt{\pi})dx \equiv \mathrm{diag}(e^{\lambda_1^2}, e^{\lambda_2^2}, \dots, e^{\lambda_n^2}).$$

This proves the identity (3). Since $U^* = U^{-1}$ we also proved identity (1).

Problem 8.   Given the nonlinear system of differential equations

$$\frac{d^2\theta}{ds^2} - \frac{2b\sin(\phi)}{a + b\cos(\phi)}\frac{d\theta}{ds}\frac{d\phi}{ds} = 0, \quad \frac{d^2\phi}{ds^2} + \frac{(a + b\cos(\phi))\sin(\phi)}{b}\left(\frac{d\theta}{ds}\right)^2 = 0 \tag{1}$$

where $a > b > 0$. Show that

$$\frac{d^2\phi}{ds^2} + \frac{C^2 \sin(\phi)}{b(a + b\cos(\phi))^3} = 0 \tag{2}$$

where C is a constant of integration.

Solution 8.   From (1) we find

$$\frac{\frac{d^2\theta}{ds^2}}{\frac{d\theta}{ds}} = \frac{2b\sin(\phi)}{a + b\cos(\phi)}\frac{d\phi}{ds} \Rightarrow \frac{d}{ds}\left(\ell n\left(\frac{d\theta}{ds}\right)\right) = \frac{2b\sin(\phi)}{a + b\cos(\phi)}\frac{d\phi}{ds}.$$

Therefore

$$\int \frac{d}{ds}\left(\ell n\left(\frac{d\theta}{ds}\right)\right)ds = \int \frac{2b\sin(\phi)}{a + b\cos(\phi)}\frac{d\phi}{ds}ds = \int \frac{2b\sin(\phi)}{a + b\cos(\phi)}d\phi$$

or

$$\ell n\left(\frac{d\theta}{ds}\right) = -2\ell n|a + b\cos(\phi)| + K$$

where K is the constant of integration. From this equation we obtain

$$\frac{d\theta}{ds} = \frac{C}{(a + b\cos(\phi))^2}$$

where $C = e^K$. Inserting this equation into (1b) yields (2).

Problem 9.   Consider a one-dimensional lattice (chain) with lattice constant a. Let k be the sum over the first Brioullin zone we have

$$\frac{1}{N}\sum_{k\in 1.BZ} F(\epsilon(k)) \to \frac{a}{2\pi}\int_{-\pi/a}^{\pi/a} F(\epsilon(k))dk = G$$

where $\epsilon(k) = \epsilon_0 - 2\epsilon_1 \cos(ka)$. Using the identity

$$\int_{-\infty}^{\infty} \delta(E - \epsilon(k))F(E)dE \equiv F(\epsilon(k))$$

we can write

$$G = \frac{a}{2\pi} \int_{-\infty}^{\infty} F(E) \left(\int_{-\pi/a}^{\pi/a} \delta(E - \epsilon(k))dk \right) dE.$$

Calculate

$$g(E) = \int_{-\pi/a}^{\pi/a} \delta(E - \epsilon(k))dk$$

where $g(E)$ is called the density of states.

Solution 9.   Since $\epsilon(k) = \epsilon_0 - 2\epsilon_1 \cos(ka)$ where $\epsilon_1 > 0$ we have

$$g(E) = \int_{-\pi/a}^{\pi/a} \delta(E - \epsilon_0 + 2\epsilon_1 \cos(ka))dk.$$

Setting $\bar{E} := (E - \epsilon_0)/(2\epsilon_1)$ we obtain

$$g(E) = \int_{-\pi/a}^{\pi/a} \delta(2\epsilon_1(\bar{E} + \cos(ka))dk = \frac{1}{2\epsilon_1} \int_{-\pi/a}^{\pi/a} \delta(\bar{E} + \cos(ka))dk.$$

The substitution $u = \cos(ka)$, $du = -a \sin(ka)dk$ provides

$$g(E) = \frac{1}{\epsilon_1} \int_{-1}^{1} \frac{\delta(\bar{E} + u)}{\sqrt{1 - u^2}} du.$$

Therefore

$$g(E) = \begin{cases} \dfrac{2}{\sqrt{4\epsilon_1^2 - (\bar{E} - \epsilon_0)^2}} & \text{for} \quad \bar{E} \in [-1, 1] \\ 0 & \text{otherwise} \end{cases}$$

Problem 10.   Calculate

$$I(\lambda) = \int_{|\mathbf{q}| < k_F} \frac{dq_1 dq_2 dq_3}{|\mathbf{k} - \mathbf{q}|^2 + \lambda^2} \tag{1}$$

where $|\mathbf{k} - \mathbf{q}|^2 := (k_1 - q_1)^2 + (k_2 - q_2)^2 + (k_3 - q_3)^2$ and $\lambda^2 > 0$. The integral (1) plays an important role in solid state physics. Here $\hbar k_f$ is the *Fermi momentum*. The absolute value of momentum of particles at zero temperature is called the Fermi momentum and the region of momentum space with momentum $\hbar k_f$ is called the Fermi surface.

Solution 10. Introducing spherical coordinates (q, ϕ, θ) we obtain

$$I(\lambda) = \int_0^{k_F} q^2 dq \int_0^\pi \sin(\theta) d\theta \int_0^{2\pi} d\phi \frac{1}{k^2 - 2qk\cos(\theta) + q^2 + \lambda^2}$$

where θ is given by

$$\mathbf{k} \cdot \mathbf{q} = kq\cos(\theta).$$

We set $u = \cos(\theta)$. Then $du = -\sin(\theta)d\theta$ and $\theta = 0 \to u = 1$, $\theta = \pi \to u = -1$. We obtain

$$I(\lambda) = -2\pi \int_0^{k_F} q^2 dq \int_1^{-1} du \frac{1}{k^2 + q^2 - 2qku + \lambda^2}.$$

The integration with respect to u can easily be performed. We find

$$I(\lambda) = 2\pi \int_0^{k_F} q^2 dq \left[\frac{1}{-2qk} \ln(k^2 + q^2 + \lambda^2 - 2qku) \right]_{u=1}^{u=-1}.$$

Thus

$$I(\lambda) = -\frac{\pi}{k} \underbrace{\int_0^{k_F} q \ln(\lambda^2 + (k+q)^2) dq}_{F_1} + \frac{\pi}{k} \underbrace{\int_0^{k_F} q \ln(\lambda^2 + (k-q)^2) dq}_{F_2}.$$

Now we have to calculate the integrals F_1 and F_2. We find

$$F_1 = \int_{q=0}^{k_F} q \ln(\lambda^2 + (k+q)^2) dq = \int_{x=k}^{k_F+k} (x-k) \ln(\lambda^2 + x^2) dx$$

or

$$F_1 = -\underbrace{\int_{x=k}^{k_F+k} k \ln(\lambda^2 + x^2) dx}_{F_1'} + \underbrace{\int_{x=k}^{k_F+k} x \ln(\lambda^2 + x^2) dx}_{F_1''}.$$

Now we have to calculate F_1' and F_1''. Since

$$\int \ln(\lambda^2 + x^2) dx = x \ln(x^2 + \lambda^2) + 2\lambda \arctan(x/\lambda) - 2x$$

we find for F_1'

$$F_1' = -k[(k_F + k)\ln((k_F + k)^2 + \lambda^2) + 2\lambda \arctan((k_F + k)/\lambda)$$
$$-2(k_F + k) - k\ln(k^2 + \lambda^2) - 2\lambda \arctan(k/\lambda)) + 2k].$$

Since $\int x f(x^2) dx = \frac{1}{2} \int f(u) du$ and with $x^2 = u$ we find

$$F_1'' = \int_{x=k}^{k_F+k} x \ln(\lambda^2 + x^2) dx = \frac{1}{2} \int_{u=k^2}^{(k_F+k)^2} \ln(\lambda^2 + u) du.$$

Thus

$$F_1'' = \frac{1}{2}(((k_F + k)^2 + \lambda^2)\ln((k_F + k)^2 + \lambda^2) - (k_F + k)^2$$
$$-(k^2 + \lambda^2)\ln(k^2 + \lambda^2) + k^2).$$

The integral F_2 is given by

$$F_2 = \int_{q=0}^{k_F} q \ln(\lambda^2 + (q - k)^2)dq.$$

It follows that

$$F_2 = \underbrace{\int_{x=-k}^{k_F-k} k \ln(\lambda^2 + x^2)dx}_{F_2'} + \underbrace{\int_{x=-k}^{k_F-k} x \ln(\lambda^2 + x^2)dx}_{F_2''}.$$

Now we have to calculate F_2'. We obtain

$$F_2' = \int_{x=-k}^{k_F-k} k \ln(\lambda^2 + x^2)dx.$$

Thus

$$F_2' = k(k_F - k)\ln((k_F - k)^2 + \lambda^2) + 2k\lambda \arctan\left(\frac{k_F - k}{\lambda}\right)$$
$$-2k(k_F - k) + k^2 \ln(k^2 + \lambda^2) - 2k\lambda \arctan\left(\frac{-k}{\lambda}\right) - 2k^2.$$

For F_2'' we obtain

$$F_2'' = \int_{x=-k}^{k_F-k} x \ln(\lambda^2 + x^2)dx = \frac{1}{2}\int_{u=k^2}^{(k_F-k)^2} \ln(\lambda^2 + u)du$$

or

$$F_2'' = \frac{1}{2}(((k_F-k)^2 + \lambda^2)\ln((k_F-k)^2 + \lambda^2) - (k_F-k)^2 - (k^2+\lambda^2)\ln(k^2+\lambda^2) + k^2)$$

where we have used $u = x^2$ and therefore $du = 2xdx$.

Problem 11.   Calculate the integral (which plays a rôle in electrostatic fields)

$$U(r) = \frac{1}{4\pi}\int_{r'}\int_{\theta=0}^{\pi}\int_{\phi=0}^{2\pi} \frac{\rho(r')}{(r^2 + r'^2 - 2rr'\cos(\theta))^{1/2}} r'^2 \sin(\theta)dr'd\theta d\phi.$$

Solution 11.   The integration with respect to ϕ can easily be performed. We find

$$U(r) = \frac{1}{2} \int_{r'} \int_{\theta=0}^{\pi} \frac{\rho(r')}{(r^2 + r'^2 - 2rr'\cos(\theta))^{1/2}} r'^2 \sin(\theta) dr' d\theta.$$

To perform the θ-integration we set $u = \cos(\theta)$. Therefore, $du = -\sin(\theta)d\theta$ and

$$U(r) = -\frac{1}{2} \int_{r'} \int_{u=1}^{-1} \frac{\rho(r')}{(r^2 + r'^2 - 2rr'u)^{1/2}} r'^2 dr' du.$$

Thus

$$U(r) = \frac{1}{2} \int_{r'} \rho(r') \left(\frac{(r^2 + r'^2 + 2rr')^{1/2}}{rr'} - \frac{(r^2 + r'^2 - 2rr')^{1/2}}{rr'} \right) r'^2 dr'.$$

Since

$$(r^2 + r'^2 - 2rr')^{1/2} = ((r - r')^2)^{1/2} = r - r' \quad \text{for} \quad r \geq r'$$

$$(r^2 + r'^2 - 2rr')^{1/2} = ((r' - r)^2)^{1/2} = r' - r \quad \text{for} \quad r' \geq r$$

we obtain

$$U(r) = \frac{1}{2} \int_{r'=0}^{r} \rho(r') \frac{(r + r') - (r - r')}{r} r' dr'$$
$$+ \frac{1}{2} \int_{r'=r}^{\infty} \rho(r') \frac{(r + r') - (r' - r)}{r} r' dr'.$$

Consequently,

$$U(r) = \frac{1}{r} \int_{r'=0}^{r} \rho(r') r'^2 dr' + \int_{r'=r}^{\infty} \rho(r') r' dr'.$$

Problem 12.   The sum

$$\lim_{n \to \infty} \frac{1}{n} \sum_{k=1}^{n} \exp\left(2 \cos\left(\frac{k\pi}{n+1} \right) \right)$$

can be cast into the integral

$$\lim_{n \to \infty} \frac{1}{n} \int_0^n \exp(2 \cos(\pi x)) dx. \tag{1}$$

Calculate this integral.

Solution 12.   We have

$$\lim_{n\to\infty} \frac{1}{n} \int_0^n \exp(2\cos(\pi x))dx = \lim_{n\to\infty} \frac{1}{n} n \int_0^1 \exp(2\cos(\pi x))dx$$

$$= \int_0^1 \exp(2\cos(\pi x))dx.$$

Using $y = \pi x$ we arrive at

$$\lim_{n\to\infty} \frac{1}{n} \int_0^n \exp(2\cos(\pi x))dx = \frac{1}{\pi} \int_0^\pi \exp(2\cos(y))dy = \sqrt{\pi}\Gamma(1/2)I_0(2)$$

where Γ is the gamma function and I_0 is the cylinder function.

Problem 13.   Two friends plan to meet at the library during a given 1-hour period. Their arrival times are independent and randomly distributed across the 1-hour period. Each agrees to wait for 15 minutes, or until the end of the hour. If the friend has not appeared during that time, the other friend will leave. Find the probability that the friends will meet.

Solution 13.   If X_1 denotes one person's arrival time in $[0,1]$, the 1-hour period, and X_2 denotes the second person's arrival time, then (X_1, X_2) can be modeled as having a two-dimensional uniform distribution over the unit square. That is,

$$f(x_1, x_2) = \begin{cases} 1 & 0 \le x_1 \le 1, \ 0 \le x_2 \le 1 \\ 0 & \text{otherwise.} \end{cases} \tag{1}$$

The event that the two friends will meet depends upon the time, U, between their arrivals, where $U = |X_1 - X_2|$. Next we find the distribution function for the random variable U. For $U \le u$ we have $|X_1 - X_2| \le u$. Therefore $-u \le X_1 - X_2 \le u$

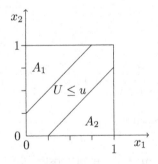

The figure shows that the region over which (X_1, X_2) has positive probability is the region defined by $U \le u$. The probability that $U \le u$ can be found by integrating f over the six-sided region shown in the figure. This can be simplified by integrating over the triangles (A_1 and A_2) and subtracting from one. We have

$$F_U(u) = P(U \le u) = \int\int_{|x_1 - x_2| \le u} f(x_1, x_2) dx_1 dx_2.$$

Thus

$$F_U(u) = 1 - \int\int_{A_1} f(x_1, x_2) dx_1 dx_2 - \int\int_{A_2} f(x_1, x_2) dx_1 dx_2.$$

Therefore

$$F_U(u) = 1 - \int_u^1 \int_0^{x_2 - u} dx_1 dx_2 - \int_0^{1-u} \int_{x_2+u}^1 dx_1 dx_2.$$

Finally, we arrive at $F_U(u) = 1 - (1 - u)^2$, $0 \le u \le 1$. Thus

$$P\left(U \le \frac{15}{60}\right) = 1 - \left(1 - \frac{15}{60}\right)^2 = \frac{7}{16}.$$

Problem 14. Consider a one-dimensional chain of length N with open end boundary conditions. The counting is from left to right starting at 0. The canonical *partition function* $Z(\beta)$ ($\beta > 0$) is given by the multiple integral

$$Z_N(\beta) = \int_{-1}^1 ds_0 \int_{-1}^1 ds_1 \cdots \int_{-1}^1 ds_{N-1} e^{\beta|s_0 - s_1|} e^{\beta|s_1 - s_2|} \ldots e^{\beta|s_{N-2} - s_{N-1}|}.$$

Show that there is a coordinate transformation which decouples the sites. Find $Z_2(\beta)$ and $Z_3(\beta)$.

Solution 14. We have the invertible transformation

$$t_0 = s_0, \quad t_1 = -s_0 + s_1, \quad t_2 = -s_1 + s_2, \quad \ldots, \quad t_j = s_j - s_{j-1}.$$

The inverse transformation is

$$s_0 = t_0, \quad s_1 = t_0 + t_1, \quad s_2 = t_0 + t_1 + t_2, \quad \ldots, \quad s_j = \sum_{k=0}^j t_k.$$

The Jacobian determinant is $+1$. The boundary conditions for the integrations are now

$$-1 \le t_0 \le 1, \quad -1 - t_0 \le t_1 \le 1 - t_0, \quad -1 - t_0 - t_1 \le t_2 \le 1 - t_0 - t_1$$

etc. For $Z_2(\beta)$ we find

$$Z_2(\beta) = -\frac{4}{\beta} + \frac{2}{\beta^2}(e^{2\beta} - 1).$$

For $Z_3(\beta)$ we have

$$Z_3(\beta) = \int_{-1}^{1} dt_0 \int_{-1-t_0}^{1-t_0} dt_1 e^{\beta|t_1|} \int_{-1-t_0-t_1}^{1-t_0-t_1} dt_2 e^{\beta|t_2|}$$
$$= \frac{1}{\beta^2}(8 + 4e^{2\beta}) + \frac{1}{\beta^3}(7 - 8e^{2\beta} + e^{4\beta}).$$

Problem 15.   Let $a, b \in \mathbb{R}$, $a, b \neq 0$ and $b > a$. Show that

$$\frac{1}{ab} = \int_0^1 \frac{dz}{(az + b(1 - z))^2}. \tag{1}$$

If a and b have opposite sign, z is to be considered as a complex variable and the path of the integration must deviate from the real axis so as to avoid the singularity at $z^* = b/(b - a)$.

Solution 15.   Identity (1) is obtained by noting that

$$\frac{1}{ab} = \frac{1}{b - a}\left(\frac{1}{a} - \frac{1}{b}\right) = \frac{1}{b - a}\int_a^b \frac{dx}{x^2}$$

and introducing in the last integral the new variable z via $x = az + b(1 - z)$. Thus $dx = (a - b)dz$ and $z = 0 \Rightarrow x = b$, $z = 1 \Rightarrow x = a$. The identity (1) holds for all values of a and b with $a, b \neq 0$. If, however, a and b are of the opposite sign, z is to be considered as a complex variable and the path of integration must deviate from the real axis so as to avoid the singularity (pole) at $z^* = b/(b - a)$. Any path joining $z = 0$ and $z = 1$ but not passing through the singularity may actually be chosen since the residue of the integrand at the singularity vanishes.

The integral given by identity (1) is a special case of the following general identity

$$\frac{1}{a_1 a_2 \cdots a_n} = (n - 1)! \int_0^1 \cdots \int_0^1 \frac{dz_1 dz_2 \cdots dz_n}{(a_1 z_1 + a_2 z_2 + \cdots + a_n z_n)^n}$$

where

$$\sum_{i=1}^{n} z_i = 1.$$

To evaluate this integral we can use

$$\frac{1}{(n-1)!}\frac{1}{a_1 a_2 \cdots a_n}$$

$$= \int_0^1 dz_1 \int_0^{z_1} dz_2 \cdots \int_0^{z_{n-2}} dz_{n-1}$$

$$\times \frac{1}{(a_n z_{n-1} + a_{n-1}(z_{n-2} - z_{n-1}) + \cdots + a_1(1 - z_1))^n}$$

$$= \int_0^1 \epsilon_1^{n-2} d\epsilon_1 \int_0^1 \epsilon_2^{n-3} d\epsilon_2 \cdots \int_0^1 d\epsilon_{n-1}$$

$$\times \frac{1}{(a_1 \epsilon_1 \epsilon_2 \cdots \epsilon_{n-1} + a_2 \epsilon_1 \cdots \epsilon_{n-2}(1 - \epsilon_{n-1}) + \cdots + a_n(1 - \epsilon_1))^n}.$$

Problem 16. Consider a right circular cone (elliptic cone) of radius r and height h with axis as z-axis and $z \geq 0$. The surface is described by

$$\frac{x^2}{r^2} + \frac{y^2}{r^2} = \frac{z^2}{h^2}. \tag{1}$$

Find the volume of the cone.

Solution 16. To find the volume we have to do the *triple integral*

$$\int_{z=0}^{z=h} \left(\int_{y=g_1(z)}^{y=g_2(z)} \left(\int_{x=f_1(y,z)}^{x=f_2(y,z)} dx \right) dy \right) dz.$$

We consider the quadrant $x \geq 0$ and $y \geq 0$ and we multiply the final result by 4 to find the volume. For the x-integration we have

$$f_1(y,z) = 0, \qquad f_2(y,z) = \frac{\sqrt{r^2 z^2 - h^2 y^2}}{h}$$

where we used (1). Thus

$$\int_{x=0}^{x=\sqrt{r^2 z^2 - h^2 y^2}/h} dx = \frac{1}{h}\sqrt{r^2 z^2 - h^2 y^2}.$$

For the y-integration we have $g_1(z) = 0$ and $g_2(z) = zr/h$ since z and y are related by the linear equation $z = my$. For $z = h$ we have $y = r$ and therefore $m = h/r$, $z = hy/r$. Thus

$$\frac{1}{h}\int_{y=0}^{y=zr/h} \sqrt{r^2 z^2 - h^2 y^2}\, dy.$$

To perform the y-integration we apply the substitution $\tilde{y} = hy$. Therefore $d\tilde{y} = hdy$. For $y = 0$ we have $\tilde{y} = 0$ and for $y = zr/h$ we obtain $\tilde{y} = zr$. It follows that

$$\frac{1}{h^2} \int_{\tilde{y}=0}^{\tilde{y}=zr} \sqrt{r^2 z^2 - \tilde{y}^2} \, d\tilde{y} = \frac{1}{2h^2} \left| \tilde{y} \sqrt{r^2 z^2 - \tilde{y}^2} + r^2 z^2 \arcsin\left(\frac{\tilde{y}}{rz}\right) \right|_{\tilde{y}=0}^{\tilde{y}=zr}$$

$$= \frac{1}{4h^2} r^2 z^2 \pi$$

where we used $\arcsin(1) = \pi/2$ and $\arcsin(0) = 0$. The last integration over z is

$$\frac{\pi r^2}{4h^2} \int_{z=0}^{z=h} z^2 dz = \frac{\pi}{4} \cdot \frac{r^2 h}{3}.$$

Since we only considered the quadrant $x \geq 0$ and $y \geq 0$ we find that the volume of the cone is $V = \pi r^2 h/3$.

Problem 17. Let

$$x(\tau) = \begin{cases} 1 & \text{for} \quad 0 \leq \tau \leq 1 \\ 0 & \text{otherwise} \end{cases}, \quad h(\tau) = \begin{cases} 1 & \text{for} \quad 0 \leq \tau \leq 1 \\ 0 & \text{otherwise} \end{cases}.$$

Find the *convolution integral*

$$y(t) = \int_{-\infty}^{\infty} x(\tau) h(t - \tau) d\tau.$$

Solution 17. The value of the function y at time t is given by the amount of overlap (i.e. the integral of the overlapping region) between $h(t - \tau)$ and $x(\tau)$. Thus we have four limits of integration, namely $t < 0$, $0 \leq t < 1$, $1 \leq t \leq 2$ and $t > 2$. For the region $t < 0$ we have $y(t) = 0$. For the region $0 \leq t < 1$ we have

$$y(t) = \int_0^t d\tau = t.$$

The third region $1 \leq t \leq 2$ provides

$$y(t) = \int_{t-1}^1 d\tau = 2 - t.$$

The fourth region $t > 2$ provides $y(t) = 0$. Thus

$$y(t) = \begin{cases} 0 & \text{if} \quad t < 0 \\ t & \text{if } 0 \leq t < 1 \\ 2 - t & \text{if } 1 \leq t \leq 2 \\ 0 & \text{if} \quad t > 2 \end{cases}.$$

Problem 18. Calculate

$$I(x) = PV \int_{-\infty}^{\infty} \frac{e^{-t^2}}{t - x} dt, \qquad x \in \mathbb{R} \tag{1}$$

where the integral is taken in the sense of the *Cauchy principal value*. This is the *Hilbert transform* of $\exp(-t^2)$.

Solution 18. By definition of the Cauchy principal value integral, we have

$$I(x) = \lim_{\epsilon \to 0} \left(\int_{-\infty}^{x-\epsilon} + \int_{x+\epsilon}^{\infty} \right) \frac{e^{-t^2}}{t - x} dt = \lim_{\epsilon \to 0} \int_{\epsilon}^{\infty} \frac{e^{-(x+t)^2} - e^{-(x-t)^2}}{t} dt$$

$$= -e^{-x^2} \int_0^{\infty} \frac{e^{2xt} - e^{-2xt}}{t} e^{-t^2} dt$$

or

$$I(x) = -4x e^{-x^2} \int_0^{\infty} \frac{\sinh(2xt)}{2xt} e^{-t^2} dt. \tag{2}$$

Using Taylor expansion of the hyperbolic sine, term-by-term integration and

$$\int_0^{\infty} t^{2k} e^{-t^2} dt = \frac{1}{2} \Gamma\left(k + \frac{1}{2}\right)$$

yields

$$\int_0^{\infty} \frac{\sinh(2xt)}{2xt} e^{-t^2} dt = \frac{1}{2} \sum_{k=0}^{\infty} \frac{\Gamma(k + \frac{1}{2})}{\Gamma(2k + 2)} (2x)^{2k}. \tag{3}$$

The duplication formula for the *Gamma function* is

$$\Gamma(2k + 2) \equiv \frac{2^{2k+3/2}}{\sqrt{2\pi}} \Gamma(k + 1) \Gamma\left(k + \frac{1}{2}\right).$$

Thus

$$\int_0^{\infty} \frac{\sinh(2xt)}{2xt} e^{-t^2} dt = \frac{\sqrt{\pi}}{4} \sum_{k=0}^{\infty} \frac{1}{k + \frac{1}{2}} \frac{x^{2k}}{k!}. \tag{4}$$

The series on the right-hand side is the *error function* of a purely imaginary argument, namely

$$\sum_{k=0}^{\infty} \frac{1}{k + \frac{1}{2}} \frac{x^{2k}}{k!} = \sqrt{\pi} \frac{\mathrm{erf}(ix)}{ix} = \frac{2}{x} \int_0^{x} e^{t^2} dt. \tag{5}$$

Substitution of (4) and (5) into (2) expresses the Hilbert transform in terms of *Dawson's integral*

$$I(x) = -2\sqrt{\pi} F(x), \qquad F(x) := e^{-x^2} \int_0^{x} e^{t^2} dt.$$

From the asymptotic expansion of the error function one finds

$$I(x) \sim -\frac{\sqrt{\pi}}{x}\left(1 + \sum_{k=1}^{\infty} \frac{1 \cdot 3 \cdots (2k-1)}{(2x^2)^k}\right) \qquad \text{as} \quad x \to \infty.$$

Problem 19. Let $n = 1, 2, \ldots$ and

$$I_n := \int_0^{\pi/4} \tan^n(x) dx.$$

Calculate $f(n) = I_n + I_{n+2}$. Note that $I_1 = \frac{1}{2} \ln(2)$.

Solution 19. Using $1 + \tan^2(x) \equiv \sec^2(x)$, $\sec(x) \equiv 1/\cos(x)$ and the substitution $u = \tan(x)$ we have

$$I_n + I_{n+2} = \int_0^{\pi/4} (\tan^n(x) + \tan^{n+2}(x)) dx = \int_0^{\pi/4} \tan^n(x) \sec^2(x) dx$$

$$= \int_0^1 u^n du = \frac{1}{n+1}.$$

Problem 20. Let f be a Riemann integrable periodic function over the interval $[0, 2\pi]$. Then f can be written as

$$f(\theta) = \sum_{n=-\infty}^{\infty} c_n e^{in\theta}, \qquad c_m = \frac{1}{2\pi} \int_0^{2\pi} e^{-im\theta} f(\theta) d\theta.$$

Let $m \geq 0$ and

$$s_m := \sum_{n=-m}^{m} c_n e^{in\theta}.$$

Calculate the *Cesáro sum* $\sigma_m(\theta)$ defined by

$$\sigma_m(\theta) := \frac{s_0 + s_1 + \cdots + s_m}{m+1}.$$

Express the result using the *Fejér kernels*

$$K_m(y) := \sum_{n=-m}^{m} \frac{m+1-|n|}{m+1} e^{iny}.$$

(ii) Show that

$$K_m(y) = \frac{1}{m+1}\left(\frac{\sin((m+1)y/2)}{\sin(y/2)}\right)^2.$$

Solution 20. (i) We have

$$\sigma_m(\theta) = \frac{1}{m+1} \sum_{j=0}^{m} \sum_{n=-j}^{j} c_n e^{in\theta} = \sum_{n=-m}^{m} \frac{m+1-|n|}{m+1} c_n e^{in\theta}$$

$$= \sum_{n=-m}^{m} \frac{m+1-|n|}{m+1} \left(\frac{1}{2\pi} \int_0^{2\pi} e^{-inx} f(x) dx \right) e^{in\theta}$$

$$= \frac{1}{2\pi} \int_0^{2\pi} f(x) \left(\sum_{n=-m}^{m} \frac{m+1-|n|}{m+1} e^{in(\theta-x)} \right) dx$$

$$= \frac{1}{2\pi} \int_0^{2\pi} f(x) K_m(\theta - x) dx.$$

The substitution $y = \theta - x$ provides

$$\frac{1}{2\pi} \int_0^{2\pi} f(x) K_m(\theta - x) dx = \frac{1}{2\pi} \int_0^{2\pi} f(\theta - y) K_m(y) dy.$$

(ii) We have

$$K_m(y) = \frac{1}{m+1} (e^{-imy} + 2e^{-i(m-1)y} + \cdots + (m+1)e^0 + \cdots + e^{imy})$$

$$= \frac{1}{m+1} (e^{-imy/2} + e^{-i(m/2-1)y} + \cdots + e^{imy/2})^2$$

$$= \frac{1}{m+1} \left(\frac{e^{i(m+1)y/2} - e^{-i(m+1)y/2}}{e^{iy/2} - e^{-iy/2}} \right)^2$$

$$= \frac{1}{m+1} \left(\frac{\sin((m+1)y/2)}{\sin(y/2)} \right)^2.$$

Thus $K_m(0) = m + 1$. The function K_m satisfies $K_m(y) \geq 0$ for all $y \in \mathbb{R}$.

Problem 21. Let f be a continuous function.
(i) Show that the double integral

$$\int_0^x d\xi \int_0^\xi f(t) dt$$

can be expressed by a single integral.
(ii) Show that $(n \geq 2)$

$$\int_0^x d\xi_1 \int_0^{\xi_1} d\xi_2 \cdots \int_0^{\xi_{n-1}} f(\xi_n) d\xi_n$$

can be expressed by a single integral.

Solution 21.   (i) We have

$$\int_0^x d\xi \int_0^\xi f(t)dt = \int_0^x dt \int_t^x f(t)d\xi = \int_0^x f(t)dt \int_t^x d\xi = \int_0^x (x-t)f(t)dt.$$

(ii) Using the result from (i) and induction we have

$$\int_0^x d\xi_1 \int_0^{\xi_1} d\xi_2 \cdots \int_0^{\xi_{n-1}} f(\xi_n)d\xi_n = \int_0^x \frac{(x-t)^{n-1}}{(n-1)!} f(t)dt.$$

For $n = 2$ we have the result of (i).

Problem 22.   Evaluate the *Lebesgue integral*

$$\int_0^1 x^2 dx. \tag{1}$$

Solution 22.   Let $E = [0, 1)$. Then E is an elementary set having Lebesgue measure 1. The function $f : \mathbb{R} \to \mathbb{R}$ defined by

$$f(x) = \begin{cases} x^2 & \text{if } x \in E \\ 0 & \text{if } x \notin E \end{cases}$$

is a measurable function since $\{x : f(x) \geq a\}$ is measurable for all $a \in \mathbb{R}$. We introduce a monotonic increasing sequence of simple functions tending to f as follows. Let

$$Q_{p,s} := \left\{ x : \frac{p-1}{2^s} \leq x < \frac{p}{2^s} \right\}$$

where $p = 1, 2, 4, \ldots, 2^s$, $s = 1, 2, \ldots$. We define

$$f_s(x) := \begin{cases} \left(\frac{p-1}{2^s}\right)^2 & \text{for } x \in Q_{p,s} \\ 0 & \text{for } x \in \mathbb{R} \setminus E \end{cases}.$$

Then for all $x \in \mathbb{R}$ we have $0 \leq f_s(x) \leq f(x)$. Moreover, for all $x \in \mathbb{R}$, we find $f_{s+1}(x) \geq f_s(x)$. This means f_s is a monotonic increasing function with increasing s. Furthermore

$$0 \leq f(x) - f_s(x) \leq \left(\frac{2^s}{2^s}\right)^2 - \left(\frac{2^s-1}{2^s}\right)^2 = \frac{2^{2s}-1}{2^s 2^s} < \frac{2}{2^s}$$

so that $f_s(x) \to f(x)$ as $s \to \infty$. Therefore

$$\int_E f_s(x)dx = \frac{1}{2^s}\left(0 + \left(\frac{1}{2^s}\right)^2 + \left(\frac{2}{2^s}\right)^2 + \cdots + \left(\frac{2^s-1}{2^s}\right)^2\right).$$

Thus

$$\int_E f_s(x)dx = \frac{1}{n^3}(1 + 2^2 + 3^2 + \cdots + (n-1)^2) = \frac{1}{n^3}\left(\frac{(n-1)n(2n-1)}{6}\right)$$

where we used $n = 2^s$. Therefore

$$\int_E f_s(x)dx \to \frac{1}{3} \quad \text{as} \quad s \to \infty.$$

It follows that $\int_0^1 x^2 dx = \frac{1}{3}$.

Problem 23. Given a smooth Hamilton function

$$H(\mathbf{p}, \mathbf{q}) = \sum_{j=1}^n \frac{p_j^2}{2} + U(\mathbf{q})$$

with n degrees of freedom ($\mathbf{p} = (p_1, \ldots, p_n)$, $\mathbf{q} = (q_1, \ldots, q_n)$). Let $V(E)$ be the classical phase space volume at energy E of a smooth Hamilton function is given by

$$V(E) = \int_{\mathbb{R}^{2n}} \Theta(E - H(\mathbf{p}, \mathbf{q}))d^n\mathbf{p}d^n\mathbf{q}$$

where Θ is the step function. Assume that $U(\epsilon\mathbf{q}) = \epsilon^m U(\mathbf{q})$.
(i) Consider the transformation

$$\mathbf{p} = E^{1/2}\mathbf{p}', \qquad \mathbf{q} = E^{1/n}\mathbf{q}'$$

with the inverse transformation

$$\mathbf{p}' = E^{-1/2}\mathbf{p}, \qquad \mathbf{q}' = E^{-1/n}\mathbf{q}.$$

Find $d^n\mathbf{p}'d^n\mathbf{q}'$ and $H(\mathbf{p}', \mathbf{q}')$.
(ii) Calculate $V(E)$ with the assumption that $E > 0$. Find the asymptotic behaviour.

Solution 23. (i) We have

$$d^n\mathbf{p}d^n\mathbf{q} = E^{n/2}d^n\mathbf{p}'E^{n/m}d^n\mathbf{q}' = E^{n(m+2)/(2m)}d^n\mathbf{p}'d^n\mathbf{q}'$$

and $H(\mathbf{p}, \mathbf{q}) = EH(\mathbf{p}', \mathbf{q}')$, where

$$H(\mathbf{p}', \mathbf{q}') = \sum_{j=1}^n \frac{p_j'^2}{2} + U(\mathbf{q}').$$

(ii) We find

$$V(E) = \int_{\mathbb{R}^{2n}} \Theta(E - H(\mathbf{p}, \mathbf{q})) d^n \mathbf{p} d^n \mathbf{q}$$

$$= \int_{\mathbb{R}^{2n}} \Theta(E - EH(\mathbf{p}', \mathbf{q}')) E^{n(m+2)/(2m)} d^n \mathbf{p}' d^n \mathbf{q}'$$

$$= E^{n(m+2)/(2m)} \int_{\mathbb{R}^{2n}} \Theta(E(1 - H(\mathbf{p}', \mathbf{q}'))) d^n \mathbf{p} d^n \mathbf{q}.$$

If $E > 0$ we obtain

$$V(E) = E^{n(m+2)/(2m)} \int_{\mathbb{R}^{2n}} \Theta(1 - H(\mathbf{p}, \mathbf{q})) d^n \mathbf{p}' d^n \mathbf{q}'$$

where the integral on the right hand side is a constant. Thus the asymptotic behaviour is

$$V(E) \propto E^{n(m+2)/(2m)}.$$

For $m = 4$ and $n = 2$ we have $V(E) \propto E^{3/2}$.

Problem 24.   Let $a > 0$. Find the area of the surface in $\mathbb{R}^3$ given by the intersection of a hyperbolic *paraboloid* $x_3 = x_1 x_2/a$ and the cylinder $x_1^2 + x_2^2 = a^2$. We have the parameter representation

$$x_1(r, \phi) = r \cos(\phi), \quad x_2(r, \phi) = r \sin(\phi), \quad x_3(r, \phi) = \frac{1}{a} r^2 \cos(\phi) \sin(\phi).$$

Note the identity $\sin(\phi) \cos(\phi) \equiv \frac{1}{2} \sin(2\phi)$.

Solution 24.   The surface element is $do = \sqrt{g} dr d\phi$ with

$$g = g_{11} g_{22} - g_{12}^2, \quad g_{jk} = \mathbf{t}_j \cdot \mathbf{t}_k$$

where $\cdot$ denotes the scalar product and

$$\mathbf{t}_1 = \begin{pmatrix} \partial x_1/\partial \phi \\ \partial x_2/\partial \phi \\ \partial x_3/\partial \phi \end{pmatrix} = \begin{pmatrix} -r \sin(\phi) \\ r \cos(\phi) \\ (r^2/a^2) \cos(2\phi) \end{pmatrix}$$

$$\mathbf{t}_2 = \begin{pmatrix} \partial x_1/\partial r \\ \partial x_2/\partial r \\ \partial x_3/\partial r \end{pmatrix} = \begin{pmatrix} \cos(\phi) \\ \sin(\phi) \\ (r/a) \sin(2\phi) \end{pmatrix}.$$

Then

$$g_{11} = \mathbf{t}_1 \cdot \mathbf{t}_1 = r^2 + \frac{r^4}{a^2} \cos^2(2\phi), \quad g_{22} = \mathbf{t}_2 \cdot \mathbf{t}_2 = 1 + \frac{r^2}{a^2} \sin^2(2\phi),$$

$$g_{12} = \mathbf{t}_1 \cdot \mathbf{t}_2 = \frac{r^3}{a^2} \cos(2\phi) \sin(2\phi).$$

For the rank with have

$$\text{rank}(\mathbf{t}_1, \mathbf{t}_2) = \begin{pmatrix} -r\sin(\phi) & \cos(\phi) \\ r\cos(\phi) & \sin(\phi) \\ (r^2/a)\cos(2\phi) & (r/a)\sin(2\phi) \end{pmatrix} = 2$$

for $r > 0$. Then

$$g = g_{11}g_{22} - g_{12}^2 = r^2 + \frac{r^4}{a^2}.$$

Then $do = r\sqrt{1 + r^2/a^2} \, dr d\phi$ and hence

$$O = \int_{\phi=0}^{2\pi} \int_{r=0}^{a} r\sqrt{1 + r^2/a^2} \, dr d\phi = \frac{2}{3}\pi a^2 (2\sqrt{2} - 1).$$

Problem 25. Consider the two curves $\mathbf{r}_1(s_1)$ and $\mathbf{r}_2(s_2)$ in $\mathbb{R}^3$, where $\mathbf{r}_1(L_1) = \mathbf{r}_1(0)$ and $\mathbf{r}_2(L_2) = \mathbf{r}_2(0)$. Find

$$\oint \oint \frac{(d\mathbf{r}_1 \times d\mathbf{r}_2) \cdot (\mathbf{r}_1 - \mathbf{r}_2)}{|\mathbf{r}_1 - \mathbf{r}_2|^3}$$

where $\times$ denotes the vector product and $\cdot$ denotes the scalar product.

Solution 25. We have

$$\oint \oint \frac{(\mathbf{r}_1 \times \mathbf{r}_2) \cdot (\mathbf{r}_1 - \mathbf{r}_2)}{|\mathbf{r}_1 - \mathbf{r}_2|^3} = \oint d\mathbf{r}_1 \cdot \oint \mathbf{r}_2 \times (\mathbf{r}_1 - \mathbf{r}_2) = \oint d\mathbf{r}_1 \cdot \text{curl}_1 \oint \frac{d\mathbf{r}_2}{|\mathbf{r}_1 - \mathbf{r}_2|}.$$

Applying *Stokes theorem* we arrive at

$$\oint \oint \frac{(\mathbf{r}_1 \times \mathbf{r}_2) \cdot (\mathbf{r}_1 - \mathbf{r}_2)}{|\mathbf{r}_1 - \mathbf{r}_2|^3} = \int d\mathbf{S}_1 \cdot \left(\text{curl} \left(\text{curl} \oint \frac{d\mathbf{r}_2}{|\mathbf{r}_1 - \mathbf{r}_2|} \right) \right).$$

Now $\text{curl}(\text{curl}(.)) \equiv \text{grad}(\text{div}(.)) - \nabla^2(.)$ and applying Green's lemma

$$\int \text{div}_1 \frac{d\mathbf{r}_2}{|\mathbf{r}_1 - \mathbf{r}_2|} = -\int \frac{d\mathbf{r}_2 \cdot (\mathbf{r}_1 - \mathbf{r}_2)}{|\mathbf{r}_1 - \mathbf{r}_2|^3} = -\int d\mathbf{S}_2 \left(\text{curl} \left(\text{grad} \frac{1}{|\mathbf{r}_1 - \mathbf{r}_2|} \right) \right) = 0.$$

Now

$$\nabla \frac{1}{|\mathbf{r}_1 - \mathbf{r}_2|} = -4\pi \delta(\mathbf{r}_1 - \mathbf{r}_2)$$

were $\delta(.)$ is the delta function and

$$\int d\mathbf{S}_1 \cdot \oint d\mathbf{r}_2 \delta(\mathbf{r}_1 - \mathbf{r}_2)$$

is the algebraic number of times n the curves $\mathbf{r}_2(s_2)$ passes through the surface $\mathbf{S}_1$ with $\mathbf{r}_1(s_1)$ as perimeter. It follows that

$$\oint \oint \frac{(d\mathbf{r}_1 \times d\mathbf{r}_2) \cdot (\mathbf{r}_1 - \mathbf{r}_2)}{|\mathbf{r}_1 - \mathbf{r}_2|^3} = 4\pi n.$$

Problem 26. The *Riemann-Liouville definition* for the *fractional derivative* of a function f is given by

$$\frac{d^\alpha f(t)}{dt^\alpha} = \frac{1}{\Gamma(n - \alpha)} \frac{d^n}{dt^n} \int_{\tau=0}^{\tau=t} \frac{f(\tau)}{(t - \tau)^{\alpha - n + 1}} d\tau$$

where $\Gamma(.)$ is the gamma function and the integer n is given by $n - 1 \le \alpha < n$. Let $f(t) = t^2$. Find the fractional derivative of f with $\alpha = 1/2$.

Solution 26. Since $\alpha = 1/2$ we have $n = 1$ and $\alpha - n + 1 = 1/2$. Now we have

$$\int_{\tau=0}^{\tau=t} \frac{\tau^2}{\sqrt{t - \tau}} d\tau = -\frac{2(3\tau^2 + 4t\tau + 8t^2)}{15} \sqrt{t - \tau} \Big|_{\tau=0}^{\tau=t} = \frac{16}{15} t^{5/2}.$$

Since $dt^{5/2}/dt = 5t^{3/2}/2$ and $\Gamma(1/2) = \sqrt{\pi}$ we obtain

$$\frac{d^{1/2} f(t)}{dt^{1/2}} = \frac{8}{3\sqrt{\pi}} t^{3/2}.$$

Supplementary Problems

Problem 1. Let $k \in \mathbb{N}$. Show that

$$\int_0^{2\pi} \cos^k(\theta) d\theta = 2(1 + (-1)^k)\pi \frac{(k - 1)!!}{k!!}.$$

Problem 2. (i) Let $k > 0$. Show that

$$\int_0^\infty e^{-kx} dx = \frac{1}{k}.$$

(ii) Let $0 < k_1 < k_2$. Show that

$$\ln(k_2/k_1) = \int_0^\infty \frac{e^{-k_1 x} - e^{-k_2 x}}{x} dx.$$

Problem 3. Let $k = 1, 2, \ldots$. Study the integrals

$$L_k(bcc) = \frac{1}{\pi^3} \int_0^\pi \int_0^\pi \int_0^\pi (\cos(x_1)\cos(x_2)\cos(x_3))^k dx_1 dx_2 dx_3$$

$$L_k(sc) = \frac{1}{\pi^3} \frac{1}{3^k} \int_0^\pi \int_0^\pi \int_0^\pi (\cos(x_1) + \cos(x_2) + \cos(x_3))^k dx_1 dx_2 dx_3.$$

Let

$$f(x_1, x_2, x_3) = (\cos(x_1)\cos(x_2) + \cos(x_2)\cos(x_3) + \cos(x_3)\cos(x_1)).$$

Study the integral

$$L_k(fcc) = \frac{1}{\pi^3} \frac{1}{3^k} \int_0^\pi \int_0^\pi \int_0^\pi (f(x_1, x_2, x_3))^k dx_1 dx_2 dx_3.$$

They play a role in solid state physics for the body-centred cubic lattice, simple cubic lattice, face-center cubic lattice.

Problem 4. The linear one-dimensional *diffusion equation* is given by

$$\frac{\partial u}{\partial t} = D \frac{\partial^2 u}{\partial x^2}, \qquad t \geq 0, \quad -\infty < x < \infty$$

where $u(x,t)$ denotes the concentration at time t and position $x \in \mathbb{R}$. D is the diffusion constant which is assumed to be independent of x and t. Given the initial condition $c(x,0) = f(x)$, $x \in \mathbb{R}$ the solution of the one-dimensional diffusion equation is given by

$$u(x,t) = \int_{-\infty}^{\infty} G(x,t|x',0) f(x') dx'$$

where

$$G(x,t|x',t') = \frac{1}{\sqrt{4\pi D(t-t')}} \exp\left(-\frac{(x-x')^2}{4D(t-t')}\right).$$

Here $G(x,t|x',t')$ is called the fundamental solution of the diffusion equation obtained for the initial data $\delta(x - x')$ at $t = t'$, where δ denotes the Dirac delta function.
(i) Let $u(x,0) = f(x) = \exp(-x^2/(2\sigma))$. Find $u(x,t)$.
(ii) Let $u(x,0) = f(x) = \exp(-|x|/\sigma)$. Find $u(x,t)$.

Problem 5. For a $\lambda/2$ antenna we obtain the expression

$$E(r,\theta) = -\frac{\omega I_0 \sin\theta}{4\pi\epsilon_0 c^2 r} \int_{-\lambda/4}^{\lambda/4} \cos(k\ell) \sin(\omega(t - c^{-1}(r - \ell\cos(\theta)))) d\ell.$$

Calculate $E(r, \theta)$.

Problem 6. The *Gauss invariant* for two given closed loops C_α and C_β in $\mathbb{R}^3$ parametrized by $\mathbf{r}_\alpha(s)$, $\mathbf{r}_\beta$ is defined by

$$G(C_\alpha, C_\beta) := \frac{1}{4\pi} \oint_{C_\alpha} ds \oint_{C_\beta} ds' \frac{d\mathbf{r}_\alpha(s)}{ds} \times \frac{d\mathbf{r}_\beta(s')}{ds'} \frac{\mathbf{r}_\alpha(s) - \mathbf{r}_\beta(s')}{|\mathbf{r}_\alpha(s) - \mathbf{r}_\beta(s')|^3}$$

where $\times$ denotes the vector product. Find $G(C_\alpha, C_\beta)$ for the two curves $C_\alpha : x_1^2 + x_3^2 = 1$, $C_\beta : (x_1 - 1)^2 + x_2^2 = 1$.

Problem 7. Let $k = 0, 1, 2, \dots$. Consider the integral

$$c_k = \int_0^1 \left(\int_0^{x_1} \left(\int_0^{x_3} e^{-2\pi i k (x_1 - 2x_2 + x_3)} dx_2 \right) dx_3 \right) dx_1.$$

Show that

$$c_0 = \frac{1}{6}, \qquad c_k = -\frac{1}{8\pi^2 k^2} \quad (k > 1).$$

Problem 8. Let $k > 0$. Show that (*Laplace integral*)

$$\int_0^\infty \frac{\cos(kx)}{k^2 + s^2} ds = \frac{\pi}{2} \frac{e^{-k|x|}}{k}.$$

Problem 9. Show that the length of the arc described by the *logarithmic spiral*

$$r(\theta) = ae^{k\theta}, \qquad a > 0, k > 0$$

from $\theta = 0$ to $\theta = \pi$ is given by

$$\frac{a}{k} (1 + k^2)^{1/2} (e^{k\pi} - 1).$$

Note that $r(0) = a$.

Problem 10. Show that

$$I(\ell) = \int_0^{2\pi} \left(\int_0^{|\cos(\theta)|(\ell/2)} dp \right) d\theta = \frac{\ell}{2} \int_0^{2\pi} |\cos(\theta)| d\theta = 2\ell.$$

Chapter 8

Inequalities

In a normed vector space V, for example $\mathbb{C}^n$, the *triangular inequality* is

$$\|\mathbf{v} + \mathbf{w}\| \leq \|\mathbf{v}\| + \|\mathbf{w}\| \quad \text{for all} \quad \mathbf{v}, \mathbf{w} \in V.$$

In $\mathbb{R}^n$ with $\mathbf{x} = (x_1, \ldots, x_n)$, $\mathbf{y} = (y_1, \ldots, y_n)$ *Cauchy-Schwarz inequality* is given by

$$(x_1 y_1 + x_2 y_2 + \cdots + x_n y_n)^2 \leq (x_1^2 + x_2^2 + \cdots + x_n^2)(y_1^2 + y_2^2 + \cdots + y_n^2).$$

Let $\mathbf{v}$, $\mathbf{w}$ be two vectors in $\mathbb{R}^n$. Then the Cauchy-Schwarz inequality states that

$$|\mathbf{v} \cdot \mathbf{w}| \leq |\mathbf{v}| \cdot |\mathbf{w}|$$

where $\mathbf{v} \cdot \mathbf{w}$ denotes the scalar product

$$\mathbf{v} \cdot \mathbf{w} = \sum_{j=1}^{n} v_j w_j.$$

The *Hölder inequality* for two vectors $\mathbf{x}, \mathbf{y} \in \mathbb{R}^n$ takes the form

$$|x_1 y_1 + x_2 y_2 + \ldots + x_n y_n| \leq (|x_1|^p + \cdots + |x_n|^p)^{1/p}(|y_1|^q + \cdots + |y_n|^q)^{1/q}$$

where $p > 1$, $q > 1$ and $1/p + 1/q = 1$.

Let $x_1, x_2, \ldots, x_n$ be positive integers and A be the arithmetic mean, G the geometric mean and H the harmonic mean, i.e.

$$A := \frac{1}{n} \sum_{j=1}^{n} x_j, \quad G := \left(\prod_{j=1}^{n} x_j \right)^{1/n}, \quad \frac{1}{H} := \frac{1}{n} \sum_{j=1}^{n} \frac{1}{x_j}.$$

Then

$$H \leq G \leq A.$$

Equality holds when $x_1 = x_2 = \cdots = x_n$.

Let $\mathbf{x} = (x_1, \ldots, x_n)$, $\mathbf{y} = (y_1, \ldots, y_n)$ be vectors in $\mathbb{R}^n$ and

$$x_1 \geq x_2 \geq \cdots \geq x_n, \quad y_1 \geq y_2 \geq \cdots \geq y_n.$$

Then (Chebyshev's inequality)

$$\left(\sum_{j=1}^{n} x_j \right) \left(\sum_{j=1}^{n} y_j \right) \leq n \sum_{j=1}^{n} x_j y_j.$$

Let $x \geq 0$ and $f(x) = x \ln(x) - x + 1$. Then $f(x) \geq 0$. This is *Gibbs inequality*.

Let A, B be $n \times n$ hermitian matrices. Then

$$\mathrm{tr}(e^{A+B}) \leq \mathrm{tr}(e^A e^B).$$

Let A, B be positive definite matrices. Then

$$\mathrm{tr}(AB)^{2^{p+1}} \leq \mathrm{tr}(A^2 B^2)^{2^p}, \quad p \text{ a non-negative integer.}$$

Let A, B be $n \times n$ positive definite matrices. Then

$$\mathrm{tr}(A \ln(A)) - \mathrm{tr}(A \ln(B)) \geq \mathrm{tr}(A - B).$$

Let $\hat{A}$, $\hat{B}$ be self-adjoint operators in a Hilbert space $\mathcal{H}$ and $|\psi\rangle$ be a normalized state ($\langle \psi | \psi \rangle = 1$) in this Hilbert space. Let $[,]$ be the commutator. Then (*uncertainty relation*)

$$(\Delta \hat{A})(\Delta \hat{B}) \geq \frac{1}{2} |\langle [\hat{A}, \hat{B}] \rangle|$$

where $\Delta \hat{A}$ is the standard deviation of the self-adjoint operator $\hat{A}$ described by $|\psi\rangle$, i.e.

$$\Delta \hat{A} := \sqrt{\langle \hat{A}^2 \rangle - \langle \hat{A} \rangle^2}$$

with $\langle \hat{A} \rangle := \langle \psi | \hat{A} | \psi \rangle$, $\langle \hat{A}^2 \rangle = \langle \psi | \hat{A}^2 | \psi \rangle$.

Problem 1. (i) Let $a, b \in \mathbb{R}$. Show that $a^2 + b^2 \geq 2ab$.
(ii) Let a, b be two non-negative numbers. Show that $\frac{1}{2}(a + b) \geq \sqrt{ab}$.

Solution 1. (i) Since $(a - b)^2 \geq 0$ we find $a^2 + b^2 - 2ab \geq 0$. It follows that $a^2 + b^2 \geq 2ab$.
(ii) Since $(\sqrt{a} - \sqrt{b})^2 \geq 0$ we have $a + b - 2\sqrt{ab} \geq 0$. Therefore $a + b \geq 2\sqrt{ab}$.

Problem 2. Consider the differentiable function $f : [0, \infty) \to \mathbb{R}$

$$f(x) = \frac{x}{1 + x}.$$

Let $a, b \in \mathbb{R}^+$. Show that $f(|a + b|) \leq f(|a| + |b|)$.

Solution 2. Obviously we have (*triangular inequality*) $|a + b| \leq |a| + |b|$. Differentiation of f gives $df/dx = 1/(1 + x)^2$ which is positive. Thus the function f is monotone increasing. Consequently the inequality follows.

Problem 3. Let $a, b \in \mathbb{R}$. Show that

$$\frac{|a + b|}{1 + |a + b|} \leq \frac{|a|}{1 + |a|} + \frac{|b|}{1 + |b|}.$$

Solution 3. From the triangular inequality $|a + b| \leq |a| + |b|$ we obtain

$$\frac{|a + b|}{1 + |a + b|} \leq \frac{|a| + |b|}{1 + |a| + |b|} = \frac{|a|}{1 + |a| + |b|} + \frac{|b|}{1 + |a| + |b|}$$
$$\leq \frac{|a|}{1 + |a|} + \frac{|b|}{1 + |b|}.$$

Problem 4. Find all $x \in \mathbb{R}$ such that $x^2 - x - 6 \leq 0$.

Solution 4. We have $(x + 2)(x - 3) \leq 0$. Thus if $x \in [-2, 3]$ the inequality is satisfied. For all other x the inequality is not satisfied.

Problem 5. Let $a_1, a_2, \ldots, a_n, b_1, b_2, \ldots, b_n \in \mathbb{R}$.
(i) Show that

$$(a_1 b_1 + a_2 b_2 + \cdots + a_n b_n)^2 \leq (a_1^2 + a_2^2 + \cdots + a_n^2)(b_1^2 + b_2^2 + \cdots + b_n^2). \quad (1)$$

(ii) Show that

$$\sqrt{(a_1 + b_1)^2 + \cdots + (a_n + b_n)^2} \leq \sqrt{a_1^2 + \cdots + a_n^2} + \sqrt{b_1^2 + \cdots + b_n^2}. \quad (2)$$

Solution 5.   (i) If $a_1 = a_2 = \cdots = a_n = 0$, then inequality (1) is satisfied. Assume that at least one of the a_j's is nonzero. Let ϵ be an arbitrary real number. Then we obviously have

$$(a_1\epsilon + b_1)^2 + (a_2\epsilon + b_2)^2 + \cdots + (a_n\epsilon + b_n)^2 \geq 0. \tag{3}$$

Let

$$A^2 := a_1^2 + a_2^2 + \cdots + a_n^2, \qquad B^2 := b_1^2 + b_2^2 + \cdots + b_n^2$$

and $C := a_1b_1 + a_2b_2 + \cdots + a_nb_n$. Then from (3) we have

$$A^2\epsilon^2 + 2C\epsilon + B^2 \geq 0.$$

The left-hand side of this inequality is a smooth function of ϵ, say

$$f(\epsilon) := A^2\epsilon^2 + 2C\epsilon + B^2.$$

Since

$$\frac{df}{d\epsilon} = 2A^2\epsilon + 2C, \qquad \frac{d^2f}{d\epsilon^2} = 2A^2 > 0$$

we find that f has a minimum at $\epsilon = -C/A^2$. Inserting ϵ into the left-hand side of the inequality yields

$$\frac{A^2C^2}{A^4} - \frac{2C^2}{A^2} + B^2 \geq 0$$

or $A^2B^2 \geq C^2$. This is inequality (1).
(ii) Obviously

$$0 \leq \sum_{j=1}^{n}(a_j + b_j)^2 \equiv \sum_{j=1}^{n}a_j^2 + \sum_{j=1}^{n}b_j^2 + 2\sum_{j=1}^{n}a_jb_j.$$

From (1) we have

$$\sum_{j=1}^{n}a_jb_j \leq \sqrt{\sum_{j=1}^{n}a_j^2}\sqrt{\sum_{j=1}^{n}b_j^2}.$$

Therefore we obtain

$$\sum_{j=1}^{n}(a_j + b_j)^2 \leq \sum_{j=1}^{n}a_j^2 + \sum_{j=1}^{n}b_j^2 + 2\sqrt{\sum_{j=1}^{n}a_j^2}\sqrt{\sum_{j=1}^{n}b_j^2}.$$

Taking the square root on both sides leads to (2).

Problem 6.   Let $x \geq 0$ and

$$f(x) = x\ln(x) - x + 1. \tag{1}$$

Show that
$$f(x) \geq 0. \tag{2}$$

From L'Hospital's rule we find that $f(0) = 1$. The inequality (2) is called *Gibbs inequality* and plays an important rôle in statistical physics.

Solution 6. From (1) we obtain $df/dx = \ln(x)$, $d^2f/dx^2 = 1/x$. Thus we find that the function f has one minimum at $x = 1$ for $x \geq 0$, since $\ln(1) = 0$ and
$$\left.\frac{d^2 f}{dx^2}\right|_{x=1} = 1 > 0.$$

It follows that $f(1) = 0$. Therefore inequality (2) follows.

Problem 7. Suppose that $f : \mathbb{R} \to \mathbb{R}$ is twice differentiable with
$$f''(x) \geq 0$$

for all x. Prove that for all a and b, $a < b$,
$$f\left(\frac{a+b}{2}\right) \leq \frac{f(a) + f(b)}{2}. \tag{1}$$

Solution 7. By the *mean-value theorem* there is a number x_1 in the interval $(a, \frac{1}{2}(a+b))$ such that
$$\frac{f(\frac{1}{2}(a+b)) - f(a)}{\frac{1}{2}(a+b) - a} = f'(x_1)$$

and a number x_2 in $(\frac{1}{2}(a+b), b)$ such that
$$\frac{f(b) - f(\frac{1}{2}(a+b))}{b - \frac{1}{2}(a+b)} = f'(x_2).$$

However $f''(x) \geq 0$ for all x in (x_1, x_2), so f' is a nondecreasing function. Thus $f'(x_2) \geq f'(x_1)$ or equivalently,
$$\frac{f(b) - f(\frac{1}{2}(a+b))}{b - a} \geq \frac{f(\frac{1}{2}(a+b)) - f(a)}{b - a}.$$

Thus (1) follows.

Problem 8. Let $a_j > 0$ for $j = 1, 2, \ldots, n$. The *arithmetic mean* of $a_1, a_2, \ldots, a_n$ is the number $(a_1 + a_2 + \cdots + a_n)/n$ and the *geometric mean* of $a_1, a_2, \ldots, a_n$ is the number $(a_1 a_2 \cdots a_n)^{1/n}$.

(i) Show that

$$(a_1 a_2 \cdots a_n)^{1/n} \leq \frac{a_1 + a_2 + \cdots + a_n}{n}. \tag{1}$$

(ii) Consider a rectangular parallelepiped. Let the lengths of the three adjacent sides be a, b and c. Let A and V be the surface area and the volume, respectively. Prove that $A \geq 6V^{2/3}$.

Solution 8.   (i) Let $n = 2$. Then we have to prove that

$$(a_1 a_2)^{1/2} \leq \frac{a_1 + a_2}{2}.$$

This inequality was proved in problem 1. The proof for larger values of n can be handled by mathematical induction.

(ii) Obviously

$$A = 2(ab + bc + ca), \qquad V = abc.$$

Thus $V^2 = a^2 b^2 c^2 = (ab)(bc)(ca)$. Using the *arithmetic mean geometric mean inequality* we obtain

$$V^2 \leq \left(\frac{ab + bc + ca}{3} \right)^3 = \left(\frac{2(ab + bc + ca)}{6} \right)^3 = \left(\frac{A}{6} \right)^3.$$

Thus the inequality follows. If $ab = bc = ca$ or equivalently $a = b = c$ we obtain $6V^{2/3} = A$.

Problem 9.   Let $x \in \mathbb{R}$ and $x > 0$. Show that $\sqrt[e]{e} \geq \sqrt[x]{x}$.

Solution 9.   Since

$$f(x) = \sqrt[x]{x} \equiv \exp\left(\frac{1}{x} \ln(x) \right) \Rightarrow \frac{df}{dx} = \frac{1 - \ln(x)}{x^2} \exp\left(\frac{1}{x} \ln(x) \right).$$

The necessary condition for an extremum is $df/dx = 0$. It follows that

$$1 - \ln(x) = 0 \Rightarrow x = e \quad \text{(critical point)}.$$

Therefore $f(e) = e^{1/e} \approx 1.44466786101....$ The second derivative is

$$\frac{d^2 f}{dx^2} = \left(\left(\frac{1 - \ln(x)}{x^2} \right)^2 + \frac{-x - (1 - \ln(x))2x}{x^4} \right) \exp\left(\frac{1}{x} \ln(x) \right).$$

Setting $x = e$ and using $\ln(e) = 1$ yields

$$\frac{d^2 f}{dx^2} \bigg|_{x=e} = \left(-\frac{1}{e^3} \right) \exp(1/e) < 0.$$

Thus f has a maximum at $x = e$. For $x = 0$ we find that f takes the form $f(x = 0) = 0$ (L'Hospital's rule). Furthermore we have

$$\lim_{x \to \infty} \sqrt[x]{x} = 1.$$

Problem 10. Let $a_1, a_2, \ldots, a_n$ be positive numbers and $b_1, b_2, \ldots, b_n$ be nonnegative numbers such that

$$\sum_{j=1}^{n} b_j > 0.$$

Show that (*log-sum inequality*)

$$\sum_{j=1}^{n} \left(a_j \ln \left(\frac{a_j}{b_j} \right) \right) \geq \left(\sum_{j=1}^{n} a_j \right) \ln \left(\frac{\left(\sum_{j=1}^{n} a_j \right)}{\left(\sum_{j=1}^{n} b_j \right)} \right)$$

with the conventions based on continuity arguments

$$0 \cdot \ln(0) = 0, \qquad 0 \cdot \ln \left(\frac{p}{0} \right) = \infty, \quad p > 0.$$

Show that equality holds iff $a_j / b_j = $ constant for all $j = 1, 2, \ldots, n$.

Solution 10. We set $A := \sum_{j=1}^{n} a_j$, $B := \sum_{j=1}^{n} b_j$. Then

$$\sum_{j=1}^{n} \left(a_j \ln \left(\frac{a_j}{b_j} \right) \right) - A \ln \left(\frac{A}{B} \right) = \sum_{j=1}^{n} \left(a_j \ln \left(\frac{a_j}{b_j} \right) \right) - A \ln \left(\frac{A}{B} \right)$$

$$= A \sum_{j=1}^{n} \frac{a_j}{A} \ln \left(\frac{a_j / A}{b_j / B} \right)$$

$$\geq A \sum_{j=1}^{n} \left(\frac{a_j}{A} \left(1 - \frac{b_j / B}{a_j / A} \right) \right)$$

$$= A \sum_{j=1}^{n} \left(\frac{a_j}{A} - \frac{b_j}{B} \right) = A(1 - 1) = 0.$$

The inequality comes from $1 - 1/x \leq \ln(x)$ ($x > 0$) with equality iff $x = 1$. For equality to hold, we need to have

$$\frac{a_j / A}{b_j / B} = 1 \quad \text{for all } j = 1, 2, \ldots, n.$$

This implies

$$\frac{a_j}{b_j} = \text{constant} \quad \text{for all } j = 1, 2, \ldots, n.$$

Problem 11. Let n be a positive integer and $a, b \geq c/2 > 0$. Show that

$$|a^{-n} - b^{-n}| \leq 4nc^{-n-1}|a - b|.$$

Solution 11. . We have

$$|a^{-n} - b^{-n}| = \left| \frac{b^n - a^n}{(ab)^n} \right| = \frac{\sum_{i=1}^n b^{n-i} a^{i-1}}{(ab)^n} |b - a|$$

$$= |a - b| \sum_{i=1}^n b^{-i} a^{i-n-1} \leq 4nc^{-n-1}|b - a|.$$

Problem 12. Let f and g be two integrable functions. Assume that

$$\int_I f(x)dx = \int_I g(x)dx \tag{1}$$

and $f(x) > 0$, $g(x) > 0$ for $x \in I$, where $I \subset \mathbb{R}$. Show that

$$\int_I f(x) \ln(f(x)) \, dx \geq \int_I f(x) \ln(g(x)) \, dx. \tag{2}$$

Solution 12. We have

$$\int_I f \ln(f) \, dx - \int_I f \ln(g) \, dx \equiv \int_I f(\ln(f) - \ln(g))dx \equiv \int_I f \ln\left(\frac{f}{g}\right) dx$$

$$\equiv \int_I g \left(\frac{f}{g} \ln\left(\frac{f}{g}\right)\right) dx.$$

Thus

$$\int_I f \ln(f)dx - \int_I f \ln(g)dx \equiv \int_I g \left(\frac{f}{g} \ln\left(\frac{f}{g}\right) - \frac{f}{g} + 1\right) dx \geq 0$$

where we have used (1) and the fact that

$$\frac{f}{g} \ln\left(\frac{f}{g}\right) - \frac{f}{g} + 1 \geq 0.$$

Problem 13.   (i) Let $0 < \mu < 1$ and $a \geq 0$, $b \geq 0$. Show that

$$a^\mu b^{1-\mu} \leq \mu a + (1 - \mu)b. \tag{1}$$

(ii) Let p and q be two positive numbers which satisfy the condition $1/p + 1/q = 1$. Let $c, d \in \mathbb{C}$. Show that

$$|cd| \leq \frac{|c|^p}{p} + \frac{|d|^q}{q}. \tag{2}$$

Solution 13.   (i) For $a = b$ the inequality (1) is obviously satisfied. Without loss of generality we can assume that $b > a > 0$. Applying the *theorem of the mean*, we obtain

$$b^{1-\mu} - a^{1-\mu} = (1 - \mu)(b - a)\xi^{-\mu}$$

with $a < \xi < b$. Since $\xi^{-\mu} < a^{-\mu}$ we obtain

$$b^{1-\mu} - a^{1-\mu} \leq (1 - \mu)(b - a)a^{-\mu}.$$

Multiplying the left and right-hand sides of this inequality by a^μ leads to inequality (1).
(ii) If we set $\mu = 1/p$, $1 - \mu = 1/q$, $a = |c|^p$, $b = |d|^q$ in inequality (1) we obtain (2).

Problem 14.   Let $a_1, \ldots, a_n$ and $b_1, \ldots, b_n$ be arbitrary complex numbers. Let p and q be two real positive numbers which satisfy $1/p + 1/q = 1$. Show that

$$\sum_{k=1}^n |a_k b_k| \leq \left[\sum_{k=1}^n |a_k|^p \right]^{1/p} \left[\sum_{k=1}^n |b_k|^q \right]^{1/q}.$$

Solution 14.   We set $A^p := \sum_{k=1}^n |a_k|^p$, $B^q := \sum_{k=1}^n |b_k|^q$. We can assume that both A and B are positive. Let $\bar{a}_k := a_k/A$, $\bar{b}_k := b_k/B$. Then

$$|\bar{a}_k \bar{b}_k| \leq \frac{|\bar{a}_k|^p}{p} + \frac{|\bar{b}_k|^q}{q}.$$

Taking the sum over k on both sides yields

$$\sum_{k=1}^n |\bar{a}_k \bar{b}_k| \leq \sum_{k=1}^n \left(\frac{|\bar{a}_k|^p}{p} + \frac{|\bar{b}_k|^q}{q} \right) = \frac{1}{p} + \frac{1}{q} = 1.$$

Consequently, $\sum_{k=1}^n |a_k b_k| \leq AB$.

Problem 15.   Let B be an $n \times n$ matrix over $\mathbb{R}$. Let B^T be the transpose. Show that

$$\text{tr}(B^{2m}) \leq \text{tr}(B^m B^{mT}) \tag{1}$$

where tr denotes the trace.

Solution 15.   Let A be an $n \times n$ matrix over $\mathbb{R}$. Then

$$(A^T A)_{jl} = \sum_{k=1}^{n} a_{kj} a_{kl} \tag{2}$$

and therefore

$$(A^T A)_{jj} = \sum_{k=1}^{n} a_{kj} a_{kj} \geq 0 \;\Rightarrow\; \text{tr}(A^T A) \geq 0. \tag{3}$$

We set $A^T := B^m - B^{mT}$. Hence $A = B^{mT} - B^m$. Using (3) we obtain

$$\text{tr}((B^m - B^{mT})(B^{mT} - B^m)) \geq 0.$$

It follows that

$$\text{tr}(B^m B^{mT}) + \text{tr}(B^{mT} B^m) - \text{tr}(B^{mT} B^{mT}) - \text{tr}(B^m B^m) \geq 0.$$

Since $\text{tr}(B^m B^{mT}) \equiv \text{tr}(B^{mT} B^m)$, $\text{tr}(B^m B^m) \equiv \text{tr}(B^{mT} B^{mT})$ inequality (1) follows.

Problem 16.   Show that for two $n \times n$ positive-semidefinite real matrices A and B and $0 \leq \alpha \leq 1$ we have

$$\text{tr}(A^\alpha B^{1-\alpha}) \leq (\text{tr}(A))^\alpha (\text{tr}(B))^{1-\alpha}. \tag{1}$$

Solution 16.   Since A and B are positive-semidefinite their eigenvalues are real and nonnegative. We order the eigenvalues of A and B in decreasing order

$$\lambda_1 \geq \lambda_2 \geq \cdots \geq \lambda_n \geq 0, \qquad \mu_1 \geq \mu_2 \geq \cdots \geq \mu_n \geq 0.$$

For an arbitrary orthonormal set of vectors $\mathbf{x}_i$ we have

$$\sum_{i=1}^{k} \langle \mathbf{x}_i, B^{1-\alpha} \mathbf{x}_i \rangle \leq \sum_{i=1}^{k} \mu_i^{1-\alpha}, \qquad k = 1, 2, \ldots, n$$

where $\langle \, , \, \rangle$ denotes the scalar product. Choosing $\mathbf{x}_i$ to be the eigenvectors of A and summing by parts gives

$$\text{tr}(A^\alpha B^{1-\alpha}) = \sum_{i=1}^{n} \lambda_i^\alpha \langle \mathbf{x}_i, B^{1-\alpha} \mathbf{x}_i \rangle.$$

Thus

$$\operatorname{tr}(A^\alpha B^{1-\alpha}) \le \sum_{i=1}^n \lambda_i^\alpha \mu_i^{1-\alpha} \le \left(\sum_{i=1}^n \lambda_i\right)^\alpha \left(\sum_{i=1}^n \mu_i\right)^{1-\alpha} = (\operatorname{tr}(A))^\alpha (\operatorname{tr}(B))^{1-\alpha}$$

where the last inequality is the *Hölder inequality* for positive real numbers. We also used that $\operatorname{tr}(A) = \sum_{i=1}^n \lambda_i$, $\operatorname{tr}(B) = \sum_{i=1}^n \mu_i$.

Problem 17. Let A be a bounded linear operator in a Banach space. Let

$$f_{n,m}(A) := \left(\sum_{k=0}^m \frac{1}{k!}\left(\frac{A}{n}\right)^k\right)^n. \tag{1}$$

Show that

$$\|e^A - f_{n,m}(A)\| \le \frac{1}{n^m(m+1)!}\|A\|^{m+1}e^{\|A\|}. \tag{2}$$

Solution 17. Using the properties of the norm we have the inequality

$$\|e^A - f_{n,m}(A)\| \le \|e^{A/n} - h\| \cdot \|(e^{A/n})^{n-1} + (e^{A/n})^{n-2}h + \cdots + h^{n-1}\|$$
$$\le n\|e^{A/n} - h\| \cdot e^{(n-1)\|A\|/n}$$

where

$$h := \sum_{k=0}^m \frac{1}{k!}\left(\frac{A}{n}\right)^k.$$

Using Taylor's theorem we find

$$\|e^{A/n} - h\| = \left\| \sum_{k=m+1}^\infty \frac{1}{k!}\left(\frac{A}{n}\right)^k \right\| \le \sum_{k=m+1}^\infty \frac{1}{k!}\left(\frac{\|A\|}{n}\right)^k$$
$$= e^{\|A\|/n} - \sum_{k=0}^m \frac{1}{k!}\left(\frac{\|A\|}{n}\right)^k$$
$$= \frac{1}{(m+1)!}\left(\frac{\|A\|}{n}\right)^{m+1} e^{\theta\|A\|/n}, \qquad 0 < \theta < 1.$$

Inserting this equation into the inequality above we obtain (1). $f_{n,m}$ converges to e^A if $n \to \infty$ or if $m \to \infty$.

Problem 18. Let A and B be two symmetric $n \times n$ matrices over $\mathbb{R}$. It can be shown that

$$\operatorname{tr}(e^{A+B}) \le \operatorname{tr}(e^A e^B) \le \frac{1}{2}\operatorname{tr}(e^{2A} + e^{2B})$$
$$\operatorname{tr}(e^{A+B}) \le \operatorname{tr}(e^A e^B) \le (\operatorname{tr}(e^{pA})^{1/p})(\operatorname{tr}(e^{qB})^{1/q})$$

where $p > 1$, $q > 1$ with $1/p + 1/q = 1$. Is

$$(\mathrm{tr} e^{pA})^{1/p}(\mathrm{tr} e^{qB})^{1/q} \le \frac{1}{2}\mathrm{tr}(e^{2A} + e^{2B})\,?$$

Prove or disprove.

Solution 18.   Since A and B are symmetric the eigenvalues are real numbers. Let $\ln(\lambda_1)$, ..., $\ln(\lambda_n)$ and $\ln(\mu_1)$, ..., $\ln(\mu_n)$ be the (real) eigenvalues of A and B, respectively. Then the conjectured inequality takes the form

$$(\lambda_1^p + \cdots + \lambda_n^p)^{1/p}(\mu_1^q + \cdots + \mu_n^q)^{1/q} \le \frac{1}{2}\left(\lambda_1^2 + \cdots + \lambda_n^2 + \mu_1^2 + \cdots + \mu_n^2\right).$$

If $n = 1$ or $p = q = 2$, this inequality is trivially true. It is generally false in all other cases. Let $n > 1$ and take $q < 2$. Choose $\lambda_1 = n^{1/4}$, $\lambda_2 = \cdots = \lambda_n = \epsilon$ and $\mu_1 = \mu_2 = \cdots = \mu_n = n^{-1/4}$. As $\epsilon \to 0$, the inequality tends to $n^{1/q} \le n^{1/2}$ which is false. Thus the desired counterexample is obtained for all sufficiently small ϵ.

Problem 19.   Let A, B be hermitian matrices. Then

$$\mathrm{tr}(e^{A+B}) \le \mathrm{tr}(e^A e^B).$$

Assume that

$$A = \begin{pmatrix} 0 & 0 \\ 0 & 1 \end{pmatrix}, \qquad B = \begin{pmatrix} 0 & 1 \\ 1 & 0 \end{pmatrix}.$$

Calculate the left and right-hand side of the inequality. Does equality hold?

Solution 19.   We note that the commutator of A and B is given by

$$[A, B] = \begin{pmatrix} 0 & -1 \\ 1 & 0 \end{pmatrix}.$$

Thus equality does not hold. We find since $A^2 = A$ and $B^2 = I_2$

$$e^A = \begin{pmatrix} 1 & 0 \\ 0 & e \end{pmatrix}, \qquad e^B = \begin{pmatrix} \cosh(1) & \sinh(1) \\ \sinh(1) & \cosh(1) \end{pmatrix}.$$

Thus $\mathrm{tr}(e^A e^B) = \cosh(1) + e\sinh(1)$. The eigenvalues of $A + B$ are $\lambda_1 = (1 + \sqrt{5})/2$ and $\lambda_2 = (1 - \sqrt{5})/2$. Thus

$$\mathrm{tr}(e^{A+B}) = e^{\lambda_1} + e^{\lambda_2} = e^{1/2}(e^{\sqrt{5}/2} + e^{-\sqrt{5}/2}).$$

Problem 20. Let $\hat{A}$, $\hat{B}$ be self-adjoint operators in a Hilbert space $\mathcal{H}$ and $|\psi\rangle$ be a normalized state ($\langle\psi|\psi\rangle = 1$) in this Hilbert space. Let $[,]$ be the commutator. Then (*Uncertainty relation*)

$$(\Delta\hat{A})(\Delta\hat{B}) \geq \frac{1}{2}|\langle[\hat{A},\hat{B}]\rangle|$$

where $\Delta\hat{A}$ is the standard deviation of the self-adjoint operator $\hat{A}$ described by $|\psi\rangle$, i.e.

$$\Delta\hat{A} := \sqrt{\langle\hat{A}^2\rangle - \langle\hat{A}\rangle^2}$$

with $\langle\hat{A}\rangle := \langle\psi|\hat{A}|\psi\rangle$, $\langle\hat{A}^2\rangle = \langle\psi|\hat{A}^2|\psi\rangle$. Since $\Delta\hat{A}$, $\Delta\hat{B}$ and $|\langle[\hat{A},\hat{B}]\rangle|$ are non-negative we can write the inequality as

$$(\Delta\hat{A})^2(\Delta\hat{B})^2 \geq \frac{1}{4}|\langle[\hat{A},\hat{B}]\rangle|^2. \tag{1}$$

A state $|\psi\rangle$ is called an *intelligent state* when equality holds. A stronger bound is given by

$$(\Delta\hat{A})^2(\Delta\hat{B})^2 \geq \frac{1}{4}|\langle[\hat{A},\hat{B}]\rangle|^2 + \frac{1}{4}|\langle[\hat{A}-\langle\hat{A}\rangle I, \hat{B}-\langle\hat{B}\rangle I]_+\rangle|^2 \tag{2}$$

where $[,]_+$ denotes the anti-commutator and I is the identity operator. If the equality holds in this inequality for some $|\psi\rangle$ we call $|\psi\rangle$ a generalized intelligent state.

(i) Consider the Hilbert space $\mathbb{C}^4$, the *Bell state*

$$|\psi^+\rangle = \frac{1}{\sqrt{2}}\begin{pmatrix}1 & 0 & 0 & 1\end{pmatrix}^T$$

and $\hat{A} = \sigma_1 \otimes \sigma_1$, $\hat{B} = \sigma_3 \otimes \sigma_3$. Do we have an intelligent state?

(ii) Consider the Hilbert space $\mathbb{C}^8$ and the *GHZ state*

$$|GHZ\rangle = \frac{1}{\sqrt{2}}\begin{pmatrix}1 & 0 & 0 & 0 & 0 & 0 & 0 & 1\end{pmatrix}^T.$$

Let $\hat{A} = \sigma_1 \otimes \sigma_1 \otimes \sigma_1$, $\hat{B} = \sigma_3 \otimes \sigma_3 \otimes I_2$. Show that $|GHZ\rangle$ is an intelligent and generalized intelligent state.

Solution 20. (i) We have $\hat{A}^2 = I_2 \otimes I_2$, $\hat{B}^2 = I_2 \otimes I_2$. Hence $\langle\psi^+|\hat{A}^2|\psi^+\rangle = 1$, $\langle\psi^+|\hat{B}^2|\psi^+\rangle = 1$. Furthermore $\langle\psi^+|\hat{A}|\psi^+\rangle = 1$, $\langle\psi^+|\hat{B}|\psi^+\rangle = 1$. It follows that $\Delta\hat{A} = \Delta\hat{B} = 0$. Obviously the right-hand side of (1) and (2) is also 0. Hence the Bell state $|\psi^+\rangle$ is an intelligent state and a generalized intelligent state.

(ii) We have $[\hat{A},\hat{B}] = 0_8$ and

$$\langle GHZ|\hat{A}|GHZ\rangle = 1, \quad \langle GHZ|\hat{B}|GHZ\rangle = 1,$$

$$\langle GHZ|\hat{A}^2|GHZ\rangle = 1, \quad \langle GHZ|\hat{B}^2|GHZ\rangle = 1.$$

Hence $|GHZ\rangle$ is an intelligent and generalized intelligent state.

Supplementary Problems

Problem 1.   Let $x_1, \ldots, x_n \geq 0$. Show that

$$\frac{1}{n}\sum_{j=1}^{n} x_j \geq \left(\prod_{j=1}^{n} x_j\right)^{1/n}.$$

Problem 2.   Let $x_1, x_2, x_3 \in \mathbb{R}$. Show that

$$(x_1 x_2 + x_2 x_3 + x_1 x_3) - 3x_1 x_2 x_3(x_1 + x_2 + x_3) \geq 0.$$

Problem 3.   Let $\mathbf{v}$ be a normalized (column) vector in $\mathbb{C}^n$ and let A be an $n \times n$ hermitian matrix. Is

$$\mathbf{v}^* e^A \mathbf{v} \geq e^{\mathbf{v}^* A \mathbf{v}}$$

for all normalized $\mathbf{v}$? Prove or disprove.

Chapter 9

Optimization

Let M be a smooth manifold and f be a real valued function of class $C^{(1)}$ on some open set containing M. We consider the problem of finding the extrema of the function $f|M$. This is called a problem of *constrained extrema*.

The *Lagrange multiplier rule* is as follows. Assume that f has a constrained relative extremum at $\mathbf{x}^* = (x_1^*, x_2^*, \ldots, x_n^*)$. Then there exist real numbers $\lambda_1, \lambda_2, \ldots, \lambda_m$ such that $\mathbf{x}^*$ is a critical point of the function

$$L(\mathbf{x}) := f(\mathbf{x}) + \lambda_1 g_1(\mathbf{x}) + \cdots + \lambda_m g_m(\mathbf{x}).$$

The numbers $\lambda_1, \lambda_2, \cdots, \lambda_m$ are called *Lagrange multipliers*. The critical points are given by the solution of the system

$$\frac{\partial L}{\partial x_k} = 0, \qquad k = 1, 2, \ldots, n.$$

L is called the *Lagrangian*.

Using differential forms to solve constraint max-min problems has the advantage over the Lagrange multiplier method that the Lagrange multipliers are eliminated. The equation

$$df|_P = \lambda dg|_P$$

between differential one forms is equivalent to the Lagrange condition

$$\nabla f|_P = \lambda \nabla g|_P.$$

201

Geometrically the differential condition says that the tangent lines to the contour and the constraint curves are identical, while the gradient condition says that the normal vectors of these lines are parallel.

A subset C of $\mathbb{R}^n$ is said to be *convex* if for any $\mathbf{a}$ and $\mathbf{b}$ in C and any θ in $\mathbb{R}$, $0 \leq \theta \leq 1$, the n-tuple $\theta\mathbf{a} + (1 - \theta)\mathbf{b}$ also belongs to C. In other words, if $\mathbf{a}$ and $\mathbf{b}$ are in C, then $\{\theta\mathbf{a} + (1 - \theta)\mathbf{b} : 0 \leq \theta \leq 1\} \subset C$. Let C be a convex subset of $\mathbb{R}^n$. Let f be a real-valued function with domain C. If for each $\mathbf{x}$ and $\mathbf{y}$ in C and each θ, $0 \leq \theta \leq 1$, the inequality

$$f(\theta\mathbf{x} + (1 - \theta)\mathbf{y}) \leq \theta f(\mathbf{x}) + (1 - \theta)f(\mathbf{y})$$

holds, then f is said to be a *convex function*. If $-f$ is convex, then f is called a *concave function*.

Many optimization problems also include inequalities. The *Karush-Kuhn-Tucker conditions* extend the Lagrange multiplier method to include inequality constraints. Given an optimization problem with convex domain $\Omega \subseteq \mathbb{R}^n$,

$$\begin{aligned}
\text{minimize} \quad & f(\mathbf{x}), \quad \mathbf{x} \in \Omega \\
\text{subject to} \quad & g_k(\mathbf{x}) \geq 0 , \quad k = 1, \ldots, K \\
& h_m(\mathbf{x}) = 0 , \quad j = 1, \ldots, M.
\end{aligned}$$

For this problem we can construct the *Lagrange function*

$$L(\mathbf{x}, \boldsymbol{\lambda}, \boldsymbol{\mu}) = f(\mathbf{x}) - \sum_{k=1}^{K} \lambda_k g_k(\mathbf{x}) - \sum_{m=1}^{M} \mu_m h_m(\mathbf{x})$$

where λ_k and μ_m are the Lagrange multipliers. If the problem has a solution

$$\mathbf{x}^* = (x_1^*, x_2^*, \ldots, x_n^*)$$

i.e. $\min_{\mathbf{x}} f(\mathbf{x}) = f(\mathbf{x}^*)$ and all constraints are satisfied, then the following Karush-Kuhn-Tucker conditions hold

$$\nabla f(\mathbf{x}^*) - \sum_{k=1}^{K} \lambda_k^* \nabla g_k(\mathbf{x}^*) - \sum_{m=1}^{M} \mu_m^* \nabla h_m(\mathbf{x}^*) = \mathbf{0}$$

and

$$\begin{aligned}
g_k(\mathbf{x}^*) &\geq 0 & k &= 1, 2, \ldots, K \\
h_m(\mathbf{x}^*) &= 0 & m &= 1, 2, \ldots, M \\
\lambda_k^* g_k(\mathbf{x}^*) &= 0 & k &= 1, 2, \ldots, K \\
\lambda_k^* &\geq 0 & k &= 1, 2, \ldots, K.
\end{aligned}$$

In convex programming problems, the Karush-Kuhn-Tucker conditions are necessary and sufficient for a global minimum.

Problem 1. Let R_i be a fixed resistor and U_0 be a fixed voltage. Consider

$$P(R_a) = R_a I^2 \equiv \frac{R_a n^2 U_0^2}{(nR_i + R_a)^2}$$

where $I = nU_0/(nR_i + R_a)$, $R_i = 0.1\, Ohm = 0.1\, kg\,.\,meter^2\,.\,sec^{-3}\,.\,A^{-2}$, $n = 10$, $U_0 = 2\,Volt = 2\,kg\,.\,meter^2\,.\,sec^{-3}\,.\,A^{-1}$. Find the maximum of $P(R_a)$.

Solution 1. We have

$$\frac{dP}{dR_a} = \frac{(nR_i + R_a)^2 n^2 U_0^2 - R_a n^2 U_0^2 2(nR_i + R_a)}{(nR_i + R_a)^4}.$$

Hence from $dP/dR_a = 0$ we obtain the equation

$$(nR_i + R_a) - 2R_a = 0, \quad \text{or} \quad R_a = nR_i = 1\, Ohm.$$

The second derivative d^2P/dR_a^2 at the critical point $R_a = nR_i$ shows that we have a maximum. Furthermore

$$P(nR_i) = \frac{nU_0^2}{4R_i} = 100\, meter^2\,.\,sec^{-3}\,.\,kg.$$

Problem 2. Let $f : \mathbb{R}^3 \to \mathbb{R}$, $f(\mathbf{x}) = x_1 - x_2 + 2x_3$. Find the maximum and minimum values of f on the *ellipsoid*

$$M := \{\, (x_1, x_2, x_3) : x_1^2 + x_2^2 + 2x_3^2 = 2 \,\}.$$

Solution 2. Let $g(\mathbf{x}) = 2 - (x_1^2 + x_2^2 + 2x_3^2)$ and $L(\mathbf{x}) = f(\mathbf{x}) + \lambda g(\mathbf{x})$. The Lagrange multiplier λ is yet to be determined. From the multiplier rule we obtain three equations

$$\frac{\partial L}{\partial x_1}\bigg|_{\mathbf{x}=\mathbf{x}^*} = 1 - 2\lambda x_1^* = 0$$

$$\frac{\partial L}{\partial x_2}\bigg|_{\mathbf{x}=\mathbf{x}^*} = -1 - 2\lambda x_2^* = 0$$

$$\frac{\partial L}{\partial x_3}\bigg|_{\mathbf{x}=\mathbf{x}^*} = 2 - 4\lambda x_3^* = 0.$$

From these and the fourth equation (constraint) $g(\mathbf{x}^*) = 0$, we obtain

$$x_1^* = \frac{1}{2\lambda}, \quad x_2^* = -\frac{1}{2\lambda}, \quad x_3^* = \frac{1}{2\lambda}, \quad \frac{1}{\lambda} = \pm\sqrt{2}.$$

Therefore $\mathbf{x}^* = \pm(\sqrt{2}/2)(\mathbf{e}_1 - \mathbf{e}_2 + \mathbf{e}_3)$ depending on which of the two possible values for λ is used. Here $\mathbf{e}_1$, $\mathbf{e}_2$ and $\mathbf{e}_3$ denote the standard basis in $\mathbb{R}^3$. Since f is continuous and M is a compact set, f has a maximum and a minimum value on M. One of the two critical points obtained by the multiplier rule must give the maximum and the other the minimum. Since

$$f[(\sqrt{2}/2)(\mathbf{e}_1 - \mathbf{e}_2 + \mathbf{e}_3)] = 2\sqrt{2}$$

and

$$f[-(\sqrt{2}/2)(\mathbf{e}_1 - \mathbf{e}_2 + \mathbf{e}_3)] = -2\sqrt{2}$$

these numbers are the maximum and minimum values, respectively.

Problem 3. Calculate the greatest volume of a rectangular box that can be securely tied up with a piece of string 360 cm in length. Let a, b, c be the width, length and height respectively of the box. Thus the volume of the box is $V = abc$. The string must pass twice around the width of the parcel and once around the length. Hence the length of the string is

$$S = 2b + 4a + 6c = 360 \text{ cm.} \tag{1}$$

This is the constraint. Apply the Lagrange multiplier method.

Solution 3. Thus we have the Lagrange function

$$L(a, b, c) = abc + \lambda(2b + 4a + 6c - 360)$$

where λ is the Lagrange multiplier. Therefore

$$\frac{\partial L}{\partial a} = bc + 4\lambda = 0, \qquad \frac{\partial L}{\partial b} = ac + 2\lambda = 0, \qquad \frac{\partial L}{\partial c} = ab + 6\lambda = 0.$$

From these three equations we obtain $2a^* = b^*$, $b^* = 3c^*$, $2a^* = 3c^*$. Inserting these equations into the constraint (1) yields

$$a^* = 30 \text{ cm,} \qquad b^* = 60 \text{ cm,} \qquad c^* = 20 \text{ cm.}$$

This is obviously a global maximum.

Problem 4. Let $f : \mathbb{R}^n \to \mathbb{R}$, $f(\mathbf{x}) = x_1 x_2 \cdots x_n$ and

$$M := \{ (x_1, x_2, \ldots, x_n) : x_1^2 + x_2^2 + \cdots + x_n^2 = 1 \}.$$

Find the extrema of $f|_M$.

Solution 4. We define $r^2 := x_1^2 + x_2^2 + \cdots + x_n^2$. From the system

$$\frac{\partial}{\partial x_k}(f(\mathbf{x}) + \lambda(1 - r^2)) = 0, \qquad k = 1, 2, \ldots, n$$

we find

$$2\lambda x_k^{*2} = x_1^* x_2^* \cdots x_n^*, \qquad k = 1, 2, \ldots, n \tag{1}$$

and

$$2\lambda = n x_1^* x_2^* \cdots x_n^* \tag{2}$$

where we have taken into account the constraint $r^2 = 1$. From (1) and (2) we obtain $x_k^{*2} = 1/n$, $(k = 1, 2, \ldots, n)$. Thus at the extrema we have $f(\mathbf{x}^*) = \pm\sqrt{1/n^n}$.

Problem 5. Show that the angles α, β, γ of a triangle maximize the function

$$f(\alpha, \beta, \gamma) = \sin(\alpha) \sin(\beta) \sin(\gamma)$$

if and only if the triangle is equilateral. The constraint is $\alpha + \beta + \gamma = \pi$.

Solution 5. Let λ be the Lagrange multiplier. The Lagrangian is

$$L(\alpha, \beta, \gamma) = f(\alpha, \beta, \gamma) + \lambda(\alpha + \beta + \gamma - \pi).$$

Then

$$\frac{\partial L}{\partial \alpha} = 0 \Rightarrow \cos(\alpha^*) \sin(\beta^*) \sin(\gamma^*) + \lambda = 0$$

$$\frac{\partial L}{\partial \beta} = 0 \Rightarrow \sin(\alpha^*) \cos(\beta^*) \sin(\gamma^*) + \lambda = 0$$

$$\frac{\partial L}{\partial \gamma} = 0 \Rightarrow \sin(\alpha^*) \sin(\beta^*) \cos(\gamma^*) + \lambda = 0.$$

We can assume (why?) that $\sin(\alpha^*) \neq 0$, $\sin(\beta^*) \neq 0$, $\sin(\gamma^*) \neq 0$. Thus

$$\cot(\alpha^*) = \cot(\beta^*), \quad \cot(\alpha^*) = \cot(\gamma^*)$$

with the constraint $0 < \alpha^*, \beta^*, \gamma^* < \pi$. We obtain $\alpha^* = \beta^* = \gamma^*$ and therefore $\alpha^* = \beta^* = \gamma^* = \pi/3$. Thus $f(\pi/3, \pi/3, \pi/3) = 3\sqrt{3}/8$. This is obviously a (global) maximum.

Problem 6. Consider two smooth non-intersecting curves

$$f(x, y) = 0, \qquad g(x, y) = 0$$

in the Euclidean plane $\mathbb{E}^2$. Find the necessary conditions for the shortest distance between the two curves. Utilize the Lagrange multiplier method. Apply the equations to the curves (circles)

$$f(x, y) = x^2 + y^2 - 1 = 0, \qquad g(x, y) = (x - 3)^2 + (y - 3)^2 - 1 = 0.$$

Solution 6.   We denote the first curve with the index 1 and the second with the index 2. Thus the square of the Euclidean distance between the two curves is

$$d^2 = (x_1 - x_2)^2 + (y_1 - y_2)^2$$

with the two constraints $f(x_1, y_1) = 0$ and $g(x_2, y_2) = 0$. Thus the Lagrangian is

$$L(x_1, y_1, x_2, y_2) = d^2 + \lambda_1(x_1^2 + y_1^2 - 1) + \lambda_2((x_2 - 3)^2 + (y_2 - 3)^2 - 1).$$

Therefore we have to solve the system of equations

$$\frac{\partial L}{\partial x_1} = 2(x_1 - x_2) + 2\lambda_1 x_1 = 0$$

$$\frac{\partial L}{\partial y_1} = 2(y_1 - y_2) + 2\lambda_1 y_1 = 0$$

$$\frac{\partial L}{\partial x_2} = -2(x_1 - x_2) + 2\lambda_2(x_2 - 3) = 0$$

$$\frac{\partial L}{\partial y_2} = -2(y_1 - y_2) + 2\lambda_2(y_2 - 3) = 0$$

together with the two constraints $x_1^2 + y_1^2 = 1$, $(x_2 - 3)^2 + (y_2 - 3)^2 = 1$. Thus we have six equations with six unknowns $x_1, y_1, x_2, y_2, \lambda_1, \lambda_2$. The six equations can be written as

$$x_1(1 + \lambda_1) = x_2, \quad y_1(1 + \lambda_1) = y_2$$
$$x_2(1 + \lambda_2) = x_1 + 3\lambda_2, \quad y_2(1 + \lambda_2) = y_1 + 3\lambda_2$$
$$x_1^2 + y_1^2 = 1, \quad (x_2 - 3)^2 + (y_2 - 3)^2 = 1.$$

We use the first two equations to eliminate x_2 and y_2. We arrive at the four equations

$$x_1(1 + \lambda_1)(1 + \lambda_2) = x_1 + 3\lambda_2$$
$$y_1(1 + \lambda_1)(1 + \lambda_2) = y_1 + 3\lambda_2$$
$$x_1^2 + y_1^2 = 1$$
$$(x_1(1 + \lambda_1) - 3)^2 + (y_1(1 + \lambda_1) - 3)^2 = 1$$

with the four unknowns $x_1, y_1, \lambda_1, \lambda_2$. We consider two cases. First $(1 + \lambda_1)(1 + \lambda_2) = 1$ yields $\lambda_1 = \lambda_2 = 0$, $x_1 = x_2$, $y_1 = y_2$ which cannot satisfy the constraints. Second $(1 + \lambda_1)(1 + \lambda_2) \neq 1$ yields $x_1 = y_1 = \pm 1/\sqrt{2}$, $x_2 = y_2 = 3 \pm 1/\sqrt{2}$. The shortest distance over the four values is

$$d = \sqrt{18} - 2 \equiv 3\sqrt{2} - 2$$

which is obviously the shortest distance between the two circles.

Problem 7. Show that the Lagrange multiplier method fails for the following problem. Maximize $f(x, y) = -y$ subject to the constraint $g(x, y) = y^3 - x^2$.

Solution 7. From

$$L(x, y, \lambda) = f(x, y) + \lambda g(x, y)$$

we obtain the system of equations $-2\lambda x = 0$, $-1 + 3\lambda y^2 = 0$, $-x^2 + y^3 = 0$. We have to do a case study. For the first equation to be satisfied, we must have $x = 0$ or $\lambda = 0$. If $\lambda = 0$, the second equation is not satisfied. If $x = 0$, the third equation implies $y = 0$, and once again the second equation cannot be satisfied. Using differential forms we find the right solution. We have

$$df \wedge dg = -dy \wedge (3y^2 dy - 2x dx) = -2x dx \wedge dy.$$

Here $\wedge$ is the *exterior product* (also called *wedge product*) which is linear and has the property $dx \wedge dy = -dy \wedge dx$. From $df \wedge dg = 0$ we obtain $x = 0$. Therefore from $y^3 - x^2 = 0$ we obtain $y = 0$ and the maximum $f(0, 0) = 0$.

Problem 8. A number of linear optimization problems may be cast in the form

$$\text{Min}(E) = c_1 x_1 + c_2 x_2 + \cdots + c_n x_n \equiv \sum_{j=1}^{n} c_j x_j$$

with linear constraints

$$\sum_{j=1}^{n} a_{ij} x_j (\leq, =, \geq) b_i, \quad i = 1, 2, \ldots, m; \qquad x_j \geq 0, \quad j = 1, 2, \ldots, n$$

where the coefficients c_j and a_{ij} are constants and the notation $(\leq, =, \geq)$ signifies that any of the three possibilities may hold in any constraint. This is the standard linear programming problem.
Find the minimum of

$$E = -5x_1 - 4x_2 - 6x_3$$

subject to the constraints

$$x_1 + x_2 + x_3 \leq 100, \quad 3x_1 + 2x_2 + 4x_3 \leq 210, \quad 3x_1 + 2x_2 \leq 150$$

with $x_1, x_2, x_3 \geq 0$.

Solution 8.   We apply a *gradient method.* We convert to equalities by introducing three nonnegative *slack variables* x_4, x_5, x_6.

$$x_1 + x_2 + x_3 + x_4 = 100$$
$$3x_1 + 2x_2 + 4x_3 + x_5 = 210 \tag{1}$$
$$3x_1 + 2x_2 + x_6 = 150$$

$$5x_1 + 4x_2 + 6x_3 + E = 0$$

where $x_1, x_2, x_3, x_4, x_5, x_6 \geq 0$. A basic feasible solution to this system is

$$x_4 = 100, \quad x_5 = 210, \quad x_6 = 150, \quad x_1 = 0, \quad x_2 = 0, \quad x_3 = 0.$$

The nonzero variables x_4, x_5, x_6 are the basic variables at this stage. Thus $E = 0$. Now, computing the gradient of E yields

$$\frac{\partial E}{\partial x_1} = -5, \qquad \frac{\partial E}{\partial x_2} = -4, \qquad \frac{\partial E}{\partial x_3} = -6$$

so that improvement in E can be obtained by increasing x_1, x_2, and/or x_3. Since the magnitude of the gradient in the x_3 direction is greatest we move in x_3-direction. We retain x_1 and x_2 as nonbasic (zero). Thus we obtain

$$x_3 + x_4 = 100, \qquad 4x_3 + x_5 = 210.$$

Either x_4 or x_5 must go to zero (since x_3 will be nonzero) while the other remains nonnegative. If x_4 goes to zero, $x_3 = 100$ and $x_5 = -190$, while if x_5 goes to zero $x_3 = 52.5$ and $x_4 = 47.5$. That is

$$x_3 = \min\left(\frac{100}{1}, \frac{210}{4}\right) = 52.5$$

and x_5 is to be eliminated (set to zero) as x_3 is introduced. The calculation is one of dividing the right-hand column by the coefficient of x_3 and comparing. The new basic variables, x_4, x_3, and x_6 each appear in only a single equation. This is done by dividing the second equation by the coefficient of x_3, then multiplying it by the coefficient of x_3 in each of the other equations, and subtracting to obtain an equivalent set of equations. The second equation is chosen because it is the one in which x_5 appears. Thus we find

$$\tfrac{1}{4}x_1 + \tfrac{1}{2}x_2 + x_4 - \tfrac{1}{4}x_5 = 47\tfrac{1}{2}$$
$$\tfrac{3}{4}x_1 + \tfrac{1}{2}x_2 + x_3 + \tfrac{1}{4}x_5 = 52\tfrac{1}{2}$$
$$3x_1 + 2x_2 + x_6 = 150$$

and

$$\frac{1}{2}x_1 + x_2 - \frac{3}{2}x_5 + E = -315.$$

The basic feasible solution is

$$x_1 = 0, \quad x_2 = 0, \quad x_5 = 0, \quad x_4 = 47\frac{1}{2}, \quad x_3 = 52\frac{1}{2}, \quad x_6 = 150.$$

From (4) we obtain $E = -315 - x_1/2 - x_2 + 3x_5/2$ so that the value of E in the basic variables is -315. Repeating the above procedure, we now find that the largest positive coefficient which is that of x_2, and so x_2 is to be the new basic (nonzero) variable. The variable to be eliminated is found from the coefficients of the basic variable x_2 in the form

$$x_2 = \min\left(\frac{47\frac{1}{2}}{\frac{1}{2}}, \frac{52\frac{1}{2}}{\frac{1}{2}}, \frac{150}{2}\right) = 75$$

which corresponds to eliminating x_6. Thus, we now use the Gauss-Jordan procedure to obtain x_2 in the third equation only. The result is

$$-\tfrac{1}{2}x_1 + x_4 - \tfrac{1}{4}x_5 - \tfrac{1}{4}x_6 = 10$$
$$x_3 + \tfrac{1}{4}x_5 - \tfrac{1}{4}x_6 = 15$$
$$\tfrac{3}{2}x_1 + x_2 + \tfrac{1}{2}x_6 = 75$$

and $-x_1 - \tfrac{3}{2}x_5 - \tfrac{1}{2}x_6 + E = -390$. The basic feasible solution is then

$$x_1 = 0, \quad x_5 = 0, \quad x_6 = 0, \quad x_4 = 10, \quad x_3 = 15, \quad x_2 = 75.$$

The corresponding value of E is $E = -390$. There are no positive coefficients in this system, and so this is the minimum.

Problem 9.   Let S be a nonempty convex set in $\mathbb{R}^n$.
Definition 1. A vector $\mathbf{x} \in S$ is called an *extreme point* of S if

$$\mathbf{x} = \lambda \mathbf{x}_1 + (1 - \lambda)\mathbf{x}_2$$

with $\mathbf{x}_1, \mathbf{x}_2 \in S$, and $\lambda \in (0, 1)$ implies $\mathbf{x} = \mathbf{x}_1 = \mathbf{x}_2$.
Definition 2. Let $f : S \to \mathbb{R}$ be convex. Then $\mathbf{y}$ is called a *subgradient* of f at $\bar{\mathbf{x}}$ if $f(\mathbf{x}) \geq f(\bar{\mathbf{x}}) + \mathbf{y}^T(\mathbf{x} - \bar{\mathbf{x}})$ for all $\mathbf{x} \in S$.
(i) Minimize the analytic function

$$f(x_1, x_2) = \left(x_1 - \frac{3}{2}\right)^2 + (x_2 - 5)^2$$

subject to the four inequalities $-x_1 + x_2 \leq 2$, $2x_1 + 3x_2 \leq 11$, $x_1 \geq 0$, $x_2 \geq 0$.
(ii) Solve the optimization problem using the Karush-Kuhn-Tucker method. The constraints $-x_1 + x_2 \leq 2$, $2x_1 + 3x_2 \leq 11$ we have to rewrite as

$$2 + x_1 - x_2 \geq 0, \qquad 11 - 2x_1 - 3x_2 \geq 0.$$

Solution 9.   (i) A convex polyhedral set S is represented by the four inequalities. The extreme points of S are $(0,0)$, $(0,2)$, $(1,3)$, $(11/2,0)$. The function f is a convex function, which gives the square of the distance from the point $(3/2, 5)$.

Let $f : \mathbb{R}^n \to \mathbb{R}$ be a convex function, and S be a nonempty convex set in $\mathbb{R}^n$. Consider the problem to minimize f subject to $\mathbf{x} \in S$. The point $\bar{\mathbf{x}} \in S$ is an optimal solution if and only if f has a subgradient $\mathbf{y}$ at $\bar{\mathbf{x}}$ such that

$$\mathbf{y}^T(\mathbf{x} - \bar{\mathbf{x}}) \geq 0$$

for all $\mathbf{x} \in S$. We have $\nabla f(x_1, x_2) = (2(x_1 - 3/2), 2(x_2 - 5))^T$, where ^{T} denotes transpose. Thus the gradient vector of f at the point $(1, 3)$ is

$$\nabla f(1, 3) = (-1, -4)^T.$$

The vector $(-1, -4)$ makes an angle of $\leq \pi/4$ with each vector of the form $(x_1 - 1, x_2 - 3)$, where $(x_1, x_2) \in S$. Thus the optimality condition given above is verified. Now, suppose that it is claimed that $(0, 0)$ is an optimal point. Note that

$$\nabla f(0, 0) = (-3, -10)^T$$

and for each nonzero $\mathbf{x} \in S$, we have $-3x_1 - 10x_2 < 0$. Hence, the origin could not be an optimal point. Moreover, we can improve f by moving from $\mathbf{0}$ in the direction $\mathbf{x} - \mathbf{0}$ for any $\mathbf{x} \in S$. In this case, the best local direction is $-\nabla f(0, 0)$, that is the direction $(3, 10)$. Thus the solution is $(x_1, x_2) = (1, 3)$.

(ii) The domain is convex and thus we can apply the Karush-Kuhn-Tucker condition. The Lagrangian is

$$L(\mathbf{x}, \boldsymbol{\lambda}) = f(\mathbf{x}) - \lambda_1 x_1 - \lambda_2 x_2 - \lambda_3(2 + x_1 - x_2) - \lambda_4(11 - 2x_1 - 3x_2).$$

Thus we obtain the conditions

$$\frac{\partial L}{\partial x_1} = 0 \Rightarrow 2(x_1 - 3/2) - \lambda_1 - \lambda_3 + 2\lambda_4 = 0$$

$$\frac{\partial L}{\partial x_2} = 0 \Rightarrow 2(x_2 - 5) - \lambda_2 + \lambda_3 + 3\lambda_4 = 0$$

$$(2 + x_1 - x_2) \geq 0, \quad (11 - 2x_1 - 3x_2) \geq 0$$

$$\lambda_3(2 + x_1 - x_2) = 0, \quad \lambda_4(11 - 2x_1 - 3x_2) = 0$$

and

$$x_1 \geq 0, \ x_2 \geq 0, \ \lambda_1 x_1 = 0, \ \lambda_2 x_2 = 0, \ \lambda_1 \geq 0, \ \lambda_2 \geq 0, \ \lambda_3 \geq 0, \ \lambda_4 \geq 0.$$

The solution is $x_1 = 1$, $x_2 = 3$, $\lambda_1 = 0$, $\lambda_2 = 0$, $\lambda_3 = 1$, $\lambda_4 = 1$.

Problem 10. Use the Karush-Kuhn-Tucker to solve the following optimization problem. Minimize

$$f(\mathbf{x}) = (x_1 - 4)^2 + (x_2 - 4)^2$$

subject to the constraints $x_1 \geq 0$, $x_2 \geq 0$, $4 - x_1 - x_2 \geq 0$, $9 - x_1 - 3x_2 \geq 0$.

Solution 10. The domain is convex and thus we can apply the Karush-Kuhn-Tucker condition. The Lagrangian is

$$L(\mathbf{x}, \lambda_1, \lambda_2, \lambda_3, \lambda_4) = f(\mathbf{x}) - \lambda_1 x_1 - \lambda_2 x_2 - \lambda_3(4 - x_1 - x_2) - \lambda_4(9 - x_1 - 3x_2).$$

Thus we obtain the conditions

$$\frac{\partial L}{\partial x_1} = 0 \Rightarrow 2(x_1 - 4) - \lambda_1 + \lambda_3 + \lambda_4 = 0$$

$$\frac{\partial L}{\partial x_2} = 0 \Rightarrow 2(x_2 - 4) - \lambda_2 + \lambda_3 + 3\lambda_4 = 0$$

$$(4 - x_1 - x_2) \geq 0, \quad (9 - x_1 - 3x_2) \geq 0,$$

$$\lambda_3(4 - x_1 - x_2) = 0, \quad \lambda_4(9 - x_1 - 3x_2) = 0$$

and $x_1 \geq 0$, $x_2 \geq 0$, $\lambda_1 x_1 = 0$, $\lambda_2 x_2 = 0$, $\lambda_1 \geq 0$, $\lambda_2 \geq 0$, $\lambda_3 \geq 0$, $\lambda_4 \geq 0$. The solution is $x_1 = 2$, $x_2 = 2$, $\lambda_1 = 0$, $\lambda_2 = 0$, $\lambda_3 = 4$, $\lambda_4 = 0$.

Problem 11. Let B^n be the unit ball in $\mathbb{R}^n$, i.e.

$$B^n := \{(a_1, a_2, \ldots, a_n) \in \mathbb{R}^n : \sum_{j=1}^{n} a_j^2 \leq 1\}.$$

Show that B^n is convex.

Solution 11. Let $\mathbf{a} \in B^n$, $\mathbf{b} \in B^n$ and $c_j := \theta a_j + (1 - \theta)b_j$, where $j = 1, 2, \ldots, n$. Then we have

$$\sum_{j=1}^{n} c_j^2 = \sum_{j=1}^{n}(\theta a_j + (1-\theta)b_j)^2 = \theta^2 \sum_{j=1}^{n} a_j^2 + (1-\theta)^2 \sum_{j=1}^{n} b_j^2 + 2\theta(1-\theta) \sum_{j=1}^{n} a_j b_j.$$

Thus

$$\sum_{j=1}^{n} c_j^2 \leq \theta^2 + (1 - \theta)^2 + 2\theta(1 - \theta) \sum_{j=1}^{n} a_j b_j.$$

Next we prove that $\sum_{j=1}^{n} a_j b_j \leq 1$. Since

$$0 \leq \sum_{j=1}^{n}(a_j - b_j)^2 = \sum_{j=1}^{n} a_j^2 + \sum_{j=1}^{n} b_j^2 - 2 \sum_{j=1}^{n} a_j b_j$$

we have $1 - \sum_{j=1}^{n} a_j b_j \geq 0$. It follows that

$$\sum_{j=1}^{n} c_j^2 \leq \theta^2 + (1-\theta)^2 + 2\theta(1-\theta) = (\theta + (1-\theta))^2 = 1.$$

Problem 12. (i) Let $b > a$. Consider the function $f : \mathbb{R} \to \mathbb{R}$, $f(x) = x^2$ in this interval. Show that f is convex.
(ii) Let A be an $n \times n$ positive semidefinite matrix over $\mathbb{R}$. Show that the positive semidefinite quadratic form $f(\mathbf{x}) = \mathbf{x}^T A \mathbf{x}$ is a convex function throughout $\mathbb{R}^n$.

Solution 12. (i) We have to evaluate the difference

$$d := \theta f(x) + (1-\theta) f(y) - f(\theta x + (1-\theta)y).$$

We have to show that $d \geq 0$ for any $\mathbf{x}$ and $\mathbf{y}$ in C and any θ with $0 \leq \theta \leq 1$. From (1) we find

$$
\begin{aligned}
d &= \theta x^2 + (1-\theta)y^2 - (\theta x + (1-\theta)y)^2 \\
&= \theta x^2 + (1-\theta)y^2 - \theta^2 x^2 - (1-\theta)^2 y^2 - 2\theta(1-\theta)xy \\
&= x^2\theta(1-\theta) + y^2\theta(1-\theta) - 2\theta(1-\theta)xy \\
&= \theta(1-\theta)(x^2 + y^2 - 2xy).
\end{aligned}
$$

Thus $d = \theta(1-\theta)(x-y)^2$. Since $\theta \geq 0$, $1-\theta \geq 0$ and $(x-y)^2 \geq 0$ it follows that $d \geq 0$.
(ii) We define $\tilde{\mathbf{x}} := \lambda \mathbf{x}_2 + (1-\lambda)\mathbf{x}_1$ for any vectors $\mathbf{x}_1, \mathbf{x}_2 \in \mathbb{R}^n$ and $\lambda \in [0,1]$. Thus we obtain

$$
\begin{aligned}
\tilde{\mathbf{x}}^T A \tilde{\mathbf{x}} &= (\lambda \mathbf{x}_2 + (1-\lambda)\mathbf{x}_1)^T A(\lambda \mathbf{x}_2 + (1-\lambda)\mathbf{x}_1) \\
&= (\mathbf{x}_1 + \lambda(\mathbf{x}_2 - \mathbf{x}_1))^T A(\mathbf{x}_1 + \lambda(\mathbf{x}_2 - \mathbf{x}_1)) \\
&= \mathbf{x}_1^T A \mathbf{x}_1 + 2\lambda(\mathbf{x}_2 - \mathbf{x}_1)^T A \mathbf{x}_1 + \lambda^2(\mathbf{x}_2 - \mathbf{x}_1)^T A(\mathbf{x}_2 - \mathbf{x}_1).
\end{aligned}
$$

Since $\mathbf{x}^T A \mathbf{x} \geq 0$ for all $\mathbf{x}$ and $\lambda \in [0,1]$, we have $\lambda^2 \mathbf{x}^T A \mathbf{x} \leq \lambda \mathbf{x}^T A \mathbf{x}$ for all $\mathbf{x} \in \mathbb{R}^n$. Thus we can write

$$
\begin{aligned}
\tilde{\mathbf{x}}^T A \tilde{\mathbf{x}} &\leq \mathbf{x}_1^T A \mathbf{x}_1 + 2\lambda(\mathbf{x}_2 - \mathbf{x}_1)^T A \mathbf{x}_1 + \lambda(\mathbf{x}_2 - \mathbf{x}_1)^T A(\mathbf{x}_2 - \mathbf{x}_1) \\
&= \mathbf{x}_1^T A \mathbf{x}_1 + \lambda(\mathbf{x}_2 - \mathbf{x}_1)^T A \mathbf{x}_1 + \lambda(\mathbf{x}_2 - \mathbf{x}_1)^T A \mathbf{x}_2 \\
&= \lambda \mathbf{x}_2^T A \mathbf{x}_2 + (1-\lambda)\mathbf{x}_1^T A \mathbf{x}_1.
\end{aligned}
$$

Problem 13. Let $\mathbb{E}^n$ be the n-dimensional Euclidean space. Let S be a closed convex set in $\mathbb{E}_n$ and $\mathbf{y} \notin S$.

(i) Show that there exists a unique point $\bar{\mathbf{x}} \in S$ with a minimum distance from $\mathbf{y}$.

(ii) Show that $\bar{\mathbf{x}}$ is the minimizing point if and only if $(\mathbf{x} - \bar{\mathbf{x}})^T(\bar{\mathbf{x}} - \mathbf{y}) \geq 0$ for all $\mathbf{x} \in S$.

Solution 13. (i) Let

$$\inf\{\,\|\mathbf{y} - \mathbf{x}\| : \mathbf{x} \in S\,\} = \gamma > 0.$$

There exists a sequence $\{\mathbf{x}_k\}$ in S such that $\|\mathbf{y} - \mathbf{x}_k\| \to \gamma$. We show that $\{\mathbf{x}_k\}$ has a limit $\bar{\mathbf{x}} \in S$ by showing that $\{\mathbf{x}_k\}$ is a *Cauchy sequence*. By the *parallelogram law*, we have

$$\|\mathbf{x}_k - \mathbf{x}_m\|^2 = 2\|\mathbf{x}_k - \mathbf{y}\|^2 + 2\|\mathbf{x}_m - \mathbf{y}\|^2 - \|\mathbf{x}_k + \mathbf{x}_m - 2\mathbf{y}\|^2.$$

It follows that

$$\|\mathbf{x}_k - \mathbf{x}_m\|^2 = 2\|\mathbf{x}_k - \mathbf{y}\|^2 + 2\|\mathbf{x}_m - \mathbf{y}\|^2 - 4\,\|(\mathbf{x}_k + \mathbf{x}_m)/2 - \mathbf{y}\|^2.$$

Note that $(\mathbf{x}_k + \mathbf{x}_m)/2 \in S$, and by definition of γ we have

$$\|(\mathbf{x}_k + \mathbf{x}_m)/2 - \mathbf{y}\|^2 \geq \gamma^2.$$

Therefore $\|\mathbf{x}_k - \mathbf{x}_m\|^2 \leq 2\|\mathbf{x}_k - \mathbf{y}\|^2 + 2\|\mathbf{x}_m - \mathbf{y}\|^2 - 4\gamma^2$. By choosing k and m sufficiently large, $\|\mathbf{x}_k - \mathbf{y}\|^2$ and $\|\mathbf{x}_m - \mathbf{y}\|^2$ can be made sufficiently close to γ^2, hence $\|\mathbf{x}_k - \mathbf{x}_m\|^2$ can be made sufficiently close to zero. Therefore $\{\mathbf{x}_k\}$ is a Cauchy sequence and has a limit $\bar{\mathbf{x}}$. Since S is closed, $\bar{\mathbf{x}} \in S$. To show uniqueness, suppose that there is an $\bar{\mathbf{x}}' \in S$ such that

$$\|\mathbf{y} - \bar{\mathbf{x}}\| = \|\mathbf{y} - \bar{\mathbf{x}}'\| = \gamma.$$

As a result of the convexity of S, we have $\frac{1}{2}(\bar{\mathbf{x}} + \bar{\mathbf{x}}') \in S$. By the *Schwarz inequality*, we obtain

$$\|\mathbf{y} - (\bar{\mathbf{x}} + \mathbf{x}')/2\| \leq \frac{1}{2}\|\mathbf{y} - \bar{\mathbf{x}}\| + \frac{1}{2}\|\mathbf{y} - \bar{\mathbf{x}}'\| = \gamma.$$

If strict inequality holds, we violate the definition of γ. Therefore equality holds, and we must have $\mathbf{y} - \bar{\mathbf{x}} = \lambda(\mathbf{y} - \bar{\mathbf{x}}')$ for some λ. Since

$$\|\mathbf{y} - \bar{\mathbf{x}}\| = \|\mathbf{y} - \bar{\mathbf{x}}'\| = \gamma$$

it follows that $|\lambda| = 1$. Clearly, $\lambda \neq -1$, because otherwise $\mathbf{y} = (\bar{\mathbf{x}} + \bar{\mathbf{x}}')/2 \in S$, contradicting the assumption that $\mathbf{y} \notin S$. So $\lambda = 1$, $\bar{\mathbf{x}}' = \bar{\mathbf{x}}$, and uniqueness is established.

(ii) We need to show that $(\mathbf{x} - \bar{\mathbf{x}})^T(\bar{\mathbf{x}} - \mathbf{y}) \geq 0$ for all $\mathbf{x} \in S$ is both a necessary and sufficient condition for $\bar{\mathbf{x}}$ to be the point in S closest to $\mathbf{y}$. To prove sufficiency, let $\mathbf{x} \in S$. Then,

$$\|\mathbf{y} - \mathbf{x}\|^2 = \|\mathbf{y} - \bar{\mathbf{x}} + \bar{\mathbf{x}} - \mathbf{x}\|^2 = \|\mathbf{y} - \bar{\mathbf{x}}\|^2 + \|\bar{\mathbf{x}} - \mathbf{x}\|^2 + 2(\bar{\mathbf{x}} - \mathbf{x})^T(\mathbf{y} - \bar{\mathbf{x}}).$$

Since $\|\bar{\mathbf{x}} - \mathbf{x}\|^2 \geq 0$ and $(\bar{\mathbf{x}} - \mathbf{x})^T(\mathbf{y} - \bar{\mathbf{x}}) \geq 0$ by assumption,

$$\|\mathbf{y} - \mathbf{x}\|^2 \geq \|\mathbf{y} - \bar{\mathbf{x}}\|^2$$

and $\bar{\mathbf{x}}$ is the minimizing point. Conversely, assume that $\|\mathbf{y} - \mathbf{x}\|^2 \geq \|\mathbf{y} - \bar{\mathbf{x}}\|^2$ for all $\mathbf{x} \in S$. Let $\mathbf{x} \in S$ and note that $\bar{\mathbf{x}} + \lambda(\mathbf{x} - \bar{\mathbf{x}}) \in S$ for $\lambda > 0$ and sufficiently small. Therefore, $\|\mathbf{y} - \bar{\mathbf{x}} - \lambda(\mathbf{x} - \bar{\mathbf{x}})\|^2 \geq \|\mathbf{y} - \bar{\mathbf{x}}\|^2$. Also

$$\|\mathbf{y} - \bar{\mathbf{x}} - \lambda(\mathbf{x} - \bar{\mathbf{x}})\|^2 = \|\mathbf{y} - \bar{\mathbf{x}}\|^2 + \lambda^2 \|\mathbf{x} - \bar{\mathbf{x}}\|^2 + 2\lambda(\mathbf{x} - \bar{\mathbf{x}})^T(\bar{\mathbf{x}} - \mathbf{y}).$$

From these two equations we obtain

$$\lambda^2 \|\mathbf{x} - \bar{\mathbf{x}}\|^2 + 2\lambda(\mathbf{x} - \bar{\mathbf{x}})^T(\bar{\mathbf{x}} - \mathbf{y}) \geq 0$$

for all $\lambda > 0$ and sufficiently small. Dividing by $\lambda > 0$ and letting $\lambda \to 0$, the result follows.

Problem 14.   (i) Show that the intersection of convex sets is convex. (ii) Show that the Cartesian product $I_1 \times \ldots \times I_n \subset \mathbb{R}^n$ of the intervals

$$I_j := \{\, x \in \mathbb{R} : a_j \leq x \leq b_j \,\}, \quad j = 1, 2, \ldots, n$$

is convex.

Solution 14.   (i) Let S and T be two convex sets and a and b be in $S \cap T$. Then for any $0 \leq \theta \leq 1$, $\theta a + (1 - \theta)b$ is in S, since S is convex, and $\theta a + (1 - \theta)b$ is in T, since T is convex. Hence $(\theta a + (1 - \theta)b) \in S \cap T$. Therefore $S \cap T$ is convex.
(ii) Let $\mathbf{x} = (x_1, \ldots, x_n)$, $\mathbf{y} = (y_1, \ldots, y_n)$ with

$$a_j \leq x_j \leq b_j, \qquad a_j \leq y_j \leq b_j, \qquad j = 1, \ldots, n.$$

Let $0 \leq \theta \leq 1$. Then if $\mathbf{z} := \theta\mathbf{x} + (1 - \theta)\mathbf{y}$ we have

$$z_j = \theta x_j + (1 - \theta)y_j \leq \theta b_j + (1 - \theta)b_j = b_j, \qquad j = 1, \ldots, n$$
$$z_j = \theta x_j + (1 - \theta)y_j \geq \theta a_j + (1 - \theta)a_j = a_j, \qquad j = 1, \ldots, n.$$

Hence $a_j \leq z_j \leq b_j$, $(j = 1, \ldots, n)$ and $\mathbf{z} \in I_1 \times \cdots \times I_n$. Thus $I_1 \times \cdots \times I_n$ is convex.

Problem 15.   Let $y = f(x)$ be a convex function. Assume that f is twice differentiable. Thus $f''(x) > 0$ for all $x \in \mathbb{R}$. The *Legendre transformation* of the function f is a new function g of a new variable p as a maximum with respect to x at the point $x(p)$. Now we define

$$g(p) := F(p, x(p)). \tag{1}$$

The point $x(p)$ is defined by the extremal condition $\partial F/\partial x = 0$ i.e. $f'(x) = p$. Since f is convex, the point $x(p)$ is unique.
Let $f(x) = x^2$. Show that $F(p, x) = px - x^2$, $x(p) = \frac{1}{2}p$, $g(p) = \frac{1}{4}p^2$.

Solution 15.   Obviously, f is convex, since $d^2f/dx^2 = 2 > 0$ for all $x \in \mathbb{R}$. Now

$$F(p, x(p)) = px - x^2$$

and $\partial F/\partial x = p - 2x$ or $p = 2x$. Therefore

$$g(p) = F(p, x(p)) = p\frac{p}{2} - \left(\frac{p}{2}\right)^2 = \frac{1}{4}p^2.$$

Problem 16.   Let $X \subset \mathbb{R}$ be an interval. A function $\psi : X \to \mathbb{R}$ is *convex* if for all $x_1, x_2 \in X$ and numbers $\alpha_1, \alpha_2 \geq 0$ with $\alpha_1 + \alpha_2 = 1$,

$$\psi(\alpha_1 x_1 + \alpha_2 x_2) \leq \alpha_1 \psi(x_1) + \alpha_2 \psi(x_2). \tag{1}$$

This means that every chord of the graph of ψ lies above the graph. Let $\psi : X \to \mathbb{R}$ be convex, let $x_1, x_2, \ldots, x_n \in X$, and let $\alpha_1, \alpha_2, \ldots, \alpha_n \geq 0$ satisfy $\sum_{j=1}^{n} \alpha_j = 1$. Show that (*Jensen's inequality*)

$$\psi\left(\sum_{i=1}^{n} \alpha_i x_i\right) \leq \sum_{i=1}^{n} \alpha_i \psi(x_i). \tag{2}$$

Solution 16.   For $n \geq 3$ and $\alpha_n \neq 1$ we use the inductive step

$$\psi\left(\sum_{i=1}^{n} \alpha_i x_i\right) = \psi\left((1 - \alpha_n)\sum_{i=1}^{n-1} \alpha_i(1 - \alpha_n)^{-1}x_i + \alpha_n x_n\right)$$

$$\leq (1 - \alpha_n)\psi\left(\sum_{i=1}^{n-1} \alpha_i(1 - \alpha_n)^{-1}x_i\right) + \alpha_n\psi(x_n)$$

by (1). Thus (2) follows from the inequality for $n - 1$, since

$$\sum_{i=1}^{n-1} \alpha_i(1 - \alpha_n)^{-1} = 1.$$

Problem 17.   Consider the one-dimensional map $f : [0, 1] \to [0, 1]$

$$x_{t+1} = f(x_t) \tag{1}$$

where $t = 0, 1, 2, \ldots$ and $x_0 \in [0, 1]$. The n-th moment of the time evolution of x_t is defined as

$$\langle x_t^n \rangle := \lim_{T \to \infty} \frac{1}{T} \sum_{t=0}^{T} x_t^n \tag{2}$$

where $n = 1, 2, \ldots$. The moments depend on the initial conditions. In the case of ergodic systems, the moments and the probability density ρ are related by

$$\langle x_t^n \rangle = \int_0^1 y^n \rho(y) dy \tag{3}$$

where $\rho > 0$ for $x \in [0, 1]$ and

$$\int_0^1 \rho(x) dx = 1. \tag{4}$$

We can calculate the Ljapunov exponent as follows

$$\lambda = \int_0^1 \rho(x) \ln \left| \frac{df}{dx} \right| dx. \tag{5}$$

The missing information function (*entropy*) of a probability density ρ is defined as

$$I = - \int_0^1 \rho(x) \ln(\rho(x)) dx. \tag{6}$$

Apply the *maximum entropy formalism* to obtain the probability density approximately using as information, N moments. Apply it to the logistic map $f(x) = 4x(1 - x)$ with $N = 2$.

Solution 17.   In the maximum entropy formalism, one maximizes the missing information subject to the constraints of the available information and to the normalization of the probability density. We assume that we have the N lowest moments. The constraints are introduced via the method of the *Lagrange multipliers* $\lambda_1, \lambda_2, \ldots, \lambda_N$. The aim is to find the approximate probability density ρ_{app} which minimizes

$$I' = - \int_0^1 \rho_{app} \ln \rho_{app} \, dx + \lambda_0 \left(1 - \int_0^1 \rho_{app} dx \right) + \sum_{n=1}^{N} \lambda_n \left(\langle x^n \rangle - \int_0^1 x^n \rho_{app} dx \right)$$

where λ_n, $n = 0, 1, \ldots, N$ are the Lagrange multipliers. Performing the minimization we obtain

$$\rho_{app}(x) = \exp \left(-1 - \sum_{n=0}^{N} \lambda_n x^n \right) \equiv \frac{1}{Z} \exp \left(- \sum_{n=1}^{N} \lambda_n x^n \right)$$

where $Z = \exp(1 + \lambda_0)$. From the normalization condition, we obtain

$$1 = \frac{1}{Z} \int_0^1 \exp\left(-\sum_{n=1}^{N} \lambda_n x^n\right) dx.$$

The remaining Lagrange multipliers are obtained by solving the following set of N coupled nonlinear equations for λ_m, $m = 1, 2, \ldots, N$,

$$\langle x^m \rangle = \frac{1}{Z} \int_0^1 x^m \exp\left(-\sum_{n=1}^{N} \lambda_n x^n\right) dx, \qquad m = 1, 2, \ldots, N.$$

For the *logistic map* $f(x) = 4x(1 - x)$, the moments are given by

$$\langle x^n \rangle = \frac{1}{2^{2n}} \binom{2n}{n}.$$

Thus $\langle x \rangle = 1/2$ and $\langle x^2 \rangle = 3/8$. Thus we have to solve

$$1 = \int_0^1 \exp(-1 - \lambda_0 - \lambda_1 x - \lambda_2 x^2) dx$$

$$\frac{1}{2} = \int_0^1 \exp(-1 - \lambda_0 - \lambda_1 x - \lambda_2 x^2) dx$$

$$\frac{3}{8} = \int_0^1 \exp(-1 - \lambda_0 - \lambda_1 x - \lambda_2 x^2) dx.$$

We solve this system numerically and find

$$\lambda_0 = 2.69242, \qquad \lambda_1 = -6.76825, \qquad \lambda_2 = -\lambda_1.$$

Obviously $\lambda_1 = -\lambda_2$.

Problem 18. Let $A = (a_{jk})_{j,k=1}^{n}$ be an $n \times n$ symmetric matrix over $\mathbb{R}$. The eigenvalue problem is given by $A\mathbf{x} = \lambda\mathbf{x}$, $\mathbf{x} \neq \mathbf{0}$, where we assume that the eigenvectors are normalized, i.e. $\mathbf{x}^*\mathbf{x} = 1$. Thus

$$\sum_{k=1}^{n} a_{jk} x_k = \lambda x_j, \qquad j = 1, \ldots, n.$$

Consider the function

$$f(\mathbf{x}) = \sum_{j,k=1}^{n} a_{jk} x_j x_k.$$

To obtain the maximum and minimum of f subject to the constraint

$$\sum_{j=1}^{n} x_j^2 = 1$$

we consider the Lagrange function

$$L(\mathbf{x}) = f(\mathbf{x}) - \lambda \sum_{j=1}^{n} x_j^2.$$

Show that the largest and smallest eigenvalue of A are the maximum and minimum of the function f subject to the constraint $\sum_{j=1}^{n} x_j^2 = 1$.

Solution 18.   From

$$\frac{\partial L}{\partial x_j} = 0 \;\Rightarrow\; 2 \sum_{k=1}^{n} a_{jk} x_k - 2\lambda x_j = 0, \quad j = 1, \ldots, n$$

we obtain

$$\sum_{k=1}^{n} a_{jk} x_{k0} = \lambda x_{j0}, \quad j = 1, \ldots, n.$$

Hence $\mathbf{x}_0$ is a constrained critical point of f subject to $\sum_{j=1}^{n} x_j^2 = 1$ iff $A\mathbf{x}_0 = \lambda \mathbf{x}_0$ for some λ, i.e. λ is an eigenvalue and $\mathbf{x}_0$ is a normalized eigenvector. It follows that

$$f(\mathbf{x}_0) = \sum_{j=1}^{n} \left(\sum_{k=1}^{n} a_{jk} x_{k0} \right) x_{j0} = \sum_{j=1}^{n} (\lambda x_{j0}) x_{j0} = \lambda \sum_{j=1}^{n} x_{j0}^2 = \lambda.$$

Consequently, the largest and smallest eigenvalues of A are the maximum and minimum of f subject to the constrain $\sum_{j=1}^{n} x_{j0}^2 = 1$.

Supplementary Problems

Problem 1.   (i) Let $k > 0$. Find the minima and maxima of the function $f : \mathbb{R} \to \mathbb{R}$, $f(x) = \ln(\cosh(kx))$.
(ii) Let $a > 0$, $b > 0$. Find the minima and maxima of the analytic function $f : \mathbb{R} \to \mathbb{R}$, $f(x) = \exp(-ax^2 - bx^4)$.

Problem 2.   Let $a_1, a_2, a_3 > 0$. The equation $x_1^2/a_1^2 - x_2^2/a_2^2 - x_3^2/a_3^2 = 1$ defines a *hyperboloid* of two sheets. Find the shortest distance between the two disconnected sheets.

Problem 3.   Consider the normalized vector $\mathbf{v}$ in $\mathbb{C}^4$ (four dimensional *spherical coordinates*)

$$\mathbf{v} = \begin{pmatrix} \cos(\theta_1) \\ \sin(\theta_1)\cos(\theta_2) \\ \sin(\theta_1)\sin(\theta_2)\cos(\theta_3) \\ \sin(\theta_1)\sin(\theta_2)\sin(\theta_3) \end{pmatrix}$$

and

$$K = \frac{1}{\sqrt{2}} \begin{pmatrix} 1 & 0 & 0 & 1 \\ 0 & 1 & 1 & 0 \\ 0 & 1 & -1 & 0 \\ 1 & 0 & 0 & -1 \end{pmatrix}.$$

Find the function $f(\theta_1, \theta_2, \theta_3) = \mathbf{v}^* K \mathbf{v}$ and minimize.

Problem 4. Minimize $\mathbf{c}^T \mathbf{x}$ subject to $A\mathbf{x} = \mathbf{b}$, where $\mathbf{c}, \mathbf{x}, \mathbf{b} \in \mathbb{R}$ and A is an $n \times n$ matrix over $\mathbb{R}$.

Problem 5. (i) Minimize

$$\sum_{j=1}^{n} x_j \quad \text{subject to} \quad \prod_{j=1}^{n} x_j = 1$$

with $x_j \geq 0$ for $j = 1, \ldots, n$.
(ii) Maximize

$$\prod_{j=1}^{n} x_j \quad \text{subject to} \quad \sum_{j=1}^{n} x_j$$

with $x_j \geq 0$ for $j = 1, \ldots, n$.

Problem 6. Let A, B be $n \times n$ matrices and $\mathbf{v}_1$, $\mathbf{v}_2$, $\ldots$, $\mathbf{v}_n$ be the normalized eigenvectors of A which should form an orthonormal basis in $\mathbb{C}^n$ and analogously $\mathbf{u}_1$, $\mathbf{u}_2$, $\ldots$, $\mathbf{u}_n$ are the normalized eigenvectors of B which should form a orthonormal basis in $\mathbb{C}^n$. The quantity

$$\max_{j,k=1,\ldots,n} |\langle v_j | u_k \rangle|^2$$

plays a role for entropic inequalities. Let $n = 4$. Find this quantity for the two matrices

$$B = \frac{1}{\sqrt{2}} \begin{pmatrix} 1 & 0 & 0 & 1 \\ 0 & 1 & 1 & 0 \\ 0 & 1 & -1 & 0 \\ 1 & 0 & 0 & -1 \end{pmatrix}, \quad C = \frac{1}{\sqrt{2}} \begin{pmatrix} 1 & 1 \\ 1 & -1 \end{pmatrix} \otimes \frac{1}{\sqrt{2}} \begin{pmatrix} 0 & 1 \\ 1 & 0 \end{pmatrix}.$$

Problem 7. Consider the Hamilton operator

$$\hat{H} = -\frac{\hbar^2}{2\mu} \sum_{j=1}^{N} \frac{\partial^2}{\partial \hat{r}_j^2} + \sum_{(j,k)} \hat{V}(\hat{r}_{jk})$$

where the summation (i, k) is over all different pairs of atoms. The potential $\hat{V}$ is the *Lennard-Jones potential*

$$\hat{V}(\hat{r}_{jk}) = 4\epsilon \left(\left(\frac{\sigma}{\hat{r}_{jk}} \right)^{12} - \left(\frac{\sigma}{\hat{r}_{jk}} \right)^{6} \right)$$

defined in terms of σ, the diameter of the potential and ϵ the well depth of the potential. Show that the minimum of the potential us given by

$$\hat{r} = 2^{1/6}\sigma.$$

Show that the Hamilton operator can be written in dimensionless form

$$\hat{H} = -\frac{\alpha}{2} \sum_{j=1}^{N} \frac{\partial^2}{\partial r_j^2} + \sum_{(j,k)} V(r_{jk}), \qquad \alpha = \frac{\hbar^2}{2^{1/3}\mu\sigma^2\epsilon}, \qquad r = 2^{1/6}\sigma\hat{r}$$

and V is the dimensionless Lennard-Jones potential

$$V(r_{jk}) = \frac{1}{r_{jk}^{12}} - \frac{2}{r_{jk}^{6}}.$$

Show that $V(r_{jk})$ has a minimum of -1 at $r_{jk} = 1$. Note that $r_{jk} = |\mathbf{r}_k - \mathbf{r}_j|$.

Problem 8.   Given the two-dimensional Euclidean plane $\mathbb{E}^2$. Find a point **p** in the plane whose sum of weighted Euclidean distances from a given set of n points $\mathbf{x}_1, \mathbf{x}_2, \ldots, \mathbf{x}_n$ is minimized. This means

$$\min \sum_{j=1}^{n} w_j \|\mathbf{p} - \mathbf{x}_j\|$$

subject to $\mathbf{p} \in \mathbb{E}^2$ with $w_1, \ldots, w_n$ are given positive numbers.

Problem 9.   Let $N \geq 2$ and $u_\tau \geq 0$. The *harvest problem* is given by

$$f(u_0, \ldots, u_{N-1}) = \max \sum_{\tau=0}^{N-1} \sqrt{u_\tau}$$

subject to the equation of growth $(a > 0)$

$$x_{\tau+1} = ax_\tau - u_\tau, \qquad \tau = 0, 1, \ldots$$

and the constraint $x_0 = x_N$. The initial value x_0 is given. Find f.

Chapter 10

Ordinary Differential Equations

Consider the system of first order differential equations

$$\frac{du_j}{dt} = f_j(u_1, \ldots, u_n, t), \quad j = 1, \ldots, n$$

where it is assumed that the functions f_j are differentiable. If the functions f_j do not depend explicitly on t, then the system is called autonomous. Second order and higher order differential equations can be cast into a system of first order differential equations. In most cases we consider an initial value problem, i.e. given $u_j(t = 0)$ for $j = 1, \ldots, n$.

The superposition principle applies to linear differential equations. If two solutions of the linear differential equation are given, then the sum of the two solutions is also a solution of the linear differential equation. A solution can also be multiplied with a constant to be a solution again.

The *Euler-Lagrange equation* with Lagrange function $L(\mathbf{x}, \dot{\mathbf{x}}, t)$ are given by

$$\frac{\partial L}{\partial x_k} - \frac{d}{dt} \frac{\partial L}{\partial \dot{x}_k} = 0, \quad k = 1, 2, 3.$$

The Hamilton equations of motion with Hamilton function $H(\mathbf{p}, \mathbf{q})$ are given by

$$\frac{dq_k}{dt} = \frac{\partial H}{\partial p_k}, \quad \frac{dp_k}{dt} = -\frac{\partial H}{\partial q_k}.$$

Problem 1.   Find the general solution for the *driven damped harmonic oscillator*

$$\frac{d^2u}{dt^2} + \alpha\frac{du}{dt} + \omega^2 u = k_1\cos(\Omega_1 t) + k_2\cos(\Omega_2 t) \tag{1}$$

where $\alpha > 0$ and $\Omega_1 \neq \Omega_2$. Discuss the dependence on Ω_1 and Ω_2.

Solution 1.   The general solution of equation (1) is given by

$$u = u_h + u_p$$

where u_h is the general solution to the homogeneous equation and u_p is a particular solution to the inhomogeneous equation. We first determine the general solution of the homogeneous differential equation

$$\frac{d^2u}{dt^2} + \alpha\frac{du}{dt} + \omega^2 u = 0. \tag{2}$$

Since (2) is a linear differential equation with constant coefficients we can solve the equation with the exponential ansatz

$$u(t) \propto e^{\lambda t}.$$

Inserting this ansatz into (2) yields the quadratic equation

$$\lambda^2 + \alpha\lambda + \omega^2 = 0.$$

The roots are

$$\lambda_\pm = -\frac{\alpha}{2} \pm \sqrt{\left(\frac{\alpha}{2}\right)^2 - \omega^2}.$$

Consequently, the general solution of the homogeneous part of the linear differential equation is given by one of the following cases

$\omega^2 = \left(\frac{\alpha}{2}\right)^2 : u_h(t) = (C + Dt)e^{-\alpha t/2}$

$\omega^2 < \left(\frac{\alpha}{2}\right)^2 : u_h(t) = Ce^{(-\alpha/2+\sqrt{(\alpha/2)^2-\omega^2})t} + De^{(-\alpha/2-\sqrt{(\alpha/2)^2-\omega^2})t}$

$\omega^2 > \left(\frac{\alpha}{2}\right)^2 : u_h(t) = (C\cos\sqrt{\omega^2 - (\alpha/2)^2}t + D\sin\sqrt{\omega^2 - (\alpha/2)^2}t)e^{-\alpha t/2}$

where C and D are the constants of integration. To find the particular solution to (1) we make the ansatz

$$u_p(t) = A_1\cos(\Omega_1 t) + B_1\sin(\Omega_1 t) + A_2\cos(\Omega_2 t) + B_2\sin(\Omega_2 t).$$

We find the following linear equations for A_1, A_2, B_1 and B_2

$$k_1 = -A_1\Omega_1^2 + \alpha B_1\Omega_1 + \omega^2 A_1, \quad 0 = -B_1\Omega_1^2 - \alpha A_1\Omega_1 + \omega^2 B_1,$$

$$k_2 = -A_2\Omega_2^2 + \alpha B_2\Omega_2 + \omega^2 A_2, \quad 0 = -B_2\Omega_2^2 - \alpha A_2\Omega_2 + \omega^2 B_2.$$

Consequently, we find that the general solution to (1) is given by

$$u(t) = u_h(t) + \sum_{j=1}^{2} \frac{k_j}{(\omega^2 - \Omega_j^2)^2 + \alpha^2 \Omega_j^2} [(\omega^2 - \Omega_j^2)\cos(\Omega_j t) + \alpha \Omega_j \sin(\Omega_j t)].$$

Problem 2. Let $m = 0.1\,kg$. Solve the initial value problem of the linear first order differential equation

$$\frac{dc}{dt} = k(m - c)$$

with $c(t = 0) = 0$. t is the time and k is a positive constant with dimension sec^{-1}.

Solution 2. We have

$$\int \frac{dc}{m - c} = \int k\,dt \;\Rightarrow\; -\ln(m - c) = kt + A.$$

It follows that

$$e^{-\ln(m-c)} = e^{kt+A} \;\Rightarrow\; \frac{1}{m - c} = e^{kt+A} \;\Rightarrow\; m - c(t) = e^{-kt-A}.$$

From the initial condition $c(0) = 0$ we find that $A = -ln(m)$. Thus we find

$$m - c(t) = e^{-kt+\ln(m)} \;\Rightarrow\; m - c(t) = me^{-kt} \;\Rightarrow\; c(t) = m(1 - e^{-kt}).$$

Problem 3. Let ω be a fixed frequency. Find the differential equation that $u(t) = A(\sin(\omega t) + \cos(\omega t))$ satisfies together with its initial conditions.

Solution 3. Since

$$\frac{du}{dt} = A\omega(\cos(\omega t) - \sin(\omega t)), \quad \frac{d^2u}{dt^2} = A\omega^2(-\sin(\omega t) - \cos(\omega t))$$

we obtain the linear second order differential equation

$$\frac{d^2u}{dt^2} = -A\omega^2 u(t)$$

together with the initial conditions $u(0) = A$, $du(0)/dt = A\omega$.

Problem 4. The *motion of a charge q* in an electromagnetic field is given by

$$m\frac{d\mathbf{v}}{dt} = q(\mathbf{E} + \mathbf{v} \times \mathbf{B}) \tag{1}$$

where m denotes the mass and $\mathbf{v}$ the velocity. $\mathbf{E}$ is the *electric field strength* and $\mathbf{B}$ is the *magnetic flux density*. Assume that

$$\mathbf{E} = \begin{pmatrix} E_1 \\ E_2 \\ E_3 \end{pmatrix}, \qquad \mathbf{B} = \begin{pmatrix} B_1 \\ B_2 \\ B_3 \end{pmatrix}$$

are constant fields. Find the solution of the initial value problem.

Solution 4.   Equation (1) can be written in the form

$$\begin{pmatrix} dv_1/dt \\ dv_2/dt \\ dv_3/dt \end{pmatrix} = \frac{q}{m} \begin{pmatrix} 0 & B_3 & -B_2 \\ -B_3 & 0 & B_1 \\ B_2 & -B_1 & 0 \end{pmatrix} \begin{pmatrix} v_1 \\ v_2 \\ v_3 \end{pmatrix} + \frac{q}{m} \begin{pmatrix} E_1 \\ E_2 \\ E_3 \end{pmatrix}.$$

We set

$$B_j \to \frac{q}{m} B_j, \qquad E_j \to \frac{q}{m} E_j.$$

Thus

$$\begin{pmatrix} dv_1/dt \\ dv_2/dt \\ dv_3/dt \end{pmatrix} = \begin{pmatrix} 0 & B_3 & -B_2 \\ -B_3 & 0 & B_1 \\ B_2 & -B_1 & 0 \end{pmatrix} \begin{pmatrix} v_1 \\ v_2 \\ v_3 \end{pmatrix} + \begin{pmatrix} E_1 \\ E_2 \\ E_3 \end{pmatrix}. \qquad (2)$$

Equation (2) is a system of nonhomogeneous linear differential equations with constant coefficients. The solution to the homogeneous equation

$$\begin{pmatrix} dv_1/dt \\ dv_2/dt \\ dv_3/dt \end{pmatrix} = \begin{pmatrix} 0 & B_3 & -B_2 \\ -B_3 & 0 & B_1 \\ B_2 & -B_1 & 0 \end{pmatrix} \begin{pmatrix} v_1 \\ v_2 \\ v_3 \end{pmatrix}$$

is given by

$$\begin{pmatrix} v_1(t) \\ v_2(t) \\ v_3(t) \end{pmatrix} = e^{tM} \begin{pmatrix} v_1(0) \\ v_2(0) \\ v_3(0) \end{pmatrix}$$

where M is the matrix of the right hand side and $v_j(0) = v_j(t = 0)$. The solution of the system of nonhomogeneous linear differential equations can be found with the help of the method called *variation of constants*. One sets

$$\mathbf{v}(t) = e^{tM} \mathbf{f}(t)$$

where $\mathbf{f} : \mathbb{R} \to \mathbb{R}^3$ is some differentiable curve. Then differentiation yields

$$\frac{d\mathbf{v}}{dt} = M e^{tM} \mathbf{f}(t) + e^{tM} \frac{d\mathbf{f}}{dt}.$$

Inserting this ansatz into (2) yields

$$Mv(t) + \mathbf{E} = Me^{tM}\mathbf{f}(t) + e^{tM}\frac{d\mathbf{f}}{dt} = Mv(t) + e^{tM}\frac{d\mathbf{f}}{dt}.$$

Consequently $d\mathbf{f}/dt = e^{-tM}\mathbf{E}$. By integration we obtain

$$\mathbf{f}(t) = \int_0^t e^{-sM}\mathbf{E}\,ds + \mathbf{K}$$

where $\mathbf{K} = (K_1\,K_2\,K_3)^T$. Therefore we obtain the general solution of the initial value problem of the nonhomogeneous system (2), namely

$$\mathbf{v}(t) = e^{tM}\left(\int_0^t e^{-sM}\mathbf{E}\,ds + \mathbf{K}\right)$$

where

$$\mathbf{K} = \begin{pmatrix} v_1(0) \\ v_2(0) \\ v_3(0) \end{pmatrix}.$$

We now have to calculate e^{tM} and e^{-sM}. We find that

$$M^2 = \begin{pmatrix} -B_2^2 - B_3^2 & B_1 B_2 & B_1 B_3 \\ B_1 B_2 & -B_1^2 - B_3^2 & B_2 B_3 \\ B_1 B_3 & B_2 B_3 & -B_1^2 - B_2^2 \end{pmatrix}$$

and $M^3 = M^2 M = -B^2 M$, where $B^2 = B_1^2 + B_2^2 + B_3^2$ and

$$B := \sqrt{B_1^2 + B_2^2 + B_3^2}.$$

Therefore $M^4 = -B^2 M^2$. Since

$$e^{tM} := \sum_{k=0}^{\infty} \frac{(tM)^k}{k!} = I_3 + \frac{tM}{1!} + \frac{t^2 M^2}{2!} + \frac{t^3 M^3}{3!} + \frac{t^4 M^4}{4!} + \cdots$$

where I_3 denotes the 3×3 unit matrix, we obtain

$$e^{tM} = I_3 + M\left(t - \frac{t^3}{3!}B^2 + \frac{t^5}{5!}B^4 - \cdots\right) + M^2\left(\frac{t^2}{2!} - \frac{t^4}{4!}B^2 + \frac{t^6}{6!}B^4 - \cdots\right).$$

Thus

$$e^{tM} = I_3 + \frac{M}{B}\sin(Bt) + \frac{M^2}{B^2}(1 - \cos(Bt)).$$

With $\sin(-\alpha) = -\sin(\alpha)$ we obtain

$$e^{-sM} = I_3 - \frac{M}{B}\sin(Bs) + \frac{M^2}{B^2}(1 - \cos(Bs)).$$

Since

$$\int_0^t e^{-sM} \mathbf{E} ds = \int_0^t \left(I_3 - \frac{M}{B} \sin(Bs) + \frac{M^2}{B^2} (1 - \cos(Bs)) \right) \mathbf{E} ds$$

$$= \mathbf{E}t + \frac{M\mathbf{E}}{B^2} \cos(Bt) - \frac{M\mathbf{E}}{B^2} + \frac{M^2 \mathbf{E}t}{B^2} - \frac{M^2 \mathbf{E}}{B^3} \sin(Bt)$$

we find as the solution of the initial value problem of system (2)

$$\mathbf{v}(t) = \left(1 + \frac{M^2}{B^2} \right) \mathbf{E}t - \frac{M^2 \mathbf{E}}{B^3} \sin(Bt) + \frac{M\mathbf{v}(0)}{B} \sin(Bt)$$

$$+ \frac{M\mathbf{E}}{B^2} (1 - \cos(Bt)) + \frac{M^2 \mathbf{v}(0)}{B^2} (1 - \cos(Bt)) + \mathbf{v}(0)$$

where $\mathbf{v}(t = 0) \equiv \mathbf{v}(0)$.

Problem 5.   Consider a ball of radius t, mass density ρ_B, volume V and mass M, i.e. $M = V\rho_B$. The ball moves down a liquid of mass density ρ_L and dynamical viscosity η ($meter^{-1} . sec^{-1} . kg$) under the gravitational force of the earth. Solve the initial value problem of the first order ordinary linear differential equation with constant coefficients

$$\frac{dv}{dt} = Mg - \rho_L Vg - \frac{6\pi\eta r}{M} v$$

where the gravitational force, *Archimede law* and *Stokes law* are taken into account. We set $v(t = 0) = 0$.

Solution 5.   We have

$$\frac{dv}{dt} = g \left(1 - \frac{\rho_L}{\rho_B} \right) - \frac{6\pi\eta r}{M} v.$$

Solving first the homogeneous part and then the inhomogeneous part and finally imposing the initial condition provides the solution

$$v(t) = \frac{mg(1 - \rho_l/\rho_B)}{6\pi\eta r} \left(1 - \exp\left(-\frac{6\pi\eta r}{M} t \right) \right).$$

Problem 6.   Solve the linear second order differential equation

$$\frac{d^2 u}{dx^2} + u = 0 \tag{1}$$

with the following boundary conditions

$$u(0) = 1, \qquad u(1) = 1 \tag{2a}$$

$$u(0) = 1, \qquad u(\pi) = -1 \tag{2b}$$

$$u(0) = 1, \qquad u(\pi) = -2. \tag{2c}$$

Discuss the result.

Solution 6. The general solution to (1) is

$$u(x) = C_1 \cos(x) + C_2 \sin(x)$$

where C_1 and C_2 are the constants of integration. Imposing the boundary condition (2a) yields

$$u(0) = 1, \; u(1) = 1 \Rightarrow C_1 = 1, \quad C_2 = \frac{1 - \cos(1)}{\sin(1)}.$$

Thus with the boundary condition (2a) we have one solution. Imposing the boundary condition (2b) yields

$$u(0) = 1, \quad u(\pi) = -1 \Rightarrow C_1 = 1, \quad C_2 = \text{arbitrary}.$$

Since C_2 is arbitrary we have infinitely-many solutions for boundary condition (2b). Imposing the boundary condition (2c) gives

$$u(0) = 1, \quad u(\pi) = -2 \Rightarrow C_1 = 1, \quad \underbrace{-2 = 1(-1) + C_2 \cdot 0}_{\text{cannot be satisfied}}.$$

Consequently, no solution exists for boundary condition (2c). *Boundary value problems* can lead to (a) unique solutions, (b) arbitrarily many solutions, (c) no solution.

Problem 7. Show that the substitution

$$u = -\frac{1}{v}\frac{dv}{dt} \tag{1}$$

reduces the nonlinear differential equation (*Riccati differential equation*)

$$\frac{du}{dt} = u^2 + t, \qquad u(0) = 1 \tag{2}$$

to the linear second-order differential equation

$$\frac{d^2 v}{dt^2} + tv = 0. \tag{3}$$

Find the inverse transformation of (1).

Solution 7.   From (1) we obtain

$$\frac{du}{dt} = \frac{1}{v^2}\left(\frac{dv}{dt}\right)^2 - \frac{1}{v}\frac{d^2v}{dt^2}. \tag{4}$$

Inserting (4) and (1) into (2) yields (3). We have

$$u(t) = -\frac{d}{dt}\ln(v(t)) \implies v(t) = \exp\left(-\int^t u(s)ds\right).$$

Equation (2) has no solution in terms of the elementary functions. *Bessel functions* are needed to solve it.

Problem 8.   Let f be an analytic function of x and u. In *Picard's method* one approximates a solution of a first order differential equation

$$\frac{du}{dx} = f(x, u)$$

with initial conditions $u(x_0) = u_0$ as follows. Integrating both sides yields

$$u(x) = u_0 + \int_{x_0}^{x} f(s, u(s))ds.$$

Now starting with u_0 this formula can be used to approach the exact solution iteratively if the series converges. The next approximation is given by

$$u_{k+1}(x) = u_0 + \int_{x_0}^{x} f(s, u_k(s))ds, \qquad k = 0, 1, 2, \dots.$$

Apply this approach to the linear differential equation $\frac{du}{dx} = x + u$, where $x_0 = 0$, $u(x_0) = 1$.

Solution 8.   Five steps in Picard's method provide

$$u(x) \approx 1 + x + x^2 + \frac{x^3}{3} + \frac{x^4}{12} + \frac{x^5}{60} + \frac{x^6}{720}.$$

What is the radius of convergence of this series?

Problem 9.   Find the solution of the initial value problem

$$\frac{d^3u}{dt^3} + u = 0$$

with $u(t = 0) = u_0$, $du(t = 0)/dt = u_{t0}$, $d^2u(t = 0)/dt^2 = u_{tt0}$ and u is a real-valued function.

Solution 9. From

$$u(t) \propto e^{\lambda t}, \quad du/dt \propto \lambda e^{\lambda t}, \quad d^2u/dt^2 \propto \lambda^2 e^{\lambda t}, \quad d^3u/dt^3 \propto \lambda^3 e^{\lambda t}$$

we obtain the characteristic equation $\lambda^3 + 1 = 0$ with the solutions

$$\lambda_1 = -1, \quad \lambda_2 = -e^{i2\pi/3} \equiv \frac{1}{2}(1 - i\sqrt{3}), \quad \lambda_3 = -e^{i4\pi/3} \equiv \frac{1}{2}(1 + i\sqrt{3}).$$

Thus we have the general solution

$$u(t) = Ae^{-t} + Be^{t/2}e^{-i\sqrt{3}t/2} + Ce^{t/2}e^{i\sqrt{3}t/2}$$

where A, B, C are arbitrary constants. We can also write with

$$B = B_1 + iB_2, \qquad C = C_1 + iC_2$$

(A, B_1, B_2, C_1, C_2 real)

$$u(t) = Ae^{-t} + ((B_1 + C_1)\cos(\sqrt{3}t/2) + (B_2 - C_2)\sin(\sqrt{3}t/2))e^{t/2}$$
$$+ i((B_2 + C_2)\cos(\sqrt{3}t/2) + (C_1 - B_1)\sin(\sqrt{3}t/2))e^{t/2}.$$

Thus we can set $B_2 + C_2 = 0$ and $C_1 - B_1 = 0$. Therefore

$$u(t) = Ae^{-t} + (2B_1\cos(\sqrt{3}t/2) - 2C_2\sin(\sqrt{3}t/2))e^{t/2}.$$

Now we can set $\widetilde{B} = 2B_1$ and $\widetilde{C} = -2C_2$ and arrive at

$$u(t) = Ae^{-t} + \widetilde{B}\cos(\sqrt{3}t/2)e^{t/2} + \widetilde{C}\sin(\sqrt{3}t/2)e^{t/2}.$$

Finally we have to insert the initial conditions which provide

$$u_0 = A + \widetilde{B}, \quad u_{t0} = -A + \frac{1}{2}\widetilde{B} + \frac{\sqrt{3}}{2}\widetilde{C}, \quad u_{tt0} = A - \frac{1}{2}\widetilde{B} + \frac{\sqrt{3}}{2}\widetilde{C}.$$

Problem 10. Consider the linear differential equation

$$\frac{d\mathbf{x}}{dt} = P(t)\mathbf{x}$$

where $P(t)$ is periodic with principal period T and differentiable. Thus T is the smallest positive number for which $P(t + T) = P(t)$ and $-\infty < t < \infty$. Can we conclude that all solutions are periodic? For example, consider

$$\frac{dx}{dt} = (1 + \sin(t))x.$$

Solution 10.   Thus $P(t) = 1 + \sin(t)$ has period 2π, but all solutions are given by

$$x(t) = C \exp(t - \cos(t))$$

where C is any constant, so only the solution $x(t) = 0$ is periodic.

Problem 11.   Solve the initial value problem for the system of linear differential equations

$$\frac{dc_0}{dt} = \frac{1}{2} i\Omega e^{-i\phi} e^{i(\omega - \nu)t} c_1, \quad \frac{dc_1}{dt} = \frac{1}{2} i\Omega e^{i\phi} e^{-i(\omega - \nu)t} c_0$$

where Ω, ω, ν are constant frequencies. Note that the system depends explicitly on the time t. Study the special case $\omega = \nu$.

Solution 11.   Differentiating the first equation with respect to t and inserting the second equation provides a second order differential equation with constant coefficients

$$\frac{d^2 c_0}{dt^2} - i(\omega - \nu)\frac{dc_0}{dt} + \frac{1}{4}\Omega^2 c_0 = 0.$$

This equation can be solved with the exponential ansatz. Using

$$\frac{dc_0(0)}{dt} = \frac{1}{2} i\Omega e^{-i\phi} c_1(0)$$

we obtain the solution

$$c_0(t) = e^{i\Delta t/2}\left(\cos(ft/2) - i\frac{\Delta}{f}\sin(ft/2)\right)c_0(0) + i\frac{\Omega}{f}e^{i\Delta t/2}e^{-i\phi}\sin(ft/2)c_1(0)$$

$$c_1(t) = i\frac{\Omega}{f}e^{-i\Delta t/2}e^{i\phi}\sin(ft/2)c_0(0) + e^{-i\Delta t/2}\left(\cos(ft/2) + i\frac{\Delta}{f}\sin(ft/2)\right)c_1(0)$$

where $\Delta := \omega - \nu$ and $f := \sqrt{\Omega^2 + (\omega - \nu)^2}$. For the special case $\omega = \nu$ we have

$$c_0(t) = \cos(ft/2)c_0(0) + ie^{-i\phi}\sin(ft/2)c_1(0)$$
$$c_1(t) = ie^{i\phi}\sin(ft/2)c_0(0) + \cos(ft/2)c_1(0).$$

Problem 12.   The *time-independent Schrödinger equation* (eigenvalue equation) for a one-particle problem is given by

$$\left(-\frac{\hbar^2}{2m}\left(\frac{\partial^2}{\partial x_1^2} + \frac{\partial^2}{\partial x_2^2} + \frac{\partial^2}{\partial x_3^2}\right) + U(\mathbf{r})\right)u = Eu. \tag{1}$$

Show that if the potential energy $U(\mathbf{r})$ can be written as a sum of functions of a single coordinate, $U(\mathbf{r}) = U_1(x_1) + U_2(x_2) + U_3(x_3)$ then the equation can be decomposed into a set of one-dimensional differential equations of the form

$$\frac{d^2 u_j(x_j)}{dx_j^2} + \frac{2m}{\hbar^2}\left(E_j - U_j(x_j)\right)u_j(x_j) = 0, \qquad j = 1, 2, 3$$

with the help of the *separation ansatz*

$$u(\mathbf{r}) = u_1(x_1)u_2(x_2)u_3(x_3) \tag{2}$$

and $E = E_1 + E_2 + E_3$.

Solution 12. Substituting the separation ansatz (2) into (1) and dividing by $u_1 u_2 u_3$, we obtain

$$\sum_{j=1}^{3}\left(\frac{1}{u_j}\frac{d^2 u_j}{dx_j^2} - \frac{2m}{\hbar^2}U_j(x_j)\right) = -\frac{2m}{\hbar^2}E.$$

Since the terms in each bracket of the sum contain independent variables, the equality can be valid for all (x_1, x_2, x_3) only if each bracket is a constant, i.e. if

$$\frac{1}{u_j}\frac{d^2 u_j}{dx_j^2} - \frac{2m}{\hbar^2}U_j = -\frac{2m}{\hbar^2}E_j, \qquad j = 1, 2, 3$$

where the E_j are constants, and $E = E_1 + E_2 + E_3$.

Problem 13. The nonlinear system of ordinary differential equations

$$\frac{du_1}{dt} = u_1 - u_1 u_2, \qquad \frac{du_2}{dt} = -u_2 + u_1 u_2$$

is a so-called *Lotka-Volterra model*, where $u_1 > 0$ and $u_2 > 0$.
(i) Give an interpretation of the system assuming that u_1 and u_2 are describing species.
(ii) Find the fixed points (time-independent solutions).
(iii) Find the variational equation. Study the stability of the fixed points.
(iv) Find the first integral of the system.
(v) Describe why the solution cannot be given explicitly.

Solution 13. (i) Since the quantities u_1 and u_2 are positive we find the following behaviour: owing to the first term u_1 on the right-hand side of the first equation u_1 is growing. The second term $-u_1 u_2$ describes the interaction of species 1 with species 2. Owing to the minus sign u_1 is decreasing.

For the second equation we have the opposite behaviour. Owing to $-u_2$ on the right-hand side we find that u_2 is decreasing and the interacting part $u_1 u_2$ leads to a growing u_2. Consequently, one expects that the quantities u_1 and u_2 are oscillating.

(ii) The equations which determine the *fixed points* are given by

$$u_1^* - u_1^* u_2^* = 0, \qquad -u_2^* + u_1^* u_2^* = 0.$$

Since $u_1 > 0$ and $u_2 > 0$ we obtain only one fixed point, namely

$$u_1^* = u_2^* = 1.$$

The fixed points are also called *time-independent solutions* or *steady-state solutions* or *equilibrium solutions*.

(iii) The variational equation is given by

$$\frac{dv_1}{dt} = v_1 - u_1 v_2 - u_2 v_1, \qquad \frac{dv_2}{dt} = -v_2 + u_1 v_2 + u_2 v_1.$$

Inserting the fixed point solution $(u_1^*, u_2^*) = (1, 1)$ into this system yields

$$\frac{dv_1}{dt} = -v_2, \quad \frac{dv_2}{dt} = v_1 \quad \Rightarrow \quad \begin{pmatrix} dv_1/dt \\ dv_2/dt \end{pmatrix} = \begin{pmatrix} 0 & -1 \\ 1 & 0 \end{pmatrix} \begin{pmatrix} v_1 \\ v_2 \end{pmatrix}.$$

Thus in a sufficiently small neighbourhood of the fixed point $\mathbf{u}^* = (1, 1)$ we find a periodic solution, since the eigenvalues of the matrix on the right-hand side of this equation are given by i, $-i$.

(iv) We have

$$\frac{du_1}{u_1(1 - u_2)} = \frac{du_2}{u_2(-1 + u_1)} = dt \quad \Rightarrow \quad \frac{(-1 + u_1)du_1}{u_1} = \frac{(1 - u_2)du_2}{u_2}.$$

Integrating this equation leads to the *first integral* $I(\mathbf{u}) = u_1 u_2 e^{-u_1 - u_2}$.

(v) To find the explicit solution we have to integrate

$$\frac{du_1}{u_1(1 - u_2)} = \frac{du_2}{u_2(-1 + u_1)} = \frac{dt}{1}.$$

Owing to the first integral the constant of motion is given by

$$u_1 u_2 e^{-u_1 - u_2} = c$$

where c is a positive constant. This equation cannot be solved with respect to u_1 or u_2. Since $c > 0$ we find that the Lotka-Volterra model oscillates around the fixed point $\mathbf{u}^* = (1, 1)$.

Problem 14. Consider the autonomous system of differential equation

$$\frac{du_1}{dt} = 2u_1^3, \qquad \frac{du_2}{dt} = -(1 + 6u_1^2)u_2. \tag{1}$$

Show that it has negative divergence and a particular solution with *exploding amplitude*. Note that $(u_1^* = 0, u_2^* = 0)$ is a fixed point. Let $t_c < \infty$. If $u_j(t) \to \infty$ as $t \to t_c$, then $u_j(t)$ is said to have an exploding amplitude.

Solution 14. The divergence of system (1) is obviously

$$\frac{\partial}{\partial u_1}(2u_1^3) - \frac{\partial}{\partial u_2}(1 + 6u_1^2)u_2 = 6u_1^2 - 1 - 6u_1^2 = -1.$$

If $u_2(t) = 0$, then $du_2/dt = -(1 + 6u_1^2)u_2$ is satisfied and we obtain

$$\frac{du_1}{dt} = 2u_1^3.$$

This equation admits the particular solution $u_1(t) = 1/(2\sqrt{1-t})$ with $u_1(0) = \frac{1}{2}$ and $0 \le t < 1$. If $t \to t_c = 1$, then $u_1(t) \to \infty$.

Problem 15. Let

$$\frac{d\mathbf{u}}{dt} = L\mathbf{u} + (\mathbf{a} \cdot \mathbf{u})\mathbf{u} \tag{1}$$

be a nonlinear autonomous system of n first-order ordinary differential equations, where L is an $n \times n$ matrix with constant coefficients and

$$\mathbf{a} \cdot \mathbf{u} := a_1 u_1 + a_2 u_2 + \cdots + a_n u_n.$$

Find the solution of the initial-value problem, where $\mathbf{u}(0) = \mathbf{u}_0$.

Solution 15. Method 1. We start from the ansatz

$$\mathbf{u}(\mathbf{u}_0, t) = f(t)\mathbf{v}(\mathbf{u}_0, t). \tag{2}$$

Differentiating (2) with respect to t gives

$$\frac{d\mathbf{u}}{dt} = \frac{df}{dt}\mathbf{v} + f\frac{d\mathbf{v}}{dt}.$$

Inserting this expression into (1) yields

$$\frac{df}{dt}\mathbf{v} + f\frac{d\mathbf{v}}{dt} = Lf\mathbf{v} + f(\mathbf{a} \cdot \mathbf{v})f\mathbf{v} = f(L\mathbf{v}) + f^2(\mathbf{a} \cdot \mathbf{v})\mathbf{v}.$$

Therefore

$$\frac{d\mathbf{v}}{dt} = L\mathbf{v}, \qquad \mathbf{v}(0) = \mathbf{u}_0 \tag{3}$$

and

$$\frac{df}{dt} = f^2 \mathbf{a} \cdot \mathbf{v}, \qquad f(0) = 1. \tag{4}$$

The method of the solution of system (1) is equivalent to the well known method of *variation of constants*. Equation (4) is a *Riccati differential equation*. Solving first the system of linear differential equations (3) and then the nonlinear differential equation (4), we obtain

$$\mathbf{u}(\mathbf{u}_0, t) = \frac{\mathbf{v}(\mathbf{u}_0, t)}{1 - \int_0^t \mathbf{a} \cdot \mathbf{v}(\mathbf{u}_0, \tau) d\tau}.$$

Method 2. The transformation

$$\mathbf{u}(\mathbf{u}_0, t) = \frac{1}{1 - \mathbf{a} \cdot \mathbf{B}} \frac{d\mathbf{B}}{dt}$$

where $\mathbf{a} \cdot \mathbf{B} = a_1 B_1 + a_2 B_2 + \cdots + a_n B_n$ with $d\mathbf{B}(0)/dt = \mathbf{u}_0$, $\mathbf{B}(0) = \mathbf{0}$ is the linearization of system (1) to the system of linear differential equations with constant coefficients

$$\frac{d^2 \mathbf{B}}{dt^2} = L \frac{d\mathbf{B}}{dt}.$$

Problem 16.   The mathematical *pendulum* is described by the nonlinear second order ordinary differential equation

$$\frac{d^2\theta}{dt^2} + \omega^2 \sin(\theta) = 0 \tag{1}$$

where $\omega^2 = g/L$ (L length of the pendulum, g acceleration due to gravity) and θ is the angular displacement of the pendulum from its position of equilibrium. Find the solution of the initial-value problem $\theta(t = 0) = \alpha$ and $d\theta(t = 0)/dt = 0$.

Solution 16.   The constant of motion of equation (1) is given by

$$\frac{1}{2}\left(\frac{d\theta}{dt}\right)^2 - \omega^2 \cos(\theta) = C \tag{2}$$

where C is a constant of integration. The constant of integration is given by the initial condition. We find

$$C = -\left(\frac{g}{L}\cos(\alpha)\right).$$

We solve (2) for $d\theta/dt$, and thus obtain

$$\frac{d\theta}{dt} = \sqrt{\frac{2g}{L}}\sqrt{\cos(\theta) - \cos(\alpha)}. \tag{3}$$

It follows that

$$dt = \sqrt{\frac{L}{2g}} \frac{d\theta}{\sqrt{\cos(\theta) - \cos(\alpha)}}. \tag{4}$$

In order to reduce the right-hand member of this equation to standard form, we introduce $\cos(\theta) =: 1 - 2k^2 \sin^2(\phi)$, $k := \sin(\alpha/2)$ and use the following equations

$$\cos(\theta) - \cos(\alpha) = 2k^2 \cos^2(\phi), \quad \sin(\theta) = 2k \sin(\phi)\sqrt{1 - k^2 \sin^2(\phi)}$$

$$\sin(\theta)d\theta = 4k^2 \sin(\phi) \cos(\phi)d\phi.$$

When these equations are substituted into (4), we obtain

$$dt = \sqrt{\frac{L}{g}} \frac{d\phi}{\sqrt{1 - k^2 \sin^2(\phi)}}. \tag{5}$$

Therefore the time T required for the pendulum to swing from its position of equilibrium at $\theta = 0$ to a displacement of $\theta = \theta_0$ is given by

$$T = \sqrt{\frac{L}{g}} \int_0^{\phi_0} \frac{d\phi}{\sqrt{1 - k^2 \sin^2(\phi)}}$$

where ϕ_0 is obtained from $\sin^2(\phi_0) = \frac{1 - \cos(\theta_0)}{2k^2} \equiv \frac{\sin^2(\theta_0/2)}{k^2}$. Thus

$$\phi_0 = \arcsin\left(\frac{\sin(\theta_0/2)}{k}\right).$$

In terms of elliptic integrals, we can write

$$T = \sqrt{\frac{L}{g}} F(\phi_0, k).$$

The period of the simple pendulum is defined to be the time required to make a complete oscillation between positions of maximum displacement. To determine this position of maximum displacement, we use (3). Consequently

$$\frac{d\theta}{dt} = 2k\sqrt{\frac{g}{L}} \cos(\phi).$$

Since the desired value of θ is that for which $d\theta/dt = 0$, we see that this corresponds to $\phi = \frac{1}{2}\pi$. If we let $P(k)$ be the period of the pendulum, we obtain

$$P(k) = 4\sqrt{\frac{L}{g}} \int_0^{\pi/2} \frac{d\phi}{\sqrt{1 - k^2 \sin^2(\phi)}} = 4\sqrt{\frac{L}{g}} K(k)$$

where $K(k)$ is the *complete elliptic integral* of the first kind. When $k = 0$, this reduces to $P = 2\pi \sqrt{L/g}$. To find the displacement θ as a function of t, we integrate (5). We find

$$t = \sqrt{\frac{L}{g}} \int_0^\phi \frac{d\phi}{\sqrt{1 - k^2 \sin^2(\phi)}}.$$

This equation can be written as

$$\operatorname{sn}\left(t\sqrt{\frac{g}{L}}, k\right) = \sin(\phi) = \frac{1}{k}\sin(\theta/2)$$

from which we obtain

$$\theta(t) = 2\arcsin\left(k\operatorname{sn}\left(t\sqrt{\frac{g}{L}}, k\right)\right), \quad k = \sin(\alpha/2).$$

Problem 17. The *Korteweg-de Vries equation* is given by

$$\frac{\partial u}{\partial t} + u\frac{\partial u}{\partial x} + \beta\frac{\partial^3 u}{\partial x^3} = 0. \tag{1}$$

Find a solution of the form (*traveling wave solution*)

$$u(x, t) = f(kx - \omega t) = f(s) \tag{2}$$

where $s = kx - \omega t$ (*similarity variable*).

Solution 17. From the ansatz (2) we obtain

$$\frac{\partial u}{\partial t} = -\omega\frac{df}{ds}, \quad \frac{\partial u}{\partial x} = k\frac{df}{ds}, \quad \frac{\partial^3 u}{\partial x^3} = k^3\frac{d^3 f}{ds^3}. \tag{3}$$

Inserting (3) into (1) yields the ordinary third order differential equation

$$-\omega\frac{df}{ds} + kf\frac{df}{ds} + k^3\beta\frac{d^3 f}{ds^3} = 0. \tag{4}$$

Integrating (4) once gives

$$-\omega f + \frac{1}{2}kf^2 + k^3\beta\frac{d^2 f}{ds^2} + C_1 = 0.$$

Multiplying the differential equation with df/ds and then integrating again we find the first order differential equation

$$-\frac{1}{2}\omega f^2 + \frac{1}{6}kf^3 + \frac{1}{2}k^3\beta\left(\frac{df}{ds}\right)^2 + C_1 f + C_2 = 0. \tag{5}$$

This differential equation can be written as

$$\left(\frac{df}{ds}\right)^2 = -\frac{1}{3\beta k^2}f^3 + \frac{\omega}{\beta k^3}f^2 - \frac{2C_1}{\beta k^3}f - \frac{2C_2}{k^2}.$$

This differential equation can be solved with the *Jacobi elliptic function* sn(z,k).

Problem 18. Consider the nonlinear autonomous system of ordinary differential equations

$$\frac{du_1}{dt} = u_2 + u_1(1 - u_1^2 - u_2^2), \qquad \frac{du_2}{dt} = -u_1 + u_2(1 - u_1^2 - u_2^2).$$

(i) Find the fixed points.
(ii) Show that the system admits a periodic solution of the form

$$u_1(t) = \sin(\omega t), \qquad u_2(t) = \cos(\omega t).$$

Determine ω.
(iii) Find the solution for $t \to \infty$.

Solution 18. (i) The fixed points are determined by

$$u_2^* + u_1^*(1 - u_1^{*2} - u_2^{*2}) = 0, \qquad -u_1^* + u_2^*(1 - u_1^{*2} - u_2^{*2}) = 0.$$

We find one fixed point, namely $u_1^* = u_2^* = 0$.
(ii) Inserting the periodic solution into the system and applying the identity $\sin^2(\omega t) + \cos^2(\omega t) \equiv 1$ yields $\omega = 1$.
(iii) Introducing the quantity $r^2 := u_1^2 + u_2^2$ and differentiating we obtain the equation of motion

$$\frac{d}{dt}r^2 = 2r^2(1 - r^2).$$

Obviously, we find the time-independent solution $r^2 = 0$ (i.e. $u_1^* = u_2^* = 0$) and the *limit cycle* $r^2 = 1$ as solutions of the differential equations. Let $r^2 \neq 1$ and $r^2 \neq 0$. For $0 < r^2 < 1$, the solution of the above equation is given by

$$r^2(t, r_0^2) = \frac{1}{1 + \left(\frac{1}{r_0^2} - 1\right)e^{-2t}}$$

with $0 < r_0^2 < 1$ and $-\infty < t < \infty$. For $1 < r^2 < \infty$, the solution of the initial value problem is given by

$$r^2(t, r_0^2) = \frac{1}{1 + \left(\frac{1}{r_0^2} - 1\right)e^{-2t}}$$

with $1 < r_0^2 < \infty$ and

$$\frac{1}{2} \ln \left(1 - \frac{1}{r_0^2}\right) < t < \infty.$$

For both cases as $t \to \infty$, we have $r^2(t \to \infty) = 1$. Consequently, the limit cycle is stable. This can also be seen from applying the *Ljapunov theory*. Let $V : \mathbb{R}^2 \to \mathbb{R}$ with $V(\mathbf{u}) = r^2/2$. Then

$$\frac{dV}{dt} = r\frac{dr}{dt} = r^2(1 - r^2).$$

Thus $dV/dt > 0$ for $0 < r^2 < 1$ and $dV/dt < 0$ for $r^2 > 1$.

Problem 19.　The nonlinear system of ordinary differential equations

$$A\frac{dp}{dt} + (C - B)qr = Mg(y_0\gamma'' - z_0\gamma')$$
$$B\frac{dq}{dt} + (A - C)rp = Mg(z_0\gamma - x_0\gamma'')$$
$$C\frac{dr}{dt} + (B - A)pq = Mg(x_0\gamma' - y_0\gamma)$$

$$\frac{d\gamma}{dt} = r\gamma' - q\gamma'', \quad \frac{d\gamma'}{dt} = p\gamma'' - r\gamma, \quad \frac{d\gamma''}{dt} = q\gamma - p\gamma'$$

describes the motion of a *heavy rigid body* about a fixed point, where M denotes the mass and A, B and C are the principle moments of inertia.
(i) Show that

$$I_1 = Ap^2 + Bq^2 + Cr^2 - 2Mg(x_0\gamma + y_0\gamma' + z_0\gamma'')$$
$$I_2 = Ap\gamma + Bq\gamma' + Cr\gamma''$$
$$I_3 = \gamma^2 + \gamma'^2 + \gamma''^2$$

are first integrals of the nonlinear dynamical system.
(ii) Find the conditions on the constants A, B, C, x_0, y_0 and z_0 under which

$$I_4 = A^2p^2 + B^2q^2 + C^2r^2$$
$$I_5 = r$$
$$I_6 = x_0p + y_0q + z_0r$$
$$I_7 = (p^2 - q^2 + c\gamma)^2 + (2pq + c\gamma')^2$$

are first integrals of the nonlinear dynamical system, where $c = Mgx_0/C$.

Solution 19. (i) We recall that I is a *first integral* if $dI/dt = 0$. Straightforward calculation yields

$$\frac{dI_1}{dt} = 2Ap\frac{dp}{dt} + 2Bq\frac{dq}{dt} + 2Cr\frac{dr}{dt} - 2Mg\left(x_0\frac{d\gamma}{dt} + y_0\frac{d\gamma'}{dt} + z_0\frac{d\gamma''}{dt}\right).$$

Inserting the dynamical system yields

$$\frac{dI_1}{dt} = 2p[(B-C)qr + Mg(y_0\gamma'' - z_0\gamma')]$$
$$+2q[(C-A)rp + Mg(z_0\gamma - x_0\gamma'')]$$
$$+2r[(A-B)pq + Mg(x_0\gamma' - y_0\gamma)]$$
$$-2Mg[x_0(r\gamma' - q\gamma'') + y_0(p\gamma'' - r\gamma) + z_0(q\gamma - p\gamma')].$$

Hence we arrive at $dI_1/dt = 0$. Analogously, for I_2 and I_3 we find $dI_2/dt = 0$, $dI_3/dt = 0$.

(ii) The condition that I_4 is a first integral of the nonlinear dynamical system, i.e. $dI_4/dt = 0$ leads to

$$\frac{dI_4}{dt} = 2Mg[\gamma(z_0Bq - y_0Cr) + \gamma'(x_0Cr - z_0Ap) + \gamma''(y_0Ap - x_0Bq)] = 0.$$

Therefore we obtain the condition $x_0 = y_0 = z_0 = 0$. We find that the condition that $I_5 = r$ is a first integral is given by $A = B$, $x_0 = y_0 = 0$. The condition that I_6 is a first integral leads to

$$\frac{dI_6}{dt} = x_0\frac{dp}{dt} + y_0\frac{dq}{dt} + z_0\frac{dr}{dt} = 0.$$

Thus

$$\frac{dI_6}{dt} = \frac{x_0Mg}{A}(y_0\gamma'' - z_0\gamma') + \frac{y_0Mg}{B}(z_0\gamma - x_0\gamma'') + \frac{z_0Mg}{C}(x_0\gamma' - y_0\gamma) = 0.$$

Thus we find $A = B = C$. Taking the time derivative of I_7 we find terms such as

$$4p^3qr\left(\frac{B-C}{A} - \frac{C-A}{B} + 2\frac{C-A}{B}\right) + 4pq^3r\left(\frac{C-B}{A} + \frac{C-A}{B} + 2\frac{B-C}{A}\right).$$

From this we find the condition $A = B = 2C$. Furthermore we find the term

$$4\frac{Mg}{A}p^3(y_0\gamma'' - z_0\gamma').$$

We conclude that $z_0 = y_0 = 0$. All other terms vanish if these conditions are satisfied.

Problem 20.   Consider the eigenvalue equation

$$\frac{d^2\psi}{dx^2} = (V(x) - E)\psi(x) \tag{1}$$

where $V(x) = V(-x)$. Let

$$\Phi(x) := x^{-s}\psi(x) \tag{2}$$

where $s = 0$ or $s = 1$ for even or odd states, respectively. The logarithmic derivative

$$f(x) := -\frac{1}{\Phi(x)}\frac{d\Phi(x)}{dx} \tag{3}$$

is regular at the origin for all eigenstates of (1). Thus we have

$$\Phi(x) = \exp\left(-\int^x f(s)ds\right).$$

(i) Find the differential equation that f satisfies.
(ii) Assume that

$$V(x) = \sum_{j=1}^{K} v_j x^{2j}, \qquad v_K > 0. \tag{4}$$

We expand f in a *Taylor series* around the origin

$$f(x) = \sum_{j=0}^{\infty} f_j x^{2j+1}. \tag{5}$$

Find the condition on the coefficients f_j.

Solution 20.   (i) Inserting (3) into (1) yields

$$\frac{df}{dx} - f^2(x) + 2s\frac{f(x)}{x} = E - V(x). \tag{6}$$

This is a *Riccati differential equation*.
(ii) Inserting the Taylor expansion (5) and the expansion (4) into (6) we find that the coefficients f_j satisfy the condition

$$f_j = \frac{1}{2j + 2s + 1}\left(\sum_{i=0}^{j-1} f_i f_{j-i-1} + E\delta_{j0} - \sum_{i=1}^{K} v_i \delta_{ij}\right)$$

where δ_{ij} denotes the Kronecker delta. The function f can be approximated by a sequence of rational functions

$$g(x) = \frac{A(x)}{B(x)}$$

where

$$A(x) = \sum_{j=0}^{M} a_j x^{2j+1}, \qquad B(x) = \sum_{j=0}^{N} b_j x^{2j}, \quad b_0 = 1.$$

If g is exactly a *Padé approximant*, then $f(x) - g(x) = O(x^{2(M+N)+3})$.

Problem 21. (i) Let $f : \mathbb{R} \to \mathbb{R}$ be an analytic function. Solve the functional equation

$$f(x + y) = f(x) + f(y) \qquad (1)$$

using differentiation.

(ii) Let $f : \mathbb{R} \to \mathbb{R}$ be an analytic function. Solve the functional equation

$$f(x + y) = f(x)f(y) \qquad (2)$$

using differentiation.

(iii) Let $f : \mathbb{R} \to \mathbb{R}$ be an analytic function. Solve the functional equation

$$f(x + y) + f(x - y) = 2f(x)f(y) \qquad (3)$$

using differentiation.

Solution 21. (i) If we set $y = 0$ in (1) we obtain $f(0) = 0$. If we differentiate (1) with respect to x we obtain

$$f'(x + y) = f'(x).$$

Thus $f'(x) = c$, $f(x) = cx + b$, where c and b are constants. Owing to $f(0) = 0$ we obtain $b = 0$ and thus $f(x) = cx$.

(ii) Inserting $y = 0$ into (2) yields $f(x) = 0$ or $f(0) = 1$. We find f with the condition $f(0) = 1$. If we differentiate (2) with respect to x we obtain

$$f'(x + y) = f'(x)f(y).$$

Setting $x = 0$ yields

$$\frac{f'(y)}{f(y)} = f'(0) = c.$$

Thus $f(x) = \exp(cx)$.

(iii) Setting $y = 0$ we find $f(x) = f(x)f(0)$ and (3) admits the trivial solution $f(x) = 0$. For a non-trivial solution of (3) we find that $f(0) = 1$. Setting $x = 0$ we obtain $f(-x) = f(x)$. Taking the second derivative of (3) with respect to x we obtain

$$f''(x + y) + f''(x - y) = 2f''(x)f(y).$$

Taking the second derivative of (3) with respect to y we obtain

$$f''(x+y) + f''(x-y) = 2f(x)f''(y).$$

Thus

$$f''(x)f(y) = f(x)f''(y).$$

It follows that $f''(x) = kf(x)$. For $k = 0$ we obtain $f(x) = 1$, for $k > 0$ we obtain $f(x) = \cosh(\sqrt{k}x)$. For $k < 0$ we obtain $f(x) = \cos(\sqrt{-k}x)$.

Problem 22.   Consider the second-order ordinary differential equation

$$\frac{d^2x}{dt^2} + f(x)\frac{dx}{dt} + g(x) = 0. \tag{1}$$

This equation will have a unique periodic solution provided that

$$\oint F(x)dy = 0, \quad y = \frac{dx}{dt} + F(x) \tag{2}$$

where
1) f is even, g is odd, $xg(x) > 0$ for all $x \neq 0$, and $f(0) < 0$
2) f and g are continuous and g is Lipschitzian
3) $F(x) \to \pm\infty$ as $x \to \pm\infty$, where

$$F(x) := \int_0^x f(s)ds$$

4) F has one single positive zero $x = a$, for $x \geq a$ the function F increases monotonically with x.

Give a physical interpretation of these conditions.

Solution 22.   Equation (1) can be written as

$$\frac{d^2x}{dt^2} = -f(x)\frac{dx}{dt} - g(x)$$

and may be thought of as representing a unit mass acted on by a restoring force $-g(x)$ and a damping force $-f(x)dx/dt$. As the unit mass moves a distance dx, the damper does an amount of work dW given by

$$dW = -f(x)\frac{dx}{dt}dx.$$

Since $f(0) < 0$, we find that $dW/dx > 0$ when the unit mass passes through the origin with a positive velocity. Thus at this point the damper is putting energy into the system. As x increases, $F(x)$ reaches a minimum, at which point $f(x) = 0$. After this point has been reached f becomes positive and the damper removes energy from the system. From physical considerations, we realize the possibility of a limit cycle. When the limit cycle is reached, the sum of all of the elements of work done by the damper in one cycle must be zero. Thus, if the system is in a steady-state oscillation

$$W_{cycle} = \int_{cycle} dW = \int_{cycle} \left(-f(x)\frac{dx}{dt} \right) dx = 0. \tag{3}$$

If we let $dx/dt = v$, then (3) can be integrated by parts

$$(vF(x))_{cycle} - \int_{cycle} F(x)dv = 0.$$

The first term of this equation is zero because the integration is over a cycle. Thus, the condition for a steady-state oscillation becomes

$$\int_{cycle} F(x)dv = 0.$$

This equation can be converted into (2) by introducing the variable y. Thus we can write

$$\int F(x)(dy - f(x)dx) = \int_{cycle} F(x)dy - \int_{cycle} F(x)f(x)dx.$$

The second integral is zero because the integration is over a cycle. Hence we obtain

$$\oint F(x)dy = 0$$

which is the usual curvilinear integral taken along a trajectory.

Problem 23. Consider the differential equation

$$\frac{du}{dt} = f(t, u(t)).$$

Let (t_0, u_0) be a particular pair of values assigned to the real variable (t, u) such that within a rectangular domain D surrounding the point (t_0, u_0) and defined by the inequalities $|t - t_0| \leq a$, $|u - u_0| \leq b$. If (t, u) and (t, U) are two points within D, then the *Lipschitz condition* is

$$|f(t, U) - f(t, u)| < K|(U - u)|$$

where K is a constant. Discuss the solution of the one-dimensional nonlinear differential equation

$$\frac{du}{dt} = -u^{1/3} \tag{1}$$

where $u(t) \geq 0$. Show that the Lipschitz condition is violated.

Solution 23.   The differential equation (1) has a fixed point at $u^* = 0$. Let $f(u) = -u^{1/3}$. Then

$$\frac{df}{du} = -\frac{1}{3}u^{-2/3}.$$

Thus $df/du \to -\infty$ as $u \to 0$. Thus the Lipschitz condition is violated. The fixed point $u^* = 0$ is an attractor with "infinite" local stability. Let $u(t = 0) = u_0 > 0$. Then the time t_f to reach the fixed point (attractor) is finite, i.e.

$$t_f = -\int_{u_0}^{0} \frac{du}{u^{1/3}} = \frac{3}{2}u_0^{3/2} < \infty.$$

Problem 24.   Consider the hypothetical chemical reaction mechanism (*Lotka-Volterra model*)

$$A + X \overset{k_1}{\to} 2X, \quad X + Y \overset{k_2}{\to} 2Y, \quad Y \overset{k_3}{\to} B$$

where X and Y are intermediaries, k_1, k_2, and k_3 are the reaction rate constants, and the concentrations of the reactants A and B are kept constant. Find the kinetic equations. Denote the concentrations of A, B, X, and Y by the same letters for convenience.

Solution 24.   The law of mass action then gives

$$\frac{dA}{dt} = k_1 AX$$

$$\frac{dX}{dt} = -k_1 AX + k_1 X^2 - k_2 XY$$

$$\frac{dY}{dt} = -k_2 XY + k_2 Y^2 - k_3 Y$$

$$\frac{dB}{dt} = k_3 Y.$$

Since A and B are kept constant, the conditions on A and B mean that the system is open and so there must be an exchange of matter with the surroundings. Under this condition the equations reduce to two equations

$$\frac{dX}{dt} = -k_1 AX + k_1 X^2 - k_2 XY, \quad \frac{dY}{dt} = -k_2 XY + k_2 Y^2 - k_3 Y$$

where A, k_1, k_2 and k_3 are constants.

Problem 25.   Let

$$H(\mathbf{p}, \mathbf{q}) = \frac{1}{2}\left(\frac{p_1^2}{m} + \frac{p_2^2}{m}\right) + \frac{1}{2}k(q_2 - q_1)^2 \tag{1}$$

be a Hamilton function, where k is a positive constant. Solve Hamilton's equations of motion for this Hamilton function. *Hamilton's equations of motion* are given by

$$\frac{dq_j}{dt} = \frac{\partial H}{\partial p_j}, \quad \frac{dp_j}{dt} = -\frac{\partial H}{\partial q_j}, \quad j = 1, 2. \tag{2}$$

Solution 25. Inserting the Hamilton function (1) into system (2) yields

$$\frac{dq_1}{dt} = \frac{p_1}{m}, \quad \frac{dq_2}{dt} = \frac{p_2}{m} \tag{3a}$$

$$\frac{dp_1}{dt} = k(-q_1 + q_2), \quad \frac{dp_2}{dt} = k(q_1 - q_2). \tag{3b}$$

This is an autonomous system of first-order linear differential equations with constant coefficients. System (3) can be expressed as

$$A := \begin{pmatrix} 0 & 0 & \frac{1}{m} & 0 \\ 0 & 0 & 0 & \frac{1}{m} \\ -k & k & 0 & 0 \\ k & -k & 0 & 0 \end{pmatrix} \Rightarrow \begin{pmatrix} dq_1/dt \\ dq_2/dt \\ dp_1/dt \\ dp_2/dt \end{pmatrix} = A \begin{pmatrix} q_1 \\ q_2 \\ p_1 \\ p_2 \end{pmatrix}.$$

Thus the general solution of the initial value problem is given by

$$\begin{pmatrix} q_1(t) \\ q_2(t) \\ p_1(t) \\ p_2(t) \end{pmatrix} = e^{tA} \begin{pmatrix} q_1 \\ q_2 \\ p_1 \\ p_2 \end{pmatrix} \Bigg|_{q_1 \to q_1(0), \dots, p_2 \to p_2(0)}.$$

The exponential function $\exp(tA)$ can be evaluated by determining the eigenvalues and normalized eigenvectors of A. The eigenvalues of A are given by $\lambda_{1,2} = 0$, $\lambda_{3,4} = \pm i\omega$, where $\omega^2 = 2k/m$. Then the solution of the initial value problem is given by

$$q_1(t) = \frac{q_{10} - q_{20}}{2} \cos(\omega t) + \frac{p_{10} - p_{20}}{2\omega m} \sin(\omega t) + \frac{p_{10} + p_{20}}{2m} t + \frac{q_{10} + q_{20}}{2}$$

$$q_2(t) = \frac{q_{20} - q_{10}}{2} \cos(\omega t) - \frac{p_{10} - p_{20}}{2\omega m} \sin(\omega t) + \frac{p_{10} + p_{20}}{2m} t + \frac{q_{10} + q_{20}}{2}$$

where $q_{j0} \equiv q_j(t = 0)$ and $p_{j0} \equiv p_j(t = 0)$ are the initial values. The momenta p_1 and p_2 are given by (3a) and (3b), respectively.

Problem 26. Consider the Hamilton function

$$H(p_{x_1}, p_{x_2}, x_1, x_2) = \frac{1}{2} \left(p_1^2 + p_2^2 + (x_1^2 x_2^2)^{1/\alpha} \right) \tag{1}$$

where $\alpha \in [0, 1]$. Discuss the equation of motion for $x_1 \gg x_2$.

Solution 26.   In the limit $\alpha \to 0$ we obtain the hyperbola billiard. Increasing the parameter α means gradual softening of the billiard walls and when $\alpha = 1$ we recover the $x_1^2 x_2^2$ potential. The symmetry group of this family is C_{4v}. For $x_1 \gg x_2$ the motion in the x_1-direction will be much slower than the motion in the x_2-direction. Thus x may be regarded as a slowly varying parameter. In the *adiabatic approximation*, the motion in the x_2-direction is described by the Hamilton function

$$H_{x_2} = \frac{1}{2} \left(p_2^2 + (x_1^2 x_2^2)^{1/\alpha} \right). \tag{2}$$

By the *adiabatic theorem* the action integral in the x_2-direction is given by

$$J_2 = \oint p_2 dx_2 = \frac{(2H_{x_2})^{(1+\alpha)/2} f(2/\alpha)}{x_1} \tag{3}$$

where

$$f(s) := \frac{1}{2\pi} \oint \sqrt{1 - |z|^s} dz. \tag{4}$$

The quantity J_2 is approximately a constant of motion. The full Hamilton function may now be written as

$$H \approx \frac{1}{2} p_1^2 + H_{x_2}(x_1) \tag{5}$$

giving (approximately) the x_1 motion for any J_{x_2}. If we transform the (x_1, p_1) pair to action-angle variables we obtain the following expression for H

$$H \approx \frac{1}{2} \left[\frac{J_1 J_2}{f(2/\alpha) f(2/(\alpha + 1))} \right]^{2/(\alpha+2)}.$$

Although this expression was derived under the asymmetric assumption $x_1 \gg x_2$, the Hamilton function (1) is symmetric in x_1 and x_2. These adiabatic expressions are not valid in the central region.

Problem 27.   Let $a > 0$ with dimension length. A particle with mass m is moving without friction on a surface described by

$$x_3 = \sqrt{x_1^2 + x_2^2 + a^2}$$

with a gravitational force acting in x_3 direction $(0 \quad 0 \quad -mg)^T$. Then the *Lagrange function* is given by

$$L(\dot{x}_1, \dot{x}_2, \dot{x}_3, x_3) = \frac{1}{2} m(\dot{x}_1^2 + \dot{x}_2^2 + \dot{x}_3^2) - mgx_3.$$

Express the Lagrange function in the coordinates ζ, ϕ

$$x_1(\zeta, \phi) = a \sinh(\zeta) \cos(\phi), \quad x_2(\zeta, \phi) = a \sinh(\zeta) \sin(\phi).$$

Give the equation of motion. Find the Hamilton function and the Hamilton equations of motion.

Solution 27.   Since $x_3 = \sqrt{x_1^2 + x_2^2 + a^2} = a \cosh(\zeta)$ it follows that

$$\frac{dx_1}{dt} = a \cosh(\zeta) \cos(\phi) \frac{d\zeta}{dt} - a \sinh(\zeta) \sin(\phi) \frac{d\phi}{dt}$$

$$\frac{dx_2}{dt} = a \cosh(\zeta) \sin(\phi) \frac{d\zeta}{dt} + a \sinh(\zeta) \sin(\phi) \frac{d\phi}{dt}$$

$$\frac{dx_3}{dt} = a \sinh(\zeta) \frac{d\zeta}{dt}$$

and

$$\left(\frac{dx_1}{dt}\right)^2 + \left(\frac{dx_2}{dt}\right)^2 + \left(\frac{dx_3}{dt}\right)^2 =$$

$$a^2 (\cosh^2(\zeta) + \sinh^2(\zeta)) \left(\frac{d\zeta}{dt}\right)^2 + a^2 \sinh^2(\zeta) \left(\frac{d\phi}{dt}\right)^2.$$

Hence the Lagrange function takes the form

$$L(\zeta, \dot{\zeta}, \dot{\phi}) =$$

$$\frac{1}{2} m a^2 \left((\cosh^2(\zeta) + \sinh^2(\zeta)) \left(\frac{d\zeta}{dt}\right)^2 + \sinh^2(\zeta) \left(\frac{d\phi}{dt}\right)^2 \right) - mga \cosh(\zeta).$$

Note that L does not depend on ϕ. Since

$$p_\zeta = \frac{\partial L}{\partial \dot{\zeta}} = m a^2 (\cosh^2(\zeta) + \sinh^2(\zeta)) \frac{d\zeta}{dt}$$

$$\frac{d\zeta}{dt} = \frac{p_\zeta}{m a^2 (\cosh^2(\zeta) + \sinh^2(\zeta))}$$

$$p_\phi = \frac{\partial L}{\partial \dot{\phi}} = m a^2 \sinh^2(\zeta) \frac{d\phi}{dt}$$

$$\frac{d\phi}{dt} = \frac{p_\phi}{m a^2 \sinh^2(\zeta)}.$$

Thus the Hamilton function $H = p_\zeta \dot{\zeta} + p_\phi \dot{\phi} - L$ takes the form

$$H(\zeta, p_\zeta, p_\phi) = \frac{p_\zeta^2}{2m a^2 (\cosh^2(\zeta) + \sinh^2(\zeta))} + \frac{p_\phi^2}{2m a^2 \sinh^2(\zeta)} + mga \cosh(\zeta)$$

and the equation of motions are $dp_\phi/dt = -\partial H/\partial \phi = 0$ and

$$\frac{dp_\zeta}{dt} = -\frac{\partial H}{\partial \zeta} = \frac{2\sinh(\zeta)\cosh(\zeta)}{ma^2(\cosh^2(\zeta)+\sinh^2(\zeta))^2}p_\zeta^2 + \frac{\operatorname{cosech}^2(\zeta)\coth(\zeta)}{ma^2}p_\phi^2$$
$$-mga\sinh(\zeta)$$

$$\frac{d\phi}{dt} = \frac{\partial H}{\partial p_\phi} = \frac{p_\phi}{ma^2\sinh^2(\zeta)}$$

$$\frac{d\zeta}{dt} = \frac{\partial H}{\partial p_\zeta} = \frac{p_\zeta}{ma^2(\cosh^2(\zeta)+\sinh^2(\zeta))}.$$

Problem 28.   Suppose that an elementary magnet is situated at the origin and that its axis corresponds to the x_3-axis. The trajectory

$$(x_1(t), x_2(t), x_3(t))$$

of an electrical particle in this magnetic field is then given as the autonomous system of second order ordinary differential equations

$$\frac{d^2x_1}{dt^2} = \frac{1}{r^5}\left(3x_2x_3\frac{dx_3}{dt} - (3x_3^2 - r^2)\frac{dx_2}{dt}\right)$$

$$\frac{d^2x_2}{dt^2} = \frac{1}{r^5}\left((3x_3^2 - r^2)\frac{dx_1}{dt} - 3x_1x_3\frac{dx_3}{dt}\right)$$

$$\frac{d^2x_3}{dt^2} = \frac{1}{r^5}\left(3x_1x_3\frac{dx_2}{dt} - 3x_2x_3\frac{dx_1}{dt}\right)$$

where $r^2 := x_1^2 + x_2^2 + x_3^2$. Show that the system can be simplified by introducing *polar coordinates* $x_1(t) = R(t)\cos(\phi(t))$, $x_2(t) = R(t)\sin(\phi(t))$.

Solution 28.   Since

$$\frac{dx_1}{dt} = \frac{dR}{dt}\cos(\phi) - R\frac{d\phi}{dt}\sin(\phi)$$

$$\frac{dx_2}{dt} = \frac{dR}{dt}\sin(\phi) + R\frac{d\phi}{dt}\cos(\phi)$$

and

$$\frac{d^2x_1}{dt^2} = \frac{d^2R}{dt^2}\cos(\phi) - 2\frac{dR}{dt}\frac{d\phi}{dt}\sin(\phi) - R\frac{d^2\phi}{dt^2}\sin(\phi) - R\left(\frac{d\phi}{dt}\right)^2\cos(\phi)$$

$$\frac{d^2x_2}{dt^2} = \frac{d^2R}{dt^2}\sin(\phi) + 2\frac{dR}{dt}\frac{d\phi}{dt}\cos(\phi) + R\frac{d^2\phi}{dt^2}\cos(\phi) - R\left(\frac{d\phi}{dt}\right)^2\cos(\phi)$$

we obtain

$$\frac{d^2R}{dt^2} = \left(\frac{2C}{R} + \frac{R}{r^3}\right)\left(\frac{2C}{R^2} + \frac{3R^2}{r^5} - \frac{1}{r^3}\right), \qquad \frac{d^2x_3}{dt^2} = \left(\frac{2C}{R} + \frac{R}{r^3}\right)\frac{3Rx_3}{r^5}$$

$$\frac{d\phi}{dt} = \left(\frac{2C}{R} + \frac{R}{r^3}\right)\frac{1}{R}$$

where $r^2 = R^2 + x_3^2$ and C is some constant originating from the integration of $d^2\phi/dt^2$.

Problem 29. Consider the pair of coupled nonlinear differential equations relevant to the quantum field theory of charged solitons

$$\frac{d^2u}{dx^2} = -u + u^3 + duv^2, \qquad \frac{d^2v}{dx^2} = fv + \lambda v^3 + dv(u^2 - 1) \qquad (1)$$

where u and v are real scalar fields and d, f, λ are constants. Try to find an exact solution of (1) with the ansatz

$$v(x) = b_1 \tanh(\lambda_0(x + c_0)), \qquad u(x) = \sum_{n=1} a_n \tanh^n(\lambda_0(x + c_0)). \qquad (2)$$

Solution 29. Inserting (2) into the first equation of (1) and equating with the terms of the same order in $\tanh(\lambda_0(x+c_0))$ we obtain a recurrence relation for the coefficients a_n, namely

$$(n+1)(n+2)\lambda_0^2 - (2n^2\lambda_0^2 - 1)a_n + ((n-1)(n-2)\lambda_0^2 - db_1^2)a_{n-2} - c_n = 0 \quad (3)$$

where

$$c_n = \sum_{i+j+k=n} a_i a_j a_k. \qquad (4)$$

Inserting (2) into the second equation of (1), we obtain

$$2\lambda_0^2 + f - d = 0, \qquad 2\lambda_0^2 - \lambda b_1^2 - dc_2' = 0, \qquad (5)$$

and $c_n' = 0$, $n \neq 2$, where

$$c_n' = \sum_{i+j=n} a_i a_j. \qquad (6)$$

Using $c_m' = 0$ ($n \neq 2$), we have $a_n = 0$ ($n \neq 1$). Let $n = 0, 1, 2 \ldots$. Then from (3), (4) and (6) we find $\lambda_0 = \sqrt{1/2}$ and

$$a_1 = \pm\sqrt{(d - \lambda)/(d^2 - \lambda)}, \qquad b_1 = \pm\sqrt{(d - 1)/(d^2 - \lambda)}.$$

Thus we find an exact static soliton solution for (1)

$$v(x) = \pm\sqrt{(d - 1)/(d^2 - \lambda)} \tanh\left(\sqrt{\frac{1}{2}}(x + c_0)\right)$$

$$u(x) = \pm\sqrt{(d-\lambda)/(d^2-\lambda)} \tanh\left(\sqrt{\frac{1}{2}}(x+c_0)\right).$$

Problem 30. In *Nambu mechanics* the equations of motion are given by

$$\frac{du_j}{dt} = \frac{\partial(u_j, I_1, ..., I_{n-1})}{\partial(u_1, u_2, ..., u_n)}, \qquad j = 1, 2, ..., n \tag{1}$$

where $\partial(u_j, I_1, ..., I_{n-1})/\partial(u_1, u_2, ..., u_n)$ denotes the *Jacobian determinant* and $I_k : \mathbb{R}^n \to \mathbb{R}$ ($k = 1, ..., n-1$) are $n-1$ smooth functions. Equation (1) can also be written in summation convention as

$$\frac{du_j}{dt} = \varepsilon_{jl_1...l_{n-1}}(\partial_{l_1} I_1) \cdots (\partial_{l_{n-1}} I_{n-1}), \qquad j = 1, 2, ..., n$$

where $\varepsilon_{jl_1...l_{n-1}}$ is the generalized Levi-Civita symbol and $\partial_j \equiv \partial/\partial u_j$. Show that I_k ($k = 1, ..., n-1$) are first integrals of system (1).

Solution 30. The proof that $I_1, ..., I_{n-1}$ are first integrals of system (1) is as follows. We have (summation convention is used)

$$\frac{dI_i}{dt} = \frac{\partial I_i}{\partial u_j}\frac{du_j}{dt} = \varepsilon_{jl_1...l_{n-1}}(\partial_j I_i)(\partial_{l_1} I_1) \cdots (\partial_{l_{n-1}} I_{n-1})$$

$$= \frac{\partial(I_i, I_1, ..., I_{n-1})}{\partial(u_1, ..., u_n)}.$$

Thus $dI_i/dt = 0$ as the Jacobian matrix has two equal rows and is therefore singular. If the first integrals are polynomials, then the dynamical system (1) is algebraically completely integrable.

For $n = 3$ the equations of motion are given by

$$\frac{du_1}{dt} = \frac{\partial I_1}{\partial u_2}\frac{\partial I_2}{\partial u_3} - \frac{\partial I_1}{\partial u_3}\frac{\partial I_2}{\partial u_2}$$

$$\frac{du_2}{dt} = \frac{\partial I_1}{\partial u_3}\frac{\partial I_2}{\partial u_1} - \frac{\partial I_1}{\partial u_1}\frac{\partial I_2}{\partial u_3}$$

$$\frac{du_3}{dt} = \frac{\partial I_1}{\partial u_1}\frac{\partial I_2}{\partial u_2} - \frac{\partial I_1}{\partial u_2}\frac{\partial I_2}{\partial u_1}.$$

Assume that

$$I_1(\mathbf{u}) = a_{11}u_1^2/2 + a_{12}u_1u_2 + \cdots + a_{33}u_3^2/2 + a_1u_1 + a_2u_2 + a_3u_3$$
$$I_2(\mathbf{u}) = b_{11}u_1^2/2 + b_{12}u_1u_2 + \cdots + b_{33}u_3^2/2 + b_1u_1 + b_2u_2 + b_3u_3.$$

Find the dynamical system.

Problem 31. Let

$$\Omega_1 \frac{du_1}{dt} = (\Omega_3 - \Omega_2)u_2 u_3 \tag{1a}$$

$$\Omega_2 \frac{du_2}{dt} = (\Omega_1 - \Omega_3)u_3 u_1 \tag{1b}$$

$$\Omega_3 \frac{du_3}{dt} = (\Omega_2 - \Omega_1)u_1 u_2 \tag{1c}$$

where Ω_1, Ω_2 and Ω_3 are positive constants. This dynamical system describes *Euler's rigid body motion*. Show that this system can be derived within Nambu mechanics.

Solution 31. Using the result from the previous problem the equations of motion can be derived within Nambu mechanics from the first integrals

$$I_1(\mathbf{u}) = \frac{1}{2}(\Omega_1 u_1^2 + \Omega_2 u_2^2 + \Omega_3 u_3^2)$$

$$I_2(\mathbf{u}) = \frac{1}{2}\left(\frac{\Omega_1}{\Omega_2 \Omega_3}u_1^2 + \frac{\Omega_2}{\Omega_3 \Omega_1}u_2^2 + \frac{\Omega_3}{\Omega_1 \Omega_2}u_3^2\right).$$

The first integral I_1 represents the total kinetic energy. The general solution of system (1) can be expressed in terms of Jacobi elliptic functions.

Problem 32. The nonrelativistic motion of a charged particle with mass m and charge q in a constant electric field strength $\mathbf{E} = (E_1, E_2, E_3)$ and constant magnetic flux density $\mathbf{B} = (B_1, B_2, B_3)$ is given by

$$\frac{dv_1}{dt} = \frac{q}{m}E_1 + \frac{q}{m}(B_3 v_2 - B_2 v_3)$$

$$\frac{dv_2}{dt} = \frac{q}{m}E_2 + \frac{q}{m}(B_1 v_3 - B_3 v_1)$$

$$\frac{dv_3}{dt} = \frac{q}{m}E_3 + \frac{q}{m}(B_2 v_1 - B_1 v_2)$$

where $\mathbf{v} = (v_1, v_2, v_3)$ denotes the velocity of the particle. If $B_1 = B_2 = 0$, $B_3 \neq 0$ and $E_2 = E_3 = 0$, $E_1 \neq 0$, then we obtain

$$\frac{dv_1}{dt} = \frac{q}{m}E_1 + \frac{q}{m}B_3 v_2, \qquad \frac{dv_2}{dt} = -\frac{q}{m}B_3 v_1, \qquad \frac{dv_3}{dt} = 0.$$

Show that this dynamical system can be derived within Nambu mechanics.

Solution 32. Using the result from problem (2) we find that the dynamical system can be derived from the first integrals

$$I_1(\mathbf{v}) = -v_3, \qquad I_2(\mathbf{v}) = \frac{1}{2}B_3^2 v_1^2 + \frac{1}{2}B_3^2 v_2^2 + E_1 B_3 v_2.$$

Problem 33. Consider the hierarchy of an autonomous system of first order ordinary differential equations

$$\frac{du_j}{dt} = cu_1 \cdots u_{j-1}\hat{u}_j u_{j+1} \cdots u_n$$

where $n \geq 2$ and $j = 1, 2, \ldots, n$, $\hat{u}_j$ indicates omission and c is a nonzero constant. Find the first integrals. Can the dynamical system be derived from these first integrals using Nambu mechanics?

Solution 33. We find $n - 1$ first integrals

$$\frac{1}{2}c(u_1^2 - u_2^2), \quad \frac{1}{2}c(u_2^2 - u_3^2), \quad \ldots, \quad \frac{1}{2}c(u_{n-1}^2 - u_n^2).$$

Using these first integrals and Nambu mechanics we obtain the dynamical system.

Problem 34. Consider the *Rössler model*

$$\frac{dx}{dt} = -y - \epsilon z, \quad \frac{dy}{dt} = x + \epsilon ry, \quad \frac{dz}{dt} = 1 + (x - C)z$$

where ϵ, r and C are positive constants. There is numerical evidence that the system shows chaotic behaviour for $\epsilon = 0.2$, $r = 1$ and $C = 4$. The system can be considered as a harmonic oscillator in x and y nonlinearly coupled to a third variable. Find a *Poincaré map* for the system with $\epsilon \ll 1$.

Solution 34. For $\epsilon = 0$ we have a harmonic oscillator

$$\frac{dx}{dt} = -y, \quad \frac{dy}{dt} = x$$

with the solution $x(t) = a_0 \cos(\omega_0 t)$, $y(t) = a_0 \sin(\omega_0 t)$, where $\omega_0 = 1$. For $\epsilon \ll 1$ we consider the expansion

$$x(t) = x_0(t) + \epsilon x_1(t) + \cdots, \qquad y(t) = y_0(t) + \epsilon y_1(t) + \cdots$$

$$z(t) = z_0(t) + \epsilon z_1(t) + \cdots$$

where the zeroth solution of x and y is a *harmonic oscillator*

$$x_0(t) = a_0 \cos(\omega t), \qquad y_0(t) = a_0 \sin(\omega t)$$

with $\omega = \omega_0 + \epsilon \omega_1 + \cdots$. We still have to find ω_1. The terms of first order are given by

$$x_1(t) = a_1(t) \cos(\omega t) + b_1(t) \sin(\omega t)$$

$$y_1(t) = a_1(t) \sin(\omega t) - b_1(t) \cos(\omega t).$$

Inserting this ansatz into (1) we find up to the zeroth order that

$$(\omega_0 - 1)x_0 = 0, \qquad (\omega_0 - 1)y_0 = 0, \qquad \frac{dz_0}{dt} = 1 + (x_0 - C)z_0.$$

The first two equations are satisfied for $\omega_0 = 1$. The third equation can be integrated. We obtain

$$z_0(t) = \frac{z_0(0) + \int_0^t f(s)ds}{f(t)}$$

where $z_0(0)$ is the initial value of z_0 and the function f is given by

$$f(t) = \exp\left(-\int_0^t (x_0(s) - C)ds\right) = \exp\left(-\frac{a_0}{\omega}\sin(\omega t) + Ct\right).$$

Since we are only interested in the attractor, we have to find $z_0(t \to \infty)$. At time $t + n\tau$ with $0 \le t \le \tau$ the quantity z_0 is given by

$$z_0(t + n\tau) = \frac{z_0 + \int_0^{t+n\tau} f(s)ds}{f(t + n\tau)}.$$

The function f has the property $f(t + n\tau) = f(t)\exp(nC\tau)$. Using this property we find

$$z_0(t + n\tau) = \frac{(z_0 - z_C(\tau))\exp(-nC\tau) + z_C(\tau) + \int_0^t f(s)ds}{f(t)}$$

where

$$z_C(\tau) = \frac{\int_0^\tau f(s)ds}{\exp(C\tau) - 1}.$$

For $n \to \infty$ the function $z_0(t + n\tau)$ approaches the asymptotic function $\tilde{z}_0(t)$

$$z_0(t + n\tau) \to \tilde{z}_0(t) = \frac{z_C(\tau) + \int_0^t f(s)ds}{f(t)}.$$

We also find $\tilde{z}_0(t + n\tau) = \tilde{z}_0(t)$. In first-order we find

$$\frac{da_1}{dt} = (\omega_1 y_0 - z_0)\cos(\omega t) + (-\omega_1 x_0 + r y_0)\sin(\omega t)$$

$$\frac{db_1}{dt} = (\omega_1 y_0 - z_0)\sin(\omega t) - (-\omega_1 x_0 + r y_0)\cos(\omega t)$$

where x_0, y_0, z_0 are given above. This system of differential equations can be integrated to give

$$a_1(\tau) = \frac{r}{2}a_0\tau - \int_0^\tau \tilde{z}_0(t)\cos(\omega t)dt$$

$$b_1(\tau) = \omega_1 a_0 \tau - \int_0^\tau \tilde{z}_0(t) \sin(\omega t) dt$$

with the initial values $a_1(0) = b_1(0) = 0$ and $x(0) = a_0$, $y(0) = 0$, $z(0) = z_C(\tau)$. After one period τ, we have

$$x(\tau) = a_0 + \epsilon a_1(\tau) + O(\epsilon^2) = a(\tau), \quad y(\tau) = -\epsilon b_1(\tau) + O(\epsilon^2) = b(\tau),$$

$$z(\tau) = z_C(\tau) + O(\epsilon).$$

The initial values a_0 and $b_0 = 0$ are mapped, after one period, into

$$a(\tau) = a_0 + \epsilon \tau \left(\frac{r}{2} a_0 - \frac{1}{\tau} \int_0^\tau \tilde{z}_0(t) \cos(\omega t) dt \right)$$

$$b(\tau) = \epsilon \tau \left(\omega_1 a_0 - \frac{1}{\tau} \int_0^\tau \tilde{z}_0(t) \sin(\omega t) dt \right).$$

This is the discrete Poincaré map.

Problem 35.   Consider the second order linear differential equation

$$\frac{d^2\phi}{dt^2} = 0. \tag{1}$$

(i) Let

$$u(t) = \frac{d}{dt}(\ln(\phi)). \tag{2a}$$

Derive the differential equation for u.
(ii) Let

$$u(t) = \frac{d^2}{dt^2}(\ln(\phi)). \tag{2b}$$

Derive the differential equation for u. Apply the following notation

$$A_1(t) := \int^t u(t_0) dt_0, \qquad A_2(t) := \int^t dt_1 \int^{t_1} u(t_0) dt_0.$$

Solution 35.   (i) Taking the derivative of A_2 and A_1 yields

$$\frac{dA_2}{dt} = A_1, \qquad \frac{dA_1}{dt} = A_0 \equiv u$$

and $de^{A_n}/dt = A_{n-1} e^{A_n}$, where $n = 1, 2$. From (2a) we obtain

$$u(t) = \frac{1}{\phi} \frac{d\phi}{dt}.$$

Therefore $\phi(t) = \exp(A_1)$. Taking the derivative we find

$$\frac{d\phi}{dt} = e^{A_1}u, \quad \frac{d^2\phi}{dt^2} = e^{A_1}u^2 + e^{A_1}\frac{du}{dt} \equiv e^{A_1}\left(\frac{du}{dt} + u^2\right).$$

Consequently (2a) leads to the nonlinear differential equation

$$\frac{du}{dt} + u^2 = 0.$$

This is a *Riccati differential equation*.

(ii) From (2b) we obtain $\phi(t) = \exp(A_2)$. The derivatives of ϕ give

$$\frac{d\phi}{dt} = e^{A_2}A_1, \quad \frac{d^2\phi}{dt^2} = e^{A_2}(A_1^2 + u).$$

Therefore $A_1^2 + u = 0$. Taking the derivative of this equation leads to

$$2A_1u + \frac{du}{dt} = 0, \quad 2u^2 + 2A_1\frac{du}{dt} + \frac{d^2u}{dt^2} = 0.$$

To eliminate A_1 we multiply this equation with u and insert $2A_1u + du/dt = 0$. This leads to the nonlinear differential equation

$$u\frac{d^2u}{dt^2} - \left(\frac{du}{dt}\right)^2 + 2u^3 = 0.$$

The technique to find nonlinear ordinary differential equations from a linear differential equation using (1) can be extended to

$$A_n(t) := \int^t dt_{n-1}\int^{t_{n-1}} dt_{n-2}\cdots\int^{t_1} u(t_0)dt_0.$$

Problem 36. Let

$$\frac{d^2w}{dz^2} + p_1(z)\frac{dw}{dz} + p_2(z)w = 0 \tag{1}$$

be a second-order linear differential equation in the complex domain. Consider the transformation

$$\tilde{z}(z) = z, \quad v(\tilde{z}(z)) = w(z)\exp\left(\frac{1}{2}\int^z p_1(s)ds\right).$$

Find the differential equation for $v(\tilde{z})$.

Solution 36. Let

$$R(z) := \frac{1}{2}\int^z p_1(s)ds.$$

Then

$$\frac{dv}{dz} = \frac{dv}{d\tilde{z}}\frac{d\tilde{z}}{dz} = \frac{dv}{d\tilde{z}} = \frac{dw}{dz}e^R + \frac{1}{2}p_1we^R \tag{1}$$

where we have used that $dR/dz = p_1/2$. Since $d\tilde{z}/dz = 1$ we obtain from (1)

$$\frac{d^2v}{d\tilde{z}^2} = \frac{d^2w}{dz^2}e^R + p_1\frac{dw}{dz}e^R + \frac{1}{2}\frac{dp_1}{dz}we^R + \frac{1}{4}p_1^2we^R.$$

Consequently

$$\frac{d^2v}{d\tilde{z}^2} + I(\tilde{z})v = 0, \quad I(\tilde{z}) := p_2(\tilde{z}) - \frac{1}{2}\frac{dp_1(\tilde{z})}{d\tilde{z}} - \frac{1}{4}p_1^2(\tilde{z}).$$

Problem 37. The nonlinear ordinary differential equation

$$\frac{d^2u}{dt^2} + 3\frac{du}{dt}u + u^3 = 0 \tag{1}$$

occurs in the investigation of univalued functions defined by second-order differential equations and in the study of the *modified Emden equation*.
(i) Show that (1) can be linearized with the help of the transformations

$$U(T) = u^2(t(T)), \qquad dT = u(t(T))dt(T) \tag{2}$$

$$u(t) = \frac{1}{\phi(t)}\frac{d\phi(t)}{dt}. \tag{3}$$

(ii) Find the general solution to (1).

Solution 37. (i) First we consider the transformation given by (2). Since

$$\frac{dU}{dT} = 2u\frac{du}{dt}\frac{dt}{dT} = 2\frac{du}{dt}, \quad \frac{d^2U}{dT^2} = 2\frac{d^2u}{dt^2}\frac{dt}{dT} = \frac{2}{u}\frac{d^2u}{dt^2}$$

we obtain

$$\frac{u}{2}\frac{d^2U}{dT^2} + \frac{3u}{2}\frac{dU}{dT} + uU = 0 \quad \Rightarrow \quad \frac{d^2U}{dT^2} + 3\frac{dU}{dT} + 2U = 0$$

with $u \neq 0$. Now we consider the transformation given by (3). Since

$$\frac{du}{dt} = \frac{1}{\phi^2}\left(\frac{d^2\phi}{dt^2}\phi - \left(\frac{d\phi}{dt}\right)^2\right)$$

and

$$\frac{d^2u}{dt^2} = \frac{1}{\phi^3}\left(\frac{d^3\phi}{dt^3}\phi^2 - 3\phi\frac{d^2\phi}{dt^2}\frac{d\phi}{dt} + 2\left(\frac{d\phi}{dt}\right)^3\right)$$

we obtain (with $\phi \neq 0$) the linear differential equation

$$\frac{d^3\phi}{dt^3} = 0. \tag{4}$$

(ii) The general solution to (1) can now easily be found with the help of (4) and transformation (2). The general solution to (4) is given by

$$\phi(t) = C_1 t^2 + C_2 t + C_3$$

where C_1, C_2 and C_3 are the constants of integration. Inserting $\phi(t)$ into (3) yields the general solution

$$u(t) = \frac{2t + (C_2/C_1)}{t^2 + (C_2/C_1)t + (C_3/C_1)}$$

where $C_1 \neq 0$. If $C_1 = 0$ and $C_2 \neq 0$, then $u(t) = 1/(t + C_3/C_2)$. If $C_1 = C_2 = 0$ and $C_3 \neq 0$, then $u(t) = 0$.

Problem 38. Consider the ordinary differential equation

$$\frac{d^2u}{dt^2} + \lambda(\phi(t))^{2m-2}u = 0 \tag{1}$$

where $m = 1, 2, \ldots$, and λ is a real parameter. The smooth function ϕ satisfies

$$\frac{d^2\phi}{dt^2} + (\phi(t))^{2m-1} = 0 \tag{2}$$

with the initial conditions $\phi(0) = 1$, $d\phi(0)/dt = 0$. Perform the transformation

$$z(t) = (\phi(t))^{2m}, \qquad \bar{u}(z(t)) = u(t).$$

Solution 38. First we notice that the integration of the differential equation (2) yields

$$\frac{1}{2}\left(\frac{d\phi}{dt}\right)^2 + \frac{1}{2m}\phi^{2m} = \frac{1}{2m}$$

where we have taken into account the initial conditions. From the transformation it follows that

$$\frac{d\bar{u}}{dt} = \frac{d\bar{u}}{dz}\frac{dz}{dt} = \frac{du}{dt}, \quad \frac{d^2\bar{u}}{dt^2} = \frac{d^2\bar{u}}{dz^2}\left(\frac{dz}{dt}\right)^2 + \frac{d\bar{u}}{dz}\frac{d^2z}{dt^2} = \frac{d^2u}{dt^2}.$$

From $z(t) = (\phi(t))^m$ we find that

$$\frac{dz}{dt} = 2m(\phi(t))^{2m-1}\frac{d\phi}{dt}.$$

Consequently,

$$\left(\frac{dz}{dt}\right)^2 = 4m^2(\phi(t))^{4m-2}\left(\frac{d\phi}{dt}\right)^2 = 4m(\phi(t))^{2m-2}z(1-z).$$

From this equation we also obtain

$$\frac{d^2z}{dt^2} = 2(\phi(t))^{2m-2}((1-3m)z + 2m - 1).$$

It follows that

$$z(1-z)\frac{d^2\bar{u}}{dz^2} + (\gamma - (\alpha + \beta + 1)z)\frac{d\bar{u}}{dz} - \alpha\beta\bar{u} = 0$$

where $\alpha + \beta = (m-1)/(2m)$, $\alpha\beta = -\lambda/(4m)$, $\gamma = (2m-1)/(2m)$. This second order linear ordinary differential equation has three singular points at $z = 0$, $z = 1$ and $z = \infty$. It is called the *hypergeometric equation*.

Problem 39.   Let m be the mass of a body. For a gravitational field with a linear gradient, the Lagrange function L is given by

$$L(x_3, \dot{x}_3) = \frac{1}{2}m\left(\frac{dx_3}{dt}\right)^2 - mg_0x_3 + \frac{1}{2}m\gamma x_3^2$$

where x_3 denotes the altitude, g_0 the gravitational acceleration at $x_3 = 0$ and γ the linear gravitational gradient. The signs of the gravitational acceleration g_0 and the linear gravitational gradient γ are chosen to be positive for normal conditions (objects accelerate downwards, magnitude of acceleration decreases with altitude). Find the *Euler-Lagrange equation*

$$\frac{d}{dt}\left(\frac{\partial L}{\partial \dot{x}_3}\right) - \frac{\partial L}{\partial x_3} = 0$$

and solve the initial value problem.

Solution 39.   With

$$\frac{\partial L}{\partial \dot{x}_3} = m\frac{dx_3}{dt}, \qquad \frac{\partial L}{\partial x_3} = -mg_0 + m\gamma x_3$$

and

$$\frac{d}{dt}\frac{\partial L}{\partial \dot{x}_3} = m\frac{d^2x_3}{dt^2}$$

we obtain the second order differential equation

$$\frac{d^2x_3}{dt^2} + mg_0 - m\gamma x_3 = 0.$$

The equation can be directly integrated and we find

$$x_3(t) = \frac{g_0}{\gamma} + \left(x_{3,0} - \frac{g_0}{\gamma} \right) \cosh(\sqrt{\gamma}t) + \frac{v_0}{\sqrt{\gamma}} \sinh(\sqrt{\gamma}t)$$

and

$$\frac{dx_3}{dt} = \sqrt{\gamma} \left(x_{3,0} - \frac{g_0}{\gamma} \right) \sinh(\sqrt{\gamma}t) + v_0 \cosh(\sqrt{\gamma}t)$$

with the initial values $x_3(t = 0) = x_{3,0}$, $dx_3(0)/dt = v_0$.

Problem 40. Consider the eigenvalue equation

$$-\frac{\hbar^2}{2m}\frac{d^2u}{dx^2} + V(x)u(x) = Eu(x). \tag{1}$$

(i) Let the potential V be given by

$$V(x) = \frac{\hbar^2}{2m}a^2x^6 - 3\frac{\hbar^2}{2m}ax^2 \tag{2}$$

where $a > 0$. Show that for this potential the ground-state wave function u_0 is

$$u_0(x) = \exp(-ax^4/4). \tag{3}$$

(ii) What happens when we analytically continue the eigenvalue $E_0(a)$ from positive to negative values of the parameter a?

Solution 40. (i) Inserting (3) into (1) yields the energy eigenvalue

$$E(a) = 0.$$

The eigenfunction corresponds to the lowest-lying energy eigenvalue because it has no nodes.

(ii) For $a > 0$ the eigenfunction is an element of the Hilbert space $L_2(\mathbb{R})$. When we analytically continue the function $E_0(a)$ from positive to negative values, the result is still $E_0(a) = 0$. However, this conclusion is wrong. The spectrum of the potential (2) for $a < 0$ is strictly positive. When we continue the wave function ψ_0 to a negative value of a we find a nonnormalizable solution of the eigenvalue equation (1), which indicates that a simple replacement of a by $-a$ is not allowed.

Problem 41. Study the boundary value problem

$$\frac{d^2u}{dx^2} + k^2u = 0$$

with $u(0) = u(-\ell)$, $du(x = 0)/dx = du(x = -\ell)/dx$.

Solution 41.   Starting with $u(x) = A \sin(kx) + B \cos(kx)$ we obtain

$$\frac{du}{dx} = Ak \cos(kx) - Bk \sin(kx), \quad \frac{d^2 u}{dx^2} = -Ak^2 \sin(kx) - Bk^2 \cos(kx).$$

Thus for non-trivial solutions we have to satisfy $\cos(k\ell) = 1$. It follows that $k_n = 2\pi n/\ell$ $(n \in \mathbb{Z})$. Hence

$$u_n(x) = A \sin(k_n x) + B \cos(k_n n), \quad n = \pm 1, \pm 2, \dots$$

and $u_0(\mathbf{x}) = B$ for $k_0 = 0$. The eigenvalue $k_0 = 0$ is not degenerate, whereas the eigenvalues k_n $(n \neq 0)$ are twice degenerate.

Problem 42.   Let $r, s > 0$. Show that

$$f(x) = \sqrt{\frac{2s}{r}} \frac{1}{\cosh(\sqrt{s}x)}$$

satisfies the second order nonlinear ordinary differential equation

$$\frac{d^2 f}{dx^2} - sf + rf^3 = 0.$$

Solution 42.   We apply the Maxima program

```
/* diffeq1.mac */
f: sqrt(2*s/r)/cosh(sqrt(s)*x);
df: diff(f,x);
d2f: diff(df,x);
d2f: trigsimp(d2f);
R: d2f - s*f + r*f*f*f;
R: trigsimp(R);
```

where `trigsimp()` will do trigonometric simplifications.

Supplementary Problems

Problem 1.   Solve the initial value problem (i.e. given $x(t = 0)$ and $y(t = 0)$) of the autonomous system of first order differential equations

$$\frac{dx}{dt} = -v_1 \frac{x}{\sqrt{x^2 + y^2}}, \quad \frac{dy}{dt} = v_2 - v_1 \frac{y}{\sqrt{x^2 + y^2}}$$

where v_1, v_2 are constant velocities with $v_1 = 2 \, meter \cdot sec^{-1}$ and $v_2 = 1 \, meter \cdot sec^{-1}$. The initial conditions are $x(t = 0) = 10 \, meter$ and $y(t = 0) = 0 \, meter$. Note that

$$y\frac{dx}{dt} - x\frac{dy}{dt} = -xv_2.$$

Problem 2. The differential equation for a particle in a gravitational field moving in $\mathbb{R}^3 \setminus \{0\}$ is given by

$$\frac{d^2\mathbf{r}}{dt^2} + \frac{\mu\mathbf{r}}{r^3} = 0$$

where $\mu > 0$ is the constant field strength. Each orbit is planar. Show that in the plane with *polar coordinates* (r, θ) we find the system of second order differential equations

$$\frac{d^2r}{dt^2} = r\left(\frac{d\theta}{dt}\right)^2 - \frac{\mu}{r}, \quad \frac{d^2\theta}{dt^2} = \frac{2}{r}\frac{dr}{dt}\frac{d\theta}{dt}.$$

Show that the conserved quantities are the energy

$$E = \frac{1}{2}((dr/dt)^2 + h^2/r^2 - \mu/r)$$

the magnitude of *angular momentum* $h = |\mathbf{L}| = r^2 d\theta/dt$ and the *Runge-Lenz vector*.

Problem 3. Consider the Hilbert space $L_2([0,1])$ and the boundary value problem

$$-\frac{d^2u}{dx^2} = \sin(\pi x/2), \quad u(0) = \frac{du(0)}{dx} = 0.$$

Show that $u(x) = \left(\frac{2}{\pi}\right)^2 \sin(\pi x/2)$ is the solution of the problem.

Problem 4. Find the Lie symmetries of the second order ordinary differential equation (Poisson-Boltzmann equation)

$$\frac{d^2u}{dr^2} + \frac{L}{r}\frac{du}{dr} = e^u - \delta e^{-u}$$

where $u = (e\Phi)/(k_B T)$ is a dimensionless potential.

Problem 5. Let $R(t) = |\mathbf{R}(t)|$. Show that the system of second order ordinary differential equations

$$\frac{d^2\mathbf{R}}{dt^2} + h(R)\mathbf{R} \times \frac{d\mathbf{R}}{dt} + q(R)\mathbf{R} = 0$$

admits the Lie symmetry vector fields

$$T = \frac{\partial}{\partial t}, \quad S_{12} = X_2\frac{\partial}{\partial X_1} - X_1\frac{\partial}{\partial X_2},$$

$$S_{31} = X_1\frac{\partial}{\partial X_3} - X_3\frac{\partial}{\partial X_1}, \quad S_{23} = X_3\frac{\partial}{\partial X_2} - X_2\frac{\partial}{\partial X_3}.$$

Problem 6. Show that the van der Pol equation can be written as

$$\frac{du_1}{dt} = u_2 = \frac{\partial S}{\partial u_1} + \frac{\partial H}{\partial u_2}$$

$$\frac{du_2}{dt} = \alpha(1 - u_1^2)u_2 - u_1 = \frac{\partial S}{\partial u_2} - \frac{\partial H}{\partial u_1}$$

where

$$S = \alpha \left(\frac{u_1^2}{2} - \frac{u_1^4}{12} \right), \quad H = \frac{1}{2}u_1^2 - \alpha \left(u_1 - \frac{u_1^3}{3} \right) u_2 + \frac{u_1^2}{2}.$$

Problem 7. Let m be the mass of a particle and $v(t)$ its velocity. The equation of motion for the average velocity of a Brownian particle in one dimension is given by

$$m\langle dv/dt \rangle + m \int_{-\infty}^{t} \gamma(t - t')v(t')dt' = K(t)$$

where $\gamma(t - t')$ describes the retarded effect of the frictional force and $K(t)$ is the external force. Show that applying the Fourier transform

$$\hat{v}(\omega) = \frac{1}{2\pi} \int_{-\infty}^{\infty} v(t)e^{i\omega t}dt, \quad \hat{K}(\omega) = \frac{1}{2\pi} \int_{-\infty}^{\infty} K(t)e^{i\omega t}dt$$

one obtains $\langle \hat{v}(\omega) \rangle = Y(\omega)\hat{K}(\omega)$ with

$$Y(\omega) = \frac{1}{m} \frac{1}{-i\omega + \hat{\gamma}(\omega)}, \quad \hat{\gamma}(\omega) = \int_{0}^{\infty} \gamma(t)e^{i\omega}dt.$$

Problem 8. Let $k > 0$ with dimension $meter^{-1}$. Show that

$$u(x) = A\sin(kx) + B\sinh(kx) + C\cos(kx) + D\cosh(kx))$$

satisfies the differential equation $d^4u/dx^4 = k^4u$. Let $\ell > 0$ (dimension length). Impose the boundary conditions $u(0) = u(\ell) = 0$.

Problem 9. Let $n \in \mathbb{N}$. Consider the time-dependent *Boltzmann equation* in the form

$$\frac{\partial f(n)}{\partial t} = \sum_{n_1=1}^{\infty} \sum_{n_2=1}^{\infty} \sum_{n_3=1}^{\infty} (f(n_2)f(n_3) - f(n)f(n_1))$$

with the constraint $n + n_1 = n_2 + n_3$. It follows that

$$\frac{\partial f(n)}{\partial t} = \sum_{n_2,n_3;n_2+n_3>n} f(n_2)f(n_3) - f(n) \sum_{n_1=1}^{\infty} (n + n_1 - 1)f(n_1).$$

Show that the total number of molecules $\sum_{n=1}^{\infty} f(n) = N$ and the total energy $\sum_{n=1}^{\infty} nf(n) = E$ remain constant. Show that

$$\frac{\partial f(n)}{\partial t} + ((n-1)N + E)f(n) = N^2 - \sum_{n_2,n_3=1}^{n_2+n_3 \leq n} f(n_2)f(n_3).$$

Problem 10. The torque-free motion of a rigid body about its centre of mass is given by

$$\frac{d\mathbf{L}}{dt} = -(I^{-1}\mathbf{L}) \times \mathbf{L}$$

where $\mathbf{L}$ is the angular momentum of the body and I its moment of inertia tensor (3×3 matrix assumed to be invertible). Solve the initial value problem.

Problem 11. Let $\omega, b > 0$. Solve the initial value problem of the *Verhulst model*

$$\frac{du}{dt} = \omega u \left(1 - \frac{u}{b}\right)$$

with $u(t = 0) = u_0 > 0$.

Problem 12. Solve the system of coupled linear second order differential equations

$$\frac{d^2u}{dx^2} + k\frac{dv}{dx} = -1, \qquad \frac{d^2v}{dx^2} + k\frac{du}{dx} = 0$$

with $u(\pm 1) = v(\pm 1) = 0$.

Problem 13. Consider the variable frequency oscillator with the Hamilton function

$$H(t) = \frac{1}{2m}p^2 + \frac{1}{2}m\omega(t)q^2.$$

Show that an invariant has the form

$$I(t) = \frac{1}{\sigma^2}q^2 + \left(\frac{d\sigma}{dt}q - \frac{\sigma}{m}p\right)^2$$

where $\sigma(t)$ is any solution of the *Pinney equation*

$$\frac{d^2\sigma}{dt^2} + w(t)\sigma = \frac{1}{\sigma^3}.$$

Problem 14.   Let $n \in \mathbb{Z}$. Consider the master equation for the linear birth-death process

$$\frac{\partial u(n,t)}{\partial t} = (n-1)\lambda u(n-1,t) + (n-1)\mu u(n+1,t) - (n\lambda + n\mu)u(n,t).$$

Here λ (μ) is the probability per unit time that a single bacterium has a birth (death). Consider the generating function

$$F(s,t) = \sum_{n \in \mathbb{Z}} u(n,t)s^n.$$

Show that $F(s,t)$ satisfies the partial differential equation

$$\frac{\partial F}{\partial t} = (s-1)(\lambda s - \mu)\frac{\partial F}{\partial s}$$

and that if we equate powers of n we find the master equation again. Show that to find F one has to integrate

$$\frac{dt}{1} = \frac{-ds}{(s-1)(\lambda s - \mu)} = \frac{dF}{0}.$$

Show that integrating the left two equations one finds

$$C_1 = \frac{(s-1)}{(\lambda s - \mu)}e^{-(\mu-\lambda)t}$$

where C_1 is an integration constant. Show that integrating the two equations on the right one finds $F(s,t) = C_2$, where C_2 is an integration constant.

Problem 15.   Let $x \in \mathbb{R}$ and

$$S(x) := x + \frac{x^3}{1 \cdot 3} + \frac{x^5}{1 \cdot 3 \cdot 5} + \frac{x^7}{1 \cdot 3 \cdot 5 \cdot 7} + \cdots$$

Show that $S(x)$ satisfies the differential equation

$$\frac{dS}{dx} = 1 + xS(x) \text{ with } S(0) = 0.$$

Show that

$$S(x) = e^{x^2/2}\int_0^x e^{-x^2/2}dx.$$

Problem 16.   Consider the Hamilton function of the time dependent damped driven one dimensional harmonic oscillator ($\alpha > 0$)

$$H(p,x,t) = \frac{1}{2m}e^{-\alpha t}p^2 + e^{\alpha t}\left(\frac{1}{2}m\omega_0^2 x^2 - xf(t)\right)$$

where $f(t)$ is an external smooth driving force. Show that the equations of motion are given by

$$\frac{dx}{dt} = \frac{1}{m}e^{-\alpha t}p, \quad \frac{dp}{dt} = -e^{\alpha t}(m\omega_0^2 x - f(t)).$$

Show that the Lagrange function is given by

$$L(x, dx/dt, t) = e^{\alpha t}\left(\frac{1}{2}m\left(\frac{dx}{dt}\right)^2 - \frac{1}{2}m\omega_0 x^2 + xf(t)\right)$$

with the equation of motion

$$\frac{d^2 x}{dt^2} + \alpha\frac{dx}{dt} + \omega_0^2 x^2 = \frac{1}{m}f(t).$$

Show that the solution is given by

$$x(t) = Ae^{-\alpha t/2}\cos(\omega t + \phi)$$

$$+ e^{-\alpha t}\cos(\omega t)\int^t \cos(2\omega t')dt'\int^{t'}\frac{f(s)}{m}e^{-\alpha s/2}\cos(\omega s)ds$$

where $\omega = (\omega_0^2 - \alpha^2/4)^{1/2}$. Study the case with $f(t) = B\sin(\Omega t)$.

Problem 17. Show that the *Lorenz model* can be written as

$$\frac{du_1}{dt} = -\sigma u_1 + \sigma u_2 = \frac{\partial S}{\partial u_1} + \frac{\partial H}{\partial u_2}$$

$$\frac{du_2}{dt} = u_1 u_3 + r u_1 - u_2 = \frac{\partial S}{\partial u_2} - \frac{\partial H}{\partial u_1}$$

$$\frac{du_3}{dt} = u_1 u_2 - b u_3 = \frac{\partial S}{\partial u_3}$$

where

$$S = -\sigma\frac{1}{2}u_1^2 - \frac{1}{2}u_2^2 - \frac{1}{2}bu_3^2 + u_1 u_2 u_3, \quad H = \frac{1}{2}(\sigma - u_3)u_2^2 + u_1^2(u_3 - r/2).$$

Problem 18. The variational equation of the *Lorenz model*

$$\frac{dX}{dt} = -\sigma X + \sigma Y \tag{1a}$$

$$\frac{dY}{dt} = -XZ + \tau X - Y \tag{1b}$$

$$\frac{dZ}{dt} = XY - bZ \tag{1c}$$

is given by

$$
\begin{pmatrix} dx_0/dt \\ dy_0/dt \\ dz_0/dt \end{pmatrix} = \begin{pmatrix} -\sigma & \sigma & 0 \\ (\tau - Z) & -1 & -X \\ Y & X & -b \end{pmatrix} \begin{pmatrix} x_0 \\ y_0 \\ z_0 \end{pmatrix}. \tag{2}
$$

(i) Show that the Lorenz model possesses the steady-state solution $X = Y = Z = 0$, representing the state of no convection.

(ii) Show that with this basic solution, the characteristic equation of the variational matrix is

$$
(\lambda + b)(\lambda^2 + (\sigma + 1)\lambda + \sigma(1 - \tau)) = 0. \tag{3}
$$

(iii) Show that this equation has three real roots when $\tau > 0$; all are negative when $\tau < 1$, but one is positive when $\tau > 1$. The criterion for the onset of convection is therefore $\tau = 1$.

(iv) Show that when $\tau > 1$, system (1) possesses two additional steady state solutions (fixed points) $X = Y = \pm\sqrt{b(\tau - 1)}$, $Z = \tau - 1$.

(v) Show that for either of these solutions, the characteristic equation of the matrix in (2) is $\lambda^3 + (\sigma + b + 1)\lambda^2 + (\tau + \sigma)b\lambda + 2\sigma b(\tau - 1) = 0$.

(vi) Show that this equation possesses one real negative root and two complex conjugate roots when $\tau > 1$. Show that the complex conjugate roots are pure imaginary if the product of the coefficients of λ^2 and λ equals the constant term, or $\tau = \sigma(\sigma + b + 3)(\sigma - b - 1)^{-1}$.

Problem 19. Let c be the speed of light. The Hamilton function of a relativistic massless particle with the potential (harmonic oscillator) takes the form

$$
H = c|\mathbf{p}| + \frac{1}{2}\kappa^2\mathbf{x}^2
$$

with $\mathbf{x}$ and $\mathbf{p}$ are the position and momentum of the particle. The kinetic energy is $c|\mathbf{p}|$. Show that the Hamilton equations of motion are given by

$$
\frac{d\mathbf{x}}{dt} = c\frac{\mathbf{p}}{|\mathbf{p}|}, \quad \frac{d\mathbf{p}}{dt} = -\kappa^2\mathbf{x}.
$$

Show that $E = c|\mathbf{p}| + \frac{\kappa^2}{2}\mathbf{x}^2$ and $\mathbf{J} = \mathbf{x} \times \mathbf{p}$ are first integrals.

Problem 20. Let Γ be the gamma function. The *Liouville-Riemann definition* for the *fractional integral operator* $_0D_x^{-q}$ is given by

$$
_0D_x^{-q}f(x) := \frac{1}{\Gamma(q)}\int_0^x (x - y)^{q-1}f(y)dy, \quad q > 0. \tag{1}
$$

The *fractional differential operator* $_0D_x^\nu$ for $\nu > 0$ is defined as

$$
_0D_x^\nu f(x) := \frac{d^n}{dx^n}(_0D_x^{\nu-n}f(x)), \quad \nu - n < 0 \tag{2}
$$

where $_0D_x^{\nu-n}$ for $\nu - n < 0$ is given in definition (1). Show that normalized one-sided Lévy-type probability densities

$$f(x) = \frac{a^\mu}{\Gamma(\mu)} x^{-\mu-1} \exp(-a/x), \quad a > 0, \quad x > 0$$

are solutions of the fractional integral equation

$$x^{2q} f(x) = a_0^q D_x^{-q} f(x)$$

when the Lévy-index μ is identified as the fractional order q of the integral operator $_0D_x^{-q}$.

Chapter 11

Partial Differential Equations

Let $(x_1, \ldots, x_m) \in \mathbb{R}^m$ $(m \geq 2)$ be the independent variables and $(u_1, \ldots, u_n)$ be the dependent variables. So one can consider the system of partial differential equations $(i = 1, \ldots, m; j = 1, \ldots, n; k = 1, \ldots, p)$.

$$f_k(x_i, u_j(x_i), \partial u_j / \partial x_i) = 0.$$

In many case boundary conditions and initial conditions have to be added.

In mathematical physics the *diffusion equation*

$$\frac{\partial^2 u}{\partial x_1^2} + \frac{\partial^2 u}{\partial x_2^2} + \frac{\partial^2 u}{\partial x_3^2} = D \frac{\partial u}{\partial t}$$

and the *wave equation*

$$\frac{\partial^2 u}{\partial x_1^2} + \frac{\partial^2 u}{\partial x_2^2} + \frac{\partial^2 u}{\partial x_3^2} = \frac{1}{c^2} \frac{\partial^2 u}{\partial t^2}$$

play a central role. *Maxwell's equations* in free space are given by

$$\nabla \times \mathbf{B} = \epsilon_0 \mu_0 \frac{\partial \mathbf{E}}{\partial t}, \quad \nabla \times \mathbf{E} = -\frac{\partial \mathbf{B}}{\partial t}, \quad \nabla \cdot \mathbf{E} = 0, \quad \nabla \cdot \mathbf{B} = 0$$

where $\mathbf{B}$ is the *magnetic induction* $\mathbf{B} = \mu_0 \mathbf{H}$, $\mathbf{E}$ is the *electric field intensity*

and

$$\nabla \times \mathbf{B} = \begin{pmatrix} \dfrac{\partial B_3}{\partial x_2} - \dfrac{\partial B_2}{\partial x_3} \\[2mm] \dfrac{\partial B_1}{\partial x_3} - \dfrac{\partial B_3}{\partial x_1} \\[2mm] \dfrac{\partial B_2}{\partial x_1} - \dfrac{\partial B_1}{\partial x_2} \end{pmatrix}.$$

The *Navier-Stokes equation* of a viscous incompressible fluid is given by

$$\left(\frac{\partial}{\partial t} + \mathbf{u} \cdot \nabla \right) \mathbf{u} = -\frac{1}{\rho} \nabla p + \nu \Delta \mathbf{u}.$$

If the fluid is incompressible one has $\nabla \cdot \mathbf{u} = \mathbf{0}$.

In soliton theory the following equations play a central role. The inviscid Burgers equation is given by

$$\frac{\partial u}{\partial t} + u \frac{\partial u}{\partial x} = 0$$

and Burgers equation takes the form

$$\frac{\partial u}{\partial t} + u \frac{\partial u}{\partial x} = D \frac{\partial^2 u}{\partial x^2}.$$

The Korteweg-de Vries equation

$$\frac{\partial u}{\partial t} + u \frac{\partial u}{\partial x} + \beta \frac{\partial^3 u}{\partial x^3} = 0$$

is one of the core equation in soliton theory. The Kadomtsev-Petviashivili equation can be considered as a two-dimensional extension of the Korteweg-de Vries equation

$$\frac{\partial}{\partial x} \left(\frac{\partial u}{\partial t} + u \frac{\partial u}{\partial x} + \epsilon^2 \frac{\partial^3 u}{\partial x^3} \right) + \lambda \frac{\partial^2 u}{\partial y^2} = 0.$$

The one-dimensional sine-Gordon equation and one-dimensional Liouville equation take the form

$$\frac{\partial^2 u}{\partial x_0^2} - \frac{\partial^2 u}{\partial x_1^2} = \sin(u), \quad \frac{\partial^2 u}{\partial x_0^2} - \frac{\partial^2 u}{\partial x_1^2} = \exp(u).$$

The one-dimensional nonlinear Schrödinger equation plays a role in fibre optics

$$i\hbar \frac{\partial \psi}{\partial t} + \frac{\hbar^2}{2m} \frac{\partial^2 \psi}{\partial x^2} = c|\psi|^2 \psi.$$

Problem 1.   Consider the one-dimensional *linear diffusion equation*

$$\frac{\partial u}{\partial t} = D\frac{\partial^2 u}{\partial x^2}$$

with $u(x,0) = f(x)$, $x \in [0,\infty)$, $u(0,t) = U$, $t \geq 0$. Let

$$T(x, x_1) = \exp(-(x - x_1)^2/(4Dt)) - \exp(-(x + x_1)^2/(4Dt)).$$

The exact solution is given by

$$u(x,t) = U\mathrm{erfc}\left(\frac{1}{2}x/(Dt)^{1/2}\right) + \frac{1}{2\sqrt{\pi Dt}}\int_0^\infty f(x_1)T(x, x_1)dx_1$$

where erfc is the *error function*. Find the asymptotic expansion of $u(x,t)$ as $t \to \infty$, $\eta = O(1)$ with the *similarity variable* η (dimensionless)

$$\eta = \frac{1}{2}x/(Dt)^{1/2}.$$

Solution 1.   The solution $u(x,t)$ can be written as

$$u(x,t) \sim U\mathrm{erfc}\left(\frac{1}{2}\frac{x}{\sqrt{Dt}}\right) + \frac{1}{\sqrt{\pi}}\left(\frac{I_2(0)\eta e^{-\eta^2}}{Dt} + \frac{I_4(0)(\eta^3 - \frac{3}{2}\eta)e^{-\eta^2}}{(Dt)^2} + \cdots\right).$$

The coefficients being proportional to *Hermite polynomials*. If we assume that the integrals converge, then the numbers $I_2(0)$, $I_4(0)$, $\ldots$, are given by

$$I_2(x) = \int_x^\infty dx_1 \int_{x_1}^\infty f(x_2)dx_2, \quad I_4(x) = \int_x^\infty dx_1 \int_{x_1}^\infty I_2(x_2)dx_2, \quad \ldots$$

at $x = 0$. The expansion depends on the initial condition only via these coefficients.

Problem 2.   The linear diffusion equation in one-space dimension is given by

$$\frac{\partial u}{\partial t} = \frac{\partial^2 u}{\partial x^2}. \tag{1}$$

Insert the ansatz

$$u(x,t) = \prod_{j=1}^n (x - a_j(t)) \tag{2}$$

into (1) and show that the time dependent functions a_j satisfy the nonlinear autonomous system of ordinary differential equations

$$\frac{da_k}{dt} = -2\sum_j{}' \frac{1}{a_k - a_j} \tag{3}$$

where $\sum'$ means that $j \neq k$. Use the identity

$$\sum_{\substack{j,k=1 \\ j \neq k}}^{n} \frac{1}{x - a_j} \frac{1}{x - a_k} \equiv 2 \sum_{\substack{j,k=1 \\ j \neq k}}^{n} \frac{1}{x - a_k} \frac{1}{a_k - a_j}. \tag{4}$$

Solution 2. Using

$$\frac{\partial u}{\partial t} = -u \sum_{j=1}^{n} \frac{1}{x - a_j} \frac{da_j}{dt}, \quad \frac{\partial u}{\partial x} = u \sum_{j=1}^{n} \frac{1}{x - a_j}$$

where u is given by ansatz (2). It follows that

$$\frac{\partial^2 u}{\partial x^2} = u \left(\sum_{k=1}^{n} \frac{1}{x - a_k} \right) \left(\sum_{j=1}^{n} \frac{1}{x - a_j} \right) - u \sum_{j=1}^{n} \frac{1}{(x - a_j)^2}$$

and therefore

$$\frac{\partial^2 u}{\partial x^2} = u \sum_{j \neq k}^{n} \frac{1}{x - a_j} \frac{1}{x - a_k} \quad \Rightarrow \quad \frac{\partial^2 u}{\partial x^2} = 2u \sum_{j \neq k}^{n} \frac{1}{x - a_k} \frac{1}{a_k - a_j}.$$

Inserting $\partial u / \partial t$ and $\partial^2 u / \partial x^2$ into the linear diffusion equation (1) gives

$$\sum_{k=1}^{n} \frac{1}{x - a_k} \frac{da_k}{dt} = -2 \sum_{j \neq k}^{n} \frac{1}{x - a_k} \frac{1}{a_k - a_j}.$$

Consequently, equation (3) follows.

Problem 3. (i) We set $t \mapsto Dt$, where D is the diffusion coefficient. The linear one-dimensional diffusion equation

$$\frac{\partial u}{\partial t} = \frac{\partial^2 u}{\partial x^2} \tag{1}$$

is invariant under the transformation group

$$t'(x, t, \epsilon) = t, \quad x'(x, t, \epsilon) = t\epsilon + x, \quad u'(x'(x, t), t'(x, t), \epsilon) = u(x, t)e^{-\frac{1}{2}(\frac{1}{2}t\epsilon^2 + x\epsilon)}$$

where ϵ is a real parameter.
(i) Find the infinitesimal generator of this transformation.
(ii) Show that the transformation given by (2) can be derived from the infinitesimal generator.

(iii) Find a *similarity ansatz* and *similarity solution* from the transformation.

Solution 3.   (i) From the transformation we obtain the mapping

$$t'(x,t,\epsilon) = t, \quad x'(x,t,\epsilon) = t\epsilon + x, \quad u'(x,t,u,\epsilon) = ue^{-\frac{1}{2}(\frac{1}{2}t\epsilon^2 + x\epsilon)}.$$

The transformation is called the *Galilean transformation*. From this equation we find

$$\frac{dt'}{d\epsilon}\bigg|_{\epsilon=0} = 0, \quad \frac{dx'}{d\epsilon}\bigg|_{\epsilon=0} = t, \quad \frac{du'}{d\epsilon}\bigg|_{\epsilon=0} = -\frac{1}{2}xu.$$

Therefore the infinitesimal generator is given by

$$G = t\frac{\partial}{\partial x} - \frac{1}{2}xu\frac{\partial}{\partial u}.$$

(ii) The autonomous system associated with the symmetry generator G is

$$\frac{dt}{d\epsilon} = 0, \quad \frac{dx}{d\epsilon} = t, \quad \frac{du}{d\epsilon} = -\frac{1}{2}xu.$$

Solving the first equation yields $t(\epsilon) = t_0$. Inserting this solution into the second equation and integrating, we find $x(\epsilon) = t_0\epsilon + x_0$. Inserting this solution into the third equation and integrating, we arrive at

$$u(\epsilon) = u_0 e^{-\frac{1}{2}(\frac{1}{2}t_0\epsilon^2 + x_0\epsilon)}$$

which is the general solution to this autonomous system. When we set

$$t \to t', \quad t_0 \to t, \quad x \to x', \quad x_0 \to x, \quad u \to u', \quad u_0 \to u$$

we obtain the mapping. To find the transformation we have to calculate

$$\begin{pmatrix} x'(x,t,\epsilon) \\ t'(x,t,\epsilon) \\ u'(x'(x,t), t'(x,t),\epsilon) \end{pmatrix} = e^{\epsilon G} \begin{pmatrix} x \\ t \\ u \end{pmatrix}\bigg|_{u \to u(x,t)}$$

where

$$e^{\epsilon G}x = x + \epsilon\left(t\frac{\partial}{\partial x} - \frac{1}{2}xu\frac{\partial}{\partial u}\right)x + \frac{\epsilon^2}{2!}\left(t\frac{\partial}{\partial x} - \frac{1}{2}xu\frac{\partial}{\partial u}\right)^2 x + \cdots = x + \epsilon t$$

$$e^{\epsilon G}t = t$$

$$e^{\epsilon G}u = u - \frac{\epsilon}{2}xu + \frac{\epsilon^2}{2!}\left(-\frac{1}{2}tu + \frac{1}{4}x^2u\right) + \cdots = ue^{-\frac{1}{2}(\frac{1}{2}t\epsilon^2 + x\epsilon)}.$$

Thus we find the transformation group.

(iii) To find a similarity ansatz we write the mapping as

$$t(x_0, t_0, \epsilon) = t_0, \quad x(x_0, t_0, \epsilon) = t_0\epsilon + x_0, \quad u(x_0, t_0, u_0, \epsilon) = u_0 e^{-\frac{1}{2}(\frac{1}{2}t_0\epsilon^2 + x_0\epsilon)}.$$

We set $t_0 = s$ and $x_0 = 0$, where s is the *similarity variable*. Then we find $s = t$, $\epsilon = x/s$. Therefore the similarity ansatz is given by

$$u(x, t) = f(s)e^{-\frac{1}{4}\frac{x^2}{s}}.$$

Inserting the similarity ansatz into (1) leads to the linear ordinary differential equation

$$\frac{df}{ds} + \frac{1}{2s}f = 0.$$

The general solution of this differential equation is given by $f(s) = C/\sqrt{s}$, where C is the constant of integration. Since $s = t$ is the similarity variable, we find an exact solution of the partial differential equation (1)

$$u(x, t) = \frac{C}{\sqrt{t}} \exp(-x^2/(4t)).$$

Problem 4. Let $a > 0$, $b > 0$ with dimension length. Consider the two-dimensional linear diffusion equation

$$\frac{\partial u}{\partial t} = D \left(\frac{\partial^2 u}{\partial x_1^2} + \frac{\partial^2 u}{\partial x_2^2} \right)$$

with the boundary conditions

$$u(0, x_2, t) = 0, \quad u(a, x_2, t) = 0, \quad 0 \le x_2 \le b, \ t \ge 0$$
$$u(x_1, 0, t) = 0, \quad u(x_1, b, t) = 0, \quad 0 \le x_1 \le a, \ t \ge 0$$

and the initial condition

$$u(x_1, x_2, 0) = f(x_1, x_2), \quad 0 \le x_1 \le a, \ 0 \le x_2 \le b$$

where f is a continuous function. Find solutions of the form (*separation ansatz*)

$$u(x_1, x_2, t) = X_1(x_1)X_2(x_2)T(t).$$

Solution 4. Inserting the separation ansatz into the differential equation yields

$$\frac{d^2X_1}{dx_1^2}X_2T + X_1\frac{d^2X_2}{dx_2^2}T = \frac{X_1X_2}{D}\frac{dT}{dt} \Rightarrow \frac{1}{X_1}\frac{d^2X_1}{dx_1^2} = -\frac{1}{X_2}\frac{d^2X_2}{dx_2^2} + \frac{1}{DT}\frac{dT}{dt}.$$

Thus we find the three ordinary differential equations

$$\frac{d^2 X_1}{dx_1^2} + \lambda X_1 = 0, \quad \frac{d^2 X_2}{dx_2^2} + \mu X_2 = 0, \quad \frac{dT}{dt} D(\lambda + \mu)T = 0$$

where λ, μ are constants and to be determined by the boundary conditions. Imposing the boundary conditions for the first two equations we have

$$X_1(x_1) = C_1 \sin(m\pi x_1/a), \quad X_2(x_2) = C_2 \sin(n\pi x_2/b)$$

with $\lambda_m = m^2\pi^2/a^2$, $\mu_n = n^2\pi^2/b^2$ and $m, n = 1, 2, \ldots$. Then

$$T(t) = C_3 \exp(-D((m\pi/a)^2 + (n\pi/b)^2)t).$$

Hence

$$u_{mn}(x_1, x_2, t) = C_{mn} \exp(-D((m\pi/a)^2 + (n\pi/b)^2)t) \sin(m\pi x_1/a) \sin(n\pi x_2/b)$$

where C_{mn} is an arbitrary constant. It follows that (superposition)

$$u(x_1, x_2, t) =$$

$$\sum_{m=1}^{\infty} \sum_{n=1}^{\infty} C_{mn} \exp(-D((m\pi/a)^2 + (n\pi/b)^2)t) \sin(m\pi x_1/a) \sin(n\pi x_2/b).$$

Next imposing the initial condition for $t = 0$ we have

$$u(x_1, x_2, 0) = f(x_1, x_2) = \sum_{m=1}^{\infty} \sum_{n=1}^{\infty} C_{mn} \sin(m\pi x_1/a) \sin(n\pi x_2/b).$$

Note that $\pi_{mn}(x_1, x_2) = \sin(m\pi x_1/a) \sin(n\pi x_2/b)$ $(m.n = 1, 2, \ldots)$ is an orthonormal basis in the Hilbert space $L_2([0, a] \times [0, b])$. Thus applying the scalar product of f and ϕ_{mn} in this Hilbert space we obtain

$$C_{mn} = \frac{4}{ab} \int_0^b \int_0^a f(x_1, x_2) \sin(m\pi x_1/a) \sin(n\pi x_2/b) dx_1 dx_2.$$

Problem 5.   (i) Show that the general solution of the one-dimensional *wave equation*

$$\frac{1}{c^2} \frac{\partial^2 u}{\partial t^2} = \frac{\partial^2 u}{\partial x^2} \tag{1}$$

is given by

$$u(x, t) = f(x - ct) + g(x + ct) \tag{2}$$

where f and g are smooth functions and c is a positive constant.

(ii) Solve the initial value problem

$$u(t = 0, x) = u_0(x), \quad \frac{\partial u}{\partial t}(t = 0, x) = u_1(x).$$

Solution 5. (i) Let $s := x - ct$ and $r := x + ct$. Then

$$\frac{\partial u}{\partial t} = \frac{df}{ds}\frac{\partial s}{\partial t} + \frac{dg}{dr}\frac{\partial r}{\partial t} = -c\frac{df}{ds} + c\frac{dg}{dr}, \quad \frac{\partial^2 u}{\partial t^2} = c^2\frac{d^2 f}{ds^2} + c^2\frac{d^2 g}{dr^2}. \quad (3)$$

Analogously,

$$\frac{\partial^2 u}{\partial x^2} = \frac{d^2 f}{ds^2} + \frac{d^2 g}{dr^2}. \quad (4)$$

Inserting (3) and (4) into (1) shows that (2) is the (general) solution. Equation (2) is the general solution of the one-dimensional wave equation since f and g are arbitrary.

(ii) From (2) we obtain $u_0(x) = f(x) + g(x)$. Since

$$\frac{\partial u(x, t)}{\partial t} = -c\frac{df(x - ct)}{ds} + c\frac{dg(x + ct)}{dr}$$

it follows that

$$u_1(x) = -c\frac{df(x)}{dx} + c\frac{dg(x)}{dx}.$$

Integrating this equation yields

$$A + \int_a^x u_1(\alpha)d\alpha = -cf(x) + cg(x)$$

where A, a are two arbitrary constants. Using $f(x) = u_0(x) - g(x)$ and $g(x) = u_0(x) - f(x)$ provides the two equations

$$2cf(x) = cu_0(x) - A - \int_a^x u_1(\alpha)d\alpha$$

$$2cg(x) = cu_0(x) + A + \int_a^x u_1(\alpha)d\alpha.$$

With the translation $x \to x - ct$, $x \to x + ct$ it follows that

$$2cf(x - ct) = cu_0(x - ct) - A - \int_a^{x-ct} u_1(\alpha)d\alpha$$

$$2cg(x + ct) = cu_0(x + ct) + A + \int_a^{x+ct} u_1(\alpha)d\alpha.$$

Adding these two equations yields

$$u(x,t) = \frac{1}{2}(u_0(x+ct) + u_0(x-ct)) + \frac{1}{2c}\int_{x-ct}^{x+ct} u_1(\alpha)d\alpha.$$

Consider the Cauchy problem of the the partial differential equation

$$\frac{\partial^2 u}{\partial t^2} - c^2 \frac{\partial^2 u}{\partial x^2} = f(t,x)$$

with

$$u(t,x)|_{t=0} = \phi(x), \qquad \frac{\partial u}{\partial t}\bigg|_{t=0} = \psi(x)$$

where $\phi \in C^2(\mathbb{R})$, $\psi \in C^1(\mathbb{R})$ and the function f is continuous together with the first derivative with respect to x in the half-plane

$$\{t \geq 0, \ -\infty < x < +\infty\}.$$

The solution is (d'Alembert's formula)

$$u(x,t) = \frac{1}{2c}\int_0^t \int_{x-c(t-\tau)}^{x+c(t-\tau)} f(\tau,\eta)d\eta d\tau$$
$$+ \frac{1}{2c}\int_{x-ct}^{x+ct} \psi(\eta)d\eta + \frac{1}{2}(\phi(x+ct) + \phi(x-ct)).$$

Problem 6. The *three-dimensional wave equation* is given by

$$\frac{1}{c^2}\frac{\partial^2 u}{\partial t^2} = \frac{\partial^2 u}{\partial x_1^2} + \frac{\partial^2 u}{\partial x_2^2} + \frac{\partial^2 u}{\partial x_3^2} \equiv \Delta u. \tag{1}$$

(i) Express the wave equation in spherical coordinates. Omit the angle part and find the wave equation for the radial part.
(ii) Find the general solution of the wave equation which only includes the radial part.

Solution 6. (i) The *spherical coordinates* are given by

$$x_1(r,\phi,\theta) = r\cos(\phi)\sin(\theta), \quad x_2(r,\phi,\theta) = r\sin(\phi)\sin(\theta), \quad x_3(r,\phi,\theta) = r\cos(\theta)$$

where $0 \leq \phi < 2\pi$, $0 < \theta < \pi$ and $r > 0$. Let

$$u(x_1(r,\phi,\theta), x_2(r,\phi,\theta), x_3(r,\phi,\theta), t) = v(r,\phi,\theta,t).$$

Applying the chain rule we find from (1) that

$$\frac{1}{r^2}\frac{\partial}{\partial r}\left(r^2\frac{\partial v}{\partial r}\right) + \frac{1}{r^2\sin(\theta)}\frac{\partial}{\partial\theta}\left(\sin(\theta)\frac{\partial v}{\partial\theta}\right) + \frac{1}{r^2\sin^2(\theta)}\frac{\partial^2 v}{\partial\phi^2} = \frac{1}{c^2}\frac{\partial^2 v}{\partial t^2}.$$

If v is a *spherically-symmetric solution* of the wave equation, i.e. $\partial v/\partial\theta = 0$ and $\partial v/\partial\phi = 0$ the wave equation becomes

$$\frac{1}{r^2}\frac{\partial}{\partial r}\left(r^2\frac{\partial v}{\partial r}\right) = \frac{1}{c^2}\frac{\partial^2 v}{\partial t^2} \quad\Rightarrow\quad \frac{1}{r^2}\left(2r\frac{\partial v}{\partial r} + r^2\frac{\partial^2 v}{\partial r^2}\right) = \frac{1}{c^2}\frac{\partial^2 v}{\partial t^2}.$$

We have the identity

$$\frac{\partial^2(rv)}{\partial r^2} \equiv \frac{\partial}{\partial r}\left(v + r\frac{\partial v}{\partial r}\right) \equiv \frac{\partial v}{\partial r} + \frac{\partial v}{\partial r} + r\frac{\partial^2 v}{\partial r^2} \equiv \frac{1}{r}\left(2r\frac{\partial v}{\partial r} + r^2\frac{\partial^2 v}{\partial r^2}\right).$$

Thus the wave equation takes the form

$$\frac{1}{r}\frac{\partial^2(rv)}{\partial r^2} = \frac{1}{c^2}\frac{\partial^2 v}{\partial t^2}, \quad\text{or}\quad \frac{\partial^2(rv)}{\partial r^2} = \frac{1}{c^2}\frac{\partial^2(rv)}{\partial t^2}.$$

We define $\psi(r,t) := rv(r,t)$. Then

$$\frac{\partial^2\psi}{\partial r^2} = \frac{1}{c^2}\frac{\partial^2\psi}{\partial t^2}.$$

We have found the general solution of this partial differential equation, namely

$$\psi(r,t) = f_1(r - ct) + f_2(r + ct).$$

It follows that

$$v(r,t) = \frac{1}{r}f_1(r - ct) + \frac{1}{r}f_2(r + ct)$$

with $r > 0$. Let $v_0(r) = v(r,0)$, $v_1(r) = \partial v(r,0)/\partial t$ be the initial distributions. Then

$$\psi_0(r) = \psi(r,0) = (rv)(r,0) = rv_0$$

$$\psi_1(r) = \frac{\partial\psi}{\partial t}(r,0) = r\frac{\partial v}{\partial t}(r,0) = rv_1(r).$$

Using the method described above the solution to the initial value problem is given by

$$v(r,t) = \frac{1}{2r}(r + ct)v_0(r + ct) + \frac{1}{2r}(r - ct)v_0(r - ct) + \frac{1}{2cr}\int_{r-ct}^{r+ct}\alpha v_1(\alpha)d\alpha.$$

Problem 7. Consider the two-dimensional wave equation

$$\frac{\partial^2 u}{\partial x_0^2} - \frac{\partial^2 u}{\partial x_1^2} - \frac{\partial^2 u}{\partial x_2^2} = F(u)$$

and

$$u(x_0, x_1, x_2) = u(f(x_0, x_1, x_2)).$$

Find the ordinary differential equation $u(f)$, where the function f satisfies

$$\frac{\partial^2 f}{\partial x_0^2} - \frac{\partial^2 f}{\partial x_1^2} - \frac{\partial^2 f}{\partial x_2^2} = -f, \quad \left(\frac{\partial f}{\partial x_0}\right)^2 - \left(\frac{\partial f}{\partial x_1}\right)^2 - \left(\frac{\partial f}{\partial x_2}\right)^2 = -f^2.$$

Solution 7.   Applying the *chain rule* we have

$$\frac{\partial u}{\partial x_0} = \frac{\partial u}{\partial f}\frac{\partial f}{\partial x_0}, \quad \frac{\partial u}{\partial x_1} = \frac{\partial u}{\partial f}\frac{\partial f}{\partial x_1}, \quad \frac{\partial u}{\partial x_2} = \frac{\partial u}{\partial f}\frac{\partial f}{\partial x_2}$$

and

$$\frac{\partial^2 u}{\partial x_0^2} = \frac{\partial^2 u}{\partial f^2}\left(\frac{\partial f}{\partial x_0}\right)^2 + \frac{\partial u}{\partial f}\frac{\partial^2 f}{\partial x_0^2},$$

$$\frac{\partial^2 u}{\partial x_1^2} = \frac{\partial^2 u}{\partial f^2}\left(\frac{\partial f}{\partial x_1}\right)^2 + \frac{\partial u}{\partial f}\frac{\partial^2 f}{\partial x_1^2},$$

$$\frac{\partial^2 u}{\partial x_2^2} = \frac{\partial^2 u}{\partial f^2}\left(\frac{\partial f}{\partial x_2}\right)^2 + \frac{\partial u}{\partial f}\frac{\partial^2 f}{\partial x_2^2}.$$

It follows that

$$\frac{\partial^2 u}{\partial f^2}\left(\left(\frac{\partial f}{\partial x_0}\right)^2 - \sum_{j=1}^{2}\left(\frac{\partial f}{\partial x_j}\right)^2\right) + \frac{\partial u}{\partial f}\left(\frac{\partial^2 f}{\partial x_0^2} - \sum_{j=1}^{2}\frac{\partial^2 f}{\partial x_j^2}\right) = F(u(f))$$

and the ordinary differential equation is

$$\frac{\partial^2 u}{\partial f^2}f^2 + \frac{\partial u}{\partial f}f + F(u(f)) = 0.$$

Problem 8.   The *telegraph equation* is given by

$$\frac{\partial^2 w}{\partial t^2} + a\frac{\partial w}{\partial t} + bw = c^2\frac{\partial^2 w}{\partial x^2}. \tag{1}$$

Show that this equation can be transformed into the canonical form

$$\frac{\partial^2 u}{\partial \eta \partial \xi} + ku = 0 \tag{2}$$

where $k = (a^2 - 4b^2)/(16c^2)$ by applying the transformation

$$\eta(x,t) = x - ct, \qquad \xi(x,t) = x + ct \tag{3a}$$

$$w(x(\eta,\xi), t(\eta,\xi)) = u(\eta,\xi)\exp(-at(\eta,\xi)/2). \tag{3b}$$

Solution 8. From (3a) and (3b) we find by applying the chain rule

$$\frac{\partial w}{\partial \eta} = \frac{\partial w}{\partial x}\frac{\partial x}{\partial \eta} + \frac{\partial w}{\partial t}\frac{\partial t}{\partial \eta}$$

$$= \frac{\partial u}{\partial \eta}\exp(-at(\eta,\xi)/2) - \frac{a}{2}u(\eta,\xi)\exp(-at(\eta,\xi)/2)\frac{\partial t}{\partial \eta}.$$

Since $\partial x/\partial \eta = 1/2$, $\partial t/\partial \eta = -1/(2c)$, $\partial x/\partial \xi = 1/2$, $\partial t/\partial \xi = 1/(2c)$ and setting $s := -at(\beta,\xi)/2$ we obtain

$$\frac{1}{2}\frac{\partial w}{\partial x} - \frac{1}{2c}\frac{\partial w}{\partial t} = \frac{\partial u}{\partial \eta}\exp(s) + \frac{a}{4c}u(\eta,\xi)\exp(s). \tag{4}$$

Analogously when we take the derivative of w with respect to ξ we find

$$\frac{1}{2}\frac{\partial w}{\partial x} + \frac{1}{2c}\frac{\partial w}{\partial t} = \frac{\partial u}{\partial \xi}\exp(s) - \frac{a}{4c}u(\eta,\xi)\exp(s). \tag{5}$$

Subtracting (4) and (5) yields

$$\frac{\partial w}{\partial t} = -c\frac{\partial u}{\partial \eta}\exp(s) + c\frac{\partial u}{\partial \xi}\exp(s) - \frac{a}{2}u\exp(s).$$

Taking the derivative of (4) with respect to ξ yields

$$\frac{\partial^2 w}{\partial x^2} - \frac{1}{c^2}\frac{\partial^2 w}{\partial t^2} = 4\frac{\partial^2 u}{\partial \eta \partial \xi}\exp(s) - \frac{a}{c}\frac{\partial u}{\partial \eta}\exp(s) + \frac{a}{c}\frac{\partial u}{\partial \xi}\exp(s) - \frac{a^2}{4c^2}u\exp(s).$$

Taking the derivative of (5) with respect to η yields

$$\frac{\partial^2 w}{\partial x^2} - \frac{1}{c^2}\frac{\partial^2 w}{\partial t^2} = 4\frac{\partial^2 u}{\partial \eta \partial \xi}\exp(s) + \frac{a}{c}\frac{\partial u}{\partial \xi}\exp(s) - \frac{a}{c}\frac{\partial u}{\partial \eta}\exp(s) - \frac{a^2}{4c^2}u\exp(s).$$

Inserting these equations into (1) yields (2).

Problem 9. Consider the linear partial differential equation

$$\frac{\partial u}{\partial t} + c_0\frac{\partial u}{\partial x} + \beta\frac{\partial^3 u}{\partial x^3} = 0 \tag{1}$$

where β and c_0 are positive constants.
(i) Find the *dispersion relation* (i.e. the relation between the frequency ω and the wave vector k) using the ansatz

$$u(x,t) = u_0\exp(i(kx - \omega(k)t))$$

(ii) Let $x'(x,t) = x - c_0t$, $t'(x,t) = t$, $u'(x'(x,t),t'(x,t)) = u(x,t)$. Find the partial differential equation for $u'(x',t')$.

(iii) Show that the solution to the initial-value problem of the partial differential equation

$$\frac{\partial u}{\partial t} + \beta \frac{\partial^3 u}{\partial x^3} = 0 \tag{2}$$

is given by

$$u(x,t) = (\pi)^{-1/2}(3\beta t)^{-1/3} \int_{-\infty}^{\infty} \text{Ai}\left(\frac{x-s}{(3\beta t)^{1/3}}\right) u(s, t=0) ds \tag{3}$$

where

$$\text{Ai}(s) := \frac{1}{\sqrt{\pi}} \int_{0}^{\infty} \cos\left(\frac{v^3}{3} + vs\right) dv$$

is the *Airy function*. The Airy functions $\text{Ai}(z)$ and $\text{Bi}(z)$ can be expressed in terms of Bessel functions. Vice versa the Bessel functions can be expressed in terms of the Airy functions $\text{Ai}(z)$ and $\text{Bi}(z)$.

Solution 9.   (i) Inserting the ansatz into (1) leads to the dispersion relation $\omega(k) = c_0 k - \beta k^3$.
(ii) Straightforward application of the chain rule yields

$$\frac{\partial u'}{\partial t'} + \beta \frac{\partial^3 u'}{\partial x'^3} = 0.$$

(iii) The Airy function has the asymptotic representation

$$\text{Ai}(s) = \begin{cases} \frac{1}{2} s^{-1/4} \exp\left(-\frac{2}{3} s^{3/2}\right) & \text{for } s \to \infty \\ |s|^{-1/4} \cos\left(\frac{2}{3}|s|^{3/2} - \frac{\pi}{4}\right) & \text{for } s \to -\infty. \end{cases}$$

If we assume that the initial distribution $u(x, t = 0)$ tends to zero sufficiently fast as $x \to \pm\infty$ we find by differentiating that (3) is the solution of the initial-value problem of the partial differential equation (2).

Problem 10.   Let u be a differentiable function of x_1, x_2, x_3 which satisfies the partial differential equation

$$(x_2 - x_3)\frac{\partial u}{\partial x_1} + (x_3 - x_1)\frac{\partial u}{\partial x_2} + (x_1 - x_2)\frac{\partial u}{\partial x_3} = 0$$

i.e. we have the smooth *vector field* in $\mathbb{R}^3$

$$V = (x_2 - x_3)\frac{\partial}{\partial x_1} + (x_3 - x_1)\frac{\partial}{\partial x_2} + (x_1 - x_2)\frac{\partial}{\partial x_3}.$$

Show that u contains x_1, x_2, x_3 only in combinations $x_1 + x_2 + x_3$ and $x_1^2 + x_2^2 + x_3^2$.

Solution 10. The auxiliary equations of the partial differential equation are

$$\frac{dx_1}{x_2 - x_3} = \frac{dx_2}{x_3 - x_1} = \frac{dx_3}{x_1 - x_2} = \frac{du}{0}.$$

They are equivalent to the three differential relations

$$du = 0, \qquad dx_1 + dx_2 + dx_3 = 0, \qquad x_1 dx_1 + x_2 dx_2 + x_3 dx_3 = 0.$$

Thus the integrals are $u = c_1$, $x_1 + x_2 + x_3 = c_2$, $x_1^2 + x_2^2 + x_3^2 = c_3$, where c_1, c_2, c_3 are constants. Therefore the general solution is given by

$$u(x_1, x_2, x_3) = f(x_1 + x_2 + x_3, x_1^2 + x_2^2 + x_3^2)$$

where f is an arbitrary differentiable function.

Problem 11. Consider an electron of mass m confined to the $x_1 - x_2$ plane and a constant *magnetic flux density* **B** parallel to the x_3-axis, i.e.

$$\mathbf{B} = \begin{pmatrix} 0 \\ 0 \\ B \end{pmatrix}.$$

The Hamilton operator for this two-dimensional electron is given by

$$\hat{H} = \frac{(\hat{\mathbf{p}} + e\mathbf{A})^2}{2m} = \frac{1}{2m}((\hat{p}_1 + eA_1)^2 + (\hat{p}_2 + eA_2)^2)$$

where **A** is the *vector potential* with $\mathbf{B} = \nabla \times \mathbf{A}$ and

$$\hat{p}_1 = -i\hbar\frac{\partial}{\partial x_1}, \qquad \hat{p}_2 = -i\hbar\frac{\partial}{\partial x_2}.$$

(i) Show that **B** can be obtained from the vector potential

$$\mathbf{A} = \begin{pmatrix} 0 \\ x_1 B \\ 0 \end{pmatrix} \quad \text{or} \quad \mathbf{A} = \begin{pmatrix} -x_2 B \\ 0 \\ 0 \end{pmatrix}.$$

(ii) Use the second choice for **A** to find the Hamilton operator $\hat{H}$.
(iii) Show that $[\hat{H}, \hat{p}_1] = 0$.
(iv) Let $k = p_1/\hbar$. Make the ansatz for the wave function

$$\psi(x_1, x_2) = e^{ikx_1}\phi(x_2)$$

and show that the eigenvalue equation $\hat{H}\psi = E\psi$ reduces to

$$\left(-\frac{\hbar^2}{2m}\frac{d^2}{dx_2^2} + \frac{m\omega_c^2}{2}(x_2 - x_{20})^2 \right)\phi(x_2) = E\phi(x_2)$$

where $\omega_c := eB/m$, $x_{20} := \hbar k/(eB)$.
(v) Show that the eigenvalues are given by

$$E_n = (n+1/2)\hbar\omega_c, \qquad n = 0, 1, 2, \ldots .$$

Solution 11.   (i) Since

$$\nabla \times \mathbf{A} := \begin{pmatrix} \partial A_3/\partial x_2 - \partial A_2/\partial x_3 \\ \partial A_1/\partial x_3 - \partial A_3/\partial x_1 \\ \partial A_2/\partial x_1 - \partial A_1/\partial x_2 \end{pmatrix}$$

we obtain the desired result.
(ii) Inserting $A_1 = -x_2 B$, $A_2 = 0$, $A_3 = 0$ into the Hamilton operator provides

$$\hat{H} = \frac{1}{2m}(\hat{p}_1 - ex_2 B)^2 + \frac{1}{2m}\hat{p}_2^2.$$

(iii) Since coordinate x_1 does not appear in the Hamilton operator $\hat{H}$ and $[\hat{x}_2, \hat{p}_1] = [\hat{p}_2, \hat{p}_1] = 0$ it follows that $[\hat{H}, \hat{p}_1] = 0$.
(iv) From (iii) we have

$$\hat{p}_1 \psi(x_1, x_2) \equiv -i\hbar \frac{\partial}{\partial 1} \psi(x_1, x_2) = \hbar k \psi(x_1, x_2), \quad \hat{H}\psi(x_1, x_2) = E\psi(x_1, x_2).$$

Inserting the ansatz $\psi(x_1, x_2) = e^{ikx_1}\psi(x_2)$ into the first equation we find

$$\psi(x_1, x_2) = e^{ikx_1}\phi(x_2).$$

For $\hat{H}\psi$ we obtain

$$\begin{aligned}
\hat{H}\psi &= \frac{1}{2m}((\hat{p}_1 - eyB)^2 + \hat{p}_2)e^{ikx}\phi(y) \\
&= \frac{1}{2m}e^{ikx}((\hbar k - eyB)^2 + \hat{p}_2^2)\phi(x_2) \\
&= e^{ikx}\left(\frac{m\omega_c^2}{2}(y - y_0)^2 + \frac{\hat{p}_2^2}{2m}\right)\phi(x_2).
\end{aligned}$$

Now the right-hand side must be equal to $E\psi(x_1, x_2) = Ee^{ikx_1}\phi(x_2)$. Thus since

$$\hat{p}_2^2 = -\hbar^2 \frac{\partial^2}{\partial x_2^2}$$

the second order ordinary differential equation follows

$$\left(-\frac{\hbar^2}{2m}\frac{d^2}{dx_2^2} + \frac{m\omega_c^2}{2}(x_2 - x_{20})^2\right)\phi(x_2) = E\phi(x_2).$$

(v) The eigenvalue problem of (iv) is essentially the one-dimensional harmonic oscillator except the term is $(x_2 - x_{20})^2$ instead of x_2^2. This means that the centre of oscillation is at $x_2 = x_{20}$ instead of 0. This has no influence on the eigenvalues which are the same as for the harmonic oscillator, namely

$$E_n = \left(n + \frac{1}{2}\right)\hbar\omega_c, \quad n = 0, 1, 2, \dots.$$

Problem 12. Consider the *Dirac-Hamilton operator*

$$\hat{H} = mc^2\beta + c\alpha \cdot \mathbf{p} \equiv mc^2\beta + c(\alpha_1 p_1 + \alpha_2 p_2 + \alpha_3 p_3) \tag{1}$$

where m is the rest mass, c is the speed of light and

$$\beta := \begin{pmatrix} 1 & 0 & 0 & 0 \\ 0 & 1 & 0 & 0 \\ 0 & 0 & -1 & 0 \\ 0 & 0 & 0 & -1 \end{pmatrix}, \quad \alpha_1 := \begin{pmatrix} 0 & 0 & 0 & 1 \\ 0 & 0 & 1 & 0 \\ 0 & 1 & 0 & 0 \\ 1 & 0 & 0 & 0 \end{pmatrix},$$

$$\alpha_2 := \begin{pmatrix} 0 & 0 & 0 & -i \\ 0 & 0 & i & 0 \\ 0 & -i & 0 & 0 \\ i & 0 & 0 & 0 \end{pmatrix}, \quad \alpha_3 := \begin{pmatrix} 0 & 0 & 1 & 0 \\ 0 & 0 & 0 & -1 \\ 1 & 0 & 0 & 0 \\ 0 & -1 & 0 & 0 \end{pmatrix}$$

$$\hat{p}_1 := -i\hbar\frac{\partial}{\partial x_1}, \quad \hat{p}_2 := -i\hbar\frac{\partial}{\partial x_2}, \quad \hat{p}_3 := -i\hbar\frac{\partial}{\partial x_3}.$$

Thus the Dirac-Hamilton operator takes the form

$$\hat{H} = c\hbar \begin{pmatrix} mc/\hbar & 0 & -i\frac{\partial}{\partial x_3} & -i\frac{\partial}{\partial x_1} - \frac{\partial}{\partial x_2} \\ 0 & mc/\hbar & -i\frac{\partial}{\partial x_1} + \frac{\partial}{\partial x_2} & i\frac{\partial}{\partial x_3} \\ -i\frac{\partial}{\partial x_3} & -i\frac{\partial}{\partial x_1} - \frac{\partial}{\partial x_2} & -mc/\hbar & 0 \\ -i\frac{\partial}{\partial x_1} + \frac{\partial}{\partial x_2} & i\frac{\partial}{\partial x_3} & 0 & -mc/\hbar \end{pmatrix}.$$

Let I_4 be the 4×4 unit matrix. Use the *Heisenberg equation of motion* to find the time evolution of β, α_j and $I_4 p_j$, where $j = 1, 2, 3$. The *Heisenberg equation of motion* is given by

$$i\hbar\frac{d\hat{A}}{dt} = [\hat{A}, \hat{H}](t).$$

Solution 12. We find for the commutator

$$[\beta, \hat{H}] = 2c\hbar \begin{pmatrix} 0 & 0 & -i\frac{\partial}{\partial x_3} & -i\frac{\partial}{\partial x_1} - \frac{\partial}{\partial x_2} \\ 0 & 0 & -i\frac{\partial}{\partial x_1} + \frac{\partial}{\partial x_2} & i\frac{\partial}{\partial x_3} \\ i\frac{\partial}{\partial x_3} & i\frac{\partial}{\partial x_1} + \frac{\partial}{\partial x_2} & 0 & 0 \\ i\frac{\partial}{\partial x_1} - \frac{\partial}{\partial x_2} & -i\frac{\partial}{\partial x_3} & 0 & 0 \end{pmatrix}.$$

Thus the Heisenberg equation of motion takes the form

$$i\hbar\frac{d\beta}{dt} = 2c\beta\alpha_1(p_1 - ip_2)(t) + 2c\alpha_3 p_3(t).$$

Since $[\alpha_1, \hat{H}] = -2\hat{H}\alpha_1 + 2cI_4p_1$ we obtain

$$i\hbar\frac{d\alpha_1}{dt} = -(2\hat{H}\alpha_1 + 2cI_4p_1)(t).$$

Analogously,

$$i\hbar\frac{d\alpha_2}{dt} = -(2\hat{H}\alpha_2 + 2cI_4p_2)(t), \quad i\hbar\frac{d\alpha_3}{dt} = -(2\hat{H}\alpha_3 + 2cI_4p_3)(t).$$

Furthermore we obviously have $[I_4p_j, \hat{H}] = 0_4$, $(j = 1, 2, 3)$. Thus

$$i\hbar\frac{d}{dt}I_4p_j = 0_4, \qquad j = 1, 2, 3.$$

From these equations we obtain

$$i\hbar\frac{d^2\alpha}{dt^2} = 2\frac{d\alpha}{dt}\hat{H} = -2\hat{H}\frac{d\alpha}{dt}.$$

This equation can be integrated once and we find

$$\frac{d\alpha}{dt} = \frac{d\alpha(0)}{dt}\exp(-2i\hat{H}t/\hbar).$$

We also have the identities $[\hat{H}, \dot{\alpha}]_+ = 0_4$, $[\hat{H}, \dot{\alpha}] = 2\hat{H}\alpha$.

Problem 13.   The nonlinear partial differential equation

$$\frac{\partial \mathbf{S}}{\partial t} = \mathbf{S} \times \frac{\partial^2 \mathbf{S}}{\partial x^2} \tag{1}$$

where (constraint) $S_1^2 + S_2^2 + S_3^2 = 1$ is called the *Heisenberg ferromagnet equation* in one-space dimension. Here $\times$ denotes the vector product and

$$\mathbf{S} = (\, S_1 \quad S_2 \quad S_3 \,)^T.$$

(i) Write equation (1) in components.
(ii) Let $Q := 1 + u^2 + v^2$ and set

$$S_1 := \frac{2u}{Q}, \qquad S_2 := \frac{2v}{Q}, \qquad S_3 := \frac{-1 + u^2 + v^2}{Q}. \tag{2}$$

Show that constraint (2) is satisfied identically. Find the time evolution of u and v. This transformation is called *stereographic projection*.

Solution 13. (i) Utilizing the properties of the vector product provides

$$\frac{\partial S_1}{\partial t} = S_2 \frac{\partial^2 S_3}{\partial x^2} - S_3 \frac{\partial^2 S_2}{\partial x^2} \tag{3a}$$

$$\frac{\partial S_2}{\partial t} = S_3 \frac{\partial^2 S_1}{\partial x^2} - S_1 \frac{\partial^2 S_3}{\partial x^2} \tag{3b}$$

$$\frac{\partial S_3}{\partial t} = S_1 \frac{\partial^2 S_2}{\partial x^2} - S_2 \frac{\partial^2 S_1}{\partial x^2}. \tag{3c}$$

(ii) From (2) we find

$$\frac{\partial S_1}{\partial t} = \frac{2}{Q} \frac{\partial u}{\partial t} - \frac{2u}{Q^2} \frac{\partial Q}{\partial t} = \frac{1}{Q^2} \left(2Q \frac{\partial u}{\partial t} - 2u \frac{\partial Q}{\partial t} \right) \tag{4a}$$

$$\frac{\partial S_2}{\partial t} = \frac{2}{Q} \frac{\partial v}{\partial t} - \frac{2v}{Q^2} \frac{\partial Q}{\partial t} = \frac{1}{Q^2} \left(2Q \frac{\partial v}{\partial t} - 2v \frac{\partial Q}{\partial t} \right). \tag{4b}$$

Since

$$S_3 = \frac{-1 + u^2 + v^2}{Q} \equiv \frac{Q - 2}{Q} = 1 - \frac{2}{Q}$$

we have

$$\frac{\partial S_3}{\partial t} = \frac{2}{Q^2} \frac{\partial Q}{\partial t}. \tag{4c}$$

Now

$$\frac{\partial^2 S_1}{\partial x^2} = \frac{2}{Q} \frac{\partial^2 u}{\partial x^2} - \frac{4}{Q^2} \frac{\partial u}{\partial x} \frac{\partial Q}{\partial x} - \frac{2u}{Q^2} \frac{\partial^2 Q}{\partial x^2} + \frac{4u}{Q^3} \left(\frac{\partial Q}{\partial x} \right)^2 \tag{5a}$$

$$\frac{\partial^2 S_2}{\partial x^2} = \frac{2}{Q} \frac{\partial^2 v}{\partial x^2} - \frac{4}{Q^2} \frac{\partial v}{\partial x} \frac{\partial Q}{\partial x} - \frac{2v}{Q^2} \frac{\partial^2 Q}{\partial x^2} + \frac{4v}{Q^3} \left(\frac{\partial Q}{\partial x} \right)^2 \tag{5b}$$

and

$$\frac{\partial^2 S_3}{\partial x^2} = \frac{2}{Q^2} \frac{\partial^2 Q}{\partial x^2} - \frac{4}{Q^3} \left(\frac{\partial Q}{\partial x} \right)^2. \tag{5c}$$

Inserting (4) through (5) into (3c) yields

$$\frac{\partial Q}{\partial t} = u \left(2 \frac{\partial^2 v}{\partial x^2} - \frac{4}{Q} \frac{\partial v}{\partial x} \frac{\partial Q}{\partial x} \right) - v \left(2 \frac{\partial^2 u}{\partial x^2} - \frac{4}{Q} \frac{\partial u}{\partial x} \frac{\partial Q}{\partial x} \right).$$

Inserting (4) through (5) and this equation into (3a) and (3b) yields

$$Q \frac{\partial v}{\partial t} + Q \frac{\partial^2 u}{\partial x^2} - 2 \left(\left(\frac{\partial u}{\partial x} \right)^2 - \left(\frac{\partial v}{\partial x} \right)^2 \right) u - 4v \frac{\partial u}{\partial x} \frac{\partial v}{\partial x} = 0$$

$$-Q \frac{\partial u}{\partial t} + Q \frac{\partial^2 v}{\partial x^2} + 2 \left(\left(\frac{\partial u}{\partial x} \right)^2 - \left(\frac{\partial v}{\partial x} \right)^2 \right) v - 4u \frac{\partial u}{\partial x} \frac{\partial v}{\partial x} = 0.$$

Problem 14.   Let

$$-\frac{\hbar}{i}\frac{\partial\psi}{\partial t} = \hat{H}\psi \tag{1}$$

be the *Schrödinger equation*, where

$$\hat{H} := -\frac{\hbar^2}{2m}\Delta + U(\mathbf{r}), \qquad \Delta := \frac{\partial^2}{\partial x_1^2} + \frac{\partial^2}{\partial x_2^2} + \frac{\partial^2}{\partial x_3^2}$$

and $\mathbf{r} = (x_1, x_2, x_3)$. Let

$$\rho(\mathbf{r}, t) := \bar{\psi}(\mathbf{r}, t)\psi(\mathbf{r}, t)$$

where $\bar{\psi}$ denotes the complex conjugate of ψ. Find $\mathbf{j}$ such that

$$\mathrm{div}(\mathbf{j}) + \frac{\partial\rho}{\partial t} = 0 \tag{2}$$

where

$$\mathrm{div}(\mathbf{j}) := \frac{\partial j_1}{\partial x_1} + \frac{\partial j_2}{\partial x_2} + \frac{\partial j_3}{\partial x_3}.$$

Equation (2) is called a *conservation law*.

Solution 14.   We have

$$-\frac{\hbar}{i}\frac{\partial\psi}{\partial t} = -\frac{\hbar^2}{2m}\Delta\psi + U\psi \;\Rightarrow\; \frac{\hbar}{i}\frac{\partial\bar{\psi}}{\partial t} = -\frac{\hbar^2}{2m}\Delta\bar{\psi} + U\bar{\psi}.$$

From ρ we obtain

$$\frac{\partial\rho}{\partial t} = \frac{\partial}{\partial t}(\bar{\psi}\psi) = \bar{\psi}\frac{\partial\psi}{\partial t} + \frac{\partial\bar{\psi}}{\partial t}\psi.$$

Inserting the Schrödinger equation gives

$$-\frac{\hbar}{i}\frac{\partial\rho}{\partial t} = \bar{\psi}\hat{H}\psi - \psi\hat{H}\bar{\psi}.$$

Now

$$\bar{\psi}\hat{H}\psi - \psi\hat{H}\bar{\psi} = -\frac{\hbar^2}{2m}(\bar{\psi}\Delta\psi - \psi\Delta\bar{\psi}) = -\frac{\hbar^2}{2m}\mathrm{div}(\bar{\psi}\nabla\psi - \psi\nabla\bar{\psi}).$$

Here ∇ denotes the gradient. Thus

$$\mathbf{j} = \frac{\hbar}{2mi}(\bar{\psi}\nabla\psi - \psi\nabla\bar{\psi}).$$

Problem 15.   (i) Solve the first order partial differential equation (*inviscid Burgers equation*)

$$F\left(u, \frac{\partial u}{\partial t}, \frac{\partial u}{\partial x}\right) \equiv \frac{\partial u}{\partial t} + u\frac{\partial u}{\partial x} = 0 \tag{1}$$

with the initial condition $u(x, t = 0) = x$.

(ii) Show that the nonlinear partial differential equation

$$\frac{\partial u}{\partial t} + (\alpha + \beta u)\frac{\partial u}{\partial x} = 0 \tag{2}$$

admits the general solution

$$u(x, t) = f(x - (\alpha + \beta u(x, t))t) \tag{3}$$

to the initial-value problem, where $u(x, t = 0) = f(x)$ and $\alpha, \beta \in \mathbb{R}$. Simplify case (ii) to case (i).

Solution 15. (i) From (1) we obtain the surface $F(u, p, q) \equiv p + uq = 0$. Then we obtain the autonomous system of first-order ordinary differential equations

$$\frac{dt}{ds} = \frac{\partial F}{\partial p} = 1, \quad \frac{dx}{ds} = \frac{\partial F}{\partial q} = u, \quad \frac{du}{ds} = p\frac{\partial F}{\partial p} + q\frac{\partial F}{\partial q} = p + uq = 0$$

where we have used $F(u, p, q) \equiv p + uq$. The corresponding vector field is

$$V = \frac{\partial}{\partial t} + u\frac{\partial}{\partial x} + 0\frac{\partial}{\partial u} = \frac{\partial}{\partial t} + u\frac{\partial}{\partial x}.$$

These autonomous system differential equations determine the *characteristic strip*. The solution of the initial-value problem of this autonomous system of differential equations is given by

$$x(s) = u_0 s + x_0, \quad t(s) = s + t_0, \quad u(s) = u_0.$$

To impose the initial condition we have to set

$$x_0(\alpha) = \alpha, \quad t_0(\alpha) = 0, \quad u_0(\alpha) = \alpha.$$

Thus we obtain as solution of the autonomous system

$$x(s, \alpha) = \alpha s + \alpha, \quad t(s, \alpha) = s, \quad u(s, \alpha) = \alpha.$$

Since

$$D := \det\begin{pmatrix} \partial t/\partial s & \partial t/\partial \alpha \\ \partial x/\partial s & \partial x/\partial \alpha \end{pmatrix} = \begin{pmatrix} 1 & 0 \\ \alpha & s+1 \end{pmatrix} = 1 + s$$

we can solve these equations with respect to s and α if $D \neq 0$. We find

$$s(x, t) = t, \quad \alpha(x, t) = \frac{x}{1+t}.$$

Inserting s and α into $u(s, \alpha) = \alpha$ gives the solution $u(x, t) = x/(1 + t)$ of the initial-value problem.

(ii) Since

$$\frac{\partial u}{\partial t} = \left(-\alpha - \beta u - \beta t \frac{\partial u}{\partial t}\right) f', \qquad \frac{\partial u}{\partial x} = \left(1 - \beta t \frac{\partial u}{\partial x}\right) f'$$

where f' is the derivative of f with respect to the argument we obtain

$$\frac{\partial u}{\partial t} = (-\alpha - \beta u)\frac{\partial u}{\partial x}$$

which is (2). Let us now simplify (2) to case (i). Let $\alpha = 0$ and $\beta = 1$. Then we obtain from (3) that $u(x, t) = f(x - u(x, t)t)$. Since $u(x, t = 0) = f(x) = x$ it follows that

$$u(x, t) = f(x - u(x, t)t) = x - u(x, t)t.$$

From this solution we see that the solution $u(x, t) = x/(1 + t)$ follows.

Problem 16.   (i) Show that the nonlinear partial differential equation

$$\frac{\partial^2 u}{\partial x_0^2} - \frac{\partial^2 u}{\partial x_1^2} = \mu^2 u - \lambda u^3 \tag{1}$$

can be derived from the *Lagrange density*

$$\mathcal{L} = \frac{1}{2}\left(\left(\frac{\partial u}{\partial x_0}\right)^2 - \left(\frac{\partial u}{\partial x_1}\right)^2\right) + \frac{1}{2}\mu^2 u^2 - \frac{1}{4}\lambda u^4 \tag{2}$$

where $\mu^2 > 0$ and $\lambda > 0$. From the Lagrange density $\mathcal{L}$ it follows that we can introduce the *Hamilton density* and therefore the *energy functional*

$$E(u) = \int_{-\infty}^{+\infty} dx_1 \left(\frac{1}{2}\left(\frac{\partial u}{\partial x_0}\right)^2 + \frac{1}{2}\left(\frac{\partial u}{\partial x_1}\right)^2 - \frac{1}{2}\mu^2 u^2 + \frac{1}{4}\lambda u^4\right).$$

(ii) Find the space-time independent solutions of (1).
(iii) Calculate the energy difference between the different solutions of (i).
(iv) Show that (1) admits the time-independent solution

$$u_{\text{kink}}(x_1) = \pm\frac{\mu}{\sqrt{\lambda}}\tanh\left(\frac{\mu}{\sqrt{2}}(x_1 - x_{1,0})\right) \tag{3}$$

and calculate the energy difference. Owing to its shape, this solution is called the *kink solution*.

Solution 16.   (i) The *Euler-Lagrange equation* is given by

$$\frac{\partial \mathcal{L}}{\partial u} - \frac{\partial}{\partial x_1}\left(\frac{\partial \mathcal{L}}{\partial \left(\frac{\partial u}{\partial x_1}\right)}\right) - \frac{\partial}{\partial x_0}\left(\frac{\partial \mathcal{L}}{\partial \left(\frac{\partial u}{\partial x_0}\right)}\right) = 0.$$

Inserting the Lagrange density (2) into (4) gives (1).

(ii) The space-time independent solutions are determined by

$$\mu^2 u - \lambda u^3 = 0.$$

We obtain the three solutions $u = 0$, $u = \pm\mu/\sqrt{\lambda}$.

(iii) Straightforward calculation shows that

$$E(u = 0) - E\left(u = \pm\frac{\mu}{\sqrt{\lambda}}\right) = \lim_{L\to\infty} \int_{-L}^{+L} dx_1 \frac{1}{4}\frac{\mu^4}{\lambda} = +\infty$$

which implies that the difference between the energy densities is positive and moreover that $u = \pm\mu/\sqrt{\lambda}$ are the minima ($u = 0$ is a maximum). The set of vacua is degenerate.

(iv) If u is independent of x_0 we obtain from (1)

$$\frac{d^2u}{dx^2} + \mu^2 u - \lambda u^3 = 0.$$

Consequently integration yields

$$\frac{1}{2}\left(\frac{du}{dx_1}\right)^2 + \mu^2\frac{u^2}{2} - \frac{\lambda u^4}{4} = \text{const.}$$

It follows that

$$x_1 - x_{1,0} = \int_0^u \frac{du'}{\sqrt{\lambda u'^4/2 - \mu^2 u'^2 + k}}, \qquad x_{1,0}, k = \text{const}$$

which represents an *elliptic integral*. Therefore $u(x_1)$ is periodic on the complex x_1-plane. Hence $E[u]$ is of infinite energy, unless the two zeroes of the square root coalesce. This happens for $k = \mu^4/2\lambda$. Then we obtain the solution (3). Straightforward calculation gives

$$E(u_{\text{kink}}) - E(u_{\text{vac}}) = \frac{2\sqrt{2}}{3}\frac{\mu^3}{\lambda} < +\infty$$

where $u_{\text{vac}} = \pm\mu/\sqrt{\lambda}$. A space-time dependent solution can be found from (3) by applying the *Lorentz transformation*.

Problem 17. The *Navier-Stokes equation* of a viscous incompressible fluid is given by

$$\left(\frac{\partial}{\partial t} + \mathbf{u}\cdot\nabla\right)\mathbf{u} = -\frac{1}{\rho}\nabla p + \nu\Delta\mathbf{u}. \tag{1}$$

We assume that the fluid is incompressible, i.e. $\nabla\cdot\mathbf{u} = \mathbf{0}$.

(i) Express the Navier-Stokes equation in cylindrical coordinates r, ϕ, x_3.

(ii) Consider a flow between concentric rotating cylinders. The cylinders are infinitely long. Let r_1, r_2 and Ω_1, Ω_2 denote the radii and angular velocities of the inner and outer cylinders, respectively. We denote the velocity components in the increasing r, ϕ and x_3 directions by u, v and w. The boundary conditions are

$$u = w = 0 \quad \text{at} \quad r = r_1 \quad \text{and} \quad r = r_2$$

$$v(r_1, \phi, x_3, t) = r_1 \Omega_1, \qquad v(r_2, \phi, x_3, t) = r_2 \Omega_2.$$

These conditions are called *no-slip conditions*. Find an exact solution of the form $u = w = 0$, $v = V(r)$.

Solution 17.   Writing the Navier-Stokes equation in *cylindrical coordinates* $x_1(\phi, r) = r \cos(\phi)$, $x_2(\phi, r) = r \sin(\phi)$, $x_3 = x_3$ yields

$$\left(\frac{\partial}{\partial t} + \mathbf{u} \cdot \nabla \right) u - \frac{v^2}{r} = -\frac{\partial}{\partial r} \frac{p}{\rho} + \nu \left(\Delta - \frac{1}{r^2} \right) u - \frac{2\nu}{r^2} \frac{\partial}{\partial \phi} v$$

$$\left(\frac{\partial}{\partial t} + \mathbf{u} \cdot \nabla \right) v + \frac{uv}{r} = -\frac{1}{r} \frac{\partial}{\partial \phi} \frac{p}{\rho} + \nu \left(\Delta - \frac{1}{r^2} \right) v + \frac{2\nu}{r^2} \frac{\partial}{\partial \phi} u$$

$$\left(\frac{\partial}{\partial t} + \mathbf{u} \cdot \nabla \right) w = -\frac{\partial}{\partial x_3} \frac{p}{\rho} + \nu \Delta w$$

$$\left(\frac{\partial}{\partial r} + \frac{1}{r} \right) u = -\frac{1}{r} \frac{\partial}{\partial \phi} v - \frac{\partial}{\partial x_3} w$$

where

$$\mathbf{u} \cdot \nabla := u \frac{\partial}{\partial r} + \frac{1}{r} v \frac{\partial}{\partial \phi} + w \frac{\partial}{\partial x_3}, \quad \Delta := \frac{\partial^2}{\partial r^2} + \frac{1}{r} \frac{\partial}{\partial r} + \frac{1}{r^2} \frac{\partial^2}{\partial \phi^2} + \frac{\partial^2}{\partial x_3^2}.$$

Inserting $u = w = 0$ and $v = V(r)$ into the first equation yields

$$-\frac{V^2(r)}{r} = -\frac{1}{\rho} \frac{\partial p}{\partial r}.$$

Thus we can assume that the pressure p depends only on r. The third equation is satisfied identically. For the second equation we find

$$\nu \left(\frac{\partial^2}{\partial r^2} + \frac{1}{r} \frac{\partial}{\partial r} - \frac{1}{r^2} \right) V(r) = 0.$$

Therefore the solution is given by $V(r) = Ar + B/r$, where A and B are the constants of integration. Imposing the boundary conditions gives

$$A = \frac{\Omega_2 r_2^2 - \Omega_1 r_1^2}{r_2^2 - r_1^2}, \qquad B = \frac{r_1^2 r_2^2 (\Omega_1 - \Omega_2)}{r_2^2 - r_1^2}.$$

Problem 18. Consider the nonlinear diffusion equation

$$\frac{\partial u}{\partial t} = D\frac{\partial}{\partial x}\left(\frac{1}{u}\frac{\partial u}{\partial x}\right)$$

where $u(x,t)$ is the density and x, t are the space, time coordinates and $x \in [0,1]$. D is a positive constant. We set at the boundaries

$$u(0,t) = u(1,t) = u_0 \geq 0.$$

We assume that $u(x,0) \geq u_0$.
(i) Show that the problem is well-posed.
(ii) Introduce the new independent and dependent variables

$$\tau(t,x) = \frac{D}{u_0}t, \qquad y(t,x) = x \qquad (1a)$$

$$v(\tau(t,x),y(t,x)) = \ln(u(x,t)/u_0). \qquad (1b)$$

Find the partial differential equation for these variables.
(iii) How do the boundary conditions change?

Solution 18. (i) Since

$$\frac{d}{dt}\int_0^1 u(x,t)dx = D\left(\frac{1}{u}\frac{\partial u}{\partial x}\right)\Big|_0^1 \qquad (2)$$

the flux (i.e. the right-hand side of (2)) will be finite and the problem is well posed.
(ii) From (1b) we find

$$u(x,t) = u_0\exp(v(\tau(t,x),y(t,x))). \qquad (3)$$

Using (3) and (1b) we find, by applying the chain rule,

$$\frac{\partial^2 v}{\partial y^2} = \exp(v)\frac{\partial v}{\partial \tau} = \frac{\partial}{\partial \tau}\exp(v).$$

(iii) The boundary conditions change to $v(0,t) = v(1,t) = 0$. The quantity v is nonnegative and the new time scale differs from the old by a factor of u_0/D.

Problem 19. Consider the complex *Ginzburg-Landau equation*

$$\frac{\partial A}{\partial t} = (1+ic_1)\frac{\partial^2 A}{\partial x^2} + A - (-1+ic_2)|A|^2A. \qquad (1)$$

Find a solution of the form

$$A(x,t) = R(x,t)\exp(i\Theta(x,t)) \tag{2}$$

where R and Θ are real-valued functions.

Solution 19. Inserting (2) into (1) we obtain two real equations

$$\frac{1}{c_1^2}\frac{\partial R}{\partial t} = \frac{1}{c_1^2}\frac{\partial^2 R}{\partial x^2} - \frac{1}{c_1^2}R\left(\frac{\partial\Theta}{\partial x}\right)^2 - \frac{2}{c_1}\frac{\partial R}{\partial x}\frac{\partial\Theta}{\partial x} - \frac{1}{c_1}R\frac{\partial^2\Theta}{\partial x^2} + \frac{1}{c_1^2}R + \frac{1}{c_1^2}R^3$$

$$\frac{1}{c_1}R\frac{\partial\Theta}{\partial t} = \frac{2}{c_1}\frac{\partial R}{\partial x}\frac{\partial\Theta}{\partial x} + \frac{1}{c_1}R\frac{\partial^2\Theta}{\partial x^2} + \frac{\partial^2 R}{\partial x^2} - R\left(\frac{\partial\Theta}{\partial x}\right)^2 - \frac{c_2}{c_1}R^3.$$

Adding the two equation provides

$$\epsilon^2\frac{\partial R}{\partial t} + \epsilon R\frac{\partial\Theta}{\partial t} = (1+\epsilon^2)\left(\frac{\partial^2 R}{\partial x^2} - \left(\frac{\partial\Theta}{\partial x}\right)^2 R\right) + \epsilon^2 R + (\beta + \epsilon^2)R^3$$

where $c_2 = -\beta c_1$ and $\epsilon = 1/c_1$. We also obtain

$$-\frac{1}{2}\epsilon\frac{\partial R^2}{\partial t} + \epsilon^2 R^2\frac{\partial\Theta}{\partial t} = (1+\epsilon^2)\frac{\partial}{\partial x}\left(R^2\frac{\partial\Theta}{\partial x}\right) - \epsilon(1 + (1-\beta)R^2)R^2.$$

For sufficiently large values of c_1 an expansion in ϵ becomes meaningful

$$R = R_0 + \epsilon^2 R_2 + \cdots, \qquad \Theta = \frac{1}{\epsilon}(\Theta_{-1} + \epsilon^2\Theta_1 + \cdots).$$

For the order ϵ^{-2} we find $R_0(\partial\Theta_{-1}/\partial x)^2 = 0$. For the order ϵ^{-1} we find

$$\frac{\partial}{\partial x}\left(R_0^2\frac{\partial\Theta_{-1}}{\partial x}\right) = 0.$$

Excluding $R_0 = 0$ we obtain $\partial\Theta_{-1}/\partial x = 0$, so that Θ_{-1} only depends on t. Setting $\gamma(t) := \partial\Theta_{-1}/\partial t$ we obtain for the ϵ^0

$$0 = \frac{\partial^2 R_0}{\partial x^2} - \gamma(t)R_0 + \beta R_0^3. \tag{3a}$$

For ϵ^1 we find

$$-\frac{\partial}{\partial x}\left(R_0^2\frac{\partial\Theta_1}{\partial x}\right) = \frac{1}{2}\frac{\partial R_0^2}{\partial t} - (1 + \gamma(t))R_0^2 - (1-\beta)R_0^4. \tag{3b}$$

For $\beta, \gamma > 0$ we obtain from (3a) the spatially periodic solution

$$R_0(x,t) = \left(\frac{2\gamma(t)}{(2-m(t))\beta}\right)^{1/2}\mathrm{dn}\left(\left(\frac{\gamma(t)}{2-m(t)}\right)^{1/2}x, m(t)\right). \tag{4}$$

Here $dn(u, m)$ is a Jacobian elliptic function that varies between $(1 - m)$ and 1 with period $2K(m)$. The parameter m is between 0 and 1, and $K(m)$ is the complete elliptic integral of the first kind. For $m \to 1$ the period of the function dn goes to infinity and (3) degenerates into the pulse

$$(2\gamma/\beta)^{1/2}\mathrm{sech}(\gamma^{1/2}x)$$

while for $m \to 0$ one has small, harmonic oscillations.

Problem 20. In *plasma physics* the differential equations for the density of the ion-fluid n and its velocity u, in dimensionless form, is given by

$$\frac{\partial n}{\partial t} + \frac{\partial(nu)}{\partial x} = 0 \qquad \text{equation of continuity}$$

$$\frac{\partial u}{\partial t} + u\frac{\partial u}{\partial x} = E \qquad \text{equation of motion}$$

$$\frac{\partial n_e}{\partial x} = -n_e E \qquad \text{balance of pressure and electric force}$$

$$\frac{\partial E}{\partial x} = n - n_e \qquad \text{Poisson equation}$$

where n_e is the electron density, E the electric field, and the inertia term is neglected because of the small mass of electrons.
(i) Eliminate n and E to find the differential equations for u and n_e.
(ii) Introduce $\xi := \epsilon^{1/2}(x - t)$, $\eta := \epsilon^{3/2}x$ and find a solution of the form

$$u = \epsilon u^{(1)} + \epsilon^2 u^{(2)} + \cdots, \qquad n_e = 1 + \epsilon n_e^{(1)} + \epsilon^2 n_e^{(2)} + \cdots. \qquad (1)$$

Solution 20. (i) Eliminating n and E yields

$$\frac{\partial u}{\partial t} + u\frac{\partial u}{\partial x} + \frac{1}{n_e}\frac{\partial n_e}{\partial x} = 0 \qquad (2a)$$

$$\frac{\partial n_e}{\partial t} + \frac{\partial(n_e u)}{\partial x} + \frac{\partial P}{\partial x} = 0 \qquad (2b)$$

where

$$P := -\left(\frac{\partial}{\partial t} + u\frac{\partial}{\partial x}\right)\left(\frac{1}{n_e}\frac{\partial n_e}{\partial x}\right).$$

(ii) Inserting expansion (1) into (2) we obtain $u^{(1)} = n_e^{(1)}$ and

$$\frac{\partial u^{(1)}}{\partial \eta} + u^{(1)}\frac{\partial u^{(1)}}{\partial \xi} + \frac{\partial^3 u^{(1)}}{\partial \xi^3} = 0, \qquad \frac{\partial n_e^{(1)}}{\partial \eta} + n_e^{(1)}\frac{\partial n_e^{(1)}}{\partial \xi} + \frac{\partial^3 n_e^{(1)}}{\partial \xi^3} = 0.$$

We see that u, the ion-fluid velocity, and n_e, the electron density, obey the same *Korteweg-de Vries equation* and move with the same phase $u^{(1)} = n_e^{(1)}$.

The nonlinear term $u^{(1)}\partial u^{(1)}/\partial\xi$ comes from the interaction of the ions with the electrons which affects the ions themselves, and $n_e^{(1)}\partial n_e^{(1)}/\partial\xi$ expresses similar effects on the electrons through the interaction with the ions. We have

$$n^{(1)}(\eta,\xi) = -\int^{\xi} E^{(1)}(\eta,s)ds = u^{(1)}(\eta,\xi) = n_e^{(1)}(\eta,\xi).$$

Problem 21.   Consider the *Korteweg-de Vries-Burgers equation*

$$\frac{\partial u}{\partial t} + a_1 u\frac{\partial u}{\partial x} + a_2\frac{\partial^2 u}{\partial x^2} + a_3\frac{\partial^3 u}{\partial x^3} = 0 \tag{1}$$

where a_1, a_2 and a_3 are non-zero constants. It contains dispersive, dissipative and nonlinear terms.
(i) Find a solution of the form

$$u(x,t) = \frac{b_1}{(1 + \exp(b_2(x + b_3 t + b_4)))^2} \tag{2}$$

where b_1, b_2, b_3 and b_4 are constants determined by a_1, a_2, a_3 and a_4.
(ii) Study the case $t \to \infty$ and $t \to -\infty$.

Solution 21.   (i) This is a typical problem for computer algebra. Inserting the ansatz (2) into (1) yields the conditions

$$b_3 = -(a_1 b_1 + a_2 b_2 + a_3 b_2^2) \tag{3}$$

and

$$(2a_1 b_1 + 3a_2 b_2 + 9a_3 b_2^2)\exp((a_1 b_1 b_2 + a_2 b_2^2 + a_3 b_2^3)t) + (a_1 b_1 + 3a_2 b_2 - 3a_3 b_2^2) = 0.$$

The quantity b_4 is arbitrary. Thus

$$2a_1 b_1 = -3a_2 b_2 - 9a_3 b_2^2, \qquad a_1 b_1 = -3a_2 b_2 + 3a_3 b_2^2. \tag{4}$$

Solving these system of equations with respect to b_1 and b_2 yields

$$b_1 = -\frac{12a_2^2}{25a_1 a_3}, \qquad b_2 = \frac{a_2}{5a_3}.$$

Inserting b_1 and b_2 into (3) gives $b_3 = 6a_2^2/(25a_3)$. Thus the solution is

$$u(x,t) = -\frac{12a_2^2}{25a_1 a_3}\left(1 + \exp\left(\frac{a_2}{5a_3}\left(x + \frac{6a_2^2}{25a_3}t + b_4\right)\right)\right)^{-2}.$$

(ii) Two asymptotic values exist which are for $t \to -\infty$, $u_I = 2v_c/a$ and for $t \to \infty$, $u_{II} = 0$.

Problem 22. For a barotropic fluid of index γ the *Navier-Stokes equation* in one space dimension is given by

$$\frac{\partial c}{\partial t} + v\frac{\partial c}{\partial x} + \frac{(\gamma - 1)}{2}c\frac{\partial v}{\partial x} = 0 \qquad \text{continuity equation} \qquad (1a)$$

$$\frac{\partial v}{\partial t} + v\frac{\partial v}{\partial x} + \frac{2}{(\gamma - 1)}c\frac{\partial c}{\partial x} - \frac{\partial^2 v}{\partial x^2} = 0 \qquad \text{Euler's equation} \qquad (1b)$$

where v represents the fluid's velocity and c the speed of sound. For the class of solutions characterized by a vanishing pressure (i.e. $c = 0$), the above system reduces to the Burgers equation. We assume for simplicity the value $\gamma = 3$ in what follows.

(i) Show that the *velocity potential* Φ exists, and it is given by the following pair of equations

$$\frac{\partial \Phi}{\partial x} = v, \qquad \frac{\partial \Phi}{\partial t} = \frac{\partial v}{\partial x} - \frac{1}{2}(v^2 + c^2). \qquad (2)$$

(ii) Consider the *similarity ansatz*

$$v(x,t) = f(s)\frac{x}{t}, \qquad c(x,t) = g(s)\frac{x}{t} \qquad (3)$$

where the *similarity variable* s is given by $s(x,t) := x/\sqrt{t}$. Show that the Navier-Stokes equation yields the following system of ordinary differential equations

$$s\frac{df}{ds}g + s\left(f - \frac{1}{2}\right)\frac{dg}{ds} + g(2f - 1) = 0$$

$$2\frac{d^2 f}{ds^2} + \frac{df}{ds}\left(\frac{4}{s} + s(1 - 2f)\right) + 2f(1 - f) = 2g\left(s\frac{dg}{ds} + g\right).$$

(iii) Show that the continuity equation (1a) admits a first integral, expressing the law of mass conservation, namely $s^2 g(2f - 1) = C$, where C is a constant.

(iv) Show that g can be eliminated, and we obtain a second-order ordinary differential equation for the function f.

Solution 22. (i) From (2) with

$$\frac{\partial^2 \Phi}{\partial t \partial x} = \frac{\partial^2 \Phi}{\partial x \partial t}$$

the Euler equation (1b) follows. Thus the condition of integrability of Φ precisely coincides with the Euler equation (1b). Conversely inserting $\partial \Phi/\partial x = v$ into (1b) yields the expression for $\partial \Phi/\partial t$.

(ii) From (3) we obtain

$$\frac{\partial v}{\partial t} = \frac{df}{ds}\frac{ds}{dt}\frac{x}{t} - f\frac{x}{t^2}, \qquad \frac{\partial v}{\partial x} = \frac{df}{ds}\frac{ds}{dx}\frac{x}{t} + f\frac{1}{t}$$

$$\frac{\partial c}{\partial t} = \frac{dg}{ds}\frac{ds}{dt}\frac{x}{t} - g\frac{x}{t^2}, \qquad \frac{\partial c}{\partial x} = \frac{dg}{ds}\frac{ds}{dx}\frac{x}{t} + g\frac{1}{t}$$

and

$$\frac{\partial^2 v}{\partial x^2} = \frac{d^2 f}{ds^2}\left(\frac{ds}{dx}\right)^2\frac{x}{t} + \frac{df}{ds}\frac{d^2 s}{dx^2}\frac{x}{t} + \frac{df}{ds}\frac{ds}{dx}\frac{1}{t} + \frac{df}{ds}\frac{ds}{dx}\frac{1}{t}.$$

Since $ds/dx = 1/\sqrt{t}$, $ds/dt = -xt^{-3/2}/2$ the system of ordinary differential equation follows.

Problem 23. For the functions $u, v : [0, \infty) \times \mathbb{R} \to \mathbb{R}$ we consider the Cauchy problem

$$\frac{\partial u}{\partial t} + \frac{\partial u}{\partial x} = v^2 - u^2, \qquad \frac{\partial v}{\partial t} - \frac{\partial v}{\partial x} = u^2 - v^2 \qquad (1a)$$

$$u(0, x) = u_0(x), \qquad v(0, x) = v_0(x). \qquad (1b)$$

This is the *Carleman model* introduced above.
(i) Define

$$S := u + v, \qquad D := u - v. \qquad (2)$$

Find the partial differential equations for S and D.
(ii) Find explicit solutions assuming that S and D are conjugate harmonic functions.

Solution 23. (i) We find that S and D satisfy the system of partial differential equations

$$\frac{\partial S}{\partial t} + \frac{\partial D}{\partial x} = 0, \qquad \frac{\partial D}{\partial t} + \frac{\partial S}{\partial x} = -2DS \qquad (3)$$

and the conditions $u \geq 0$, $v \geq 0$ take the form $S \geq 0$ and $S^2 - D^2 \geq 0$.
(ii) Assume that S and D conjugate harmonic functions, i.e.

$$\frac{\partial S}{\partial t} + \frac{\partial D}{\partial x} = 0, \qquad \frac{\partial S}{\partial x} - \frac{\partial D}{\partial t} = 0. \qquad (4)$$

Let $z := x + it$ and $f(z) := S + iD$. Then

$$\frac{df}{dz} = -\frac{i}{2}(f^2(z) + c), \qquad c \in \mathbb{R} \qquad (5)$$

since

$$\frac{df}{dz} = \frac{\partial S}{\partial x} + i\frac{\partial D}{\partial x} = -\frac{i}{2}(S^2 - D^2 + 2iDS + c), \qquad \frac{\partial S}{\partial x} = DS.$$

Owing to $\partial S/\partial x = \partial D/\partial t$ the second equation is also satisfied. Let

$$g(z) := \alpha f(\alpha z), \qquad \alpha \in \mathbb{R}$$

then

$$\frac{dg(z)}{dz} = \alpha^2 f'(\alpha z) = -\frac{i\alpha^2}{2}(f^2(\alpha z) + c) = -\frac{i}{2}(g^2(z) + \alpha^2 c).$$

For the solution

$$f(z) = \begin{cases} \dfrac{2f(z_0)}{2 + if(z_0)z_0 - if(z_0)z} & c = 0 \\[3mm] id\dfrac{(f(z_0) + id)e^{d(z-z_0)} + f(z_0) - id}{(f(z_0) + id)e^{d(z-z_0)} - f(z_0) + id} & d = \sqrt{c} \neq 0 \end{cases}$$

of (5) we only have to study the three cases $c = -1, 0, 1$. Since the problem is analytic we can consider

$$c = -1, \quad z_0 = 0, \quad f(z_0) = \frac{1 + \theta}{1 - \theta}, \qquad \theta \in \mathbb{R}, \quad \theta \neq 1$$

$$f(z) = -\frac{\theta e^{iz} + 1}{\theta e^{iz} - 1} = \frac{1 - \theta^2 e^{-2t} + i2\theta e^{-t}\sin(x)}{1 - 2\theta e^{-t}\cos(x) + \theta^2 e^{-2t}}.$$

We can check that for $|\theta| \leq \sqrt{2} - 1$ the conditions $S \geq 0$ and $S^2 - D^2 \geq 0$ are satisfied. Thus a particular solution of (1) is given by

$$u(t, x) = -\frac{1}{2}\frac{\mathrm{sgn}(\theta)\sinh(t - \log|\theta|) + \sin(x)}{\mathrm{sgn}(\theta)\cosh(t - \log|\theta|) - \cos(x)}$$

$$v(t, x) = \frac{1}{2}\frac{\mathrm{sgn}(\theta)\sinh(t - \log|\theta|) - \sin(x)}{\mathrm{sgn}(\theta)\cosh(t - \log|\theta|) - \cos(x)}.$$

Problem 24. Consider the partial differential equation

$$\frac{\partial u}{\partial t} + u\frac{\partial u}{\partial x} + \delta\frac{\partial^3 u}{\partial x^3} + \frac{\partial^2 u}{\partial x^2} + \frac{\partial^4 u}{\partial x^4} = 0 \tag{1}$$

subject to periodic boundary conditions in the interval $[0, L]$, with initial conditions $u(x, 0) = u_0(x)$. We only consider solutions with zero spatial average. We recall that for $L \leq 2\pi$ all initial conditions evolve into $u(x, t) = 0$. We expand the solution for u in the Fourier series

$$u(x, t) = \sum_{n=-\infty}^{\infty} a_n(t)\exp(ik_n x) \tag{2}$$

where $k_n := 2n\pi/L$ and the complex expansion coefficients satisfy

$$a_{-n}(t) = \bar{a}_n(t). \tag{3}$$

Here $\bar{a}$ denotes the complex conjugate of a. Since we choose solutions with zero average we have $a_0 = 0$.

(i) Show that inserting the series expansion (2) into (1) we obtain the following system for the time evolution of the Fourier amplitudes $a_n(t)$

$$\frac{da_n}{dt} + (k_n^4 - k_n^2 - i\delta k_n^3)a_n + \frac{1}{2}ik_n \sum_{m=0}^{\infty}(a_m a_{n-m} + \bar{a}_m a_{n+m}) = 0. \tag{4}$$

(ii) Find the autonomous system when we only keeping the first five modes, where $k := 2\pi/L$, $\mu_n := k_n^4 - k_n^2$.

Solution 24.   (i) Since

$$\frac{\partial u}{\partial x} = \sum_{m=-\infty}^{\infty} a_m(t)ik_m e^{ik_m x}$$

we have

$$u\frac{\partial u}{\partial x} = \sum_{q=-\infty}^{\infty}\sum_{p=-\infty}^{\infty} a_p(t)a_q(t)ik_q e^{i(k_q+k_p)x}.$$

Furthermore

$$\frac{\partial u}{\partial t} = \sum_{n=-\infty}^{\infty} \frac{da_n}{dt}e^{ik_n x}$$

and the higher order derivatives with respect to x are given by

$$\frac{\partial^2 u}{\partial x^2} = -\sum_{m=-\infty}^{\infty} a_m(t)k_m^2 e^{ik_m x}, \quad \frac{\partial^3 u}{\partial x^3} = -i\sum_{m=-\infty}^{\infty} a_m(t)k_m^3 e^{ik_m x}$$

and

$$\frac{\partial^4 u}{\partial x^4} = \sum_{m=-\infty}^{\infty} a_m(t)k_m^4 e^{ik_m x}.$$

Thus

$$\sum_{m=-\infty}^{\infty}\left(\frac{da_m}{dt} + (k_m^4 - k_m^2 - i\delta k_m^3)a_m\right)e^{ik_m t} + \sum_{\substack{q=-\infty\\p=-\infty}}^{\infty} a_p(t)a_q(t)ik_q e^{i(k_q+k_p)x} = 0.$$

Applying the Kronecker delta $\delta_{m,n}$ and $\delta_{p+q,n}$ to this equation we obtain (4) where we used (3).

(ii) We obtain the autonomous system of first order

$$\frac{da_1}{dt} + (\mu_1 - i\delta k^3)a_1 + ik(\bar{a}_1 a_2 + \bar{a}_2 a_3 + \bar{a}_3 a_4 + \bar{a}_4 a_5) = 0$$

$$\frac{da_2}{dt} + (\mu_2 - 8i\delta k^3)a_2 + ik(a_1^2 + 2\bar{a}_1 a_3 + 2\bar{a}_2 a_4 + 2\bar{a}_3 a_5) = 0$$

$$\frac{da_3}{dt} + (\mu_3 - 27i\delta k^3)a_3 + 3ik(a_1 a_2 + \bar{a}_1 a_4 + \bar{a}_2 a_5) = 0$$

$$\frac{da_4}{dt} + (\mu_4 - 64i\delta k^3)a_4 + 2ik(a_2^2 + 2a_1 a_3 + 2\bar{a}_1 a_5) = 0$$

$$\frac{da_5}{dt} + (\mu_5 - 125i\delta k^3)a_5 + 5ik(a_1 a_4 + a_2 a_3) = 0.$$

Problem 25. The spherically symmetric $SU(2)$ *Yang-Mills equations* can be written in the form

$$\frac{\partial \phi_1}{\partial x_0} - \frac{\partial \phi_2}{\partial r} = -A_0 \phi_2 - A_1 \phi_1 \tag{1a}$$

$$\frac{\partial \phi_2}{\partial x_0} + \frac{\partial \phi_1}{\partial r} = -A_1 \phi_2 + A_0 \phi_1 \tag{1b}$$

$$r^2 \left(\frac{\partial A_1}{\partial x_0} - \frac{\partial A_0}{\partial r} \right) = 1 - (\phi_1^2 + \phi_2^2) \tag{1c}$$

where r is the spatial radius vector and $x_0 = ct$ with t the time. Introduce the new variables

$$w(r, x_0) = \phi_1(r, x_0) + i\phi_2(r, x_0) \tag{2a}$$

$$w(r, x_0) = R(r, x_0) \exp(i\theta(r, x_0)) \tag{2b}$$

to find particular solutions of system (1). Find the differential equation for R.

Solution 25. Using (2a) we obtain from (1a) and (1b) that

$$\frac{\partial w}{\partial x_0} + i\frac{\partial w}{\partial r} = (iA_0 - A_1)w. \tag{3}$$

Using (2b) and (1a), (1b) we obtain after separating the imaginary and real parts

$$A_0 = \frac{\partial \theta}{\partial x_0} + \frac{\partial \ln(R)}{\partial r}, \qquad A_1 = \frac{\partial \theta}{\partial r} - \frac{\partial \ln(R)}{\partial x_0}. \tag{4}$$

Substitution of this equation into (1c) yields

$$r^2 \left(\frac{\partial^2}{\partial r^2} + \frac{\partial^2}{\partial x_0^2} \right) \ln(R) = R^2 - 1.$$

Thus θ does not arise in this differential equation. Changing the variables

$$R(r, x_0) = r \exp(g(r, x_0))$$

we find the *Liouville equation*

$$\left(\frac{\partial^2}{\partial r^2} + \frac{\partial^2}{\partial x_0^2} \right) g = \exp(2g)$$

which has the general solution in terms of two harmonic functions $a(r, x_0)$ and $b(r, x_0)$ related by the *Cauchy-Riemann condition*

$$\exp(2g) = 4 \left(\left(\frac{\partial a}{\partial r} \right)^2 + \left(\frac{\partial a}{\partial t} \right)^2 \right) \frac{1}{(1 - a^2 - b^2)^2}$$

$$\left(\frac{\partial^2}{\partial r^2} + \frac{\partial^2}{\partial x_0^2} \right) a = 0, \quad \left(\frac{\partial^2}{\partial r^2} + \frac{\partial^2}{\partial x_0^2} \right) b = 0$$

$$\frac{\partial a}{\partial r} = \frac{\partial b}{\partial x_0}, \quad \frac{\partial a}{\partial x_0} = -\frac{\partial b}{\partial r}.$$

Equation (3) was obtained with an arbitrary smooth function $\theta(r, t)$.

Problem 26. If a partial differential equation admits continuous symmetries (Lie symmetries) we can reduce it to an ordinary differential equation using a similarity variable which is constructed from the Lie symmetries. The motion of an incompressible constant-property fluid in two-space dimension is described by the *stream function equation*

$$\nabla^2 \frac{\partial \psi}{\partial t} + \frac{\partial \psi}{\partial x_2} \nabla^2 \frac{\partial \psi}{\partial x_1} - \frac{\partial \psi}{\partial x_1} \nabla^2 \frac{\partial \psi}{\partial x_2} = \nu \nabla^4 \psi \tag{1}$$

where $\nabla^2 := \partial^2/\partial x_1^2 + \partial^2/\partial x_2^2$ and ν is a positive constant. The *stream function* ψ is defined by $u = \partial \psi / \partial x_2$, $v = -\partial \psi / \partial x_1$ in order to satisfy the *continuity equation* identically, i.e. $\partial u / \partial x_1 + \partial v / \partial x_2 = 0$.
(i) Show that (1) is invariant under the transformation

$$x_1'(x_1, x_2, t, \epsilon) = x_1 \cos(\epsilon t) + x_2 \sin(\epsilon t)$$
$$x_2'(x_1, x_2, t, \epsilon) = -x_1 \sin(\epsilon t) + x_2 \cos(\epsilon t)$$
$$t'(x_1, x_2, t, \epsilon) = t$$

$$\psi'(x_1'(x_1, x_2, t), x_2'(x_1, x_2, t), t'(x_1, x_2, t), \epsilon) = \psi(x_1, x_2, t) + \frac{1}{2} \epsilon (x_1^2 + x_2^2)$$

where ϵ is a real parameter.
(ii) Find the infinitesimal generator of this transformation.

Solution 26. (i) Applying the chain rule we find

$$\frac{\partial \psi'}{\partial x'}\cos(\epsilon t) - \frac{\partial \psi'}{\partial x'_2}\sin(\epsilon t) = \frac{\partial \psi}{\partial x_1} + \epsilon x_1 \tag{2a}$$

$$\frac{\partial \psi'}{\partial x'_1}\sin(\epsilon t) + \frac{\partial \psi'}{\partial x'_2}\cos(\epsilon t) = \frac{\partial \psi}{\partial x_2} + \epsilon x_2 \tag{2b}$$

and

$$\frac{\partial \psi'}{\partial x'}(-\epsilon x_1 \sin(\epsilon t)+\epsilon x_2 \cos(\epsilon t))+\frac{\partial \psi'}{\partial y'}(-\epsilon x_1 \cos(\epsilon t)-\epsilon x_2 \sin(\epsilon t))+\frac{\partial \psi'}{\partial t'} = \frac{\partial \psi}{\partial t}. \tag{2c}$$

From (2a) and (2b) it follows that

$$\frac{\partial^2 \psi'}{\partial x'^2}\cos^2(\epsilon t) - 2\frac{\partial^2 \psi'}{\partial x'\partial x'_2}\cos(\epsilon t)\sin(\epsilon t) + \frac{\partial^2 \psi'}{\partial x_2'^2}\sin^2(\epsilon t) = \frac{\partial^2 \psi}{\partial x^2} + \epsilon \tag{3a}$$

$$\frac{\partial^2 \psi'}{\partial x'^2}\sin^2(\epsilon t) + 2\frac{\partial^2 \psi'}{\partial x'_1\partial x'_2}\cos(\epsilon t)\sin(\epsilon t) + \frac{\partial^2 \psi'}{\partial x_2'^2}\cos^2(\epsilon t) = \frac{\partial^2 \psi}{\partial x_2^2} + \epsilon. \tag{3b}$$

Therefore

$$\frac{\partial^2 \psi'}{\partial x_1'^2} + \frac{\partial^2 \psi'}{\partial x_2'^2} = \frac{\partial^2 \psi}{\partial x_1^2} + \frac{\partial^2 \psi}{\partial x^2} + 2\epsilon.$$

Thus it is obvious that $\nu\nabla'^4\psi' = \nu\nabla^4\psi$. From (2c) we find that

$$\nabla'^2\frac{\partial \psi'}{\partial t'} = \nabla^2\frac{\partial \psi}{\partial t} + \epsilon x_1\nabla^2\frac{\partial \psi}{\partial x_2} - \epsilon x_2\nabla^2\frac{\partial \psi}{\partial x_1}.$$

From (2a) and (2b) we also find that

$$\frac{\partial \psi'}{\partial x'_2}\nabla'^2\frac{\partial \psi'}{\partial x'_1} - \frac{\partial \psi'}{\partial x'_1}\nabla'^2\frac{\partial \psi'}{\partial x'_2} = \frac{\partial \psi}{\partial x_2}\nabla^2\frac{\partial \psi}{\partial x_1} - \frac{\partial \psi}{\partial x_1}\nabla^2\frac{\partial \psi}{\partial x_2} - \epsilon x_1\nabla^2\frac{\partial \psi}{\partial x_2} + \epsilon x_2\nabla^2\frac{\partial \psi}{\partial x_1}.$$

Thus we arrive at

$$\nabla'^2\frac{\partial \psi'}{\partial t'} + \frac{\partial \psi'}{\partial y'}\nabla'^2\frac{\partial \psi'}{\partial x'} - \frac{\partial \psi'}{\partial x'}\nabla'^2\frac{\partial \psi'}{\partial y'} - \nu\nabla'^4\psi' = 0.$$

(ii) From the transformation we obtain the mapping

$$x'_1(x,y,t,\epsilon) = x\cos(\epsilon t) + x_2\sin(\epsilon t)$$
$$x'_2(x,y,t,\epsilon) = -x\sin(\epsilon t) + x_2\cos(\epsilon t)$$
$$t'(x,y,t,\epsilon) = t$$
$$\psi'(x_1,x_2,t,\psi,\epsilon) = \psi + \frac{1}{2}\epsilon(x_1^2 + x_2^2).$$

It follows that

$$\left.\frac{dx'}{d\epsilon}\right|_{\epsilon=0} = x_2t, \quad \left.\frac{dy'}{d\epsilon}\right|_{\epsilon=0} = -x_1t, \quad \left.\frac{dt'}{d\epsilon}\right|_{\epsilon=0} = 0, \quad \left.\frac{d\psi'}{d\epsilon}\right|_{\epsilon=0} = \frac{1}{2}(x_1^2 + x_2^2).$$

Thus the infinitesimal symmetry generator is given by

$$S = yt\frac{\partial}{\partial x_1} - xt\frac{\partial}{\partial x_2} + \frac{1}{2}(x_1^2 + x_2^2)\frac{\partial}{\partial \psi}.$$

We find the transformation from the symmetry generator S when we apply the exponential map, i.e.

$$\begin{pmatrix} x'(x_1, x_2, t, \epsilon) \\ y'(x, y, t, \epsilon) \\ t'(x, y, t, \epsilon) \\ \psi'(x'(x, y, t), y'(x, y, t), t'(x, y, t), \epsilon) \end{pmatrix} = e^{\epsilon S} \begin{pmatrix} x_1 \\ x_2 \\ t \\ \psi \end{pmatrix}\Bigg|_{\psi \to \psi(x_1, x_2, t)}.$$

Problem 27. The nonlinear partial differential equation

$$\frac{\partial^2 u}{\partial x_1 \partial x_2} + \frac{\partial u}{\partial x_1} + u^2 = 0 \tag{1}$$

describes the relaxation to a *Maxwell distribution*. The symmetry vector fields are given by

$$Z_1 = \frac{\partial}{\partial x_1}, \quad Z_2 = \frac{\partial}{\partial x_2}, \quad Z_3 = -x_1\frac{\partial}{\partial x_1} + u\frac{\partial}{\partial u}, \quad Z_4 = e^{x_2}\frac{\partial}{\partial x_2} - e^{x_2}u\frac{\partial}{\partial u}.$$

Construct a similarity ansatz from the symmetry vector field

$$Z = c_1\frac{\partial}{\partial x_1} + c_2\frac{\partial}{\partial x_2} + c_3\left(-x_1\frac{\partial}{\partial x_1} + u\frac{\partial}{\partial u}\right)$$

and find the corresponding ordinary differential equation. Here $c_1, c_2, c_3 \in \mathbb{R}$.

Solution 27. The corresponding initial value problem of the symmetry vector field Z is given by

$$\frac{dx_1'}{d\epsilon} = c_1 - c_3x_1', \quad \frac{dx_2'}{d\epsilon} = c_2, \quad \frac{du'}{d\epsilon} = c_3u'.$$

The solution to this system provides the transformation group

$$x_1'(\mathbf{x}, u, \epsilon) = \frac{c_1}{c_3} - \frac{c_1 - c_3x_1}{c_3}e^{-c_3\epsilon}, \quad x_2'(\mathbf{x}, u, \epsilon) = c_2\epsilon + x_2, \quad x_3'(\mathbf{x}, u, \epsilon) = ue^{c_3\epsilon}$$

where $c_3 \neq 0$. Now let $x_2 = s/c$ and $x_1 = 1$ with the constant $c \neq 0$. The *similarity variable s* follows as

$$s = cx_2' + \frac{c_2 c}{c_3} \ln\left(\frac{c_3 x_1' - c_1}{c_3 - c_1}\right)$$

and the *similarity ansatz* is

$$u'(x_1', x_2') = v(s)\frac{c_1 - c_3}{c_1 - c_3 x_1'}.$$

Inserting this equation into

$$\frac{\partial^2 u'}{\partial x_1' \partial x_2'} + \frac{\partial u'}{\partial x_1'} + u'^2 = 0$$

leads to the ordinary differential equation

$$c\frac{d^2 v}{ds^2} + (1 - c)\frac{dv}{ds} - (1 - v)v = 0.$$

Problem 28. Let

$$\bar{t}(x,t) = t, \quad \bar{x}(x,t) = \int^x \frac{1}{u(s,t)}ds, \quad \bar{u}(\bar{x}(x,t), \bar{t}(x,t)) = u(x,t).$$

Let

$$\frac{\partial u}{\partial t} = u^2 \frac{\partial^2 u}{\partial x^2}.$$

Find the partial differential equation for $\bar{u}(\bar{x}, \bar{t})$.

Solution 28. We obtain

$$\frac{\partial \bar{t}}{\partial t} = 1, \quad \frac{\partial \bar{t}}{\partial x} = 0, \quad \frac{\partial \bar{x}}{\partial x} = \frac{1}{u(x,t)}$$

and

$$\frac{\partial \bar{x}}{\partial t} = -\int^x \frac{u_t(s,t)}{u^2(s,t)}ds = -\int^x \frac{\partial^2 u(s,t)}{\partial s^2}ds = -\frac{\partial u}{\partial x}$$

where $u_t \equiv \partial u/\partial t$. Applying the chain rule we have

$$\frac{\partial u}{\partial x} = \frac{\partial \bar{u}}{\partial \bar{x}}\frac{\partial \bar{x}}{\partial x} + \frac{\partial \bar{u}}{\partial \bar{t}}\frac{\partial \bar{t}}{\partial x} = \frac{\partial u}{\partial x} \quad \Rightarrow \quad \frac{\partial \bar{u}}{\partial \bar{x}} = u\frac{\partial u}{\partial x}.$$

Analogously,

$$\frac{\partial u}{\partial t} = \frac{\partial \bar{u}}{\partial \bar{x}}\frac{\partial \bar{x}}{\partial t} + \frac{\partial \bar{u}}{\partial \bar{t}}\frac{\partial \bar{t}}{\partial t} = \frac{\partial u}{\partial t} \quad \Rightarrow \quad -\frac{\partial \bar{u}}{\partial \bar{x}}\frac{\partial u}{\partial x} + \frac{\partial \bar{u}}{\partial \bar{t}} = \frac{\partial u}{\partial t}.$$

Furthermore we obtain

$$\frac{\partial}{\partial x}\left(\frac{\partial \bar{u}}{\partial \bar{x}}\right) = \left(\frac{\partial u}{\partial x}\right)^2 + u\frac{\partial^2 u}{\partial x^2}.$$

It follows that

$$\left(\frac{\partial^2 \bar{u}}{\partial \bar{x}^2}\frac{\partial \bar{x}}{\partial x} + \frac{\partial^2 \bar{u}}{\partial \bar{x}\partial \bar{t}}\frac{\partial \bar{t}}{\partial x}\right) = \left(\frac{\partial u}{\partial x}\right)^2 + u\frac{\partial^2 u}{\partial x^2}.$$

Since $\partial \bar{t}/\partial x = 0$ and $\partial \bar{x}/\partial x = u^{-1}$ we obtain

$$\frac{\partial^2 \bar{u}}{\partial \bar{x}^2} = u\left(\frac{\partial u}{\partial x}\right)^2 + u^2\frac{\partial^2 u}{\partial x^2}.$$

Finally we obtain $\partial \bar{u}/\partial \bar{t} = \partial^2 \bar{u}/\partial \bar{x}^2$. This partial differential equation is the *linear diffusion equation*.

Problem 29.   Consider the transformation

$$\bar{t}(x,t) = t, \quad \bar{x}(x,t) = u(x,t), \quad \bar{u}(\bar{x}(x,t),\bar{t}(x,t)) = x$$

and the nonlinear partial differential equation

$$\frac{\partial u}{\partial t} = \frac{\partial^2 u}{\partial x^2} + u\frac{\partial u}{\partial x}.$$

Find the partial differential equation for $\bar{u}(\bar{x},\bar{t})$.

Solution 29.   Taking the derivatives with respect to t and x provides

$$\frac{\partial \bar{t}}{\partial t} = 1, \quad \frac{\partial \bar{t}}{\partial x} = 0, \quad \frac{\partial \bar{x}}{\partial t} = \frac{\partial u}{\partial t}, \quad \frac{\partial \bar{x}}{\partial x} = \frac{\partial u}{\partial x}.$$

and

$$\frac{\partial \bar{u}}{\partial x} = \frac{\partial \bar{u}}{\partial \bar{x}}\frac{\partial \bar{x}}{\partial x} + \frac{\partial \bar{u}}{\partial \bar{t}}\frac{\partial \bar{t}}{\partial x} = 1.$$

Therefore $(\partial \bar{u}/\partial \bar{x})(\partial u/\partial x) = 1$. Analogously,

$$\frac{\partial \bar{u}}{\partial t} = \frac{\partial \bar{u}}{\partial \bar{x}}\frac{\partial \bar{x}}{\partial t} + \frac{\partial \bar{u}}{\partial \bar{t}}\frac{\partial \bar{t}}{\partial t} = 0, \quad \frac{\partial \bar{u}}{\partial \bar{x}}\frac{\partial u}{\partial t} + \frac{\partial \bar{u}}{\partial t} = 0.$$

It follows that

$$\frac{\partial}{\partial x}\left(\frac{\partial \bar{u}}{\partial \bar{x}}\frac{\partial u}{\partial x}\right) = 0, \quad \left(\frac{\partial^2 \bar{u}}{\partial \bar{x}^2}\frac{\partial \bar{x}}{\partial x} + \frac{\partial^2 \bar{u}}{\partial \bar{x}\partial \bar{t}}\frac{\partial \bar{t}}{\partial x}\right)\frac{\partial u}{\partial x} + \frac{\partial \bar{u}}{\partial \bar{x}}\frac{\partial^2 u}{\partial x^2} = 0.$$

Consequently

$$\frac{\partial^2 \bar{u}}{\partial \bar{x}^2}\left(\frac{\partial u}{\partial x}\right)^2 + \frac{\partial \bar{u}}{\partial \bar{x}}\frac{\partial^2 u}{\partial x^2} = 0.$$

Thus we have

$$u = \bar{x}, \quad \frac{\partial u}{\partial t} = -\frac{\bar{u}_{\bar{t}}}{\bar{u}_{\bar{x}}}, \quad \frac{\partial u}{\partial x} = \frac{1}{\bar{u}_{\bar{x}}}, \quad \frac{\partial^2 u}{\partial x^2} = -\frac{\bar{u}_{\bar{x}\bar{x}}}{\bar{u}_{\bar{x}}^3}$$

where $\bar{u}_{\bar{t}} := \partial \bar{u}/\partial \bar{t}$ etc. Inserting these results into the nonlinear partial differential equation provides the nonlinear partial differential equation

$$\left(\frac{\partial \bar{u}}{\partial \bar{x}}\right)^2 \frac{\partial \bar{u}}{\partial \bar{t}} = \frac{\partial^2 \bar{u}}{\partial \bar{x}^2} - \left(\frac{\partial \bar{u}}{\partial \bar{x}}\right)^2 \bar{x}.$$

Problem 30. Consider the Hamilton operator

$$\hat{H} = -\frac{\hbar^2}{2m}\frac{d^2}{dx^2} + D(1 - e^{-\alpha x})^2 + eEx\cos(\omega t)$$

where $\alpha > 0$. So the third term is a driving force. Find the quantum Liouville equation for this Hamilton operator.

Solution 30. From the definition of the Wigner function we find that

$$\dot{\rho}(p, q, t) = \frac{1}{\pi\hbar}\int dx e^{2ipx/\hbar}[\dot{\psi}^*(q + x)\psi(q - x) + \psi^*(q + x)\dot{\psi}(q - x)].$$

Using the Schrödinger equation we arrive at

$$\dot{\rho}(p, q, t) = \frac{1}{\pi\hbar}\int dx e^{2ipx/\hbar}[(\hat{H}\psi^*)(q + x)\psi(q - x) - \psi^*(q + x)(\hat{H}\psi)(q - x)].$$

This expression is linear in $\hat{H}$. Thus each part of the Hamilton operator can be considered separately. The kinetic part $\hat{H} = -(\hbar^2/2m)d^2/dx^2$ becomes, after integration by parts

$$\dot{\rho} = -\frac{p}{m}\frac{\partial\rho}{\partial q}$$

which is the classical Liouville operator for a free particle. The next contribution is from the exponential potential and are both of the form $Ce^{-\beta x}$. The expression for $\dot{\rho}$ can be written compactly if it is assumed that ρ can be analytically continued. We obtain

$$\dot{\rho} = e^{-\beta q}\frac{i}{\hbar}[\rho(q, p + i\beta\hbar/2) - \rho(q, p - i\beta\hbar/2)].$$

For the linear driving force we find

$$\dot{\rho} = eE\cos(\omega t)\frac{\partial\rho}{\partial p}.$$

Thus the *quantum Liouville equation* for the Hamilton operator $\hat{H}$ is given by

$$\dot{\rho} = -\frac{p}{m}\frac{\partial\rho}{\partial q} + eE\cos(\omega t)\frac{\partial\rho}{\partial p} + 2D\alpha e^{-\alpha q}\frac{i}{\hbar}[\rho(q,p+i\alpha\hbar/2) - \rho(q,p-i\alpha\hbar/2)]$$

$$-2D\alpha e^{-2\alpha q}\frac{i}{\hbar}[\rho(q,p+i\alpha h) - \rho(q,p-i\alpha\hbar)].$$

Problem 31.   Consider *Maxwell's equations*

$$\nabla \times \mathbf{B} = \frac{1}{c^2}\frac{\partial\mathbf{E}}{\partial t}, \quad \nabla \times \mathbf{E} = -\frac{\partial\mathbf{B}}{\partial t}, \quad \mathrm{div}(\mathbf{E}) = 0, \quad \mathrm{div}(\mathbf{B}) = 0$$

with $\mathbf{B} = \mu_0\mathbf{H}$.
(i) Assume that $E_2 = E_3 = 0$ and $B_1 = B_3 = 0$. Simplify Maxwell's equations.
(ii) Now assume that E_1 and B_2 only depends on x_3 and t with

$$E_1(x_3,t) = f(t)\sin(k_3x_3), \qquad B_2(x_3,t) = g(t)\cos(k_3x_3)$$

where k_3 is the third component of the *wave vector* $\mathbf{k}$. Find the system of ordinary differential equations for $f(t)$ and $g(t)$ and solve it and thus find the *dispersion relation*. Note that

$$\nabla \times \mathbf{E} = \begin{pmatrix} \partial E_3/\partial x_2 - \partial E_2/\partial x_3 \\ \partial E_1/\partial x_3 - \partial E_3/\partial x_1 \\ \partial E_2/\partial x_1 - \partial E_1/\partial x_2 \end{pmatrix}.$$

Solution 31.   (i) Maxwell's equation written down in components take the form

$$\frac{\partial B_1}{\partial t} - \frac{\partial E_2}{\partial x_3} + \frac{\partial E_3}{\partial x_2} = 0$$

$$\frac{\partial B_2}{\partial t} - \frac{\partial E_3}{\partial x_1} + \frac{\partial E_1}{\partial x_3} = 0$$

$$\frac{\partial B_3}{\partial t} - \frac{\partial E_1}{\partial x_2} + \frac{\partial E_2}{\partial x_1} = 0$$

$$\frac{1}{c^2}\frac{\partial E_1}{\partial t} - \frac{\partial B_3}{\partial x_2} + \frac{\partial B_2}{\partial x_3} = 0$$

$$\frac{1}{c^2}\frac{\partial E_2}{\partial t} - \frac{\partial B_1}{\partial x_3} + \frac{\partial B_3}{\partial x_1} = 0$$

$$\frac{1}{c^2}\frac{\partial E_3}{\partial t} - \frac{\partial B_2}{\partial x_1} + \frac{\partial B_1}{\partial x_2} = 0$$

and

$$\frac{\partial E_1}{\partial x_1} + \frac{\partial E_2}{\partial x_2} + \frac{\partial E_3}{\partial x_3} = 0, \quad \frac{\partial B_1}{\partial x_1} + \frac{\partial B_2}{\partial x_2} + \frac{\partial B_3}{\partial x_3} = 0.$$

With the simplification $E_2 = E_3 = 0$ and $B_1 = B_3 = 0$ we arrive at

$$\frac{\partial E_1}{\partial x_2} = 0, \quad \frac{\partial B_2}{\partial t} + \frac{\partial E_1}{\partial x_3} = 0, \quad \frac{\partial B_2}{\partial x_1} = 0, \quad \frac{\partial E_1}{\partial t} + c^2 \frac{\partial B_2}{\partial x_3} = 0$$

and $\partial E_1/\partial x_1 = 0$, $\partial B_2/\partial x_2 = 0$.

(ii) With the assumption that E_1 and B_2 only depend on x_3 and t the equations reduce to two equations

$$\frac{\partial B_2}{\partial t} + \frac{\partial E_1}{\partial x_3} = 0, \quad \frac{\partial E_1}{\partial t} + c^2 \frac{\partial B_2}{\partial x_3} = 0.$$

Inserting the ansatz for E_1 and B_2 yields the system of differential equations for f and g

$$\frac{df}{dt} - c^2 k_3 g = 0, \quad \frac{dg}{dt} + k_3 f = 0.$$

These are the equations for the harmonic oscillator. Starting from the ansatz for the solution

$$f(t) = A \sin(\omega t) + B \cos(\omega t)$$

where ω is the frequency, yields

$$g(t) = \frac{A\omega}{c^2 k_3} \cos(\omega t) - \frac{B\omega}{c^2 k_3} \sin(\omega t)$$

and the *dispersion relation* is $\omega^2 = c^2 k_3^2$.

Problem 32. Consider the one-dimensional nonlinear Schrödinger equation

$$\frac{\partial w}{\partial t} + i a_1 \frac{\partial^2 w}{\partial x^2} + a_2 \frac{\partial^3 w}{\partial x^3} = i b_1 |w|^2 w + b_2 \frac{\partial |w|^2 w}{\partial x}$$

where w is a complex valued function and a_1, a_2, b_1, b_2 are real constants with $a_2 b_2 < 0$. Show that

$$w(x,t) = \sqrt{-\frac{2 r a_2}{b_2}} \exp(i(kx + \omega t + c_0)) \frac{1}{\cosh(\sqrt{r}(x - vt - d))}$$

where r has dimension meter^{-2} and

$$k = \frac{a_2 b_1 - a_1 b_2}{2 a_2 b_2}, \quad \omega = -r(a_1 + 3 a_2 k) + a_1 k^2 + a_2 k^3,$$

$$v = a_2 r - 2 a_1 k - 3 a_2 k^2.$$

Solution 32. We apply the Maxima program

```
/* partial1.mac */
w: sqrt(-2*r*a2/b2)*exp(%i*(k*x+omega*t+c0))*1/cosh(sqrt(r)*(x-v*t-d));
wb: sqrt(-2*r*a2/b2)*exp(-%i*(k*x+omega*t+c0))*1/cosh(sqrt(r)*(x-v*t-d));
T1: diff(w,t) + %i*a1*diff(w,x,2) + a2*diff(w,x,3);
T1: trigsimp(T1);
T2: %i*b1*w*wb*w + b2*diff(w*wb*w,x);
T2: trigsimp(T2);
T3: T1-T2;
T3: trigsimp(T3);
T3: substitute(-r*(a1+3*a2*k)+a1*k*k+a2*k*k*k,omega,T3);
T3: substitute(a2*r-2*a1*k-3*a2*k*k,v,T3);
T3: substitute((a2*b1-a1*b2)/(2*a2*b2),k,T3);
T3: trigsimp(T3);
```

where `substitute()` will make the substitution and `trigsimp()` will do the trigonometric simplification.

Supplementary Problems

Problem 1.   Show that the Laplace operator $\Delta \equiv \nabla^2$ can be written in *spherical coordinates* as

$$\nabla^2 = \frac{\partial^2}{\partial r^2} + \frac{2}{r}\frac{\partial}{\partial r} + \frac{1}{r^2}\left(\frac{\partial^2}{\partial \theta^2} + \cot(\theta)\frac{\partial}{\partial \theta} + \frac{1}{\sin^2(\theta)}\frac{\partial^2}{\partial \phi^2}\right) \equiv \frac{\partial^2}{\partial r^2} + \frac{2}{r}\frac{\partial}{\partial r} - \frac{\mathbf{L}^2}{\hbar^2 r^2}$$

where $\mathbf{L} = \mathbf{r} \times \hat{\mathbf{P}} = -i\hbar\mathbf{r} \times \nabla$.

Problem 2.   Consider the initial value problem

$$\frac{\partial^2 u}{\partial x^2} + \frac{1}{c^2}\frac{\partial^2 u}{\partial t^2} = 0, \qquad -\infty < x < +\infty, \quad t \geq 0$$

$$u(x,0) = f(x), \qquad -\infty < x < +\infty$$

$$\frac{\partial u(x,0)}{\partial x} = g(x), \qquad -\infty < x < +\infty$$

where c denotes a speed. Fix down for dimension.
(i) Show that $u(x,t) = A\sin(kx)\sinh(\omega t)$ is a solution of the partial differential equation with $f(x) = g(x) = 0$ and the dispersion relation $c^2 k^2 = \omega^2$.
(ii) Show that

$$u(x,t) = A\sin(kx)\cosh(\omega t)$$

is a solution of the partial differential equation with $f(x) = A\sin(kx)$ and $g(x) = Ak\cos(kx)$ and the dispersion relation $c^2 k^2 = \omega^2$.

Problem 3.   Let $\ell_1 > 0$, $\ell_2 > 0$, $\ell_3 > 0$ with dimension length. Consider the Hilbert space $L_2([0,\ell_1] \times [0,\ell_2] \times [0,\ell_3])$. Let $n_1, n_2, n_3 \in \mathbb{N}_0$.

(i) Show that the functions

$$u(x_1, x_2, x_3) = \cos(k_1 x_1) \sin(k_2 x_2) \sin(k_3 x_3)$$
$$v(x_1, x_2, x_3) = \sin(k_1 x_1) \cos(k_2 x_2) \sin(k_3 x_3)$$
$$w(x_1, x_2, x_3) = \sin(k_1 x_1) \sin(k_2 x_2) \cos(k_3 x_3)$$

in this Hilbert space, where $k_1 = n_1 \pi / \ell_1$, $k_2 = n_2 \pi / \ell_2$, $k_3 = n_3 \pi / \ell_3$.
(ii) Show that $u(x_1, x_2, x_3)$, $v(x_1, x_2, x_3)$, $w(x_1, x_2, x_3)$ are 0 at the boundaries of the box (rectangular cavity) $[0, \ell_1] \times [0, \ell_2] \times [0, \ell_3]$.
(iii) Show that the functions (A, B, C are constants)

$$\tilde{u}(x_1, x_2, x_3, t) = Au(x_1, x_2, x_3)e^{i\omega t}$$
$$\tilde{v}(x_1, x_2, x_3, t) = Bv(x_1, x_2, x_3)e^{i\omega t}$$
$$\tilde{w}(x_1, x_2, x_3, t) = Cw(x_1, x_2, x_3)e^{i\omega t}$$

satisfy the wave equation

$$\frac{1}{c^2} \frac{\partial^2 \psi}{\partial t^2} = \frac{\partial^2 \psi}{\partial x_1^2} + \frac{\partial^2 \psi}{\partial x_2^2} + \frac{\partial^2 \psi}{\partial x_3^2}$$

with (*dispersion relation*) $\omega_{\mathbf{k}} = c(k_1^2 + k_2^2 + k_3^2)$.

Problem 4. Let $\mathbf{A}$ be the *magnetic vector potential* ($kg \cdot m \cdot s^{-2} \cdot A^{-1}$) and ϕ the *electric potential* ($kg \cdot m^2 \cdot s^{-3} \cdot A^{-1}$). The Hamilton function of a non-relativistic electron (without spin) moving in an electromagnetic field $(\mathbf{A}, \phi)$ is given by

$$H = \frac{1}{2m_e}(\mathbf{p} - e\mathbf{A}(\mathbf{x}, t))^2 + e\phi(\mathbf{x}, t).$$

Here m_e is the mass of the electron, $e = -e_0$ is the *charge* of the electron and

$$\mathbf{p} = m_e \frac{d\mathbf{x}}{dt} + e\mathbf{A}(\mathbf{x}, t)$$

is its canonical momentum. Show that if there is a constant $\mathbf{p}_0$ such that the electric potential is given by

$$\phi(\mathbf{x}, t) = \frac{1}{m_e}\mathbf{p}_0 \cdot \mathbf{A}(\mathbf{x}, t) - \frac{e}{2m_e}\mathbf{A}^2(\mathbf{x}, t)$$

then the Hamilton function can be written as

$$H = \frac{1}{2m_e}\mathbf{p}^2 - \frac{e}{m_e}(\mathbf{p} - \mathbf{p}_0) \cdot \mathbf{A}(\mathbf{x}, t)$$

where $\cdot$ denotes the scalar product.

Problem 5.   Consider the eigenvalue problem (Schrödinger equation for the three-body harmonic oscillator)

$$-\frac{\hbar^2}{2m}\sum_{j=1}^{3}\Delta_{\mathbf{r}_j}u(\mathbf{r}_1,\mathbf{r}_2,\mathbf{r}_3)+\frac{k}{2}\sum_{j>k=1}^{3}(\mathbf{r}_k-\mathbf{r}_j)^2u(\mathbf{r}_1,\mathbf{r}_2,\mathbf{r}_3)=Eu(\mathbf{r}_1,\mathbf{r}_2,\mathbf{r}_3)$$

where m denotes the mass of the identical particle, k denotes the elastic constant between each pair of particles and $\Delta_{\mathbf{r}_j}$ is the Laplace operator with respect to the position vector $\mathbf{r}_j$ ($j=1,2,3$). Consider the transformation

$$\mathbf{R}=\frac{1}{\sqrt{3}}(\mathbf{r}_1+\mathbf{r}_2+\mathbf{r}_3)=0,\quad \mathbf{x}=\frac{1}{\sqrt{6}}(2\mathbf{r}_1-\mathbf{r}_2-\mathbf{r}_3)=\sqrt{\frac{3}{2}}\mathbf{r}_1,\quad \mathbf{y}=\frac{1}{\sqrt{2}}(\mathbf{r}_2-\mathbf{r}_3)$$

$$\tilde{u}(\mathbf{R}(\mathbf{r}_1,\mathbf{r}_2,\mathbf{r}_3),\mathbf{x}(\mathbf{r}_1,\mathbf{r}_2,\mathbf{r}_3),\mathbf{y}(\mathbf{r}_1,\mathbf{r}_2,\mathbf{r}_3))=u(\mathbf{r}_1,\mathbf{r}_2,\mathbf{r}_3)$$

(center of mass frame, Jacobi coordinate vectors). Show that the eigenvalue equation takes the form

$$-\frac{\hbar^2}{2m}(\Delta_{\mathbf{x}}+\Delta_{\mathbf{y}})\tilde{u}(\mathbf{R},\mathbf{x},\mathbf{y})+\frac{3k}{2}(\mathbf{x}^2+\mathbf{y}^2)\tilde{u}(\mathbf{R},\mathbf{x},\mathbf{y})=E\tilde{u}(\mathbf{R},\mathbf{x},\mathbf{y}).$$

Problem 6.   *Darcy's law* tells us that in a porous medium the fluid flux U is proportional to the gradient of the pressure p

$$U=-\frac{k}{\mu}\nabla p$$

where μ is the viscosity of the fluid and k is the permeability of the medium. Consider incompressible fluids $\nabla\cdot U\equiv\mathrm{div}(U)=0$. Show that

$$\nabla\cdot(k\nabla p)=0.$$

For a uniform system the quantity k is constant. Show that the pressure p satisfies the Laplace equation

Problem 7.   Consider the Schrödinger equation in two space dimensions

$$-\frac{\hbar^2}{2m}\left(\frac{\partial^2 u}{\partial x_1^2}+\frac{\partial^2 u}{\partial x_2^2}\right)+V(x_1,x_2)u=Eu.$$

Let (r,θ) be *polar coordinates*, i.e. $x_1=r\cos(\theta)$, $x_2=r\sin(\theta)$. Then we can write (*Fourier expansion*)

$$V(r,\theta)=\sum_{\mu=-\infty}^{\infty}V_\mu(r)\exp(i\mu\theta),\quad u(r,\theta)=\sum_{\lambda=-\infty}^{\infty}R_\lambda(r)\exp(i\lambda\theta).$$

Assume that for $V_\mu(r) = 0$ with $\mu \neq 0$ (circular potential) the Schrödinger equation can be written as

$$-\frac{\hbar^2}{2m}\left(\frac{d^2}{dr^2} + \frac{1}{r}\frac{d}{dr} - \frac{\lambda^2}{r^2}\right) R_\lambda(r) = E R_\lambda(r).$$

Defining the quantities (dimension $1/length^2$)

$$U(r) = \frac{2m}{\hbar}V(r), \qquad k^2 = -\kappa^2 = \frac{2m}{\hbar^2}E = E'$$

where k^2 refers to scattering states and κ^2 refers to bound states. Show that the differential equation takes the form

$$\frac{d^2 R_\lambda}{dr^2} + \frac{1}{r}\frac{dR_\lambda}{dr} + \left(k^2 - U(r) - \frac{\lambda^2}{r^2}\right) R_\lambda = 0.$$

Applying the transformation

$$R_\lambda(r) = \frac{\phi_\lambda(r)}{\sqrt{r}}$$

show that the differential equation then takes the form

$$\frac{d^2\phi_\lambda}{dr^2} + \left(k^2 - U(r) - \frac{(\lambda - 1/2)(\lambda + 1/2)}{r^2}\right)\phi_\lambda = 0.$$

Problem 8. Let $\sigma_1, \sigma_2, \sigma_3$ be the Pauli spin matrices. Consider the 4×4 matrices α_j $(j = 1, 2, 3)$ and β

$$\alpha_j = \begin{pmatrix} \sigma_j & 0_2 \\ 0_2 & -\sigma_j \end{pmatrix}, \qquad \beta = \begin{pmatrix} 0_2 & I_2 \\ I_2 & 0_2 \end{pmatrix}$$

with I_2 the 2×2 identity matrix and 0_2 the 2×2 zero matrix. The (Euclidean) Dirac's equation in four space-time dimensions

$$\hbar\frac{\partial\psi}{\partial t} = -\hat{H}\psi$$

where

$$\hat{H} = c\alpha \cdot (-i\hbar\nabla - e\mathbf{A}) + mc^2\beta + eA_0 I_4$$

where $\cdot$ denotes the scalar product and $\mathbf{A}$ and A_0 are the (time independent) electromagnetic potentials in the gauge $\nabla \cdot \mathbf{A} = 0$. Setting

$$\psi = \begin{pmatrix} \phi & \chi \end{pmatrix}^T = \begin{pmatrix} \phi_1 & \phi_2 & \chi_1 & \chi_2 \end{pmatrix}^T$$

show that the Dirac equation can be written as

$$\hbar\frac{\partial\phi}{\partial t} = c\boldsymbol{\sigma}\cdot(i\hbar\nabla + e\mathbf{A})\phi - mc^2\chi - eA_0\phi$$

$$\hbar\frac{\partial\chi}{\partial t} = -c\boldsymbol{\sigma}\cdot(i\hbar\nabla + e\mathbf{A})\chi - mc^2\phi - eA_0\chi$$

where

$$\chi = -\frac{1}{mc^2}\left(\hbar I_2\frac{\partial}{\partial t} - c\boldsymbol{\sigma}\cdot(i\hbar\nabla + e\mathbf{A}) + eA_0\right)\phi.$$

Problem 9.   Show that the *Helmholtz equation*

$$(\nabla^2 + k^2)u(\mathbf{r}) = -f(\mathbf{r})$$

admits the *Green's function*

$$G(\mathbf{r}, \mathbf{r}') = \frac{\exp(ik|\mathbf{r} - \mathbf{r}'|)}{4\pi|\mathbf{r} - \mathbf{r}'|}.$$

Show that $G(\mathbf{r}, \mathbf{r}')$ can be expanded into the series

$$G(\mathbf{r}, \mathbf{r}') = ik\sum_{\ell=0}^{\infty}\left(h_\ell(kr)j_\ell(kr')\left(\sum_{m=-\ell}^{\ell} Y_{\ell m}(\theta, \phi)Y_{\ell m}^*(\theta', \phi')\right)\right)$$

where $r, \theta, \phi)$ and (r', θ', ϕ') are the spherical coordinates of the radius vectors $\mathbf{r}$ and $\mathbf{r}'$ with $r > r'$ and h_ℓ, j_ℓ represent, respectively the spherical Hankel function of the first type (representing outgoing waves) and the spherical Bessel function.

Problem 10.   Consider the one-dimensional scalar conservation law

$$\frac{\partial u}{\partial t} + \frac{\partial}{\partial x}f(u(x, t)) = 0$$

with the initial condition $u(x, 0) = u_0(x)$. Assume that the flux $f(u)$ is a convex function (an example would be $f(u) = u^2/2$), i.e. $d^2f(u)/du^2 > 0$ and that the initial data are piecewise smooth functions which are either periodic or of compact support. Show that under this assumption the solution $u(x, t)$ of the initial value problem becomes discontinuous at some finite time $t = t_c$.

Problem 11.   Find solutions to the partial differential equation

$$\frac{\partial^2 u}{\partial x_0\partial x_1} - \frac{\partial u}{\partial x_0}\frac{\partial u}{\partial x_1} = 0.$$

First try the special cases

$$u(x_0, x_1) = f(x_0) + f(x_1), \quad u(x_0, x_1) = f(x_0)f(x_1), \quad u(x_0, x_1) = f(kx_1 - wx_0).$$

Problem 12. Consider the generalized *Korteweg-de Vries equation*

$$\frac{\partial u}{\partial t} + (a + bu^c)u^c\frac{\partial u}{\partial x} + d\frac{\partial^3 u}{\partial x^3} = 0.$$

Show that a travelling wave solution ansatz

$$u(x, t) = f(x - vt) = f(s)$$

provides the nonlinear ordinary differential equation

$$-v\frac{df}{ds} + (a + bf^c)f^c\frac{df}{ds} + d\frac{d^3 f}{ds^3} = 0.$$

Show that this differential equation can be integrated once

$$-vf + \left(\frac{a}{c+1} + \frac{b}{2c+1}f^c\right)f^{c+1} + d\frac{d^2 f}{ds^2} = C_1.$$

Show that this differential equation can be integrated once more with df/ds as an integrating factors

$$-\frac{v}{2}f^2 + \left(\frac{a}{(c+1)(c+2)} + \frac{b}{(2c+1)(2c+2)}f^c\right)f^{c+2} + \frac{d}{2}\left(\frac{df}{ds}\right)^2 = C_1 f + C_2.$$

Problem 13. The wave *group velocity* is given by $v_g = \partial\omega/\partial k$. If $v_g \neq v_p$ (v_p is the *phase velocity*) the wave is said to be *dispersive*. Show that the Monge-Ampére equation

$$\frac{\partial u}{\partial t} + u\frac{\partial u}{\partial x} = D\frac{\partial^2 u}{\partial x^2}$$

known as *Burgers equation* incorporates both amplitude dispersion and diffusion effects.

Problem 14. Consider the *Whitham equation*

$$\frac{\partial u}{\partial t} + Cu\frac{\partial u}{\partial x} + \frac{\partial}{\partial x}\int_{\mathbb{R}} u(x', t)G(x' - x)dx' = 0$$

with

$$G(x) = \frac{1}{2\pi}\int_{\mathbb{R}} c(k)e^{ikx}dk$$

and the boundary conditions $u \to 0$ for $|x| \to \infty$. Here $c(k)$ is the infinitesimal wave phase speed dispersion. Assume that

$$c(k) = c_0 \left(1 - \frac{1}{2}kd \left(\coth(kD) - \frac{1}{kD} \right) \right).$$

Show that if $D \to 0$ and $c(k) - c_0 \sim k^2$ the partial differential equation reduces to the Korteweg-de Vries equation. Show that for $D \to \infty$, $c(k) - c_0 \sim |k|$ the partial differential equation reduces to the Benjamin-Ono equation.

Problem 15. Consider the Schrödinger equation

$$i\hbar \frac{\partial \psi(\mathbf{x}, t)}{\partial t} = \hat{H}\psi(\mathbf{x}, t)$$

where the Hamilton operator is given by

$$\hat{H} = -\frac{\hbar^2}{2m}\nabla^2 + V(\mathbf{x}).$$

Let

$$\psi(\mathbf{x}, t) = \phi(\mathbf{x}, t)\exp(iS(\mathbf{x}, t)/\hbar), \quad \rho(\mathbf{x}, t) = \psi^*(\mathbf{x}, t)\psi(\mathbf{x}, t) = \phi^2(\mathbf{x}, t)$$

$$\mathbf{J}(\mathbf{x}, t) = \phi^2(\mathbf{x}, t)\frac{1}{m}\nabla_{\mathbf{x}}S(\mathbf{x}, t) = \rho(\mathbf{x}, t)\mathbf{v}(\mathbf{x}, t).$$

Show that the Schrödinger equation can be expressed as

$$\frac{\partial \rho}{\partial t} = -\nabla \cdot \mathbf{J}, \quad \frac{\partial \mathbf{v}}{\partial t} + (\mathbf{v} \cdot \nabla)\mathbf{v} = -\frac{1}{m}\nabla(V + Q)$$

where

$$Q = -\frac{\hbar^2}{2m}\frac{1}{\rho^{1/2}}\nabla^2\rho^{1/2}.$$

Problem 16. The *Gross-Pitaevskii equation* for the condensate wave function ψ is given by

$$i\hbar \frac{\partial \psi}{\partial t} = -\frac{\hbar^2}{2m}\nabla^2\psi + g|\psi|^2\psi$$

where m is the mass of the particle and g is a positive interaction coefficient. Find the time evolution (transport equation) for the number density $\rho = |\psi|^2$ and the time evolution (transport equation) for the mass current $\mathbf{J}$ given by

$$J_\mu = \frac{i\hbar}{2}\left(\psi \frac{\partial \psi^*}{\partial x_\mu} - \psi^* \frac{\partial \psi}{\partial x_\mu} \right), \quad \mu = 1, 2, 3.$$

Show that (*momentum conservation*)

$$\frac{\partial J_\mu}{\partial x_\mu} + \sum_{\nu=1}^{3} \frac{T_{\mu\nu}}{\partial x_\nu} = 0, \quad \mu = 1, 2, 3$$

where $T_{\mu\nu}$ is the *tensor flux* of the momentum

$$T_{\mu\nu} = -\frac{\hbar^2}{4m} \left(\psi \frac{\partial^2 \psi^*}{\partial x_\mu \partial x_\nu} + \psi^* \frac{\partial^2 \psi}{\partial x_\mu \partial x_\nu} - \frac{\partial \psi^*}{\partial x_\mu} \frac{\partial \psi}{\partial x_\nu} - \frac{\partial \psi}{\partial x_\mu} \frac{\partial \psi^*}{\partial x_\nu} \right) + \frac{g\hbar}{2} \delta_{\nu\mu} |\psi|^4$$

where $\mu, \nu = 1, 2, 3$ and $\delta_{\nu\mu}$ is the Kronecker delta.

Problem 17. Consider the Schrödinger equation

$$i\hbar \frac{\partial \psi}{\partial t} = -\frac{\hbar^2}{2m} \sum_{j=1}^{3} \frac{\partial^2 \psi}{\partial q_j^2} + V(\mathbf{q}).$$

Show that substituting $\psi(\mathbf{q}, t) = R(\mathbf{q}, t) \exp(iS(\mathbf{q}, t)/\hbar)$ (where R, S are real) into the Schrödinger equation and separating out the real and imaginary parts provides

$$\frac{\partial S}{\partial t} + \frac{1}{2m}(\nabla S)^2 + V(\mathbf{q}) - \frac{\hbar^2}{2m} \frac{(\nabla^2 R)}{R} = 0, \quad \frac{\partial R^2}{\partial t} + \nabla \left(\frac{1}{m} R^2 \nabla S \right) = 0.$$

Set

$$S(\mathbf{q}, t) = \hbar \arctan \left(i \frac{\psi^* - \psi}{\psi^* + \psi} \right).$$

Show that

$$\mathbf{p} = \nabla S = \frac{1}{|\psi|^2} \Re(\psi^*(-i\hbar\nabla)\psi).$$

Problem 18. Let G be the gravitational constant. The *Schrödinger-Newton equations* are given by

$$-\frac{\hbar^2}{2m} \Delta\psi + U(\mathbf{x})\psi = \mu\psi, \quad \Delta U = 4\pi G m^2 |\psi|^2$$

with

$$\int_{\mathbb{R}^3} |\psi|^2 d\mathbf{x} = 1.$$

Show that by multiplying the first equation with ψ^* and integrating we find μ as a function of the eigenstate ψ

$$\mu[\psi] = \int_{\mathbb{R}^3} \left(\frac{\hbar^2}{2m} |\nabla\psi(\mathbf{x})|^2 + U(\mathbf{x})|\psi(\mathbf{x})|^2 \right) d\mathbf{x}.$$

The energy functional is defined as

$$E[\psi] = \int_{\mathbb{R}^3} \left(\frac{\hbar^2}{2m} |\nabla \psi(\mathbf{x})|^2 + \frac{U(\mathbf{x})}{2} |\psi(\mathbf{x})|^2 \right) d\mathbf{x}.$$

Problem 19. The Schrödinger equation with the electromagnetic vector (magnetic vector potential) potential and electric potential ϕ is given by

$$i\hbar \frac{\partial \psi}{\partial t} = \left(\frac{1}{2m} \left(\frac{\hbar}{i} \nabla - e\mathbf{A}(\mathbf{x}) \right)^2 + e\phi(\mathbf{x}) + V(\mathbf{x}) \right) \psi.$$

Let $\psi(\mathbf{x}, t) = R(\mathbf{x}, t) \exp(iS(\mathbf{x}, t))$ with $R(\mathbf{x}, t)$, $S(\mathbf{x}, t)$ be real functions. Show that

$$i\hbar e^{iS} \left(\frac{\partial R}{\partial t} + iR \frac{\partial S}{\partial t} \right) = -\frac{\hbar^2}{2m} e^{iS} (2i(\nabla S) \cdot (\nabla R) + \nabla^2 R - R(\nabla S)^2 + iR\nabla^2 S)$$

$$+ ie\hbar m e^{iS} (\mathbf{A} \cdot (\nabla R) + iR\mathbf{A} \cdot (\nabla S))$$

$$+ Re^{iS} \left(\frac{ie\hbar}{2m} \nabla \cdot \mathbf{A} + \frac{e^2}{2m} \mathbf{A}^2 + e\phi + V \right).$$

Setting $\rho(\mathbf{x}, t) = |\psi(\mathbf{x}, t)|^2$ and $R = \rho^{1/2}$ show that this equation can be written as a pair of partial differential equations

$$\frac{\partial \rho}{\partial t} + \frac{\hbar}{m} \nabla \cdot (\rho \nabla S) - \frac{e}{m} \nabla \cdot (\rho \mathbf{A}) = 0$$

$$\hbar \frac{\partial S}{\partial t} + \frac{\hbar^2}{2m} \left(\left(\frac{\nabla \rho}{2\rho} \right)^2 - \frac{\nabla^2 \rho}{2\rho} + (\nabla S)^2 \right) - \frac{e\hbar}{m} \mathbf{A} \cdot (\nabla S) + \frac{e^2}{2m} \mathbf{A}^2 = -e\phi - V.$$

Problem 20. Consider the equation of motion of the flow of a liquid

$$\rho \frac{\partial \mathbf{u}}{\partial t} + \rho(\mathbf{u} \cdot \nabla)\mathbf{u} = -\nabla p + \rho \mathbf{g} - 2\rho(\mathbf{\Omega} \times \mathbf{u}) \tag{1}$$

where ρ is the mass density, $\mathbf{u} = (u_1, u_2, u_3)$ is the velocity field, p is the pressure and $\mathbf{g}$ is the gravitational acceleration in x_3 direction. Show that introducing the *vorticity* of the flow $\zeta := \nabla \times \mathbf{u}$ and taking the curl of (1) one obtains

$$\frac{d\omega}{dt} = (\omega \cdot \nabla)\mathbf{u} - \omega \nabla \cdot \mathbf{u} + \frac{\nabla \rho \times \nabla p}{\rho^2}$$

where d/dt is the substantial derivative and ω is the absolute vorticity given by $\omega = \nabla \times (\mathbf{u} + \mathbf{\Omega} \times \mathbf{x}) = \zeta + 2\mathbf{\Omega}$.

Chapter 12

Functional Analysis, Hilbert Spaces and Wavelets

A *Hilbert space* is a set, $\mathcal{H}$ of elements, or vectors, $(f, g, h, \ldots)$ which satisfies the following conditions (1) to (5).

(1) If f and g belong to $\mathcal{H}$, then there is a unique element of $\mathcal{H}$, denoted by $f + g$, the operation of addition $(+)$ being invertible, commutative and associative.

(2) If c is a complex number, then for any f in $\mathcal{H}$, there is an element cf of $\mathcal{H}$; and the multiplication of vectors by complex numbers thereby defined satisfies the distributive conditions

$$c(f + g) = cf + cg, \qquad (c_1 + c_2)f = c_1 f + c_2 f.$$

(3) Hilbert spaces $\mathcal{H}$ possess a zero element, 0, characterized by the property that $0 + f = f$ for all vectors f in $\mathcal{H}$.

(4) For each pair of vectors f, g in $\mathcal{H}$, there is a complex number $\langle f|g \rangle$, termed the inner product or scalar product of f with g, such that

$$\langle f|g \rangle = \overline{\langle g|f \rangle}$$

$$\langle f|g + h \rangle = \langle f|g \rangle + \langle f|h \rangle$$

$$\langle f|cg\rangle = c\langle f|g\rangle$$

and

$$\langle f|f\rangle \geq 0.$$

Equality in the last formula occurs only if $f = 0$. The scalar product defines the norm $\|f\| = \langle f|f\rangle^{1/2}$.

(5) If $\{f_n\}$ is a sequence in $\mathcal{H}$ satisfying the Cauchy condition that

$$\|f_m - f_n\| \to 0$$

as m and n tend independently to infinity, then there is a unique element f of $\mathcal{H}$ such that $\|f_n - f\| \to 0$ as $n \to \infty$.

Let $B = \{\phi_n : n \in I\}$ be an orthonormal basis in the Hilbert space $\mathcal{H}$, where I is the countable index set. Then

$$(1) \qquad \bigwedge_{f \in \mathcal{H}} \quad f = \sum_{n \in I}\langle f|\phi_n\rangle\phi_n$$

$$(2) \qquad \bigwedge_{f,g \in \mathcal{H}} \quad \langle f|g\rangle = \sum_{n \in I}\overline{\langle f|\phi_n\rangle}\langle g|\phi_n\rangle.$$

Let $L_2(\mathbb{R})$ be the Hilbert space of the square integrable function over $\mathbb{R}$. Let $f \in L_2(\mathbb{R})$ and $f \in L_1(\mathbb{R})$. The *Fourier transform* $\hat{f}(k)$ of $f(x)$ is defined as

$$\hat{f}(k) := \int_{\mathbb{R}} f(x)e^{ikx}dx.$$

The inverse Fourier transform is given by

$$f(x) = \frac{1}{2\pi}\int_{\mathbb{R}} \hat{f}(k)e^{-ikx}dk.$$

For $L_2(\mathbb{R}^n)$ we have

$$\hat{f}(\mathbf{k}) := \int_{\mathbb{R}^n} f(\mathbf{x})e^{i\mathbf{k}\cdot\mathbf{x}}dx$$

and the inverse

$$f(\mathbf{x}) = \frac{1}{(2\pi)^n}\int_{\mathbb{R}^n} \hat{f}(\mathbf{k})e^{-i\mathbf{k}\cdot\mathbf{x}}d\mathbf{k}$$

where $d\mathbf{x} = dx_1\cdots dx_n$, $d\mathbf{k} = dk_1\cdots dk_n$ and

$$\mathbf{k}\cdot\mathbf{x} := k_1x_1 + \cdots + k_nx_n$$

is the scalar product. In some one-dimensional applications x would be the time t and k would then be the frequency ω. In applications in three dimensions $\mathbf{k}$ would be the *wave vector* and $\mathbf{x}$ would be the space coordinates. The Fourier transform and its inverse are linear transformations.

Problem 1. Let $d \geq 1$ and $|0\rangle$, $|1\rangle$, ..., $|d\rangle$ be an orthonormal basis in the Hilbert space $\mathbb{C}^{d+1}$. Define the linear operators $((d+1) \times (d+1)$ matrices)

$$b = \sum_{j=1}^{d} \sqrt{j}|j-1\rangle\langle j|, \quad b^\dagger = \sum_{k=1}^{d} \sqrt{k}|k\rangle\langle k-1|.$$

Are the operators hermitian? Are the operators unitary? Find the commutator $[b, b^\dagger]$. Is the commutator hermitian? Is the commutator unitary? Study the case with $d = 1$ the standard basis and the Hadamard basis.

Solution 1. The operators are neither hermitian nor unitary. Utilizing $\langle j|k\rangle = \delta_{jk}$ we have

$$[b, b^\dagger] = \sum_{j=0}^{d-1} |j\rangle\langle j| - d|d\rangle\langle d| = I_{d+1} - (d+1)|d\rangle\langle d|$$

where we used the *completeness relation* $\sum_{j=0}^{d} |j\rangle\langle j| = I_{d+1}$. For $d = 1$ the standard basis is given by

$$|0\rangle = \begin{pmatrix} 1 \\ 0 \end{pmatrix}, \quad |1\rangle = \begin{pmatrix} 0 \\ 1 \end{pmatrix}.$$

It follows that

$$b = \begin{pmatrix} 0 & 1 \\ 0 & 0 \end{pmatrix}, \quad b^\dagger = \begin{pmatrix} 0 & 0 \\ 1 & 0 \end{pmatrix} \quad \Rightarrow \quad [b, b^\dagger] = \sigma_3.$$

The Hadamard basis is given

$$|0\rangle = \frac{1}{\sqrt{2}} \begin{pmatrix} 1 \\ 1 \end{pmatrix}, \quad |1\rangle = \frac{1}{\sqrt{2}} \begin{pmatrix} 1 \\ -1 \end{pmatrix}.$$

Then

$$b = \frac{1}{2} \begin{pmatrix} 1 & -1 \\ 1 & -1 \end{pmatrix}, \quad b^\dagger = \frac{1}{2} \begin{pmatrix} 1 & 1 \\ -1 & -1 \end{pmatrix} \quad \Rightarrow \quad [b, b^\dagger] = \sigma_1.$$

Problem 2. Let A and B be two arbitrary $n \times n$ matrices over $\mathbb{R}$. We define the scalar product

$$\langle A, B \rangle := \operatorname{tr}(AB^T) \tag{1}$$

where B^T denotes the transpose of B.
(i) Let $c \in \mathbb{R}$. Show that

$$\langle A, A \rangle \geq 0 \tag{2a}$$

$$\langle A, B \rangle = \langle B, A \rangle \tag{2b}$$

$$\langle cA, B \rangle = c \langle A, B \rangle \tag{2c}$$

$$\langle A_1 + A_2, B \rangle = \langle A_1, B \rangle + \langle A_2, B \rangle. \tag{2d}$$

(ii) Let O be an orthogonal $n \times n$ matrix over $\mathbb{R}$, i.e. $O^{-1} = O^T$. Find the scalar product $\langle O, O \rangle$.

(iii) Can the result be extended to infinite-dimensional matrices?

Solution 2. (i) We find

$$\langle A, A \rangle = \text{tr}(AA^T) = \sum_{j=1}^{n} \sum_{k=1}^{n} a_{jk} a_{jk}.$$

Since $a_{jk} a_{jk} \geq 0$ we obtain inequality (2a). Owing to

$$\langle A, B \rangle = \text{tr}(AB^T) = \sum_{j=1}^{n} \sum_{k=1}^{n} a_{jk} b_{jk}, \quad \langle B, A \rangle = \text{tr}(BA^T) = \sum_{j=1}^{n} \sum_{k=1}^{n} b_{jk} a_{jk}$$

we obtain (2b). Since $\langle cA, B \rangle = \text{tr}(cAB^T) = c\,\text{tr}(AB^T) = c \langle A, B \rangle$ we find (2c). To prove (2d) we recall that $\text{tr}(X + Y) = \text{tr}(X) + \text{tr}(Y)$ for any $n \times n$ matrices X and Y. Therefore

$$\langle A_1 + A_2, B \rangle = \text{tr}((A_1 + A_2)B^T) = \text{tr}(A_1 B^T) + \text{tr}(A_2 B^T) = \langle A_1, B \rangle + \langle A_2, B \rangle.$$

Consequently, definition (1) defines a scalar product for the vector space of $n \times n$ matrices over the field of real numbers.

(ii) Since $O^T = O^{-1}$ for an orthogonal matrix we find

$$\langle O, O \rangle = \text{tr}(OO^T) = \text{tr}(OO^{-1}) = \text{tr}(I_n) = n.$$

Note that $\langle OA, OB \rangle = \text{tr}(OAB^T O^T) = \text{tr}(AB^T)$.

(iii) The results can be extended when we impose the condition

$$\sum_{j=1}^{\infty} \sum_{k=1}^{\infty} |a_{jk}|^2 < \infty.$$

If an infinite-dimensional matrix A satisfies this condition we call A a *Hilbert-Schmidt operator*. The norm of a Hilbert-Schmidt operator A is given by

$$\|A\| := \sqrt{\sum_{j=1}^{\infty} \sum_{k=1}^{\infty} |a_{jk}|^2}.$$

Problem 3. Consider the Hilbert space $\mathcal{H}$ of the 2×2 matrices over the complex numbers with the scalar product

$$\langle A, B \rangle := \operatorname{tr}(AB^*), \qquad A, B \in \mathcal{H}.$$

Show that the rescaled Pauli matrices $\mu_j := \frac{1}{\sqrt{2}}\sigma_j$, $j = 1, 2, 3$

$$\mu_1 = \frac{1}{\sqrt{2}} \begin{pmatrix} 0 & 1 \\ 1 & 0 \end{pmatrix}, \quad \mu_2 = \frac{1}{\sqrt{2}} \begin{pmatrix} 0 & -i \\ i & 0 \end{pmatrix}, \quad \mu_3 = \frac{1}{\sqrt{2}} \begin{pmatrix} 1 & 0 \\ 0 & -1 \end{pmatrix}$$

plus the rescaled 2×2 identity matrix

$$\mu_0 = \frac{1}{\sqrt{2}} \begin{pmatrix} 1 & 0 \\ 0 & 1 \end{pmatrix}$$

form an orthonormal basis in the Hilbert space $\mathcal{H}$.

Solution 3. Let $c_0, c_1, c_2, c_3 \in \mathbb{C}$. Then the equation

$$c_0\mu_0 + c_1\mu_1 + c_2\mu_2 + c_3\mu_3 = \begin{pmatrix} 0 & 0 \\ 0 & 0 \end{pmatrix}$$

only admits the zero solution, i.e. $c_0 = c_1 = c_2 = c_3 = 0$. Thus the four matrices are linearly independent. Since $\operatorname{tr}(\mu_j\mu_k) = 0$, $j \neq k$ and $\operatorname{tr}(\mu_j\mu_j) = 1$ we have an orthonormal basis in this Hilbert space.

Problem 4. Consider a complex Hilbert space $\mathcal{H}$ and $|\phi_1\rangle, |\phi_2\rangle \in \mathcal{H}$. Let $c_1, c_2 \in \mathbb{C}$. An *antilinear operator* K in this Hilbert space $\mathcal{H}$ is characterized by

$$K(c_1|\phi_1\rangle + c_2|\phi_2\rangle) = c_1^* K|\phi_1\rangle + c_2^* K|\phi_2\rangle.$$

A *comb* is an antilinear operator K with zero expectation value for all states $|\psi\rangle$ of a certain complex Hilbert space $\mathcal{H}$. This means

$$\langle \psi|K|\psi \rangle = \langle \psi|LC|\psi \rangle = \langle \psi|L|\psi^* \rangle = 0$$

for all states $|\psi\rangle \in \mathcal{H}$, where L is a linear operator and C is the complex conjugation.
(i) Consider the two-dimensional Hilbert space $\mathcal{H} = \mathbb{C}^2$. Find a unitary 2×2 matrix such that $\langle \psi|UC|\psi \rangle = 0$.
(ii) Consider the Pauli spin matrices σ_1, σ_2, σ_3 and $\sigma_0 = I_2$. Find

$$\sum_{\mu=0}^{3}\sum_{\nu=0}^{3} \langle \psi|\sigma_\mu C|\psi \rangle g^{\mu,\nu} \langle \psi|\sigma_\nu C|\psi \rangle$$

where $g^{\mu,\nu} = \operatorname{diag}(-1, 1, 0, 1)$.

Solution 4.   (i) We find $U = \sigma_2$ since

$$\langle\psi|\sigma_2 C|\psi\rangle = \langle\psi|\sigma_2|\psi^*\rangle = (\,\psi_1^* \quad \psi_2^*\,) \begin{pmatrix} 0 & -i \\ i & 0 \end{pmatrix} \begin{pmatrix} \psi_1^* \\ \psi_2^* \end{pmatrix}$$

$$= (\,\psi_1^* \quad \psi_2^*\,) \begin{pmatrix} -i\psi_2^* \\ i\psi_1^* \end{pmatrix} = 0.$$

(ii) We have

$$\sum_{\mu=0}^{3}\sum_{\nu=0}^{3}\langle\psi|\sigma_\mu C|\psi\rangle g^{\mu,\nu}\langle\psi|\sigma_\nu C|\psi\rangle = 0.$$

Problem 5.   Show that if $a < b$, and $n \in \mathbb{Z}$ the functions $f_n(x) = e^{inx}$ are a linearly independent set on the interval $[a, b]$.

Solution 5.   If $(b - a) \geq 2\pi$ and

$$\sum_{n=p}^{q} c_n e^{inx} = 0$$

on $[a, b]$, then the orthogonality relationships among the functions f_n, i.e.

$$\int_0^{2\pi} e^{i(n-m)x}dx = 2\pi\delta_{mn}$$

imply that all c_n are equal to zero. Here δ_{mn} denotes the Kronecker delta. If $0 < b - a < 2\pi$ and

$$g(x) = \sum_{n=p}^{q} c_n e^{inx} = 0$$

on $[a, b]$, then g has an extension to an entire function on $\mathbb{C}$. Thus

$$g(x) = 0, \qquad \sum_{n=p}^{q} c_n e^{inx} = 0$$

on $[0, 2\pi]$ and the previous argument shows all c_n are equal to zero.

Problem 6.   (i) Let $I = [0, 1]$ with Lebesgue measure and $x \in I$. Let the nth *Rademacher function* r_n be defined as

$$r_n(x) := \text{sgn}(\sin(2^n\pi x))$$

where $n = 0, 1, 2, \ldots$ and sgn denotes the *signum function*. Show that the Rademacher functions r_n constitute a linearly independent set.

(ii) For $m \in \mathbb{N}$ there is in $\mathbb{N} \cup \{0\}$ a unique finite set $\{n_1, n_2, \dots, n_{K_m}\}$ such that

$$m = \sum_{k=1}^{K_m} 2^{n_k}.$$

The m-th *Walsh function* is defined by

$$W_m(x) := \prod_{k=1}^{K_m} r_{n_k}(x)$$

where $x \in I \equiv [0,1]$. Show that $\{W_m\}_{m=1}^{\infty}$ is a complete orthogonal set in the Hilbert space $L_2(I)$. This is the Hilbert space of the square-integrable functions.

Solution 6. (i) The function r_n partitions $[0,1]$ into three sets, $S_n^{\pm}$ and S_n^0, where

$$S_n^{\pm} = r_n^{-1}(\pm 1), \qquad S_n^0 = r_n^{-1}(0).$$

Furthermore, for $n = 0$ we have $S_0^+ = (0,1)$, $S_0^- = \emptyset$, $S_0^0 = \{0,1\}$. If $n \geq 1$ each of $S_n^{\pm}$ consists of 2^{n-1} disjoint intervals, each of length 2^{-n} and S_n^0 consists of their $2^n + 1$ endpoints. The intervals of S_n^+ alternate with those of S_n^-. If $m > n$ the intervals of $S_m^{\pm}$ equipartition those of $S_n^{\pm}$ from which the linear independence of the Rademacher functions r_n, $n = 1, 2, \dots$ follows. Obviously, $\{r_n\}_n$ is an orthonormal set in the Hilbert space $L_2(I)$. However, the Rademacher functions do not form a basis in the Hilbert space $L_2(I)$.

(ii) The result of (i) shows that $\{W_m\}_{m=1}^{\infty}$ is an orthonormal set in the Hilbert space $L_2(I)$. Let

$$f \in (\{W_m\}_{m=1}^{\infty})^{\perp}$$

where $M^{\perp}$ is the *orthogonal complement* of M. The set $M^{\perp}$ denotes all vectors which are orthogonal to every vector in M. Then for all $N \in \mathbb{N}$ and all x,

$$F(x) = \int_0^1 f(t) \prod_{m=0}^{N} (1 + r_m(x)r_m(t)) dt = 0.$$

Induction shows that if $N > 1$ and $k/2^m \leq x \leq (k+1)/2^m$ then

$$\prod_{m=0}^{N} (1 + r_m(x)r_m(t)) \neq 0$$

if and only if $k/2^N \leq t \leq (k+1)/2^N$. Obviously $1 + r_2(x)r_2(t) \neq 0$ if and only if x and t are in the same half of I,

$$(1 + r_2(x)r_2(t))(1 + r_3(x)r_3(t)) \neq 0$$

if and only if x and t are in the same quarter of I, etc. Hence

$$\int_{k/2^N}^{(k+1)/2^N} f(t)dt = 0$$

for all k, $N \in \mathbb{N}$. Consequently, $f(t) = 0$ almost everywhere. It follows that $\{W_m\}_{m=1}^{\infty}$ is a complete orthonormal system. In other words the Walsh functions form an orthonormal basis in the Hilbert space $L_2(I)$.

Problem 7.   Let P and Q be two linear operators in a normed space. Let a be a nonzero real or complex number and let I be the identity operator. Show that if all iterated operators Q^m exist then the relation

$$PQ - QP = aI \tag{1}$$

cannot be satisfied by two bounded operators P, Q in a normed space.

Solution 7.   We assume that the iterated operators Q^m $(m = 0, 1, 2 \ldots)$ are meaningful. From (1) it follows that P and Q can neither be the null operator nor a constant. Consequently, $\|P\| \neq 0$, $\|Q\| \neq 0$, where $\|.\|$ denotes the norm. From (1) we find by induction that

$$PQ^n - Q^n P = anQ^{n-1}. \tag{2}$$

For $n = 1$, (2) is obviously true. It follows then that

$$PQ^{n+1} - Q^{n+1}P = (PQ^n - Q^n P)Q + Q^n(PQ - QP)$$
$$= anQ^{n-1}Q + Q^n aI = a(n+1)Q^n$$

which completes the induction. We assume now that P and Q are bounded operators with the norms $\|P\|$ and $\|Q\|$. Taking norms on both sides of this equation for $n = 1, 2, \ldots$ we find

$$|a|n\|Q^{n-1}\| = \|PQ^n - Q^n P\| \leq \|P\|\,\|Q^n\| + \|Q^n\|\,\|P\| = 2\|P\|\,\|Q^n\|.$$

Thus $|a|n\|Q^{n-1}\| \leq 2\|P\|\,\|Q^{n-1}\|\,\|Q\|$. Let N be an integer with

$$N > \frac{2}{|a|}\|P\|\,\|Q\|.$$

We have to study two cases:
Case 1. Let $\|Q^{N-1}\| \neq 0$. This result is in contradiction for $n = N$.
Case 2: Let $\|Q^{N-1}\| = 0$. Owing to $|a|\,n\,\|Q^{n-1}\| \leq 2\|P\|\,\|Q^n\|$ for all $n = 1, 2, \ldots$ we obtain

$$\|Q^{N-2}\| = 0, \quad \|Q^{N-3}\| = 0, \quad \ldots, \quad \|Q^1\| = 0.$$

Again we find a contradiction.

Owing to this result there are no two finite dimensional matrices P and Q which satisfy (1) with $a \neq 0$. This also follows from $\text{tr}([A, B]) = 0$ for $n \times n$ matrices A and B over the field of the complex numbers. There are *unbounded operators* which satisfy (1). Consider the infinite-dimensional matrix

$$b := \begin{pmatrix} 0 & \sqrt{1} & 0 & 0 & \\ 0 & 0 & \sqrt{2} & 0 & \\ 0 & 0 & 0 & \sqrt{3} & \\ & \ddots & & & \ddots \\ & & \ddots & & \\ & & & \ddots & \end{pmatrix}.$$

Let b^T be the transpose of b. Then $[b, b^T] = I$, where I is the infinite-dimensional unit matrix. The operator $b^T b$ is given by the infinite-dimensional diagonal matrix (unbounded operator) $b^T b = \text{diag}(0, 1, 2, \dots)$.

Problem 8. Show that if $M \in C[a, b]$ and

$$\int_a^b M(x) h(x) dx = 0 \tag{1}$$

for all $h \in \mathcal{H}$, then it is necessary that $M(x) = 0$ for all $x \in [a, b]$, where

$$\mathcal{H} := \{\, h : h \in C^1[a, b], \ h(a) = h(b) = 0 \,\}.$$

Solution 8. We prove it by contradiction. Suppose that there exists an $x_0 \in (a, b)$ such that $M(x_0) \neq 0$. Then we may assume without loss of generality that $M(x_0) > 0$. Since $M \in C[a, b]$ there exists a neighbourhood $N^\delta(x_0) \subset (a, b)$ such that $M(x) > 0$ for all $x \in N^\delta(x_0)$. Let

$$h(x) = \begin{cases} (x - x_0 - \delta)^2 (x - x_0 + \delta)^2 & \text{for } x \in N^\delta(x_0) \\ 0 & \text{for } x \notin N^\delta(x_0) \end{cases}$$

Clearly, $h \in \mathcal{H}$. With this choice of h we find

$$\int_a^b M(x) h(x) dx = \int_{x_0 - \delta}^{x_0 + \delta} M(x)(x - x_0 - \delta)^2 (x - x_0 + \delta)^2 dx > 0$$

with contradicts the hypothesis. Thus $M(x) = 0$ for all $x \in (a, b)$. That $M(x)$ also has to vanish at the endpoints of the interval follows from the continuity of M.

Problem 9.   Let f_1, f_2, ..., f_n be continuous real-valued functions on the interval $[a, b]$. Show that the set $\{\, f_1, f_2, ..., f_n \,\}$ is linearly dependent on $[a, b]$ if and only if

$$\det \left(\int_a^b f_i(x) f_j(x) dx \right) = 0.$$

Solution 9.   Let A be the matrix with the entries

$$A_{ij} := \int_a^b f_i(x) f_j(x) dx.$$

If the determinant of the matrix A vanishes, the matrix A is singular. Let **a** be a nonzero (column) n-vector with $A\mathbf{a} = \mathbf{0}$. Then

$$0 = \mathbf{a}^T A \mathbf{a} = \sum_{i=1}^{n} \sum_{j=1}^{n} \int_a^b a_i f_i(x) a_j f_j(x) dx = \int_a^b \left(\sum_{i=1}^{n} a_i f_i(x) \right)^2 dx.$$

Since the f_i's are continuous functions, the linear combinations

$$\sum_{j=1}^{n} a_j f_j$$

must vanish identically. Hence, the set $\{\, f_i \,\}$ is linearly dependent on $[a, b]$. Conversely, if $\{\, f_i \,\}$ is linearly dependent, some f_i can be expressed as a linear combination of the rest, so some row of A is a linear combination of the rest and A is singular.

Problem 10.   Consider the inner product space

$$C[a, b] = \{\, f(x) : f \text{ is continuous on } x \in [a, b] \,\}$$

with the inner product

$$\langle f, g \rangle := \int_a^b f(x) g^*(x) dx \;\; \Rightarrow \;\; \langle f, f \rangle = \int_a^b f(x) f^*(x) dx = \|f\|^2.$$

Show that $C[a, b]$ is incomplete. This means find a *Cauchy sequence* in the space $C[a, b]$ which converges to an element which is not in the space $C[a, b]$.

Solution 10.   Consider the sequence of continuous functions

$$g_k(x) = \frac{1}{2} + \frac{1}{\pi} \arctan(kx), \qquad -1 \leq x \leq 1$$

i.e. $a = -1$ and $b = 1$. We find

$$\lim_{k,p\to\infty} \|g_k - g_p\|^2 = \lim_{k,p\to\infty} \int_{-1}^{+1} (g_k(x) - g_p(x))^2 dx = 0.$$

In other words, the sequence is a Cauchy sequence. For fixed x we have

$$\lim_{k\to\infty} g_k(x) = g(x) = \begin{cases} 1 & 0 < x \le 1 \\ 1/2 & x = 0 \\ 0 & -1 \le x < 0 \end{cases}$$

which is a discontinuous function, i.e. $g \notin C[-1,1]$. We say that the sequence of functions $\{\, g_k \,:\, k = 1, 2, \dots \}$ converges pointwise to the function g. We also have

$$\lim_{k\to\infty} \|g - g_k\|^2 = \lim_{k\to\infty} \int_{-1}^{+1} (g(x) - g_k(x))^2 dx = 0.$$

Thus the limit of the Cauchy sequence does not lie in the inner product space $C[-1,1]$. Thus $C[a,b]$ is incomplete.

Problem 11.   Consider the Hilbert space $L_2[0,1]$. Find a non-trivial polynomial p

$$p(x) = ax^3 + bx^2 + cx + d$$

such that $\langle p, 1 \rangle = 0$, $\langle p, x \rangle = 0$, $\langle p, x^2 \rangle = 0$.

Solution 11.   Using the ansatz for a polynomial p we obtain

$$\langle p, 1 \rangle = \int_0^1 p(x) dx = \frac{a}{4} + \frac{b}{3} + \frac{c}{2} + d = 0$$

$$\langle p, x \rangle = \int_0^1 p(x)x\, dx = \frac{a}{5} + \frac{b}{4} + \frac{c}{3} + \frac{d}{2} = 0$$

$$\langle p, x^2 \rangle = \int_0^1 p(x)x^2\, dx = \frac{a}{6} + \frac{b}{5} + \frac{c}{4} + \frac{d}{3} = 0.$$

This system can be written in matrix form

$$\begin{pmatrix} 1/4 & 1/3 & 1/2 \\ 1/5 & 1/4 & 1/3 \\ 1/6 & 1/5 & 1/4 \end{pmatrix} \begin{pmatrix} a \\ b \\ c \end{pmatrix} = \begin{pmatrix} -d \\ -d/2 \\ -d/3 \end{pmatrix}.$$

The inverse of the matrix on the left-hand side exists. If we assume that $d \ne 0$ we find a non-trivial polynomial.

Problem 12.   Let $C^m[a, b]$ be the vector space of real-valued m-times differentiable functions and the m-th derivative is continuous over the interval $[a, b]$ ($b > a$). We define an inner product (scalar product) of two such functions f and g as

$$\langle f, g \rangle_m := \int_a^b \left(fg + \frac{df}{dx}\frac{dg}{dx} + \cdots + \frac{d^m f}{dx^m}\frac{d^m g}{dx^m} \right) dx.$$

Given (Legendre polynomials)

$$f(x) = \frac{1}{2}(3x^2 - 1), \qquad g(x) = \frac{1}{2}(5x^3 - 3x)$$

and the interval $[-1, 1]$, i.e. $a = -1$ and $b = 1$. Show that f and g are orthogonal with respect to the inner product $\langle f, g \rangle_0$. Are they orthogonal with respect to $\langle f, g \rangle_1$?

Solution 12.   Straightforward calculation yields

$$\langle f, g \rangle_0 = \frac{1}{4}\int_{-1}^1 (3x^2 - 1)(5x^3 - 3x)dx = \frac{1}{4}\int_{-1}^1 (15x^5 - 14x^3 + 3x)dx = 0.$$

We have $df/dx = 3x$, $dg/dx = 15x^2/2 - 3/2$. Then since $\langle f, g \rangle_0 = 0$ we have

$$\langle f, g \rangle_1 = \int_{-1}^1 3x\left(\frac{15}{2}x^2 - \frac{3}{2} \right) dx = \int_{-1}^1 \left(\frac{45}{2}x^3 - \frac{9}{2}x \right) dx = 0.$$

Problem 13.   Let $L_2[0, 1]$ be the Hilbert space of the square-integrable functions in the Lebesgue sense over the unit interval $[0, 1]$. Let $f : [0, 1] \mapsto [0, 1]$ be

$$f(x) = \begin{cases} 2x & \text{for} \quad 0 \le x \le 1/2 \\ 2(1 - x) & \text{for} \quad 1/2 < x \le 1. \end{cases} \tag{1}$$

An orthonormal basis in $L_2[0, 1]$ is given by $\{ u_n(x) := e^{2\pi i x n} : n \in \mathbb{Z} \}$. Obviously, $f \in L_2[0, 1]$. Then f can be expanded in a Fourier series with respect to u_n. Find the Fourier coefficients.

Solution 13.   The *Fourier expansion* of f is given by

$$f(x) = \sum_{n \in \mathbb{Z}} \langle f, u_n \rangle u_n \equiv \sum_{n \in \mathbb{Z}} c_n u_n.$$

Hence the expansion coefficients c_n are given by

$$c_n = \langle f, u_n \rangle = 2\int_0^{1/2} x e^{-2\pi i x n}dx + 2\int_{1/2}^1 e^{-2\pi i x n}dx - 2\int_{1/2}^1 x e^{-2\pi i x n}dx.$$

For $n = 0$ we find $c_0 = 1/2$. For $a \neq 0$ we have

$$\int e^{ax} dx = \frac{1}{a} e^{ax}, \qquad \int x e^{ax} dx = \frac{1}{a} e^{ax} x - \frac{1}{a^2} e^{ax}.$$

Using these two integrals we find for $n \neq 0$ that $c_n = 4(1 - (-1)^n)/a^2$, where $a = -2\pi i n$.

Problem 14. (i) Let $a > 0$. We define

$$f_a(x) = \begin{cases} \frac{1}{2a} & \text{for} \quad |x| \leq a \\ 0 & \text{for} \quad |x| > a \end{cases}. \tag{1}$$

Thus $f_a(x) \in L_2(\mathbb{R})$. Calculate

$$\int_{\mathbb{R}} f_a(x) dx \tag{2}$$

and the *Fourier transform* of f_a. Discuss the cases: a large and a small.
(ii) Let $N \in \mathbb{N}$ and let

$$V(t) = \begin{cases} V_0 e^{i\omega_0 t} & \text{if} \quad nT \leq t \leq (nT + \tau) \text{ for } n = 0, 1, \ldots, N-1 \\ 0 & \text{otherwise} \end{cases}$$

where V_0 and τ are positive constants. Calculate the Fourier transform.

Solution 14. (i) We find

$$\int_{\mathbb{R}} f_a(x) dx = \frac{1}{2a} \int_{-a}^{a} dx = 1.$$

Thus the integral is independent of a. We have

$$\hat{f}_a(k) = \int_{-a}^{a} \frac{e^{ikx}}{2a} dx = \frac{1}{2a} \int_{-a}^{a} e^{ikx} dx = \frac{1}{2aik} \left. e^{ikx} \right|_{-a}^{a} = \frac{\sin(ak)}{ak}.$$

For $k = 0$ we find (L'Hospital) $\hat{f}_a(0) = 1$.
(ii) For general N the Fourier transform is given by

$$S(\omega) = V_0 \sum_{n=0}^{N-1} \int_{nT}^{nT+\tau} e^{-i\omega_0 t} e^{i\omega t} dt. = V_0 \frac{e^{i(\omega - \omega_0)\tau}}{i(\omega - \omega_0)} \sum_{n=0}^{N-1} e^{in(\omega - \omega_0)T}.$$

The sum over n is a geometric series, which can be evaluated as follows

$$\sum_{n=0}^{N-1} e^{in\alpha} = \frac{1 - e^{iN\alpha}}{1 - e^{i\alpha}} = e^{i(N-1)\alpha/2} \frac{\sin(N\alpha/2)}{\sin(\alpha/2)}.$$

With $\alpha := (\omega - \omega_0)T$ we obtain

$$S(\omega) = V_0 \frac{e^{i(\omega - \omega_0)\tau}}{i(\omega - \omega_0)} e^{i(N-1)(\omega - \omega_0)/2} \frac{\sin(N(\omega - \omega_0)T/2)}{\sin((\omega - \omega_0)T/2)}.$$

Problem 15. Calculate the Fourier transform of

$$f(\mathbf{r}) = \frac{1}{r^2 + \lambda^2}$$

where $\mathbf{r} = (x_1, x_2, x_3)$, $\mathbf{r}^2 \equiv r^2 = x_1^2 + x_2^2 + x_3^2$ and $\lambda^2 > 0$.

Solution 15. The Fourier transform is defined as

$$\hat{f}(\mathbf{k}) := \int_{-\infty}^{+\infty} \int_{-\infty}^{+\infty} \int_{-\infty}^{+\infty} \frac{e^{i\mathbf{k}\cdot\mathbf{r}}}{r^2 + \lambda^2} dx_1 dx_2 dx_3$$

where $\mathbf{k} \cdot \mathbf{r} = k_1 x_1 + k_2 x_2 + k_3 x_3$. Introducing *spherical coordinates*

$$x_1(r, \phi, \theta) = r\cos(\phi)\sin(\theta), \quad x_2(r, \phi, \theta) = r\sin(\phi)\sin(\theta),$$

$$x_3(r, \phi, \theta) = r\cos(\theta)$$

we obtain the *volume element*

$$dx_1 dx_2 dx_3 = r^2 \sin(\theta) dr d\theta d\phi$$

where $0 \le \phi < 2\pi$, $0 \le \theta < \pi$, and $r \ge 0$. We can set

$$\mathbf{k} \cdot \mathbf{r} = kr\cos(\theta)$$

where $k = \|\mathbf{k}\| = \sqrt{k_1^2 + k_2^2 + k_3^2}$. It follows that

$$\hat{f}(\mathbf{k}) = \int_{r=0}^{\infty} \int_{\theta=0}^{\pi} \int_{\phi=0}^{2\pi} \frac{e^{ikr\cos(\theta)}}{r^2 + \lambda^2} r^2 \sin(\theta) dr d\theta d\phi.$$

The integration over ϕ can be easily performed and we find

$$\hat{f}(\mathbf{k}) = 2\pi \int_{r=0}^{\infty} \int_{\theta=0}^{\pi} \frac{e^{ikr\cos(\theta)}}{r^2 + \lambda^2} r^2 \sin(\theta) dr d\theta.$$

We set $u = \cos(\theta)$ and therefore $du = -\sin(\theta)d\theta$. It follows that $\theta = 0 \to u = 1$ and $\theta = \pi \to u = -1$. Then $\hat{f}(\mathbf{k})$ takes the form

$$\hat{f}(\mathbf{k}) = -2\pi \int_{r=0}^{\infty} \frac{r^2}{r^2 + \lambda^2} \left(\int_{u=1}^{u=-1} e^{ikru} du \right) dr.$$

The integration over u can easily be performed. We find

$$\int_1^{-1} e^{ikru}du = \frac{-2\sin(kr)}{kr}.$$

Thus we arrive at

$$\hat{f}(\mathbf{k}) = \frac{4\pi}{k}\int_{r=0}^{\infty}\frac{r\sin(kr)}{r^2+\lambda^2}dr.$$

The integral on the right-hand side can be solved by applying the integration in the complex plane. We obtain

$$\hat{f}(\mathbf{k}) = \frac{2\pi^2}{k}e^{-\lambda k}.$$

Problem 16. Let $f, g \in L_2(\mathbb{R})$ and $f, g \in L_1(\mathbb{R})$. Let

$$\hat{f}(k) = \int_{\mathbb{R}} f(x)e^{ikx}dx, \qquad \hat{g}(k) = \int_{\mathbb{R}} g(x)e^{ikx}dx.$$

Show that

$$\langle f, g\rangle = \frac{1}{2\pi}\langle \hat{f}, \hat{g}\rangle \tag{1}$$

where $\langle\,,\,\rangle$ denotes the scalar product in the Hilbert space $L_2(\mathbb{R})$.

Solution 16. We have

$$\langle f, g\rangle := \int_{\mathbb{R}} \bar{f}(x)g(x)dx = \frac{1}{2\pi}\int_{\mathbb{R}}\bar{f}(x)\left(\int_{\mathbb{R}}\hat{g}(k)e^{-ikx}dk\right)dx.$$

Thus

$$\langle f, g\rangle = \frac{1}{2\pi}\int_{\mathbb{R}}\hat{g}(k)\left(\int_{\mathbb{R}}\bar{f}(x)e^{-ikx}dx\right)dk = \frac{1}{2\pi}\int_{\mathbb{R}}\overline{\hat{f}(k)}\hat{g}(k)dk.$$

Thus (1) follows. Equation (1) is called *Parseval's equation*. We have used that

$$\overline{\hat{f}}(k) = \int_{\mathbb{R}}\bar{f}(x)e^{-ikx}dx.$$

Problem 17. (i) The mother *Haar wavelet* is given by

$$f(t) = \begin{cases} -1 & \text{for} & 0 \le t < 1/2 \\ +1 & \text{for} & 1/2 \le t < 1 \\ 0 & \text{otherwise} \end{cases}.$$

Find the Fourier transform.

(ii) The *Poisson wavelet* is given by

$$f(t) = \left(t\frac{d}{dt} + 1 \right) P(t), \quad P(t) = \frac{1}{\pi}\frac{1}{1+t^2}.$$

Find the Fourier transform of f.

Solution 17.   (i) We obtain

$$\hat{f}(\omega) = \int_{-\infty}^{\infty} f(t)e^{i\omega t}dt = 2\frac{1-\cos(\omega)}{\omega}e^{i(\omega+\pi)/2}$$

with $\hat{f}(0) = 0$ (L'Hospital rule).
(ii) We obtain $\hat{f}(\omega) = |\omega|e^{-|\omega|}$. Note that $\hat{f}$ is not differentiable at $\omega = 0$, but continuous.

Problem 18.   Given the *Mexican hat wavelet*

$$\psi(t) = \frac{2}{\sqrt{3}\sqrt[4]{\pi}}(1-t^2)e^{-t^2/2} = \frac{2}{\sqrt{3}\sqrt[4]{\pi}}\frac{d}{dt}\left(te^{-t^2/2}\right).$$

Calculate the wavelet transform for the signal

$$f(t) = \begin{cases} 1 & |t| \leq \frac{1}{2} \\ 0 & \text{otherwise} \end{cases}$$

The *wavelet transform* is given by

$$W(f(t))(a,b) = \frac{1}{\sqrt{|a|}}\int_{-\infty}^{\infty} f(t)\psi\left(\frac{t-b}{a}\right)dt.$$

Solution 18.   Using the substitution $\tau = (t-b)/a$ and therefore $d\tau = dt/a$ we find

$$W(f(t))(a,b) = \frac{1}{\sqrt{|a|}}\int_{-\infty}^{\infty} f(t)\psi\left(\frac{t-b}{a}\right)dt = \frac{1}{\sqrt{|a|}}\int_{-1/2}^{1/2}\psi\left(\frac{t-b}{a}\right)dt$$

$$= \frac{a}{\sqrt{|a|}}\int_{(-1-2b)/(2a)}^{(1-2b)/(2a)}\psi(\tau)d\tau = \frac{a}{\sqrt{|a|}}\frac{2}{\sqrt{3}\sqrt[4]{\pi}}\tau e^{-\tau^2/2}\Big|_{(-1-2b)/(2a)}^{(1-2b)/(2a)}$$

$$= \frac{1}{\sqrt{3}\sqrt[4]{\pi}}\frac{1}{\sqrt{|a|}}\left((1-2b)e^{-(1-2b)^2/(8a^2)} + (1+2b)e^{-(1+2b)^2/(8a^2)}\right).$$

Problem 19.   Consider the Hilbert space $L_2[0,\pi]$ and the function $f(x) = \sin(x)$. Find a and b $(a,b \in \mathbb{R})$ such that

$$\|f(x) - (ax^2 + bx)\|$$

is a minimum. The norm in the Hilbert space $L_2[0, \pi]$ is induced by the scalar product. Hence

$$\|f(x) - (ax^2 + bx)\|^2 = \int_0^\pi (f(x) - (ax^2 + bx))^2 dx.$$

Solution 19. We have

$$h(a, b) := \int_0^\pi (\sin(x) - (ax^2 + bx))^2 dx.$$

From the conditions $\partial h/\partial a = 0$, $\partial h/\partial b = 0$ we obtain

$$\int_0^\pi x^2 (\sin(x) - (ax^2 + bx)) dx = 0, \quad \int_0^\pi x(\sin(x) - (ax^2 + bx)) dx = 0.$$

From these two equations we obtain

$$a \int_0^\pi x^4 dx + b \int_0^\pi x^3 dx = \int_0^\pi x^2 \sin(x) dx$$

$$a \int_0^\pi x^3 dx + b \int_0^\pi x^2 dx = \int_0^\pi x \sin(x) dx.$$

Integrating yields

$$\frac{\pi^5 a}{5} + \frac{\pi^4 b}{4} = \pi^2 - 4, \qquad \frac{\pi^4 a}{4} + \frac{\pi^3 b}{3} = \pi.$$

This is a system of linear equations for a and b. Solving for a and b we find

$$a = \frac{20}{\pi^3} - \frac{320}{\pi^5}, \qquad b = \frac{240}{\pi^4} - \frac{12}{\pi^2}.$$

Problem 20. Consider the Hilbert space $L_2(\mathbb{R})$. Let $\phi \in L_2(\mathbb{R})$ be a real-valued function with

$$\int_{\mathbb{R}} \phi(x) dx = 1$$

and

$$\int_{\mathbb{R}} \phi(x)\phi(x - n) dx = \delta_{n0}, \quad n \in \mathbb{Z}.$$

Assume that

$$\phi(x) = \sum_{k=0}^{M-1} c_k \phi(2x - k)$$

where $c_k \in \mathbb{R}$.

(i) Show that
$$\sum_{k=0}^{M-1} c_k = 2, \qquad \sum_{k=0}^{M-1} c_k^2 = 2.$$

(ii) Give a function that satisfies these conditions.

Solution 20.   (i) We have
$$1 = \int_{\mathbb{R}} \phi(x)dx = \int_{\mathbb{R}} \sum_{k=0}^{M-1} c_k \phi(2x - k)dx = \sum_{k=0}^{M-1} c_k \int_{\mathbb{R}} \phi(2x - k)dx.$$

Using the transformation $y = 2x - k$ and $dy = 2dx$ we find
$$1 = \frac{1}{2} \sum_{k=0}^{M-1} c_k \int_{\mathbb{R}} \phi(y)dy = \frac{1}{2} \sum_{k=0}^{M-1} c_k.$$

Thus the first equation follows. To prove the second equation we start from
$$1 = \int_{\mathbb{R}} \phi(x)\phi(x)dx = \int_{\mathbb{R}} \left(\sum_{k=0}^{M-1} c_k \phi(2x - k)\right)\left(\sum_{j=0}^{M-1} c_j \phi(2x - j)\right) dx.$$

Thus
$$1 = \sum_{k=0}^{M-1}\sum_{j=0}^{M-1} c_k c_j \int_{\mathbb{R}} \phi(2x - k)\phi(2x - j)dx.$$

Using $y = 2x - k$ and $dy = 2dx$ we obtain
$$1 = \frac{1}{2}\sum_{k=0}^{M-1}\sum_{j=0}^{M-1} c_k c_j \int_{\mathbb{R}} \phi(y)\phi(y + k - j)dy = \frac{1}{2}\sum_{k=0}^{M-1} c_k^2.$$

(iii) The function
$$\phi(x) = \begin{cases} 1 & \text{for} \quad 0 \le x \le 1 \\ 0 & \text{otherwise} \end{cases}$$

satisfies the conditions given above. The function ϕ plays an important role in wavelet theory. The function ϕ is called the *scaling function*. What is desired is a function ψ which is also orthogonal to its dilations, or scales, i.e.
$$\int_{\mathbb{R}} \psi(x)\psi(2x - k)dx = 0.$$

Such a function ψ (the so-called associated *wavelet function*) does exist and is given by
$$\psi(x) = \sum_{k=1}^{M} (-1)^k c_{1-k}\phi(2x - k)$$

which is dependent on the solution of ϕ. Periodic boundary conditions are used $c_k \equiv c_{k+nM}$.

Problem 21. Consider the function $H \in L_2(\mathbb{R})$

$$H(x) := \begin{cases} 1 & 0 \leq x < 1/2 \\ -1 & 1/2 \leq x \leq 1 \\ 0 & \text{otherwise} \end{cases} . \tag{1}$$

Let

$$H_{mn}(x) := 2^{-m/2} H(2^{-m}x - n)$$

where $m, n \in \mathbb{Z}$ and n is the translation parameter and m the dilation parameter.
(i) Find H_{11}, H_{12}.
(ii) Show that $\langle H_{mn}(x), H_{kl}(x) \rangle = \delta_{mk}\delta_{nl}$ $(k, l \in \mathbb{Z})$.
(iii) Expand the function $(f \in L_2(\mathbb{R}))$, $f(x) = \exp(-|x|)$ with respect to H_{mn}. The functions H_{mn} form an orthonormal basis in $L_2(\mathbb{R})$.

Solution 21. (i) We have

$$H_{mn}(x) = 2^{-\frac{m}{2}} H(2^{-m}x - n) = \begin{cases} 2^{-\frac{m}{2}} & 0 \leq 2^{-m}x - n < \frac{1}{2} \\ -2^{-\frac{m}{2}} & \frac{1}{2} \leq 2^{-m}x - n \leq 1 \\ 0 \text{ otherwise} \end{cases} .$$

Thus

$$H_{mn}(x) = \begin{cases} 2^{-\frac{m}{2}} & 2^m n \leq x < 2^m(n + \frac{1}{2}) \\ -2^{-\frac{m}{2}} & 2^m(n + \frac{1}{2}) \leq x \leq 2^m(n + 1) \\ 0 \text{ otherwise} \end{cases} .$$

It follows that

$$H_{11}(x) = \begin{cases} \frac{1}{\sqrt{2}} & 2 \leq x < 3 \\ -\frac{1}{\sqrt{2}} & 3 \leq x \leq 4 \\ 0 \text{ otherwise} \end{cases} \qquad H_{12}(x) = \begin{cases} \frac{1}{\sqrt{2}} & 4 \leq x < 5 \\ -\frac{1}{\sqrt{2}} & 5 \leq x \leq 6 \\ 0 \text{ otherwise} \end{cases} .$$

(ii) We have

$$I_{mnkl} := \langle H_{mn}(x), H_{kl}(x) \rangle = \int_{-\infty}^{\infty} H_{mn}(x) H_{kl}(x) dx.$$

The intervals on which H_{mn} and H_{kl} are non-zero are

$$I_{mn} := (2^m n, 2^m(n + 1)), \quad I_{kl} := (2^k l, 2^k(l + 1)).$$

We consider the different cases:
Case 1. $m = k$, $n = l$ with

$$I_{mnkl} = \int_{2^m n}^{2^m(n+1)} 2^{-m} dx = 1.$$

Case 2. $m = k, n \neq l$ with $I_{mnkl} = 0$ since $I_{mn} \cap I_{kl} = \emptyset$.

Case 3. $m \neq k$. Suppose without loss of generality that $m < k$. Either $I_{mn} \cap I_{kl} = \emptyset$ $(I_{mnkl} = 0)$, or $I_{mn} \subset I_{kl}$ (as shown below). We have

$$2^k l \leq 2^m n < 2^k \left(l + \frac{1}{2}\right) \Rightarrow 2^{k-m} l \leq n < 2^{k-m} \left(l + \frac{1}{2}\right) \Rightarrow$$

$$\Rightarrow 2^{k-m} l \leq n + 1 \leq 2^{k-m} \left(l + \frac{1}{2}\right) \Rightarrow 2^k l \leq 2^m (n+1) \leq 2^k \left(l + \frac{1}{2}\right).$$

Also

$$2^k \left(l + \frac{1}{2}\right) \leq 2^m n < 2^k (l+1) \Rightarrow 2^k \left(l + \frac{1}{2}\right) \leq 2^m (n+1) \leq 2^k (l+1)$$

$$2^k l < 2^m (n+1) \leq 2^k \left(l + \frac{1}{2}\right) \Rightarrow 2^k l \leq 2^m n \leq 2^k \left(l + \frac{1}{2}\right)$$

$$2^k \left(l + \frac{1}{2}\right) < 2^m (n+1) \leq 2^k (l+1) \Rightarrow 2^k \left(l + \frac{1}{2}\right) \leq 2^m n \leq 2^k (l+1)$$

which gives

$$\langle H_{mn}(x), H_{kl}(x)\rangle = \pm \int_{2^m n}^{2^m (n+\frac{1}{2})} 2^{-\frac{1}{2}(m+k)} dx \mp \int_{2^m (n+\frac{1}{2})}^{2^m (n+1)} 2^{-\frac{1}{2}(m+k)} dx = 0.$$

Thus $I_{mnkl} = \delta_{mk}\delta_{nl}$.

(iii) The scalar product $\langle f(x), H_{mn}(x)\rangle$ is given by

$$\langle f(x), H_{mn}(x)\rangle = \int_{-\infty}^{\infty} f(x) H_{mn}(x) dx$$

$$= \int_{2^m n}^{2^m (n+\frac{1}{2})} 2^{-\frac{m}{2}} e^{-|x|} dx - \int_{2^m (n+\frac{1}{2})}^{2^m (n+1)} 2^{-\frac{m}{2}} e^{-|x|} dx$$

$$= 2^{-\frac{m}{2}} \begin{cases} -e^{-x}\Big|_{2^m n}^{2^m (n+\frac{1}{2})} + e^{-x}\Big|_{2^m (n+\frac{1}{2})}^{2^m (n+1)} & n \geq 0 \\ e^x\Big|_{2^m n}^{2^m (n+\frac{1}{2})} - e^x\Big|_{2^m (n+\frac{1}{2})}^{2^m (n+1)} & n < 0 \end{cases}$$

$$= -\frac{n + \delta_{n,0}}{|n| + \delta_{n,0}} 2^{-\frac{m}{2}} (2e^{-2^m |n+\frac{1}{2}|} - e^{-2^m |n|} - e^{-2^m |n+1|}).$$

The expansion is given by

$$f(x) = \sum_{m,n \in \mathbf{Z}} \langle f(x), H_{mn}(x)\rangle H_{mn}(x).$$

Problem 22.   Let $\mathbb{C}^n$ denote the complex Euclidean space. Let $\mathbf{z} = (z_1, \ldots, z_n) \in \mathbb{C}^n$ and $\mathbf{w} = (w_1, \ldots, w_n) \in \mathbb{C}^n$ then the scalar product (inner product) is given by

$$\mathbf{z} \cdot \mathbf{w} := \mathbf{z}\mathbf{w}^* = \mathbf{z}\overline{\mathbf{w}}^T$$

where $\bar{\mathbf{z}} = (\bar{z}_1, \ldots, \bar{z}_n)$. Let E_n denote the set of entire functions in $\mathbb{C}^n$. Let F_n denote the set of $f \in E_n$ such that

$$\|f\|^2 := \frac{1}{\pi^n} \int_{\mathbb{C}^n} |f(\mathbf{z})|^2 \exp(-|\mathbf{z}|^2) dV$$

is finite. Here dV is the volume element (Lebesgue measure)

$$dV = \prod_{j=1}^{n} dx_j dy_j = \prod_{j=1}^{n} r_j dr_j d\theta_j$$

with $z_j = r_j e^{i\theta_j}$. The norm follows from the scalar product of two functions $f, g \in F_n$

$$\langle f, g \rangle := \frac{1}{\pi^n} \int_{\mathbb{C}^n} f(\mathbf{z}) \overline{g(\mathbf{z})} \exp(-|\mathbf{z}|^2) dV.$$

Let

$$\mathbf{z}^m := z_1^{m_1} \cdots z_n^{m_n}$$

where the multi-index m is defined by $m! = m_1! \cdots m_n!$ and $|m| = \sum_{j=1}^{n} m_j$. Find the scalar product $\langle \mathbf{z}^m, \mathbf{z}^p \rangle$.

Solution 22. We obtain

$$\langle \mathbf{z}^m, \mathbf{z}^p \rangle = \langle z_1^{m_1} \cdots z_n^{m_n}, z_1^{p_1} \cdots z_n^{p_n} \rangle = 0 \quad \text{for} \ m \neq p$$

and

$$\langle \mathbf{z}^m, \mathbf{z}^m \rangle = \langle z_1^{m_1} \cdots z_n^{m_n}, z_1^{m_1} \cdots z_n^{m_n} \rangle m! = m_1! \cdots m_n!.$$

Thus the *monomials* $z^m / \sqrt{m!}$ are orthonormal in this Hilbert space.

Problem 23. Let Ψ be a complex-valued differentiable function of ϕ in the interval $[0, 2\pi]$ and $\Psi(0) = \Psi(2\pi)$, i.e. Ψ is an element of the Hilbert space $L_2([0, 2\pi])$. Assume that (normalization condition)

$$\int_0^{2\pi} \Psi^*(\phi) \Psi(\phi) d\phi = 1.$$

Calculate

$$\Im \left(\frac{\hbar}{i} \int_0^{2\pi} \Psi^*(\phi) \phi \frac{d}{d\phi} \Psi(\phi) d\phi \right)$$

where $\Im$ denotes the imaginary part.

Solution 23. Using integration by parts, $\Psi(2\pi) = \Psi(0)$ and the normalization condition we find

$$\int_0^{2\pi} \Psi^*(\phi)\phi \frac{d}{d\phi}\Psi(\phi)d\phi = \Psi^*(\phi)\phi\Psi(\phi)\big|_0^{2\pi} - \int_0^{2\pi} \Psi(\phi)\left(\frac{d}{d\phi}(\Psi^*(\phi)\phi)\right)d\phi$$

$$= 2\pi\Psi^*(2\pi)\Psi(2\pi) - \int_0^{2\pi}\Psi(\phi)\phi\frac{d}{d\phi}\Psi^*(\phi)d\phi$$

$$- \int_0^{2\pi}\Psi(\phi)\Psi^*(\phi)d\phi$$

$$= 2\pi\Psi^*(0)\Psi(0) - 1 - \int_0^{2\pi}\Psi(\phi)\phi\frac{d}{d\phi}\Psi^*(\phi)d\phi.$$

Thus

$$\int_0^{2\pi}\left(\Psi^*(\phi)\phi\frac{d}{d\phi}\Psi(\phi) + \Psi(\phi)\phi\frac{d}{d\phi}\Psi^*(\phi)\right)d\phi = 2\pi|\Psi(0)|^2 - 1.$$

It follows that

$$\Im\left(\frac{\hbar}{i}\int_0^{2\pi}\Psi^*(\phi)\phi\frac{d}{d\phi}\Psi(\phi)d\phi\right) = \frac{\hbar}{2}(1 - 2\pi|\Psi(0)|^2).$$

Problem 24. Consider the Hilbert space $L_2([0,1])$. The *shifted Legendre polynomials*, defined on the interval $[0,1]$, are obtained from the Legendre polynomial by the transformation $y = 2x - 1$. The shifted Legendre polynomials are given by the recurrence formula

$$P_j(x) = \frac{(2j+1)(2x-1)}{j+1}P_j(x) - \frac{j}{j+1}P_{j-1}(x) \qquad j = 1, 2, \dots$$

and $P_0(x) = 1$, $P_1(x) = 2x - 1$. They are elements of the Hilbert space $L_2([0,1])$. A function u in the Hilbert space $L_2([0,1])$ can be approximated in the form of a series with $n+1$ terms

$$u(x) = \sum_{j=0}^n c_j P_j(x)$$

where the coefficients $c_j \in \mathbb{R}$, $j = 0, 1, \dots, n$. Consider the *Volterra integral equation* of first kind

$$\lambda \int_0^x \frac{y(t)}{(x-t)^\alpha}dt = f(x), \qquad 0 \le t \le x \le 1$$

with $0 < \alpha < 1$ and $f \in L_2([0,1])$. Consider the ansatz

$$y_n(x) = a_0 x^\alpha + \sum_{j=0}^n c_j P_j(x)$$

to find an approximate solution to the Volterra integral equation of first kind ($\alpha = 1/2$)

$$\lambda \int_0^x \frac{y(t)}{\sqrt{x-t}} dt = f(x), \quad f(x) = \frac{2}{105} \sqrt{x}(105 - 56x^2 + 48x^3).$$

Solution 24.   Inserting the ansatz in the Volterra integral equation yields

$$\lambda a_0 \int_0^x \frac{t^\alpha}{(x-t)^\alpha} dt + \lambda \sum_{j=0}^n c_j \int_0^x \frac{P_j(t)}{(x-t)^\alpha} dt = f(x)$$

where $\alpha = 1/2$. Since

$$\int_0^x \frac{t^n}{(x-t)^\alpha} dt = \frac{\Gamma(n+1)\Gamma(1-\alpha)}{\Gamma(n+2-\alpha)} x^{n+1-\alpha}, \quad \int_0^x \frac{t^\alpha}{(x-t)^\alpha} = \frac{\pi\alpha}{\sin(\pi\alpha)} x$$

we find

$$\int_0^x \frac{P_j(t)}{(x-t)^{(\alpha)}} dt = \sum_{k=0}^j a_{jk}^\alpha x^{k+1-\alpha}$$

where

$$a_{jk}^{(\alpha)} = (-1)^{j+k} \frac{(j+k)!\,\Gamma(1-\alpha)}{k!(j-k)!\,\Gamma(k+2-\alpha)}.$$

Thus we find

$$\lambda a_0 \frac{\pi\alpha}{\sin(\pi\alpha)} x + \lambda \sum_{j=0}^n \sum_{k=0}^j c_j a_{jk}^{(\alpha)} x^{k+1-\alpha} = f(x).$$

Since $\alpha = 1/2$ we have $\sin(\pi/2) = 1$, $\Gamma(1/2) = \sqrt{\pi}$ and the equation simplifies to

$$\frac{1}{2}\lambda a_0 \pi x + \lambda \sum_{j=0}^n \sum_{k=0}^j c_j a_{jk}^{(\alpha)} x^{k+1-\alpha} = f(x).$$

We consider this equation at the roots $y_j : (j = 1, 2, \ldots, n+1)$ of the shifted Legendre polynomials and $y_0 = 0$. Thus we find $n + 2$ equations for the coefficients c_j $(j = 0, 1, \ldots, n)$ and a_0. We have

$$P_2(x) = 6x^2 - 6x + 1, \qquad P_3(x) = 20x^3 - 30x^2 + 12x - 1.$$

We find $a_0 = 0$ and the expansion coefficients

$$c_0 = \frac{11}{12}, \quad c_1 = -\frac{1}{20}, \quad c_2 = \frac{1}{12}, \quad c_3 = \frac{1}{20}.$$

Problem 25.   (i) Consider the Hilbert space $L_2[0,1]$ with the scalar product $\langle \cdot, \cdot \rangle$. Let $f : [0,1] \to [0,1]$

$$f(x) := \begin{cases} 2x & \text{if } x \in [0, 1/2) \\ 2(1-x) & \text{if } x \in [1/2, 1] \end{cases}.$$

Thus $f \in L_2[0,1]$. Calculate the moments μ_k, $k = 0, 1, 2, \ldots$ defined by

$$\mu_k := \langle f(x), x^k \rangle \equiv \int_0^1 f(x) x^k dx.$$

(ii) Show that

$$\sum_{k=0}^{\infty} |\mu_k|^2 < \pi \int_0^1 |f(x)|^2 dx.$$

Solution 25.   (i) For $k = 0$ we have

$$\mu_0 = \int_0^1 f(x) dx = \frac{1}{2}.$$

For $k \geq 1$ we obtain

$$\mu_k = \int_0^{1/2} 2x x^k dx + \int_{1/2}^1 (2 - 2x) x^k dx$$

$$= 2 \int_0^{1/2} x^{k+1} dx + 2 \int_{1/2}^1 x^k dx - 2 \int_{1/2}^1 x^{k+1} dx$$

$$= 2\frac{(1/2)^{k+2}}{k+2} + 2\frac{1}{k+1} - 2\frac{(1/2)^{k+1}}{k+1} - 2\frac{1}{k+2} + 2\frac{(1/2)^{k+2}}{k+2}$$

$$= -\left(\frac{1}{2}\right)^{k+1} \frac{1}{(k+2)(k+1)} + \frac{2}{(k+1)(k+2)}.$$

(ii) For the right-hand side we find

$$\int_0^1 |f(x)|^2 dx = \int_0^{1/2} 4x^2 dx + \int_0^1 2(1-x)2(1-x) dx = \frac{1}{3}.$$

Problem 26.   Let $L_2(\mathbb{R}^2)$ be the Hilbert space of the square-integrable functions over $\mathbb{R}^2$. Let $C^2(\mathbb{R}^2, \mathbb{R})$ be the set of $\mathbb{R}$ valued functions having two continuous derivatives on $\mathbb{R}^2$. Assume that

$$f, g \in C^2(\mathbb{R}^2, \mathbb{R}) \cap L_2(\mathbb{R}^2) \qquad \Delta f, \Delta g \in L_2(\mathbb{R}^2)$$

and that $\nabla f, \nabla g \in L_2(\mathbb{R}^2)$. This means each component of each vector is in $L_2(\mathbb{R}^2)$. Show that

$$\int_{\mathbb{R}^2} f(x,y) \Delta g(x,y) dx dy = \int_{\mathbb{R}^2} g(x,y) \Delta f(x,y) dx dy. \tag{1}$$

Solution 26. Owing to the assumptions we know that all functions are in the Hilbert space $L_2(\mathbb{R}^2)$. It follows that all integrals exist. We apply *Green's theorem* to solve the problem. Let P and Q be two real valued functions that are continuous on a region $E \subset \mathbb{R}^2$ bounded by a rectifiable Jordan curve Γ and such that $\partial P/\partial y$ and $\partial Q/\partial x$ exist and are bounded in their interior of E and that

$$\int\int_E \frac{\partial Q}{\partial x} dx dy \qquad \text{and} \qquad \int\int_E \frac{\partial P}{\partial y} dx dy$$

exist. Then, if Γ is positively orientated, the line integral exists and we have

$$\int_\Gamma (P dx + Q dy) = \int\int_E \left(\frac{\partial Q}{\partial x} - \frac{\partial P}{\partial y} \right) dx dy.$$

Green's theorem implies that for all positive R,

$$\int_{B(0,R)} (f \Delta g - g \Delta f) dx dy = \int_{\partial B(0,R)} \left(f \frac{\partial g}{\partial x} - g \frac{\partial f}{\partial x} \right) dy$$
$$+ \int_{\partial B(0,R)} \left(g \frac{\partial f}{\partial y} - f \frac{\partial g}{\partial y} \right) dx.$$

An estimate for the integrals in these equations is

$$\left| \int_{\partial B(0,R)} f \frac{\partial g}{\partial x} dy \right| \leq \int_0^{2\pi} |f| \left| \frac{\partial g}{\partial x} \right| |R \sin(\phi)| d\phi = A(R)$$

where we have introduced polar coordinates. Let $\|.\|_2$ be the norm of the Hilbert space $L_2(\mathbb{R}^2)$. Since

$$\int_{B(0,R)} \left| f(r\cos(\phi), r\sin(\phi)) \frac{\partial g(r\cos(\phi), r\sin(\phi))}{\partial x} \right| r dr d\phi = \int_0^R A(r) r dr$$
$$\leq \|f\|_2 \|\partial g/\partial x\|_2$$

it follows that there is a sequence $\{R_n\}_{n=1}^\infty$ such that $R_n \to \infty$ and

$$A(R_n) \to 0$$

as $n \to \infty$. In a similar manner the other three integrals may be estimated. Thus

$$\left| \int_{B(0,R_n)} (f \Delta g - g \Delta f) dx dy \right|$$

is dominated by the sum of four quantities, each of which tends to zero as $n \to \infty$. Consequently, (1) follows.

Problem 27. Consider the Hilbert space $L_2(\mathbb{R})$. Let

$$\hat{\psi}(\omega) = \begin{cases} 1 & \text{if} \quad 1/2 \le |\omega| \le 1 \\ 0 & \text{otherwise} \end{cases}$$

and $\hat{\phi}(\omega) = e^{-\alpha|\omega|}$, $\alpha > 0$.

(i) Calculate the inverse Fourier transforms of $\hat{\psi}(\omega)$ and $\hat{\phi}(\omega)$, i.e.

$$\psi(t) = \frac{1}{2\pi} \int_{\mathbb{R}} e^{-i\omega t} \hat{\psi}(\omega) d\omega, \quad \phi(t) = \frac{1}{2\pi} \int_{\mathbb{R}} e^{-i\omega t} \hat{\phi}(\omega) d\omega.$$

(ii) Calculate the scalar product $\langle \psi(t)|\phi(t)\rangle$ by utilizing the identity

$$2\pi \langle \psi(t)|\phi(t)\rangle = \langle \hat{\psi}(\omega)|\hat{\phi}(\omega)\rangle.$$

Solution 27.   (i) For the inverse Fourier transforms we find

$$\psi(t) = \frac{1}{2\pi} \left(\int_{-1}^{-1/2} e^{-i\omega t} d\omega + \int_{1/2}^{1} e^{-i\omega t} d\omega \right) = \frac{1}{\pi t}(\sin(t) - \sin(t/2)).$$

$$\phi(t) = \frac{1}{2\pi} \left(\int_{-\infty}^{0} e^{-i\omega t} e^{\alpha\omega} d\omega + \int_{0}^{\infty} e^{-i\omega t} e^{-\alpha\omega} d\omega \right) = \frac{1}{\pi} \frac{\alpha}{t^2 + \alpha^2}.$$

(ii) We find

$$\langle \psi(t)|\phi(t)\rangle = \frac{1}{2\pi} \langle \hat{\psi}(\omega)|\hat{\phi}(\omega)\rangle = \frac{1}{2\pi} \left(\int_{-1}^{-1/2} e^{\alpha\omega} d\omega + \int_{1/2}^{1} e^{-\alpha\omega} d\omega \right) = 0.$$

Problem 28.   Consider the two-body problem in quantum mechanics with the coordinates $\mathbf{r}_1 = (x_1, y_1, z_1)^T$, $\mathbf{r}_2 = (x_2, y_2, z_2)^T$ and the operators

$$\Delta_1 = \frac{\partial^2}{\partial x_1^2} + \frac{\partial^2}{\partial y_1^2} + \frac{\partial^2}{\partial z_1^2}, \qquad \Delta_2 = \frac{\partial^2}{\partial x_2^2} + \frac{\partial^2}{\partial y_2^2} + \frac{\partial^2}{\partial z_2^2}$$

where $\mathbf{r}_1$ is the position vector of particle 1 and $\mathbf{r}_2$ is the position vector of particle 2. The eigenvalue equation takes the form

$$\hat{H}u(\mathbf{r}_1, \mathbf{r}_2) \equiv \left(-\frac{\hbar^2}{2m_1}\Delta_1 - \frac{\hbar^2}{2m_2}\Delta_2 + V(\mathbf{r}_1 - \mathbf{r}_2) \right) u(\mathbf{r}_1, \mathbf{r}_2) = Eu(\mathbf{r}_1, \mathbf{r}_2)$$

where, on physical grounds, only potentials depending on $\mathbf{r}_1 - \mathbf{r}_2$ are considered. The function u is an element of the Hilbert space $L_2(\mathbb{R}^6)$. Consider the linear invertible transformation $(\mathbf{r}_1, \mathbf{r}_2) \to (\mathbf{r}, \mathbf{R})$

$$\mathbf{r}(\mathbf{r}_1, \mathbf{r}_2) = \mathbf{r}_2 - \mathbf{r}_1$$

$$\mathbf{R}(\mathbf{r}_1, \mathbf{r}_2) = \frac{m_1\mathbf{r}_1 + m_2\mathbf{r}_2}{m_1 + m_2}$$

$$U(\mathbf{r}(\mathbf{r}_1, \mathbf{r}_2), \mathbf{R}(\mathbf{r}_1, \mathbf{r}_2)) = u(\mathbf{r}_1, \mathbf{r}_2)$$

with $\mathbf{r} = (x, y, z)^T$, $\mathbf{R} = (X, Y, Z)^T$, where $\mathbf{R}$ is the centre of mass position vector and $\mathbf{r}$ is the relative position vector.
(i) Find the inverse transformation.
(ii) Express the Hamilton operator in terms these new coordinates.

Solution 28. (i) Straightforward calculation yields

$$\mathbf{r_1} = \frac{(m_1 + m_2)\mathbf{R} - m_2\mathbf{r}}{m_1 + m_2}, \quad \mathbf{r_2} = \frac{(m_1 + m_2)\mathbf{R} + m_1\mathbf{r}}{m_1 + m_2}$$

(ii) We set $M := m_1 + m_2$. Using the chain rule we find

$$\frac{\partial U}{\partial x_1} = -\frac{\partial U}{\partial x} + \frac{m_1}{M}\frac{\partial U}{\partial X} = \frac{\partial u}{\partial x_1}$$

$$\frac{\partial U}{\partial y_1} = -\frac{\partial U}{\partial y} + \frac{m_1}{M}\frac{\partial U}{\partial Y} = \frac{\partial u}{\partial y_1}$$

$$\frac{\partial U}{\partial z_1} = -\frac{\partial U}{\partial z} + \frac{m_1}{M}\frac{\partial U}{\partial Z} = \frac{\partial u}{\partial z_1}$$

$$\frac{\partial U}{\partial x_2} = \frac{\partial U}{\partial x} + \frac{m_2}{M}\frac{\partial U}{\partial X} = \frac{\partial u}{\partial x_2}$$

$$\frac{\partial U}{\partial y_2} = \frac{\partial U}{\partial y} + \frac{m_2}{M}\frac{\partial U}{\partial Y} = \frac{\partial u}{\partial y_2}$$

$$\frac{\partial U}{\partial z_2} = \frac{\partial U}{\partial z} + \frac{m_2}{M}\frac{\partial U}{\partial Z} = \frac{\partial u}{\partial z_2}.$$

For the second order derivatives we have

$$\frac{\partial^2 U}{\partial x_1^2} = \frac{\partial^2 U}{\partial x^2} - 2\frac{m_1}{M}\frac{\partial^2 U}{\partial x \partial X} + \frac{m_1^2}{M^2}\frac{\partial^2 U}{\partial X^2} = \frac{\partial^2 u}{\partial x_1^2}$$

$$\frac{\partial^2 U}{\partial x_2^2} = \frac{\partial^2 U}{\partial x^2} + 2\frac{m_2}{M}\frac{\partial^2 U}{\partial x \partial X} + \frac{m_2^2}{M^2}\frac{\partial^2 U}{\partial X^2} = \frac{\partial^2 u}{\partial x_2^2}$$

and analogously for y_1, z_1, y_2 and z_2. Thus we obtain the new Hamilton operator

$$\hat{H}_S = -\frac{\hbar^2}{2M}\Delta_R - \frac{\hbar^2}{2m}\Delta + V(\mathbf{r})$$

where

$$\Delta_R := \frac{\partial^2}{\partial X^2} + \frac{\partial^2}{\partial Y^2} + \frac{\partial^2}{\partial Z^2}, \quad \Delta := \frac{\partial^2}{\partial x^2} + \frac{\partial^2}{\partial y^2} + \frac{\partial^2}{\partial z^2}$$

with the *reduced mass* $m = (m_1 m_2)/(m_1 + m_2)$.

Problem 29. Let $|\psi\rangle$, $|s\rangle$, $|\phi\rangle$ be normalized states in a Hilbert space $\mathcal{H}$. Let U be a unitary operator, i.e. $U^{-1} = U^*$ in the Hilbert space $\mathcal{H} \otimes \mathcal{H}$

such that

$$U(|\psi\rangle \otimes |s\rangle) = |\psi\rangle \otimes |\psi\rangle, \quad U(|\phi\rangle \otimes |s\rangle) = |\phi\rangle \otimes |\phi\rangle.$$

Show that $\langle\phi|\psi\rangle = \langle\phi|\psi\rangle^2$. Find solutions to this equation.

Solution 29.   Taking the scalar product of these two equations with $U^* = U^{-1}$ and $\langle s|s\rangle = 1$ we obtain

$$((\langle\psi| \otimes \langle s|)U^*U(|\phi\rangle \otimes |s\rangle) = ((\langle\psi| \otimes \langle\psi|)(|\phi\rangle \otimes |\phi\rangle))$$
$$((\langle\psi| \otimes \langle s|)(|\phi\rangle \otimes |s\rangle)) = \langle\psi|\phi\rangle\langle\psi|\phi\rangle$$
$$\langle\psi|\phi\rangle = \langle\psi|\phi\rangle^2.$$

The equation can be satisfied if $\langle\psi|\phi\rangle = 0$ ($|\psi\rangle$ and $|\phi\rangle$ are orthonormal to each other) or $\langle\psi|\phi\rangle = 1$, i.e. $|\psi\rangle = |\phi\rangle$.

Supplementary Problems

Problem 1.   Let $a > 0$. Consider the Hilbert space $L_2([0, a])$.
(i) Show that the functions

$$\phi_n(x) = \frac{1}{\sqrt{a}}e^{2\pi inx/a}, \quad n \in \mathbb{Z}$$

form an orthonormal basis in $L_2([0, a])$.
(ii) Show that the functions

$$1/\sqrt{a}, \quad \sqrt{2/a}\cos(2\pi nx/a), \quad \sqrt{2/a}\sin(2\pi nx/a), \quad n = 1, 2, 3, \ldots$$

form an orthonormal basis in $L_2([0, a])$.
(iii) Show that the functions

$$\phi_n(x) = \sqrt{2/a}\sin(\pi nx/a), \quad n = 1, 2, 3, \ldots$$

form an orthonormal basis in $L_2([0, a])$.
(iv) Show that the functions

$$1/\sqrt{a}, \quad \sqrt{2/a}\cos(\pi nx/a), \quad n = 1, 2, 3, \ldots$$

form an orthonormal basis in $L_2([0, a])$.

Problem 2.   Find the Fourier transform of $f(x) = \cos(x)e^{-x^2/2}$.

Problem 3. Let $f(x) \in S(\mathbb{R})$. Its Fourier transform is defined by

$$\hat{f}(k) = \int_{\mathbb{R}} f(x)e^{-ikx}dx.$$

Show that

$$\frac{d^n \hat{f}(k)}{dk^n} = (-i)^n \int_{\mathbb{R}} x^n e^{-ikx} f(x)dx$$

and for setting $k = 0$ one obtains

$$\frac{d^n \hat{f}(0)}{dk^n} = (-i)^n \int_{\mathbb{R}} x^n f(x)dx.$$

Apply it to the Gauss function $f(x) = \exp(-x^2/2)$.

Problem 4. Consider the Hilbert space $L_2(\mathbb{R})$. The *Littlewood-Paley basis* is generated from the *mother wavelet*

$$g(x) = \frac{1}{\pi x}(\sin(2\pi x) - \sin(\pi x))$$

via $g_{mn}(x) = 2^{-m/2}g(2^{-m}x - n)$ with $m, n \in \mathbb{Z}$. Show that the mother wavelet is well localized in momentum space, (i.e. it has compact support)

$$\hat{g}(k) = \begin{cases} (2\pi)^{-1/2} & \text{for} \quad \pi \leq k \leq 2\pi \\ 0 & \text{otherwise} \end{cases}.$$

Problem 5. Let $\mathcal{H}$ be a Hilbert space and $f, g \in \mathcal{H}$. Show that

$$\|f\|^2 + \|g\|^2 \geq 2|\langle f, g \rangle|.$$

Problem 6. Consider the Hilbert space $L_2(0, \infty)$ and the *Breit-Wigner functions* defined by

$$f_j(x) = \frac{1}{x + \mu_j}, \quad j = 1, 2, \ldots$$

where $x \in (0, \infty)$, $\mu_j \in \mathbb{C} \setminus \mathbb{R}^-$ with $\mathbb{R}^- = (-\infty, 0)$. Show that for certain sequences μ_j in the complex plane that linear combinations of f_j approximate in the mean square any function in $L_2(0, \infty)$.

Problem 7. Let $f \in L_2([0, \infty))$. The holomorphic function for $\Re(z) \geq 0$

$$F(z) = \frac{1}{\sqrt{2\pi}} \int_0^\infty e^{-tz} f(t)d\mu(t)$$

is called the *Laplace transform* of f. Let $f(t) = \exp(-|t|)$. Find the Laplace transform.

Problem 8.   Consider the Hilbert spaces $L_2(\mathbb{R})$ and the $\ell_2(\mathbb{N}_0)$ and the Hamilton operator

$$\hat{H} = \frac{p^2}{2\mu} + \frac{1}{2}\mu\omega^2 q^2 + aq^4.$$

Show that applying the normalized harmonic oscillator eigenfunctions as a basis of expansion one finds the (symmetric) matrix representation ($n = 0, 1, 2, \ldots$)

$$H_{n,n} = 3\rho(2n^2 + 2n + 1) + (2n + 1)e$$
$$H_{n,n-2} = H_{n-2,n} = 2\rho(2n - 1)(n(n - 1))^{1/2}, \quad n \geq 2$$
$$H_{n,n-4} = H_{n-4,n} = \rho(n(n - 1)(n - 2)(n - 3))^{1/2}, \quad n \geq 4.$$

All other entries in the matrix are equal to 0 and the quantities ρ, e given by

$$\rho = a\left(\frac{\hbar}{2\mu\omega}\right)^2, \qquad e = \frac{1}{2}\hbar\omega$$

have dimension energy.

Problem 9.   Let $\hat{p}$, $\hat{x}$ be the momentum and position operators, respectively. Consider the *characteristic operator* defined as

$$\hat{M}(\hat{x}, \hat{p}, \tau, \phi) := \exp(i(\phi\hat{x} + \tau\hat{p})/\hbar)$$

where τ and ϕ are real scalar parameters with dimension position and momentum, respectively. Let $|\psi\rangle$ be a normalized state. The characteristic function is defined as

$$M(\tau, \phi) = \langle\psi|\hat{M}(\hat{x}, \hat{p}, \tau, \phi)|\psi\rangle \equiv \langle\psi|\exp(i(\phi\hat{x} + \tau\hat{p})/\hbar)|\psi\rangle.$$

Show that $M(\tau, \phi) = \langle\psi|e^{i\tau\hat{p}/(2\hbar)}e^{i\phi\hat{x}/\hbar}e^{i\tau\hat{p}/(2\hbar)}|\psi\rangle$ applying the Campbell-Baker-Hausdorff identity. Show that

$$M(\tau, \phi) = \int_{-\infty}^{\infty} dq e^{i\phi x/\hbar}\langle\psi|x - \tau/2\rangle\langle x + \tau/2|\psi\rangle$$

utilizing that $e^{-i\tau\hat{p}/(2\hbar)}|x\rangle = |x + \tau/2\rangle$, where $|x\rangle$ is the position eigenstate with eigenvalue x. Furthermore apply the completeness relation

$$I = \int_{-\infty}^{\infty} dx\,|x\rangle\langle x|.$$

Problem 10. Consider the Hamilton operator

$$\hat{H} = \frac{1}{2m}(\hat{p}_{x_1}^2 + \hat{p}_{x_2}^2) + \frac{1}{2}m\Omega(x_1^2 + x_2^2) + \omega(x_1\hat{p}_{x_2} - x_2\hat{p}_{x_1}).$$

Hence we have a two-dimensional *harmonic oscillator* of frequency Ω. The oscillator itself is rotating about the negative x_3-axis with frequency ω. Show that the last term in the Hamilton operator commutes with the harmonic oscillator part. Apply this fact to show that spectrum of the Hamilton operator can be written as

$$E_{n_r,\ell} = (2n_r + |\ell| + 1)\hbar\Omega + \ell\hbar\omega$$

where n_r is the radial quantum number ($n_r = 0, 1, 2, \dots$) and $\hbar\ell$ is the momentum along the x_3 axis, where $\ell = 0, \pm 1, \pm 2, \dots$.

Problem 11. Let $\mathbb{Z}$ be the set of integers and $n, m \in \mathbb{Z}$. We consider the Hilbert space $\ell_2(\mathbb{Z})$ and denote the standard basis by $|n\rangle$ ($n \in \mathbb{Z}$) and the dual one by $\langle m|$ ($m \in \mathbb{Z}$) with the scalar product $\langle m|n\rangle = \delta_{mn}$. Let $\mathbb{C}^2$ be the two-dimensional Hilbert space with the standard basis

$$|+\rangle = \begin{pmatrix} 1 \\ 0 \end{pmatrix}, \quad |-\rangle = \begin{pmatrix} 0 \\ 1 \end{pmatrix}, \quad \langle +| = (1 \quad 0), \quad \langle -| = (0 \quad 1).$$

We consider the product Hilbert space $\ell_2(\mathbb{Z}) \otimes \mathbb{C}^2$ and define the linear operator $\hat{S}$ acting in $\ell_2(\mathbb{Z}) \otimes \mathbb{C}^2$ as

$$\hat{S}(|n\rangle \otimes |\pm\rangle) = |n \pm 1\rangle \otimes |\pm\rangle$$

and the unitary operator U_{θ_n} ($\theta_n \in \mathbb{R}$) acting in $\ell_2(\mathbb{Z}) \otimes \mathbb{C}^2$

$$\hat{U}_{\theta_n} =$$

$$\sum_{n\in\mathbb{Z}}(|n\rangle\langle n|\otimes(\cos(\theta_n)|+\rangle\langle +|+\sin(\theta_n)|+\rangle\langle -|+\sin(\theta_n)|-\rangle\langle +|-\cos(\theta_n)|-\rangle\langle -|)).$$

The underlying matrix

$$\begin{pmatrix} \cos(\theta_n) & \sin(\theta_n) \\ \sin(\theta_n) & -\cos(\theta_n) \end{pmatrix}$$

has determinant -1. Setting $\theta_n = \pi/4$ we obtain the Hadamard matrix. We can consider now the linear operators $\hat{S}\hat{U}_{\theta_n}$. Assume now that θ_n does not depend on n and we set $\theta_n = \theta$. Consider the normalized state $|0\rangle \otimes |+\rangle$. Find

$$(\hat{S}\hat{U}_\theta)(|0\rangle \otimes |+\rangle), \quad (\langle 0| \otimes \langle +|)(\hat{S}\hat{U}_\theta)(|0\rangle \otimes |+\rangle)),$$

$$(\langle 0| \otimes \langle -|)(\hat{S}\hat{U}_\theta)(|0\rangle \otimes |+\rangle)).$$

Is $[S, U_\theta] = 0 \otimes 0_2$?

Chapter 13

Special Functions

Legendre, Hermite, Laguerre and Chebyshev polynomials play an important role in mathematical physics, for example as an orthonormal basis in the Hilbert space of square integrable functions. Jacobi elliptic functions and Weierstrass elliptic functions play a role in the solution of nonlinear differential equations. The *Legendre polynomials* are given by the *Rodrigue's formula*

$$P_n(x) := \frac{1}{2^n n!} \frac{d^n}{dx^n} (x^2 - 1)^n, \qquad n = 0, 1, 2, \ldots$$

where $P_0(x) = 1$, $P_1(x) = x$, $P_2(x) = (3x^2 - 1)/2$. The *Hermite polynomials* are given by Rodrigue's formula

$$H_n(x) := (-1)^n e^{x^2} \frac{d^n}{dx^n} (e^{-x^2}), \qquad n = 0, 1, 2, \ldots$$

where $H_0(x) = 1$, $H_1(x) = 2x$, $H_2(x) = 4x^2 - 2$. The *Laguerre polynomials* are given by Rodrigue's formula

$$L_n(x) := e^x \frac{d^n}{dx^n} (x^n e^{-x}), \qquad n = 0, 1, 2, \ldots$$

where $L_0(x) = 1$, $L_1(x) = -x + 1$, $L_2(x) = x^2 - 4x + 2$. The *Chebyshev polynomials* are given by

$$T_n(x) = \cos(n \arccos(x)) = x^n - \binom{n}{2} x^{n-2}(1 - x^2) + \binom{n}{4} x^{n-4}(1 - x^2)^2 - \cdots$$

where $T_0(x) = 1$, $T_1(x) = x$, $T_2(x) = 2x^2 - 1$.

Consider the generating function

$$U(x,t) = \exp(x(t-1/t)/2).$$

Expansion with respect to t provides

$$U(x,t) = \sum_{n=-\infty}^{\infty} t^n \sum_{\mu=0}^{\infty} \frac{(-1)^\mu}{\mu!(n+\mu)!} \left(\frac{x}{2}\right)^{n+2\mu} = \sum_{n=-\infty}^{\infty} t^n J_n(x)$$

where $J_n(x)$ are the *Bessel functions* of first kind.

For the Jacobi elliptic functions $\mathrm{sn}(s,k)$, $\mathrm{cn}(s,k)$ we have

$$\mathrm{sn}^{-1}(s,k) := \int_0^{\arcsin(s)} \frac{dx}{\sqrt{1-k^2\sin^2(x)}}$$

$$\mathrm{cn}^{-1}(s,k) := \int_0^{\arccos(s)} \frac{dx}{\sqrt{1-k^2\sin^2(x)}}$$

where $k \in [0,1]$. Furthermore $\mathrm{sn}(0,k) = 0$, $\mathrm{cn}(0,k) = 1$ and the expansion

$$\mathrm{sn}(x,k) = x - (1+k^2)\frac{x^3}{3!} + (1+14k^2+k^4)\frac{x^5}{5!} - \cdots$$

$$\mathrm{cn}(x,k) = 1 - \frac{x^2}{2!} + (1+4k^2)\frac{x^4}{4!} - \cdots$$

$$\mathrm{dn}(x,k) = 1 - k^2\frac{x^2}{2!} + k^2(4+k^2)\frac{x^4}{4!} - \cdots .$$

The *addition theorems* are

$$\mathrm{sn}(u+v,k) = \frac{\mathrm{sn}(u,k)\mathrm{cn}(v,k)\mathrm{dn}(v,k) + \mathrm{cn}(u,k)\mathrm{sn}(v,k)\mathrm{dn}(u,k)}{1-k^2\mathrm{sn}^2(u,k)\mathrm{sn}^2(v,k)}$$

$$\mathrm{cn}(u+v,k) = \frac{\mathrm{cn}(u,k)\mathrm{cn}(v,k) - \mathrm{sn}(u,k)\mathrm{sn}(v,k)\mathrm{dn}(u,k)\mathrm{dn}(v,k)}{1-k^2\mathrm{sn}^2(u,k)\mathrm{sn}^2(v,k)}$$

$$\mathrm{dn}(u+v,k) = \frac{\mathrm{dn}(u,k)\mathrm{dn}(v,k) - k^2\mathrm{sn}(u,k)\mathrm{sn}(v,k)\mathrm{cn}(u,k)\mathrm{cn}(v,k)}{1-k^2\mathrm{sn}^2(u,k)\mathrm{sn}^2(v,k)}$$

The *Jacobi theta function* is defined by

$$\Theta(z,\tau) = \sum_{n=-\infty}^{\infty} \exp(\pi i n^2 \tau + 2\pi i n z) \equiv 1 + 2\sum_{n=1}^{\infty} (e^{\pi i \tau})^{n^2} \cos(2\pi n z)$$

where $z \in \mathbb{C}$ and $\Im(\tau) > 0$.

Problem 1.   (i) Consider the Legendre polynomials given by the *formula of Rodrigues*. Let f be an analytic function. Show that the integral

$$I = \int_{-1}^{1} f(x) P_n(x) dx$$

can be brought into the form

$$I = \frac{(-1)^n}{2^n n!} \int_{-1}^{1} (x^2 - 1)^n \frac{d^n}{dx^n} f(x) dx. \tag{1}$$

(ii) Show that

$$\int_{-1}^{1} P_m(x) P_n(x) dx = 0, \qquad m \neq n. \tag{2}$$

Solution 1.   (i) Applying integration by parts to (1) we obtain

$$I = \frac{1}{2^n n!} \left| f(x) \frac{d^{n-1}}{dx^{n-1}} (x^2 - 1)^n \right|_{-1}^{+1} - \frac{1}{2^n n!} \int_{-1}^{1} \frac{df}{dx} \left(\frac{d^{n-1}}{dx^{n-1}} (x^2 - 1)^n dx \right). \tag{3}$$

Since $x^2 - 1$ is equal to zero at 1 and -1 we find that the first term on the right-hand side of (3) vanishes. Repeating this process of integration by parts we obtain (3).
(ii) Let $f(x) = P_m(x)$ with $m < n$. Then $d^n f/dx^n = 0$ and therefore (2) follows.

Problem 2.   A generating function for the *Legendre polynomials* $P_l(x)$ is given by

$$G(x, r) := \frac{1}{(1 - 2xr + r^2)^{1/2}} = \sum_{l=0}^{\infty} r^l P_l(x)$$

where $x = \cos(\theta)$ and $|r| < 1$. Prove that $P_l(x)$ satisfies the linear differential relation

$$x \frac{dP_l(x)}{dx} = \frac{P'_{l-1}(x)}{dx} + l P_l(x). \tag{1}$$

Solution 2.   Differentiating G with respect to r yields

$$\frac{(x - r)}{(1 - 2xr + r^2)^{3/2}} = \sum_{l=0}^{\infty} l r^{l-1} P_l(x). \tag{2}$$

Differentiating G with respect to x gives

$$\frac{r}{(1 - 2xr + r^2)^{3/2}} = \sum_{l=0}^{\infty} r^l \frac{dP_l(x)}{dx}. \tag{3}$$

Eliminating $(1 - 2xr + r^2)^{-3/2}$ between (2) and (3) leads to

$$\sum_{l=0}^{\infty} (x - r)r^l \frac{dP_l(x)}{dx} = \sum_{l=0}^{\infty} lr^l P_l(x)$$

which, upon equating coefficients of equal powers of r, gives (1). The first three Legendre polynomials are given by

$$P_0(x) = 1, \qquad P_1(x) = x, \qquad P_2(x) = \frac{1}{2}(3x^2 - 1).$$

Problem 3. The *dominant tidal potential* at position (r, ϕ, λ) due to the moon or sun is given by

$$U(\mathbf{r}) = \frac{GM^* r^2}{r^{*3}} P_2^0(\cos(\psi))$$

where M^* is the mass of the moon or sun located at (r^*, ϕ^*, λ^*). Moreover, ψ is the angle between mass M^* and the observation point at (r, ϕ, λ), where ϕ is the latitude and λ is the longitude. By the *spherical cosine theorem* we have

$$\cos(\psi) = \sin(\phi)\sin(\phi^*) + \cos(\phi)\cos(\phi^*)\cos(\lambda - \lambda^*).$$

The *Legendre polynomials* are defined by the Rodrigue's formula given above. The *associated Legendre polynomials* are defined as

$$P_n^m(x) := (1 - x^2)^{m/2} \frac{d^m}{dx^m} P_n(x) = \frac{(1-x^2)^{m/2}}{2^n n!} \frac{d^{m+n}}{dx^{m+n}} (x^2 - 1)^n$$

with $P_n^0(x) = P_n(x)$ and $P_n^m(x) = 0$ if $m > n$.

(i) Show that $U(\mathbf{r})$ can be written as

$$U(\mathbf{r}) = \frac{GM^* r^2}{r^{*3}} \Big(P_2^0(\sin(\phi)) P_2^0(\sin(\phi^*)) + \frac{1}{3} P_2^1(\sin(\phi)) P_2^1(\sin(\phi^*)) \cos(\lambda - \lambda^*)$$

$$+ \frac{1}{12} P_2^2(\sin(\phi)) P_2^2(\sin(\phi^*)) \cos(2(\lambda - \lambda^*)) \Big).$$

(ii) Give an interpretation (maxima and nodes) of the terms in the parenthesis.

Solution 3. (i) Since

$$\cos^2(\psi) = \sin^2(\phi)\sin^2(\phi^*) + \cos^2(\phi)\cos^2(\phi^*)\cos^2(\lambda - \lambda^*)$$
$$+ 2\sin(\phi)\cos(\phi)\sin(\phi^*)\cos(\phi^*)\cos(\lambda - \lambda^*)$$

and $\cos^2(\alpha) \equiv \frac{1}{2} + \frac{1}{2}\cos(2\alpha)$ we arrive at

$$
\cos^2(\psi) = \frac{3}{2}\sin^2(\phi)\sin^2(\phi^*) - \frac{1}{2}(\sin^2(\phi) + \sin^2(\phi^*)) + \frac{1}{2}
$$
$$
+ \frac{1}{2}\cos^2(\phi)\cos^2(\phi^*)\cos(2(\lambda - \lambda^*))
$$
$$
+ 2\sin(\phi)\cos(\phi)\sin(\phi^*)\cos(\phi^*)\cos(\lambda - \lambda^*).
$$

Thus

$$
P_2^0(\cos\psi) = \frac{1}{2}(3\cos^2(\psi) - 1)
$$
$$
= \frac{9}{4}\sin^2(\phi)\sin^2(\phi^*) - \frac{3}{4}(\sin^2(\phi) + \sin^2(\phi^*)) - \frac{1}{4}
$$
$$
+ 3\sin(\phi)\cos(\phi)\sin(\phi^*)\cos(\phi^*)\cos(\lambda - \lambda^*)
$$
$$
+ \frac{3}{4}\cos^2(\phi)\cos^2(\phi^*)\cos(2(\lambda - \lambda^*)).
$$

Since

$$
P_2^0(\sin\phi)P_2^0(\sin\phi^*) = \frac{1}{4}(9\sin^2(\phi)\sin^2(\phi^*) - 3\sin^2(\phi) - 3\sin^2(\phi^*) + 1)
$$
$$
P_2^1(\sin\phi)P_2^1(\sin\phi^*) = 9\sin(\phi)\cos(\phi)\sin(\phi^*)\cos(\phi^*)
$$
$$
P_2^2(\sin(\phi))P_2^2(\sin(\phi^*)) = 9\cos^2(\phi)\cos^2(\phi^*)
$$

we find the expression for $U(\mathbf{r})$.

(ii) The first term in the parenthesis has a maximum at the poles and 2 nodes in the latitude but no longitude dependence. Thus it represents long periodic zonal tides. The second term has a maximum at $\pm 45^\circ$ latitude, one nodal line along the equator and two nodal longitudes. Thus it represents diurnal tesseral tides. The third term has a maximum at the equator and no nodes in the latitude other than those at the poles, but 4 nodal longitudes. Thus it represents semi-diurnal sectorial tides. Since the location of M^*, i.e. (r^*, ϕ^*, λ^*) are time dependent, the terms

$$
P_2^m(\sin(\phi^*))\cos(m(\lambda - \lambda^*))
$$

for the sun or moon can be expanded in trigonometric series in t, m and λ.

Problem 4.   A generating function $F(x,t)$ of the *Hermite polynomials* $H_n(x)$ is given by

$$
F(x,t) := e^{x^2 - (t-x)^2} = \sum_{k=0}^{\infty} H_k(x)\frac{t^k}{k!}
$$

where $n = 0, 1, 2, \ldots$ and $t, x \in \mathbb{R}$.

(i) Express $H_n(x)$ as a contour integral.

(ii) Prove that $H_n(x)$ satisfies *Hermite's differential equation*

$$\frac{d^2 H_n}{dx^2} - 2x\frac{dH_n}{dx} + 2nH_n = 0.$$

(iii) Show that $dH_n/dx = 2nH_{n-1}$, $n = 1, 2, \ldots$.

Solution 4. (i) We know that

$$\oint \frac{1}{z^n}dz = \begin{cases} 2\pi i & \text{if } n = 1 \\ 0 & \text{otherwise} \end{cases} \tag{1}$$

where the integration is performed over a closed contour in the complex z-plane enclosing the origin and $n \in \mathbb{Z}$. Dividing the generating equation $F(x, t)$ by t^{n+1} and integrating over a closed contour in the complex t-plane enclosing the origin, we obtain

$$H_n(x) = \frac{n!}{2\pi i} \oint \frac{\exp(x^2 - (t - x)^2)}{t^{n+1}}dt \tag{2}$$

where we have used (1).

(ii) One forms the quantity

$$\frac{\partial^2 F}{\partial x^2} + 2t\frac{\partial F}{\partial t} - 2x\frac{\partial F}{\partial x}$$

and verifies that it is identically zero. Using the expansion for F in terms of H_n, this identity takes the form

$$\sum_{n=0}^{\infty} \frac{(H_n''(x) - 2xH_n'(x) + 2nH_n(x))t^n}{n!} = 0$$

for all t, where $H_n' \equiv dH_n/dx$. Therefore $H_n''(x) - 2xH_n'(x) + 2nH_n(x) = 0$.

(iii) Differentiating the integral representation for $H_n(x)$ given by (2) with respect to x we obtain

$$\frac{dH_n}{dx} = \frac{2n((n - 1)!)}{2\pi i} \oint \frac{\exp(x^2 - (t - x)^2)}{t^n}dt.$$

Taking into account (2) we find that

$$\frac{dH_n}{dx} = 2nH_{n-1}.$$

Problem 5. Let $L_n(x)$, $H_n(x)$ be the Laguerre and Hermite polynomials, where $n = 0, 1, \ldots$. Let

$$L_n^{(\alpha)}(x) := \frac{x^{-\alpha}e^x}{n!}\frac{d^n}{dx^n}(e^{-x}x^{n+\alpha})$$

be the associated Laguerre polynomials with $\alpha > -1$ and $n = 0, 1, \ldots$. The Laguerre polynomials are recovered by setting $\alpha = 0$. We have

$$H_{2n}(x) = (-4)^n n! L_n^{(-1/2)}(x^2) \tag{1}$$

and the following addition formula for the associated Laguerre polynomials $L_n^\alpha(x)$

$$L_n^{(\alpha+\beta+1)}(x+y) = \sum_{k=0}^n L_{n-k}^{(\alpha)}(x) L_k^{(\beta)}(y). \tag{2}$$

(i) Find a new sum rule by inserting (1) into (2).
(ii) Consider the *sum rule*

$$\frac{1}{n! 2^n} H_n(\sqrt{2}x) H_n(\sqrt{2}y) = \sum_{k=1}^n (-1)^k L_{n-k}^{(-1/2)}((x+y)^2) L_k^{(-1/2)}((x-y)^2). \tag{3}$$

Insert (1) into (3) to find a sum rule for Hermite polynomials.

Solution 5.   (i) Setting $\alpha = \beta = -1/2$, $x \to x^2$, $y \to y^2$ we obtain

$$(-4)^n L_n(x^2 + y^2) = \sum_{k=0}^n \frac{1}{(n-k)! k!} H_{2n-2k}(x) H_{2k}(y).$$

(ii) Setting $x + y \to x$ and $x - y \to y$ we obtain

$$\frac{(-2)^n}{n!} H_n\left(\frac{x+y}{\sqrt{2}}\right) H_n\left(\frac{x-y}{\sqrt{2}}\right) = \sum_{k=0}^n \frac{(-1)^k}{(n-k)! k!} H_{2n-2k}(x) H_{2k}(y).$$

Problem 6.   Consider the Hilbert space $L_2[0, \infty)$. An orthonormal basis $\{\phi_n(x) : n = 0, 1, 2, \ldots\}$ is given by $\phi_n(x) = e^{-x/2} L_n(x)$, where the *Laguerre polynomials* are defined by

$$L_n(x) := \frac{1}{n!} e^x \frac{d^n}{dx^n}(x^n e^{-x}).$$

(i) Find ϕ_0, ϕ_1 and ϕ_2.
(ii) Calculate the scalar product $\langle \phi_2(x) | \phi_2(x) \rangle$. We note that

$$\int_0^\infty x^n e^{-x} dx = n!.$$

(iii) Consider the function $f : [0, \infty) \to \mathbb{R}$

$$f(x) = \begin{cases} 1 & \text{for } 0 \le x \le 1 \\ 0 & \text{for } \quad x > 1 \end{cases}.$$

Calculate the coefficients $c_0 = \langle f, \phi_0 \rangle$, $c_1 = \langle f, \phi_1 \rangle$ of the expansion

$$f(x) = \sum_{n=0}^{\infty} \langle f, \phi_n \rangle \phi_n(x)$$

i.e. $c_n = \langle f, \phi_n \rangle$.

Solution 6. (i) We have

$$\phi_0(x) = e^{-x/2}, \quad \phi_1(x) = (-x+1)e^{-x/2}, \quad \phi_2(x) = \frac{1}{2}(x^2 - 4x + 2)e^{-x/2}.$$

(ii) For the scalar product we find

$$\langle \phi_2(x) | \phi_2(x) \rangle = \frac{1}{4} \int_0^{\infty} (x^4 - 8x^3 + 20x^2 - 16x + 4)e^{-x} dx$$

$$= \frac{1}{4}(4! - 8 \cdot 3! + 20 \cdot 2! - 16 \cdot 1! + 4) = 1.$$

This is obvious since the functions ϕ_n form an orthonormal basis in the Hilbert space $L_2[0, \infty)$.

(iii) We have

$$c_0 = \int_0^{\infty} f(x)\phi_0(x)dx = \int_0^1 e^{-x/2}dx = -2e^{-1/2} + 2$$

$$c_1 = \int_0^{\infty} f(x)\phi_1(x)dx = \int_0^1 e^{-x/2}(-x+1)dx = 4e^{-1/2} - 2.$$

Problem 7. Consider

$$T_{n+1}(x) - 2xT_n(x) + T_{n-1}(x) = 0 \tag{1}$$

where $n = 1, 2, \ldots$,

$$T_0(x) = 1, \qquad T_1(x) = x \tag{2}$$

and x is a fixed parameter. Solve the linear difference equation (1) where the initial values are given by (2). The quantities $T_n(x)$ are called *Chebyshev polynomials* of the first kind.

Solution 7. Since (1) is a linear difference equation with constant coefficients it can be solved with the ansatz

$$T_n(x) := A(r(x))^n$$

where A is a constant. Inserting this ansatz into (1) yields

$$r^{n+1} - 2xr^n + r^{n-1} = 0$$

or $r^2 - 2xr + 1 = 0$. The solution to this quadratic equation is given by

$$r_{1,2}(x) = x \pm \sqrt{x^2 - 1}.$$

Thus the solution of the difference equation (1) takes the form

$$T_n(x) = A(x + \sqrt{x^2 - 1})^n + B(x - \sqrt{x^2 - 1})^n$$

where A and B are the constants of integration and $n = 0, 1, 2, \ldots$. Imposing the initial conditions (2) yields

$$T_n(x) = \frac{1}{2}(x + \sqrt{x^2 - 1})^n + \frac{1}{2}(x - \sqrt{x^2 - 1})^n.$$

The solution can be written as $T_n(x) = \cos(n \arccos(x))$, $n = 0, 1, 2, \ldots$.

Problem 8.   (i) Let n be a positive integer. Let $f : [a, b] \rightarrow \mathbb{R}$ be a continuous function. The *Bernstein polynomials* of degree n associated with the continuous function f are given by

$$B_n(f, x) := \frac{1}{(b-a)^n} \sum_{j=0}^{n} \binom{n}{j} (x-a)^j (b-x)^{n-j} f(x_j)$$

where

$$x_j = a + j \frac{b-a}{n}, \qquad j = 0, 1, \ldots, n.$$

Consider the function $f : [0, 1] \rightarrow \mathbb{R}$ given by $f(x) = \sin(4x)$. Show that $B_2(f, x)$ is not a "good approximation" for f.
(ii) The *Bernstein basis polynomials* are defined as

$$B_{n,j}(x) := \binom{n}{j} x^j (1-x)^{n-j}, \quad j = 0, 1, \ldots, n, \quad x \in [0, 1].$$

Show that

$$\sum_{j=0}^{n} B_{n,j}(x) = 1.$$

Show that $B_{n,j}$ satisfies the recursion relations

$$B_{n,j}(x) = (1-x)B_{n-1,j}(x) + xB_{n,j-1}(x), \qquad j = 1, 2, \ldots, n-1$$
$$B_{n,0}(x) = (1-x)B_{n-1,0}(x)$$
$$B_{n,n}(x) = xB_{n-1,n-1}(x).$$

Solution 8.   (i) For $x = \pi/8$ we have $\sin(\pi/2) = 1$. Using $\sin(0) = 0$ and

$$\sin(2) \equiv 2\sin(1)\cos(1), \quad \sin(4) \equiv 4\sin(1)\cos(1) - 8\sin^3(1)\cos(1)$$

we have $B_2(f(x = \pi/8), x = \pi/8) = 4x \sin(1) \cos(1)(1 - 2x \sin^2(1))$. Thus $1 - B_2(f(x = \pi/8), x = \pi/8) \approx 0.65$.

(ii) We have

$$\sum_{j=0}^{n} B_{n,j}(x) = \sum_{j=0}^{n} \binom{n}{j} x^j (1 - x)^{n-j} = ((1 - x) + x)^n = 1.$$

To prove the recursion relation, we note that

$$(1 - x)B_{n-1,j}(x) + xB_{n,j-1}(x) = \frac{(n-1)!}{j!(n-1-j)!} x^j (1 - x)^{n-j-1}(1 - x)$$

$$+ \frac{(n-1)!}{(j-1)!(n-1-j+1)!} x^{j-1}(1 - x)^{n-j} x$$

$$= x^j (1 - x)^{n-j} \frac{n!}{j!(n-j)!} \left(\frac{n-j}{n} + \frac{j}{n} \right)$$

$$= B_{n,j}(x)$$

Problem 9. Let $a > b > 0$. Then the equation of the *ellipse* in parametric form is given by $x(\phi) = a \sin(\phi)$, $y(\phi) = b \cos(\phi)$, where $0 \leq \phi < 2\pi$. The *metric tensor field* of the two-dimensional Euclidean space is given by

$$g = dx \otimes dx + dy \otimes dy. \tag{1}$$

Calculate the *arc length* of an ellipse.

Solution 9. Since $dx = a \cos(\phi)d\phi$, $dy = -b \sin(\phi)d\phi$ we obtain

$$dx \otimes dx = a^2 \cos^2(\phi)d\phi \otimes d\phi, \qquad dy \otimes dy = b^2 \sin^2(\phi)d\phi \otimes d\phi.$$

Consequently, we obtain from (1)

$$g = (a^2 \cos^2(\phi) + b^2 \sin^2(\phi))d\phi \otimes d\phi.$$

Therefore the *line element* is given by

$$\left(\frac{ds}{d\phi} \right)^2 = a^2 \cos^2(\phi) + b^2 \sin^2(\phi).$$

Using the identity $\sin^2(\phi) + \cos^2(\phi) \equiv 1$ we arrive at

$$\int_0^{2\pi} ds = a \int_0^{2\pi} \sqrt{1 - \frac{a^2 - b^2}{a^2} \sin^2(\phi)} d\phi.$$

This is a *complete elliptic integral of the second kind*, where

$$k^2 = \frac{a^2 - b^2}{a^2} =: e^2.$$

Here e is the *eccentricity* of the ellipse. We find

$$\int_0^{2\pi} ds = 4aE(k, \pi/2).$$

The complete elliptic integral of the second kind can be given as

$$E(k, \pi/2) = \int_0^{\pi/2} \sqrt{1 - k^2 \sin^2(\phi)} d\phi.$$

Thus

$$E(k, \pi/2) = \frac{\pi}{2}\left(1 - \left(\frac{1}{2}\right)^2 k^2 - \left(\frac{1\cdot 3}{2\cdot 4}\right)^2 \frac{k^4}{3} - \left(\frac{1\cdot 3\cdot 5}{2\cdot 4\cdot 6}\right)^2 \frac{k^6}{5} - \cdots\right)$$

where $|k^2| < 1$. We have used the *Taylor series expansion* around $x = 0$ of the function $f(x) = \sqrt{1 - x^2}$ with $x^2 < 1$, i.e.

$$\sqrt{1 - x^2} = 1 - \frac{1}{2}x^2 - \frac{1}{2\cdot 4}x^4 - \frac{1\cdot 3}{2\cdot 4\cdot 6}x^6 - \cdots, \qquad -1 < x \le 1.$$

Problem 10. (i) Let

$$\operatorname{sn}^{-1}(s, k) := \int_0^{\arcsin(s)} \frac{dx}{\sqrt{1 - k^2 \sin^2(x)}}$$

where $k \in [0, 1]$. Calculate the derivative $d(\operatorname{sn}^{-1}(s, k))/ds$.
(ii) Let

$$\operatorname{cn}^{-1}(s, k) := \int_0^{\arccos(s)} \frac{dx}{\sqrt{1 - k^2 \sin^2(x)}}.$$

Calculate the derivative $d(\operatorname{cn}^{-1}(s, k))/ds$.

Solution 10.   We use the *Leibniz rule* for differentiation of integrals

$$\frac{d}{ds}\int_{f(s)}^{g(s)} F(x, s)dx = \int_{f(s)}^{g(s)} \frac{\partial F}{\partial s}dx + F(g(s), s)\frac{dg}{ds} - F(f(s), s)\frac{df}{ds}. \qquad (1)$$

If F is independent of s, then (1) simplifies to

$$\frac{d}{ds}\int_{f(s)}^{g(s)} F(x)dx = F(g(s))\frac{dg}{ds} - F(f(s))\frac{df}{ds}.$$

(i) Applying this rule we obtain

$$\frac{d}{ds} \text{sn}^{-1}(s,k) = \frac{1}{\sqrt{1 - k^2(\sin(\arcsin(s)))^2}} \frac{d}{ds} \arcsin(s).$$

Since $\sin(\arcsin(s)) = s$ and $d(\arcsin(s))/ds = 1/\sqrt{1 - s^2}$ we obtain

$$\frac{d}{ds} \text{sn}^{-1}(s,k) = \frac{1}{\sqrt{1 - k^2 s^2}} \frac{1}{\sqrt{1 - s^2}}.$$

(ii) We find

$$\frac{d}{ds} \text{cn}^{-1}(s,k) = \frac{1}{\sqrt{1 - k^2(\sin(\arccos(s)))^2}} \frac{d}{ds} \arccos(s).$$

Since $\sin(\arccos(s)) = \sqrt{1 - s^2}$ and $d(\arccos(s))/ds = -1/\sqrt{1 - s^2}$ we obtain

$$\frac{d}{ds} \text{cn}^{-1}(s,k) = -\frac{1}{\sqrt{1 - k^2(1 - s^2)}} \frac{1}{\sqrt{1 - s^2}}.$$

Problem 11. Consider the function

$$\sigma(z; \omega_1, \omega_2) = z \prod_k{}' \left(1 - \frac{z}{\Omega_k}\right) \exp\left(\frac{z}{\Omega_k} + \frac{1}{2}\left(\frac{z}{\Omega_k}\right)^2\right)$$

with $\Omega_k = m_k \omega_1 + n_k \omega_2$, where m_k, n_k is the sequence of all pairs of integers. The prime after the product sign indicates that the pair $(0,0)$ should be omitted. ω_1 and ω_2 are two complex numbers with $\Im(\omega_1/\omega_2) \neq 0$. In the following the dependence on ω_1 and ω_2 will be omitted. The logarithmic derivative $\sigma'(z)/\sigma(z)$ where $\sigma'(z) \equiv d\sigma/dz$ is the meromorphic function $\zeta(z)$ of Weierstrass. The function

$$\wp(z) = -\zeta'(z)$$

is a meromorphic, doubly periodic (or elliptic) function with periods ω_1, ω_2 whose only singularities are double poles $m\omega_1 + n\omega_2$. We find

$$\wp(z; \omega_1, \omega_2) = \frac{1}{z^2} + \sum_k{}' \left(\frac{1}{(z - \Omega_k)^2} - \frac{1}{(\Omega_k)^2}\right).$$

The function $\wp$ is called *Weierstrass function $\wp$*.
(i) Show that
$$\sigma(z) = z + c_5 z^5 + c_7 z^7 + \cdots \tag{1}$$

(ii) Show that
$$\frac{\sigma(2u)}{\sigma^4(u)} = -\wp'(u) \tag{2a}$$

$$2\zeta(2u) - 4\zeta(u) = \frac{\wp''(u)}{\wp'(u)}. \tag{2b}$$

Solution 11.   (i) Grouping the factors corresponding to the pairs (m, n), $(-m, -n)$ we find

$$\sigma(z) = z\prod_k{}' \left(1 - \frac{z}{\Omega_k^2}\right) \exp(z^2/\Omega_k^2).$$

Since every factor has the form $1 - z^4/\Omega_k^2$ we find (1).

(ii) The function $\sigma(2u)/\sigma^4(u)$ is doubly periodic and has poles of order three at $m\omega_1 + n\omega_2$. Furthermore the function is equal to

$$2\frac{1}{u^3} + Au + \cdots$$

near the origin. Consequently,

$$\wp'(u) + \frac{\sigma(2u)}{\sigma^4(u)} = 0$$

since the left hand side has only removable singularities. Calculating the *logarithmic derivative*

$$(\ln(f))' = \frac{f'}{f}$$

of both sides of (2a) and $\zeta(z) = \sigma'(z)/\sigma(z)$ we find (2b).

Problem 12.   Let $f(q)$ be a meromorphic function. Let r, k, $l = 1, 2, \ldots, N$. Assume that f satisfies the following equation

$$\frac{f(q_k - q_r)f'(q_r - q_l) - f'(q_k - q_r)f(q_r - q_l)}{f(q_k - q_l)} = g(q_k - q_r) - g(q_r - q_l)$$

where f' denotes differentiation with respect to the arguments and $k \neq l$. Let $U(q) = f^2(q)$. Show that

$$[U(x)U'(y) - U'(x)U(y)] + [U(y)U'(z) - U'(y)U(z)]$$
$$+[U(z)U'(x) - U'(z)U(x)] = 0$$

where $x + y + z = (q_k - q_r) + (q_r - q_l) + (q_l - q_k) = 0$.

Solution 12.   The six permutations of (k, r, l) are given by

$$(k, r, l), \quad (k, l, r), \quad (r, k, l), \quad (r, l, k), \quad (l, k, r), \quad (l, r, k).$$

Adding all equations (1) corresponding to all permutations of (k, r, l) we find

$$\frac{f(q_r - q_l)f'(q_l - q_k) - f'(q_r - q_l)f(q_l - q_k)}{f(q_k - q_r)}$$

$$+ \frac{f(q_l - q_k)f'(q_k - q_r) - f'(q_l - q_k)f(q_k - q_r)}{f(q_r - q_l)}$$

$$+ \frac{f(q_k - q_r)f'(q_r - q_l) - f'(q_k - q_r)f(q_r - q_l)}{f(q_l - q_k)} = 0.$$

Therefore the equation for U follows. This equation is the *addition formula* for the Weierstrass $\wp$ function.

Problem 13. A symmetric $(r \times r)$ matrix $B = (B_{jk})$ with negative definite real part $\Re(B) = (\Re(B_{jk}))$ is called a *Riemann matrix*. The *Riemann theta function* is defined by

$$\theta(\mathbf{z}|B) := \sum_{\mathbf{N} \in \mathbb{Z}^r} \exp\left(\frac{1}{2}\langle B\mathbf{N}, \mathbf{N}\rangle + \langle \mathbf{N}, \mathbf{z}\rangle\right). \tag{1}$$

Here $\mathbf{z} = (z_1, \ldots, z_r) \in \mathbb{C}^r$ is a complex vector. The triangular brackets denote the Euclidean scalar product

$$\langle \mathbf{N}, \mathbf{z}\rangle := \sum_{j=1}^{r} N_j z_j, \qquad \langle B\mathbf{N}, \mathbf{N}\rangle := \sum_{i,j=1}^{r} B_{ij} N_i N_j.$$

The summation in (1) is taken over the lattice of integer vectors $\mathbf{N} = (N_1, \ldots, N_r)$. The general term of this series depends only on the symmetric part of the $r \times r$ matrix B. From the estimate

$$\Re\langle B\mathbf{N}, \mathbf{N}\rangle \leq -b\langle \mathbf{N}, \mathbf{N}\rangle, \quad b > 0$$

where $-b$ is the largest eigenvalue of the matrix $\Re(B)$ one finds that the series (1) is absolutely convergent, uniformly on compact sets. Thus the function $\theta(\mathbf{z}|B)$ is analytic in the whole space vector space $\mathbb{C}^r$. Let $\mathbf{e}_1, \ldots,$ $\mathbf{e}_r$ be the basis vectors in $\mathbb{C}^r$ with the coordinates $(e_k)_j = \delta_{kj}$. We also introduce vectors $\mathbf{f}_1, \ldots, \mathbf{f}_r$, setting $(f_k)_j = B_{kj}$, $k, j = 1, \ldots, r$. The vector $\mathbf{f}_k$ can also be written in the form $\mathbf{f}_k = B\mathbf{e}_k$.
(i) Show that

$$\theta(\mathbf{z} + 2\pi i \mathbf{e}_k) = \theta(\mathbf{z}). \tag{2}$$

(ii) Show that

$$\theta(\mathbf{z} + \mathbf{f}_k) = \exp\left(-\frac{1}{2}B_{kk} - z_k\right)\theta(\mathbf{z}). \tag{3}$$

Solution 13.   (i) The periodicity of (2) is obvious. The general term of (1) does not change under the shift $\mathbf{z} \to \mathbf{z} + 2\pi i\mathbf{e}_k$.
(ii) We have

$$\theta(\mathbf{z} + \mathbf{f}_k) = \sum_{\mathbf{N} \in \mathbb{Z}^r} \exp\left(\frac{1}{2}\langle B\mathbf{N}, \mathbf{N}\rangle + \langle \mathbf{N}, \mathbf{z} + \mathbf{f}_k\rangle\right).$$

To prove (3) we change the summation index $\mathbf{N}$ in (1) setting $\mathbf{N} = \mathbf{M} - \mathbf{e}_k$, $\mathbf{M} \in \mathbb{Z}^r$. Thus

$$\theta(\mathbf{z} + \mathbf{f}_k) = \sum_{\mathbf{M} \in \mathbb{Z}^r} \exp\left(\frac{1}{2}\langle B\mathbf{M}, \mathbf{M}\rangle - \langle \mathbf{M}, B\mathbf{e}_k\rangle + \frac{1}{2}\langle B\mathbf{e}_k, \mathbf{e}_k\rangle\right.$$
$$\left. + \langle \mathbf{M}, \mathbf{z}\rangle + \langle \mathbf{M}, \mathbf{f}_k\rangle - \langle \mathbf{e}_k, \mathbf{z}\rangle - \langle \mathbf{e}_k, \mathbf{f}_k\rangle\right).$$

It follows that

$$\theta(\mathbf{z} + \mathbf{f}_k) = \exp\left(-\frac{1}{2}B_{kk} - z_k\right) \sum_{\mathbf{M} \in \mathbb{Z}^r} \exp\left(\frac{1}{2}\langle B\mathbf{M}, \mathbf{M}\rangle + \langle \mathbf{M}, \mathbf{z}\rangle\right).$$

Finally $\theta(\mathbf{z} + \mathbf{f}_k) = \exp\left(-\frac{1}{2}B_{kk} - z_k\right)\theta(\mathbf{z})$.

Supplementary Problems

Problem 1.   Let j_ℓ ($\ell = 0, 1, 2, \ldots$) be the spherical Bessel functions. Show that (*Bauer's expansion*)

$$\exp(-i k\mathbf{s} \cdot \mathbf{r}) = 4\pi \sum_{\ell=0}^{\infty} \sum_{m=-\ell}^{m=+\ell} (-i)^\ell Y_\ell^m(\alpha, \beta) Y_\ell^{m*}(\theta, \phi) j_\ell(kr)$$

where α, β denotes the polar angles of the vector $\mathbf{s}$ and θ, ϕ those of $\mathbf{r}$.

Problem 2.   Consider the function

$$f(z) = \sum_{n=1}^{\infty} \frac{x^n}{n^2}.$$

Show that this function admits the integral transform

$$f(x) = -\int_0^x \frac{\ell n(1-s)}{s}ds$$

and satisfies $df/dx = -(1-x)/x$.

Problem 3. Let $\mu > 0$ and

$$\mathbf{R} = \begin{pmatrix} x_1 \\ x_2 \\ x_3 \end{pmatrix}, \qquad \mathbf{R}' = \begin{pmatrix} x_1' \\ x_2' \\ x_3' \end{pmatrix}.$$

(i) Show that

$$\frac{\exp(-|\mathbf{R} - \mathbf{R}'|/\mu)}{|\mathbf{R} - \mathbf{R}'|}$$

can be expressed with Bessel function.

(ii) Consider the functions

$$f_{s,k,n}(\mathbf{R}) = \frac{\sqrt{k}}{2\pi} J_n(kr) e^{in\phi + isx_3}$$

where $0 \le k < \infty$, $-\infty < s < \infty$, and $n = 0, \pm 1, \pm 2, \ldots$. Show that

$$\int_0^{2\pi} \int_{-\infty}^{+\infty} dx_3 \int_0^\infty f_{s,k,n}(\mathbf{R}) \bar{f}_{s',k',n'}(\mathbf{R}) r \, dr = \delta_{nn'} \delta(s - s') \delta(k - k').$$

Problem 4. Consider the eigenvalue problem

$$-\frac{\hbar^2}{2m} \left(\frac{d^2}{dr^2} + \frac{2}{r} \frac{d}{dr} - \frac{L^2}{r^2} \right) u(r) + \frac{1}{2} m\omega^2 r^2 u(r) = E u(r)$$

where $L^2 = \ell(\ell+1)$, $\ell = 0, 1, \ldots$. The boundary condition is $u(r = R) = 0$, where R is the radius of the box. Show that introducing the dimensionless quantities $\xi = r/b$, $\epsilon = E/(\hbar\omega)$, $b^2 = \hbar/(m\omega)$ we arrive at

$$\left(\frac{d^2}{d\xi^2} + \frac{2}{\xi} \frac{d}{d\xi} - \frac{L^2}{\xi^2} - \xi^2 + 2\epsilon \right) u(\xi) = 0.$$

Consider the ansatz $u(\xi) = \xi^\ell \exp(-\xi^2/2) f(\xi)$. Setting $x = \xi^2$ show that f satisfies *Kummer's equation*

$$x\frac{d^2 f}{dx^2} + (\ell + 3/2 - x)\frac{df}{dx} + \frac{1}{2}(\epsilon - \ell - 3/2) f(x) = 0.$$

Show that the solution can be given with the confluent hypergeometric function

$$f(x) = {}_1F_1((\ell + 3/2 - \epsilon)/2; \ell + 3/2; x).$$

Setting $\eta = \frac{1}{2}(\epsilon - \ell - 3/2)$ or $\epsilon = 2\eta + \ell + 3/2$ show that the boundary condition can be written as

$${}_1F_1(-\eta; \ell + 3/2; x_0) = 0, \quad x_0 = (R/b)^2.$$

The solutions of this equation for the discrete η's then provide the eigenvalues of the problem.

Problem 5.   The *Gegenbauer polynomials* $C_n^{(\alpha)}(x)$ $(n = 0, 1, 2, \ldots)$ can be defined with the *generating function*

$$\frac{1}{(1 - 2xt + t^2)^{(\alpha)}} = \sum_{n=0}^{\infty} C_n^{(\alpha)}(x) t^n.$$

(i) Show that the Gegenbauer polynomials satisfy the recurrence relation $C_0^{(\alpha)}(x) = 1$, $C_1^{(\alpha)}(x) = 2\alpha x$

$$C_n^{(\alpha)}(x) = \frac{1}{n}\left(2x(n + \alpha - 1)C_{n-1}^{(\alpha)}(x) - (n + 2\alpha - 2)C_{n-2}^{(\alpha)}(x)\right), \quad n \geq 2.$$

(ii) Show that the Gegenbauer polynomials satisfy the linear second order differential equation

$$(1 - x^2)\frac{d^2 u}{dx^2} - (2\alpha + 1)x\frac{du}{dx} + n(n + 2\alpha)u = 0.$$

Chapter 14

Generalized Functions

The vector space $D(\mathbb{R}^n)$ denotes all infinitely-differentiable functions in $\mathbb{R}$ with compact support. The functions $\phi \in D(\mathbb{R}^n)$ are called test functions. Each linear continuous functional over the vector space of test functions $D(\mathbb{R}^n)$ is called a generalized function.

Instead of considering the space $D(\mathbb{R}^n)$ for the test functions we can also consider the vector space $S(\mathbb{R}^n)$. The vector space $S(\mathbb{R}^n)$ is the set of all infinitely differentiable functions which decrease as $|\mathbf{x}| \to \infty$, together with all their derivatives, faster than any power of $|\mathbf{x}|^{-1}$. These functions are called test functions. Each linear continuous functional over the vector space $S(\mathbb{R}^n$ is called a generalized function (of slow growth). Note that $S(\mathbb{R}^n) \subset L_2(\mathbb{R}^n)$ and the vector space $S(\mathbb{R}^n)$ is dense in the Hilbert space $L_2(\mathbb{R}^n)$.

The *delta function* is defined by

$$(\delta(\mathbf{x}), \phi(\mathbf{x})) := \phi(\mathbf{0})$$
$$(\delta(\mathbf{x} - \mathbf{x}_0), \phi(\mathbf{x})) := \phi(\mathbf{x}_0).$$

The delta function δ is a singular generalized function.

Let ϕ be a test function. Then the derivative of a generalized function T is defined as

$$\left(\frac{\partial T}{\partial x_j}, \phi \right) := - \left(T, \frac{\partial \phi}{\partial x_j} \right).$$

Problem 1. Let $a > 0$. We define

$$f_a(x) := \begin{cases} \dfrac{1}{2a} & \text{for} \quad |x| \leq a \\ 0 & \text{for} \quad |x| > a. \end{cases}$$

(i) Calculate the derivative of f_a in the sense of generalized functions.
(ii) What happens if $a \to +0$?

Solution 1. (i) Using the definition given in the introduction we find

$$\left(\frac{df_a(x)}{dx}, \phi(x) \right) = -\left(f_a(x), \frac{d\phi(x)}{dx} \right).$$

Thus

$$\left(\frac{df_a(x)}{dx}, \phi(x) \right) = -\int_{\mathbb{R}} f_a(x) \frac{d\phi}{dx} dx = -\frac{1}{2a} \int_{-a}^{a} \frac{d\phi}{dx} dx.$$

Finally

$$\left(\frac{df_a(x)}{dx}, \phi(x) \right) = \frac{1}{2a} \left(-\phi(a) + \phi(-a) \right) = \frac{1}{2a} ((\delta(x+a), \phi(x)) - (\delta(x-a), \phi(x))).$$

We can write

$$\frac{df_a(x)}{dx} = \frac{1}{2a} \left(\delta(x + a) - \delta(x - a) \right)$$

where δ denotes the *delta function*.
(ii) In the sense of generalized functions the function f_a tends to the delta function for $a \to +0$.

Problem 2. Show that

$$\frac{1}{2\pi} \sum_{k=-\infty}^{\infty} e^{ikx} \equiv \sum_{k=-\infty}^{\infty} \delta(x - 2k\pi) \tag{1}$$

in the sense of generalized functions. Hint. Expand the 2π-periodic function

$$f(x) = \frac{1}{2} - \frac{x}{2\pi} \tag{2}$$

into a Fourier series. An orthonormal basis in the Hilbert space $L_2(0, 2\pi)$ is given by

$$\left\{ \phi_k(x) := \frac{1}{\sqrt{2\pi}} \exp(ikx) : k \in \mathbb{Z} \right\}.$$

Solution 2. Since we assume that the function f is periodic with respect to 2π we can expand it into a *Fourier expansion*

$$f(x) = \sum_{k \in \mathbb{Z}} \langle f, \phi_k \rangle \phi_k.$$

Calculating the scalar product $\langle f, \phi_k \rangle$ in the Hilbert space $L_2(0, 2\pi)$ we obtain

$$f(x) = -\frac{i}{2\pi} \sum_{\substack{k=-\infty \\ k \neq 0}}^{\infty} \frac{1}{k} e^{ikx}$$

where

$$\langle f, \phi_0 \rangle = \int_0^{2\pi} \left(\frac{1}{2} - \frac{x}{2\pi} \right) \frac{1}{\sqrt{2\pi}} dx = 0.$$

In the sense of generalized functions we can differentiate the right-hand side of (2) term by term. We obtain

$$f' = -\frac{1}{2\pi} + \sum_{k=-\infty}^{\infty} \delta(x - 2k\pi) = \frac{1}{2\pi} \sum_{\substack{k=-\infty \\ k \neq 0}}^{\infty} e^{ikx}.$$

Thus we obtain identity (1). The identity

$$\delta(f(x)) \equiv \sum_n \frac{1}{|f'(x_n)|} \delta(x - x_n)$$

can also be used to find identity (1). Here the sum runs over all the zeros of f and $f'(x_n) \equiv df(x = x_n)/dx$.

Problem 3. Calculate the Fourier transform of $f(t) = \sin(\Omega t)$, $g(t) = \cos(\Omega t)$ in the sense of generalized functions. The Fourier transform for generalized functions is given by

$$(F[f](\omega), \psi(\omega)) := 2\pi(f(t), \phi(t)), \qquad \psi(\omega) := \int_{-\infty}^{+\infty} e^{i\omega t} \phi(t) \, dt.$$

Solution 3. Inserting $f(t) = \sin(\Omega t)$ into the equation for the Fourier transform for generalized function we obtain

$$(F[f](\omega), \psi(\omega)) = 2\pi \int_{-\infty}^{+\infty} \sin(\Omega t) \phi(t) dt.$$

Using the identity $\sin(\Omega t) \equiv (e^{i\Omega t} - e^{-i\Omega t})/(2i)$ we find

$$(F[f](\omega), \psi(\omega)) = \frac{\pi}{i} \int_{-\infty}^{+\infty} e^{i\Omega t} \phi(t) dt - \frac{\pi}{i} \int_{-\infty}^{+\infty} e^{-i\Omega t} \phi(t) dt.$$

From $\psi(\omega)$ we find

$$\psi(-\omega) = \int_{-\infty}^{+\infty} e^{-i\omega t}\phi(t)dt.$$

Therefore

$$(F[f](\omega), \psi(\omega)) = \frac{\pi}{i}[\psi(\Omega) - \psi(-\Omega)].$$

Consequently,

$$(F[f](\omega), \psi(\omega)) = \frac{\pi}{i}((\delta(\omega - \Omega), \psi(\omega)) - (\delta(\omega + \Omega), \psi(\omega))).$$

Thus

$$F[f](\omega) = \frac{\pi}{i}(\delta(\omega - \Omega) - \delta(\omega + \Omega)).$$

Analogously we find the Fourier transform of g

$$F[g](\omega) = \pi(\delta(\omega + \Omega) + \delta(\omega - \Omega)).$$

The Fourier transform of $\exp(bx)$ in the sense of generalized functions is

$$F[\exp(bx)](\omega) = 2\pi\delta(\omega - ib).$$

Using the identities

$$\exp(i\Omega t) \equiv \cos(\Omega t) + i\sin(\Omega t), \quad \exp(-i\Omega t) \equiv \cos(\Omega t) - i\sin(\Omega t)$$

and $b = i\Omega$ and $b = -i\Omega$, we can also find $F[f](\omega)$ and $F[g](\omega)$ from $F[\exp(bx)](\omega)$.

Problem 4.   Let $\mathbf{x} = (x_1, x_2, \ldots, x_n)$. Show that

$$E(\mathbf{x}, t) := \frac{1}{(4\pi a^2 t)^{n/2}} \exp\left(-\frac{|\mathbf{x}|^2}{4a^2 t}\right) \to \delta(\mathbf{x}), \qquad t \to +0 \qquad (1)$$

in the sense of generalized functions, where $n = 1, 2, \ldots$.

Solution 4.   Let $\phi \in S(\mathbb{R}^n)$. Then we have the estimates

$$\left| \int_{\mathbb{R}^n} E(\mathbf{x}, t)[\phi(\mathbf{x}) - \phi(0)]d\mathbf{x} \right| \leq \frac{K}{(4\pi a^2 t)^{n/2}} \int_{\mathbb{R}^n} \exp\left(-\frac{|\mathbf{x}|^2}{4a^2 t}\right) |\mathbf{x}|d\mathbf{x}$$

$$= \frac{K\sigma_n}{(4\pi a^2 t)^{n/2}} \int_0^\infty \exp\left(-\frac{r^2}{4a^2 t}\right) r^n dr$$

$$= K'\sqrt{t} \int_0^\infty \exp(-u^2)u^n du = C\sqrt{t}$$

where $\sigma_n := 2\pi^{n/2}/\Gamma(n/2)$ is the area of the surface of a unit sphere in $\mathbb{R}^n$ and

$$\Gamma(z) := \int_0^\infty t^{z-1}e^{-t}dt, \qquad \Re(z) > 0$$

where Γ denotes the *Gamma function*. For $t > 0$ we have

$$\int_{\mathbb{R}^n} E(\mathbf{x}, t)d\mathbf{x} = \frac{1}{(2a\sqrt{\pi t})^n} \int_{\mathbb{R}^n} \exp\left(-\frac{|\mathbf{x}|^2}{4a^2t}\right) d\mathbf{x}$$

$$= \prod_{j=1}^n \frac{1}{\sqrt{\pi}} \int_{-\infty}^\infty \exp(-\xi_j^2)d\xi_j^2 = 1.$$

Thus we obtain

$$(E(\mathbf{x}, t), \phi(\mathbf{x})) = \int_{\mathbb{R}^n} E(\mathbf{x}, t)\phi(\mathbf{x})d\mathbf{x}$$

$$= \phi(0) \int_{\mathbb{R}^n} E(\mathbf{x}, t)d\mathbf{x} + \int_{\mathbb{R}^n} E(\mathbf{x}, t)[\phi(\mathbf{x}) - \phi(0)]d\mathbf{x} \to \phi(0)$$

$$= (\delta, \phi).$$

Problem 5.   Show that

$$E(\mathbf{x}) = -\frac{e^{ik|\mathbf{x}|}}{4\pi|\mathbf{x}|}$$

satisfies the linear partial differential equation $\Delta E + k^2 E = \delta$ in the sense of generalized functions, where $\mathbf{x} = (x_1, x_2, x_3)$ and

$$|\mathbf{x}| \equiv r = \sqrt{x_1^2 + x_2^2 + x_3^2}$$

and δ is the delta function.

Solution 5.   First we have to show that

$$\Delta \frac{1}{|\mathbf{x}|} = -4\pi\delta(\mathbf{x})$$

in the sense of generalized functions. Let ϕ be a test function. Without loss of generality we can assume that ϕ is spherically symmetric. From the definition of the derivative of a generalized function and using spherical coordinates we obtain

$$\left(\Delta\frac{1}{r}, \phi(r)\right) := \left(\frac{1}{r}, \Delta\phi(r)\right) = \int_{\mathbb{R}^3} \frac{\Delta\phi}{r}r^2 \sin(\theta)drd\theta d\phi. \tag{1}$$

Since ϕ depends only on r, the angle part of Δ can be omitted. We find that the Laplace operator takes the form

$$\Delta := \frac{1}{r^2} \frac{\partial}{\partial r} \left(r^2 \frac{\partial}{\partial r} \right).$$

Inserting this equation into (1) gives

$$\left(\frac{1}{r}, \Delta\phi \right) = 4\pi \int_{r=0}^{\infty} \frac{1}{r} \left(2r \frac{d}{dr}\phi + r^2 \frac{d^2}{dr^2}\phi \right) dr.$$

Thus

$$\left(\frac{1}{r}, \Delta\phi \right) = -4\pi\phi(0) = -4\pi(\delta, \phi)$$

where we have used integration by parts and that $\phi(r) \to 0$ as $r \to \infty$. We can write

$$\Delta \frac{1}{r} = -4\pi\delta.$$

For $r > 0$ we have

$$\frac{\partial}{\partial x_j} \frac{1}{|\mathbf{x}|} = -\frac{x_j}{|\mathbf{x}|^3}, \qquad \frac{\partial}{\partial x_j} e^{ik|\mathbf{x}|} = ik \frac{x_j}{|\mathbf{x}|} e^{ik|\mathbf{x}|}$$

and

$$\Delta e^{ik|\mathbf{x}|} = \left(\frac{2ik}{|\mathbf{x}|} - k^2 \right) e^{ik|\mathbf{x}|}.$$

It follows that

$$(\Delta + k^2) \frac{1}{|\mathbf{x}|} e^{ik|\mathbf{x}|} = -4\pi\delta(r).$$

Problem 6.   Consider the function $H : \mathbb{R} \to \mathbb{R}$

$$H(x) := \begin{cases} 1 & 0 \le x < 1/2 \\ -1 & 1/2 \le x \le 1 \\ 0 & \text{otherwise} \end{cases} .$$

(i) Find the derivative of H in the sense of generalized functions. Obviously H can be considered as a regular functional

$$\int_{\mathbb{R}} H(x)\phi(x)dx.$$

(ii) Find the Fourier transform of H.

Solution 6. (i) We have

$$
\left(\frac{dH}{dx}, \phi\right) := -\left(H, \frac{d\phi}{dx}\right) = -\int_0^{\frac{1}{2}} \frac{d\phi}{dx} dx + \int_{\frac{1}{2}}^1 \frac{d\phi}{dx} dx
$$
$$
= -\phi(1/2) + \phi(0) + \phi(1) - \phi(1/2)
$$
$$
= (\delta(x) + \delta(x-1) - 2\delta(x-1/2), \phi).
$$

(ii) For the Fourier transform we find

$$
\int_{-\infty}^{\infty} e^{ikx} H dx = \int_0^{\frac{1}{2}} e^{ikx} dx - \int_{\frac{1}{2}}^1 e^{ikx} dx = \frac{e^{ikx}}{ik}\Big|_0^{\frac{1}{2}} - \frac{e^{ikx}}{ik}\Big|_{\frac{1}{2}}^1
$$
$$
= \frac{1}{ik}(2e^{\frac{ik}{2}} - 1 - e^{ik}) = \frac{e^{\frac{ik}{2}}}{ik}(2 - (e^{\frac{ik}{2}} + e^{-\frac{ik}{2}}))
$$
$$
= \frac{2ie^{ik/2}}{k}(\cos(k/2) - 1).
$$

Problem 7. Find a particular solution to the linear differential equation

$$
\frac{d^2u}{dt^2} + a\frac{du}{dt} + bu = \delta(t) \qquad a, b \in \mathbb{R} \tag{1}
$$

in the sense of generalized functions.

Solution 7. Consider the ansatz

$$
u(t) = \theta(t)f(t) \tag{2}
$$

where

$$
\theta(t) := \begin{cases} 1 & t \geq 0 \\ 0 \text{ otherwise} \end{cases} \tag{3}
$$

and f is a differentiable function. Let ϕ be a test function $\phi \in S(\mathbb{R})$. Then from (1) we obtain

$$
\left(\frac{d^2u}{dt^2}, \phi\right) + a\left(\frac{du}{dt}, \phi\right) + b(u, \phi) = (\delta(t), \phi) = \phi(0). \tag{4}
$$

Applying differentiation in the sense of generalized functions we obtain from (4)

$$
\left(u, \frac{d^2\phi}{dt^2}\right) - a\left(u, \frac{d\phi}{dt}\right) + b(u, \phi) = \phi(0). \tag{5}
$$

Inserting the ansatz (2) into (5) it follows that

$$
\int_{-\infty}^{\infty} \theta(t)f(t)\frac{d^2\phi}{dt^2}dt - a\int_{-\infty}^{+\infty} \theta(t)f(t)\frac{d\phi}{dt}dt + b\int_{-\infty}^{\infty} \theta(t)f(t)\phi(t)dt = \phi(0).
$$

Applying the property of the step-function (3) leads to

$$\int_0^\infty f(t)\frac{d^2\phi}{dt^2}dt - a\int_0^\infty f(t)\frac{d\phi}{dt}dt + b\int_0^\infty f(t)\phi(t)dt = \phi(0). \qquad (6)$$

Applying integration by parts and taking into account the boundary conditions for ϕ, $\phi(|\infty|) = 0$, $\phi'(|\infty|) = 0$ we find

$$\int_0^\infty f(t)\frac{d\phi}{dt}dt = -f(0)\phi(0) - \int_0^\infty \frac{df}{dt}\phi(t)dt$$

and

$$\int_0^\infty f(t)\frac{d^2\phi}{dt^2}dt = -f(0)\frac{d\phi(0)}{dt} - \int_0^\infty \frac{df}{dt}\frac{d\phi}{dt}dt$$

$$= -f(0)\frac{d\phi(0)}{dt} + \frac{df(0)}{dt}\phi(0) + \int_0^\infty \frac{d^2f}{dt^2}\phi(t)dt.$$

Inserting these two equations into (6) we find

$$\int_0^\infty \left(\frac{d^2f}{dt^2} + a\frac{df}{dt} + bf\right)\phi(t)dt - f(0)\frac{d\phi(0)}{dt} + \frac{df(0)}{dt}\phi(0) + af(0)\phi(0) = \phi(0).$$

To satisfy this equation we have to fulfill the following three conditions

$$\frac{d^2f}{dt^2} + a\frac{df}{dt} + bf = 0, \quad f(0) = 0, \quad \frac{df(0)}{dt} + af(0) = 1.$$

Problem 8. Use the identity

$$\sum_{k=-\infty}^{\infty} \delta(x - 2\pi k) \equiv \frac{1}{2\pi}\sum_{k=-\infty}^{\infty} e^{ikx} \qquad (1)$$

to show that

$$2\pi\sum_{k=-\infty}^{\infty} \phi(2\pi k) = \sum_{k=-\infty}^{\infty} F[\phi](k) \qquad (2)$$

where F denotes the Fourier transform and ϕ is a test function. Equation (2) is called *Poisson's summation formula*.

Solution 8. We rewrite (1) in the form

$$2\pi\sum_{k=-\infty}^{\infty} \delta(x - 2\pi k) = \sum_{k=-\infty}^{\infty} F[\delta(x - k)].$$

Applying this equation to $\phi \in S(\mathbb{R})$, we obtain

$$2\pi \left(\sum_{k=-\infty}^{\infty} \delta(x - 2\pi k), \phi \right) = 2\pi \sum_{k=-\infty}^{\infty} \phi(2\pi k) = \left(\sum_{k=-\infty}^{\infty} F[\delta(x - k)], \phi \right).$$

Finally we arrive at

$$2\pi \left(\sum_{k=-\infty}^{\infty} \delta(x - 2\pi k), \phi \right) = \sum_{k=-\infty}^{\infty} (\delta(x - k), F[\phi]) = \sum_{k=-\infty}^{\infty} F[\phi](k).$$

This proves Poisson's summation formula.

Problem 9. The *Hermite polynomials* are defined by

$$H_n(x) := (-1)^n e^{x^2} \frac{d^n}{dx^n} e^{-x^2}, \qquad n = 0, 1, 2, \ldots \tag{1}$$

where $H_0(x) = 1$, $H_1(x) = 2x$, $H_2(x) = 4x^2 - 2$ etc. An orthonormal basis in the Hilbert space $L_2(\mathbb{R})$ is given by

$$B := \left\{ \frac{(-1)^n}{2^{n/2}\sqrt{n!}\sqrt[4]{\pi}} e^{x^2/2} \frac{d^n}{dx^n} e^{-x^2} \; : \; n = 0, 1, 2, \ldots \right\}. \tag{2}$$

Use the completeness relation to find an expansion of a function $f \in L_2(\mathbb{R})$.

Solution 9. Using the basis (2) and the Hermite polynomials (1) the *completeness relation* is given by

$$\sum_{n=0}^{\infty} \frac{1}{2^n n!\sqrt{\pi}} e^{-(x^2 + x'^2)/2} H_n(x) H_n(x') = \delta(x - x'), \qquad x, x' \in \mathbb{R}.$$

Using the completeness relation and the properties of the delta function we obtain

$$f(x) = \sum_{n=0}^{\infty} \frac{1}{2^n n!\sqrt{\pi}} H_n(x) e^{-x^2/2} \int_{-\infty}^{\infty} f(x') H_n(x') e^{-x'^2/2} dx'. \tag{4}$$

Problem 10. The probability amplitude for the three-dimensional transition of a particle with mass m from point $\mathbf{r}_0$ at the time t_0 to the point $\mathbf{r}_N$ at time t_N is given by the kernel

$$K(\mathbf{r}_N, t_N; \mathbf{r}_0, t_0) =$$

$$\lim_{N \to \infty} \prod_{n=1}^{N} \left(\frac{m}{2\pi i \hbar (t_n - t_{n-1})} \right)^{3/2} \int \cdots \int d^3\mathbf{r}_1 \ldots d^3\mathbf{r}_{N-1} \exp((i/\hbar)S(\mathbf{r}_N, \mathbf{r}_0)). \tag{1}$$

In a compact form one writes

$$K(\mathbf{r}_N, t_N; \mathbf{r}_0, t_0) = \int_0^N \exp((i/\hbar)S(\mathbf{r}_N, \mathbf{r}_0)D\mathbf{r}(t). \tag{2}$$

The integrals extend over all paths in three-dimensional coordinate space and S is the *action* of the corresponding classical system. Thus

$$S(\mathbf{r}_N, \mathbf{r}_0) = \int_{t_0}^{t_N} L(\dot{\mathbf{r}}, \mathbf{r}, t)dt \tag{3}$$

where $L(\dot{\mathbf{r}}, \mathbf{r}, t)$ represents the *Lagrange function*. We can also write (3) as

$$S(\mathbf{r}_N, \mathbf{r}_0) = \sum_{n=1}^{N} (t_n - t_{n-1}) L\left(\frac{\mathbf{r}_n - \mathbf{r}_{n-1}}{t_n - t_{n-1}}, \mathbf{r}_n, t_n \right). \tag{4}$$

The wave function ψ satisfies the integral equation

$$\psi(\mathbf{r}, t) = \int K(\mathbf{r}, t; \mathbf{r}_0, t_0)\psi(\mathbf{r}_0, t_0)d^3\mathbf{r}_0. \tag{5}$$

The momentum amplitude is given by the Fourier transform of the wave function $(\mathbf{p} = \hbar\mathbf{k})$

$$\phi(\mathbf{k}, t) = \int \exp(-2\pi i \mathbf{k} \cdot \mathbf{r})\psi(\mathbf{r}, t)d^3\mathbf{r}. \tag{6}$$

The momentum amplitude satisfies a similar integral equation as the wave function

$$\phi(\mathbf{k}, t) = \int \kappa(\mathbf{k}, t; \mathbf{k}_0, t_0)\phi(\mathbf{k}_0, t_0)d^3\mathbf{k}_0.$$

The momentum kernel κ represents the probability amplitude of the transition from the wave vector $\mathbf{k}_0$ at the time t_0 to the wave vector $\mathbf{k}$ at the time t. We have

$$\kappa(\mathbf{k}, t; \mathbf{k}_0, t_0) = \int \int \exp(-2\pi i \mathbf{k} \cdot \mathbf{r})K(\mathbf{r}, t; \mathbf{r}_0, t_0) \exp(2\pi \mathbf{k}_0 \cdot \mathbf{r}_0)d^3\mathbf{r}_0 d^3\mathbf{r}.$$

Describe the motion of a particle in a local potential $V(\mathbf{r}, t)$. The Lagrange function is given by

$$L(\dot{\mathbf{r}}, \mathbf{r}, t) = \frac{1}{2}m\dot{\mathbf{r}}^2 - V(\mathbf{r}, t)$$

where $V(\mathbf{r}, t)$ represents the potential energy.

Solution 10. The action S becomes

$$S(\mathbf{r}_N, \mathbf{r}_0) = \sum_{n=1}^{N} (t_n - t_{n-1}) \left(\frac{1}{2} m \left(\frac{\mathbf{r}_n - \mathbf{r}_{n-1}}{t_n - t_{n-1}} \right)^2 - V(\mathbf{r}_n, t_n) \right).$$

Inserting this expression into (1) yields

$$K(\mathbf{r}_N, t_N; \mathbf{r}_0, t_0) = \lim_{N \to \infty} \prod_{n=1}^{N} \left(\frac{m}{ih(t_n - t_{n-1})} \right)^{3/2} \int \cdots \int d^3\mathbf{r}_1 \ldots d^3\mathbf{r}_{N-1}$$

$$\times \exp\left((i\pi m/h) \sum_{n=1}^{N} \frac{(\mathbf{r}_n - \mathbf{r}_{n-1})^2}{(t_n - t_{n-1})} \right)$$

$$\times \exp\left(-(2\pi i/h) \sum_{n=1}^{N} (t_n - t_{n-1}) V(\mathbf{r}_n, t_n) \right).$$

Using the identity

$$\left(\frac{m}{iht} \right)^{3/2} \exp(i\pi m r^2/ht) \equiv \int d^3\mathbf{k} \exp(2\pi i(\mathbf{k} \cdot \mathbf{r} - ht\mathbf{k}^2/2m))$$

where the integration extends over the three-dimensional momentum space, $K(\mathbf{r}_N, t_N; \mathbf{r}_0, t_0)$ can be written as

$$K(\mathbf{r}_N, t_N; \mathbf{r}_0, t_0) = \lim_{N \to \infty} \int \cdots \int d^3\mathbf{r}_1 \cdots d^3\mathbf{r}_{N-1} d^3\mathbf{k}_0 \cdots d^3\mathbf{k}_{N-1}$$

$$\times \exp\left(2\pi i \sum_{n=1}^{N} \mathbf{k}_{n-1} \cdot (\mathbf{r}_n - \mathbf{r}_{n-1}) - \left(\frac{h}{2m} \right) (t_n - t_{n-1}) \mathbf{k}_{n-1}^2 \right)$$

$$\times \exp\left(-2\pi i/h \sum_{n=1}^{N} (t_n - t_{n-1}) V(\mathbf{r}_n, t_n) \right).$$

From (3) we obtain

$$\kappa(\mathbf{k}_N, t_N; \mathbf{k}_0', t_0) = \lim_{N \to \infty} \int \cdots \int d^3\mathbf{r}_0 \cdots d^3\mathbf{r}_N d^3\mathbf{k}_0 \cdots d^3\mathbf{k}_{N-1}$$

$$\times \exp(-2\pi i(\mathbf{k}_0 - \mathbf{k}_0') \cdot \mathbf{r}_0) \exp\left(-2\pi i \sum_{n=1}^{N} (\mathbf{k}_n - \mathbf{k}_{n-1}) \cdot \mathbf{r}_n \right)$$

$$\times \exp\left(-(2\pi i/h) \sum_{n=1}^{N} (t_n - t_{n-1}) V(\mathbf{r}_n, t_n) \right)$$

$$\times \exp\left(-(\pi i h/m) \sum_{n=1}^{N} (t_n - t_{n-1}) \mathbf{k}_{n-1}^2 \right).$$

The integration over $\mathbf{r}_0$ can be carried out giving the delta function

$$\int d^3 \mathbf{r}_0 \exp(-2\pi i (\mathbf{k}_0 - \mathbf{k}'_0) \cdot \mathbf{r}_0) = \delta(\mathbf{k}_0 - \mathbf{k}'_0).$$

Using the property of the delta function we can also calculate the integral over $\mathbf{k}_0$. We obtain

$$\kappa(\mathbf{k}_N, t_N; \mathbf{k}_0, t_0) = \lim_{N \to \infty} \int \cdots \int d^3 \mathbf{r}_1 \cdots d^3 \mathbf{r}_N d^3 \mathbf{k}_1 \cdots \mathbf{k}_{N-1}$$

$$\times \exp\left(-2\pi i \sum_{n=1}^{N} (\mathbf{k}_n - \mathbf{k}_{n-1}) \cdot \mathbf{r}_n\right)$$

$$\times \exp\left(-(2\pi i/h) \sum_{n=1}^{N} (t_n - t_{n-1}) V(\mathbf{r}_n, t_n)\right)$$

$$\times \exp\left(-(i\pi h/m) \sum_{n=1}^{N} (t_n - t_{n-1}) \mathbf{k}_{n-1}^2\right).$$

Defining

$$F(\mathbf{k}_n - \mathbf{k}_{n-1}) :=$$

$$\int d^3 \mathbf{r}_n \exp(-2\pi i (\mathbf{k}_n - \mathbf{k}_{n-1}) \cdot \mathbf{r}_n) \exp(-(2\pi i/h)(t_n - t_{n-1}) V(\mathbf{r}_n, t_n))$$

we can write

$$\kappa(\mathbf{k}_N, t_N; \mathbf{k}_0, t_0) =$$

$$\lim_{N \to \infty} \int \cdots \int d^3 \mathbf{k}_1 \cdots d^3 \mathbf{k}_{N-1} \prod_{n=1}^{N} F(\mathbf{k}_n - \mathbf{k}_{n-1}) e^{-(i\pi h/m)((t_n - t_{n-1})\mathbf{k}_{n-1}^2)}.$$

The physical interpretation of this formula is that the transmission of the wave, with wave vector $\mathbf{k}_{n-1}$ through the n-th slice can be described by the propagation factor exp multiplied by the probability amplitude

$$F(\mathbf{k}_n - \mathbf{k}_{n-1})$$

for the scattering into the new wave vector $\mathbf{k}_n$. Integration over all possible wave vectors and time slices gives the expression for the momentum kernel.

Problem 11. Given a function (signal) $f(\mathbf{t}) = f(t_1, t_2, \ldots, t_n) \in L_2(\mathbb{R}^n)$ of n real variables $\mathbf{t} = (t_1, t_2, \ldots, t_n)$. We define the *symplectic tomogram* associated with the square integrable function f

$$w(\mathbf{X}, \boldsymbol{\mu}, \boldsymbol{\nu}) = \prod_{k=1}^{n} \frac{1}{2\pi |\nu_k|} \left| \int_{\mathbb{R}^n} dt_1 \cdots dt_n f(\mathbf{t}) \exp\left(\sum_{j=1}^{n} \left(\frac{i\mu_j}{2\nu_j} t_j^2 - \frac{iX_j}{\nu_j} t_j\right)\right) \right|^2$$

where ($\nu_j \neq 0$ for $j = 1, 2, \ldots, n$)

$$\mathbf{X} = (X_1, X_2, \ldots, X_n), \quad \boldsymbol{\mu} = (\mu_1, \mu_2, \ldots, \mu_n), \quad \boldsymbol{\nu} = (\nu_1, \nu_2, \ldots, \nu_n).$$

(i) Prove the equality

$$\int_{\mathbb{R}^n} w(\mathbf{X}, \boldsymbol{\mu}, \boldsymbol{\nu}) d\mathbf{X} = \int_{\mathbb{R}^n} |f(\mathbf{t})|^2 d\mathbf{t} \tag{1}$$

for the special case $n = 1$. The tomogram is the probability distribution function of the random variable $\mathbf{X}$. This probability distribution function depends on $2n$ extra real parameters $\boldsymbol{\mu}$ and $\boldsymbol{\nu}$.

(ii) The map of the function $f(\mathbf{t})$ onto the tomogram $w(\mathbf{X}, \boldsymbol{\mu}, \boldsymbol{\nu})$ is invertible. The square integrable function $f(\mathbf{t})$ can be associated with the density matrix

$$\rho_f(\mathbf{t}, \mathbf{t}') = f(\mathbf{t}) f^*(\mathbf{t}').$$

This density matrix can be mapped onto the *Ville-Wigner function*

$$W(\mathbf{q}, \mathbf{p}) = \int_{\mathbb{R}^n} \rho_f \left(\mathbf{q} + \frac{\mathbf{u}}{2}, \mathbf{q} - \frac{\mathbf{u}}{2} \right) e^{-i\mathbf{p} \cdot \mathbf{u}} d\mathbf{u}.$$

Show that this map is invertible.

(iii) How is the tomogram $w(\mathbf{X}, \boldsymbol{\mu}, \boldsymbol{\nu})$ related to the Ville-Wigner function?

(iv) Show that the Ville-Wigner function can be reconstructed from the function $w(\mathbf{X}, \boldsymbol{\mu}, \boldsymbol{\nu})$.

(v) Show that the density matrix $f(\mathbf{t}) f^*(\mathbf{t}')$ can be found from $w(\mathbf{X}, \boldsymbol{\mu}, \boldsymbol{\nu})$.

Solution 11. (i) Since

$$\frac{1}{2\pi\nu} \left| \int_{\mathbb{R}} dt f(t) e^{i\mu t^2 / (2\nu) - iXt/\nu} \right|^2$$

$$= \frac{1}{2\pi\nu} \int_{\mathbb{R}} \int_{\mathbb{R}} dt dt' f(t) f^*(t') e^{i\mu(t^2 - (t')^2)/(2\nu)} e^{iX(t' - t)/\nu}$$

and in the sense of generalized functions ($\nu \neq 0$)

$$\int_{\mathbb{R}} e^{iX(t' - t)/\nu} dX = 2\pi \delta((t' - t)/\nu) = 2\pi\nu \delta(t' - t)$$

we find, applying the property of the delta function, that equality (1) holds.

(ii) Since we have a Fourier transform the inverse transform is given by

$$\rho(\mathbf{t}, \mathbf{t}') = \frac{1}{(2\pi)^n} \int_{\mathbb{R}^n} W \left(\frac{\mathbf{t} + \mathbf{t}'}{2}, \mathbf{p} \right) e^{i\mathbf{p} \cdot (\mathbf{t} - \mathbf{t}')} d\mathbf{p}.$$

(iii) Using a delta function we can write

$$w(\mathbf{X}, \boldsymbol{\mu}, \boldsymbol{\nu}) = \int_{\mathbb{R}^n} W(\mathbf{q}, \mathbf{p}) \prod_{k=1}^{n} \delta(X_k - \mu_k q_k - \nu_k p_k) \frac{dp_k dq_k}{2\pi}.$$

(iv) The Ville-Wigner function can be reconstructed from the function $w(\mathbf{X}, \boldsymbol{\mu}, \boldsymbol{\nu})$ via

$$W(\mathbf{p}, \mathbf{q}) = \frac{1}{(2\pi)^n} \int_{\mathbb{R}^n} w(\mathbf{X}, \boldsymbol{\mu}, \boldsymbol{\nu}) \prod_{k=1}^{n} e^{i(X_k - \mu_k q_k - \nu_k p_k)} dX_k d\mu_k d\nu_k.$$

(v) The density matrix $f(\mathbf{t})f^*(\mathbf{t}')$ can be found from $w(\mathbf{X}, \boldsymbol{\mu}, \boldsymbol{\nu})$ as

$$f(\mathbf{t})f^*(\mathbf{t}') = \frac{1}{(2\pi)^n} \int_{\mathbb{R}^n} w(\mathbf{X}, \boldsymbol{\mu}, \mathbf{t}-\mathbf{t}') \prod_{k=1}^{n} \exp\left(i\left(X_k - \mu_k \frac{t_k + t'_k}{2}\right)\right) dX_k d\mu_k.$$

Problem 12.   Consider the one-dimensional Schrödinger equation ($c > 0$)

$$-\frac{\hbar^2}{2m} \frac{d^2\psi}{dx^2} + c\delta^{(n)}(x)\psi = E\psi$$

where $\delta^{(n)}$ ($n = 0, 1, 2, \dots$) denotes the n-th derivative of the delta function. Derive the joining conditions on the wave function ψ.

Solution 12.   Let $\epsilon > 0$. We define

$$\overline{f}(0) := \frac{1}{2}(f(0^+) + f(0^-)).$$

For functions f that are discontinuous at the origin the expression

$$\int_{-\epsilon}^{\epsilon} \delta(x)f(x)dx$$

is in general not well-defined. If we assume that the delta function is the limit of a sequence of even functions, then

$$\int_{-\epsilon}^{\epsilon} \delta(x)f(x)dx = \overline{f}(0).$$

This we assume in the following. Integrating from $-\epsilon$ to ϵ the eigenvalue equation yields

$$-\frac{\hbar^2}{2m}[\psi'(\epsilon) - \psi'(-\epsilon)) + c\int_{-\epsilon}^{\epsilon} \left(\left(\frac{d}{dx}\right)^n \delta(x)\right) \psi(x)dx = E\int_{-\epsilon}^{\epsilon} \psi(x)dx.$$

Using integration by parts n times and using the fact that all boundary terms vanish yields

$$\int_{-\epsilon}^{\epsilon} \left(\left(\frac{d}{dx}\right)^n \delta(x) \right) \psi(x) dx = (-1)^n \int_{-\epsilon}^{\epsilon} \delta(x) \left(\left(\frac{d}{dx}\right)^n \psi(x) \right) dx$$

$$= (-1)^n \overline{\psi}^{(n)}(0).$$

In the limit $\epsilon \to 0$ we have

$$\lim_{\epsilon \to 0} \int_{-\epsilon}^{\epsilon} \psi(\epsilon) dx = 0.$$

Thus we have the first boundary condition

$$\Delta \psi' = (-1)^n \frac{2mc}{\hbar^2} \overline{\psi}^{(n)}(0).$$

Furthermore integrating the eigenvalue equation from $-L$ (L positive) to x yields

$$-\frac{\hbar^2}{2m}(\psi'(x) - \psi'(-L)) + c \int_{-L}^{x} \left(\left(\frac{d}{ds}\right)^n \delta(s) \right) \psi(s) ds = E \int_{-L}^{x} \psi(s) ds. \quad (1)$$

We integrate by parts and find

$$\int_{-L}^{x} \left(\left(\frac{d}{ds}\right)^n \delta(s) \right) \psi(s) ds = \delta^{(n-1)}(x)\psi(x) - \delta^{(n-2)}(x)\psi'(x)$$

$$+ \delta^{(n-3)}(x)\psi^{(2)}(x) - \cdots + (-1)^{n-1}\delta(x)\psi^{(n-1)}(x) + (-1)^n \theta(x)\overline{\psi}^{(n)}(0)$$

where now the upper boundary terms are nonzero. Integrating (1) from $-\epsilon$ to ϵ and taking the limit $\epsilon \to 0$ we obtain

$$-\frac{\hbar^2}{2m}\Delta\psi + c \left(\int_{-\epsilon}^{\epsilon} \delta^{(n-1)}(x)\psi(x) dx - \int_{-\epsilon}^{\epsilon} \delta^{(n-2)}(x)\psi'(x) dx \right.$$

$$\left. + \int_{-\epsilon}^{\epsilon} \delta^{(n-3)}(x)\psi^{(2)}(x) dx - \cdots + (-1)^{n-1} \int_{-\epsilon}^{\epsilon} \delta(x)\psi^{(n-1)}(x) dx \right) = 0.$$

Using once more integration by parts we obtain for the expression in the parenthesis $(-1)^{n-1} n \overline{\psi}^{(n-1)}(0)$. Thus we have the second boundary condition

$$\Delta\psi = (-1)^{n-1} \frac{2mc}{\hbar^2} n \overline{\psi}^{(n-1)}(0).$$

For $n = 0$ we have

$$\Delta\psi = 0, \qquad \Delta\psi' = \frac{2mc}{\hbar^2}\psi(0)$$

and for $n = 1$ we have

$$\Delta\psi = \frac{2mc}{\hbar^2}\overline{\psi}(0), \qquad \Delta\psi' = -\frac{2mc}{\hbar^2}\overline{\psi}'(0).$$

Problem 13.   Any $SU(2)$ matrix A can be written as $(x_0, x_1, x_2, x_3 \in \mathbb{R})$

$$A = \begin{pmatrix} x_0 - ix_3 & -ix_1 - x_2 \\ x_2 - ix_1 & x_0 + ix_3 \end{pmatrix}, \qquad x_0^2 + x_1^2 + x_2^2 + x_3^2 = 1 \qquad (1)$$

i.e. $\det(A) = 1$. Using *Euler angles* α, β, γ the matrix can also be written as

$$A = \begin{pmatrix} \cos(\beta/2)e^{i(\alpha+\gamma)/2} & -\sin(\beta/2)e^{i(\alpha-\gamma)/2} \\ \sin(\beta/2)e^{-i(\alpha-\gamma)/2} & \cos(\beta/2)e^{-i(\alpha+\gamma)/2} \end{pmatrix}. \qquad (2)$$

(i) Show that the invariant measure dg of $SU(2)$ can be written as

$$dg = \frac{1}{\pi^2}\delta(x_0^2 + x_1^2 + x_2^2 + x_3^2 - 1)dx_0 dx_1 dx_2 dx_3$$

where δ is the Dirac delta function.
(ii) Show that dg is normalized, i.e.

$$\int dg = 1.$$

(iii) Using (1) and (2) find $x_1(\alpha, \beta, \gamma)$, $x_2(\alpha, \beta, \gamma)$, $x_3(\alpha, \beta, \gamma)$. Find the Jacobian determinant.
(iv) Using the results from (iii) show that the invariant measure can be written as

$$\frac{1}{16\pi^2}\sin(\beta)d\alpha d\beta d\gamma.$$

Solution 13.   (i) Let $B \in SU(2)$. Under left multiplication $A' = BA$ is expressed in terms of the new variables x_0', x_1', x_2' and x_3'. By regarding $x_0 + ix_3$ and $x_2 - ix_1$ as two independent complex variables, we find that the Jacobian determinant of changing variables from (x_0, x_1, x_2, x_3) to (x_0', x_1', x_2', x_3') is simply $\det(B) = 1$. It follows that dg is invariant.
(ii) Let $r^2 := x_1^2 + x_2^2 + x_3^2$. We have

$$\int dg = \frac{1}{2\pi^2}\int_{|r|\leq 1} \frac{dx_1 dx_2 dx_3}{\sqrt{1 - r^2}} = \frac{2}{\pi}\int_0^1 \frac{r^2 dr}{\sqrt{1 - r^2}} = 1$$

where we used spherical coordinates for the angle integration which provides 4π and

$$\int \frac{x^2 dx}{\sqrt{a^2 - x^2}} = -\frac{x\sqrt{a^2 - x^2}}{2} + \frac{a^2}{2}\arcsin\left(\frac{x}{a}\right).$$

(iii) We find

$$x_1(\alpha, \beta, \gamma) = \sin(\beta/2)\sin((\alpha - \gamma)/2)$$
$$x_2(\alpha, \beta, \gamma) = \sin(\beta/2)\cos((\alpha - \gamma)/2)$$
$$x_3(\alpha, \beta, \gamma) = -\cos(\beta/2)\sin((\alpha + \gamma)/2).$$

For the Jacobian determinant we obtain

$$\left| \frac{\partial(x_1, x_2, x_3)}{\partial(\alpha, \beta, \gamma)} \right| = -\frac{1}{4}\sin(\beta/2)\cos^2(\beta/2)\cos((\alpha + \gamma)/2).$$

(iv) Since $r^2 = x_1^2 + x_2^2 + x_3^2 = 1 - \cos^2((\alpha + \gamma)/2)\cos^2(\beta/2)$ we have

$$\sqrt{1 - r^2} = \sqrt{1 - (x_1^2 + x_2^2 + x_3^2)} = \cos((\alpha + \gamma)/2)\cos(\beta/2).$$

Furthermore $\sin(\beta/2)\cos(\beta/2) \equiv \frac{1}{2}\sin(\beta)$. Using these results and the results from (ii) and (iii) we obtain

$$dg = \frac{1}{2\pi^2}\frac{dx_1 dx_2 dx_3}{\sqrt{1 - r^2}} = \frac{1}{16\pi^2}\sin(\beta)d\alpha d\beta d\gamma.$$

Problem 14. What charge distribution $\rho(r)$ does the spherical symmetric potential

$$V(r) = \frac{e^{-\mu r}}{r}$$

give? For $r \neq 0$ *Poisson's equation* in spherical coordinates is given by

$$\Delta V(\mathbf{r}) = \frac{1}{r}\frac{d^2}{dr^2}(rV(\mathbf{r})) + R(\theta, \phi)V(\mathbf{r}) = -4\pi\rho(\mathbf{r})$$

where $R(\theta, \phi)$ is the differential operator depending on the angles θ, ϕ.

Solution 14. Since $V(r)$ is spherical symmetric we have $R(\theta, \phi)V(r) = 0$ and we arrive at

$$\rho(r) = -\frac{1}{4\pi r}\frac{d^2}{dr^2}e^{-\mu r} = -\frac{\mu^2 e^{-\mu r}}{4\pi r}$$

for $r \neq 0$. For $r \to 0$ we have $V(r) \to \frac{1}{r}$. Since

$$\Delta\frac{1}{r} = -4\pi\delta(r)$$

we obtain

$$\rho(r) = \delta(r) - \frac{\mu^2}{4\pi}\frac{e^{-\mu r}}{r}.$$

Problem 15. Find the solution of the one-dimensional diffusion equation

$$\frac{\partial u(x,t)}{\partial t} = D\frac{\partial^2 u(x,t)}{\partial x^2}$$

together with the initial condition $u(x,0) = \delta(x)$.

Solution 15. Applying the *Fourier transform*

$$\tilde{u}(k,t) = \int_{\mathbb{R}} u(x,t)e^{ikx}dx$$

the diffusion equation takes the form

$$\frac{\partial \tilde{u}(k,t)}{\partial t} = -Dk^2\tilde{u}(k,t)$$

with the solution $\tilde{u}(k,t) = Ae^{-Dk^2 t}$. The integration constant is found from the initial condition $u(x,0) = \delta(x)$. We obtain $A = 1$. Taking the inverse Fourier transform we arrive at

$$u(x,t) = \frac{1}{2\pi}\int_{\mathbb{R}} e^{-ikx}e^{-Dk^2 t}dk = \frac{1}{\sqrt{4\pi Dt}}e^{-x^2/(4Dt)}.$$

Supplementary Problems

Problem 1. Let $r \geq 0$. Consider the transformation (Kustaanheimo and Stiefel)

$$u_1(r,\theta,\alpha,\phi) = \sqrt{r}\sin(\theta/2)\cos((\alpha+\phi)/2)$$
$$u_2(r,\theta,\alpha,\phi) = \sqrt{r}\cos(\theta/2)\sin((\alpha-\phi)/2)$$
$$u_3(r,\theta,\alpha,\phi) = \sqrt{r}\cos(\theta/2)\cos((\alpha-\phi)/2)$$
$$u_4(r,\theta,\alpha,\phi) = \sqrt{r}\sin(\theta/2)\sin((\alpha+\phi)/2)$$

and the three-dimensional *spherical coordinates*

$$x_1(r,\theta,\phi)=r\sin(\theta)\cos(\phi), \quad x_2(r,\theta,\phi)=r\sin(\theta)\sin(\phi), \quad x_3(r,\theta,\phi)=r\cos(\theta).$$

The range of α is from 0 to 4π. Note that $x_1^2 + x_2^2 + x_3^2 = r^2$. Show that

$$r = \sum_{j=1}^{4} u_j^2 = u^2.$$

Show that the Jacobian determinant of the transformation

$$(r,\theta,\phi,\alpha) \rightarrow (u_1,u_2,u_3,u_4)$$

is given by $r\sin(\theta)/16$. Show that

$$\int_0^R dr\, r^2 \int_{-1}^{+1} d(\cos(\theta)) \int_0^{2\pi} d\phi \int_0^{4\pi} d\alpha = 16.$$

Show that the Dirac delta functions are related by

$$\delta^3(\mathbf{r} - \mathbf{r}')\delta(\alpha - \alpha') = \frac{1}{16u^2}\delta^4(\mathbf{u} - \mathbf{u}').$$

Show that

$$\delta^3(\mathbf{r} - \mathbf{r}') = \frac{1}{16}\int_0^{4\pi} d\alpha \frac{\delta^4(\mathbf{u} - \mathbf{u}')}{u^2}.$$

Show that

$$\nabla^2 + \frac{1}{r^2\sin^2(\theta)}\frac{\partial^2}{\partial\alpha^2} = \frac{1}{4u^2}\sum_{j=1}^{4}\frac{\partial^2}{\partial u_j^2}$$

where ∇^2 is the three dimensional Laplacian expressed in spherical coordinates.

Problem 2. Show that utilizing *spherical coordinates*

$$\delta(x_1 - x_1', x_2 - x_2', x_3 - x_3') = \frac{\delta(r - r')\delta(\theta - \theta')\delta(\phi - \phi')}{r^2\sin(\theta)}.$$

Problem 3. Let $\delta(x)$ be the Dirac delta function. Consider the eigenvalue problem

$$\frac{d^2u}{dx^2} + c(\delta(x - a) + \delta(x + a))u = K^2u$$

where $K^2 = -2mE/\hbar^2$ and $E < 0$. Show that the solution is given by

$$u(x) = A\exp(-K|(x + a)|) + B\exp(-K|(x - a)|).$$

Show that the wave function $u(x)$ is continuous everywhere. Show that the derivative of the wave function is not continuous.

Problem 4. Show that the one-dimensional *Fokker-Planck equation*

$$\frac{\partial u}{\partial t} = \frac{\partial}{\partial x}\left(\left(\alpha a - \frac{\beta}{2x}\right)u\right) + \frac{\beta}{2}\frac{\partial^2 u}{\partial x^2}$$

can be solved with the separation ansatz $u(x, t) = X(x)T(t)$. Find the solution and first moment as a function of t. Assume that the initial condition $u(x, 0) = \delta(x)$ and the boundary condition $u(\pm\infty, t) = 0$.

Problem 5.   Show that in the sense of generalized functions

$$\delta(t) = \frac{1}{2\pi} \int_{\mathbb{R}} e^{-i\omega t} d\omega$$

$$\lim_{\epsilon \to 0} \frac{1}{\omega' - \omega \mp i\epsilon} = P\left(\frac{1}{\omega' - \omega}\right) \pm \delta(\omega' - \omega)$$

$$\delta(\omega) = \lim_{\epsilon \to 0} \frac{\epsilon}{\omega^2 + \epsilon^2}.$$

Problem 6.   (i) Show that the Dirac delta function can be defined as

$$\delta(x) = \lim_{\epsilon \to 0} \frac{1}{2\pi i}\left(\left(\frac{1}{x - i\epsilon}\right) - \left(\frac{1}{x + i\epsilon}\right)\right).$$

(ii) Let $\lambda \in \mathbb{R}$. Show that $x\delta(x - \lambda) = \lambda\delta(x - \lambda)$.

(iii) Study the equation

$$\frac{1}{i}\frac{df}{dx} = \lambda f \;\Rightarrow\; f(x) = e^{i\lambda x}$$

in the sense of generalized functions.

(iv) Show that in the sense of generalized functions

$$\delta(x) = \frac{1}{\pi}\lim_{n \to \infty}\frac{n}{1 + n^2(x - a)^2}.$$

(v) Show that

$$\delta(\sin(x)) = \sum_{n=-\infty}^{\infty}\frac{\delta(x - n\pi)}{|\cos(n\pi)|} = \sum_{n=-\infty}^{\infty}\delta(x - n\pi).$$

(vi) Let $a \neq 0$. Show that

$$\delta(ax + b) = \frac{1}{|a|}\delta(x + b/|a|).$$

(vii) Show that

$$\frac{1}{2\pi}\sum_{n=-\infty}^{\infty}\left(e^{in(\sigma - \bar{\sigma})} + e^{in(\sigma + \bar{\sigma})}\right) =$$

$$\sum_{n=-\infty}^{\infty}\left(\delta(\sigma - \bar{\sigma} + 2n\pi) + \delta(\sigma + \bar{\sigma} + 2n\pi)\right).$$

Chapter 15

Groups and Symmetries

A *group* G is a set of objects $\{\,a, b, c, \ldots\,\}$ (not necessarily countable) together with a binary operation which associates with any ordered pair of elements a, b in G a third element ab in G (closure). The binary operation (called group multiplication) is subject to the following requirements:

1) There exists an element e in G called the *identity element* such that $eg = ge = g$ for all $g \in G$.
2) For every $g \in G$ there exists an *inverse element* g^{-1} in G such that $gg^{-1} = g^{-1}g = e$.
3) *Associative law.* The identity $(ab)c = a(bc)$ is satisfied for all $a, b, c \in G$.

If $ab = ba$ for all $a, b \in G$ we call the group *commutative*. If G has a finite number of elements it has *finite order* $n(G)$, where $n(G)$ is the number of elements. Otherwise, G has infinite order.

Groups have matrix representations with invertible $n \times n$ matrices and matrix multiplication as group multiplication. The identity element is the identity matrix. The inverse element is the inverse matrix. An important subgroup is the set of unitary matrices U, where $U^* = U^{-1}$.

Let $(G_1, *)$ and $(G_2, \circ)$ be groups. A function $f : G_1 \rightarrow G_2$ with

$$f(a * b) = f(a) \circ f(b), \qquad \text{for all } a, b \in G_1$$

is called a *homomorphism*.

Let G be a finite group. A group element h is said to be *conjugate* to the group element k, $h \sim k$, if there exists a $g \in G$ such that

$$k = ghg^{-1}.$$

For G the number of conjugacy classes is equal to the number of irreducible matrix representations. In a *character table* the rows correspond to irreducible group representations and columns to classes of group elements. If the irreducible matrix representation is given by an $n \times n$ matrix $(n \geq 2)$ then the trace of this square matrix is the element in the table.

The *order of a finite group* is the number of elements of the group. *Lagrange's theorem* tells us that the order of a subgroup of a finite group divides the order of the group.

A *permutation matrix* contains exactly one 1 in each row and column. The inverse of a permutation matrix P is given by $P^{-1} = P^T$. The set of all $n \times n$ permutation matrices form a group under matrix multiplication. The order of the group is $n!$. *Cayley's theorem* tells us that every finite group is isomorphic to a subgroup (or the group itself) of these permutation matrices. The *order of an element* $g \in G$ is the order of the cyclic subgroup generated by $\{g\}$, i.e. the smallest positive integer m such that $g^m = e$, where e is the identity element of the group.

Let $GL(n, \mathbb{F})$ be the group of invertible $n \times n$ matrices with entries in the field $\mathbb{F}$, where $\mathbb{F}$ is $\mathbb{R}$ or $\mathbb{C}$. Let G be a group. A *matrix representation* of G over the field $\mathbb{F}$ is a homomorphism ρ from G to $GL(n, \mathbb{F})$. The degree of ρ is the integer n. Let $\rho : G \to GL(n, \mathbb{F})$. Then ρ is a representation if and only if

$$\rho(g \circ h) = \rho(g)\rho(h)$$

for all $g, h \in G$.

The $n \times n$ unitary matrices form a group under matrix multiplication called $U(n)$. The $n \times n$ unitary matrices with the constraint that the determinant is equal to 1 form a subgroup of $U(n)$ called $SU(n)$.

The $n \times n$ orthogonal matrices over $\mathbb{R}$ (called $O(n, \mathbb{R})$) form a group under matrix multiplication. A subgroup is $SO(n, \mathbb{R})$ with the condition that the determinant is equal to 1.

The $n \times n$ matrices over $\mathbb{R}$ with determinant equal to 1 form a group under matrix multiplication called $SL(n, \mathbb{R})$. The entries of the group $SL(n, \mathbb{Z})$ are integers.

Problem 1.   Consider the six 3×3 permutation matrices

$$A = \begin{pmatrix} 1 & 0 & 0 \\ 0 & 1 & 0 \\ 0 & 0 & 1 \end{pmatrix}, \qquad B = \begin{pmatrix} 1 & 0 & 0 \\ 0 & 0 & 1 \\ 0 & 1 & 0 \end{pmatrix}, \qquad C = \begin{pmatrix} 0 & 1 & 0 \\ 1 & 0 & 0 \\ 0 & 0 & 1 \end{pmatrix},$$

$$D = \begin{pmatrix} 0 & 1 & 0 \\ 0 & 0 & 1 \\ 1 & 0 & 0 \end{pmatrix}, \qquad E = \begin{pmatrix} 0 & 0 & 1 \\ 1 & 0 & 0 \\ 0 & 1 & 0 \end{pmatrix}, \qquad F = \begin{pmatrix} 0 & 0 & 1 \\ 0 & 1 & 0 \\ 1 & 0 & 0 \end{pmatrix}.$$

(i) Show that these matrices form a group under matrix multiplication.
(ii) Find all subgroups.
(iii) Find the eigenvalues of D and show that they form a group under matrix multiplication.

Solution 1.   (i) Since

$$AA = A \quad AB = B \quad AC = C \quad AD = D \quad AE = E \quad AF = F$$

$$BA = B \quad BB = A \quad BC = D \quad BD = C \quad BE = F \quad BF = E$$

$$CA = C \quad CB = E \quad CC = A \quad CD = F \quad CE = B \quad CF = D$$

$$DA = D \quad DB = F \quad DC = B \quad DD = E \quad DE = A \quad DF = C$$

$$EA = E \quad EB = C \quad EC = F \quad ED = A \quad EE = D \quad EF = B$$

$$FA = F \quad FB = D \quad FC = E \quad FD = B \quad FE = C \quad FF = A$$

we find the following. The set of matrices given above is closed under matrix multiplication. The neutral element is the matrix A, i.e. the unit matrix. Each element has an inverse. From the table given above we find that

$$A^{-1} = A, \quad B^{-1} = B, \quad C^{-1} = C, \quad D^{-1} = E, \quad E^{-1} = D, \quad F^{-1} = F.$$

Since the associative law holds for matrices, we find that the matrices given above form a finite group under matrix multiplication.
(ii) Thus our group has order 6. Applying Lagrange's theorem the subgroups must have order 3, 2, 1. From the group table we find (besides the group itself) the five subgroups

$$\{ A, D, E \}, \quad \{ A, B \}, \quad \{ A, C \}, \quad \{ A, F \}, \quad \{ A \}.$$

The set of all $n \times n$ permutation matrices form a group under matrix multiplication. *Cayley's theorem* tells us that every finite group is isomorphic to a subgroup (or the group itself) of these permutation matrices. The *order of an element* $g \in G$ is the order of the cyclic subgroup generated by $\{g\}$, i.e. the smallest positive integer m such that $g^m = e$, where e is the identity element of the group. The integer m divides the order of G. Consider, for example, the element D of our group. Then

$$D^2 = E, \qquad D^3 = A, \qquad A \quad \text{identity element.}$$

Thus $m = 3$.
(iii) The eigenvalues are $\lambda_1 = 1$, $\lambda_2 = e^{2\pi i/3}$, $\lambda_3 = e^{4\pi i/3}$ with $\lambda_2^2 = \lambda_3$, $\lambda_3^2 = \lambda_2$. Hence we have a group.

Problem 2.   Consider the set of Pauli spin matrices σ_1, σ_2, σ_3 and the 2×2 identity matrix I_2. Can we extend this set so that we obtain a group under matrix multiplication?

Solution 2.   We have $\sigma_1^2 = \sigma_2^2 = \sigma_3^2 = I_2$ and

$$\sigma_1\sigma_2 = i\sigma_3, \qquad \sigma_2\sigma_1 = -i\sigma_3, \qquad \sigma_2\sigma_3 = i\sigma_1,$$

$$\sigma_3\sigma_2 = -i\sigma_1, \qquad \sigma_3\sigma_1 = i\sigma_2, \qquad \sigma_1\sigma_3 = -i\sigma_2.$$

Thus we have to extend the set to the set of 16 elements

$$\{ \pm I_2, \pm\sigma_1, \pm\sigma_2, \pm\sigma_3, \pm iI_2, \pm i\sigma_1, \pm i\sigma_2, \pm i\sigma_3 \}$$

to obtain a group under matrix multiplication. Note that all matrices are unitary.

Problem 3.   Find the *invariance group* of the function $f : \mathbb{R} \to \mathbb{R}$

$$f(x) = \sin(x).$$

Solution 3.   We have $f(x) = f(x + 2\pi n)$ $(n \in \mathbb{Z})$. This means the function is invariant under a transformation of its domain by an element of the infinite cyclic subgroup $\langle 2\pi \rangle$ of the translation group of $\mathbb{R}$.

Problem 4.   Let G be the *dihedral group* defined by

$$D_8 := \langle a, b \ : \ a^4 = b^2 = 1, \ b^{-1}ab = a^{-1} \rangle$$

where 1 is the identity element in the group. Define the invertible 2×2 matrices

$$A = \begin{pmatrix} 0 & 1 \\ -1 & 0 \end{pmatrix}, \qquad B = \begin{pmatrix} 1 & 0 \\ 0 & -1 \end{pmatrix}.$$

Let I_2 be the 2×2 identity matrix. Show that $A^4 = B^2 = I_2$, $B^{-1}AB = A^{-1}$ and thus show that we have a matrix representation of the dihedral group.

Solution 4. We have $A^2 = -I_2$. Thus $A^4 = I_2$. Furthermore $B^2 = I_2$. The inverse of B is given by $B^{-1} = B$. The inverse of A is $-A$. Thus

$$
B^{-1}AB = BAB = \begin{pmatrix} 1 & 0 \\ 0 & -1 \end{pmatrix} \begin{pmatrix} 0 & 1 \\ -1 & 0 \end{pmatrix} \begin{pmatrix} 1 & 0 \\ 0 & -1 \end{pmatrix} = A^{-1}.
$$

It follows that the map $\rho : a^j b^k \rightarrow A^j B^k$ $(j = 0, 1, 2, 3; k = 0, 1)$ provides a matrix representation of D_8 over $\mathbb{R}$.

Problem 5. Let D_n be the *dihedral group*. This is the group of rigid motions of an n-gon $(n \geq 3)$

$$
D_n := \{ a, b \ : \ a^n = b^2 = 1, \ ba = a^{-1}b \} \tag{1}
$$

where a is a rotation by $2\pi/n$, b is a flip and 1 is the identity element. It is a noncommutative group of order $2n$. Find the *centre* of D_n. The *centre* C is defined as

$$
C := \{ c \in D_n \ : \ cx = xc \text{ for all } x \in D_n \}.
$$

Solution 5. To determine the centre of the group D_n it suffices to find those elements which commute with the generators a and b. Since $n \geq 3$, we have $a^{-1} \neq a$. Therefore

$$
a^{r+1}b = a(a^r b) = (a^r b)a = a^{r-1}b.
$$

It follows that $a^2 = 1$ a contradiction. Thus, no element of the form $a^r b$ is in the centre. Analogously, if for $1 \leq s < n$, $a^s b = ba^s = a^{-s}b$ then $a^{2s} = 1$, which is only possible if $2s = n$. Hence, a^s commutes with b if and only if $n = 2s$. Thus, if $n = 2s$ the centre of D_n is $\{ 1, a^s \}$. If n is odd the centre is $\{ 1 \}$.

Problem 6. Consider the group $G = \{ a, b, c \}$ with group operation $\cdot : G \times G \rightarrow G$ defined by the *group table*

$\cdot$	a	b	c
a	a	b	c
b	b	c	a
c	c	a	b

(i) Let $M(n)$ denote the vector space of $n \times n$ matrices over $\mathbb{C}$. Show that $f : G \rightarrow M(3)$ is a matrix representation for $(G, \cdot)$ where

$$
f(a) := \begin{pmatrix} 1 & 0 & 0 \\ 0 & 1 & 0 \\ 0 & 0 & 1 \end{pmatrix} = I_3
$$

$$f(b) := \begin{pmatrix} \frac{\sqrt{3}}{2}i - \frac{1}{2} & 0 & 0 \\ 0 & -\frac{1}{2} & \frac{\sqrt{3}}{2} \\ 0 & -\frac{\sqrt{3}}{2} & -\frac{1}{2} \end{pmatrix} \qquad f(c) := \begin{pmatrix} -\frac{\sqrt{3}}{2}i - \frac{1}{2} & 0 & 0 \\ 0 & -\frac{1}{2} & -\frac{\sqrt{3}}{2} \\ 0 & \frac{\sqrt{3}}{2} & -\frac{1}{2} \end{pmatrix}.$$

(ii) Is the matrix representation in (i) reducible? Prove or disprove.

(iii) Let $f : G \to M(m)$ be a representation of a group $(G, \cdot)$ and let $g : G \to M(n)$ also be a representation of the same group $(G, \cdot)$. Show that the maps $h : G \to M(mn)$ and $k : G \to M(m+n)$ defined by

$$h(x) := f(x) \otimes g(x), \qquad k(x) := f(x) \oplus g(x)$$

for all $x \in G$, is also a group representation of $(G, \cdot)$. Here $\otimes$ denotes the Kronecker product of matrices and $\oplus$ denotes the direct sum of matrices.

Solution 6.   (i) We obtain

$$\begin{aligned}
f(a)f(a) &= f(a), & f(a)f(b) &= f(b), & f(a)f(c) &= f(c), \\
f(b)f(a) &= f(b), & f(b)f(b) &= f(c), & f(b)f(c) &= f(a), \\
f(c)f(a) &= f(c), & f(c)f(b) &= f(a), & f(c)f(c) &= f(b).
\end{aligned}$$

Thus f is a matrix representation of the group.

(ii) We can write all three matrices as direct sums

$$f(a) = (1) \oplus \begin{pmatrix} 1 & 0 \\ 0 & 1 \end{pmatrix}$$

$$f(b) = (e^{2\pi i/3}) \oplus \begin{pmatrix} \cos(2\pi i/3) & \sin(2\pi i/3) \\ -\sin(2\pi i/3) & \cos(2\pi i/3) \end{pmatrix}$$

$$f(c) = (e^{4\pi i/3}) \oplus \begin{pmatrix} \cos(4\pi i/3) & \sin(4\pi i/3) \\ -\sin(4\pi i/3) & \cos(4\pi i/3) \end{pmatrix}.$$

Thus we find the invariant vector subspace

$$\left\{ (t \quad 0 \quad 0)^T \; : \; t \in \mathbb{C} \right\}.$$

Consequently f is reducible. Another invariant vector subspace is

$$\left\{ (0 \quad u \quad v)^T \; : \; u, v \in \mathbb{C} \right\}.$$

(iii) Using the properties of the Kronecker product and the direct sum we obtain

$$\begin{aligned}
h(a \cdot b) &= f(a \cdot b) \otimes g(a \cdot b) = (f(a)f(b)) \otimes (g(a)g(b)) \\
&= (f(a) \otimes g(a))(f(b) \otimes g(b)) = h(a)h(b)
\end{aligned}$$

and

$$k(a \cdot b) = f(a \cdot b) \oplus g(a \cdot b) = (f(a)f(b)) \oplus (g(a)g(b))$$
$$= (f(a) \oplus g(a))(f(b) \oplus g(b)) = k(a)k(b).$$

It follows that h and k are also matrix representations.

Problem 7. Consider the set of all 2×2 matrices over the set of integers $\mathbb{Z}$ with determinant equal to 1. Show that these matrices form a group under matrix multiplication. This group is called $SL(2, \mathbb{Z})$.

Solution 7. Let $A, B \in SL(2, \mathbb{Z})$. We have

$$AB = \begin{pmatrix} a_{11} & a_{12} \\ a_{21} & a_{22} \end{pmatrix} \begin{pmatrix} b_{11} & b_{12} \\ b_{21} & b_{22} \end{pmatrix} = \begin{pmatrix} a_{11}b_{11} + a_{12}b_{21} & a_{11}b_{12} + a_{12}b_{22} \\ a_{21}b_{11} + a_{22}b_{21} & a_{21}b_{12} + a_{22}b_{22} \end{pmatrix}.$$

Since $a_{ij}b_{kl} \in \mathbb{Z}$, and $\det(A) = a_{11}a_{22} - a_{12}a_{21} = 1$, $\det(B) = b_{11}b_{22} - b_{12}b_{21} = 1$ with $\det(AB) = \det(A)\det(B) = 1$ we obtain that $AB \in SL(2, \mathbb{Z})$. The neutral element (identity) is the 2×2 identity matrix. The inverse of A is given by

$$A^{-1} = \begin{pmatrix} a_{22} & -a_{12} \\ -a_{21} & a_{11} \end{pmatrix}$$

with $\det(A^{-1}) = a_{11}a_{22} - a_{12}a_{21} = 1$ and $-a_{12}, -a_{21} \in \mathbb{Z}$. Matrix multiplication is associative. Thus we have a group.

Problem 8. The underlying field is the real numbers $\mathbb{R}$. Given any *rotation matrix*, $R \in SO(n, \mathbb{R})$. If R does not admit -1 as an eigenvalue, then there is a unique skew symmetric matrix, S, $(S^T = -S)$ so that

$$R = (I_n - S)(I_n + S)^{-1}.$$

The matrix R is called the *Cayley transform* of S. Let

$$S = \begin{pmatrix} 0 & 1 \\ -1 & 0 \end{pmatrix}.$$

Find R.

Solution 8. We have

$$I_2 - S = \begin{pmatrix} 1 & -1 \\ 1 & 1 \end{pmatrix}, \quad I_2 + S = \begin{pmatrix} 1 & 1 \\ -1 & 1 \end{pmatrix}, \quad (I_2 + S)^{-1} = \frac{1}{2}\begin{pmatrix} 1 & -1 \\ 1 & 1 \end{pmatrix}.$$

It follows that

$$(I_2 - S)(I_2 + S)^{-1} = \begin{pmatrix} 1 & -1 \\ 1 & 1 \end{pmatrix} \frac{1}{2}\begin{pmatrix} 1 & -1 \\ 1 & 1 \end{pmatrix} = \begin{pmatrix} 0 & -1 \\ 1 & 0 \end{pmatrix} = R.$$

Thus $R = S^T$.

Problem 9. Let $\alpha \in \mathbb{R}$.
(i) Consider the *rotation matrix*

$$A(\alpha) := \begin{pmatrix} \cos(\alpha) & \sin(\alpha) \\ -\sin(\alpha) & \cos(\alpha) \end{pmatrix}$$

where $\alpha \in \mathbb{R}$. Show that the matrices $A(\alpha)$ form a group under matrix multiplication. Let

$$X := \left. \frac{dA(\alpha)}{d\alpha} \right|_{\alpha=0}.$$

Find X. Calculate $e^{\alpha X}$. Discuss.
(ii) Let

$$B(\alpha) := \begin{pmatrix} \cosh(\alpha) & \sinh(\alpha) \\ \sinh(\alpha) & \cosh(\alpha) \end{pmatrix}.$$

Show that the matrices $B(\alpha)$ form a group under matrix multiplication. Let

$$X := \left. \frac{dB(\alpha)}{d\alpha} \right|_{\alpha=0}.$$

Find X. Calculate $e^{\alpha X}$. Discuss.

Solution 9. (i) Since

$$A(\alpha)A(\beta) = \begin{pmatrix} \cos(\alpha + \beta) & \sin(\alpha + \beta) \\ -\sin(\alpha + \beta) & \cos(\alpha + \beta) \end{pmatrix} \tag{1}$$

the set is closed under multiplication. To find (1) we have used the identities

$$\cos(\alpha)\cos(\beta) - \sin(\alpha)\sin(\beta) \equiv \cos(\alpha + \beta)$$

$$\sin(\alpha)\cos(\beta) + \cos(\alpha)\sin(\beta) \equiv \sin(\alpha + \beta).$$

For $\alpha = 0$ we obtain the neutral element of the group (i.e. the unit matrix) $A(\alpha = 0) = I_2$. Since $\det(A(\alpha)) = 1$ the inverse exists and is given by

$$A^{-1}(\alpha) = A(-\alpha) = \begin{pmatrix} \cos(\alpha) & -\sin(\alpha) \\ \sin(\alpha) & \cos(\alpha) \end{pmatrix}.$$

For arbitrary $n \times n$ matrices the associative law holds. Consequently the matrices $A(\alpha)$ form a group. The group is called $SO(2)$. We find

$$X = \begin{pmatrix} 0 & 1 \\ -1 & 0 \end{pmatrix}$$

owing to $\sin(0) = 0$ and $\cos(0) = 1$. Since

$$e^{\alpha X} := \sum_{k=0}^{\infty} \frac{(\alpha X)^k}{k!}$$

we find $e^{\alpha X} = A(\alpha)$ because $X^2 = -I_2$. In physics X is called the *generator* of the Lie group $SO(2)$.

(ii) Since

$$B(\alpha)B(\beta) = \begin{pmatrix} \cosh(\alpha + \beta) & \sinh(\alpha + \beta) \\ \sinh(\alpha + \beta) & \cosh(\alpha + \beta) \end{pmatrix} \tag{2}$$

the set is closed under multiplication. To find (2) we have used the identities

$$\cosh(\alpha)\cosh(\beta) + \sinh(\alpha)\sinh(\beta) \equiv \cosh(\alpha + \beta)$$

$$\sinh(\alpha)\cosh(\beta) + \cosh(\alpha)\sinh(\beta) \equiv \sinh(\alpha + \beta).$$

For $\alpha = 0$ we obtain the neutral element $B(\alpha = 0) = I_2$. Since $\det(B(\alpha)) = 1$ the inverse matrix exists and is given by

$$B^{-1}(\alpha) = B(-\alpha) = \begin{pmatrix} \cosh(\alpha) & -\sinh(\alpha) \\ -\sinh(\alpha) & \cosh(\alpha) \end{pmatrix}.$$

For arbitrary $n \times n$ matrices the associative law holds. Consequently, the matrices $B(\alpha)$ form a group. The group is called $SO(1,1)$. We obtain

$$X = \begin{pmatrix} 0 & 1 \\ 1 & 0 \end{pmatrix}$$

owing to $\sinh(0) = 0$ and $\cosh(0) = 1$. Since

$$e^{\alpha X} := \sum_{k=0}^{\infty} \frac{(\alpha X)^k}{k!}$$

we find $e^{\alpha X} = B(\alpha)$ because $X^2 = I_2$.

Problem 10.   Consider the 4×4 matrix

$$A = \begin{pmatrix} 0 & 0 & 1 & 0 \\ 0 & 0 & 0 & 1 \\ -1 & 0 & 0 & 0 \\ 0 & -1 & 0 & 0 \end{pmatrix}.$$

Obviously $\text{rank}(A) = 4$ and therefore the inverse exists.

(i) Find the inverse of A.

(ii) Does the set $\{A, A^{-1}, I_4\}$ form a group under matrix multiplication? If not can we find a finite extension to the set to obtain a group?

Solution 10.   (i) We find

$$
A^{-1} = \begin{pmatrix} 0 & 0 & -1 & 0 \\ 0 & 0 & 0 & -1 \\ 1 & 0 & 0 & 0 \\ 0 & 1 & 0 & 0 \end{pmatrix} = -A.
$$

(ii) Since $A^2 = (A^{-1})^2 = -I_4$ and $A^{-1} = -A$ we find that the set

$$
\{ A, -A, I_4, -I_4 \}
$$

forms a group under matrix multiplication.

Problem 11.   Consider the three 2×2 matrices

$$
A = \begin{pmatrix} 2 & 1 \\ 1 & 2 \end{pmatrix}, \qquad I_2 = \begin{pmatrix} 1 & 0 \\ 0 & 1 \end{pmatrix}, \qquad C = \begin{pmatrix} 0 & 1 \\ 1 & 0 \end{pmatrix}.
$$

Obviously $[A, I_2] = 0_2$, $[A, C] = 0_2$.
(i) Show that the matrices I_2 and C form a group under the matrix multiplication. Find the conjugacy classes.
(ii) Find the irreducible representations of I_2 and C. Give the *character table*.
(iii) Find the projection matrices from the character table. Use the projection matrices to find the invariant subspaces. Give the eigenvalues and eigenvectors of A.

Solution 11.   (i) We find $I_2 I_2 = I_2$, $I_2 C = C$, $C I_2 = C$, $CC = I_2$. Obviously the group properties are satisfied with $C^{-1} = C$. Since $I_2 I_2 I_2 = I_2$, $C I_2 C = I_2$, $I_2 C I_2 = C$, $CCC = C$ we obtain the two classes $\{I_2\}$ and $\{C\}$.
(ii) Obviously, $I_2 \to 1$, $C \to 1$ and $I_2 \to 1$ $C \to -1$ are representations. The second one is a faithful representation. Since the representations are one-dimensional they are irreducible. The number of irreducible representations is equal to the number of classes. Taking into account the result from (i) we find that the character table is given by

	$\{I_2\}$	$\{C\}$
A_1	1	1
A_2	1	-1

(iii) For the representation A_1 we find the projection matrix

$$
\Pi_1 = \frac{1}{2} \sum_{g \in G} \chi_1(g^{-1}) g = \frac{1}{2} \sum_{g \in G} \chi_1(g) g = \frac{1}{2}(I_2 + C) = \frac{1}{2} \begin{pmatrix} 1 & 1 \\ 1 & 1 \end{pmatrix}.
$$

Analogously

$$\Pi_2 = \frac{1}{2}(I_2 - C) = \frac{1}{2}\begin{pmatrix} 1 & -1 \\ -1 & 1 \end{pmatrix}.$$

Applying the projection matrices to the standard basis in $\mathbb{R}^2$ gives

$$\Pi_1 \begin{pmatrix} 1 \\ 0 \end{pmatrix} = \frac{1}{2}\begin{pmatrix} 1 \\ 1 \end{pmatrix}, \qquad \Pi_1 \begin{pmatrix} 0 \\ 1 \end{pmatrix} = \frac{1}{2}\begin{pmatrix} 1 \\ 1 \end{pmatrix},$$

$$\Pi_2 \begin{pmatrix} 1 \\ 0 \end{pmatrix} = \frac{1}{2}\begin{pmatrix} 1 \\ -1 \end{pmatrix}, \qquad \Pi_2 \begin{pmatrix} 0 \\ 1 \end{pmatrix} = \frac{1}{2}\begin{pmatrix} -1 \\ 1 \end{pmatrix}.$$

When we normalize the right hand side we obtain the new basis

$$\frac{1}{\sqrt{2}}\begin{pmatrix} 1 \\ 1 \end{pmatrix}, \qquad \frac{1}{\sqrt{2}}\begin{pmatrix} 1 \\ -1 \end{pmatrix}$$

which is called the *Hadamard basis*. In this new basis the matrix A takes the diagonal form $\mathrm{diag}(3,1)$. It follows that the eigenvalues of the matrix A are given by 3 and 1 and the normalized eigenvectors of A are given by the Hadamard basis.

Problem 12. (i) Consider the three 4×4 permutation matrices

$$U_1 = \begin{pmatrix} 0 & 0 & 0 & 1 \\ 0 & 0 & 1 & 0 \\ 0 & 1 & 0 & 0 \\ 1 & 0 & 0 & 0 \end{pmatrix}, \quad U_2 = \begin{pmatrix} 0 & 1 & 0 & 0 \\ 1 & 0 & 0 & 0 \\ 0 & 0 & 0 & 1 \\ 0 & 0 & 1 & 0 \end{pmatrix}, \quad U_3 = \begin{pmatrix} 0 & 0 & 1 & 0 \\ 0 & 0 & 0 & 1 \\ 1 & 0 & 0 & 0 \\ 0 & 1 & 0 & 0 \end{pmatrix}.$$

Show that the set $\{ I_4, U_1, U_2, U_3 \}$ forms a group under matrix multiplication. Is the group commutative? Find the classes. Find the irreducible representations. Give the character table.
(ii) Let

$$A_1 = \begin{pmatrix} 0 & 1 & 1 & 0 \\ 1 & 0 & 0 & 1 \\ 1 & 0 & 0 & 1 \\ 0 & 1 & 1 & 0 \end{pmatrix}, \qquad A_2 = \begin{pmatrix} 0 & -1 & -1 & 0 \\ -1 & 0 & 0 & -1 \\ -1 & 0 & 0 & -1 \\ 0 & -1 & -1 & 0 \end{pmatrix},$$

$$A_3 = \begin{pmatrix} 0 & -1 & 1 & 0 \\ -1 & 0 & 0 & 1 \\ 1 & 0 & 0 & -1 \\ 0 & 1 & -1 & 0 \end{pmatrix}, \qquad A_4 = \begin{pmatrix} 0 & 1 & -1 & 0 \\ 1 & 0 & 0 & -1 \\ -1 & 0 & 0 & 1 \\ 0 & -1 & 1 & 0 \end{pmatrix}.$$

Show that $U_j A_k U_j = A_k$ for $j = 1, 2, 3$ and $k = 1, 2, 3, 4$. Find the eigenvalues and eigenvectors of A_k.

Solution 12.   (i) We find

$$I_4U_1 = U_1, \qquad I_4U_2 = U_2, \qquad I_4U_3 = U_3, \qquad U_1U_2 = U_3,$$

$$U_2U_1 = U_3, \qquad U_1U_3 = U_2, \qquad U_3U_1 = U_2, \qquad U_2U_3 = U_1,$$

$$U_3U_2 = U_1, \qquad U_1U_1 = I_4, \qquad U_2U_2 = I_4, \qquad U_3U_3 = I_4.$$

Thus the group is commutative. Since

$$I_4I_4I_4 = I_4, \quad U_1I_4U_1 = I_4, \quad U_2I_4U_2 = I_4, \quad U_3I_4U_3 = I_4,$$

$$I_4U_1I_4 = U_1, \quad U_1U_1U_1 = U_1, \quad U_2U_1U_2 = U_1, \quad U_3U_1U_3 = U_1,$$

$$I_4U_2I_4 = U_2, \quad U_1U_2U_1 = U_2, \quad U_2U_2U_2 = U_2, \quad U_3U_2U_3 = U_2,$$

$$I_4U_3I_4 = U_3, \quad U_1U_3U_1 = U_3, \quad U_2U_3U_2 = U_3, \quad U_3U_3U_3 = U_3$$

we obtain four classes, namely $\{\{I_4\}, \{U_1\}, \{U_2\}, \{U_3\}\}$. Thus we have four one-dimensional irreducible representations. We find

$$\begin{aligned}
\Gamma_1: \quad & I_4 \mapsto 1, \quad U_1 \mapsto 1, \quad U_2 \mapsto 1, \quad U_3 \mapsto 1 \\
\Gamma_2: \quad & I_4 \mapsto 1, \quad U_1 \mapsto -1 \quad U_2 \mapsto -1, \quad U_3 \mapsto 1 \\
\Gamma_3: \quad & I_4 \mapsto 1, \quad U_1 \mapsto -1, \quad U_2 \mapsto 1 \quad U_3 \mapsto -1 \\
\Gamma_4: \quad & I_4 \mapsto 1, \quad U_1 \mapsto 1 \quad U_2 \mapsto -1 \quad U_3 \mapsto -1.
\end{aligned}$$

Consequently, the character table is given by

	$\{I_4\}$	$\{U_1\}$	$\{U_2\}$	$\{U_3\}$
Γ_1	1	1	1	1
Γ_2	1	-1	-1	1
Γ_3	1	-1	1	-1
Γ_4	1	1	-1	-1

(ii) From the representation Γ_1 we find the projection matrix

$$\Pi_1 = \frac{1}{4} \sum_{g \in G} \chi_1(g^{-1})g = \frac{1}{4}(I_4 + U_1 + U_2 + U_3) = \frac{1}{4}\begin{pmatrix} 1 & 1 & 1 & 1 \\ 1 & 1 & 1 & 1 \\ 1 & 1 & 1 & 1 \\ 1 & 1 & 1 & 1 \end{pmatrix}$$

where we have used that $g^{-1} = g$ for all $g \in G$. Analogously

$$\Pi_2 = \frac{1}{4}(I_4 - U_1 - U_2 + U_3), \qquad \Pi_3 = \frac{1}{4}(I_4 - U_1 + U_2 - U_3),$$

$$\Pi_4 = \frac{1}{4}(I_4 + U_1 - U_2 - U_3).$$

Applying the projection matrix Π_1 to the standard basis $\mathbf{e}_j$ ($j = 1, 2, 3, 4$) in $\mathbb{R}^4$ gives

$$\Pi_1 \mathbf{e}_j = \frac{1}{4} \begin{pmatrix} 1 \\ 1 \\ 1 \\ 1 \end{pmatrix}, \quad j = 1, 2, 3, 4.$$

Analogously, we apply the other projection operators to the standard basis. Then the new orthonormal basis is given by

$$\frac{1}{2} \begin{pmatrix} 1 \\ 1 \\ 1 \\ 1 \end{pmatrix}, \quad \frac{1}{2} \begin{pmatrix} 1 \\ -1 \\ 1 \\ -1 \end{pmatrix}, \quad \frac{1}{2} \begin{pmatrix} 1 \\ 1 \\ -1 \\ -1 \end{pmatrix}, \quad \frac{1}{2} \begin{pmatrix} 1 \\ -1 \\ -1 \\ 1 \end{pmatrix}.$$

In this new basis the matrices A_j take the diagonal form

$$\begin{pmatrix} 2 & 0 & 0 & 0 \\ 0 & 0 & 0 & 0 \\ 0 & 0 & 0 & 0 \\ 0 & 0 & 0 & -2 \end{pmatrix}, \quad \begin{pmatrix} -2 & 0 & 0 & 0 \\ 0 & 0 & 0 & 0 \\ 0 & 0 & 0 & 0 \\ 0 & 0 & 0 & 2 \end{pmatrix},$$

$$\begin{pmatrix} 0 & 0 & 0 & 0 \\ 0 & 2 & 0 & 0 \\ 0 & 0 & -2 & 0 \\ 0 & 0 & 0 & 0 \end{pmatrix}, \quad \begin{pmatrix} 0 & 0 & 0 & 0 \\ 0 & -2 & 0 & 0 \\ 0 & 0 & 2 & 0 \\ 0 & 0 & 0 & 0 \end{pmatrix}.$$

Consequently, the eigenvalues of the matrices A_k are given by 2, -2, 0, 0 and the normalized eigenvectors are given above. We have $A_1 = U_2 + U_3$, $A_2 = -A_1$, $A_3 = -U_2 + U_3$, $A_4 = -A_3$.

Problem 13. (i) The mapping ($z \in \mathbb{C}$)

$$w(z) = \frac{az + b}{cz + d}, \quad ad - bc \neq 0 \tag{1}$$

is called the *fractional linear transformation*, where a, b, c, $d \in \mathbb{R}$. Show that these transformations form a group under the composition of mappings. (ii) The *Schwarzian derivative* of a function f is defined as

$$D(f)_z := \frac{2f'(z)f'''(z) - 3[f''(z)]^2}{2[f''(z)]^2} \tag{2}$$

where $f'(z) \equiv df/dz$. Calculate $D(w)_z$.

Solution 13. (i) First we have to show that the transformation is closed under the composition. Let

$$W(w) = \frac{Aw + B}{Cw + D}, \quad AD - BC \neq 0. \tag{3}$$

Inserting (1) yields

$$W(w(z)) = \frac{A(az+b) + B(cz+d)}{C(az+b) + D(cz+d)} = \frac{(Aa+Bc)z + (Ab+Bd)}{(Ca+Dc)z + (Cb+Dd)}. \quad (4)$$

Since $ad - bc \neq 0$ and $AD - BC \neq 0$ we find that

$$(Aa+Bc)(Cb+Dd) - (Ab+Bd)(Ca+Dc) \equiv (AD-BC)(ad-bc) \neq 0.$$

Thus mapping (1) is closed under composition. For $a = 1$, $b = 0$, $c = 0$, $d = 1$ we find the unit element $w(z) = z$. The inverse of mapping (1) is given by

$$z(w) = \frac{-dw + b}{cw - a}$$

since

$$z(w(z)) = \frac{(-d(az+b) + b(cz+d))(cz+d)}{(cz+d)(c(az+b) - a(cz+d))}$$

$$= \frac{-adz - db + bd + bcz}{bc - ad + acz - acz} = \frac{(-ad + bc)z}{(-ad + bc)} = z.$$

Finally, it can be proved that the associative law is satisfied.

(ii) We find $D(w)_z = 0$. This means $w(z)$ is the general solution of the ordinary differential equation $D(f)_z = 0$ since the solution includes three constants of integration.

Problem 14. Let S_n be the set of all $n \times n$ permutation matrices. Then

$$\Pi = \frac{1}{n!} \sum_{P \in S_n} P$$

is a *projection matrix*. Find Π for $n = 2$ and $n = 3$.

Solution 14. For $n = 2$ we have

$$\Pi = \frac{1}{2}\left(\begin{pmatrix} 1 & 0 \\ 0 & 1 \end{pmatrix} + \begin{pmatrix} 0 & 1 \\ 1 & 0 \end{pmatrix}\right) = \frac{1}{2}\begin{pmatrix} 1 & 1 \\ 1 & 1 \end{pmatrix}.$$

For $n = 3$ we have

$$\Pi = \frac{1}{6}\left(\begin{pmatrix} 1 & 0 & 0 \\ 0 & 1 & 0 \\ 0 & 0 & 1 \end{pmatrix} + \begin{pmatrix} 1 & 0 & 0 \\ 0 & 0 & 1 \\ 0 & 1 & 0 \end{pmatrix} + \begin{pmatrix} 0 & 1 & 0 \\ 1 & 0 & 0 \\ 0 & 0 & 1 \end{pmatrix}\right.$$

$$\left. + \begin{pmatrix} 0 & 1 & 0 \\ 0 & 0 & 1 \\ 1 & 0 & 0 \end{pmatrix} + \begin{pmatrix} 0 & 0 & 1 \\ 1 & 0 & 0 \\ 0 & 1 & 0 \end{pmatrix} + \begin{pmatrix} 0 & 0 & 1 \\ 0 & 1 & 0 \\ 1 & 0 & 0 \end{pmatrix}\right)$$

$$= \frac{1}{3}\begin{pmatrix} 1 & 1 & 1 \\ 1 & 1 & 1 \\ 1 & 1 & 1 \end{pmatrix}.$$

Problem 15. Let $|\psi\rangle$ be a normalized state in the Hilbert space $\mathbb{C}^n$. Show that the matrix

$$U = I_n - |\psi\rangle\langle\psi|$$

is unitary. Find the group generated by U under matrix multiplication.

Solution 15. We obtain $U^* = I_n - |\psi\rangle\langle\psi|$ and $UU^* = I_n$ since $\langle\psi|\psi\rangle = 1$. Since $U^* = U$ the group is given by U, I_n. Find the group generated by $U \otimes U$.

Problem 16. Let P be the *parity operator*, i.e.

$$P\mathbf{r} := -\mathbf{r}. \tag{1}$$

Obviously, $P = P^{-1}$. We define

$$\mathbf{O}_P u(\mathbf{r}) := u(P^{-1}\mathbf{r}) \equiv u(-\mathbf{r}). \tag{2}$$

Now $\mathbf{r}$ can be expressed in *spherical coordinates*

$$\mathbf{r} = r(\sin(\theta)\cos(\phi), \sin(\theta)\sin(\phi), \cos(\theta)) \tag{3}$$

where $0 \le \phi < 2\pi$, $0 \le \theta < \pi$ and $0 \le r < \infty$.
(i) Calculate $P(r, \theta, \phi)$.
(ii) Consider the *spherical harmonics* defined by

$$Y_{lm}(\theta, \phi) := \frac{(-1)^{l+m}}{2^l l!}\left(\frac{2l+1}{4\pi}\frac{(l-m)!}{(l+m)!}\right)^{1/2}$$
$$\times (\sin(\theta))^m \frac{d^{l+m}}{d(\cos(\theta))^{l+m}}(\sin(\theta))^{2l}e^{im\phi}$$

where $l = 0, 1, 2, \ldots$ and $m = -l, -l+1, \ldots, +l$. Find $\mathbf{O}_P Y_{lm}$.
(iii) Calculate

$$(\mathbf{O}_P \hat{L}_z)Y_{lm}, \qquad (\hat{L}_z \mathbf{O}_P)Y_{lm}$$

where

$$\hat{L}_z := -i\hbar\frac{\partial}{\partial\phi}.$$

Use these results to find the commutator $[\hat{L}_z, \mathbf{O}_P]Y_{lm}$.

Solution 16. (i) From (1) we find that

$$P(r, \theta, \phi) = (r, \pi - \theta, \pi + \phi)$$

where $\sin(\pi - \theta) \equiv \sin(\theta)$, $\cos(\pi - \theta) \equiv -\cos(\theta)$, $\sin(\pi + \phi) \equiv -\sin(\phi)$, $\cos(\pi + \phi) \equiv -\cos(\phi)$.

(ii) We find $\mathbf{O}_P Y_{lm}(\theta, \phi) = Y_{lm}(\pi - \theta, \pi + \phi) = (-1)^l Y_{lm}(\theta, \phi)$.

(iii) Since $\hat{L}_z Y_{lm} = m\hbar Y_{lm}$ we obtain

$$\mathbf{O}_P \hat{L}_z Y_{lm} = \mathbf{O}_P(\hat{L}_z Y_{lm}) = m\hbar \mathbf{O}_P Y_{lm} = m\hbar(-1)^l Y_{lm}$$

and $(\hat{L}_z \mathbf{O}_P) Y_{lm} = \hat{L}_z(\mathbf{O}_P Y_{lm}) = (-1)^l \hat{L}_z Y_{lm} = (-1)^l m\hbar Y_{lm}$. Consequently, the commutator is given by $[\hat{L}_z, \mathbf{O}_P] Y_{lm} = 0$.

Problem 17. Consider a wave function $u(x_1, x_2)$ which is confined within a regular triangle. The regular triangle is bound by the three lines

$$x_1 = -A, \qquad x_1 = \sqrt{3}x_2 + 2A, \qquad x_1 = -\sqrt{3}x_2 + 2A$$

where $A > 0$. The eigenvalue problem is given by

$$\Delta u(x_1, x_2) \equiv \frac{\partial^2 u}{\partial x_1^2} + \frac{\partial^2 u}{\partial x_2^2} = -\frac{mE}{\hbar^2} u(x_1, x_2) \tag{1}$$

where at the boundary

$$u|_B = 0. \tag{2}$$

(i) Find the symmetry group of the system.

(ii) Find the eigenvalues and eigenfunctions for an invariant subspace.

Solution 17. (i) Diffraction in a corner of a polygon does not occur whenever the angle of the corner is π divided by an integer. In our case we have $\pi/3$. Therefore it is possible to express the wave function in terms of elementary functions, i.e. sin and cos. The triangle is invariant under the following six symmetry operations:

> identity, labelled I
>
> rotation by $\frac{2}{3}\pi$ around the origin, labelled C_3
>
> rotation by $\frac{4}{3}\pi$ around the origin, labelled C_3^2
>
> reflection in the line $x_1 = (1/\sqrt{3})x_2$, labelled σ
>
> reflection in the x_1 axis, labelled σ'
>
> reflection in the line $x_1 = -(1/\sqrt{3})y$, labelled σ''.

These operations form the group C_{3v}. The *character table* is given by

	$\{I\}$	$\{2C_3\}$	$\{3\sigma\}$
A1	1	1	1
A2	1	1	-1
E	2	-1	0

(ii) To apply group theory we first have to find the projection operators. From the character table we find that the projection operators are given by

$$\Pi_{A1} = I + C_3 + C_3^2 + \sigma + \sigma' + \sigma'', \quad \Pi_{A2} = I + C_3 + C_3^2 - \sigma - \sigma' - \sigma'',$$

$$\Pi_E = 2I - C_3 - C_3^2.$$

We construct the eigenfunctions as linear combinations of

$$f(x_1, x_2) := \sin(P(x_1 + x_{10})) \sin(Qx_2), \quad g(x_1, x_2) := \sin(P(x_1 + x_{10})) \cos(Qx_2).$$

Inserting these functions into (1) yields

$$E = \frac{\hbar^2}{m}(P^2 + Q^2).$$

Now we have to apply the projection operators to f and g and then impose the boundary condition. We find $\Pi_{A1} f = 0$ and the eigenfunctions of the symmetry type $A1$ are given by

$$\phi(x_1, x_2) := \frac{1}{2} \Pi_{A1} g(x_1, x_2).$$

Thus

$$\phi(x_1, x_2) = \sin(P(x_1 + x_{10})) \cos(Qx_2)$$

$$+ \sin\left(P\left(-\frac{1}{2}x_1 - \frac{\sqrt{3}}{2}x_2 + x_{10}\right)\right) \cos\left(Q\left(-\frac{1}{2}x_2 + \frac{\sqrt{3}}{2}x_1\right)\right)$$

$$+ \sin\left(P\left(-\frac{1}{2}x_1 + \frac{\sqrt{3}}{2}x_2 + x_{10}\right)\right) \cos\left(Q\left(-\frac{1}{2}x_2 - \frac{\sqrt{3}}{2}x_1\right)\right).$$

Analogously, we find $\Pi_{A2} g = 0$ and the eigenfunctions of the symmetry type $A2$ are given by

$$\zeta(x_1, x_2) := \frac{1}{2} \Pi_{A2} f(x_1, x_2)$$

$$= \sin(P(x_1 + x_{10})) \sin(Qx_2)$$

$$+ \sin\left(P\left(-\frac{1}{2}x_1 - \frac{\sqrt{3}}{2}x_2 + x_{10}\right)\right) \sin\left(Q\left(-\frac{1}{2}x_2 + \frac{\sqrt{3}}{2}x_1\right)\right)$$

$$+ \sin\left(P\left(-\frac{1}{2}x_1 + \frac{\sqrt{3}}{2}x_2 + x_{10}\right)\right) \sin\left(Q\left(-\frac{1}{2}x_2 - \frac{\sqrt{3}}{2}x_1\right)\right).$$

Since the functions of the symmetry class $A1$ are symmetric under C_3 it is sufficient to require that $\phi(x_1, x_2) = 0$ for $x_1 = -A$ and any x_2. Then the functions automatically satisfy the boundary conditions along the two other sides of the triangle. Thus we obtain

$$P = \frac{k\pi}{3A}, \qquad Q = \frac{n\pi}{\sqrt{3}A}$$

where k and n are integers which are either both even or both odd. We can restrict the values of k and n as follows: k and n are positive integers with $k \leq n$ and are either both even or both odd. For the symmetry class $A2$ it is sufficient to require that $\zeta(x_1, x_2) = 0$ for $x_1 = -A$ and any x_2. The function ζ is identically zero for $k = n$. Therefore we can restrict the values of k as follows: k and n are positive integers with $k < n$ and are either both even or both odd. Consequently, we have an infinite countable number of eigenfunctions of the symmetry class $A1$ and $A2$. Inserting these eigenfunctions into the eigenvalue equation we find the eigenvalues

$$E_{n,k} = \frac{\pi^2 \hbar^2}{mA^2} \left(\frac{k^2}{9} + \frac{n^2}{3} \right).$$

Supplementary Problems

Problem 1.   (i) Find the group generated by the 3×3 matrices

$$A = \begin{pmatrix} 0 & 0 & -1 \\ 0 & -1 & 0 \\ -1 & 0 & 0 \end{pmatrix}, \quad B = \begin{pmatrix} 0 & 0 & 1 \\ 0 & -1 & 0 \\ 1 & 0 & 0 \end{pmatrix}, \quad C = \begin{pmatrix} -1 & 0 & 0 \\ 0 & 1 & 0 \\ 0 & 0 & -1 \end{pmatrix}$$

under matrix multiplication.   Note that $AB = C$, $BC = A$, $CA = B$, $[A, B] = 0_3$.
(ii) Consider the 4×4 matrices and matrix multiplication

$$D = \begin{pmatrix} 1 & 0 & 0 & 0 \\ 0 & i & 0 & 0 \\ 0 & 0 & -1 & 0 \\ 0 & 0 & 0 & -i \end{pmatrix}, \quad P = \begin{pmatrix} 0 & 0 & 0 & 1 \\ 1 & 0 & 0 & 0 \\ 0 & 1 & 0 & 0 \\ 0 & 0 & 1 & 0 \end{pmatrix}.$$

Find the group generated by D. Find the group generated by P. Find the group generated by D and P.
(iii) Find the group generated by

$$U = \begin{pmatrix} 0 & e^{i\pi/3} \\ 1 & 0 \end{pmatrix}.$$

Find the group generated by $U \otimes U$.
(iv) Find the group generated by the 4×4 matrix

$$X = \begin{pmatrix} 0 & 0 & 0 & 1 \\ 0 & 0 & i & 0 \\ 0 & -i & 0 & 0 \\ -1 & 0 & 0 & 0 \end{pmatrix}.$$

Find the group generated by $X \otimes X$. Let $z \in \mathbb{C}$. Find $\exp(zX)$.

Problem 2. Find the group generated by the three permutation matrices

$$
P_1 = \begin{pmatrix} 0 & 1 & 0 & 0 \\ 1 & 0 & 0 & 0 \\ 0 & 0 & 1 & 0 \\ 0 & 0 & 0 & 1 \end{pmatrix}, \quad
P_2 = \begin{pmatrix} 1 & 0 & 0 & 0 \\ 0 & 0 & 1 & 0 \\ 0 & 1 & 0 & 0 \\ 0 & 0 & 0 & 1 \end{pmatrix},
$$

$$
P_3 = \begin{pmatrix} 1 & 0 & 0 & 0 \\ 0 & 1 & 0 & 0 \\ 0 & 0 & 0 & 1 \\ 0 & 0 & 1 & 0 \end{pmatrix}
$$

under matrix multiplication. Find the group generated by the matrices $P_1 \otimes P_1$, $P_2 \otimes P_2$, $P_3 \otimes P_3$ under matrix multiplication.

Problem 3. Consider the Hamilton operators

$$
\widetilde{H} = \frac{\hat{H}}{\hbar\omega} = \begin{pmatrix} 0 & 1 & 0 \\ 1 & 0 & 1 \\ 0 & 1 & 0 \end{pmatrix}, \quad
\widetilde{K} = \frac{\hat{K}}{\hbar\omega} = \begin{pmatrix} 1 & 0 & 1 \\ 0 & 1 & 0 \\ 1 & 0 & 1 \end{pmatrix}.
$$

Both admit the same symmetry, namely the permutation matrix

$$
P = \begin{pmatrix} 0 & 0 & 1 \\ 0 & 1 & 0 \\ 1 & 0 & 0 \end{pmatrix}
$$

i.e. $P\widetilde{H}P^T = \widetilde{H}$, $P\widetilde{K}P^T = \widetilde{K}$. The eigenvalues of $\widetilde{H}$ are $0, \pm 2$ and the eigenvalues of $\widetilde{K}$ are $0, 1, 2$. Thus $\widetilde{H}$ and $\widetilde{K}$ have the eigenvalue 0 in common. For this eigenvalue both matrices admit the normalized eigenvector

$$
\mathbf{v} = \frac{1}{\sqrt{2}} (1 \quad 0 \quad -1)^T.
$$

Explain. Show that $\Pi_+ = \frac{1}{2}(I_3 + P)$, $\Pi_- = \frac{1}{2}(I_3 - P)$ are projection matrices. Find $\Pi_+ \mathbf{v}$ and $\Pi_- \mathbf{v}$. Explain.

Problem 4. Show that the *Korteweg-de Vries equation*

$$
\frac{\partial u}{\partial t} = u\frac{\partial u}{\partial x} + \beta \frac{\partial^3 u}{\partial x^3}
$$

is invariant under the *dilation symmetry (scaling symmetry)*

$$
(t, x, u) \mapsto (t/\epsilon^3, x/\epsilon, \epsilon^2 u)
$$

i.e. one can cancel out a factor ϵ^5 in the Korteweg-de Vries equation.

Chapter 16

Combinatorics

In combinatorics we study the enumeration, combination, and permutations of sets of elements. Furthermore finite or countable discrete structures (for example finite sets, graphs) are studied. Enumerative combinatorics focuses on counting the number of certain combinatorial objects. A bijection of a finite set S onto itself is called a permutation of S. If S consist of n elements then there are $n!$ possible permutations. The fundamental principle of counting tells us that if one event has $j \in \mathbb{N}$ possible outcomes and a second independent event has $k \in \mathbb{N}$ possible outcomes, then the number of possible outcomes for the combined events is given by $j \cdot k$. Let S be a finite set with n elements. Then the number of subsets is 2^n which includes the set S itself and the empty set. Let $n \geq 1$, S be a set containing n elements and $n_1, \ldots, n_r$ be positive integers with $n_1 + n_2 + \cdots + n_r = n$. Then there exist

$$\frac{n!}{n_1! n_2! \cdots n_r!}$$

different ordered partitions of S of the form $(S_1, S_2, \ldots, S_r)$, where S_1 contains n_1 elements, S_2 contains n_2 elements etc. Let A, B be two finite sets. The product set $A \times B$ of A and B is given by all ordered pairs (a, b), where $a \in A$ and $b \in B$. If A has m elements and B has n elements, then $A \times B$ has $m \cdot n$ elements. Let $n \geq 1$. The number of combinations $C(n, r)$ of n objects taken r at the time is given by

$$C(n, r) = \binom{n}{r} \equiv \frac{n!}{r!(n-r)!}.$$

The number of bitstrings of length $n \in \mathbb{N}$ is given by 2^n. The number of permutations of n distinct objects is $n!$.

Problem 1. (i) In how many ways f_n can a group of n persons arrange themselves in a row of n chairs?

(ii) In how many ways f_n can a group of n persons arrange themselves around a circular table.

Solution 1. (i) The n persons can arrange themselves in a row in

$$n(n-1)(n-2)\cdots 2\cdot 1 = n! \tag{1}$$

ways. In other words: one person can sit in n different chairs. Therefore

$$f_n = nf_{n-1} \tag{2}$$

with $n = 2, 3, \ldots$ and the initial condition is given by $f_1 = 1$. The solution of the difference equation (2) with the initial condition is given by (1).

(ii) One person can sit in any place at the circular table. The other $n - 1$ persons can then arrange themselves in

$$f_n = (n-1)(n-2)\cdots 2\cdot 1 = (n-1)! \text{ ways.}$$

In other words n objects can be arranged in a circle in $(n-1)!$ ways.

Problem 2. Consider the n-dimensional unit cube in $\mathbb{R}^n$. How many k-dimensional surfaces $(1 \leq k < n)$ does the n-dimensional unit cube have?

Solution 2. We have

$$\binom{n}{k} 2^{n-k}.$$

For example for $n = 3$ and $k = 2$ we have 6 surfaces. For $n = 3$ and $k = 1$ we have 12 (lines).

Problem 3. What is the number A_n of ways of going up n steps, if we may take one or two steps at a time? Determine

$$\sum_{n=0}^{\infty} A_n x^n.$$

Solution 3. We may begin by taking one or two steps. In the first case, we have A_{n-1} possibilities to continue; in the second, we have A_{n-2} possibilities. Thus $A_n = A_{n-1} + A_{n-2}$, where $A_1 = 1$ and $A_2 = 2$. The sequence A_n is the sequence of the *Fibonacci numbers*. We set $A_0 = 1$ and

$$f(x) = \sum_{n=0}^{\infty} A_n x^n$$

as the generating function. Then

$$xf(x) = \sum_{n=1}^{\infty} A_{n-1}x^n, \qquad x^2 f(x) = \sum_{n=2}^{\infty} A_{n-2}x^n$$

and using (1) we find

$$f(x) - xf(x) - x^2 f(x) = A_0 + (A_1 - A_0)x + \sum_{n=2}^{\infty} (A_n - A_{n-1} - A_{n-2})x^n.$$

Thus $f(x) - xf(x) - x^2 f(x) = A_0 = 1$. Consequently, the *generating function* is given by

$$f(x) = \frac{1}{1 - x - x^2}.$$

To obtain an explicit formula for A_n we write this as

$$f(x) = \frac{1}{\sqrt{5}x} \sum_{k=0}^{\infty} \left(\left(\frac{1 + \sqrt{5}}{2}x \right)^k - \left(\frac{1 - \sqrt{5}}{2}x \right)^k \right)$$

$$= \frac{1}{\sqrt{5}} \sum_{n=0}^{\infty} \left(\left(\frac{1 + \sqrt{5}}{2} \right)^{n+1} - \left(\frac{1 - \sqrt{5}}{2} \right)^{n+1} \right) x^n.$$

Therefore

$$A_n = \frac{1}{\sqrt{5}} \left(\left(\frac{1 + \sqrt{5}}{2} \right)^{n+1} - \left(\frac{1 - \sqrt{5}}{2} \right)^{n+1} \right)$$

where $n = 0, 1, 2, \ldots$.

Problem 4.   We have n Rand. Every day we buy exactly one of the following products: pretzel (1 Rand), candy (2 Rand), ice-cream (2 Rand). What is the number B_n of possible ways of spending all the money?

Solution 4.   On the first day, we have three choices: we may buy a pretzel, in which case we have B_{n-1} further possible ways to spend the remaining $n - 1$ Rand; or we may buy candy for 2 Rand, and then we can spend the rest in B_{n-2} ways; similarly we have B_{n-2} possibilities if we first buy an ice-cream. Thus we find the linear difference equation with constant coefficients

$$B_n = B_{n-1} + 2B_{n-2} \qquad (n \geq 3). \tag{1}$$

The difference equation (1) is of second order. Thus we need two initial conditions. Obviously, $B_1 = 1$, $B_2 = 3$ are the initial values of the linear difference equation (1). The first few terms are $B_3 = 5$, $B_4 = 11$, $B_5 = 21$, $B_6 = 43$. Since (1) is a linear difference equation with constant coefficients we can solve it with the ansatz $B_n = ar^n$. We obtain the quadratic equation $r^2 = r + 2$. Consequently, the solution of (1) with the initial conditions is given by

$$B_n = \frac{1}{3}(2^{n+1} + (-1)^n).$$

Problem 5. Let A_1, A_2, ... A_p be finite sets. Let $|A_k|$ be the number of elements of A_k. Show that

$$\left| \bigcup_{j=1}^{p} A_j \right| = \sum_{j=1}^{p} |A_j| - \sum_{1 \leq j < k \leq p} |A_j \cap A_k| + \cdots + (-1)^{p+1} \left| \bigcap_{j=1}^{p} A_j \right|. \qquad (1)$$

Solution 5. The proof applies induction on $p \geq 2$. For $p = 2$ we find that (1) takes the form

$$|A_1 \cup A_2| = |A_1| + |A_2| - |A_1 \cap A_2|$$

which is obviously true. Suppose the equation is true for each union of at most $p - 1$ sets. It follows that

$$|A_1 \cup A_2 \cdots \cup A_p| = |A_1 \cup A_2 \cdots \cup A_{p-1}| + |A_p| - |(A_1 \cup A_2 \cdots \cup A_{p-1}) \cap A_p|.$$

Applying the *distributive law for intersections* of sets we find

$$(A_1 \cup A_2 \cdots \cup A_{p-1}) \cap A_p = (A_1 \cap A_p) \cup (A_2 \cap A_p) \cup \cdots \cup (A_{p-1} \cap A_p)$$

and from the inductive hypothesis it follows that

$$|A_1 \cup A_2 \cdots \cup A_p| = \sum_{1 \leq j < p} |A_j| - \sum_{1 \leq j < k < p} |A_j \cap A_k| + \cdots (-1)^p \left| \bigcap_{j=1}^{p-1} A_j \right|$$

$$+ |A_p| - \sum_{1 \leq j < p} |A_j \cap A_p| + \cdots (-1)^{p+1} \left| \bigcap_{j=1}^{p} A_j \right|$$

where we have used the *idempotent law of intersection* in the form

$$(A_j \cap A_p) \cap (A_k \cap A_p) = A_j \cap A_k \cap A_p, \quad \cdots \quad \bigcap_{j=1}^{p-1} (A_j \cap A_p) = \bigcap_{j=1}^{p} A_j.$$

By regrouping terms, we obtain (1). Equation (1) is called the *principle of inclusion and exclusion*.

Problem 6.   Show that the number of arrangements N of a set of n objects in p boxes such that the jth box contains n_j objects, for $j = 1, \ldots, p$ is equal to

$$N = \frac{n!}{n_1! n_2! \cdots n_p!}$$

where $n_j \geq 0$ and $n_1 + n_2 + \cdots + n_p = n$.

Solution 6.   The n_1 objects in the first box can be chosen in

$$\binom{n}{n_1}$$

ways, the n_2 objects in the second box can be chosen from the $n - n_1$ remaining objects in

$$\binom{n - n_1}{n_2}$$

ways, etc. The total number of arrangements is equal to

$$\binom{n}{n_1}\binom{n - n_1}{n_2}\binom{n - n_1 - n_2}{n_3} \cdots \binom{n - n_1 - \ldots - n_{p-1}}{n_p} \equiv \frac{n!}{n_1! n_2! \ldots n_p!}.$$

Problem 7.   Prove that the set of n (different) elements has exactly 2^n (different) subsets.

Solution 7.   For each n, let X_n denote the number of (different) subsets of a set with n (different) elements. Let S be a set with $n+1$ elements, and designate one of its elements by s. There is a one-to-one correspondence between those subsets of S which do not contain s and those subsets that do contain s (namely, a subset T of the former type corresponds to $T \cup \{s\}$). The former types are all subsets of $S \setminus \{s\}$, a set with n elements, and therefore, it must be the case that $X_{n+1} = 2X_n$, where $X_0 = 1$. Hence $X_n = 2^n$.

Problem 8.   A derangement (or fixed-point-free permutation) of $\{1, \ldots, n\}$ is a permutation f such that $f(j) \neq j$ for all $j = 1, \ldots, n$. What is the number of derangements of n objects?

Solution 8.   There are $n!$ permutations. Using

$$\binom{n}{n} = 1, \qquad \binom{n}{1} = (n - 1)!$$

we find that the number of derangements of n objects is given by

$$n! - n(n-1)! + \binom{n}{2}(n-2)! - \cdots + (-1)^n = n! \sum_{j=2}^{n}(-1)^j \frac{1}{j!}.$$

Problem 9. *Bell numbers* are the sequence

$$\{\, 1, 1, 2, 5, 15, 52, 203, 877, 4140, \ldots \,\}.$$

The numbers count (starting from 0) the ways that n distinguishable objects can be grouped into sets if no set can be empty. For example the letters $A\ B\ C$ can be grouped into sets so that

1) A, B, C are in three separate sets: $\{A\}$, $\{B\}$, $\{C\}$
2) A and B are together and C is separate: $\{A, B\}$, $\{C\}$
3) A and C together and B separate: $\{A, C\}$, $\{B\}$
4) B and C together and A separate: $\{B, C\}$, $\{A\}$
5) A, B, C are all together in a single set: $\{A, B, C\}$

Thus for $n = 3$ there are five partitions. Thus the third Bell number is 5.
(i) Let P_n denote the nth Bell number, i.e. the number of all partitions of n objects. Then we have

$$P_n = \frac{1}{e} \sum_{k=0}^{\infty} \frac{k^n}{k!}.$$

Find a recurrence relation for P_n.
(ii) Find the MacLaurin expansion (expansion around 0) of $\exp(\exp(x))$ and establish a connection with the Bell numbers.

Solution 9. (i) Let S be the set to be partitioned and $x \in S$. If the class containing x has k elements, it can be chosen in

$$\binom{n-1}{k-1}$$

ways and the remaining $n - k$ elements can be partitioned in P_{n-k} ways. Thus the number of partitions in which the class containing x has k elements is

$$\binom{n-1}{k-1} P_{n-k}.$$

This remains true for $k = n$ if we set $P_0 = 1$. Thus

$$P_n = \sum_{k=1}^{n} \binom{n-1}{k-1} P_{n-k} = \sum_{k=0}^{n-1} \binom{n-1}{k} P_k.$$

(ii) Since $d(\exp(\exp(x)))/dx = \exp(x)\exp(\exp(x))$ and $\exp(0) = 1$ we have

$$\exp(\exp(x)) = \sum_{j=0}^{\infty} \frac{e^{jx}}{j!} = \sum_{k=0}^{\infty}\sum_{j=0}^{\infty} \frac{(jx)^k}{j!k!} = \sum_{k=0}^{\infty} \left(\sum_{j=0}^{\infty} \frac{j^k}{j!} \right) \frac{x^k}{k!}$$

$$= \exp\left(\sum_{k=0}^{\infty} P_k \frac{x^k}{k!} \right).$$

Thus we find

$$\exp(\exp(x)) = e\left(1 + \frac{1x}{1!} + \frac{2x^2}{2!} + \frac{5x^3}{3!} + \cdots \right).$$

Thus the coefficients 1, 1, 2, 5, ... are the Bell numbers.

Problem 10.   Let $d \geq 1$. Consider a bitstring of length d. Then we can form 2^d bitstrings of length d.
(i) Let $d \geq 2$ and even. How many bitstrings can be formed with $d/2$ 1's and $d/2$ 0's? Consider first the case with $d = 4$.
(ii) Let $|0\rangle$, $|1\rangle$ be an orthonormal basis in $\mathbb{C}^2$. Find an orthonormal basis in $\mathbb{C}^4$. Find an orthonormal basis in $\mathbb{C}^8$.

Solution 10.   (i) For $d = 4$ we have 0011, 0101, 0110, 1001, 1010, 1100. In general we can form

$$\frac{d!}{(d/2)!(d/2)!}$$

bitstrings.
(ii) For $\mathbb{C}^4$ we have $4 = 2^2$ vectors

$$|0\rangle \otimes |0\rangle, \quad |0\rangle \otimes |1\rangle, \quad |1\rangle \otimes |0\rangle, \quad |1\rangle \otimes |1\rangle$$

and for $\mathbb{C}^8$ we have $8 = 2^3$ vectors

$$|0\rangle \otimes |0\rangle \otimes |0\rangle, \quad |0\rangle \otimes |0\rangle \otimes |1\rangle, \quad |0\rangle \otimes |1\rangle \otimes |0\rangle, \quad |0\rangle \otimes |1\rangle \otimes |1\rangle,$$

$$|1\rangle \otimes |0\rangle \otimes |0\rangle, \quad |1\rangle \otimes |0\rangle \otimes |1\rangle, \quad |1\rangle \otimes |1\rangle \otimes |0\rangle, \quad |1\rangle \otimes |1\rangle \otimes |1\rangle.$$

These eight vectors form an orthonormal basis in $\mathbb{C}^8$.

Problem 11.   (i) Consider N boxes and n particles with $n \leq N$. In every box we can put a maximum of one particle (*Pauli principle, Fermi particles*). In how many ways can we put the n particles (which are identical) in the N boxes?

(ii) Consider N boxes and an arbitrary number of particles n. We can put an arbitrary number of particles in every box (*Bose-particles*). Again the particles are identical. In how many ways can we put the n particles in the N boxes?

(iii) The same as case (ii) but now we have distinguishable particles.

Solution 11. (i) The problem is equivalent to that of determining the total number of ways in which N objects of which n are indistinguishable from type I (say particles) and $N - n$ are indistinguishable from type II (say empty spaces, holes) can be arranged in N possible places. The total number of permutations of N objects is $N!$. From each of these $N!$ arrangements we obtain $n!$ mutually indistinguishable arrangements by permutation of the $n!$ identical particles among themselves. From each of these $n!$ we obtain again $(N - n)!$ mutually indistinguishable arrangements by permutating the $(N - n)!$ identical holes among themselves. The $N!$ possible arrangements can therefore be partitioned into disjoint sets, each containing

$$n!(N - n)!$$

indistinguishable arrangements. The total number of distinguishable arrangements is therefore

$$\frac{N!}{n!(N - n)!} \equiv \binom{N}{n}.$$

We have the recursion relation

$$\binom{N + 1}{n + 1} = \binom{N}{n} + \binom{N}{n + 1}, \qquad \binom{1}{1} := 1.$$

(ii) We select a box to begin with. The number of ways we can choose the $N - 1$ boxes and the n-particles are $(N - 1 + n)!$. However the boxes and particles are indistinguishable. Thus we find

$$\binom{n + N - 1}{n} \equiv \frac{(n + N - 1)!}{n!(n + N - 1 - n)!} \equiv \frac{(n + N - 1)!}{n!(N - 1)!}.$$

(iii) For distinguishable particles the first particle can be put in any of the N boxes, the second one in any of N boxes etc. Thus we find the desired value to be N^n.

Problem 12. A man, who lives at location A of the city street plan shown in the figure, walks daily to the home of his friend, who lives m blocks east and n blocks north of A, at location B. He can only walk north or east. In how many different ways can he go from A to B? Consider first the case $n = m = 2$.

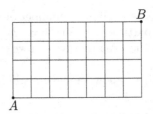

Solution 12.   We consider first the case $n = m = 2$. Each of the paths can be written as vector

$$(u, u, r, r) \quad (u, r, u, r) \quad (u, r, r, u) \quad (r, u, u, r) \quad (r, u, r, u) \quad (r, r, u, u)$$

where u stands for one step north ("up") and r stands for one step east ("right"). Thus the number of paths for $n = m = 2$ is given by 6.
The general case is as follows: At each corner the man may walk either up or to the right. Consequently, a particular path is specified by a sequence of the form $(u, u, r, u, \ldots, r)$, where the total number of $u's$ and $r's$ are n and m, respectively. The number of different ways of writing such a sequence and therefore the total number of paths is given by

$$\frac{(m+n)!}{n!\, m!}.$$

Problem 13.   The *Stirling number* of the second kind $S(n, m)$ is the number of partitions of a set with n elements into m classes.
(i) Let $X := \{a, b, c, d\}$. Thus $n = 4$. Let $m = 2$. Find the partitions.
(ii) Show that $S(n, m)$ satisfies the linear difference equation

$$S(n+1, m) = S(n, m-1) + mS(n, m) \tag{1}$$

with the initial conditions $S(n, 1) = S(n, n) = 1$.

Solution 13.   (i) We have seven partitions

$$\{\, \{a\}, \{b, c, d\}\, \}; \ \{\, \{b\}, \{a, c, d\}\, \}; \ \{\, \{c\}, \{a, b, d\}\, \}; \ \{\, \{d\}, \{a, b, c\}\, \};$$

$$\{\, \{a, b\}, \{c, d\}\, \}; \ \{\, \{a, c\}, \{b, d\}\, \}; \ \{\, \{a, d\}, \{b, c\}\, \} \, \}.$$

(ii) Consider a set with n elements, i.e.

$$X := \{\, x_1, x_2, \ldots, x_n \,\}.$$

Let $S(n, m-1)$ be the number of partitions into $m-1$ classes. One can obtain $S(n, m-1)$ partitions into m classes of a set with $n+1$ elements

$x_1, \ldots, x_{n+1}$ by adding to each partition a new class consisting of only the element x_{n+1}. The element x_{n+1} can be added to each of the already existing m classes of a partition of X in m distinct ways. These two procedures yield, without repetitions, all the partitions of the set X into m classes. Equation (1) follows. The solution to (1), taking into account the initial conditions, is given by

$$S(n, m) = \frac{1}{m!} \sum_{k=0}^{m-1} (-1)^k \binom{m}{k} (m-k)^n.$$

Problem 14. (i) Let $\mathbb{Z}$ be the set of integers. Then $\mathbb{Z}^d$ $(d = 1, 2, 3, \cdots)$ denotes the d-tuple of the set of integers, i.e.

$$\mathbb{Z}^d := \mathbb{Z} \times \mathbb{Z} \times \cdots \times \mathbb{Z} \qquad d - \text{times}.$$

Find the number of all closed paths A_n with n-steps starting from $(0, \ldots, 0)$ $\in \mathbb{Z}^d$ and returning to $(0, \ldots, 0) \in \mathbb{Z}^d$ (not necessarily the first time).
(ii) Calculate A_4 for $d = 2$.

Solution 14. (i) Obviously A_n is given by

$$A_n = \sum_{\{\Delta_1, \Delta_2, \cdots, \Delta_n\}} \delta(\Delta_1 + \Delta_2 + \cdots + \Delta_n)$$

where

$$\delta(\Delta_1 + \Delta_2 + \cdots + \Delta_n) := \begin{cases} 1 & \text{for} \quad \Delta_1 + \Delta_2 + \cdots + \Delta_n = 0 \\ 0 & \text{otherwise} \end{cases}.$$

If n is odd we find $A_n = 0$. Now we apply the *completeness relation*

$$\delta(\Delta_1 + \Delta_2 + \cdots + \Delta_n) := \frac{1}{N} \sum_{k \in 1.BZ} e^{ik(\Delta_1 + \Delta_2 + \cdots + \Delta_n)}$$

where N denotes the lattice sites in the first Brillouin zone (1.BZ). It follows that

$$A_n = \sum_{\{\Delta_1, \ldots, \Delta_n\}} \frac{1}{N} \sum_{k \in 1.BZ} e^{ik(\Delta_1 + \cdots + \Delta_n)}$$

$$= \frac{1}{N} \sum_{k \in 1.BZ} \sum_{\{\Delta_1, \ldots, \Delta_n\}} e^{ik(\Delta_1 + \cdots + \Delta_n)}.$$

Thus we can write

$$A_n = \frac{1}{N} \sum_{k \in 1.BZ} \sum_{\{\Delta_1\}} e^{ik\Delta_1} \sum_{\{\Delta_2\}} e^{ik\Delta_2} \cdots \sum_{\{\Delta_n\}} e^{ik\Delta_n} = \frac{1}{N} \sum_{k \in 1.BZ} (\epsilon(k))^n$$

where

$$\epsilon(\mathbf{k}) := \sum_{\Delta} e^{i\mathbf{k}\Delta}.$$

(ii) In two dimensions $(d = 2)$, we have $\epsilon(\mathbf{k}) = 2(\cos(k_1) + \cos(k_2))$ since $\Delta \in \{(1,0), (0,1), (-1,0), (0,-1)\}$. Thus we find

$$A_4 = \frac{1}{N} \sum_{\mathbf{k} \in 1.BZ} (\epsilon(\mathbf{k}))^4 = \frac{2^4}{4\pi^2} \int_{-\pi}^{\pi} \int_{-\pi}^{\pi} (\cos(k_1) + \cos(k_2))^4 dk_1 dk_2 = 36.$$

The *first Brillouin zone* in three dimensions is constructed as follows: Let $\mathbf{a}_1$, $\mathbf{a}_2$ and $\mathbf{a}_3$ be a set of primitive vectors for the direct lattice. Then the *reciprocal lattice* can be generated by the three primitive vectors

$$\mathbf{b}_1 = 2\pi \frac{\mathbf{a}_2 \times \mathbf{a}_3}{\mathbf{a}_1 \cdot (\mathbf{a}_2 \times \mathbf{a}_3)}, \quad \mathbf{b}_2 = 2\pi \frac{\mathbf{a}_3 \times \mathbf{a}_1}{\mathbf{a}_1 \cdot (\mathbf{a}_2 \times \mathbf{a}_3)}, \quad \mathbf{b}_3 = 2\pi \frac{\mathbf{a}_1 \times \mathbf{a}_2}{\mathbf{a}_1 \cdot (\mathbf{a}_2 \times \mathbf{a}_3)}.$$

The simple cubic Bravais lattice, with cubic primitive cell of side a, has as its reciprocal a simple cubic lattice with cubic primitive cell of side $2\pi/a$. The construction of the first Brillouin zone is as follows: choose any reciprocal lattice point as the origin. Draw the vectors connecting this point with (all) other lattice points. Next, construct a set of planes that are perpendicular bisectors of these vectors. The smallest solid figure containing the origin is the first Brillouin zone. Points $\mathbf{k}$ on the surface must satisfy the condition $\mathbf{k}^2 = (\mathbf{k} - \mathbf{K}_n)^2$ or $\mathbf{K}_n^2 - 2\mathbf{k} \cdot \mathbf{K}_n = 0$ for some reciprocal lattice vector $\mathbf{K}_n$. We need only consider those $\mathbf{k}$ values lying within the zone. Any $\mathbf{k}$ vector may be written as

$$\mathbf{k} = \left(\frac{h_1}{2N_1 + 1}\right) \mathbf{b}_1 + \left(\frac{h_2}{2N_2 + 1}\right) \mathbf{b}_2 + \left(\frac{h_3}{2N_3 + 1}\right) \mathbf{b}_3$$

where the h_i are integers in the range $-N_i \leq h_i \leq N_i$. The values of $\mathbf{k}$ that are given by the above formula may not all be within the first Brillouin zone. However, it is always possible to bring any such $\mathbf{k}$ outside the zone back into it by translation by a reciprocal lattice vector. The values of $\mathbf{k}$ form a uniform and dense distribution. In the limit in which the N_i are allowed to become infinite, we may convert sums over possible $\mathbf{k}$ values into integrals over the first Brillouin zone through the relation

$$\sum_{\mathbf{k} \in 1.BZ} \to \frac{N}{V} \int_{1.BZ} d\mathbf{k}$$

where $N = (2N_1 + 1)(2N_2 + 1)(2N_3 + 1)$ is the number of unit cells and V is the volume of the first Brillouin zone.

Problem 15. Find the probability $P_N(\mathbf{R})$ that a random walker reaches a lattice position $\mathbf{R}$,

$$\mathbf{R} = N_1 \mathbf{a}_1 + N_2 \mathbf{a}_2 + N_3 \mathbf{a}_3$$

after $N = N_1 + N_2 + N_3$ steps on a lattice with unit vectors $\mathbf{a}_i$. The lattice is the simple cubic lattice and it is periodic with a periodicity L. The transition probability $W(\mathbf{R})$ for a unit displacement is given by

$$W(\mathbf{R}) := \begin{cases} \frac{1}{6} & \text{for} \quad \mathbf{R} = (0,0,\pm1), (0,\pm1,0), (\pm1,0,0) \\ 0 & \text{otherwise} \end{cases} . \tag{1}$$

Solution 15. The probability satisfies the recurrence formula

$$P_{N+1}(\mathbf{R}) = \sum_{\mathbf{R}'} W(\mathbf{R} - \mathbf{R}') P_N(\mathbf{R}')$$

where $W(\mathbf{R} - \mathbf{R}')$ is the probability of transition from $\mathbf{R}'$ to $\mathbf{R}$. We introduce a *generating function*

$$\Phi(\mathbf{R}, z) := \sum_{N=0}^{\infty} z^N P_N(\mathbf{R}).$$

Obviously

$$\Phi(\mathbf{R}, z) - z \sum_{\mathbf{R}'} W(\mathbf{R} - \mathbf{R}') \Phi(\mathbf{R}', z) = \delta_{\mathbf{R},0} \tag{2}$$

which expresses the fact that the walker starts walking from the origin

$$P_0(\mathbf{R}) = \delta_{\mathbf{R},0} = \begin{cases} 1, & \text{for} \quad \mathbf{R} = 0 \\ 0, & \text{otherwise} \end{cases}$$

where δ is the Kronecker delta. We solve (2) by using the *Fourier transform method*. Multiplying both sides by $\exp(2\pi i \mathbf{k} \cdot \mathbf{R}/L)$ and summing over $\mathbf{R}$, we find

$$\widetilde{\Phi}(\mathbf{k}, z) - z\lambda(\mathbf{k})\widetilde{\Phi}(\mathbf{k}, z) = 1$$

with

$$\widetilde{\Phi}(\mathbf{k}, z) := \sum_{\mathbf{R}} \exp\left(\frac{2\pi i \mathbf{k} \cdot \mathbf{R}}{L}\right) \Phi(\mathbf{R}, z)$$

$$\lambda(\mathbf{k}) := \sum_{\mathbf{R}} W(\mathbf{R}) \exp\left(\frac{(2\pi i \mathbf{k} \cdot \mathbf{R})}{L}\right).$$

The Fourier inverse of $\widetilde{\Phi}(\mathbf{k}, z)$ is

$$\Phi(\mathbf{R}, z) = \frac{1}{L^3} \sum_{\mathbf{k}} \frac{\exp(-2\pi i \mathbf{k} \cdot \mathbf{R}/L)}{1 - z\lambda(\mathbf{k})}$$

which, in the limit of $L \to \infty$, gives

$$\lim_{L \to \infty} \Phi(\mathbf{R}, z) = \frac{1}{(2\pi)^3} \int_{-\pi}^{\pi} \int_{-\pi}^{\pi} \int_{-\pi}^{\pi} \frac{\exp(-i\boldsymbol{\Theta} \cdot \mathbf{R})}{1 - z\lambda(\boldsymbol{\Theta})} d\boldsymbol{\Theta}$$

where $d\Theta \equiv d\Theta_1 d\Theta_2 d\Theta_3$ and $\Theta \cdot \mathbf{R} := \Theta_1 R_1 + \Theta_2 R_2 + \Theta_3 R_3$. The probability $P_N(\mathbf{R})$ is then given by

$$P_N(\mathbf{R}) = \frac{1}{(2\pi)^3} \int_{-\pi}^{\pi} \int_{-\pi}^{\pi} \int_{-\pi}^{\pi} (\lambda(\Theta))^N \exp(-i\Theta \cdot \mathbf{R}) d\Theta.$$

Owing to (1) we find

$$\lambda(\Theta) = \frac{1}{6} \sum_{j=1}^{3} \left(e^{i\Theta_j} + e^{-i\Theta_j} \right) = \frac{1}{3} \sum_{j=1}^{3} \cos(\Theta_j).$$

Thus

$$P_N(\mathbf{R}) = \frac{1}{(2\pi)^3} \int_{-\pi}^{\pi} \int_{-\pi}^{\pi} \int_{-\pi}^{\pi} \left(\frac{1}{3} (\cos(\Theta_1) + \cos(\Theta_2) + \cos(\Theta_3)) \right)^N e^{-i\Theta \cdot \mathbf{R}} d\Theta$$

and

$$\Phi(\mathbf{R}, z) = \frac{1}{(2\pi)^3} \int_{-\pi}^{\pi} \int_{-\pi}^{\pi} \int_{-\pi}^{\pi} \frac{\exp(-i\Theta \cdot \mathbf{R})}{1 - z(\cos(\Theta_1) + \cos(\Theta_2) + \cos(\Theta_3))/3} d\Theta.$$

Problem 16.   Consider a lattice with N lattice sites. Each lattice site can only be occupied by two electrons: one with spin up and one with spin down (*Pauli principle*). Let N_e be the number of electrons we put in the lattice. Obviously, $0 \le N_e \le 2N$.
(i) Given N and N_e, find the number of ways to occupy the lattice with electrons.
(ii) Consider the case $N_e = N$.
(iii) Consider the case $N_e = N$ ($N =$ even) and $S_z = 0$, where S_z denotes the total spin.

Solution 16.   (i) Let a be the number of lattice sites occupied by one electron with spin up, b the number of lattice sites occupied by one electron with spin down, c the number of lattice sites occupied by two electrons (one spin up, one spin down) and d the unoccupied lattice sites. Obviously we have the conditions $a + b + 2c = N_e$, $a + b + c + d = N$. Thus the number of ways to occupy the lattice is

$$\frac{N!}{a! \, b! \, c! \, d!}$$

with the constraints given by (1).
(ii) If $N = N_e$ we have $d = c$ and thus

$$\frac{N!}{a! \, b! \, c! \, c!}.$$

(iii) If $N = N_e$ and if the total spin S_z is equal to 0 ($N = $ even) we find $a = b$ and therefore $a = N/2 - c$. Thus

$$\frac{N!}{\left(\frac{N}{2} - c\right)! \left(\frac{N}{2} - c\right)! \, c! \, c!}.$$

Consequently the total number of states with $N = N_e$ and $S_z = 0$ is given by

$$\sum_{c=0}^{N/2} \frac{N!}{\left(\frac{N}{2} - c\right)! \left(\frac{N}{2} - c\right)! \, c! \, c!}.$$

For example, for $N = N_e = 4$ and $S_z = 0$ we find 36.

Problem 17. Consider the problem of randomly placing r balls into n cells. The balls are indistinguishable. Let r_k be the number of balls in the k-th cell. Every n-tuple of integers satisfying

$$r_1 + r_2 + \cdots + r_n = r \tag{1}$$

describes a possible configuration of occupancy numbers. With indistinguishable balls two distributions are distinguishable only if the corresponding n-tuples are not identical.

(i) Show that the number of distinguishable distributions (i.e. the number of different solutions of (1)) is

$$A_{r,n} = \binom{n+r-1}{r} \equiv \frac{(n+r-1)!}{r!(n-1)!}.$$

(ii) Show that the number of distinguishable distributions, in which no cell remains empty, is

$$\binom{r-1}{n-1}.$$

Solution 17. (i) We represent the balls by stars and indicate the n cells by the n spaces between $n + 1$ bars. Thus, for example,

$$|\ast\ast\ast|\ast||||\ast\ast\ast\ast|$$

is used as a symbol for a distribution of $r = 8$ balls in $n = 6$ cells with occupancy number $3, 1, 0, 0, 0, 4$. Such a symbol necessarily starts and ends with a bar, but the remaining $n - 1$ bars and r stars can appear in an arbitrary order. Thus the number of distinguishable distributions equals the number of ways of selecting r places out of $n + r - 1$, namely $A_{r,n}$.

(ii) The condition that no cell be empty imposes the restriction that no two bars are adjacent and $r \geq n$. The r stars leave $r - 1$ spaces of which $n - 1$ are to be occupied by bars. Thus we have

$$\binom{r-1}{n-1} \equiv \frac{(r-1)!}{(n-1)!(r-n)!}$$

choices.

Problem 18.   (i) A numerical partition of a positive integer n is a sequence

$$p_1 \geq p_2 \geq \cdots \geq p_k \geq 1$$

such that

$$p_1 + p_2 + \cdots + p_k = n.$$

Each p_j is called a part. For example, $18 = 7+4+4+1+1+1$ is a partition of 18 into 6 parts. The number of partitions of n into k parts is denoted by $p(n, k)$. Find $p(7, 3)$.
(ii) Show that the recurrence for $p(n, k)$ is given by

$$p(n, k) = p(n - 1, k - 1) + p(n - k, k)$$

with the initial conditions $p(n, 0) = 0$, $p(k, k) = 1$. Obviously, $p(n, 1) = 1$.
(iii) Every numerical partition of a positive integer n corresponds to a unique *Ferrer's diagram*. A Ferrer's diagram of a partition is an arrangement of n dots on a square grid, where a part j in the partition is represented by placing p_j dots in a row. This means we represent each term of the partition by a row of dots, the terms in descending order with the largest at the top. Sometimes it is more convenient to use squares instead of dots (in this case the diagram is called a *Young diagram*). The partition we obtain by reading the Ferrer's diagram by columns instead of rows is called the conjugate of the original partition. Find the conjugate of the partition

$$18 = 7 + 4 + 4 + 1 + 1 + 1.$$

Solution 18.   (i) We have

$$7 = 5 + 1 + 1 = 4 + 2 + 1 = 3 + 3 + 1 = 3 + 2 + 2.$$

Thus $p(7, 3) = 4$.
(ii) Consider the cases $k = 1$ and $k > 1$. For $k = 1$ we have $p(n, 1) = 1$. For $k > 1$ we use induction.
(iii) We have $18 = 6 + 3 + 3 + 3 + 1 + 1 + 1$.

Supplementary Problems

Problem 1. Let $n, k \in \mathbb{N}$. Show that the number of ways one can form n digit ternary sequences (i.e. only 0,1,2) with k 1's is given by

$$\binom{n}{k} \cdot 2^{n-k}.$$

For example for $n = 3$ and $k = 1$ we have in lexicographical order

001, 010, 012, 021, 100, 102, 120, 122, 201, 210, 212, 221

i.e. 12 three digit ternary sequences.

Problem 2. Let $b^\dagger$, b be Bose creation and annihilation operators, i.e. $[b, b^\dagger] = I$, where I is the identity operator. Let $n \geq 1$. Show that the *Stirling numbers* of second kind are given by

$$(b^\dagger b)^n = \sum_{k=1}^{n} S(n, k)(b^\dagger)^k b^k.$$

Problem 3. How many differential two-forms $dx_j \wedge dx_k$ can we form in $\mathbb{R}^4$? Note that $dx_j \wedge dx_k = -dx_k \wedge dx_j$ and hence $dx_j \wedge dx_j = 0$.

Chapter 17

Number Theory

In number theory we are dealing with positive integers $\mathbb{N}$, non-negative integers $\mathbb{N}_0$, integers $\mathbb{Z}$, rational numbers $\mathbb{Q}$, real numbers $\mathbb{R}$ and complex numbers $\mathbb{C}$, where $\mathbb{N}$, $\mathbb{N}_0$, $\mathbb{Z}$, $\mathbb{Q}$ are countable. We have

$$\mathbb{N} \subset \mathbb{N}_0 \subset \mathbb{Z} \subset \mathbb{Q} \subset \mathbb{R} \subset \mathbb{C}.$$

Note that $\mathbb{N}_0 \times \mathbb{N}_0$, $\mathbb{Z} \times \mathbb{Z}$ are also countable. The set of integers form an abelian group under addition.

Consider an associative algebra of rank 4 with the basis elements 1, I, J, K, where 1 is the identity element, i.e. $1 * I = I$, $1 * J = J$, $1 * K = K$. The compositions are $I * I = J * J = K * K = -1$, $I * J = K$, $J * K = I$, $K * I = J$, $J * I = -K$, $K * J = -I$, $I * K = -J$. This is the so-called *quaternion algebra*. A matrix representation is

$$1 \mapsto \begin{pmatrix} 1 & 0 \\ 0 & 1 \end{pmatrix}, \quad I \mapsto -i \begin{pmatrix} 0 & 1 \\ 1 & 0 \end{pmatrix} = -i\sigma_1$$

$$J \mapsto -i \begin{pmatrix} 0 & -i \\ i & 0 \end{pmatrix} = -i\sigma_2, \quad K \mapsto -i \begin{pmatrix} 1 & 0 \\ 0 & -1 \end{pmatrix} = -i\sigma_3$$

where σ_1, σ_2, σ_3 are the Pauli spin matrices.

In mathematical physics the numbers e, π, φ (*golden mean number*) play a central role with e given by

$$e^x = \sum_{j=0}^{\infty} \frac{x^j}{j!} = \lim_{n \to \infty} \left(1 + \frac{x}{n}\right)^n, \quad x = 1$$

with (Euler identity) $e^{2\pi i} = 1$ and $e^{i\alpha} = \cos(\alpha) + i\sin(\alpha)$. Note that π can be defined as

$$\pi = 4\arctan(1) = 4\arctan(1/2) + 4\arctan(1/3)$$

and the golden mean number is

$$\varphi = \frac{1}{2}(1 + \sqrt{5}), \qquad \frac{1}{\varphi} = \frac{1}{2}(-1 + \sqrt{5}).$$

φ satisfies the quadratic equation $\varphi^2 - \varphi - 1 = 0$. The golden mean number can also be found from the recursion $\varphi_{\tau+1} = 1 + 1/\varphi_\tau$ ($\tau = 0, 1, \ldots$) and $\varphi_0 = 3/2$.

In the foundation of theoretical physics the Planck units, i.e. Planck length, Planck time, Planck mass, Planck charge and Planck temperature are of central importance.

Dimensionless quantities are the fine structure constant

$$\alpha = \frac{e^2}{4\pi\epsilon_0\hbar c} \approx \frac{1}{137.035999} = 0.0072973525664$$

and the gravitational fine structure constant

$$\alpha_\gamma = \frac{Gm_e}{\hbar c} \approx 1.7518 \cdot 10^{-45}$$

where m_e is the mass of the electron. The fine structure constant describes the strength of the electromagnetic interaction between charged particles. The gravitational coupling constant describes the gravitational attraction between a given pair of elementary particles. The electrostatic force between an electron and proton $|\mathbf{F}_c| = e^2/(4\pi\epsilon_0 r^2)$ and the gravitational force $|\mathbf{F}_G| = Gm_e m_p/r^2$ have a ratio independent of distance

$$|\mathbf{F}_c|/|\mathbf{F}_G| = e^2/(4\pi\epsilon_0 Gm_e m_p) \approx 2.27 \cdot 10^{39}.$$

The *Eddington-Dirac number* is

$$\frac{e^2}{4\pi\epsilon_0 Gm_e^2} = 4.16 \cdot 10^{42}.$$

The large number hypothesis predicts that the gravitational constant G should vary with time. The dimensionless proton to electron ration is given by

$$\frac{m_p}{m_e} = 1836.152672(80).$$

The *impedance of vacuum* Z_0 is a physical constant relating the electric and magnetic fields of electromagnetic radiation traveling through free space

$$Z_0 = \sqrt{\frac{\mu_0}{\epsilon_0}} = 376.730\,\text{Ohm}.$$

Problem 1.   (i) Show that

$$\frac{\sqrt{5}-1}{2} = \cfrac{1}{1+\cfrac{1}{1+\cfrac{1}{1+\cfrac{1}{\cdots}}}} \tag{1}$$

The irrational number $(\sqrt{5}-1)/2$ is called the *golden mean number*. This is the *worst irrational number* due to the representation (1).

(ii) The resistances of the resistors of the following network

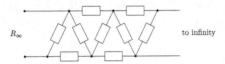

R_∞ to infinity

are all equal (say R). Calculate R_∞.

Solution 1.   (i) We have the identity

$$\frac{\sqrt{5}-1}{2} \equiv \frac{(\sqrt{5}-1)(\sqrt{5}+1)}{2(\sqrt{5}+1)} \equiv \frac{4}{2(\sqrt{5}+1)} \equiv \frac{2}{\sqrt{5}+1} \equiv \frac{1}{(\sqrt{5}+1)/2}.$$

Now we also have the identity

$$\frac{\sqrt{5}+1}{2} \equiv \frac{2+\sqrt{5}-2+1}{2} \equiv 1 + \frac{\sqrt{5}-1}{2} \equiv 1 + \frac{(\sqrt{5}-1)(\sqrt{5}+1)}{2(\sqrt{5}+1)}$$

$$\equiv 1 + \frac{4}{2(\sqrt{5}+1)}.$$

Thus the continued fraction (1) follows.

(ii) Since

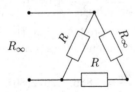

R_∞

we obtain

$$R_\infty = \cfrac{1}{\cfrac{1}{R} + \cfrac{1}{R+R_\infty}} = \frac{1}{\cfrac{R_\infty+2R}{R(R+R_\infty)}} = \frac{R(R+R_\infty)}{R_\infty+2R}.$$

The quadratic equation $R_\infty^2 + R_\infty R - R^2 = 0$ follows. The solution of the quadratic equation is given by

$$R_\infty = \frac{(\sqrt{5}-1)}{2}R.$$

Problem 2. (i) Show that

$$\sqrt{3} - 1 \equiv \cfrac{1}{1 + \cfrac{1}{2 + \cfrac{1}{1 + \cfrac{1}{2 + \cfrac{1}{\cdots}}}}}$$

(ii) The resistances of the resistors in the following network

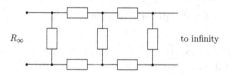

R_∞ to infinity

are all equal (say R). Calculate R_∞.

Solution 2. (i) We have

$$\sqrt{3} - 1 \equiv \frac{(\sqrt{3}-1)(\sqrt{3}+1)}{\sqrt{3}+1} \equiv \frac{2}{\sqrt{3}+1} = \frac{1}{(\sqrt{3}+1)/2}. \tag{1}$$

The denominator can be written as

$$\frac{\sqrt{3}+1}{2} \equiv \frac{1+(\sqrt{3}-1)+1}{2} \equiv 1 + \frac{\sqrt{3}-1}{2}.$$

Now

$$\frac{\sqrt{3}-1}{2} \equiv \frac{(\sqrt{3}-1)(\sqrt{3}+1)}{2\sqrt{3}+1} \equiv \frac{1}{\sqrt{3}+1}$$

and

$$\sqrt{3}+1 \equiv 2 + (\sqrt{3}-1). \tag{2}$$

Combining (1) and (2) gives

$$\sqrt{3} - 1 \equiv \cfrac{1}{1 + \cfrac{\sqrt{3}-1}{2}} \equiv \cfrac{1}{1 + \cfrac{1}{\sqrt{3}+1}} \equiv \cfrac{1}{1 + \cfrac{1}{2 + (\sqrt{3}-1)}}.$$

Thus

$$\sqrt{3} - 1 = \cfrac{1}{1 + \cfrac{1}{(2 + \sqrt{3} - 1)}}.$$

(ii) Since

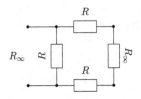

we have

$$R_\infty = \cfrac{1}{\cfrac{1}{R} + \cfrac{1}{2R + R_\infty}} \equiv \frac{R(2R + R_\infty)}{3R + R_\infty}.$$

Thus we obtain the quadratic equation $R_\infty^2 + 2RR_\infty - 2R^2 = 0$. The solution of this quadratic equation is given by $R_\infty = (\sqrt{3} - 1)R$.

Problem 3.   Let $x \in \mathbb{R}$. Let $N > 1$ be an integer. Show that there is an integer p and an integer q such that $1 \le q \le N$ and

$$\left| x - \frac{p}{q} \right| \le \frac{1}{qN}.$$

Solution 3.   The proof relies on the *pigeon hole principle* (also called the *Dirichlet box principle*). This principle says that if we put $N + 1$ objects into N boxes, then at least one of the boxes must contain more than one of the objects. We rewrite our statement as follows: There is an integer p and an integer q such that $1 \le q \le N$ and

$$|qx - p| \le \frac{1}{N}.$$

We denote by $[x]$ the largest integer $\le x$. For example $[\pi] = 3$. Consider the numbers $0 - 0 \cdot x$, $x - [x]$, $2x - [2x]$, $\ldots$ $Nx - [Nx]$. This means we consider the numbers $kx - [kx]$ for $k = 0, 1, \ldots, N$. These are $N + 1$ numbers. They will lie between 0 and 1. We divide the interval $[0, 1]$ into N equal subintervals

$$[0, 1/N), \quad [1/N, 2/N), \quad [2/N, 3/N), \quad \ldots, \quad [(N-1)/N, 1].$$

These N subintervals are the boxes. There are N of them, and so two of the numbers $kx - [kx]$ ($0 \le k \le N$) must lie in the same subinterval; that is, they can be no further apart than $1/N$. Say these two numbers are $nx - [nx]$ and $mx - [mx]$, where $0 \le n < m \le N$. Set $q = m - n$ and $p = [mx] - [nx]$. Then

$$|qx - p| = |(nx - [nx]) - (mx - [mx])| \le \frac{1}{N}.$$

Moreover, $q = m - n \le m \le N$, and $q \ge 1$, since $m - n > 0$.

Problem 4. The *Farey sequence* F_N is the set of all fractions in lowest terms between 0 and 1 whose denominators do not exceed N, arranged in order of magnitude. For example, F_6 is given by

$$\frac{0}{1} \quad \frac{1}{6} \quad \frac{1}{5} \quad \frac{1}{4} \quad \frac{1}{3} \quad \frac{2}{5} \quad \frac{1}{2} \quad \frac{3}{5} \quad \frac{2}{3} \quad \frac{3}{4} \quad \frac{4}{5} \quad \frac{5}{6} \quad \frac{1}{1}.$$

N is known as the order of the series. Let m_1/n_1 and m_2/n_2 be two successive terms in F_N. Then

$$m_2 n_1 - m_1 n_2 = 1. \tag{1}$$

For three successive terms m_1/n_1, m_2/n_2 and m_3/n_3, the middle term is the *mediant* of the other two

$$m_2/n_2 = (m_1 + m_3)/(n_1 + n_3). \tag{2}$$

(i) Find F_4.
(ii) Prove that (1) implies (2). Prove that (2) implies (1).

Solution 4. (i) We have

$$\frac{0}{1} \quad \frac{1}{4} \quad \frac{1}{3} \quad \frac{1}{2} \quad \frac{2}{3} \quad \frac{3}{4} \quad \frac{1}{1}.$$

(ii) To prove that (1) implies (2), assume we have two identities

$$m_2 n_1 - m_1 n_2 = 1, \qquad m_3 n_2 - m_2 n_3 = 1.$$

Subtract one from the other and recombine the terms yields

$$(m_3 + m_1)n_2 = m_2(n_3 + n_1)$$

which leads to (2). To prove that (2) implies (1) we use mathematical induction. Assume that we are given a series of fractions m_i/n_i $(i = 1, 2, \ldots)$ with the condition that any three consecutive terms satisfy

$$\frac{m_i}{n_i} = \frac{m_{i-1} + m_{i+1}}{n_{i-1} + n_{i+1}}. \tag{3}$$

We show that for such a series (1) also holds. However (1) only contains two fractions. Therefore, let us assume that for $i = 2$, equation (1) indeed holds. This is clearly true for F_N which starts with $0/1$, $1/N$. Assume that for some $i > 2$

$$m_i n_{i-1} - m_{i-1} n_i = 1. \tag{4}$$

We rewrite (3) as

$$(m_{i+1} + m_{i-1}) n_i = m_i (n_{i+1} + n_{i-1}). \tag{5}$$

Subtracting (5) from (3) gives $m_{i+1} n_i - m_i n_{i+1} = 1$. Equation (5) implies (3) only if $n_i \neq 0$.

Problem 5.   A real-valued function f, defined on the rational numbers $\mathbb{Q}$, satisfies $f(x + y) = f(x) + f(y)$ for all rational x and y. Prove that $f(x) = f(1) \cdot x$ for all rational x.

Solution 5.   Let $n \in \mathbb{N}$. Then

$$f(1) = f\underbrace{\left(\frac{1}{n} + \cdots + \frac{1}{n}\right)}_{n\times} \quad \Rightarrow \quad f\left(\frac{1}{n}\right) = \frac{1}{n} f(1).$$

Any rational number can be represented as m/n with m an integer and $n \in \mathbb{N}$. Then

$$f\left(\frac{m}{n}\right) = f\underbrace{\left(\frac{1}{n} + \cdots + \frac{1}{n}\right)}_{m\times} \quad \Rightarrow \quad f\left(\frac{m}{n}\right) = f\underbrace{\left(\frac{1}{n}\right) + \cdots + f\left(\frac{1}{n}\right)}_{m\times}.$$

Finally

$$f\left(\frac{m}{n}\right) = mf\left(\frac{1}{n}\right) = \frac{m}{n} f(1).$$

Problem 6.   Solve for $x \in \mathbb{Z}$ the congruence $9x + 5 \equiv 10 \bmod 11$.

Solution 6.   Subtract 5 from both sides of the congruence

$$9x \equiv 5 \bmod 11.$$

Since $\gcd(9, 11) = 1$, division by 9 is valid. Before we can perform the division, we need to choose a suitable multiple of 11 to add to the right-hand side in order to obtain a multiple of 9. We choose 22. Thus

$$9x \equiv 27 \bmod 11.$$

Now we divide by 9 providing $x \equiv 3 \bmod 11$. Thus the solution consists of all integers of the form $11n + 3$, where $n \in \mathbb{Z}$.

Problem 7. The *star-triangle transformation* is given by

$$R_u = \frac{R_2 R_3}{R_1 + R_2 + R_3}, \qquad R_1 = \frac{R_u R_v + R_v R_w + R_w R_u}{R_u}$$

$$R_v = \frac{R_1 R_3}{R_1 + R_2 + R_3}, \qquad R_2 = \frac{R_u R_v + R_v R_w + R_w R_u}{R_v}$$

$$R_w = \frac{R_1 R_2}{R_1 + R_2 + R_3}, \qquad R_3 = \frac{R_u R_v + R_v R_w + R_w R_u}{R_w}.$$

Using the star-triangle transformation find the total resistance of a circuit of equal resistors R on a unit cube between two opposite corners (say 1 and 7). The edges of a cube consist of equal resistors of resistance R, which are joined at the corners. The battery be connected to two opposite corners of a face of the cube (say 1 and 7). What is the resistance R_{17}? Drawing the circuit in the plane we have

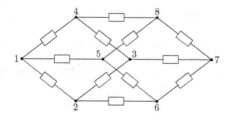

Solution 7. We solve the problem by successively applying the star-triangle transformation. With the star-triangle transformation we can simplify the circuit by using the standard rules for circuits in series

$$R_s = R_1 + R_2 + \cdots + R_n$$

and parallel

$$\frac{1}{R_s} = \frac{1}{R_1} + \frac{1}{R_2} + \cdots + \frac{1}{R_n}.$$

First we note from the star-triangle transformation that if

$$R_u = R_v = R_w = R$$

we have $R_1 = R_2 = R_3 = 3R$. In the first step, we convert the star 1, 5, 8, 6 to a triangle. Thus the node 5 disappears. In the second step we convert the star 1, 4, 8, 3 to a triangle. Thus the node 4 disappears. We repeat the step of eliminating the stars. At each step we apply the standard rules for circuits in series and parallel given above. Finally we arrive at

$$R_{17} = \frac{5}{6}R.$$

Since all the resistors are the same the problem can also be solved by symmetry considerations and Kirchhoff's laws. For example, if we want to find the resistance between 1 and 3 we could argue as follows. Conservation of the current at the corners requires

$$I = 2I_{14} + I_{15}, \qquad I_{15} = 2I_{56} = 2I_{58}$$

where I is the input current and $I_{14} = I_{12}$. The requirement that the voltage between 1 and 3 be independent of the path yields the additional equation $2I_{14}R = 2(I_{15} + I_{58})R$, where $I_{58} = I_{56}$. These three equations have the solution

$$I_{14} = \frac{3I}{8}, \qquad I_{15} = \frac{I}{4}, \qquad I_{58} = \frac{I}{8}.$$

Thus the resistance between 1 and 3 is $2I_{14}R/I$. Therefore $R_{13} = 3R/4$.

Problem 8. The recursion relation

$$F_{n+2} = F_n + F_{n+1}$$

with the initial values $F_0 = F_1 = 1$ provides the *Fibonacci sequence*

$$1, 1, 2, 3, 5, 8, 13, 21, 34, \ldots .$$

A generalization of the Fibonacci sequence is the q-analogue of the sequence defined by the recursion relation

$$F_{n+2}(q) = F_n(q) + qF_{n+1}(q)$$

and the initial conditions $F_0(q) = 1$, $F_1(q) = q$. Here, q is a real or complex number.
(i) Give the first five terms of the sequence.
(ii) Find a generating function of $F_n(q)$.
(iii) Find an explicit expression for $F_n(q)$.

Solution 8. (i) The first five terms are

$$1, \; q, \; q^2 + 1, \; q^3 + 2q, \; q^4 + 3q^2 + 1, \; \dots .$$

(ii) The generating function of $F_n(q)$ is

$$\frac{1}{1 - qs - s^2} = \sum_{n=0}^{\infty} F_n(q) s^n .$$

(iii) Writing $q = x - x^{-1}$ and partial fractioning the left-hand side of this equation yields

$$F_n(q) = \frac{x^{n+1} + (-1)^n x^{-(n+1)}}{x + x^{-1}}, \qquad n = 0, 1, 2, \dots .$$

Problem 9. Let G be the gravitational constant, $\hbar$ be the Planck constant h divided by 2π, ϵ_0 be the *permittivity* of free space, μ_0 be the *permeability* of free space, c be the *speed of light* in vacuum ($c = 1/\sqrt{\epsilon_0 \mu_0}$), $Z_0 = \sqrt{\mu_0/\epsilon_0}$ be the *impedance of vacuum*, e the elementary charge, m_e the electron mass, m_p the proton mass, k_B the Boltzmann constant. We have in the MKSA-system (where A stands for Ampere)

$$G = 6.6732 \cdot 10^{-11} \, meter^3 . sec^{-2} . kg^{-1}$$
$$\hbar = 1.0545919 \cdot 10^{-34} \, meter^2 . sec^{-1} . kg$$
$$\epsilon_0 = 8.854187817 \cdot 10^{-12} \, meter^{-3} . sec^4 . kg^{-1} . A^2$$
$$\mu_0 = 4\pi \cdot 10^{-7} meter . sec^{-2} . kg . A^{-2}$$
$$c = 1/\sqrt{\epsilon_0 \mu_0} = 2.9979250 \cdot 10^8 \, meter . sec^{-1}$$
$$Z_0 = \sqrt{\mu_0/\epsilon_0} \approx 377 \, meter^2 . sec^{-3} . kg . A^{-2}$$
$$e = 1.602176565 \cdot 10^{-19} \, sec . A$$
$$m_e = 9.10938291 \cdot 10^{-31} \, kg$$
$$m_p = 1.672621777 \cdot 10^{-27} \, kg$$
$$k_B = 1.3806488 \cdot 10^{-23} \, meter^2 . sec^{-2} . kg . K^{-1}$$

The *Planck length* is defined by

$$\ell_P := \sqrt{\frac{G\hbar}{c^3}} = 1.616229 \cdot 10^{-35} \, meter.$$

The *Planck time* is defined by

$$t_P := \sqrt{\frac{G\hbar}{c^5}} = \frac{\ell_P}{c} = 5.39116 \cdot 10^{-44} \, sec.$$

The *Planck mass* is defined by

$$m_P := \sqrt{\frac{\hbar c}{G}} = 2.176470 \cdot 10^{-8} \, kg.$$

The *Planck charge* is defined by

$$q_P := \sqrt{4\pi\epsilon_0 \hbar c} = \frac{e}{\sqrt{\alpha}} = 1.875545 \cdot 10^{-18} \, sec \cdot A$$

where α is the fine structure constant. The *Planck temperature* is defined by

$$T_p := \sqrt{\frac{\hbar c^5}{G k_B^2}} = 1.416808 \cdot 10^{32} \, K.$$

Thus the square of the *Planck length* is given by $\ell_P^2 := \frac{G\hbar}{c^3}$. Dimensionless quantities are the fine structure constant

$$\alpha = \frac{e^2}{4\pi\epsilon_0 \hbar c} = \frac{1}{137.03599911} = 0.007297352$$

and the gravitational fine structure constant

$$\alpha_g = \frac{G m_e}{\hbar c} = 1.7518 \cdot 10^{-45}.$$

Write a C++ program that calculates these quantities. Is the data type double sufficient?

Solution 9. The data type double is in the range $\pm 1.7E308$. Thus it is sufficient for calculating these quantities. The C++ program is

```
// Planck.cpp
#include <iostream>
#include <cmath>
using namespace std;

int main(void)
{
  double pi = 4.0*atan(1.0);
  double G = 6.6732E-11;
  double hbar = 1.0545919E-34;
  double c = 2.9979250E8;
  double eps0 = 8.854187817E-12;
  double kB = 1.3806488E-23;
  double e = 1.602176565E-19;
  double me = 9.10938291E-31;
  double alpha = (e*e)/(4.0*pi*eps0*hbar*c);
  cout << "alpha = " << alpha << endl;
  double alphag = (G*me*me)/(hbar*c);
  cout << "alphag = " << alphag << endl;
  double PL = sqrt((G*hbar)/(c*c*c));
```

```
    cout << "PL = " << PL << endl;
    double PT = sqrt((G*hbar)/(c*c*c*c*c));
    cout << "PT = " << PT << endl;
    double PM = sqrt((hbar*c)/G);
    cout << "PM = " << PM << endl;
    double PC = sqrt(4.0*pi*eps0*hbar*c);
    cout << "PC = " << PC << endl;
    double PK = sqrt((hbar*c*c*c*c*c)/(kB*kB*G));
    cout << "PK = " << PK << endl;
    return 0;
}
```

Problem 10. (i) Let m be a mass, E an energy, a a length and $\hbar$ the Planck constant (divided by 2π). Show that

$$\epsilon := \frac{\sqrt{2m|E|}a}{\hbar}$$

is dimensionless.

(ii) Let e be the charge, E the electric field, m the mass, ω the frequency and c the speed of light (all in SI units). Is $\eta = eE/(m\omega c)$ a dimensionless quantity so that cases such as $\eta \ll 1$ can be studied?

Solution 10. (i) In the MKSA system we have

$$[m] = kg, \quad [E] = meter^2 \cdot sec^{-2} \cdot kg, \quad [a] = meter, \quad [\hbar] = meter^2 \cdot sec^{-1} \cdot kg$$

Thus ϵ is a dimensionless quantity.

(ii) Here A stands for Ampere. We have

$$[e] = sec \cdot A, \quad [E] = meter \cdot sec^{-3} \cdot kg \cdot A^{-1}, \quad [m] = kg, \quad [\omega] = sec^{-1}$$

Thus η is a dimensionless quantity.

Supplementary Problems

Problem 1. Show that integrating over the unit sphere $\mathbb{S}^2$ in $\mathbb{R}^3$

$$\mathbb{S}^2 = \{ (x_1, x_2, x_3) \in \mathbb{R}^3 : x_1^2 + x_2^2 + x_3 = 1 \}$$

provides 4π.

Problem 2. Show that $\varphi^2 - \varphi - 1 = 0$ is invariant under the transformation $\varphi \mapsto -1/\varphi$.

Chapter 18

Kronecker and Tensor Product

Let A be an $m \times n$ matrix and let B be a $p \times q$ matrix. The *Kronecker product* of A and B is defined as

$$A \otimes B := \begin{pmatrix} a_{11}B & a_{12}B & \cdots & a_{1n}B \\ a_{21}B & a_{22}B & \cdots & a_{2n}B \\ \vdots & \vdots & \ddots & \vdots \\ a_{m1}B & a_{m2}B & \cdots & a_{mn}B \end{pmatrix}.$$

Thus $A \otimes B$ is an $mp \times nq$ matrix.

The Kronecker product is associative

$$(A \otimes B) \otimes C = A \otimes (B \otimes C).$$

Let A and B be $m \times n$ matrices and C a $p \times q$ matrix. Then

$$(A + B) \otimes C = A \otimes C + B \otimes C.$$

Let $c \in \mathbb{C}$. Then $(cA) \otimes B = c(A \otimes B) = A \otimes (cB)$. We have

$$(A \otimes B)^T = A^T \otimes B^T, \qquad (A \otimes B)^* = A^* \otimes B^*.$$

Let A be an $m \times n$ matrix, B be a $p \times q$ matrix, C be an $n \times r$ matrix and D be a $q \times s$ matrix. Then

$$(A \otimes B)(C \otimes D) = (AC) \otimes (BD).$$

The *Kronecker sum* is defined as

$$A \oplus_K B := A \otimes I_m + I_n \otimes B.$$

Assume that the matrix A is of order $m \times n$, B of order $p \times q$, C of order $n \times r$ and D of order $q \times s$. Then

$$(A \otimes B)(C \otimes D) = (AC) \otimes (BD).$$

Let A be an $m \times m$ matrix and let B be an $n \times n$ matrix. Then

$$\text{tr}(A \otimes B) \equiv (\text{tr}A)(\text{tr}B)$$

and

$$\det(A \otimes B) = (\det(A))^n(\det(B))^m.$$

If A has an inverse A^{-1} and B has an inverse B^{-1}, then

$$(A \otimes B)^{-1} = A^{-1} \otimes B^{-1}.$$

If A and B are hermitian matrices, then $A \otimes B$ is hermitian.
If A and B are skew-hermitian matrices, then $A \otimes B$ is hermitian.
If U and V are unitary matrices, then $U \otimes V$ is a unitary matrix.
If P_1, P_2 are permutation matrices, then $P_1 \otimes P_2$ is a permutation matrix.
If Π_1, Π_2 are projection matrices, then $\Pi_1 \otimes \Pi_2$ is a projection matrix.

If λ is an eigenvalue of A with eigenvector $\mathbf{v}$ and μ is an eigenvalue of B with eigenvector $\mathbf{u}$. Then $\lambda\mu$ is an eigenvalue of $A \otimes B$ with eigenvector $\mathbf{v} \otimes \mathbf{u}$. If λ is an eigenvalue of the $n \times n$ matrix A and μ is an eigenvalue of the $n \times n$ matrix B, then $\lambda + \mu$ is an eigenvalue of $A \otimes I_n + I_n \otimes B$.

Let A be an $m \times n$ matrix. The *vec-operator* $vec(A)$ stacks the columns of a matrix A one under the other to form a single column. Consider the Sylvester equation

$$AX + XB = C$$

where A is an $n \times n$ matrix, B an $m \times m$, X an $n \times m$ matrix. Then the vec operator applied on both sides of the Sylvester equation provides

$$(I_n \otimes A + B^T \otimes I_n)vec(X) = vec(C).$$

Given A, B, C we have a linear equation for X.

Problem 1. Consider the four normalized vectors in the vector space $\mathbb{R}^2$

$$\mathbf{u}_1 = \begin{pmatrix} 1 \\ 0 \end{pmatrix}, \qquad \mathbf{u}_2 = \begin{pmatrix} 0 \\ 1 \end{pmatrix}, \qquad \mathbf{v}_1 = \frac{1}{\sqrt{2}} \begin{pmatrix} 1 \\ 1 \end{pmatrix}, \qquad \mathbf{v}_2 = \frac{1}{\sqrt{2}} \begin{pmatrix} 1 \\ -1 \end{pmatrix}.$$

Calculate the vectors in the vector space $\mathbb{R}^4$

$$\mathbf{u}_1 \otimes \mathbf{u}_1, \qquad \mathbf{u}_1 \otimes \mathbf{u}_2, \qquad \mathbf{u}_2 \otimes \mathbf{u}_1, \qquad \mathbf{u}_2 \otimes \mathbf{u}_2$$

$$\mathbf{v}_1 \otimes \mathbf{v}_1, \qquad \mathbf{v}_1 \otimes \mathbf{v}_2, \qquad \mathbf{v}_2 \otimes \mathbf{v}_1, \qquad \mathbf{v}_2 \otimes \mathbf{v}_2$$

and the 2×2 matrices

$$\mathbf{u}_1^T \otimes \mathbf{u}_1, \qquad \mathbf{u}_1^T \otimes \mathbf{u}_2, \qquad \mathbf{u}_2^T \otimes \mathbf{u}_1, \qquad \mathbf{u}_2^T \otimes \mathbf{u}_2$$

$$\mathbf{v}_1^T \otimes \mathbf{v}_1, \qquad \mathbf{v}_1^T \otimes \mathbf{v}_2, \qquad \mathbf{v}_2^T \otimes \mathbf{v}_1, \qquad \mathbf{v}_2^T \otimes \mathbf{v}_2.$$

Discuss!

Solution 1. Using the definition of the Kronecker product we find

$$\mathbf{u}_1 \otimes \mathbf{u}_1 = \begin{pmatrix} 1 \\ 0 \\ 0 \\ 0 \end{pmatrix}, \qquad \mathbf{u}_1 \otimes \mathbf{u}_2 = \begin{pmatrix} 0 \\ 1 \\ 0 \\ 0 \end{pmatrix},$$

$$\mathbf{u}_2 \otimes \mathbf{u}_1 = \begin{pmatrix} 0 \\ 0 \\ 1 \\ 0 \end{pmatrix}, \qquad \mathbf{u}_2 \otimes \mathbf{u}_2 = \begin{pmatrix} 0 \\ 0 \\ 0 \\ 1 \end{pmatrix}.$$

The set $\{\mathbf{u}_1, \mathbf{u}_2\}$ is the standard basis of $\mathbb{R}^2$. We find that the set

$$\{\, \mathbf{u}_1 \otimes \mathbf{u}_1, \quad \mathbf{u}_1 \otimes \mathbf{u}_2, \quad \mathbf{u}_2 \otimes \mathbf{u}_1, \quad \mathbf{u}_2 \otimes \mathbf{u}_2 \,\}$$

is the standard basis in $\mathbb{R}^4$. Using the definition of the Kronecker product we obtain

$$\mathbf{v}_1 \otimes \mathbf{v}_1 = \frac{1}{2} \begin{pmatrix} 1 \\ 1 \\ 1 \\ 1 \end{pmatrix}, \qquad \mathbf{v}_1 \otimes \mathbf{v}_2 = \frac{1}{2} \begin{pmatrix} 1 \\ -1 \\ 1 \\ -1 \end{pmatrix},$$

$$\mathbf{v}_2 \otimes \mathbf{v}_1 = \frac{1}{2} \begin{pmatrix} 1 \\ 1 \\ -1 \\ -1 \end{pmatrix}, \qquad \mathbf{v}_2 \otimes \mathbf{v}_2 = \frac{1}{2} \begin{pmatrix} 1 \\ -1 \\ -1 \\ 1 \end{pmatrix}.$$

The set $\{\mathbf{v}_1, \mathbf{v}_2\}$ is an orthonormal basis of $\mathbb{R}^2$. Thus we find that the set

$$\{\, \mathbf{v}_1 \otimes \mathbf{v}_1, \quad \mathbf{v}_1 \otimes \mathbf{v}_2, \quad \mathbf{v}_2 \otimes \mathbf{v}_1, \quad \mathbf{v}_2 \otimes \mathbf{v}_2 \,\}$$

is an orthonormal basis in the vector space $\mathbb{R}^4$.
We find

$$\mathbf{u}_1^T \otimes \mathbf{u}_1 = (1,0) \otimes \begin{pmatrix} 1 \\ 0 \end{pmatrix} = \begin{pmatrix} 1 & 0 \\ 0 & 0 \end{pmatrix}, \quad \mathbf{u}_1^T \otimes \mathbf{u}_2 = (1,0) \otimes \begin{pmatrix} 0 \\ 1 \end{pmatrix} = \begin{pmatrix} 0 & 0 \\ 1 & 0 \end{pmatrix}$$

$$\mathbf{u}_2^T \otimes \mathbf{u}_1 = (0,1) \otimes \begin{pmatrix} 1 \\ 0 \end{pmatrix} = \begin{pmatrix} 0 & 1 \\ 0 & 0 \end{pmatrix}, \quad \mathbf{u}_2^T \otimes \mathbf{u}_2 = (0,1) \otimes \begin{pmatrix} 0 \\ 1 \end{pmatrix} = \begin{pmatrix} 0 & 0 \\ 0 & 1 \end{pmatrix}.$$

Consequently, $\mathbf{u}_1^T \otimes \mathbf{u}_1 + \mathbf{u}_2^T \otimes \mathbf{u}_2$ is the 2×2 unit matrix. We obtain

$$\mathbf{v}_1^T \otimes \mathbf{v}_1 = \frac{1}{2}(1,1) \otimes \begin{pmatrix} 1 \\ 1 \end{pmatrix} = \frac{1}{2} \begin{pmatrix} 1 & 1 \\ 1 & 1 \end{pmatrix}$$

$$\mathbf{v}_1^T \otimes \mathbf{v}_2 = \frac{1}{2}(1,1) \otimes \begin{pmatrix} 1 \\ -1 \end{pmatrix} = \frac{1}{2} \begin{pmatrix} 1 & 1 \\ -1 & -1 \end{pmatrix}$$

$$\mathbf{v}_2^T \otimes \mathbf{v}_1 = \frac{1}{2}(1,-1) \otimes \begin{pmatrix} 1 \\ 1 \end{pmatrix} = \frac{1}{2} \begin{pmatrix} 1 & -1 \\ 1 & -1 \end{pmatrix}$$

$$\mathbf{v}_2^T \otimes \mathbf{v}_2 = \frac{1}{2}(1,-1) \otimes \begin{pmatrix} 1 \\ -1 \end{pmatrix} = \frac{1}{2} \begin{pmatrix} 1 & -1 \\ -1 & 1 \end{pmatrix}$$

and $\mathbf{v}_1^T \otimes \mathbf{v}_1 + \mathbf{v}_2^T \otimes \mathbf{v}_2$ is the 2×2 unit matrix.

Problem 2. Let P and Q be two $n \times n$ matrices and I_n the $n \times n$ unit matrix. Assume that $P^2 = P$ and $Q^2 = Q$.
(i) Show that $(P \otimes Q)^2 = P \otimes Q$.
(ii) Show that $(P \otimes (I_n - P))^2 = P \otimes (I_n - P)$.
(iii) Assume that $QP = 0_n$. Calculate $(Q \otimes Q)(P \otimes P)$.

Solution 2. (i) We have $(P \otimes Q)(P \otimes Q) = (P^2 \otimes Q^2) = P \otimes Q$.
(ii) We obtain

$$(P \otimes (I_n - P))(P \otimes (I_n - P)) = (P^2 \otimes (I_n - P)^2) = P \otimes (I_n - P - P + P^2).$$

Thus

$$(P \otimes (I_n - P))(P \otimes (I_n - P)) = P \otimes (I_n - P - P + P) = P \otimes (I_n - P).$$

(iii) We find
$$(Q \otimes Q)(P \otimes P) = (QP) \otimes (QP) = 0_{n^2}.$$

The equation $(P \otimes Q)^2 = P \otimes Q$ can be extended to N-factors. Let $P_1, \ldots, P_N$ be matrices with $P_j^2 = P_j$, where $j = 1, 2, \ldots, N$. Then

$$(P_1 \otimes P_2 \otimes \cdots \otimes P_N)^2 = P_1 \otimes P_2 \otimes \cdots \otimes P_N.$$

Problem 3. (i) Let A be an $n \times n$ matrix and B be an $m \times m$ matrix. I_n is the $n \times n$ unit matrix and I_m the $m \times m$ unit matrix. Calculate the commutator $[A \otimes I_m, I_n \otimes B]$.
(ii) Let A, B, C, D be $n \times n$ matrices. Assume that $[A, B] = 0_n$, $[C, D] = 0_n$. Calculate the commutator $[A \otimes C, B \otimes D]$.

Solution 3. (i) Straightforward calculation yields

$$[A \otimes I_m, I_n \otimes B] = (A \otimes I_m)(I_n \otimes B) - (I_n \otimes B)(A \otimes I_m)$$
$$= A \otimes B - A \otimes B = 0_{n \cdot m}.$$

Consequently, the matrices $A \otimes I_m$ and $I_n \otimes B$ commute.
(ii) We find

$$[A \otimes C, B \otimes D] = (A \otimes C)(B \otimes D) - (B \otimes D)(A \otimes C)$$
$$= (AB) \otimes (CD) - (BA) \otimes (DC)$$
$$= (AB) \otimes (CD) - (AB) \otimes (CD) = 0_{n \cdot m}$$

where we have used (1).

Problem 4. (i) Let A be an $m \times m$ matrix with eigenvalue λ and corresponding eigenvector $\mathbf{a}$. Let B be an $n \times n$ matrix with eigenvalue μ corresponding eigenvector $\mathbf{b}$. Show that the matrix $A \otimes B$ has the eigenvalue $\lambda\mu$ with the corresponding eigenvector $\mathbf{a} \otimes \mathbf{b}$.
(ii) Consider the *Pauli spin matrices* σ_2 and σ_3. Find the eigenvalues of $\sigma_2 \otimes \sigma_3$.

Solution 4. (i) From the eigenvalue equations $A\mathbf{a} = \lambda\mathbf{a}$, $B\mathbf{b} = \mu\mathbf{b}$ we find

$$(A\mathbf{a}) \otimes (B\mathbf{b}) = \lambda\mu(\mathbf{a} \otimes \mathbf{b}).$$

Since $(A\mathbf{a}) \otimes (B\mathbf{b}) \equiv (A \otimes B)(\mathbf{a} \otimes \mathbf{b})$ we obtain

$$(A \otimes B)(\mathbf{a} \otimes \mathbf{b}) = \lambda\mu(\mathbf{a} \otimes \mathbf{b}).$$

This is an eigenvalue equation. Consequently, $\mathbf{a} \otimes \mathbf{b}$ is an eigenvector of $A \otimes B$ with eigenvalue $\lambda\mu$.
(ii) Both matrices are hermitian, unitary and $\text{tr}(\sigma_2) = 0$, $\text{tr}(\sigma_3) = 0$. Thus the eigenvalues of σ_2 and σ_3 are given by $\{1, -1\}$. Since the eigenvalues of σ_2 and σ_3 are given by $\{1, -1\}$, we find that the eigenvalues of $\sigma_2 \otimes \sigma_3$ are given by $\{1, 1, -1, -1\}$.

Problem 5. (i) Let A be an $m \times m$ matrix and B be an $n \times n$ matrix and $A\mathbf{a} = \lambda\mathbf{a}$, $B\mathbf{b} = \mu\mathbf{b}$ be their respective eigenvalue equations. Show that the matrix

$$A \otimes I_n + I_m \otimes B$$

where I_n is the $n \times n$ unit matrix and I_m is the $m \times m$ unit matrix, has the eigenvalue $\lambda + \mu$ with the corresponding eigenvector $\mathbf{a} \otimes \mathbf{b}$.

(ii) Consider the Pauli spin matrices σ_2 and σ_3. Find the eigenvalues of $\sigma_2 \otimes I_2 + I_2 \otimes \sigma_3$.

Solution 5.   (i) We have

$$(A \otimes I_n + I_m \otimes B)(\mathbf{a} \otimes \mathbf{b}) = (A \otimes I_n)(\mathbf{a} \otimes \mathbf{b}) + (I_m \otimes B)(\mathbf{a} \otimes \mathbf{b})$$
$$= (A\mathbf{a}) \otimes (I_n\mathbf{b}) + (I_m\mathbf{a}) \otimes (B\mathbf{b})$$
$$= (\lambda\mathbf{a}) \otimes \mathbf{b} + \mathbf{a} \otimes (\mu\mathbf{b}) = (\lambda + \mu)(\mathbf{a} \otimes \mathbf{b}).$$

Consequently, the eigenvectors and eigenvalues of $A \otimes I_n + I_m \otimes B$ admits the eigenvalue $\lambda + \mu$ and the corresponding eigenvector $\mathbf{a} \otimes \mathbf{b}$.

(ii) Since the eigenvalues of σ_2 and σ_3 are given by $\{1, -1\}$, we find that the eigenvalues of $\sigma_2 \otimes I_2 + I_2 \otimes \sigma_3$ take the form $\{2, 0, 0, -2\}$.

Problem 6.   Let A, B be two arbitrary $n \times n$ matrices and I_n be the $n \times n$ unit matrix. Prove that $\mathrm{tr}(e^{A \otimes I_n + I_n \otimes B}) \equiv (\mathrm{tr}(e^A))(\mathrm{tr}(e^B))$.

Solution 6.   Since $[A \otimes I_n, I_n \otimes B] = 0_{n \cdot n}$ we have

$$\mathrm{tr}(e^{(A \otimes I_n + I_n \otimes B)}) = \mathrm{tr}\left(e^{A \otimes I} e^{I_n \otimes B}\right).$$

Now

$$e^{A \otimes I_n} = \sum_{k=0}^{\infty} \frac{(A \otimes I_n)^k}{k!}, \qquad e^{I_n \otimes B} = \sum_{k=0}^{\infty} \frac{(I_n \otimes B)^k}{k!}.$$

An arbitrary term in the expansion of $e^{A \otimes I_n} e^{I_n \otimes B}$ is given by

$$\frac{1}{j!}\frac{1}{k!}(A \otimes I_n)^j (I_n \otimes B)^k.$$

Now we have

$$\frac{1}{j!}\frac{1}{k!}(A \otimes I_n)^j (I_n \otimes B)^k \equiv \frac{1}{j!}\frac{1}{k!}(A^j \otimes I_n^j)(I_n^k \otimes B^k)$$
$$\equiv \frac{1}{j!}\frac{1}{k!}(A^j \otimes I_n)(I_n \otimes B^k)$$
$$\equiv \frac{1}{j!}\frac{1}{k!}(A^j \otimes B^k).$$

Therefore

$$\mathrm{tr}\left(\frac{1}{j!} \cdot \frac{1}{k!}(A \otimes I_n)^j (I_n \otimes B)^k\right) = \frac{1}{j!} \cdot \frac{1}{k!}(\mathrm{tr}A^j)(\mathrm{tr}B^k).$$

Since

$$(\text{tr}e^A)(\text{tr}e^B) \equiv \text{tr}\left(\sum_{k=0}^{\infty}\frac{A^k}{k!}\right)\text{tr}\left(\sum_{k=0}^{\infty}\frac{B^k}{k!}\right) = \left(\sum_{k=0}^{\infty}\frac{1}{k!}\text{tr}A^k\right)\left(\sum_{k=0}^{\infty}\frac{1}{k!}\text{tr}B^k\right)$$

we have proved the identity.

Problem 7.   Let A be an $n \times n$ matrix and I_n be the $n \times n$ identity matrix. We define

$$A_j := \overbrace{I_n \otimes \cdots \otimes I_n \otimes A \otimes I_n \otimes \cdots \otimes I_n}^{N-factors}$$

where A is at the j-th place. Let $X := A_1 + A_2 + \cdots + A_N$. Calculate

$$\frac{\text{tr}\left(A_j e^X\right)}{\text{tr}(e^X)}.$$

Solution 7.   Since $[A_j, A_k] = 0$ for $j = 1, 2, \ldots, N$ and $k = 1, 2, \ldots, N$ we have

$$e^X \equiv e^{A_1 + A_2 + \cdots + A_N} \equiv e^{A_1}e^{A_2}\cdots e^{A_N}.$$

Now

$$e^{A_j} = \overbrace{I_n \otimes \cdots \otimes I_n \otimes e^A \otimes I_n \otimes \cdots \otimes I_n}^{N-factors}$$

where e^A is at the j-th place. It follows that

$$e^{A_1}e^{A_2}\cdots e^{A_N} = e^A \otimes e^A \otimes \cdots \otimes e^A.$$

Therefore $\text{tr}(e^X) \equiv (\text{tr}(e^A))^N$. Now

$$A_j e^X = (I_n \otimes \cdots \otimes I_n \otimes A \otimes I_n \otimes \cdots \otimes I_n)(e^A \otimes e^A \otimes \cdots \otimes e^A)$$
$$= (e^A \otimes \cdots \otimes e^A \otimes Ae^A \otimes e^A \otimes \cdots \otimes e^A).$$

Thus $\text{tr}(A_j e^X) = (\text{tr}(e^A))^{N-1}(\text{tr}(Ae^A))$. Consequently

$$\frac{\text{tr}(A_j e^X)}{\text{tr}(e^X)} = \frac{\text{tr}(Ae^A)}{\text{tr}(e^A)}.$$

Problem 8.   Let A be a symmetric 4×4 matrix over $\mathbb{R}$. Assume that the eigenvalues are given by $\lambda_1 = 0$, $\lambda_2 = 1$, $\lambda_3 = 2$ and $\lambda_4 = 3$ with the corresponding normalized eigenvectors (*Bell basis*)

$$\mathbf{u}_1 = \frac{1}{\sqrt{2}}\begin{pmatrix} 1 \\ 0 \\ 0 \\ 1 \end{pmatrix}, \qquad \mathbf{u}_2 = \frac{1}{\sqrt{2}}\begin{pmatrix} 1 \\ 0 \\ 0 \\ -1 \end{pmatrix},$$

$$\mathbf{u}_3 = \frac{1}{\sqrt{2}} \begin{pmatrix} 0 \\ 1 \\ 1 \\ 0 \end{pmatrix}, \qquad \mathbf{u}_4 = \frac{1}{\sqrt{2}} \begin{pmatrix} 0 \\ 1 \\ -1 \\ 0 \end{pmatrix}.$$

Find the symmetric matrix A with the help of the *spectral theorem*. Using the spectral theorem the matrix A we can reconstructed from the eigenvalues and normalized eigenvectors.

Solution 8. The eigenvectors to different eigenvalues are orthogonal. Since the normalized eigenvectors are pairwise orthogonal we find from the spectral theorem

$$A = \sum_{j=1}^{4} \lambda_j \mathbf{u}_j \otimes \mathbf{u}_j^T.$$

Inserting the eigenvalues yields $A = \mathbf{u}_2^T \otimes \mathbf{u}_2 + 2\mathbf{u}_3^T \otimes \mathbf{u}_3 + 3\mathbf{u}_4^T \otimes \mathbf{u}_4$. Using the definition of the Kronecker product we find that

$$A = \frac{1}{2} \begin{pmatrix} 1 & 0 & 0 & -1 \\ 0 & 5 & -1 & 0 \\ 0 & -1 & 5 & 0 \\ -1 & 0 & 0 & 1 \end{pmatrix}.$$

Problem 9. Let U be a 4×4 unitary matrix. Assume that U can be written as

$$U = V \otimes W \tag{1}$$

where V and W are 2×2 unitary matrices. Show that U can be written as

$$U = \exp(iA \otimes I_2 + iI_2 \otimes B) \tag{2}$$

where A and B are 2×2 hermitian matrices and I_2 is the 2×2 unit matrix.

Solution 9. Any unitary matrix can be written as $\exp(iH)$, where H is a hermitian matrix. Thus $V = \exp(iA)$, $W = \exp(iB)$, where A and B are 2×2 hermitian matrices. Consequently

$$U = V \otimes W = \exp(iA) \otimes \exp(iB).$$

Using the identity $\exp(iA) \otimes \exp(iB) \equiv \exp(iA \otimes I_2) \exp(iI_2 \otimes B)$ and $[A \otimes I_2, I_2 \otimes B] = 0_4$ we find equation (2).

Problem 10. Let A_i and B_j be $n \times n$ matrices over $\mathbb{C}$, where $i, j = 1, 2, \dots, p$. We define

$$r_{12} := \sum_{i=1}^{p} A_i \otimes B_i \otimes I_n, \quad r_{13} := \sum_{j=1}^{p} A_j \otimes I_n \otimes B_j, \quad r_{23} := \sum_{k=1}^{p} I_n \otimes A_k \otimes B_k$$

where I_n is the $n \times n$ unit matrix. Find $[r_{12}, r_{13}]$, $[r_{12}, r_{23}]$, $[r_{13}, r_{23}]$.

Solution 10. Since

$$(A_i \otimes B_i \otimes I_n)(A_j \otimes I_n \otimes B_j) - (A_j \otimes I_n \otimes B_j)(A_i \otimes B_i \otimes I_n)$$
$$= (A_i A_j) \otimes B_i \otimes B_j - (A_j A_i) \otimes B_i \otimes B_j$$
$$= (A_i A_j - A_j A_i) \otimes B_i \otimes B_j = [A_i, A_j] \otimes B_i \otimes B_j$$

we have

$$[r_{12}, r_{13}] = \sum_{i=1}^{p} \sum_{j=1}^{p} [A_i, A_j] \otimes B_i \otimes B_j.$$

Analogously

$$[r_{12}, r_{23}] = \sum_{i=1}^{p} \sum_{k=1}^{p} A_i \otimes [B_i, A_k] \otimes B_k$$

and

$$[r_{13}, r_{23}] = \sum_{j=1}^{p} \sum_{k=1}^{p} A_j \otimes A_k \otimes [B_j, B_k].$$

Assume that A_i and B_j are elements of a Lie algebra. The *Classical Yang-Baxter Equation* is given by $[r_{12}, r_{13}] + [r_{12}, r_{23}] + [r_{13}, r_{23}] = 0$.

Problem 11. Let $X = (x_{ij})$ be an $m \times n$ matrix. We define the vector (*vec operator*)

$$vec(X) := (x_{11}, x_{12}, \dots, x_{1n}, x_{21}, \dots, x_{2n}, \dots, x_{mn})^T \qquad (1)$$

where ^{T} denotes transpose. Thus $vec(X)$ is a column vector. Let C be a $p \times m$ matrix and let D be a $q \times n$ matrix. Then we have

$$vec(CXD^T) = (C \otimes D)vec(X). \qquad (2)$$

Show that

$$D \otimes C = P(C \otimes D)Q \qquad (3)$$

where P and Q are permutation matrices depending only on p, q and m, n respectively.

Solution 11. Consider the equation

$$CXD^T = E. \qquad (4)$$

Using (2) we find

$$(C \otimes D)vec(X) = vec(E). \qquad (5)$$

Similarly, application of (2) to the transpose of (4) yields

$$(D \otimes C)vec(X^T) = vec(E^T). \tag{6}$$

In view of (1) it is clear that

$$vec(E^T) = Pvec(E), \qquad vec(X) = Qvec(X^T) \tag{7}$$

for suitable permutation matrices P and Q which depend only on the dimensions of E and X, respectively. Equation(6), together with that resulting from substitution of (7) into (5), verifies (3).

Problem 12. The starting point in the construction of the *Yang-Baxter equation* is the 2×2 matrix

$$T := \begin{pmatrix} a & b \\ c & d \end{pmatrix} \tag{1}$$

where a, b, c, and d are $n \times n$ linear operators of an algebra over the complex numbers. In other words, T is an operator-valued matrix. Let I_2 be the 2×2 identity matrix. We define the 4×4 matrices

$$T_1 = T \otimes I_2, \qquad T_2 = I_2 \otimes T. \tag{2}$$

Thus T_1 and T_2 are operator-valued 4×4 matrices. Applying the rules for the Kronecker product we find

$$T_1 = T \otimes I_2 = \begin{pmatrix} a & 0 & b & 0 \\ 0 & a & 0 & b \\ c & 0 & d & 0 \\ 0 & c & 0 & d \end{pmatrix}, \qquad T_2 = I_2 \otimes T = \begin{pmatrix} a & b & 0 & 0 \\ c & d & 0 & 0 \\ 0 & 0 & a & b \\ 0 & 0 & c & d \end{pmatrix}.$$

The Yang-Baxter equation is given by $R_q T_1 T_2 = T_2 T_1 R_q$, where R_q is a 4×4 matrix and q a nonzero complex number. Let

$$R_q := \begin{pmatrix} 1 & 0 & 0 & 0 \\ 0 & -1 & 0 & 0 \\ 0 & 1+q & q & 0 \\ 0 & 0 & 0 & 1 \end{pmatrix}.$$

Find the condition on a, b, c and d such that condition (1) is satisfied.

Solution 12. Altogether we find 16 relations which have to be satisfied. Equation (1) gives rise to the relations of the algebra elements a, b, c and d

$$ab = q^{-1}ba, \quad dc = qcd, \quad bc = -qcb,$$

$$bd = -db, \quad ac = -ca, \quad [a,d] = (1+q^{-1})bc$$

where all the commutative relations have been omitted and $[\,,\,]$ denotes the commutator.

Problem 13.   Let $\phi \in \mathbb{R}$. Consider the normalized vector (*Bell state*)

$$|\psi\rangle = \frac{1}{\sqrt{2}} \begin{pmatrix} e^{i\phi} \\ 0 \\ 0 \\ 1 \end{pmatrix}$$

in the Hilbert space $\mathbb{C}^4$. Can the normalized vector $|\psi\rangle$ be written as the Kronecker product of two vectors $|u\rangle$, $|v\rangle$ in the Hilbert space $\mathbb{C}^2$?

Solution 13.   Let

$$|u\rangle = \begin{pmatrix} u_1 \\ u_2 \end{pmatrix}, \qquad |v\rangle = \begin{pmatrix} v_1 \\ v_2 \end{pmatrix}.$$

Then from $|\psi\rangle = |u\rangle \otimes |v\rangle$ we obtain the four equations

$$u_1 v_1 = e^{i\phi}/\sqrt{2}, \quad u_1 v_2 = 0, \quad u_2 v_1 = 0, \quad u_2 v_2 = 1/\sqrt{2}.$$

These four equations cannot be satisfied simultaneously. Thus $|\psi\rangle$ cannot be written as a Kronecker product of two vectors in $\mathbb{C}^2$. We call $|\psi\rangle$ *entangled*.

Problem 14.   Let A be an $n \times n$ matrix over $\mathbb{C}$. Consider the map

$$\rho(A) \to A \otimes I_n + I_n \otimes A.$$

Calculate the commutator $[\rho(A), \rho(B)]$.

Solution 14.   We find

$$\begin{aligned}
[\rho(A), \rho(B)] &= [A \otimes I_n + I_n \otimes A, B \otimes I_n + I_n \otimes B] \\
&= (AB) \otimes I_n - (BA) \otimes I_n + I_n \otimes (AB) - I_n \otimes (BA) \\
&= [A, B] \otimes I_n + I_n \otimes [A, B] = \rho([A, B]).
\end{aligned}$$

Problem 15.   Let

$$B(1) = \begin{pmatrix} 1 & 1 \\ 0 & 1 \end{pmatrix}$$

and $B(k)$ the $2^k \times 2^k$ matrix recursively defined by

$$B(k) := \begin{pmatrix} B(k-1) & B(k-1) \\ 0 & B(k-1) \end{pmatrix} = B(1) \otimes B(k-1), \qquad k = 2, 3, \dots$$

(i) Find $B(1)^{-1}$.

(ii) Find $B(k)^{-1}$ for $k = 2, 3, \ldots$.

(iii) Let $A(k)$ be the $2^k \times 2^k$ anti-diagonal matrix defined by $A_{ij}(k) = 1$ if $i = 2^k + 1 - j$ and 0 otherwise with $j = 1, 2, \ldots, 2^k$. Let

$$C(k) = A(k)B(k)^{-1}A(k).$$

Show that the matrices $B(k)$ and $C(k)$ satisfy the *braid-like relation*

$$B(k)C(k)B(k) = C(k)B(k)C(k).$$

Solution 15. (i) We have

$$B(1)^{-1} = \begin{pmatrix} 1 & -1 \\ 0 & 1 \end{pmatrix}.$$

(ii) We find $B(k)^{-1} = (B(1)^{-1})^{\otimes k}$.

(iii) We find

$$C(1) = \begin{pmatrix} 1 & 0 \\ -1 & 1 \end{pmatrix}.$$

The relation is true for $k = 1$ by a direct calculation. Since

$$B(k) = B(1)^{\otimes k}, \qquad C(k) = C(1)^{\otimes k}$$

and using the properties of the Kronecker product we find that the relation is true for every k.

Problem 16. Let σ_1, σ_2, σ_3 be the Pauli spin matrices and

$$R := \sigma_1 \otimes \sigma_1 + \sigma_2 \otimes \sigma_2 + \sigma_3 \otimes \sigma_3.$$

(i) Find $\text{tr}(R)$. Using this result what can be said about the eigenvalues of R?

(ii) Find R^2. Using this result and the result from (i) derive the eigenvalues of the matrix R.

(iii) Find $\frac{1}{4}(I_4 + R)^2$.

Solution 16. (i) Since $\text{tr}(\sigma_1) = \text{tr}(\sigma_2) = \text{tr}(\sigma_3) = 0$ and $\text{tr}(A \otimes B) = \text{tr}(A)\text{tr}(B)$ for any $n \times n$ matrices A and B we find $\text{tr}(R) = 0$. Thus for the four eigenvalues of R we have $\lambda_1 + \lambda_2 + \lambda_3 + \lambda_4 = 0$.

(ii) Since $\sigma_1^2 = \sigma_2^2 = \sigma_3^2 = I_2$ we find $R^2 = 3I_4 - 2R$. From the eigenvalue equation $R\mathbf{u} = \lambda\mathbf{u}$ we obtain $R^2\mathbf{u} = \lambda R\mathbf{u} = \lambda^2\mathbf{u}$. Using $R^2 = 3I_4 - 2R$ we have

$$(3I_4 - 2R)\mathbf{u} = (3 - 2\lambda)\mathbf{u} = \lambda^2\mathbf{u}$$

or $(\lambda^2 + 2\lambda - 3)\mathbf{u} = \mathbf{0}$. Thus $\lambda_+ = 1$ and $\lambda_- = -3$. Together with $\lambda_1 + \lambda_2 + \lambda_3 + \lambda_4 = 0$ we obtain the four eigenvalues $\lambda_1 = 1$, $\lambda_2 = 1$, $\lambda_3 = 1$, $\lambda_4 = -3$.

(iii) Using the result from (ii) we find

$$\frac{1}{4}(I_4 + R)^2 = \frac{1}{4}(I_4 + 2R + R^2) = I_4.$$

Problem 17.   (i) Consider the 4×4 matrix

$$T = \begin{pmatrix} 1 & 0 & 0 & 1 \\ 1 & 0 & 0 & -1 \\ 0 & 1 & 1 & 0 \\ 0 & -i & i & 0 \end{pmatrix}.$$

Show that the inverse T^{-1} of T exists and find the inverse.

(ii) Consider the *rotation matrix*

$$R_\alpha = \begin{pmatrix} \cos(\alpha) & -\sin(\alpha) \\ \sin(\alpha) & \cos(\alpha) \end{pmatrix}.$$

Calculate $T(R_\alpha \otimes R_\beta^T)T^{-1}$.

Solution 17.   (i) The row vectors (or column) vectors in the matrix T are linearly independent. Thus $\text{rank}(T) = 4$. Thus the inverse exists and is given by

$$T^{-1} = \frac{1}{2} \begin{pmatrix} 1 & 1 & 0 & 0 \\ 0 & 0 & 1 & i \\ 0 & 0 & 1 & -i \\ 1 & -1 & 0 & 0 \end{pmatrix}.$$

(ii) We find

$$T(R_\alpha \otimes R_\beta^T)T^{-1} = \begin{pmatrix} \cos(\alpha+\beta) & 0 & 0 & i\sin(\alpha+\beta) \\ 0 & \cos(\alpha-\beta) & -\sin(\alpha-\beta) & 0 \\ 0 & \sin(\alpha-\beta) & \cos(\alpha-\beta) & 0 \\ i\sin(\alpha+\beta) & 0 & 0 & \cos(\alpha+\beta) \end{pmatrix}.$$

Problem 18.   Let A, B be $n \times n$ matrices such that $A^2 = I_n$ and $B^2 = I_n$. Calculate $\exp(z(A \otimes B))$, where $z \in \mathbb{C}$.

Solution 18.   Since $(A \otimes B)^2 = I_n \otimes I_n$ we obtain

$$\exp(z(A \otimes B)) = (I_n \otimes I_n)\cosh(z) + (A \otimes B)\sinh(z).$$

Problem 19. Let A be an $m \times n$ matrix over $\mathbb{R}$, B be an $s \times t$ matrix over $\mathbb{R}$ and M be a $ms \times nt$ matrix over $\mathbb{R}$. Find A and B which minimizes

$$\|M - A \otimes B\|$$

by using the fact that the $mn \times st$ matrix $\mathrm{vec}A(\mathrm{vec}B)^T$ has the same entries as $A \otimes B$. The norm $\| \ \|$ is the Hilbert-Schmidt norm over $\mathbb{R}$ given by

$$\|X\| = \mathrm{tr}(XX^T) = \sqrt{\sum_{i=1}^{ms} \sum_{j=1}^{nt} (X)_{ij}^2}.$$

The order of the entries in the sum does not change the result. Thus we minimize

$$\|R(M) - \mathrm{vec}(A)(\mathrm{vec}(B))^T\|$$

where $R(M)$ is the $mn \times st$ matrix with the same entries as M rearranged by R where $R(A \otimes B) = \mathrm{vec}(A)(\mathrm{vec}(B))^T$. As an example consider $m = n = s = t = 2$ and the matrix

$$M = \begin{pmatrix} 1 & 0 & 0 & 1 \\ 0 & 0 & 0 & 0 \\ 0 & 0 & 0 & 0 \\ 1 & 0 & 0 & 1 \end{pmatrix}.$$

Solution 19. Since $\mathrm{vec}(A)(\mathrm{vec}(B))^T$ has rank 1, we find the nearest rank 1 approximation of $R(M)$. Let

$$R(M) = U\Sigma V^T$$

be the *singular value decomposition* of $R(M)$, where $\Sigma = \mathrm{diag}(\sigma_1, \sigma_2, \ldots)$ are the decreasing singular values of $R(M)$. Then the nearest rank 1 approximation is $U\mathrm{diag}(\sigma_1, 0, \ldots)V^T$. Consequently we solve

$$U^T \mathrm{vec}(A)(\mathrm{vec}(B))^T V = \mathrm{diag}(\sigma_1, 0, \ldots).$$

The solution is given by $\mathrm{vec}(A) = U(t\sigma_1, 0, \ldots)^T$ and $\mathrm{vec}(B) = V(1/t, 0, \ldots)^T$ for $t \in \mathbb{R} \setminus \{0\}$. For $m = n = s = t = 2$ and $R(A \otimes B) = \mathrm{vec}(A)(\mathrm{vec}(B))^T$ we find

$$R\begin{pmatrix} a_{11}b_{11} & a_{11}b_{12} & a_{12}b_{11} & a_{12}b_{12} \\ a_{11}b_{21} & a_{11}b_{22} & a_{12}b_{21} & a_{12}b_{22} \\ a_{21}b_{11} & a_{21}b_{12} & a_{22}b_{11} & a_{22}b_{12} \\ a_{21}b_{21} & a_{21}b_{22} & a_{22}b_{21} & a_{22}b_{22} \end{pmatrix} = \begin{pmatrix} a_{11}b_{11} & a_{11}b_{12} & a_{11}b_{21} & a_{11}b_{22} \\ a_{12}b_{11} & a_{12}b_{12} & a_{12}b_{21} & a_{12}b_{22} \\ a_{21}b_{11} & a_{21}b_{12} & a_{21}b_{21} & a_{21}b_{22} \\ a_{22}b_{11} & a_{22}b_{12} & a_{22}b_{21} & a_{22}b_{22} \end{pmatrix}.$$

Thus $R(M) = I_4$ with a singular value decomposition given by $U = V = \Sigma = I_4$. It follows that $\text{vec}(A) = I_4(t,0,0,0)^T$ and $\text{vec}(B) = I_4(1/t,0,0,0)^T$, i.e.

$$A = \begin{pmatrix} t & 0 \\ 0 & 0 \end{pmatrix}, \qquad B = \begin{pmatrix} 1/t & 0 \\ 0 & 0 \end{pmatrix}.$$

More generally $V = U$ for an arbitrary 4×4 orthogonal matrix U and $\text{vec}(A) = U(t,0,0,0)^T$ and $\text{vec}(B) = U(1/t,0,0,0)^T$.

Problem 20.   Let $d \geq 2$ and $|0\rangle$, $|1\rangle$, ..., $|d-1\rangle$ be an orthonormal basis in the Hilbert space $\mathbb{C}^d$. Let $|\psi\rangle$, $|\phi\rangle$ be two normalized states in $\mathbb{C}^d$. Let

$$S = \sum_{j,k=0}^{d-1} ((|j\rangle\langle k|) \otimes (|k\rangle\langle j|)).$$

Show that $S(|\psi\rangle \otimes |\phi\rangle) = |\phi\rangle \otimes |\psi\rangle$. Thus S is a *swap operator*.

Solution 20.   We have

$$S(|\psi\rangle \otimes |\phi\rangle) = \sum_{j,k=0}^{d-1} ((|j\rangle\langle k|) \otimes (|k\rangle\langle j|))(|\psi\rangle \otimes |\phi\rangle)$$

$$= \sum_{j,k=0}^{d-1} \langle k|\psi\rangle\langle j|\phi\rangle(|j\rangle \otimes |k\rangle) = \sum_{j,k=0}^{d-1} \langle j|\psi\rangle\langle k|\phi\rangle(|k\rangle \otimes |j\rangle)$$

$$= |\phi\rangle \otimes |\psi\rangle.$$

Problem 21.   Calculate the eigenvalues of the Hamilton operator

$$\hat{H} := \lambda(S_1 \otimes \hat{L}_1 + S_2 \otimes \hat{L}_2 + S_3 \otimes \hat{L}_3) \tag{1}$$

where we consider a subspace G_1 of the Hilbert space $L_2(\mathbb{S}^2)$ with

$$\mathbb{S}^2 := \{(x_1, x_2, x_3) \ x_1^2 + x_2^2 + x_3^2 = 1\}.$$

Here $\otimes$ denotes the *tensor product*. The linear operators $\hat{L}_1$, $\hat{L}_2$, $\hat{L}_3$ act in the subspace G_1. A basis of subspace G_1 is

$$Y_{1,0} = \sqrt{\frac{3}{4\pi}}\cos(\theta), \quad Y_{1,1} = -\sqrt{\frac{3}{8\pi}}\sin(\theta)e^{i\phi}, \quad Y_{1,-1} = \sqrt{\frac{3}{8\pi}}\sin(\theta)e^{-i\phi}.$$

The operators (matrices) S_1, S_2, S_3 act in the Hilbert space $\mathbb{C}^2$. The standard basis is given by

$$\mathbf{e}_1 = \begin{pmatrix} 1 \\ 0 \end{pmatrix}, \qquad \mathbf{e}_2 = \begin{pmatrix} 0 \\ 1 \end{pmatrix}.$$

The *spin-$\frac{1}{2}$ matrices* S_3, S_+ and S_- are given by

$$S_3 := \frac{1}{2}\hbar \begin{pmatrix} 1 & 0 \\ 0 & -1 \end{pmatrix}, \qquad S_+ := \hbar \begin{pmatrix} 0 & 1 \\ 0 & 0 \end{pmatrix}, \qquad S_- := \hbar \begin{pmatrix} 0 & 0 \\ 1 & 0 \end{pmatrix}$$

where $S_\pm := S_1 \pm iS_2$. The operators $\hat{L}_3$, $\hat{L}_+$ and $\hat{L}_-$ take the form

$$\hat{L}_3 := -i\hbar \frac{\partial}{\partial \phi}$$

$$\hat{L}_+ := \hbar e^{i\phi} \left(\frac{\partial}{\partial \theta} + i\cot(\theta) \frac{\partial}{\partial \phi} \right), \qquad \hat{L}_- := \hbar e^{-i\phi} \left(-\frac{\partial}{\partial \theta} + i\cot(\theta) \frac{\partial}{\partial \phi} \right)$$

where $\hat{L}_\pm := \hat{L}_1 \pm i\hat{L}_2$. Give an interpretation of the Hamilton operator $\hat{H}$ and of the subspace G_1. In some textbooks we find the notation $\hat{H} = \lambda \mathbf{S} \cdot \mathbf{L}$.

Solution 21. The Hamilton operator (1) can be written as

$$\hat{H} = \lambda(S_3 \otimes \hat{L}_3) + \frac{\lambda}{2}(S_+ \otimes \hat{L}_- + S_- \otimes \hat{L}_+).$$

In the tensor product space $\mathbb{C}^2 \otimes G_1$ a basis is given by

$$|1\rangle = \begin{pmatrix} 1 \\ 0 \end{pmatrix} \otimes Y_{1,0}, \qquad |2\rangle = \begin{pmatrix} 1 \\ 0 \end{pmatrix} \otimes Y_{1,-1}, \qquad |3\rangle = \begin{pmatrix} 1 \\ 0 \end{pmatrix} \otimes Y_{1,1},$$

$$|4\rangle = \begin{pmatrix} 0 \\ 1 \end{pmatrix} \otimes Y_{1,0}, \qquad |5\rangle = \begin{pmatrix} 0 \\ 1 \end{pmatrix} \otimes Y_{1,-1}, \qquad |6\rangle = \begin{pmatrix} 0 \\ 1 \end{pmatrix} \otimes Y_{1,1}.$$

In the following we use

$$\hat{L}_+ Y_{1,1} = 0, \qquad \hat{L}_+ Y_{1,0} = \hbar\sqrt{2} Y_{1,1}, \qquad \hat{L}_+ Y_{1,-1} = \hbar\sqrt{2} Y_{1,0},$$

$$\hat{L}_- Y_{1,1} = \hbar\sqrt{2} Y_{1,0}, \qquad \hat{L}_- Y_{1,0} = \hbar\sqrt{2} Y_{1,-1}, \qquad \hat{L}_- Y_{1,-1} = 0$$

$$\hat{L}_3 Y_{1,1} = \hbar Y_{1,1}, \qquad \hat{L}_3 Y_{1,0} = 0, \qquad \hat{L}_3 Y_{1,-1} = -\hbar Y_{1,-1}.$$

For the state $|1\rangle$ we find

$$\hat{H}|1\rangle = \left[\lambda(S_3 \otimes \hat{L}_3) + \frac{\lambda}{2}(S_+ \otimes \hat{L}_- + S_- \otimes \hat{L}_+) \right] \begin{pmatrix} 1 \\ 0 \end{pmatrix} \otimes Y_{1,0}.$$

Thus

$$\hat{H}|1\rangle = \lambda \left[S_3 \begin{pmatrix} 1 \\ 0 \end{pmatrix} \otimes \hat{L}_3 Y_{1,0} \right] + \frac{\lambda}{2} \left[S_+ \begin{pmatrix} 1 \\ 0 \end{pmatrix} \otimes \hat{L}_- Y_{1,0} + S_- \begin{pmatrix} 1 \\ 0 \end{pmatrix} \otimes \hat{L}_+ Y_{1,0} \right].$$

Finally

$$\hat{H}|1\rangle = \frac{\lambda}{2} S_- \begin{pmatrix} 1 \\ 0 \end{pmatrix} \otimes \hat{L}_+ Y_{1,0} = \frac{\lambda}{\sqrt{2}} \hbar^2 |6\rangle.$$

Analogously, we find

$$\hat{H}|2\rangle = -\frac{\lambda\hbar^2}{2}|2\rangle + \frac{\lambda\hbar^2}{\sqrt{2}}|4\rangle, \quad \hat{H}|3\rangle = \frac{\lambda\hbar^2}{2}|3\rangle, \quad \hat{H}|4\rangle = \frac{\lambda\hbar^2}{2}|2\rangle,$$

$$\hat{H}|5\rangle = \frac{\lambda\hbar^2}{2}|5\rangle, \quad \hat{H}|6\rangle = -\frac{\lambda\hbar^2}{2}|6\rangle + \frac{\lambda\hbar^2}{\sqrt{2}}|1\rangle.$$

Hence the states $|3\rangle$ and $|5\rangle$ are eigenstates with the eigenvalues $E_{1,2} = \lambda\hbar^2/2$. The states $|1\rangle$ and $|6\rangle$ form a two-dimensional subspace. The matrix representation is given by

$$\begin{pmatrix} 0 & \dfrac{\lambda\hbar^2}{\sqrt{2}} \\ \dfrac{\lambda\hbar^2}{\sqrt{2}} & -\dfrac{\lambda\hbar^2}{2} \end{pmatrix}.$$

The eigenvalues are

$$E_{3,4} = -\frac{\lambda\hbar^2}{2} \pm \frac{3\lambda\hbar^2}{4}.$$

Analogously, the states $|2\rangle$ and $|4\rangle$ form a two-dimensional subspace. The matrix representation is given by

$$\begin{pmatrix} -\dfrac{\lambda\hbar^2}{2} & \dfrac{\lambda\hbar^2}{\sqrt{2}} \\ \dfrac{\lambda\hbar^2}{\sqrt{2}} & 0 \end{pmatrix}.$$

The eigenvalues are

$$E_{5,6} = -\frac{\lambda\hbar^2}{2} \pm \frac{3\lambda\hbar^2}{4}.$$

The Hamilton operator (1) describes the *spin-orbit coupling*.

Problem 22. Let $\mathbf{v}_1, \mathbf{v}_2, \mathbf{v}_3 \in \mathbb{C}^2$. Find an 8×8 permutation matrix P such that $P(\mathbf{v}_1 \otimes \mathbf{v}_2 \otimes \mathbf{v}_3) = \mathbf{v}_2 \otimes \mathbf{v}_3 \otimes \mathbf{v}_1$.

Solution 22. We find

$$P = (1) \oplus \begin{pmatrix} 0 & 0 & 0 & 1 & 0 & 0 \\ 1 & 0 & 0 & 0 & 0 & 0 \\ 0 & 0 & 0 & 0 & 1 & 0 \\ 0 & 1 & 0 & 0 & 0 & 0 \\ 0 & 0 & 0 & 0 & 0 & 1 \\ 0 & 0 & 1 & 0 & 0 & 0 \end{pmatrix} \oplus (1).$$

Hence P is the direct sum of a 1×1 matrix, 6×6 matrix and 1×1 matrix.

Problem 23. Let $x \in \{0,1\}$ and $|0\rangle$, $|1\rangle$ be the standard basis in $\mathbb{C}^2$. Consider the boolean function $f(x) = \overline{x}$. Find a 4×4 permutation matrix P such that

$$|x\rangle \otimes |0\rangle \to |x\rangle \otimes |f(x)\rangle) \equiv |x\rangle \otimes |\overline{x}\rangle.$$

Solution 23. We have to satisfy

$$P(|0\rangle \otimes |0\rangle) = |0\rangle \otimes |1\rangle \Rightarrow P \begin{pmatrix} 1 \\ 0 \\ 0 \\ 0 \end{pmatrix} = \begin{pmatrix} 0 \\ 1 \\ 0 \\ 0 \end{pmatrix}$$

$$P(|1\rangle \otimes |0\rangle) = |1\rangle \otimes |1\rangle \Rightarrow P \begin{pmatrix} 0 \\ 0 \\ 1 \\ 0 \end{pmatrix} = \begin{pmatrix} 0 \\ 0 \\ 0 \\ 1 \end{pmatrix}.$$

A possible solution is the permutation matrix

$$P = \begin{pmatrix} 0 & 0 & 0 & 1 \\ 1 & 0 & 0 & 0 \\ 0 & 1 & 0 & 0 \\ 0 & 0 & 1 & 0 \end{pmatrix}.$$

Problem 24. Consider the *momentum operator* $\hat{P}_j = -i\hbar\partial/\partial q_j$ and the *position operator* $\hat{Q}_k$ with $j, k = 1, 2, 3$. Then we have the commutation relation

$$[\hat{Q}_k, \hat{P}_j] = i\hbar\delta_{jk}I$$

where I is the identity operator. Let S be an $n \times n$ matrix over $\mathbb{C}$ with $S^2 = I_n$. Find the commutator $[\hat{Q}_k \otimes S, \hat{P}_j \otimes S]$.

Solution 24. We obtain

$$[\hat{Q}_k \otimes S, \hat{P}_j \otimes S] = Q_k P_j \otimes I_n - P_j Q_k \otimes I_n = [Q_k, P_j] \otimes I_n = i\hbar\delta_{jk}I \otimes I_n.$$

Problem 25. Let I_n be the $n \times n$ identity matrix. A nonsingular $n^2 \times n^2$ matrix A over $\mathbb{C}$ is called an *R-matrix* when it satisfies the *Yang-Baxter equation*

$$(A \otimes I_n)(I_n \otimes A)(A \otimes I_n) = (I_n \otimes A)(A \otimes I_n)(I_n \otimes A).$$

Let $n = 2$. Show that

$$A = \begin{pmatrix} 1 & 0 & 0 & 1 \\ 0 & 1 & 1 & 0 \\ 0 & -1 & 1 & 0 \\ -1 & 0 & 0 & 1 \end{pmatrix}$$

is an R-matrix. Apply Computer Algebra.

Solution 25. The following Maxima program will do the job

```
/* R_Matrix.mac */
I2: matrix([1,0],[0,1]);
A: matrix([1,0,0,1],[0,1,1,0],[0,-1,1,0],[-1,0,0,1]);
T1: kronecker_product(A,I2);
T2: kronecker_product(I2,A);
L: T1 . T2 . T1; R: T2 . T1 . T2; F: L-R;
```

Problem 26. Let I_n be the $n \times n$ identity matrix. An invertible matrix $X \in \mathbb{C}^{n^2 \times n^2}$ satisfies the *Yang-Baxter equation* if

$$(X \otimes I_n)(I_n \otimes X)(X \otimes I_n) = (I_n \otimes X)(X \otimes I_n)(I_n \otimes).$$

If X satisfies the Yang-Baxter equation, then X^* satisfies the Yang-Baxter equation. If X satisfies the Yang-Baxter equation, then X^{-1} satisfies the Yang-Baxter equation. If X satisfies the Yang-Baxter equation and $Q \in \mathbb{C}^{n \times n}$ is an arbitrary invertible matrix. Then

$$\tilde{X} = (Q \otimes Q)X(Q \otimes Q)^{-1}$$

also satisfies the Yang-Baxter equation. Show that

$$X = \frac{(1+i)}{2} \begin{pmatrix} 1 & 0 & 0 & 1 \\ 0 & 1 & 1 & 0 \\ 0 & -1 & 1 & 0 \\ -1 & 0 & 0 & 1 \end{pmatrix}$$

satisfies the Yang-Baxter equation.

Solution 26. The following Maxima program provides the proof

```
/* YBBell.mac */
I2: matrix([1,0],[0,1]);
X: ((1+%i)/2)*matrix([1,0,0,1],[0,1,1,0],[0,-1,1,0],[-1,0,0,1]);
T1: kronecker_product(X,I2);
T2: kronecker_product(I2,X);
F: (T1 . T2 . T1) - (T2 . T1 . T2);
```

Supplementary Problems

Problem 1. Let $\mathbf{x}, \mathbf{y} \in \mathbb{R}^n$. We define

$$\mathbf{x} \wedge \mathbf{y} := \mathbf{x} \otimes \mathbf{y} - \mathbf{y} \otimes \mathbf{x}.$$

Show that $((\mathbf{x} \wedge \mathbf{y}) \wedge \mathbf{z}) + ((\mathbf{z} \wedge \mathbf{x}) \wedge \mathbf{y}) + ((\mathbf{y} \wedge \mathbf{z}) \wedge \mathbf{x}) = \mathbf{0}$.

Problem 2. Can the $\mathbb{Z}_4$ *Fourier matrix*

$$\begin{pmatrix} 1 & 1 & 1 & 1 \\ 1 & i & -1 & -i \\ 1 & -1 & 1 & -1 \\ 1 & -i & -1 & i \end{pmatrix}$$

be written as the Kronecker product of two 2×2 unitary matrices?

Problem 3. Consider the $2^5 \times 2^5$ matrices

$$A = I_2 \otimes I_2 \otimes \sigma_1 \otimes \sigma_3 \otimes \sigma_3, \quad B = I_2 \otimes I_2 \otimes \sigma_2 \otimes \sigma_3 \otimes \sigma_3,$$

$$C = \sigma_3 \otimes \sigma_3 \otimes \sigma_3 \otimes \sigma_3 \otimes \sigma_3.$$

Find A^2, B^2, C^2, $[A, B]$, $[B, C]$, $[C, A]$, $[A, B]_+$, $[B, C]_+$, $[C, A]_+$.

Problem 4. Consider the Pauli spin matrices $\sigma_1, \sigma_2, \sigma_3$ which are unitary and hermitian matrices. Consider the normalized state in $\mathbb{C}^2$

$$|\psi\rangle = \begin{pmatrix} e^{i\phi} \cos(\theta) \\ \sin(\theta) \end{pmatrix}.$$

Show that

$$\mathbf{v}(\phi, \theta) = \begin{pmatrix} \langle\psi|\sigma_1|\psi\rangle \\ \langle\psi|\sigma_2|\psi\rangle \\ \langle\psi|\sigma_3|\psi\rangle \end{pmatrix} = \begin{pmatrix} \cos(\phi)\sin(2\theta) \\ -\sin(\phi)\sin(2\theta) \\ \cos(2\theta) \end{pmatrix}.$$

Show that the vector is normalized. Study also the tensor

$$T_{jk} = ((\langle\psi| \otimes \langle\psi|)(\sigma_j \otimes \sigma_k)(|\psi\rangle \otimes |\psi\rangle)$$

where $j, k = 1, 2, 3$.

Problem 5. Let $|a\rangle$, $|b\rangle$ be normalized states in $\mathbb{C}^n$ and X, Y be $n \times n$ matrices over $\mathbb{C}$. Show that

$$((\langle a| \otimes \langle b|)((X \otimes I_n)(I_n \otimes Y))(|a\rangle \otimes |b\rangle) = \langle a|X|a\rangle\langle b|Y|b\rangle.$$

Note that $(|a\rangle \otimes |b\rangle)^* = \langle a| \otimes \langle b|$.

Problem 6.   Let $n \geq 1$ and $|0\rangle, |1\rangle, \ldots, |n\rangle$ be an orthonormal basis in $\mathbb{C}^{n+1}$. Consider the states

$$|\psi_0\rangle = \frac{1}{\sqrt{2}}|0\rangle \otimes |0\rangle + \frac{1}{\sqrt{2n}} \sum_{j=1}^{n} |j\rangle \otimes |j\rangle, \quad |\psi_1\rangle = \frac{1}{\sqrt{2}}|0\rangle \otimes |0\rangle - \frac{1}{\sqrt{2n}} \sum_{j=1}^{n} |j\rangle \otimes |j\rangle.$$

Find $\langle \psi_0|\psi_1 \rangle$ and $|\langle \psi_0|\psi_1 \rangle|^2$.

Problem 7.   The *free group* with two generators g_1 and g_2 has the matrix representation

$$g_1 = \begin{pmatrix} 1 & 0 \\ 2 & 1 \end{pmatrix}, \qquad g_2 = \begin{pmatrix} 1 & 2 \\ 0 & 1 \end{pmatrix}.$$

Find g_1^{-1}, g_2^{-1}, $g_1 \otimes g_2$, $(g_1 \otimes g_2)^{-1}$.

Problem 8.   Consider the *Bell matrix*

$$B(\phi_1, \phi_2, \phi_3, \phi_4) = \frac{1}{\sqrt{2}} \begin{pmatrix} e^{i\phi_1} & 0 & 0 & e^{i\phi_1} \\ 0 & e^{i\phi_3} & e^{i\phi_3} & 0 \\ 0 & e^{i\phi_4} & -e^{i\phi_4} & 0 \\ e^{i\phi_2} & 0 & 0 & -e^{i\phi_2} \end{pmatrix}.$$

Find the conditions on ϕ_1, ϕ_2, ϕ_3, ϕ_4 such that (Yang-Baxter equation)

$$(B \otimes I_2)(I_2 \otimes B)(B \otimes I_2) = (I_2 \otimes B)(B \otimes I_2)(I_2 \otimes B).$$

Problem 9.   Consider the Pauli spin matrices σ_2 and σ_3. Find a unitary matrix U such that

$$\sigma_2 \otimes \sigma_3 = U \begin{pmatrix} -1 & 0 & 0 & 0 \\ 0 & 0 & 0 & 1 \\ 0 & 0 & 1 & 0 \\ 0 & 1 & 0 & 0 \end{pmatrix} U^* \Rightarrow (\sigma_2 \otimes \sigma_3) U - U \begin{pmatrix} -1 & 0 & 0 & 0 \\ 0 & 0 & 0 & 1 \\ 0 & 0 & 1 & 0 \\ 0 & 1 & 0 & 0 \end{pmatrix} = 0_4.$$

Apply the vec-operator.

Problem 10.   An orthonormal basis in $\mathbb{C}^2$ is given by

$$\mathbf{v}_1 = \frac{1}{\sqrt{2}} \begin{pmatrix} 1 \\ 1 \end{pmatrix}, \quad \mathbf{v}_2 = \frac{1}{\sqrt{2}} \begin{pmatrix} 1 \\ -1 \end{pmatrix}$$

and an orthonormal basis in $\mathbb{C}^3$ is given by

$$\mathbf{u}_1 = \frac{1}{\sqrt{2}} \begin{pmatrix} 1 \\ 0 \\ 1 \end{pmatrix}, \quad \mathbf{u}_2 = \begin{pmatrix} 0 \\ 1 \\ 0 \end{pmatrix}, \quad \mathbf{u}_3 = \frac{1}{\sqrt{2}} \begin{pmatrix} 1 \\ 0 \\ -1 \end{pmatrix}.$$

Use these two bases to construct two orthonormal bases in $\mathbb{C}^6$, i.e. calculate

$$\mathbf{v}_j \otimes \mathbf{u}_k, \ j = 1, 2; \ k = 1, 2, 3 \quad \text{and} \quad \mathbf{u}_k \otimes \mathbf{v}_j, \quad k = 1, 2, 3; \ j = 1, 2.$$

Find the unitary matrix which connects the two bases.

Problem 11. Find the group generated by

$$A = \begin{pmatrix} 0 & 0 & i \\ 0 & -1 & 0 \\ -i & 0 & 0 \end{pmatrix}$$

under matrix multiplication. Find the group generated by $A \otimes A$ under matrix multiplication.

Problem 12. Let

$$x = \begin{pmatrix} 0 & 1 \\ 0 & 0 \end{pmatrix}, \quad y = \begin{pmatrix} 0 & 0 \\ 1 & 0 \end{pmatrix}, \quad h = \begin{pmatrix} 1 & 0 \\ 0 & -1 \end{pmatrix}$$

be the standard basis of the Lie algebra $s\ell(2, \mathbb{R})$. Is

$$x \otimes x, \ x \otimes y, \ x \otimes h, \ y \otimes x, \ y \otimes y, \ y \otimes h, \ h \otimes x, \ h \otimes y, \ h \otimes h$$

a basis of the Lie algebra $s\ell(4, \mathbb{R})$?

Problem 13. The *Pauli spin matrices* σ_1, σ_2, σ_3 are not elements of the Lie group $SU(2)$ since $\det(\sigma_j) = -1$, but elements of the Lie group $U(2)$. The matrices $\tau_1 = i\sigma_1$, $\tau_2 = i\sigma_2$, $\tau_3 = i\sigma_3$ are elements of the Lie group $SU(2)$. Find the group generated by

$$\tau_1 = i\sigma_1 = \begin{pmatrix} 0 & i \\ i & 0 \end{pmatrix}, \quad \tau_2 = i\sigma_2 = \begin{pmatrix} 0 & 1 \\ -1 & 0 \end{pmatrix}, \quad \tau_3 = i\sigma_3 = \begin{pmatrix} i & 0 \\ 0 & -i \end{pmatrix}$$

under matrix multiplication. Find the group generated by $\tau_1 \otimes \tau_1$, $\tau_2 \otimes \tau_2$, $\tau_3 \otimes \tau_3$ under matrix multiplication.

Problem 14. Let A be an invertible $m \times m$ matrix and B be an $n \times n$ invertible matrix. Show that 1 is an eigenvalue of $A \otimes A^{-1} \otimes B \otimes B^{-1}$ and $A \otimes B \otimes A^{-1} \otimes B^{-1}$.

Problem 15. Let A be an 2×2 matrix over $\mathbb{C}$. Find all solutions of the equation $\text{tr}(A + A^*) = \text{tr}(AA^*)$. Find all solutions of the equation

$$\text{tr}(A \otimes I_2 + I_2 \otimes A^*) = \text{tr}(A \otimes A^*).$$

Problem 16.   Let **n** and **m** ne normalized vectors in $\mathbb{R}^3$ and

$$\mathbf{n} \cdot \boldsymbol{\sigma} = n_1\sigma_1 + n_2\sigma_2 + n_3\sigma_3, \quad \mathbf{m} \cdot \boldsymbol{\sigma} = m_1\sigma_1 + m_2\sigma_2 + m_3\sigma_3.$$

Let $|\psi\rangle$ be the (normalized) Bell state

$$|\psi\rangle = \frac{1}{\sqrt{2}} \left(\begin{pmatrix} 0 \\ 1 \end{pmatrix} \otimes \begin{pmatrix} 1 \\ 0 \end{pmatrix} - \begin{pmatrix} 1 \\ 0 \end{pmatrix} \otimes \begin{pmatrix} 0 \\ 1 \end{pmatrix} \right).$$

Show that $\langle\psi|(\boldsymbol{\sigma} \cdot \mathbf{n}) \otimes (\boldsymbol{\sigma} \cdot \mathbf{m})|\psi\rangle = -\mathbf{n} \cdot \mathbf{m} = -n_1m_1 - n_2m_2 - n_3m_3$.

Problem 17.   Let **a**, **b** be unit vectors in $\mathbb{R}^3$. Consider the 2×2 matrices

$$M_{\mathbf{a}} = \frac{1}{2}(I_2 + a_1\sigma_1 + a_2\sigma_2 + a_3\sigma_3), \quad M_{\mathbf{b}} = \frac{1}{2}(I_2 + b_1\sigma_1 + b_2\sigma_2 + b_3\sigma_3)$$

which are projection matrices. Let $|\psi_j\rangle$ $(j = 1, 2, 3, 4)$ be the four Bell states. Find $\langle\psi_j|(M_{\mathbf{a}} \otimes M_{\mathbf{b}})|\psi_j\rangle$, $j = 1, 2, 3, 4$.

Problem 18.   Let $|0\rangle$, $|1\rangle$ be the standard basis in $\mathbb{C}^2$. Consider the Bell state

$$|\psi\rangle = \frac{1}{\sqrt{2}}(|0\rangle_A \otimes |0\rangle_B + |1\rangle_A \otimes |1\rangle_B) \equiv \frac{1}{\sqrt{2}} \begin{pmatrix} 1 \\ 0 \\ 0 \\ 1 \end{pmatrix}$$

where A refers to Alice and B refers to Bob. Let

$$\Pi_0 = |0\rangle\langle 0| \equiv \begin{pmatrix} 1 & 0 \\ 0 & 0 \end{pmatrix}, \quad \Pi_1 = |1\rangle\langle 1| \equiv \begin{pmatrix} 0 & 0 \\ 0 & 1 \end{pmatrix}$$

be projection matrices with $\Pi_0 + \Pi_1 = I_2$. Show that measurement of the first qubit (Alice) provide

$$p_1(0) = \langle\psi|(\Pi_0 \otimes I_2)^*(\Pi_0 \otimes I_2)|\psi\rangle = \frac{1}{2}.$$

Show that the post-measurement state $|\phi\rangle$ is given by

$$|\phi\rangle = \frac{1}{\sqrt{p_1(0)}}(\Pi_0 \otimes I_2)|\psi\rangle = |0\rangle \otimes |0\rangle.$$

Show that $\langle\phi|(I_2 \otimes \Pi_0)^*(I_2 \otimes \Pi_0)|\phi\rangle = 1$.

Chapter 19

Variational Calculus

Let W_1 be some topological vector space and f a (nonlinear) mapping $f : W \to W_1$. We call f *Gateaux differentiable* in $u \in W$ if there exists a mapping $\theta \in L(W, W_1)$ such that for all $v \in W$

$$\lim_{\epsilon \to 0} \frac{1}{\epsilon} (f(u + \epsilon v) - f(u) - \epsilon \theta v) = 0$$

in the topology of W_1. The linear mapping $\theta \in L(W, W_1)$ is the called *Gateaux derivative* of f in u and is written as $\theta = f'(u)$. The Gateaux derivative is a linear operation.

Let V and W be two Banach spaces with the norms $\| \cdot \|_V$ and $\| \cdot \|_W$, respectively. Let $U \subset V$ be an open subset of V. A function $f : U \to W$ is called *Fréchet differentiable* at $p \in U$ if there exists a bounded operator $A_p : V \to W$ such that

$$\lim_{h \to \infty} \frac{\| f(p + h) - f(p) - A_p(h) \|_W}{\| h \|_V} = 0.$$

The Fréchet derivative is a linear operation. The chain rule also applies.

If the Fréchet derivative of a mapping exists, then the Gateaux derivative of this mapping also exists.

Problem 1.   The difference equation

$$u_{t+1} = 4u_t(1 - u_t), \qquad t = 0, 1, 2, \ldots \qquad (1)$$

is called the *logistic equation*, where $u_0 \in [0, 1]$. It can be shown that $u_t \in [0, 1]$ for all $t \in \mathbb{N}$. Equation (1) can also be written as the map $f : [0, 1] \to [0, 1]$

$$f(u) = 4u(1 - u). \qquad (2)$$

Then u_t is the t-th iterate of f. Calculate the Gateaux derivative of the map f. Find the *variational equation* of (1).

Solution 1.   We obtain $f(u + \epsilon v) - f(u) = 4\epsilon v - 8\epsilon u v - 4\epsilon^2 v^2$. Therefore

$$\frac{d}{d\epsilon}(f(u + \epsilon v) - f(u))\Big|_{\epsilon=0} = 4v - 8uv = 4(1 - 2u)v.$$

Thus we find that the Gateaux derivative of the the map is given by

$$\theta v = 4v - 8uv = 4(1 - 2u)v.$$

The variational equation of the nonlinear difference equation (1) is then

$$v_{t+1} = 4v_t - 8u_t v_t = 4(1 - 2u_t)v_t.$$

The stability of fixed points and periodic solutions is studied with the help of the variational equation. For $u_t = 0$ we have $v_{t+1} = 4v_t$.

Problem 2.   Consider the system of differential equations

$$\frac{du_1}{dt} - \sigma(u_2 - u_1) = 0$$

$$\frac{du_2}{dt} + u_1 u_3 - r u_1 + u_2 = 0$$

$$\frac{du_3}{dt} - u_1 u_2 + b u_3 = 0$$

where σ, b and r are positive constants. The left-hand side defines a map f. Calculate the Gateaux derivative. This system of differential equations is called the *Lorenz model*. The equation $\theta \mathbf{v} = \mathbf{0}$ is called the corresponding *variational equation* (or *linearized equation*). Calculate the variational equation of the Lorenz model.

Solution 2.   Since

$$f(\mathbf{u} + \epsilon \mathbf{v}) - f(\mathbf{u}) = \begin{pmatrix} \epsilon dv_1/dt - \epsilon\sigma(v_2 - v_1) \\ \epsilon dv_2/dt + \epsilon u_1 v_3 + \epsilon u_3 v_1 + \epsilon^2 v_1 v_3 - \epsilon r v_1 + \epsilon v_2 \\ \epsilon dv_3/dt - \epsilon u_1 v_2 - \epsilon u_2 v_1 - \epsilon^2 v_1 v_2 + \epsilon b v_3 \end{pmatrix}$$

we obtain

$$\theta\mathbf{v} = \begin{pmatrix} dv_1/dt - \sigma(v_2 - v_1) \\ dv_2/dt + u_1v_3 + u_3v_1 - rv_1 + v_2 \\ dv_3/dt - u_1v_2 - u_2v_1 + bv_3 \end{pmatrix}.$$

Then the variational equation $\theta\mathbf{v} = \mathbf{0}$ takes the form

$$\frac{dv_1}{dt} = \sigma(v_2 - v_1)$$

$$\frac{dv_2}{dt} = -u_1v_3 - u_3v_1 + rv_1 - v_2$$

$$\frac{dv_3}{dt} = u_1v_2 + u_2v_1 - bv_3.$$

To solve this system we have find a solution of (1) to solve system (2).

Problem 3. Consider the map

$$f(u) := \frac{\partial u}{\partial t} + u\frac{\partial u}{\partial x} + \beta\frac{\partial^3 u}{\partial x^3}.$$

Calculate the Gateaux derivative. Calculate the variational equation. The nonlinear partial differential equation $f(u) = 0$, i.e.

$$\frac{\partial u}{\partial t} + u\frac{\partial u}{\partial x} + \beta\frac{\partial^3 u}{\partial x^3} = 0$$

is called the *Korteweg-de Vries equation*.

Solution 3. We obtain

$$f(u + \epsilon v) - f(u) = \epsilon\frac{\partial v}{\partial t} + \epsilon v\frac{\partial u}{\partial x} + \epsilon u\frac{\partial v}{\partial x} + \epsilon^2 v\frac{\partial v}{\partial x} + \epsilon\beta\frac{\partial^3 v}{\partial x^3}.$$

It follows that

$$\lim_{\epsilon \to 0}\frac{1}{\epsilon}(f(u + \epsilon v) - f(u)) = \frac{\partial v}{\partial t} + \frac{\partial u}{\partial x}v + u\frac{\partial v}{\partial x} + \beta\frac{\partial^3 v}{\partial x^3}.$$

Therefore

$$\theta v \equiv f'(u)v = \frac{\partial v}{\partial t} + \frac{\partial u}{\partial x}v + u\frac{\partial v}{\partial x} + \beta\frac{\partial^3 v}{\partial x^3}.$$

From this equation we find the variational equation $\theta v = 0$ as

$$\frac{\partial v}{\partial t} + \frac{\partial u}{\partial x}v + u\frac{\partial v}{\partial x} + \beta\frac{\partial^3 v}{\partial x^3} = 0.$$

This is a linear partial differential equation in v. A solution u from the Korteweg-de Vries equation has to be inserted. An example is the trivial solution $u(x, t) = 0$. Then we obtain the linear partial differential equation

$$\frac{\partial v}{\partial t} + \beta\frac{\partial^3 v}{\partial x^3} = 0.$$

Using this equation we can study the stability of the solution $u(x,t) = 0$.

Problem 4.   The Gateaux derivative of an operator-valued function $R(u)$ is defined as

$$R'(u)[v]w := \left.\frac{\partial[R(u+\epsilon v)w]}{\partial\epsilon}\right|_{\epsilon=0} \tag{1}$$

where ϵ is a real parameter. We say that $R'(u)[v]w$ is the derivative of $R(u)$ evaluated at v and then applied to w, where w, v and u are smooth functions of $x_1,\ldots,x_n$ and t. Let $n = 1$ and

$$R(u) := \frac{\partial}{\partial x} + u + \frac{\partial u}{\partial x}D_x^{-1}, \quad D_x^{-1}f(x) := \int^x f(s)ds$$

where u is a smooth function of x and t. Calculate the Gateaux derivative of $R(u)$.

Solution 4.   From (1) we find

$$R(u+\epsilon v)w = \frac{\partial w}{\partial x} + (u+\epsilon v)w + \frac{\partial}{\partial x}(u+\epsilon v)D_x^{-1}w$$

$$= \frac{\partial w}{\partial x} + uw + \frac{\partial u}{\partial x}D_x^{-1}w + \epsilon vw + \epsilon\frac{\partial v}{\partial x}D_x^{-1}w.$$

Therefore

$$R'(u)[v]w = vw + \frac{\partial v}{\partial x}D_x^{-1}w \equiv \left(v + \frac{\partial v}{\partial x}D_x^{-1}\right)w.$$

Problem 5.   Let $f,g : W \longrightarrow W$ be two maps, where W is a topological vector space ($u \in W$). Assume that the Gateaux derivative of f and g exists, i.e.

$$f'(u)[v] := \left.\frac{\partial f(u+\epsilon v)}{\partial\epsilon}\right|_{\epsilon=0}, \quad g'(u)[v] := \left.\frac{\partial g(u+\epsilon v)}{\partial\epsilon}\right|_{\epsilon=0}. \tag{1}$$

The *Lie product* (or *commutator*) of f and g is defined by

$$[f,g] := f'(u)[g] - g'(u)[f]. \tag{2}$$

Let

$$f(u) := \frac{\partial u}{\partial t} - u\frac{\partial u}{\partial x} - \frac{\partial^3 u}{\partial x^3}, \quad g(u) := \frac{\partial u}{\partial t} - \frac{\partial^2 u}{\partial x^2}. \tag{3}$$

Calculate the commutator $[f,g]$.

Solution 5. Applying the definition (1) we find

$$f'(u)[g] \equiv \frac{\partial}{\partial \epsilon} \left[\frac{\partial}{\partial t}(u + \epsilon g(u)) \right]_{\epsilon=0} - \frac{\partial}{\partial \epsilon} \left[(u + \epsilon g(u)) \frac{\partial}{\partial x}(u + \epsilon g(u)) \right]_{\epsilon=0}$$
$$- \frac{\partial}{\partial \epsilon} \left[\frac{\partial^3}{\partial x^3}(u + \epsilon g(u)) \right]_{\epsilon=0}.$$

Consequently,

$$f'(u)[g] = \frac{\partial g(u)}{\partial t} - g(u)\frac{\partial u}{\partial x} - u\frac{\partial g(u)}{\partial x} - \frac{\partial^3 g(u)}{\partial x^3}. \tag{4}$$

For the second term on the right-hand side of (2) we find

$$g'(u)[f] = \frac{\partial}{\partial \epsilon} \left[\frac{\partial}{\partial t}(u + \epsilon f(u)) \right]_{\epsilon=0} - \frac{\partial}{\partial \epsilon} \left[\frac{\partial^2}{\partial x^2}(u + \epsilon f(u)) \right]_{\epsilon=0}.$$

Consequently

$$g'(u)[f] = \frac{\partial f(u)}{\partial t} - \frac{\partial^2 f(u)}{\partial x^2}.$$

Hence

$$[f, g] = \frac{\partial}{\partial t}(g(u) - f(u)) - g(u)\frac{\partial u}{\partial x} - u\frac{\partial g(u)}{\partial x} - \frac{\partial^3 g(u)}{\partial x^3} + \frac{\partial^2 f(u)}{\partial x^2}.$$

Inserting f and g we find

$$[f, g] = \frac{\partial}{\partial t} \left(-\frac{\partial^2 u}{\partial x^2} + u\frac{\partial u}{\partial x} + \frac{\partial^3 u}{\partial x^3} \right) - \left(\frac{\partial u}{\partial t} - \frac{\partial^2 u}{\partial x^2} \right) \frac{\partial u}{\partial x}$$
$$- u\left(\frac{\partial^2 u}{\partial t \partial x} - \frac{\partial^3 u}{\partial x^3} \right) - \frac{\partial^4 u}{\partial t \partial x^3} + \frac{\partial^5 u}{\partial x^5} + \frac{\partial^3 u}{\partial x^2 \partial t}$$
$$- 3\frac{\partial u}{\partial x}\frac{\partial^2 u}{\partial x^2} - u\frac{\partial^3 u}{\partial x^3} - \frac{\partial^5 u}{\partial x^5}.$$

It follows that

$$[f, g] = -2\frac{\partial u}{\partial x}\frac{\partial^2 u}{\partial x^2}.$$

Problem 6. Let

$$\frac{\partial u}{\partial t} = A(u) \tag{1}$$

be an evolution equation, where A is a differential expression, i.e. A depends on u and its partial derivatives with respect to $x_1, \ldots, x_n$ and on the variables $x_1, \ldots, x_n$ and t. A function

$$\sigma\left(x_1, \ldots, x_n, t, u(\mathbf{x}, t), \frac{\partial u}{\partial x_1}, \ldots \right) \tag{2}$$

is called a *symmetry generator* of (1) if

$$\frac{\partial\sigma}{\partial t} = A'(u)[\sigma] \tag{3}$$

where $A'(u)[\sigma]$ is the Gateaux derivative of A, i.e.

$$A'(u)[\sigma] = \frac{\partial}{\partial\epsilon}A(u + \epsilon\sigma)|_{\epsilon=0}. \tag{4}$$

Let $n = 1$ and consider the *Korteweg-de Vries equation*

$$\frac{\partial u}{\partial t} = 6u\frac{\partial u}{\partial x} + \frac{\partial^3 u}{\partial x^3}.$$

Show that

$$\sigma = 3t\frac{\partial u}{\partial x} + \frac{1}{2}$$

is a symmetry generator.

Solution 6.   First we calculate the Gateaux derivative of A, i.e.

$$A'(u)[\sigma] = \frac{\partial}{\partial\epsilon}A(u + \epsilon\sigma)|_{\epsilon=0}$$

$$= \frac{\partial}{\partial\epsilon}\left(6(u + \epsilon\sigma)\left(\frac{\partial(u + \epsilon\sigma)}{\partial x}\right) + \frac{\partial^3}{\partial x^3}(u + \epsilon\sigma)\right)\Bigg|_{\epsilon=0}.$$

Therefore

$$A'(u)[\sigma] = 6\frac{\partial u}{\partial x}\sigma + 6u\frac{\partial\sigma}{\partial x} + \frac{\partial^3\sigma}{\partial x^3}.$$

From σ we find using differentiation with respect to t

$$\frac{\partial\sigma}{\partial t} = 3\frac{\partial u}{\partial x} + 3t\frac{\partial^2 u}{\partial x\partial t}.$$

Inserting σ into $A'(u)[\sigma]$ gives

$$A'(u)[\sigma] = 18tu\frac{\partial^2 u}{\partial x^2} + 18t\left(\frac{\partial u}{\partial x}\right)^2 + 3\frac{\partial u}{\partial x} + 3t\frac{\partial^4 u}{\partial x^4}.$$

From the Korteweg-de Vries equation we obtain

$$\frac{\partial^2 u}{\partial x\partial t} = 6\left(\frac{\partial u}{\partial x}\right)^2 + 6u\frac{\partial^2 u}{\partial x^2} + \frac{\partial^4 u}{\partial x^4}.$$

Inserting the right-hand side of this equation into the equation for $\partial\sigma/\partial t$ leads to

$$\frac{\partial\sigma}{\partial t} = 3\frac{\partial u}{\partial x} + 18t\left(\frac{\partial u}{\partial x}\right)^2 + 18tu\frac{\partial^2 u}{\partial x^2} + 3t\frac{\partial^4 u}{\partial x^4}.$$

This proves (3).

Problem 7. Let

$$\frac{\partial u}{\partial t} = A(u)$$

be an evolution equation, where A is a nonlinear differential expression, i.e. A depends on u and its partial derivatives with respect to $x_1, \ldots, x_n$ and on the variables $x_1, \ldots, x_n$ and t. An operator $R(u)$ is called a *recursion operator* if

$$R'(u)[A(u)]v = [A'(u)v, R(u)v] \tag{1}$$

where

$$R'(u)[A(u)]v := \frac{\partial}{\partial \epsilon}(R(u + \epsilon A(u))v)|_{\epsilon=0} \tag{2}$$

and the commutator is defined as

$$[A'(u)v, R(u)v] := \frac{\partial}{\partial \epsilon}A'(u)(v+\epsilon R(u)v)|_{\epsilon=0} - \frac{\partial}{\partial \epsilon}R(u)(v+\epsilon A'(u)v)|_{\epsilon=0}. \tag{3}$$

Let $n = 1$. Consider the nonlinear diffusion equation

$$\frac{\partial u}{\partial t} = \frac{\partial^2 u}{\partial x^2} + 2u\frac{\partial u}{\partial x}, \quad A(u) \equiv \frac{\partial^2 u}{\partial x^2} + 2u\frac{\partial u}{\partial x}. \tag{4}$$

Show that

$$R(u) = \frac{\partial}{\partial x} + u + \frac{\partial u}{\partial x}D_x^{-1}, \quad D_x^{-1}f(x) := \int^x f(s)ds$$

is a recursion operator of (4).

Solution 7. First we calculate the left-hand side of (1). When we apply the operator $R(u)$ to v we obtain

$$R(u)v = \frac{\partial v}{\partial x} + uv + \frac{\partial u}{\partial x}D_x^{-1}v.$$

From definition (2) we find that the left-hand side is given by

$$R'(u)[A(u)]v = \frac{\partial}{\partial \epsilon}\left(\frac{\partial}{\partial x} + u + \epsilon A(u) + \frac{\partial(u + \epsilon A(u))}{\partial x}D_x^{-1}\right)v$$

$$= A(u)v + \frac{\partial A(u)}{\partial x}D_x^{-1}v$$

$$= \left(\frac{\partial^2 u}{\partial x^2} + 2u\frac{\partial u}{\partial x}\right)v + \left(\frac{\partial^3 u}{\partial x^3} + 2\left(\frac{\partial u}{\partial x}\right)^2 + 2u\frac{\partial^2 u}{\partial x^2}\right)D_x^{-1}v.$$

To calculate the commutator we first have to find the Gateaux derivative of A. Since

$$A'(u)[v] := \frac{\partial A(u + \epsilon v)}{\partial \epsilon}\bigg|_{\epsilon=0}$$

we find

$$A'(u)[v] = \frac{\partial}{\partial \epsilon}\left(\frac{\partial^2}{\partial x^2}(u + \epsilon v) + 2(u + \epsilon v)\frac{\partial}{\partial x}(u + \epsilon v)\right)\bigg|_{\epsilon=0}$$

$$= \frac{\partial^2 v}{\partial x^2} + 2u\frac{\partial v}{\partial x} + 2\frac{\partial u}{\partial x}v.$$

Therefore

$$A'(u)v = \frac{\partial^2 v}{\partial x^2} + 2u\frac{\partial v}{\partial x} + 2\frac{\partial u}{\partial x}v.$$

For the first term of the commutator we find

$$\frac{\partial}{\partial \epsilon}A'(u)(v + \epsilon R(u)v)|_{\epsilon=0} = \frac{\partial}{\partial \epsilon}\left[\frac{\partial^2}{\partial x^2}(v + \epsilon R(u)v) + 2u\frac{\partial}{\partial x}(v + \epsilon R(u)v)\right.$$

$$\left. + 2\frac{\partial u}{\partial x}(v + \epsilon R(u)v)\right]_{\epsilon=0}$$

$$= \frac{\partial^3 v}{\partial x^3} + 3\frac{\partial^2 u}{\partial x^2}v + 5\frac{\partial u}{\partial x}\frac{\partial v}{\partial x} + 3u\frac{\partial^2 v}{\partial x^2} + 6u\frac{\partial u}{\partial x}v$$

$$+ 2u^2\frac{\partial v}{\partial x} + \left(\frac{\partial^3 u}{\partial x^3} + 2\left(\frac{\partial u}{\partial x}\right)^2 + 2u\frac{\partial^2 u}{\partial x^2}\right)D_x^{-1}v.$$

For the second term of the commutator we obtain

$$\frac{\partial}{\partial \epsilon}R(u)(v + \epsilon A'(u)v)|_{\epsilon=0} = \frac{\partial A'(u)v}{\partial x} + uA'(u)v + \frac{\partial u}{\partial x}D_x^{-1}A'(u)v.$$

Using the identity $D_x^{-1}(v\partial u/\partial x + u\partial v/\partial x) = uv$, where D_x^{-1} is defined above we find

$$\frac{\partial}{\partial \epsilon}R(u)(v + \epsilon A'(u)v)|_{\epsilon=0} = \frac{\partial^3 v}{\partial x^3} + 2\frac{\partial^2 u}{\partial x^2}v + 5\frac{\partial u}{\partial x}\frac{\partial v}{\partial x} + 3u\frac{\partial^2 v}{\partial x^2} + 4u\frac{\partial u}{\partial x}v + 2u^2\frac{\partial v}{\partial x}.$$

Combining these equations we find that the commutator $[A'(u)v, R(u)v]$ is given by the right-hand side of $R'(u)[A(u)]v$ which proves (1).

Problem 8. Let f be a continuously differentiable real function on the interval $I = [a, b]$, i.e. $f : C^1[a, b] \to \mathbb{R}$. The norm is defined as

$$\|f\| := \max_{a \leq t \leq b}(|f(t)|, |f'(t)|)$$

where $f'(t) \equiv df/dt$. Let $\Phi(t, y, z)$ be a given real function for $t \in I$ and y, z arbitrary. Φ is assumed to possess partial derivatives up to second order with respect to all arguments. Find the *Fréchet derivative* of the functional

$$Tf(t) := \int_a^b \Phi(t, f, f')dt.$$

Solution 8. Taylor's theorem yields

$$T(f + g) - T(f) = \int_a^b [\Phi(t, f + g, f' + g') - \Phi(t, f, f')]dt$$

$$= \int_a^b [g\Phi_f + g'\Phi_{f'} + g^2\tilde{\Phi}_{ff} + gg'\tilde{\Phi}_{ff'} + g'^2\tilde{\Phi}_{f'f'}]dt$$

where g is continuously differentiable in I and

$$\frac{\partial \Phi}{\partial f} \equiv \Phi_f := \left.\frac{\partial \Phi(t, y, z)}{\partial y}\right|_{y \to f(t), z \to f'(t)}$$

$$\frac{\partial \Phi}{\partial f'} \equiv \Phi_{f'} := \left.\frac{\partial \Phi(t, y, z)}{\partial z}\right|_{y \to f(t), z \to f'(t)}.$$

The tilde indicates that the function has to be taken at some intermediate argument. With the norm we have $|g|, |g'| \leq \|g\|$. Therefore the contributions of the second derivatives under the integral sign may be estimated by $\|g\|\, \epsilon(\|g\|)$ where ϵ is a null function. Thus the Fréchet derivative of $T(f)$ becomes

$$T'_{(f)}g = \int_a^b \left[\frac{\partial \Phi}{\partial f}g(t) + \frac{\partial \Phi}{\partial f'}g'(t)\right]dt.$$

If $f(a) = f_a$ and $f(b) = f_b$, then $g(a) = g(b) = 0$. *Partial integration* in this case yields

$$\int_a^b \frac{\partial \Phi}{\partial f'}g'(t)dt = -\int_a^b \left(\frac{d}{dt}\frac{\partial \Phi}{\partial f'}\right)g(t)dt.$$

Consequently, the Fréchet derivative of T is given by

$$T'_{(f)}g = \int_a^b \left[\frac{\partial \Phi}{\partial f} - \frac{d}{dt}\frac{\partial \Phi}{\partial f'}\right]g(t)dt.$$

In *variational calculus* we find that an extremum of Tf with the function $f \in C^2[a, b]$ and $f(a) = f_a$, $f(b) = f_b$ is given by

$$\frac{\partial \Phi}{\partial f} - \frac{d}{dt}\frac{\partial \Phi}{\partial f'} = 0.$$

This equation is called the *Euler-Lagrange equation*. Using the Gateaux derivative we start from

$$I(t, f(t), f'(t)) = \int_a^b F(t, f(t), f'(t))dt.$$

Then

$$I(t, f(t) + \epsilon g(t), f'(t) + \epsilon g'(t)) = \int_a^b F(t, f(t) + \epsilon g(t), f'(t) + \epsilon g'(t))dt.$$

We assume that F admits a Taylor expansion around f and f'. Then

$$\frac{d}{d\epsilon} I(t, f(t) + \epsilon g(t), f'(t) + \epsilon g'(t))\Big|_{\epsilon=0} = \int_a^b \left(\frac{\partial F}{\partial f} g + \frac{\partial F}{\partial f'} g' \right) dt.$$

From the boundary conditions $g(a) = g(b) = 0$ and integration by parts we obtain

$$\int_a^b \left(\frac{\partial F}{\partial f} - \frac{d}{dt} \frac{\partial F}{\partial f'} \right) g(t)dt = 0.$$

Problem 9. The *Euler-Lagrange equation* can be written as

$$\frac{d}{dt} \frac{\partial F(t, f(t), f'(t))}{\partial f'} - \frac{\partial F}{\partial f} = 0. \tag{1}$$

Show that if F does not depend explicitly on t the Euler-Lagrange equation can be written as

$$\frac{d}{dt} \left(f' \frac{\partial F}{\partial f'} - F \right) = 0. \tag{2}$$

Solution 9. From (1) we find

$$f'' \frac{\partial^2 F}{\partial f'^2} + f' \frac{\partial^2 F}{\partial f \partial f'} + \frac{\partial^2 F}{\partial f' \partial t} - \frac{\partial F}{\partial f} = 0.$$

Since F does not depend explicitly on t, this differential equation simplifies to

$$f'' \frac{\partial^2 F}{\partial f'^2} + f' \frac{\partial^2 F}{\partial f \partial f'} - \frac{\partial F}{\partial f} = 0.$$

On the other hand from (2) we obtain

$$f' \left(f'' \frac{\partial^2 F}{\partial f'^2} + f' \frac{\partial^2 F}{\partial f \partial f'} - \frac{\partial F}{\partial f} \right) = 0$$

where we assume that $f'(t) \neq 0$. Thus this proves that the Euler-Lagrange equation can be written in the form (2). From (2) it follows that

$$f' \frac{\partial F}{\partial f'} - F = \text{const.}$$

Problem 10. Let $\rho \in C[a, b]$. Consider the function

$$J(u(x)) = \int_a^b \rho(x)\sqrt{1 + u'(x)^2}dx \tag{1}$$

which is defined for all $u \in C^1[a, b]$. Find the Gateaux derivative.

Solution 10. From

$$J(u + \epsilon v) = \int_a^b \rho(x)\sqrt{1 + (u + \epsilon v)'(x)^2}dx$$

it follows that

$$\frac{\partial}{\partial \epsilon} J(u + \epsilon v) = \int_a^b \frac{\rho(x)(u + \epsilon v)'(x)v'(x)}{\sqrt{1 + (u + \epsilon v)'(x)^2}}dx.$$

This is justified by the continuity of this last integrand on $[a, b] \times \mathbb{R}$. Setting $\epsilon = 0$ we obtain

$$\frac{\partial}{\partial \epsilon} J(u + \epsilon v)\bigg|_{\epsilon=0} = \int_a^b \frac{\rho(x)u'(x)v'(x)}{\sqrt{1 + u'(x)^2}}dx.$$

Problem 11. The *Fréchet derivative* of a matrix function $f : \mathbb{C}^{n \times n}$ at a point $X \in \mathbb{C}^{n \times n}$ is a linear mapping $L_X : \mathbb{C}^{n \times n} \to \mathbb{C}^{n \times n}$ such that for all $Y \in \mathbb{C}^{n \times n}$

$$f(X + Y) - f(X) - L_X(Y) = o(\|Y\|).$$

Calculate the Fréchet derivative of $f(X) = X^2$.

Solution 11. We have $f(X + Y) - f(X) = XY + YX + Y^2$. Thus

$$L_X(Y) = XY + YX \equiv [X, Y]_+.$$

The right-hand side is the anti-commutator of X and Y.

Supplementary Problems

Problem 1.   Let $\mathcal{H}$ be the Hilbert space of $n \times n$ matrices over $\mathbb{C}$ with scalar product

$$\langle A, B \rangle := \text{tr}(AB^*).$$

(i) Consider the map $f : \mathcal{H} \to \mathcal{H}$, $f(A) = A^2$. Show that

$$\frac{d}{d\epsilon}(f(A + \epsilon B))\bigg|_{\epsilon=0} = AB + BA \equiv [A, B]_+.$$

(ii) Consider the map $g : \mathcal{H} \to \mathcal{H}$, $g(A) = A^3$. Show that

$$\frac{d}{d\epsilon}(g(A + \epsilon B))\bigg|_{\epsilon=0} = A[A, B]_+ + BAA.$$

(iii) Consider the map $h : \mathcal{H} \to \mathcal{H} \otimes \mathcal{H}$, $h(A) = A \otimes A$. Show that

$$\frac{d}{d\epsilon}(h(A + \epsilon B))\bigg|_{\epsilon=0} = A \otimes B + B \otimes A.$$

(iv) Consider the map $d : \mathcal{H} \to \mathbb{C}$

$$d(A) := \det(A).$$

Find

$$\frac{d}{d\epsilon}d(A + \epsilon B)\bigg|_{\epsilon=0}.$$

Problem 2.   Let A be a fixed element of the Lie group $SL(2, \mathbb{R})$. Consider the vector space of all $\mathbb{R}^{2 \times 2}$ of all 2×2 matrices over $\mathbb{R}$. Consider the map $f : \mathbb{R}^{2 \times 2} \to \mathbb{R}$

$$f(X) = \text{tr}(X^T A X).$$

Show that

$$\frac{d}{d\epsilon}f(X + \epsilon Y)\bigg|_{\epsilon=0} = \text{tr}(Y^T A X + X^T A Y).$$

Problem 3.   (i) Let A be an $n \times n$ matrix over $\mathbb{C}$. Find

$$\frac{d}{d\epsilon}(e^{A+\epsilon B})\bigg|_{\epsilon=0}.$$

(ii) Find

$$\frac{d}{d\epsilon}\det(e^{A+\epsilon B})\bigg|_{\epsilon=0}.$$

Chapter 20

Lax Representation, Solitons and Bäcklund Transformations

Soliton equations are special partial differential equations. They have a number of interesting properties such as a Lax representation, solvable by the inverse scattering method, infinite many conservation laws. They also admit Bäcklund transformations and Lie-Bäcklund symmetries. Most of them also pass the Painlevé test. Lax representations can also be found for some system of ordinary differential equations.

The most studied soliton equation is the *Korteweg-de Vries equation*.

$$\frac{\partial u}{\partial t} + u\frac{\partial u}{\partial x} + \beta\frac{\partial^3 u}{\partial x^3} = 0.$$

Other important soliton equations are the Kadomtsev-Petviashvili equation

$$\frac{\partial u}{\partial x}\left(\frac{\partial u}{\partial t} + u\frac{\partial u}{\partial x} + \epsilon^2\frac{\partial^2 u}{\partial x^2}\right) + \lambda\frac{\partial^2 u}{\partial y^2} = 0$$

the one-dimensional cubic Schrödinger equation

$$i\hbar\frac{\partial \psi}{\partial t} + \frac{\hbar^2}{2m}\frac{\partial^2 \psi}{\partial x^2} + k|\psi|^2\psi = 0$$

467

the sine-Gordon equation

$$\frac{\partial^2 u}{\partial x_0^2} - \frac{\partial^2 u}{\partial x_1^2} = \sin(u).$$

Let L and M be two linear differential operators. Assume that

$$Lv = \lambda v$$

$$\frac{\partial v}{\partial t} = Mv$$

where L is the linear differential operator of the *spectral problem*, M is the linear differential operator of an associated *time-evolution equation* and λ is a parameter. The differential operators L and M are called a *Lax pair*. For the Lax representation we have

$$L_t = [M, L](t)$$

where $[L, M] := LM - ML$ is the commutator. This equation is called the *Lax representation*. It contains a nonlinear evolution equation if L and M are correctly chosen. From $Lv = \lambda v$ we obtain

$$\frac{\partial}{\partial t}(Lv) = \frac{\partial}{\partial t}(\lambda v).$$

Since λ does not depend on t it follows that

$$L_t v + L\frac{\partial v}{\partial t} = \lambda \frac{\partial v}{\partial t}$$

where we assume that λ does not depend on t. Inserting $Lv = \lambda v$ and $\partial b/\partial t = Mv$ into this equation provides

$$L_t v + LMv = \lambda Mv.$$

From $Lv = \lambda v$ we obtain $MLv = \lambda Mv$. Inserting this expression into this equation yields

$$L_t v = -LMv + MLv.$$

Consequently, $L_t v = [M, L]v$. Since the smooth function v is arbitrary we obtain $L_t = [M, L](t)$.

Bäcklund transformations relate differential equations (both ordinary and partial) and their solutions. A Bäcklund transformation that relates solutions of the same differential equation is called an auto-Bäcklund transformation. The Painlevé test can be utilized to find auto-Bäcklund transformations.

Problem 1. Let

$$\frac{d\mathbf{u}}{dt} = \mathbf{V}(\mathbf{u}), \qquad \mathbf{u} \equiv (u_1, u_2, \ldots, u_m)^T$$

be an autonomous system of first-order ordinary differential equations. Assume that the functions $V_k : \mathbb{R}^m \to \mathbb{R}$ are smooth. Assume that this system can be written in the form (the so-called *Lax representation*)

$$\frac{dL}{dt} = [A, L](t) \equiv [A(t), L(t)]$$

where A and L are $n \times n$ matrices and $[A, L] \equiv AL - LA$. Then the $n \times n$ matrices L and A are called a *Lax pair*.
(i) Show that

$$\frac{dL^k}{dt} = [A, L^k](t). \tag{1}$$

(ii) Show that $\mathrm{tr}(L^k)$ ($k = 1, 2, \ldots$) are first integrals, where $\mathrm{tr}(.)$ denotes the trace.
(iii) Assume that L^{-1} exists. Show that $\mathrm{tr}(L^{-1})$ is a first integral.

Solution 1. (i) The formula (1) is true by assumption for $k = 1$. If it is true for k, then

$$\begin{aligned}
\frac{dL^{k+1}}{dt} &= \frac{d}{dt}(L^k L) = \frac{dL^k}{dt}L + L^k\frac{dL}{dt} \\
&= ([A, L^k]L + L^k[A, L])(t) \qquad \text{by the induction hypothesis} \\
&= (AL^k L - L^k AL + L^k AL - L^k LA)(t) = (AL^{k+1} - L^{k+1}A)(t) \\
&= [A, L^{k+1}](t).
\end{aligned}$$

(ii) We have

$$\frac{d}{dt}\mathrm{tr}(L^k) = \mathrm{tr}\left(\frac{dL^k}{dt}\right) = \mathrm{tr}([A, L^k]) = \mathrm{tr}(AL^k - L^k A) = \mathrm{tr}(AL^k) - \mathrm{tr}(L^k A) = 0$$

since $\mathrm{tr}(XY) = \mathrm{tr}(YX)$ for arbitrary $n \times n$ matrices X and Y. Consequently, $\mathrm{tr}(L^k)$ ($k = 1, 2, ..$) are first integrals of system (1).
(iii) From $L^{-1}L = I$ where I is the unit matrix it follows that

$$\frac{dL^{-1}}{dt} = -L^{-1}\frac{dL}{dt}L^{-1}.$$

Then

$$\begin{aligned}
\frac{d}{dt}\mathrm{tr}(L^{-1}) &= \mathrm{tr}\left(\frac{d}{dt}L^{-1}\right) = -\mathrm{tr}\left(L^{-1}\frac{dL}{dt}L^{-1}\right) \\
&= -\mathrm{tr}(L^{-1}[A, L]L^{-1}) = -\mathrm{tr}(L^{-1}ALL^{-1}) + \mathrm{tr}(L^{-1}LAL^{-1}) \\
&= -\mathrm{tr}(L^{-1}A) + \mathrm{tr}(AL^{-1}) = 0.
\end{aligned}$$

Problem 2.   Consider the nonlinear system of ordinary differential equations

$$\frac{du_1}{dt} = (\lambda_3 - \lambda_2)u_2u_3 \tag{1a}$$

$$\frac{du_2}{dt} = (\lambda_1 - \lambda_3)u_3u_1 \tag{1b}$$

$$\frac{du_3}{dt} = (\lambda_2 - \lambda_1)u_1u_2 \tag{1c}$$

where $\lambda_j \in \mathbb{R}$. This dynamical system describes *Euler's rigid body motion*.
(i) Show that the first integrals are given by

$$I_1(\mathbf{u}) = u_1^2 + u_2^2 + u_3^2, \qquad I_2(\mathbf{u}) = \lambda_1 u_1^2 + \lambda_2 u_2^2 + \lambda_3 u_3^2. \tag{2}$$

(ii) A Lax representation is given by

$$\frac{dL}{dt} = [L, \lambda L](t) \tag{3}$$

where

$$L := \begin{pmatrix} 0 & -u_3 & u_2 \\ u_3 & 0 & -u_1 \\ -u_2 & u_1 & 0 \end{pmatrix}, \qquad \lambda L := \begin{pmatrix} 0 & -\lambda_3 u_3 & \lambda_2 u_2 \\ \lambda_3 u_3 & 0 & -\lambda_1 u_1 \\ -\lambda_2 u_2 & \lambda_1 u_1 & 0 \end{pmatrix}. \tag{4}$$

Show that $\operatorname{tr}(L^k)$ $(k = 1, 2, \dots)$ gives only one first integral of system (1).
(iii) Instead of (3) we consider now

$$\frac{d(L + Ay)}{dt} = [L + Ay, \lambda L + By](t) \tag{5}$$

where y is a dummy variable and A and B are time-independent diagonal matrices, i.e., $A = \operatorname{diag}(A_1, A_2, A_3)$ and $B = \operatorname{diag}(B_1, B_2, B_3)$ with $A_j, B_j \in \mathbb{R}$. Equation (5) decomposes into various powers of y, namely

$$y^0 : \quad \frac{dL}{dt} = [L, \lambda L](t), \quad y^1 : \quad 0 = [L, B] + [A, \lambda L], \quad y^2 : \quad [A, B] = 0. \tag{6}$$

Equation (6c) is satisfied identically since A and B are diagonal matrices. Equation (6b) leads to

$$\lambda_i = \frac{B_j - B_k}{A_j - A_k}$$

where (i, j, k) are permutations of (1, 2, 3). This equation can be satisfied by setting $B_j = A_j^2$, $\lambda_i = A_j + A_k$. Consequently the original Lax pair L, λL satisfies the *extended Lax pair* $L + Ay$, $\lambda L + By$. Show that $\operatorname{tr}((L + Ay)^2)$ and $\operatorname{tr}((L + Ay)^3)$ provide the first integrals (2).

Solution 2. (i) Straightforward calculation yields

$$\frac{d}{dt}I_1(\mathbf{u}) = 2u_1\frac{du_1}{dt} + 2u_2\frac{du_2}{dt} + 2u_3\frac{du_3}{dt}.$$

Thus

$$\frac{d}{dt}I_1(\mathbf{u}) = 2u_1(\lambda_3 - \lambda_2)u_2u_3 + 2u_2(\lambda_1 - \lambda_3)u_3u_1 + 2u_3(\lambda_2 - \lambda_1)u_1u_2 = 0.$$

Analogously, $dI_2(\mathbf{u})/dt = 0$.

(ii) We obtain $\mathrm{tr}(L) = 0$ and $\mathrm{tr}(L^2) = -2(u_1^2 + u_2^2 + u_3^2) = -2I_1(\mathbf{u})$. Since L does not depend on λ we cannot find I_2.

(iii) Straightforward calculation yields

$$\mathrm{tr}((L + Ay)^2) = -2I_1(\mathbf{u}) + C_1, \qquad \mathrm{tr}((L + Ay)^3) = -3yI_2(\mathbf{u}) + C_2$$

where C_1 and C_2 are constants. Thus the extended Lax pair provides both first integrals.

Problem 3. Let $H : \mathbb{R}^4 \to \mathbb{R}$ be the Hamilton function

$$H(\mathbf{p}, \mathbf{q}) = \frac{1}{2}(p_1^2 + p_2^2) + e^{q_2 - q_1}. \tag{1}$$

(i) Find the equations of motion. The Hamilton function (1) is a first integral, i.e. $dH/dt = 0$. Find the second first integral on inspection of the equations of motion.

(ii) Define

$$a := \frac{1}{2}e^{(q_2 - q_1)/2}, \quad b_1 := \frac{1}{2}p_1, \quad b_2 := \frac{1}{2}p_2.$$

Give the equations of motion for a, b_1 and b_2.

(iii) Show that the equations of motion for a, b_1 and b_2 can be written in Lax form, i.e. $dL/dt = [A, L](t)$, where

$$L := \begin{pmatrix} b_1 & a \\ a & b_2 \end{pmatrix}, \qquad A := \begin{pmatrix} 0 & a \\ -a & 0 \end{pmatrix}.$$

Solution 3. (i) The *Hamilton equations of motion* are given by

$$\frac{dq_1}{dt} = \frac{\partial H}{\partial p_1} = p_1, \qquad \frac{dp_1}{dt} = -\frac{\partial H}{\partial q_1} = e^{q_2 - q_1}$$

$$\frac{dq_2}{dt} = \frac{\partial H}{\partial p_2} = p_2, \qquad \frac{dp_2}{dt} = -\frac{\partial H}{\partial q_2} = -e^{q_2 - q_1}.$$

Adding the last two equations provides the first integral $I(\mathbf{p}, \mathbf{q}) = p_1 + p_2$ since

$$\frac{dI}{dt} = \frac{dp_1}{dt} + \frac{dp_2}{dt} = e^{q_2 - q_1} - e^{q_2 - q_1} = 0.$$

Obviously, the Hamilton function H is the other first integral.
(ii) The time derivative of $a(t)$ yields

$$\frac{da}{dt} = \frac{1}{4}e^{(q_2-q_1)/2}\left(\frac{dq_2}{dt} - \frac{dq_1}{dt}\right) = \frac{1}{4}e^{(q_2-q_1)/2}(p_2-p_1) = \frac{1}{2}e^{(q_2-q_1)/2}(b_2-b_1).$$

Thus $da/dt = a(b_2 - b_1)$. The time derivative of $b_1(t)$ yields

$$\frac{db_1}{dt} = \frac{1}{2}\frac{dp_1}{dt} = \frac{1}{2}e^{q_2-q_1} = 2a^2.$$

Analogously,

$$\frac{db_2}{dt} = \frac{1}{2}\frac{dp_2}{dt} = -\frac{1}{2}e^{q_2-q_1} = -2a^2.$$

To summarize: the equations of motion for a, b_1 and b_2 are given by

$$\frac{da}{dt} = a(b_2 - b_1), \qquad \frac{db_1}{dt} = 2a^2, \qquad \frac{db_2}{dt} = -2a^2.$$

(iii) The Lax representation is given by L and A since

$$[A, L] = \begin{pmatrix} 2a^2 & a(b_2 - b_1) \\ a(b_2 - b_1) & -2a^2 \end{pmatrix}, \qquad \frac{dL}{dt} = \begin{pmatrix} db_1/dt & da/dt \\ da/dt & db_2/dt \end{pmatrix}.$$

The first integrals follow as

$$I_1(a, b_1, b_2) = \text{tr}(L) = b_1 + b_2, \qquad I_2(a, b_1, b_2) = \text{tr}(L^2) = 2a^2 + b_1^2 + b_2^2.$$

Problem 4. The semisimple Lie algebra $s\ell(2, \mathbb{R})$ is defined by the commutator rules

$$[X_0, X_+] = 2X_+ \qquad [X_0, X_-] = -2X_- \qquad [X_+, X_-] = X_0 \qquad (1)$$

where X_0, X_+, X_- denote a basis. Let

$$\{A(\mathbf{p}, \mathbf{q}), B(\mathbf{p}, \mathbf{q})\} := \sum_{j=1}^{N}\left(\frac{\partial A}{\partial q_j}\frac{\partial B}{\partial p_j} - \frac{\partial B}{\partial q_j}\frac{\partial A}{\partial p_j}\right) \qquad (2)$$

be the *Poisson bracket*. Let

$$t_+ := \frac{1}{2}\sum_{k=1}^{N}p_k^2, \qquad t_- := -\frac{1}{2}\sum_{k=1}^{N}q_k^2, \qquad t_0 := -\sum_{k=1}^{N}p_k q_k.$$

(i) Show that the functions $-t_+$, $-t_-$, $-t_0$ form a basis of a Lie algebra under the Poisson bracket and that the Lie algebra is isomorphic to $s\ell(2, \mathbb{R})$.

(ii) Let

$$H(\mathbf{p}, \mathbf{q}) = \frac{1}{2} \sum_{k=1}^{N} p_k^2 + U(\mathbf{q}) \equiv t_+ + U(\mathbf{q}). \tag{4a}$$

Find the condition on U such that $\{-H, -t_-, -t_0\}$ forms a basis of a Lie algebra which is isomorphic to $sl(2, \mathbb{R})$.

(iii) The Hamilton system

$$H(\mathbf{p}, \mathbf{q}) = \frac{1}{2} \sum_{k=1}^{N} p_k^2 + \sum_{j=2}^{N} \sum_{k=1}^{j-1} \frac{a^2}{(q_j - q_k)^2}$$

admits a Lax representation with

$$L := \begin{pmatrix} p_1 & \dfrac{ia}{(q_1 - q_2)} & \cdots & \dfrac{ia}{(q_1 - q_N)} \\ \dfrac{ia}{(q_2 - q_1)} & p_2 & \cdots & \dfrac{ia}{(q_2 - q_N)} \\ \vdots & & & \vdots \\ \dfrac{ia}{(q_N - q_1)} & & \cdots & p_N \end{pmatrix}$$

where a is a nonzero real constant. Let $N = 3$. Find the first integrals.

Solution 4. (i) Straightforward calculation yields

$$\{-t_0, -t_+\} = \{t_0, t_+\} = -2t_+, \quad \{-t_0, -t_-\} = \{t_0, t_-\} = 2t_-,$$

$$\{-t_+, -t_-\} = \{t_+, t_-\} = -t_0.$$

Consider the map $X_0 \to -t_0$, $X_- \to -t_-$, $X_+ \to -t_+$. Then the two Lie algebras are isomorphic.

(ii) Since

$$\{-t_0, -H\} = -\sum_{j=1}^{N} \left(p_j^2 - \frac{\partial U}{\partial q_j} q_j \right), \quad \{-H, -t_-\} = \sum_{j=1}^{N} p_j q_j = -t_0$$

we find from the condition $\{-t_0, -H\} = -2H$ that the potential U satisfies the linear partial differential equation

$$\sum_{j=1}^{N} q_j \frac{\partial U}{\partial q_j} = -2U.$$

This is a homogeneous equation (of rank -2) for the potential U. This equation admits the two solutions

$$U(\mathbf{q}) = \sum_{j=2}^{N} \sum_{k=1}^{j-1} \frac{a^2}{(q_j - q_k)^2}, \qquad U(\mathbf{q}) = \frac{a^2}{q_1^2 + q_2^2 + \cdots + q_N^2}.$$

(iii) The first integrals are the invariants of L. From L we find (with $N = 3$) that $\text{tr}(L) = p_1 + p_2 + p_3$ and

$$\text{tr}(L^2) = p_1^2 + p_2^2 + p_3^2 + 2a^2 \left(\frac{1}{(q_1 - q_2)^2} + \frac{1}{(q_3 - q_1)^2} + \frac{1}{(q_2 - q_3)^2} \right).$$

Obviously, $2H = \text{tr}(L^2)$. The calculation of $\text{tr}(L^3)$ is very lengthy. To find the third first integral we use

$$\det(L) = \lambda_1 \lambda_2 \lambda_3 = p_1 p_2 p_3 - a^2 \left(\frac{p_1}{(q_2 - q_3)^2} + \frac{p_2}{(q_3 - q_1)^2} + \frac{p_3}{(q_1 - q_2)^2} \right)$$

where λ_1, λ_2 and λ_3 denote the eigenvalues of L.

Problem 5.   Assume that

$$
L := \begin{pmatrix}
p_1 & if_{12} & 0 & \cdots & 0 & 0 & -if_{N1} \\
-if_{12} & p_2 & if_{23} & \cdots & 0 & 0 & 0 \\
0 & -if_{23} & p_3 & \cdots & & & \\
\vdots & & & & & & if_{N-1,N} \\
if_{N1} & 0 & 0 & \cdots & 0 & -if_{N-1,N} & p_N
\end{pmatrix}
\tag{1}
$$

where $f_{n,n+1} := f(q_n - q_{n+1})$ with $f(\mathbf{q})$ a certain given smooth real function. Assume that the Hamilton function H is given by

$$H(\mathbf{p}, \mathbf{q}) = \frac{1}{2}\text{tr}(L^2). \tag{2}$$

Let

$$
A := i \begin{pmatrix}
h_1 & g_{12} & 0 & \cdots & 0 & g_{N1} \\
g_{12} & h_2 & g_{23} & \cdots & 0 & 0 \\
0 & g_{23} & h_3 & & & \\
& & & & & \\
0 & \cdots & & & & g_{N-1,N} \\
g_{N1} & 0 & & \cdots & g_{N-1,N} & h_N
\end{pmatrix}
\tag{3}
$$

where $g(\mathbf{q})$ and $h_k(\mathbf{p}, \mathbf{q})$ are real-valued smooth functions. Find the condition on f, g and h such that L, A are a Lax pair of the Hamilton function H.

Solution 5.   From the Hamilton function we obtain

$$H(\mathbf{p}, \mathbf{q}) = \frac{1}{2}\text{tr}(L^2) = \frac{1}{2}\sum_{k=1}^{N} p_k^2 + \sum_{k=1}^{N} f_{k,k+1}^2 \tag{4}$$

with $N + 1 \equiv 1$, i.e. modulo N. Then the Hamilton equations of motion are given by

$$\frac{dq_k}{dt} = \frac{\partial H}{\partial p_k} = p_k, \qquad \frac{dp_k}{dt} = -\frac{\partial H}{\partial q_k} = 2\left(-f'_{k,k+1}f_{k,k+1} + f'_{k-1,k}f_{k-1,k}\right)$$

where $k = 1, \ldots, N$ and f' means differentiation with respect to the argument. For f we obtain

$$\frac{df_{k,k+1}}{dt} = (p_k - p_{k+1})f'_{k,k+1} \tag{6}$$

where we used $f_{k,k+1} = f(q_k - q_{k+1})$. From the requirement that L, A are a Lax pair, i.e. L, A satisfies $dL/dt = [L, A](t)$ we obtain

$$\frac{dp_k}{dt} = 2(f_{k-1,k}g_{k-1,k} - f_{k,k+1}g_{k,k+1})$$

$$\frac{df_{k,k+1}}{dt} = g_{k,k+1}(p_k - p_{k+1}) + if_{k,k+1}(h_{k+1} - h_k) \quad \text{modulo} \quad N$$

$$0 = f_{k,k+1}g_{k,k+2} - g_{k,k+1}f_{k,k+2}.$$

The choice $h_k = 0$ and

$$g_{k,k+1} = f'_{k,k+1}, \qquad 0 = f_{k,k+1}g_{k,k+2} - g_{k,k+1}f_{k,k+2}$$

provides the consistency of this system and the Hamilton equations of motion. From the last two equations we can eliminate $g_{k,k+1}$ and obtain

$$\frac{f'_{k,k+1}}{f_{k,k+1}} = \frac{f'_{k,k+2}}{f_{k,k+2}} \quad \Rightarrow \quad \frac{f'_{k,k+1}}{f_{k,k+1}} = c \equiv \text{const.}$$

Therefore, the solution of this equation is given by

$$f_{k,k+1} \equiv f(q_k - q_{k+1}) = A\exp(c(q_k - q_{k+1}))$$

where A is a constant.

Problem 6. We consider systems of $(2n + 1)$ ordinary nonlinear differential equations. These are the multiple three-wave interaction system describing triads (a_j, b_j, u), $j = 1, \ldots, n$, evolving in time alone and interacting with each other through the single common member u. These systems can be derived from a Hamilton function

$$H(\mathbf{b}, \mathbf{c}, u) = \frac{1}{2}i\sum_{j=1}^{n} \epsilon_j(c_j c_j^* - b_j b_j^*) + i\sum_{j=1}^{n} \alpha_j(ub_j^* c_j + u^* b_j c_j^*) \tag{1}$$

and with *Poisson bracket* defined as

$$\{f, g\} := \frac{\partial f}{\partial u}\frac{\partial g}{\partial u^*} - \frac{\partial f}{\partial u^*}\frac{\partial g}{\partial u} + \sum_{j=1}^{n}\left(\frac{\partial f}{\partial b_j}\frac{\partial g}{\partial b_j^*} - \frac{\partial f}{\partial b_j^*}\frac{\partial g}{\partial b_j} + \frac{\partial f}{\partial c_j}\frac{\partial g}{\partial c_j^*} - \frac{\partial f}{\partial c_j^*}\frac{\partial g}{\partial c_j}\right). \tag{2}$$

Thus (u, u^*), (b_j, b_j^*), (c_j, c_j^*) are pairs of canonical variables (where $*$ means complex conjugate). The arbitrary real parameters α_j, ϵ_j play the role of frequencies.

(i) Let $\alpha_j = 1$ for $j = 1, 2, \ldots, n$. Find the Hamilton's equation of motion from (1) and (2).

(ii) Find the Lax representation.

Solution 6.   From the Poisson bracket we find

$$\frac{du}{dt} = \{u, H\} = \frac{\partial H}{\partial u^*}$$

$$\frac{db_j}{dt} = \{b_j, H\} = \frac{\partial H}{\partial b_j^*}, \qquad \frac{dc_j}{dt} = \{c_j, H\} = \frac{\partial H}{\partial c_j^*}$$

and *c.c.*, where *c.c.* stands for complex conjugate. Thus the Hamilton's equations of motion are

$$\frac{du}{dt} = i\sum_{j=1}^{n} b_j c_j^*, \quad \frac{db_j}{dt} = -\frac{1}{2}i\epsilon_j b_j + iuc_j, \quad \frac{dc_j}{dt} = \frac{1}{2}i\epsilon_j c_j + iu^* b_j, \ \text{ and } \ \text{c.c.}$$

(ii) We have

$$\frac{dL}{dt} = \{L, H\}(t) \equiv [A, L](t)$$

in which L and A are the $(2n+2) \times (2n+2)$ matrices

$$L := \frac{1}{2}\begin{pmatrix} \pi & \sigma_1 & \cdots & \sigma_n \\ \tau_1 & \epsilon_1 I_2 & & 0 \\ \vdots & & \ddots & \\ \tau_n & 0 & & \epsilon_n I_2 \end{pmatrix}, \qquad A := \frac{1}{2}i\begin{pmatrix} 0 & \omega\sigma_1 & \cdots & \omega\sigma_n \\ \tau_1\omega & & & \\ \vdots & & 0 & \\ \tau_n\omega & & & \end{pmatrix}.$$

The π, σ_j and τ_j are the 2×2 matrices

$$\pi := \begin{pmatrix} 0 & -2u \\ 2u^* & 0 \end{pmatrix}, \qquad \tau_j := \begin{pmatrix} b_j^* & -c_j^* \\ -c_j & -b_j \end{pmatrix}, \qquad \sigma_j := \begin{pmatrix} -b_j & -c_j^* \\ -c_j & b_j^* \end{pmatrix}$$

and $\omega = \mathrm{diag}(-1, 1)$. I_2 is the 2×2 identity matrix. The Hamilton function (1) is given by

$$H = \frac{2}{3}i\mathrm{tr}(L^3) + \text{ const.}$$

Problem 7.   Consider the autonomous first order system

$$\frac{du_1}{dt} = u_1(u_2 - u_n), \quad \frac{du_2}{dt} = u_2(u_3 - u_1), \quad \ldots, \quad \frac{du_n}{dt} = u_n(u_1 - u_{n-1}).$$

(i) Find a Lax representation for the case $n = 3$ and the first integrals using $\text{tr}(L^n)$.

(ii) Find the Lax representation for the case $n = 4$ and the first integrals.

(iii) Find the Lax representation for arbitrary n.

Solution 7. (i) For $n = 3$ we find the Lax pair

$$
L = \begin{pmatrix} 0 & 1 & u_1 \\ u_2 & 0 & 1 \\ 1 & u_3 & 0 \end{pmatrix}, \quad A = \begin{pmatrix} u_1 + u_2 & 0 & 1 \\ 1 & u_2 + u_3 & 0 \\ 0 & 1 & u_1 + u_2 \end{pmatrix}.
$$

The first integrals are given by $\text{tr}(L^2) = u_1 + u_2 + u_3$, $\text{tr}(L^3) = u_1 u_2 u_3$.

(ii) For $n = 4$ we have the Lax pair

$$
L = \begin{pmatrix} 0 & 1 & 0 & u_1 \\ u_2 & 0 & 1 & 0 \\ 0 & u_3 & 0 & 1 \\ 1 & 0 & u_4 & 0 \end{pmatrix}, \quad A = \begin{pmatrix} u_1 + u_2 & 0 & 1 & 0 \\ 0 & u_2 + u_3 & 0 & 1 \\ 1 & 0 & u_3 + u_4 & 0 \\ 0 & 1 & 0 & u_4 + u_1 \end{pmatrix}.
$$

The first integrals are given by

$$
I_1(\mathbf{u}) = \sum_{j=1}^{4} u_j, \quad I_2(\mathbf{u}) = \prod_{j=1}^{4} u_j, \quad I_3(\mathbf{u}) = u_2 u_4 + u_1 u_3
$$

where $\text{tr}(L) = 0$, $\text{tr}(L^2) = 2I_1$, $\text{tr}(L^3) = 0$, $\text{tr}(L^4) = 4 + 4I_3 + 2I_1^2 - 4I_2$.

(iii) For arbitrary n the Lax representation is given by

$$
\phi_{j+1} + u_j \phi_{j-1} = \lambda \phi_j, \quad \frac{d\phi_j}{dt} = \phi_{j+2} + (u_{j+1} + u_j)\phi_j
$$

where $j = 1, \ldots, n$ and $0 \equiv n$, $1 \equiv n + 1$ etc. The parameter λ is time-independent.

Problem 8. Let

$$
\psi_{m+1} = L_m \psi_m \tag{1}
$$

$$
\frac{d\psi_m}{dt} = M_m \psi_m \tag{2}
$$

where the entries of the square matrices L_m and M_m depend on time t, λ is the so-called spectral parameter which does not depend on time and $m \in \mathbb{Z}$ or $m \in \mathbb{Z}_N$. Show that

$$
\frac{dL_m}{dt} = M_{m+1} L_m - L_m M_m. \tag{3}
$$

L_m and M_m are called a *Lax pair*. We can find conservation laws from L_m.

Solution 8.   From (1) we obtain

$$\frac{d\psi_{m+1}}{dt} = \frac{dL_m}{dt}\psi_m + L_m\frac{d\psi_m}{dt}. \tag{4}$$

Equation (2) yields

$$\frac{d\psi_{m+1}}{dt} = M_{m+1}\psi_{m+1}. \tag{5}$$

Inserting (5) and (2) into (4) gives

$$M_{m+1}\psi_{m+1} = \frac{dL_m}{dt}\psi_m + L_m M_m\psi_m.$$

Inserting (1) into this equation leads to

$$M_{m+1}L_m\psi_m = \frac{dL_m}{dt}\psi_m + L_m M_m\psi_m.$$

Consequently,

$$\frac{dL_m}{dt}\psi_m = (M_{m+1}L_m - L_m M_m)\psi_m.$$

Since ψ_m is arbitrary we obtain (3).

Problem 9.   Consider the 2^N dimensional Hilbert space

$$\mathcal{H}_N = \prod_{n=1}^{N} \otimes \mathbb{C}^2$$

and the Hamilton operator

$$\hat{H}_N = \frac{J}{4}\sum_{n=1}^{N}(\sigma_{1,n}\sigma_{1,n+1} + \sigma_{2,n}\sigma_{2,n+1} + \sigma_{3,n}\sigma_{3,n+1} - I_N) \tag{1}$$

in this Hilbert space. Here J is a real constant and I_N is the identity operator ($2^N \times 2^N$ unit matrix) in the space $\mathcal{H}_N$ (dim $\mathcal{H}_N = 2^N$). The operators $\sigma_{j,n}$ with ($j = 1, 2, 3$) have the following form

$$\sigma_{j,n} := I_2 \otimes \cdots \otimes I_2 \otimes \sigma_j \otimes I_2 \otimes \cdots \otimes I_2 \quad (N - \text{factors})$$
$$\downarrow$$
$$n - th \text{ place}$$

where I_2 is the 2×2 identity matrix, $n = 1, \ldots, N$ and σ_1, σ_2, σ_3 are the Pauli matrices. Here $\otimes$ denotes the Kronecker product. Let $\sigma_{j,N+1} \equiv \sigma_{j,1}$ (periodic boundary condition). Let

$$L_n(\lambda) := \begin{pmatrix} \lambda I_n + \frac{i}{2}\sigma_{3,n} & \frac{i}{2}\sigma_{-,n} \\ \frac{i}{2}\sigma_{+,n} & \lambda I_n - \frac{i}{2}\sigma_{3,n} \end{pmatrix}$$

be an operator valued matrix of order 2×2 with

$$\sigma_{+,n} := \sigma_{1,n} + i\sigma_{2,n}, \qquad \sigma_{-,n} := \sigma_{1,n} - i\sigma_{2,n}.$$

Here λ is a complex parameter. Thus $L_n(\lambda)$ is a $2^{N+1} \times 2^{N+1}$ matrix. L_n is called a *Lax operator*. The matrix $L_n(\lambda)$ can also be represented in the form

$$L_n(\lambda) = \lambda I_2 \otimes I_N + \frac{i}{2} \sum_{j=1}^{3} \sigma_j \otimes \sigma_{j,n}.$$

Show that (*Yang-Baxter relation*)

$$R(\lambda - \mu)(L_n(\lambda) \otimes L_n(\mu)) \equiv (L_n(\mu) \otimes L_n(\lambda))R(\lambda - \mu)$$

where

$$R(\lambda) = \frac{1}{\lambda + i} \left(\left(\frac{\lambda}{2} + i \right) I_2 \otimes I_2 + \frac{\lambda}{2} \sum_{a=1}^{3} \sigma^a \otimes \sigma^a \right).$$

We define

$$(\sigma_1 \otimes \sigma_1)(L_n(\lambda) \otimes L_n(\mu)) := (\sigma_1 L_n(\lambda)) \otimes (\sigma_1 L_n(\mu))$$

and

$$\begin{pmatrix} 0 & 1 \\ 1 & 0 \end{pmatrix} \begin{pmatrix} \lambda I_N + \frac{i}{2}\sigma_n^3 & \frac{i}{2}\sigma_n^- \\ \frac{i}{2}\sigma_n^+ & \lambda I_N - \frac{i}{2}\sigma_n^3 \end{pmatrix} = \begin{pmatrix} \frac{i}{2}\sigma_n^+ & \lambda I_N - \frac{i}{2}\sigma_n^3 \\ \lambda I_N + \frac{i}{2}\sigma_n^3 & \frac{i}{2}\sigma_n^- \end{pmatrix}$$

and so on. Owing to this definition, one says that the 2×2 identity matrix I_2 and the Pauli matrices σ_j act in the auxiliary space $\mathbb{C}^2$.

Solution 9. We put

$$L_n(\lambda) = \begin{pmatrix} a & b \\ c & d \end{pmatrix}, \qquad L_n(\mu) = \begin{pmatrix} e & f \\ g & h \end{pmatrix}$$

and $D := 1/(\lambda - \mu + i)$. Obviously $b = f$ and $c = g$ with

$$b = f = \frac{i}{2}\sigma_n^-, \qquad c = g = \frac{i}{2}\sigma_n^+.$$

Then we find

$$R(\lambda - \mu)(L_n(\lambda) \otimes L_n(\mu))$$

$$= D \begin{pmatrix} \lambda - \mu + i & 0 & 0 & 0 \\ 0 & i & \lambda - \mu & 0 \\ 0 & \lambda - \mu & i & 0 \\ 0 & 0 & 0 & \lambda - \mu + i \end{pmatrix} \begin{pmatrix} ae & af & be & bf \\ ag & ah & bg & bh \\ ce & cf & de & df \\ cg & ch & dg & dh \end{pmatrix}$$

$$= D \begin{pmatrix} (\lambda - \mu + i)ae & (\lambda - \mu + i)af & (\lambda - \mu + i)be & (\lambda - \mu + i)bf \\ iag + (\lambda - \mu)ce & iah + (\lambda - \mu)cf & ibg + (\lambda - \mu)de & ibh + (\lambda - \mu)df \\ (\lambda - \mu)ag + ice & (\lambda - \mu)ah + icf & (\lambda - \mu)bg + ide & (\lambda - \mu)bh + idf \\ (\lambda - \mu + i)cg & (\lambda - \mu + i)ch & (\lambda - \mu + i)dg & (\lambda - \mu + i)dh \end{pmatrix}$$

and

$$(L_n(\mu) \otimes L_n(\lambda))R(\lambda - \mu)$$

$$= D \begin{pmatrix} ea & eb & fa & fb \\ ec & ed & fc & fd \\ ga & gb & ha & hb \\ gc & gd & hc & hd \end{pmatrix} \begin{pmatrix} \lambda - \mu + i & 0 & 0 & 0 \\ 0 & i & \lambda - \mu & 0 \\ 0 & \lambda - \mu & i & 0 \\ 0 & 0 & 0 & \lambda - \mu + i \end{pmatrix}$$

$$= D \begin{pmatrix} (\lambda - \mu + i)ea & ieb + (\lambda - \mu)fa & (\lambda - \mu)eb + ifa & (\lambda - \mu + i)fb \\ (\lambda - \mu + i)ec & ied + (\lambda - \mu)fc & (\lambda - \mu)ed + ifc & (\lambda - \mu + i)fd \\ (\lambda - \mu + i)ga & igb + (\lambda - \mu)ha & (\lambda - \mu)gb + iha & (\lambda - \mu + i)hb \\ (\lambda - \mu + i)gc & igd + (\lambda - \mu)hc & (\lambda - \mu)gd + ihc & (\lambda - \mu + i)hd \end{pmatrix}.$$

For identity Yang-Baxter relation to be true, the matrices on the right-hand sides of these equations must be equal. We number the entries as follows

$$\begin{pmatrix} 1 & 2 & 5 & 6 \\ 3 & 4 & 7 & 8 \\ 9 & 10 & 13 & 14 \\ 11 & 12 & 15 & 16 \end{pmatrix}.$$

Therefore we have to prove 16 identities. The first identity to prove is $ea = ae$. From $ea - ae$ we find

$$(\mu I_N + \frac{i}{2}\sigma_n^3)(\lambda I_N + \frac{i}{2}\sigma_n^3) - (\lambda I_N + \frac{i}{2}\sigma_n^3)(\mu I_N + \frac{i}{2}\sigma_n^3) = 0$$

since λ and μ are complex parameters. Analogously, we can prove identity 16, i.e. $dh = hd$. Identity 6 is given by $fb = bf$. This is obviously true, because $f = b$. Analogously, identity 11 is satisfied. Identity 2 is given by

$$(\lambda - \mu)(af - fa) + i(af - eb) = 0.$$

Using the commutation relation $[\sigma_{-,n}, \sigma_{3,n}] = 2\sigma_{-,n}$ we find that the identity is satisfied. By using the same identity, we can similarly prove 5, 8 and 14. Applying the commutation relation $[\sigma_{+,n}, \sigma_{3,n}] = -2\sigma_{+,n}$ we can prove 3, 9, 12 and 15. The technique is the same as that used for identity 2. Identity 4 is given by

$$ied + (\lambda - \mu)fc = iah + (\lambda - \mu)cf.$$

Using the commutation relation $[\sigma_{+,n}, \sigma_{-,n}] = 4\sigma_{3,n}$ we find that identity 4 holds. By using the same identity, we can prove identity 13 in the same way. Identity 7 is given by

$$(\lambda - \mu)ed + ifc = ibg + (\lambda - \mu)de.$$

Thus we have to prove that

$$(\lambda - \mu)(ed - de) + i(fc - bg) = 0.$$

For $ed - de$ we find

$$ed - de = (\mu I_N - \frac{i}{2}\sigma_{3,n})(\lambda I_N - \frac{i}{2}\sigma_{3,n}) - (\lambda I_N - \frac{i}{2}\sigma_{3,n})(\mu I_N - \frac{i}{2}\sigma_{3,n}) = 0.$$

For $fc - bg$ we obtain

$$fc - bg = \frac{i}{2}\sigma_n^- \frac{i}{2}\sigma_n^+ - \frac{i}{2}\sigma_n^- \frac{i}{2}\sigma_n^+ = 0.$$

Identity 11 is proved in the same way. Consequently, we have shown that the corresponding entries of the two matrices are equal. Thus the Yang-Baxter relation follows.

Problem 10. The classical massless *Thirring model* in (1+1) dimensions is given by

$$\frac{\partial \phi}{\partial t} + v\frac{\partial \phi}{\partial x} = -2iJ\chi^*\chi\phi \tag{1a}$$

$$\frac{\partial \chi}{\partial t} - v\frac{\partial \chi}{\partial x} = -2iJ\phi^*\phi\chi \tag{1b}$$

where v is the constant velocity and J is the coupling constant.
(i) Find the general solution of the initial-value problem

$$\phi(x,0) = a(x)\exp(ib(x)), \qquad \chi(x,0) = c(x)\exp(id(x)) \tag{2}$$

where a, b, c and d are real-valued functions.
(ii) Find the Lax pair for (1). To find the Lax pair we consider

$$\frac{\partial \Psi}{\partial t} = U\Psi, \qquad \frac{\partial \Psi}{\partial x} = V\Psi.$$

Here U and V are 4×4 matrices. The consistency condition $\partial^2\Psi/\partial x\partial t = \partial^2\Psi/\partial t\partial x$ for this system yields

$$\frac{\partial U}{\partial t} - \frac{\partial V}{\partial x} + [U,V] = 0.$$

Solution 10. (i) Using the new independent variables $\xi := x - vt$, $\eta := x + vt$ we can write (1) in the form

$$v\frac{\partial \phi}{\partial \eta} = -iJ\chi^*\chi\phi, \qquad v\frac{\partial \chi}{\partial \xi} = iJ\phi^*\phi\chi. \tag{3}$$

Since

$$\frac{\partial(\phi^*\phi)}{\partial\eta} = 0, \qquad \frac{\partial(\chi^*\chi)}{\partial\xi} = 0$$

we set

$$\phi(\xi,\eta) = f(\xi)\exp(ik(\xi,\eta)), \qquad \chi(\xi,\eta) = g(\eta)\exp(i\ell(\xi,\eta)) \qquad (4)$$

where f and g are real arbitrary functions. The real functions k and ℓ are determined as follows. Substituting the expressions (4) into (3) yields

$$v\frac{\partial k}{\partial\eta} = -Jg^2(\eta), \qquad v\frac{\partial\ell}{\partial\xi} = Jf^2(\xi). \qquad (5)$$

Integrating this system of partial differential equations yields

$$k(\xi,\eta) = -\frac{J}{v}\int^\eta g^2(\eta)d\eta + \theta(\xi), \qquad \ell(\xi,\eta) = \frac{J}{v}\int^\xi f^2(\xi)d\xi + \epsilon(\eta) \qquad (6)$$

where θ and ϵ are real arbitrary functions. With (4), (6) and (3) we arrive at the general solutions for (1)

$$\phi(x,t) = f(x-vt)\exp\left(-i\frac{J}{v}\int^{x+vt} g^2(\eta)d\eta + i\theta(x-vt)\right)$$

$$\chi(x,t) = g(x+vt)\exp\left(i\frac{J}{v}\int^{x-vt} f^2(\xi)d\xi + i\epsilon(x+vt)\right).$$

Suppose that the initial conditions are given by (2) we obtain

$$f(x) = a(x), \qquad \theta(x) = \frac{J}{v}\int^x c^2(\eta)d\eta + b(x)$$

$$g(x) = c(x), \qquad \epsilon(x) = -\frac{J}{v}\int^x a^2(\xi)d\xi + d(x).$$

Thus we find

$$\phi(x,t) = a(x-vt)\exp\left(-i\frac{J}{v}\int_{x-vt}^{x+vt} c^2(\eta)d\eta + ib(x-vt)\right)$$

$$\chi(x,t) = c(x+vt)\exp\left(-i\frac{J}{v}\int_{x-vt}^{x+vt} a^2(\xi)d\xi + id(x+vt)\right).$$

(ii) We assume that the 4×4 matrices U and V take the following form

$$U = \lambda U^{(1)} + U^{(0)}, \qquad V = \lambda V^{(1)} + V^{(0)}.$$

Substituting the this expression into the consistency condition and equating the terms with the same powers of λ yields

$$[U^{(1)}, V^{(1)}] = 0$$

$$\frac{\partial U^{(1)}}{\partial t} - \frac{\partial V^{(1)}}{\partial x} + [U^{(1)}, V^{(0)}] + [U^{(0)}, V^{(1)}] = 0$$

$$\frac{\partial U^{(0)}}{\partial t} - \frac{\partial V^{(0)}}{\partial x} + [U^{(0)}, V^{(0)}] = 0.$$

We set

$$U^{(1)} := \begin{pmatrix} A & 0 \\ 0 & B \end{pmatrix}, \qquad U^{(0)} := \begin{pmatrix} C & 0 \\ 0 & D \end{pmatrix},$$

$$V^{(1)} := \begin{pmatrix} -vA & 0 \\ 0 & vB \end{pmatrix}, \qquad V^{(0)} := \begin{pmatrix} vC & 0 \\ 0 & -vD \end{pmatrix}$$

where A, B, C and D are 2×2 matrices. With this choice, $[U^{(1)}, V^{(1)}] = 0$ is satisfied identically. For the other two equations we obtain

$$\frac{\partial A}{\partial t} + v\frac{\partial A}{\partial x} + 2v[A, C] = 0, \qquad \frac{\partial B}{\partial t} - v\frac{\partial B}{\partial x} - 2v[B, D] = 0$$

and

$$\frac{\partial C}{\partial t} - v\frac{\partial C}{\partial x} = 0, \qquad \frac{\partial D}{\partial t} + v\frac{\partial D}{\partial x} = 0.$$

Thus we can set

$$A = \begin{pmatrix} 0 & \phi \\ \phi^* & 0 \end{pmatrix}, \qquad B = \begin{pmatrix} 0 & \chi \\ \chi^* & 0 \end{pmatrix},$$

$$C = -\frac{iJ}{2v}\begin{pmatrix} \chi^*\chi & 0 \\ 0 & -\chi^*\chi \end{pmatrix}, \qquad D = \frac{iJ}{2v}\begin{pmatrix} \phi^*\phi & 0 \\ 0 & -\phi^*\phi \end{pmatrix}.$$

These partial differential equations are the massless Thirring model (1).

Problem 11. The *Korteweg-de Vries equation* is given by

$$\frac{\partial u}{\partial t} + u\frac{\partial u}{\partial x} + \beta\frac{\partial^3 u}{\partial x^3} = 0. \tag{1}$$

A *conservation law* is given by

$$\frac{\partial Q}{\partial t} + \frac{\partial P}{\partial x} = 0.$$

(i) Show that (1) can be written as a conservation law.
(ii) Show that

$$\frac{\partial}{\partial t}\left(\frac{u^2}{2}\right) + \frac{\partial}{\partial x}\left(\frac{u^3}{3} + \beta\left(u\frac{\partial^2 u}{\partial x^2} - \frac{1}{2}\left(\frac{\partial u}{\partial x}\right)^2\right)\right) = 0 \tag{2}$$

and

$$\frac{\partial}{\partial t}\left(\frac{u^3}{3} - \beta\frac{\partial^2 u}{\partial x^2}\right) + \frac{\partial}{\partial x}\left(\frac{u^4}{4} + \beta\left(u^2\frac{\partial^2 u}{\partial x^2} + 2\frac{\partial u}{\partial t}\frac{\partial u}{\partial x}\right) + \beta^2\left(\frac{\partial^2 u}{\partial x^2}\right)^2\right) = 0$$

$$(3)$$

are conservation laws of the Korteweg-de Vries equation (1).

Solution 11.   (i) Since $\partial u^2/\partial x \equiv 2u\partial u/\partial x$ we find that the Korteweg-de Vries equation (1) can be written as

$$\frac{\partial u}{\partial t} + \frac{\partial}{\partial x}\left(\frac{u^2}{2} + \beta\frac{\partial^2 u}{\partial x^2}\right) = 0.$$

(ii) Multiplying (1) by u and using the identity

$$\frac{\partial}{\partial x}\left(u\frac{\partial^2 u}{\partial x^2} - \frac{1}{2}\left(\frac{\partial u}{\partial x}\right)^2\right) \equiv u\frac{\partial^3 u}{\partial x^3}$$

we obtain (2). Multiplying (1) by u^2 we obtain (3).

Problem 12.   Consider the *Schrödinger eigenvalue equation*

$$\frac{d^2 y}{dx^2} + (k^2 - u(x))y = 0 \tag{1}$$

where k is an arbitrary real parameter.
(i) Find a solution of the form

$$y(x) = e^{ikx}(2k + if(x)) \tag{2}$$

i.e. determine f and u.
(ii) Find the condition that $u(ax+bt)$ satisfies the Korteweg-de Vries equation

$$\frac{\partial u}{\partial t} - 6u\frac{\partial u}{\partial x} + \frac{\partial^3 u}{\partial x^3} = 0. \tag{3}$$

Solution 12.   (i) Since

$$\frac{d^2 y}{dx^2} = -k^2 e^{ikx}(2k + if(x)) - 2ke^{ikx}\frac{df}{dx} + ie^{ikx}\frac{d^2 f}{dx^2} \tag{4}$$

we obtain

$$i\frac{d^2 f}{dx^2} - iu(x)f(x) + k\left(-2\frac{df}{dx} - 2u(x)\right) = 0. \tag{5}$$

Separating terms according to powers of k, we arrive at

$$\frac{d^2 f}{dx^2} = fu, \quad \frac{df}{dx} = -u.$$

Elimination of u and integration yields

$$\frac{df}{dx} + \frac{1}{2}f^2 = 2c^2$$

where $2c^2$ is the constant of integration. The substitution

$$f = \frac{2}{w}\frac{dw}{dx}$$

leads to the linear second order differential equation $d^2w/dx^2 = c^2w$ with the general solution $w(x) = Ae^{cx} + Be^{-cx}$, where A, B are the constants of integration. Using $df/du = -u$ and $dw/dx = fw$ we obtain

$$u(x) = -2\frac{d^2}{dx^2}\ln(w(x)).$$

Inserting the $w(x)$ into this expression leads to

$$u(x) = -2c^2\text{sech}^2(cx - \phi)$$

where $\phi = \frac{1}{2}\ln(B/A)$ and $\text{sech}(\alpha):=1/\cosh(\alpha)$.
(ii) Inserting the ansatz

$$u(x,t) = -2c^2\text{sech}^2(cx - \phi(t))$$

into the Korteweg-de Vries equation provides $d\phi/dt = 4c^3$. Therefore

$$u(x,t) = -2c^2\text{sech}^2(cx - 4c^3t).$$

Problem 13. The function

$$u(x,t) = -12\frac{4\cosh(2x - 8t) + \cosh(4x - 64t) + 3}{(3\cosh(x - 28t) + \cosh(3x - 36t))^2} \tag{1}$$

is a so-called *two-soliton solution* of the Korteweg-de Vries equation

$$\frac{\partial u}{\partial t} - 6u\frac{\partial u}{\partial x} + \frac{\partial^3 u}{\partial x^3} = 0. \tag{2}$$

(i) Find the solution for $t = 0$.
(ii) Find the solution for $|2x - 32t| \gg 0$ and $|x - 4t| < 1$.
(iii) Find the solution for $|x - 4t| \gg 0$ and $|2x - 32t| < 1$.

Solution 13.   For $t = 0$ the function (1) simplifies to

$$u(x, t = 0) = -12\frac{4\cosh(2x) + \cosh(4x) + 3}{(3\cosh(x) + \cosh(3x))^2} = -6\text{sech}^2(x) \qquad (3)$$

where we have used the identities

$$\cosh(2x) \equiv \sinh^2(x) + \cosh^2(x), \quad \cosh(4x) \equiv 8\cosh^4(x) - 8\cosh^2(x) + 1,$$

$$\sinh(3x) \equiv 3\sinh(x) + 4\sinh^3(x).$$

Thus the two solitons completely overlap.
(ii) For $|2x - 32t| \gg 1$ and $|x - 4t| < 1$ we find

$$u(x, t) = -2\text{sech}^2\left(x - 4t \mp \tanh^{-1}(1/2)\right).$$

(iii) For $|x - 4t| \gg 1$ and $|2x - 32t| < 1$ we find

$$u(x, t) = -8\text{sech}^2\left(2x - 32t \mp \tanh^{-1}\frac{1}{2}\right).$$

Problem 14.   Consider the *Korteweg-de Vries-Burgers equation*

$$\frac{\partial u}{\partial t} + u\frac{\partial u}{\partial x} + b\frac{\partial^3 u}{\partial x^3} - a\frac{\partial^2 u}{\partial x^2} = 0 \qquad (1)$$

where u is a *conserved quantity*

$$\frac{d}{dt}\int_{-\infty}^{\infty} u(x, t)dx = 0$$

i.e. the area under u is conserved for all t. Find solutions of the form

$$u(x, t) = U(\xi(x, t)), \qquad \xi(x, t) := c(x - vt)$$

where

$$U(\xi) = S(Y(\xi)) = \sum_{n=0}^{N} a_n Y(\xi)^n, \quad Y(\xi) = \tanh(\xi(x, t)) = \tanh(c(x - vt)). \qquad (2)$$

Solution 14.   Inserting (2) into (1) yields the ordinary differential equation

$$-cvU(\xi) + \frac{1}{2}cU(\xi)^2 + bc^3\frac{d^2U}{d\xi^2} - ac^2\frac{dU(\xi)}{d\xi} = C.$$

Requiring that

$$U(\xi) \to 0, \quad \frac{dU(\xi)}{d\xi} \to 0, \quad \frac{d^2U(\xi)}{d\xi^2} \to 0 \quad \text{as} \quad \xi \to \infty \qquad (3)$$

we find that the integration constant C must be set to zero. Then the ordinary differential equation can be expressed in the variable Y as

$$-vS + \frac{1}{2}S^2 + bc^2(1 - Y^2)\left(-2Y\frac{dS}{dY} + (1 - Y^2)\frac{d^2S}{dY^2}\right) - ac(1 - Y^2)\frac{dS}{dY} = 0.$$
$$(4)$$

Substitution of the expansion (2) into (4) and balancing the highest degree in Y, yields $N = 2$. The boundary condition $U(\xi) \to 0$ for $\xi \to +\infty$ or $\xi \to -\infty$ implies that $S(Y) \to 0$ for $Y \to +1$ or $Y \to -1$. Without loss of generality we only consider the limit $Y \to 1$. Two possible solutions arise

$$S(Y) = F(Y) = b_0(1 - Y)(1 + b_1 Y)$$

or

$$S(Y) = G(Y) = d_0(1 - Y)^2.$$

In the first case $S(Y)$ decays as $\exp(2\xi)$, whereas in the second case, $S(Y)$ decays as $\exp(-4\xi)$ as $\xi \to +\infty$. We first take

$$F(Y) = (1 - Y)b_0(1 + b_1 Y).$$

Upon substituting this ansatz into (4) and subsequent cancellation of a common factor $(1 - Y)$, we take the limit $Y \to 1$, which gives

$$v = 4bc^2 + 2ac.$$

The remaining constants b_0 and b_1 can be found through algebra. We obtain $c = a/(10b)$, $v = 24bc^2$. Thus

$$F(Y) = 36bc^2(1 - Y)(1 + \frac{1}{3}Y).$$

The second possible solution $G(Y) = d_0(1-Y)^2$ also leads to a real solution.

Problem 15. The equation which describes small amplitude waves in a dispersive medium with a slight deviation from one-dimensionality is

$$\frac{\partial}{\partial x}\left(4\frac{\partial u}{\partial t} + 6u\frac{\partial u}{\partial x} + \frac{\partial^2 u}{\partial x^2}\right) \pm 3\frac{\partial^2 u}{\partial y^2} = 0. \qquad (1)$$

The $+$ refers to the two-dimensional Korteweg-de Vries equation. The $-$ refers to the two-dimensional Kadomtsev-Petviashvili equation. The formulation of the *inverse scattering transform* is as follows. Let

$$u(x, y, t) := 2\frac{\partial}{\partial x}K(x, x; y, t) \qquad (2)$$

where

$$K(x, z; y, t) + F(x, z; y, t) + \int_x^\infty K(x, s; y, t) F(s, z; y, t) ds = 0 \quad (3)$$

and F satisfies the system of linear partial differential equations

$$\frac{\partial^3 F}{\partial x^3} + \frac{\partial^3 F}{\partial z^3} + \frac{\partial F}{\partial t} = 0, \qquad \frac{\partial^2 F}{\partial x^2} - \frac{\partial^2 F}{\partial z^2} + \sigma \frac{\partial F}{\partial y} = 0 \quad (4)$$

with $\sigma = 1$ for the two-dimensional Korteweg-de Vries equation and $\sigma = i$ for the Kadomtsev-Petviashvili equation. Find solutions of the product form

$$F(x, z; y, t) = \alpha(x; y, t) \beta(z; y, t) \quad (5)$$

and

$$K(x, z; y, t) = L(x; y, t) \beta(z; y, t). \quad (6)$$

Solution 15. Inserting (5) and (6) into (3) yields

$$L(x; y, t) = -\frac{\alpha(x; y, t)}{\left(1 + \int_x^\infty \alpha(s; y, t) \beta(s; y, t) ds\right)}. \quad (7)$$

From (2) we then have the solution of (1)

$$u(x, y, t) = 2 \frac{\partial^2}{\partial x^2} \ln \left(1 + \int_x^\infty \alpha(s; y, t) \beta(s; y, t) ds\right) \quad (8)$$

provided functions α and β can be found. From (4) we obtain

$$\frac{\partial \alpha}{\partial t} + \frac{\partial^3 \alpha}{\partial x^3} = 0, \qquad \frac{\partial \beta}{\partial t} + \frac{\partial^3 \beta}{\partial z^3} = 0$$

$$\sigma \frac{\partial \alpha}{\partial y} + \frac{\partial^2 \alpha}{\partial x^2} = 0, \qquad \sigma \frac{\partial \beta}{\partial y} - \frac{\partial^2 \beta}{\partial z^2} = 0.$$

Then α and β admit the solution

$$\alpha(x, y, t) = \exp(-lx - (l^2/\sigma)y + l^3 t + \delta)$$

and similarly

$$\beta(x, y, t) = \exp(-Lz + (L^2/\sigma)y + L^3 t + \Delta)$$

where δ, Δ are arbitrary shifts and l, L are constants. This form gives the oblique solitary wave solution of the two-dimensional Korteweg-de Vries equation,

$$u(x, y, t) = 2a^2 \text{sech}^2(a(x + 2my - (a^2 + 3m^2)t)) \quad (9)$$

where $a = (l + L)/2$, $m = (l - L)/2$. For the Kadomtsev-Petviashvili equation we have $\sigma = i$, and so if we regard l, L as complex constants a real solution is just (9) with $m \to -im$ and $l = \bar{L}$.

Problem 16. Let

$$
L := \frac{\partial^2}{\partial x^2} + bu(x, y, t) + \frac{\partial}{\partial y}
\tag{1}
$$

and

$$
T := -4\frac{\partial^3}{\partial x^3} - 6bu(x, y, t)\frac{\partial}{\partial x} - 3b\frac{\partial u}{\partial x} - 3b\partial_x^{-1}\frac{\partial u}{\partial y} + \frac{\partial}{\partial t}
\tag{2}
$$

where $b \in \mathbb{R}$ and

$$
\partial_x^{-1} f := \int_{-\infty}^x f(s)\,ds.
\tag{3}
$$

Calculate the nonlinear evolution equation which follows from the condition $[T, L]v = 0$, where v is a smooth function of x, y, t. Use the property that

$$
\frac{\partial}{\partial x}\left(\partial_x^{-1} f\right) = f(x).
$$

Solution 16. Since $[T, L]v = (TL)v - (LT)v = T(Lv) - L(Tv)$ we obtain after a lengthy calculation

$$
[T, L]v = b\frac{\partial u}{\partial t}v - b\frac{\partial^3 u}{\partial x^3}v - 6b^2 u\frac{\partial u}{\partial x}v - 3b\left(\partial_x^{-1}\frac{\partial^2 u}{\partial y^2}\right)v.
$$

Since v is arbitrary, the condition $[T, L]v = 0$ yields

$$
\frac{\partial u}{\partial t} = \frac{\partial^3 u}{\partial x^3} + 6bu\frac{\partial u}{\partial x} + 3\partial_x^{-1}\frac{\partial^2 u}{\partial y^2}.
$$

To find the local form we differentiate this equation with respect to x and find

$$
\frac{\partial^2 u}{\partial t \partial x} = \frac{\partial^4 u}{\partial x^4} + 6b\left(\frac{\partial u}{\partial x}\right)^2 + 6bu\frac{\partial^2 u}{\partial x^2} + 3\frac{\partial^2 u}{\partial y^2}.
$$

This equation is called the *Kadomtsev-Petviashvili equation*.

Problem 17. Consider the system of partial differential equations

$$
\frac{\partial \psi_1}{\partial x} = \lambda\psi_1 + q\psi_2
\tag{1a}
$$

$$
\frac{\partial \psi_2}{\partial x} = r\psi_1 - \lambda\psi_2
\tag{1b}
$$

$$\frac{\partial \psi_1}{\partial t} = A\psi_1 + B\psi_2 \qquad (1c)$$

$$\frac{\partial \psi_2}{\partial t} = C\psi_1 - A\psi_2 \qquad (1d)$$

where λ is a real parameter and q and r are smooth functions of the independent variables x and t. The coefficients A, B and C are one-parameter (λ) families of x, t and q and r with their derivatives. This system is called the *AKNS system*.

(i) Find the equation which follows from the *compatibility condition* (also called *integrability condition*)

$$\frac{\partial^2 \psi_1}{\partial x \partial t} = \frac{\partial^2 \psi_1}{\partial t \partial x}, \qquad \frac{\partial^2 \psi_2}{\partial x \partial t} = \frac{\partial^2 \psi_2}{\partial t \partial x}. \qquad (2)$$

(ii) Let w be a complex-valued function of x and t. Assume that

$$A := 2i\lambda^2 + i|w|^2, \qquad B := i\frac{\partial w}{\partial x} + 2i\lambda w, \qquad C := i\frac{\partial \bar{w}}{\partial x} - 2i\lambda \bar{w}$$

$$q := w, \qquad r := -\bar{q} \equiv -\bar{w}. \qquad (3)$$

Find the partial differential equation of w.

Solution 17. Taking the derivative of (1a) and (1b) with respect to t yields

$$\frac{\partial^2 \psi_1}{\partial t \partial x} = \lambda \frac{\partial \psi_1}{\partial t} + \frac{\partial q}{\partial t}\psi_2 + q\frac{\partial \psi_2}{\partial t} = (\lambda A + qC)\psi_1 + \left(\lambda B - qA + \frac{\partial q}{\partial t}\right)\psi_2$$

and

$$\frac{\partial^2 \psi_2}{\partial t \partial x} = -\lambda \frac{\partial \psi_2}{\partial t} + \frac{\partial r}{\partial t}\psi_1 + r\frac{\partial \psi_1}{\partial t} = \left(-\lambda C + \frac{\partial r}{\partial t} + rA\right)\psi_1 + (\lambda A + rB)\psi_2$$

where we have used (1a) through (1d). Taking the derivative of (1c) and (1d) with respect to x yields

$$\frac{\partial^2 \psi_1}{\partial x \partial t} = \left(\frac{\partial A}{\partial x} + \lambda A + rB\right)\psi_1 + \left(Aq + \frac{\partial B}{\partial x} - \lambda B\right)\psi_2$$

and

$$\frac{\partial^2 \psi_2}{\partial x \partial t} == \left(\frac{\partial C}{\partial x} + \lambda C - rA\right)\psi_1 + \left(qC - \frac{\partial A}{\partial x} + \lambda A\right)\psi_2$$

where we have used (1a) through (1d). From these equations we obtain

$$\frac{\partial A}{\partial x} = qC - rB, \qquad \frac{\partial B}{\partial x} = 2\lambda B + \frac{\partial q}{\partial t} - 2Aq$$

and

$$\frac{\partial C}{\partial x} = -2\lambda C + \frac{\partial r}{\partial t} + 2rA, \qquad \frac{\partial A}{\partial x} = qC - rB.$$

(ii) Using (3) and these equations we obtain the nonlinear partial differential equation

$$i\frac{\partial w}{\partial t} + \frac{\partial^2 w}{\partial x^2} + 2|w|^2 w = 0.$$

The differential equation $\partial A/\partial x = qC - rB$ is satisfied identically. We used that $w = w_1 + iw_2$, where w_1 and w_2 are real-valued functions and $|w|^2 = w_1^2 + w_2^2$. This nonlinear partial differential equation is the so-called one-dimensional *cubic Schrödinger equation*.

Problem 18. Consider a system consisting of a linear equation for an eigenfunction and an evolution equation

$$L(x, \partial)\psi(x, \lambda) = \lambda\psi(x, \lambda) \tag{1}$$

$$\frac{\partial\psi(x, \lambda)}{\partial x_n} = B_n(x, \partial)\psi(x, \lambda). \tag{2}$$

(i) Show that

$$\frac{\partial L}{\partial x_n} = [B_n, L] \equiv B_n L - L B_n \tag{3}$$

and

$$\frac{\partial B_m}{\partial x_n} - \frac{\partial B_n}{\partial x_m} = [B_n, B_m]. \tag{4}$$

(ii) Let

$$L := \partial + u_2(x)\partial^{-1} + u_3(x)\partial^{-2} + u_4(x)\partial^{-3} + \cdots$$

where $x = (x_1, x_2, x_3, \ldots)$, $\partial \equiv \partial/\partial x_1$ and

$$\partial^{-1} f(x) := \int^{x_1} f(s, x_2, x_3, \ldots) ds.$$

Define $B_n(x, \partial)$ as the differential part of $(L(x, \partial))^n$. Show that

$$B_1 = \partial$$
$$B_2 = \partial^2 + 2u_2$$
$$B_3 = \partial^3 + 3u_2\partial + 3u_3 + 3\frac{\partial u_2}{\partial x_1}$$
$$B_4 = \partial^4 + 4u_2\partial^2 + \left(4u_3 + 6\frac{\partial u_2}{\partial x_1}\right)\partial + 4u_4 + 6\frac{\partial u_3}{\partial x_1} + 4\frac{\partial^2 u_2}{\partial x_1^2} + 6u_2^2.$$

(iii) Find the equations of motion from (4) for $n = 2$ and $m = 3$.

Solution 18.   (i) From (1) we obtain

$$\frac{\partial}{\partial x_n}(L\psi) = \frac{\partial L}{\partial x_n}\psi + L\frac{\partial \psi}{\partial x_n} = \frac{\partial}{\partial x_n}(\lambda\psi) = \lambda\frac{\partial \psi}{\partial x_n}. \tag{5}$$

Inserting (2) into (5) yields

$$\frac{\partial L}{\partial x_n}\psi + LB_n\psi = \lambda B_n\psi = B_n\lambda\psi$$

or

$$\frac{\partial L}{\partial x_n}\psi = -LB_n\psi + B_nL\psi = [B_n, L]\psi. \tag{6}$$

(ii) From this equation we obtain

$$\frac{\partial}{\partial x_n}(L\psi) - L\frac{\partial \psi}{\partial x_n} = B_nL\psi - LB_n\psi \tag{7}$$

and

$$\frac{\partial}{\partial x_m}(L\psi) - L\frac{\partial \psi}{\partial x_m} = B_mL\psi - LB_m\psi. \tag{8}$$

Taking the derivatives of (7) with respect to x_m and of (8) with respect to x_n gives

$$\frac{\partial^2}{\partial x_m \partial x_n}(L\psi) - \frac{\partial}{\partial x_m}\left(L\frac{\partial \psi}{\partial x_n}\right) = \frac{\partial}{\partial x_m}(B_nL\psi) - \frac{\partial}{\partial x_m}(LB_n\psi)$$

$$\frac{\partial^2}{\partial x_n \partial x_m}(L\psi) - \frac{\partial}{\partial x_n}\left(L\frac{\partial \psi}{\partial x_m}\right) = \frac{\partial}{\partial x_n}(B_mL\psi) - \frac{\partial}{\partial x_n}(LB_m\psi).$$

Subtracting these two equations we obtain

$$-\frac{\partial}{\partial x_m}\left(L\frac{\partial \psi}{\partial x_n}\right) + \frac{\partial}{\partial x_n}\left(L\frac{\partial \psi}{\partial x_m}\right) = \frac{\partial}{\partial x_m}(B_nL\psi) - \frac{\partial}{\partial x_m}(LB_n\psi)$$

$$+ \frac{\partial}{\partial x_n}(LB_m\psi) - \frac{\partial}{\partial x_m}(LB_n\psi).$$

Inserting $\partial\psi/\partial x_n = B_n\psi$, $\partial\psi/\partial x_m = B_m\psi$ into this equation and taking into account (1) gives

$$\frac{\partial}{\partial x_m}(B_nL\psi) - \frac{\partial}{\partial x_n}(LB_m\psi) = 0.$$

From this equation we obtain

$$\frac{\partial B_n}{\partial x_m}\psi + B_n\frac{\partial \psi}{\partial x_m} - \frac{\partial B_m}{\partial x_n}\psi - B_m\frac{\partial \psi}{\partial x_n} = 0.$$

Inserting $\partial\psi/\partial x_n = B_n\psi$ and $\partial\psi/px_m = B_m\psi$ into this equation gives

$$\frac{\partial B_n}{\partial x_m}\psi + B_n B_m\psi - \frac{\partial B_m}{\partial x_n}\psi - B_m B_n\psi = 0.$$

Since ψ is arbitrary, equation (4) follows.

(ii) From L it is obvious that $B_1 = \partial$. Let f be a smooth function. Then

$$\begin{aligned}
L^2 f &= L(Lf)\\
&= L(\partial f + u_2\partial^{-1}f + u_3\partial^{-2}f + u_4\partial^{-3}f + u_5\partial^{-4}f + \cdots)\\
&= \partial^2 f + u_2 f + (\partial u_2)\partial^{-1}f + u_3(\partial^{-1}f) + (\partial u_3)\partial^{-2}f + u_4\partial^{-2}f\\
&\quad +(\partial u_4)\partial^{-2}f + \cdots
\end{aligned}$$

where

$$\partial^{-2}f := \int^{x_1}\left(\int^{s_1} f(s_2)ds_2\right)ds_1.$$

Since f is an arbitrary smooth function we have $B_2 = \partial^2 + u_2$. In the same manner we find B_3 and B_4.

(iii) Since

$$\begin{aligned}
\frac{\partial B_3}{\partial x_2}\psi &= \frac{\partial}{\partial x_2}(\partial^3\psi) + 3\frac{\partial u_2}{\partial x_2}(\partial\psi) + 3u_2\partial\left(\frac{\partial\psi}{\partial x_2}\right) + 3\frac{\partial u_3}{\partial x_2}\psi + 3u_3\frac{\partial\psi}{\partial x_2}\\
&\quad +3\frac{\partial^2 u_2}{\partial x_1\partial x_2}\psi + 3\frac{\partial u_2}{\partial x_1}\frac{\partial\psi}{\partial x_2}
\end{aligned}$$

and

$$-\frac{\partial B_2}{\partial x_3}\psi = -\partial^2\frac{\partial\psi}{\partial x_3} - 2\frac{\partial u_2}{\partial x_3}\psi - 2u_2\frac{\partial\psi}{\partial x_3}$$

we find

$$\frac{\partial}{\partial x_1}\left(\frac{\partial u_2}{\partial x_3} - \frac{1}{4}\frac{\partial^3 u_2}{\partial x_1^3} - 3u_2\frac{\partial u_2}{\partial x_1}\right) - \frac{3}{4}\frac{\partial^2 u_2}{\partial x_2^2} = 0.$$

This partial differential equation is the called the *Kadomtsev-Petviashvili equation*.

Problem 19. The nonlinear partial differential equation

$$\frac{\partial^2 u}{\partial x\partial y} + \alpha\frac{\partial u}{\partial x} + \beta\frac{\partial u}{\partial y} + \gamma\frac{\partial u}{\partial x}\frac{\partial u}{\partial y} = 0 \tag{1}$$

is called *Thomas equation*, where α, β and γ are real constants with $\gamma \neq 0$. Show that

$$\frac{\partial\phi}{\partial x} = -\frac{\partial u}{\partial x} - \left(2\beta + \gamma\frac{\partial u}{\partial x}\right)\phi \tag{2a}$$

$$\frac{\partial\phi}{\partial y} = \frac{\partial u}{\partial y} - \left(2\alpha + \gamma\frac{\partial u}{\partial y}\right)\phi \tag{2b}$$

provides a Lax representation for the Thomas equation.

Solution 19.   Taking the derivative of equation (2b) with respect to x gives

$$\frac{\partial^2 \phi}{\partial x \partial y} = \frac{\partial^2 u}{\partial x \partial y} - 2\alpha \frac{\partial \phi}{\partial x} - \gamma \frac{\partial^2 u}{\partial x \partial y} \phi - \gamma \frac{\partial u}{\partial y} \frac{\partial \phi}{\partial x}. \tag{3}$$

Taking the derivative of (2a) with respect to y gives

$$\frac{\partial^2 \phi}{\partial y \partial x} = -\frac{\partial^2 u}{\partial y \partial x} - 2\beta \frac{\partial \phi}{\partial y} - \gamma \frac{\partial^2 u}{\partial y \partial x} \phi - \gamma \frac{\partial u}{\partial x} \frac{\partial \phi}{\partial y}. \tag{4}$$

Subtracting (3) and (4) leads to

$$0 = 2 \frac{\partial^2 u}{\partial y \partial x} - \left(2\alpha + \gamma \frac{\partial u}{\partial y} \right) \frac{\partial \phi}{\partial x} + \left(2\beta + \gamma \frac{\partial u}{\partial x} \right) \frac{\partial \phi}{\partial y}.$$

Inserting equations (2a) and (2b) into this equation gives the Thomas equation (1). Equation (1) can be linearized by applying the transformation

$$u(x, y) = \frac{1}{\gamma} \ln(v(x, y)).$$

Problem 20.   Consider the system of quasi-linear partial differential equations

$$\frac{\partial u}{\partial t} + \frac{\partial v}{\partial y} + u \frac{\partial v}{\partial x} - v \frac{\partial u}{\partial x} = 0, \qquad \frac{\partial u}{\partial y} + \frac{\partial v}{\partial x} = 0.$$

Show that this system arise as compatibility conditions $[L, M]\Psi = 0$ of an overdetermined system of linear equations $L\Psi = 0$, $M\Psi = 0$, where $\Psi(x, y, t, \lambda)$ is a function, λ is a spectral parameter, and the *Lax pair* is given by

$$L = \frac{\partial}{\partial t} - v \frac{\partial}{\partial x} - \lambda \frac{\partial}{\partial y}, \qquad M = \frac{\partial}{\partial y} + u \frac{\partial}{\partial x} - \lambda \frac{\partial}{\partial x}.$$

Solution 20.   Using the product rule we have

$$[L, M]\Psi = L(M\Psi) - M(L\Psi)$$

$$= \left(\frac{\partial}{\partial t} - v \frac{\partial}{\partial x} - \lambda \frac{\partial}{\partial y} \right) \left(\frac{\partial}{\partial y} + u \frac{\partial}{\partial x} - \lambda \frac{\partial}{\partial x} \right) \Psi$$

$$- \left(\frac{\partial}{\partial y} + u \frac{\partial}{\partial x} - \lambda \frac{\partial}{\partial x} \right) \left(\frac{\partial}{\partial t} - v \frac{\partial}{\partial x} - \lambda \frac{\partial}{\partial y} \right) \Psi$$

$$= \left(\frac{\partial u}{\partial t} - v \frac{\partial u}{\partial x} - \lambda \frac{\partial u}{\partial y} + \frac{\partial v}{\partial y} + u \frac{\partial v}{\partial x} - \lambda \frac{\partial v}{\partial x} \right) \frac{\partial \Psi}{\partial x}$$

$$= \left(\frac{\partial u}{\partial t} - v \frac{\partial u}{\partial x} + \frac{\partial v}{\partial y} + u \frac{\partial v}{\partial x} - \lambda \left(\frac{\partial v}{\partial x} + \frac{\partial u}{\partial y} \right) \right) \frac{\partial \Psi}{\partial x}.$$

Thus from $[L, M]\Psi = 0$ the system of partial differential equations follows.

Problem 21. The one-dimensional *sine-Gordon equation* in light-cone coordinates ξ, η is given by

$$\frac{\partial^2 u}{\partial \xi \partial \eta} = \sin(u).$$
(1)

Show that the transformation

$$\xi'(\xi, \eta) = \xi$$
(2a)

$$\eta'(\xi, \eta) = \eta$$
(2b)

$$\frac{\partial u'(\xi'(\xi, \eta), \eta'(\xi, \eta))}{\partial \xi'} = \frac{\partial u}{\partial \xi} - 2\lambda \sin\left(\frac{u(\xi, \eta) + u'(\xi'(\xi, \eta), \eta'(\xi, \eta))}{2}\right)$$
(2c)

$$\frac{\partial u'(\xi'(\xi, \eta), \eta'(\xi, \eta))}{\partial \eta'} = -\frac{\partial u}{\partial \eta} + \frac{2}{\lambda} \sin\left(\frac{u(\xi, \eta) - u'(\xi'(\xi, \eta), \eta'(\xi, \eta))}{2}\right)$$
(2d)

defines an *auto-Bäcklund transformation*, where λ is a nonzero real parameter. This means one has to show that

$$\frac{\partial^2 u'}{\partial \xi' \partial \eta'} = \sin(u')$$
(3)

follows from (1) and (2). Vice versa one has to show that (1) follows from (3) and (2).

Solution 21. Differentiating (2c) with respect to η, taking into account (2a) and (2b) yields

$$\frac{\partial^2 u'}{\partial \eta' \partial \xi'} = \frac{\partial^2 u}{\partial \eta \partial \xi} - \lambda \cos\left(\frac{u + u'}{2}\right)\left(\frac{\partial u}{\partial \eta} + \frac{\partial u'}{\partial \eta'}\right).$$
(4)

Differentiating (2d) with respect to ξ, taking into account (2a) and (2b) yields

$$\frac{\partial^2 u'}{\partial \xi' \partial \eta'} = -\frac{\partial^2 u}{\partial \xi \partial \eta} + \frac{1}{\lambda} \cos\left(\frac{u - u'}{2}\right)\left(\frac{\partial u}{\partial \xi} - \frac{\partial u'}{\partial \xi'}\right).$$

For the sake of simplicity we have omitted the arguments. Adding the last two equations gives

$$2\frac{\partial^2 u'}{\partial \eta' \partial \xi'} = -\lambda \cos\left(\frac{u + u'}{2}\right)\left(\frac{\partial u}{\partial \eta} + \frac{\partial u'}{\partial \eta'}\right) + \frac{1}{\lambda} \cos\left(\frac{u - u'}{2}\right)\left(\frac{\partial u}{\partial \xi} - \frac{\partial u'}{\partial \xi'}\right).$$

Inserting $\partial u/\partial \xi$ from (2c) and $\partial u/\partial \eta$ from (2d) into this equation gives

$$2\frac{\partial^2 u'}{\partial \eta' \partial \xi'} = -\lambda \cos\left(\frac{u + u'}{2}\right)\frac{2}{\lambda}\sin\left(\frac{u - u'}{2}\right) + \frac{1}{\lambda}\cos\left(\frac{u - u'}{2}\right)2\lambda\sin\left(\frac{u + u'}{2}\right).$$

Using the identity $\sin(\alpha)\cos(\beta) \equiv \frac{1}{2}\sin(\alpha+\beta) + \frac{1}{2}\sin(\alpha-\beta)$ we obtain

$$\frac{\partial^2 u'}{\partial \eta' \partial \xi'} = -\frac{1}{2}\sin(u) + \frac{1}{2}\sin(u') + \frac{1}{2}\sin(u) + \frac{1}{2}\sin(u').$$

Consequently

$$\frac{\partial^2 u'}{\partial \eta' \partial \xi'} = \sin(u').$$

In the same manner we can show that (1) follows from (3).

Problem 22. We showed that the one-dimensional sine-Gordon equation in light-cone coordinates ξ, η

$$\frac{\partial^2 u}{\partial \xi \partial \eta} = \sin(u) \tag{1}$$

admits an auto-Bäcklund transformation. Obviously

$$u(\xi, \eta) = 0 \tag{2}$$

is a trivial solution of (1). Apply the auto-Bäcklund transformation given in problem 1 to find a nontrivial solution of (1). This solution is also called the *vacuum solution* of (1).

Solution 22. Inserting $u(\xi, \eta) = 0$ into the auto-Bäcklund transformation yields

$$\frac{\partial u'(\xi', \eta')}{\partial \xi'} = -2\lambda \sin\left(\frac{u'(\xi', \eta')}{2}\right), \quad \frac{\partial u'(\xi', \eta')}{\partial \eta'} = \frac{2}{\lambda}\sin\left(\frac{-u'(\xi', \eta')}{2}\right).$$

Since

$$\int \frac{du}{\sin(u/2)} = 2\ln(\tan(u/4))$$

we obtain a solution of the one-dimensional sine-Gordon equation

$$u'(\xi', \eta') = 4\tan^{-1}(\exp(\lambda \xi' + \lambda^{-1}\eta' + C))$$

where C is a constant of integration.

Problem 23. Consider the two nonlinear ordinary differential equations

$$\frac{d^2 u}{dt^2} = \sin(u), \qquad \frac{d^2 v}{dt^2} = \sinh(v) \tag{1}$$

where u and v are real-valued functions. Show that

$$\frac{du}{dt} - i\frac{dv}{dt} = 2e^{i\lambda}\sin\left(\frac{1}{2}(u+iv)\right) \tag{2}$$

defines a Bäcklund transformation, where λ is a real parameter.

Solution 23. Taking the time derivative of (2) gives

$$\frac{d^2u}{dt^2} - i\frac{d^2v}{dt^2} = e^{i\lambda}\cos\left(\frac{1}{2}(u+iv)\right)\left(\frac{du}{dt} + i\frac{dv}{dt}\right). \tag{3}$$

The complex conjugate of (2) is

$$\frac{du}{dt} + i\frac{dv}{dt} = 2e^{-i\lambda}\sin\left(\frac{1}{2}(u-iv)\right)$$

where we have used $\overline{\sin z} = \sin(\bar{z})$. Inserting this equation into (3) yields

$$\frac{d^2u}{dt^2} - i\frac{d^2v}{dt^2} = 2\cos\left(\frac{1}{2}(u+iv)\right)\sin\left(\frac{1}{2}(u-iv)\right).$$

Using the identity

$$\cos\left(\frac{1}{2}(u+iv)\right)\sin\left(\frac{1}{2}(u-iv)\right) \equiv \frac{1}{2}\sin(u) + \frac{1}{2}\sin(-iv)$$

$$\equiv \frac{1}{2}\sin(u) - \frac{1}{2}i\sinh(v)$$

we find $d^2u/dt^2 - id^2v/dt^2 = \sin(u) - i\sinh(v)$. Therefore (1) follows.

Problem 24. The nonlinear partial differential equation

$$\frac{\partial u}{\partial t} + u\frac{\partial u}{\partial x} = D\frac{\partial^2 u}{\partial x^2} \tag{1}$$

is called *Burgers equation*. Let

$$u(x,t) = -2D\frac{\partial}{\partial x}\ln(v(x,t)). \tag{2}$$

Show that v satisfies the linear diffusion equation $\partial v/\partial t = D\partial^2 v/\partial x^2$.

Solution 24. From (2) we find

$$u = -2D\frac{1}{v}\frac{\partial v}{\partial x}. \tag{3}$$

It follows that

$$\frac{\partial u}{\partial t} = 2D\frac{1}{v^2}\frac{\partial v}{\partial t}\frac{\partial v}{\partial x} - 2D\frac{1}{v}\frac{\partial^2 v}{\partial t\partial x}, \qquad \frac{\partial u}{\partial x} = 2D\frac{1}{v^2}\left(\frac{\partial v}{\partial x}\right)^2 - 2D\frac{1}{v}\frac{\partial^2 v}{\partial x^2}$$

$$\frac{\partial^2 u}{\partial x^2} = -4D\frac{1}{v^3}\left(\frac{\partial v}{\partial x}\right)^3 + 6D\frac{1}{v^2}\frac{\partial v}{\partial x}\frac{\partial^2 v}{\partial x^2} - 2D\frac{1}{v}\frac{\partial^3 v}{\partial x^3}.$$

Inserting these three equations into (1) yields

$$\frac{\partial v}{\partial x}\left(\frac{\partial v}{\partial t} - D\frac{\partial^2 v}{\partial x^2}\right) = v\frac{\partial}{\partial x}\left(\frac{\partial v}{\partial t} - D\frac{\partial^2 v}{\partial x^2}\right).$$

This partial differential equation is satisfied if v satisfies the linear diffusion equation $\partial v/\partial t = D\partial^2 v/\partial x^2$.

Problem 25. The *Korteweg-de Vries equation* is given by

$$\frac{\partial u}{\partial t} + 6u\frac{\partial u}{\partial x} + \frac{\partial^3 u}{\partial x^3} = 0 \tag{1}$$

and the *modified Korteweg-de Vries equation* takes the form

$$\frac{\partial v}{\partial t} - 6v^2\frac{\partial v}{\partial x} + \frac{\partial^3 v}{\partial x^3} = 0. \tag{2}$$

Show that the Korteweg-de Vries equation (1) and the modified Korteweg-de Vries equation (2) are related by the Bäcklund transformation

$$\frac{\partial v}{\partial x} = u + v^2 \tag{3a}$$

$$\frac{\partial v}{\partial t} = -\frac{\partial^2 u}{\partial x^2} - 2\left(v\frac{\partial u}{\partial x} + u\frac{\partial v}{\partial x}\right). \tag{3b}$$

Equation (3a) is called the *Miura transformation*.

Solution 25. First we show that the Korteweg-de Vries equation follows from system (3). Taking the derivative of (3a) with respect to t yields

$$\frac{\partial^2 v}{\partial x\partial t} = \frac{\partial u}{\partial t} + 2v\frac{\partial v}{\partial t}. \tag{4}$$

Inserting (3b) gives

$$\frac{\partial^2 v}{\partial x\partial t} = \frac{\partial u}{\partial t} - 2v\frac{\partial^2 u}{\partial x^2} - 4v^2\frac{\partial u}{\partial x} - 4uv(u + v^2) \tag{5}$$

where we used $\partial v/\partial x = u + v^2$. Taking the derivative of (3b) with respect to x yields

$$\frac{\partial^2 v}{\partial x\partial t} = -\frac{\partial^3 u}{\partial x^3} - 2\left(2\frac{\partial u}{\partial x}\frac{\partial v}{\partial x} + v\frac{\partial^2 u}{\partial x^2} + u\frac{\partial^2 v}{\partial x^2}\right). \tag{6}$$

Taking the derivative of (3a) with respect to x yields

$$\frac{\partial^2 v}{\partial x^2} = \frac{\partial u}{\partial x} + 2v\frac{\partial v}{\partial x} = \frac{\partial u}{\partial x} + 2v(u + v^2). \tag{7}$$

Inserting (3a) and (7) into (6) leads to

$$\frac{\partial^2 v}{\partial x \partial t} = -\frac{\partial^3 u}{\partial x^3} - 2\left(2\frac{\partial u}{\partial x}(u + v^2) + v\frac{\partial^2 u}{\partial x^2} + u\left(\frac{\partial u}{\partial x} + 2uv + 2v^3\right)\right).$$

Subtracting this equation from (5) gives the Korteweg-de Vries equation. Now we have to show that the modified Korteweg-de Vries equation follows from system (3). Differentiating (3a) with respect to x yields

$$\frac{\partial^2 v}{\partial x^2} = \frac{\partial u}{\partial x} + 2v\frac{\partial v}{\partial x} = \frac{\partial u}{\partial x} + 2v(u + v^2)$$

and

$$\frac{\partial^3 v}{\partial x^3} = \frac{\partial^2 u}{\partial x^2} + 2u\frac{\partial v}{\partial x} + 2v\frac{\partial u}{\partial x} + 6v^2\frac{\partial v}{\partial x} = \frac{\partial^2 u}{\partial x^2} + 2v\frac{\partial u}{\partial x} + (2u + 6v^2)(u + v^2).$$

Adding this equation and (3b) gives

$$\frac{\partial v}{\partial t} + \frac{\partial^3 v}{\partial x^3} = -2u\frac{\partial v}{\partial x} + (2u + 6v^2)(u + v^2).$$

From (3) we obtain $u = \partial v/\partial x - v^2$. Inserting it into this partial differential equation we obtain the modified Korteweg-de Vries equation (2).

Problem 26. Consider the partial differential equations

$$\frac{\partial^2 u}{\partial x \partial y} = \exp(u), \qquad \frac{\partial^2 v}{\partial x \partial y} = 0. \tag{1}$$

The first equation is called the *Liouville equation*. Show that

$$\frac{\partial v}{\partial x} = \frac{\partial u}{\partial x} + \lambda \exp\left(\frac{1}{2}(v + u)\right), \qquad \frac{\partial v}{\partial y} = -\frac{\partial u}{\partial y} - \frac{2}{\lambda}\exp\left(\frac{1}{2}(u - v)\right) \tag{2}$$

provides a Bäcklund transformation of these two equations, where λ is a nonzero parameter.

Solution 26. Differentiating (2a) with respect to y yields

$$\frac{\partial^2 v}{\partial x \partial y} = \frac{\partial^2 u}{\partial x \partial y} + \lambda \exp\left(\frac{1}{2}(u + v)\right)\frac{1}{2}\left(\frac{\partial v}{\partial y} + \frac{\partial u}{\partial y}\right). \tag{3}$$

Inserting (2b) into (3) yields

$$\frac{\partial^2 v}{\partial x \partial y} = \frac{\partial^2 u}{\partial x \partial y} - \exp\left(\frac{1}{2}(u+v)\right)\exp\left(\frac{1}{2}(u-v)\right) = \frac{\partial^2 u}{\partial x \partial y} - \exp(u). \quad (4)$$

Differentiating (2b) with respect to x leads to

$$\frac{\partial^2 v}{\partial x \partial y} = -\frac{\partial^2 u}{\partial x \partial y} + \frac{1}{\lambda}\exp\left(\frac{1}{2}(u-v)\right)\left(\frac{\partial v}{\partial x} - \frac{\partial u}{\partial x}\right). \quad (5)$$

Inserting (2a) into (5) gives

$$\frac{\partial^2 v}{\partial x \partial y} = -\frac{\partial^2 u}{\partial x \partial y} + \exp\left(\frac{1}{2}(u-v)\right)\exp\left(\frac{1}{2}(u+v)\right) = -\frac{\partial^2 u}{\partial x \partial y} + \exp(u).$$

Adding (4) and this equation leads to (1b). Subtracting (4) from this equation leads to (1a).

Problem 27. Let

$$L := \frac{\partial^2}{\partial x^2} + u(x,t), \qquad M := 4\frac{\partial^3}{\partial x^3} + 6u(x,t)\frac{\partial}{\partial x} + 3\frac{\partial u}{\partial x}.$$

Find the evolution equation for u.

Solution 27. We have

$$L_t v = \frac{\partial}{\partial t}(Lv) - L\left(\frac{\partial v}{\partial t}\right) \Rightarrow L_t v = \frac{\partial}{\partial t}\left(\left(\frac{\partial^2}{\partial x^2} + u\right)v\right) - \left(\frac{\partial^2}{\partial x^2} + u\right)\frac{\partial v}{\partial t} = \frac{\partial u}{\partial t}v.$$

Now $[M,L]v = (ML)v - (LM)v = M(Lv) - L(Mv)$. Thus

$$[M,L]v = 6u\frac{\partial u}{\partial x}v + \frac{\partial^3 u}{\partial x^3}v = \left(6u\frac{\partial u}{\partial x} + \frac{\partial^3 u}{\partial x^3}\right)v.$$

Since v is arbitrary we obtain the *Korteweg-de Vries equation*

$$\frac{\partial u}{\partial t} = 6u\frac{\partial u}{\partial x} + \frac{\partial^3 u}{\partial x^3}.$$

The following Computer Algebra program will to the job. We calculate the right-hand side of $L_t = [M,L](t)$.

```
/* Lax.mac */
depends(u,x); depends(v,x);
L: diff(v,x,2) + u*v;
M: 4*diff(v,x,3) + 6*u*diff(v,x) + 3*diff(u,x)*v;
A: subst(v=L,M); B: subst(v=M,L);
commutator: A-B;
commutator: ev(commutator,'diff);
commutator: expand(commutator);
print("commutator=",commutator);
```

The output is

$$\left(6u\frac{\partial u}{\partial x} + \frac{\partial^3 u}{\partial x^3}\right)v.$$

Supplementary Problems

Problem 1. Consider the autonomous system of first order differential equations

$$\frac{du_k}{dt} = u_k(v_k - v_{k-1}), \qquad \frac{dv_k}{dt} = v_k(u_{k+1} + v_{k+1} - u_k - v_k), \qquad 1 \le k \le N$$

with the boundary conditions $u_{N+1} = v_0 = v_N = 0$. Show that this system admits a Lax representation.

Problem 2. Let $x_1(\tau)$, $x_2(\tau)$, $x_3(\tau)$. Consider the matrix valued differential equation

$$\frac{dL}{d\tau} = [B(t), L(t)],$$

$$L(t) = \begin{pmatrix} x_3 & \frac{1}{\sqrt{2}}(x_2 - \frac{1}{2}x_1^2) \\ \frac{1}{\sqrt{2}}(x_2 + \frac{1}{2}x_1^2) & \frac{1}{2}x_1^2 \end{pmatrix}, \qquad B(t) = \begin{pmatrix} -x_1 & -\frac{1}{\sqrt{2}}x_1 \\ \frac{1}{\sqrt{2}}x_1 & 0 \end{pmatrix}.$$

Show that

$$I_1 = \mathrm{tr}(L) = x_3 + \frac{1}{2}x_1^2, \qquad I_2 = \mathrm{tr}(L^2) = \frac{1}{2}x_2^2 + \frac{1}{2}x_3^2$$

are first integrals of the matrix valued differential equation.

Problem 3. Let $f : \mathbb{R}^3 \to \mathbb{R}$, $g : \mathbb{R}^3 \to \mathbb{R}$ be smooth functions. Consider the system of autonomous first order differential equations

$$\begin{pmatrix} du_1/dt \\ du_2/dt \\ du_3/dt \end{pmatrix} = \omega(\nabla f) \times (\nabla g).$$

Are the functions f and g are first integrals? Find the solution of the initial value problem with $f(u_1, u_2, u_3) = u_1 u_2 + u_2 u_3 + u_1 u_3$ and $g(u_1, u_2, u_3) = u_1 u_2 u_3$.

Problem 4. Show that $\partial u/\partial t = u\partial u/\partial x$ admits an infinite number of conservation laws of the form

$$\frac{\partial}{\partial t}(u^n) = \frac{\partial}{\partial x}\left(\frac{n}{n+1}u^{n+1}\right), \qquad n = 1, 2, \ldots$$

Chapter 21

Hirota Technique and Painlevé Test

The Hirota technique plays a central role in finding solutions to soliton equations. Let f, g be smooth functions. The *Hirota bilinear operators* D_x and D_t are defined as

$$D_t^n D_x^m (f \circ g) := \left(\frac{\partial}{\partial t} - \frac{\partial}{\partial t'} \right)^n \left(\frac{\partial}{\partial x} - \frac{\partial}{\partial x'} \right)^m f(t,x) g(t',x') \Big|_{x'=x, t'=t}$$

where m, $n = 0, 1, 2, \ldots$. The Hirota bilinear operator is linear. From this definition it follows that

$$D_x^m (f \circ f) = 0 \quad \text{for odd } m$$

$$D_x^m (f \circ g) = (-1)^m D_x^m g \circ f$$

$$D_x (f \circ g) = \frac{\partial f}{\partial x} g - f \frac{\partial g}{\partial x}$$

$$D_t (f \circ g) = \frac{\partial f}{\partial t} g - f \frac{\partial g}{\partial t}$$

$$\frac{\partial^2}{\partial x^2} \ln(f) = \frac{1}{2f^2} (D_x^2 f \circ f)$$

$$\frac{\partial^2}{\partial x \partial t} \ln(f) = \frac{1}{2f^2} D_x D_t f \circ f$$

$$\frac{\partial^4}{\partial x^4} \ln(f) = \frac{1}{2f^2} D_x^4 (f \circ f) - 6 \left(\frac{1}{2f^2} D_x^2 (f \circ f) \right)^2.$$

An ordinary differential equation considered in the complex domain is said to have the Painlevé property when every solution is single valued, except at the fixed singularities of the coefficients. Hence the Painlevé property requires that the movable singularities are no worse than poles.

A necessary condition that an n-th order differential equation of the form

$$\frac{d^n w}{dz^n} = H\left(z, w(z), \ldots, \frac{d^{n-1}w}{dz^{n-1}}\right)$$

where H is rational in $w, \ldots, d^{n-1}w/dz^{n-1}$ and analytic in z, has the Painlevé property is that there is a *Laurent expansion*

$$w(z) = (z - z_1)^k \sum_{j=0}^{\infty} a_j (z - z_j)^j$$

with $n - 1$ arbitrary expansion coefficients (besides the pole which is arbitrary). More than one branch may arise. These subbranches may have less than $n - 1$ arbitrary expansion coefficients. The first Painlevé transcendent is given by

$$\frac{d^2 w}{dz^2} = 6w^2 + \lambda w$$

where λ is an arbitrary parameter. The Laurent expansion is given by

$$w(z) = \frac{a_2}{(z - z_0)^2} + \frac{a_{-1}}{(z - z_0)} + a_0 + a_1(z - z_0) + a_2(z - z_0)^2 + \cdots$$

where the expansion coefficient a_4 is arbitrary as well as z_0. Furthermore $a_{-2} = 1$.

A partial differential equation has the *Painlevé property* when the solution of the partial differential equation considered in the complex domain is single-valued about the movable, singularity manifold. One requires that the solution be a single-valued functional of the data, i.e. arbitrary functions. To prove that a partial differential equation has the Painlevé property one expands a solution about a movable singular manifold

$$\phi(z_1, \ldots, z_n) = 0.$$

Let $u = u(z_1, \ldots, z_n)$ be a solution of the partial differential equation and assume that

$$u = \phi^n \sum_{j=0}^{\infty} u_j \phi^j$$

where ϕ and u_j are analytic functions of $z_1, \ldots, z_n$ in a neighbourhood of the manifold $\phi = 0$.

Problem 1.   The *Boussinesq equation* is given by

$$\frac{\partial^2 u}{\partial t^2} - \frac{\partial^2 u}{\partial x^2} - 3\frac{\partial^2}{\partial x^2}u^2 - \frac{\partial^4 u}{\partial x^4} = 0. \tag{1}$$

Express the Boussinesq equation using Hirota bilinear operators setting

$$u = 2\frac{\partial^2}{\partial x^2}\ln(f) \tag{2}$$

where $f > 0$.

Solution 1.   Obviously

$$\frac{\partial^2}{\partial x^2}(\ln(f)) = \frac{1}{2f^2}D_x^2(f \circ f)$$

and

$$\frac{\partial^4}{\partial x^4}(\ln(f)) = \frac{1}{2f^2}D_x^4 f \circ f - 6\left(\frac{1}{2f^2}D_x^2(f \circ f)\right)^2.$$

Then we obtain

$$u = \frac{1}{f^2}D_x^2 f \circ f, \qquad \frac{\partial^2 u}{\partial x^2} = \frac{1}{f^2}D_x^4(f \circ f) - 3u^2.$$

Furthermore

$$2\frac{\partial^2 \ln(f)}{\partial t^2} = \frac{1}{f^2}D_t^2(f \circ f).$$

It follows that

$$2\frac{\partial^2 \ln(f)}{\partial t^2} - u - 3u^2 - \frac{\partial^2 u}{\partial x^2} = \frac{1}{f^2}((D_t^2 - D_x^2 - D_x^4)(f \circ f)).$$

Taking the second derivative of this equation with respect to x we find

$$2\frac{\partial^4 \ln(f)}{\partial t^2 \partial x^2} - \frac{\partial^2 u}{\partial x^2} - 3\frac{\partial^2}{\partial x^2}u^2 - \frac{\partial^4 u}{\partial x^4} = \frac{\partial^2}{\partial x^2}\left(\frac{1}{f^2}(D_t^2 - D_x^2 - D_x^4)(f \circ f)\right).$$

Hence

$$\frac{\partial^2 u}{\partial t^2} - \frac{\partial^2 u}{\partial x^2} - 3\frac{\partial^2}{\partial x^2}u^2 - \frac{\partial^4 u}{\partial x^4} = \frac{\partial^2}{\partial x^2}\left(\frac{1}{f^2}(D_t^2 - D_x^2 - D_x^4)(f \circ f)\right).$$

Thus, if f is a solution of

$$(D_t^2 - D_x^2 - D_x^4)(f \circ f) = 0$$

then u as given by (2) is a solution of the Boussinesq equation. The Boussinesq equation admits a Lax representation and can be solved by the inverse scattering method. Furthermore, one finds auto-Bäcklund transformations, infinite hierarchies of conservation laws and infinite hierarchies of Lie-Bäcklund vector fields.

Problem 2. The system of partial differential equations

$$\left(\frac{\partial}{\partial t} + c_1 \frac{\partial}{\partial x}\right) u_1 = -u_1 u_2 \tag{1a}$$

$$\left(\frac{\partial}{\partial t} + c_2 \frac{\partial}{\partial x}\right) u_2 = u_1 u_2 \tag{1b}$$

describes the *interaction of two waves* u_1 and u_2 propagating with constant velocities c_1 and c_2, respectively.
(i) Find a conservation law.
(ii) Let

$$u_1 := \frac{G_1}{F}, \qquad u_2 := \frac{G_2}{F}. \tag{2}$$

Show that system (1) can be written as

$$D_1(G_1 \circ F) = -G_1 G_2 \tag{3a}$$

$$D_2(G_2 \circ F) = G_1 G_2 \tag{3b}$$

where $D_j := D_t + c_j D_x$, $j = 1, 2$.
(iii) Expand F, G_1 and G_2 in power series, i.e.

$$F = 1 + \epsilon f_1 + \epsilon^2 f_2 + \cdots$$

$$G_1 = \epsilon g_1 + \epsilon^2 g_2 + \cdots, \qquad G_2 = \epsilon h_1 + \epsilon^2 h_2 + \cdots$$

and construct a solution to system (1).

Solution 2. (i) Adding (1a) and (1b) leads to the conservation law

$$\frac{\partial}{\partial t}(u_1 + u_2) + \frac{\partial}{\partial x}(c_1 u_1 + c_2 u_2) = 0.$$

(ii) Since

$$\frac{\partial u_j}{\partial t} = \frac{1}{F^2}\left(\frac{\partial G_j}{\partial t} F - \frac{\partial F}{\partial t} G_j\right), \qquad \frac{\partial u_j}{\partial x} = \frac{1}{F^2}\left(\frac{\partial G_j}{\partial x} F - \frac{\partial F}{\partial x} G_j\right)$$

where $j = 1, 2$, we find (3a) and (3b).

(iii) Substituting the power series for F, G_1, G_2 into system (3) and collecting terms with the same power in ϵ, we find

$$
\begin{aligned}
D_1(g_1 \circ 1) &= 0 \\
D_2(h_1 \circ 1) &= 0 \\
D_1((g_1 \circ f_1) + (g_2 \circ 1)) &= -g_1 h_1 \\
D_2((h_1 \circ f_1) + (h_2 \circ 1)) &= g_1 h_1
\end{aligned}
\tag{4}
$$

and so on. From the first and second of these equations we have

$$
g_1(x, t) = g_1(x - c_1 t)
\tag{5a}
$$

$$
h_1(x, t) = h_1(x - c_2 t)
\tag{5b}
$$

where g_1 and h_1 are arbitrary functions. Inserting (5a) and (5b) into (4c) and (4d) we find that all higher terms can be chosen to be zero if f_1 satisfies the differential equations

$$
\left(\frac{\partial}{\partial t} + c_1 \frac{\partial}{\partial x} \right) f_1 = h_1, \qquad \left(\frac{\partial}{\partial t} + c_2 \frac{\partial}{\partial x} \right) f_1 = -g_1.
$$

The general solution to this system of partial differential equations is given by

$$
f_1(x, t) = F_1(x - c_1 t) + F_2(x - c_2 t)
$$

where F_1 and F_2 are related to g_1 and h_1 by

$$
\left(\frac{\partial}{\partial t} + c_1 \frac{\partial}{\partial x} \right) F_2(x - c_2 t) = h_1(x - c_2 t)
$$

$$
\left(\frac{\partial}{\partial t} + c_2 \frac{\partial}{\partial x} \right) F_1(x - c_1 t) = -g_1(x - c_1 t).
$$

Therefore we have an exact solution of system (1)

$$
u_1(x, t) = \frac{-\left(\dfrac{\partial}{\partial t} + c_2 \dfrac{\partial}{\partial x} \right) F_1(x - c_1 t)}{F_1(x - c_1 t) + F_2(x - c_2 t)}
$$

$$
u_2(x, t) = \frac{\left(\dfrac{\partial}{\partial t} + c_1 \dfrac{\partial}{\partial x} \right) F_2(x - c_2 t)}{F_1(x - c_1 t) + F_2(x - c_2 t)}.
$$

Problem 3.   Show that Hirota's operators $D_x^n(f \circ g)$ and $D_x^m(f \circ g)$ given in the introduction can be written as

$$
D_x^n(f \circ g) = \sum_{j=0}^{n} \frac{(-1)^{(n-j)} n!}{j!(n-j)!} \frac{\partial^j f}{\partial x^j} \frac{\partial^{n-j} g}{\partial x^{n-j}},
\tag{1}
$$

$$D_x^m D_t^n (f \circ g) = \sum_{j=0}^{m} \sum_{i=0}^{n} \frac{(-1)^{(m+n-j-i)} m!}{j!(m-j)!} \frac{n!}{i!(n-i)!} \frac{\partial^{i+j} f}{\partial t^i \partial x^j} \frac{\partial^{n+m-i-j} g}{\partial t^{n-i} \partial x^{m-j}}.$$

$$(2)$$

Solution 3. We prove (1) by mathematical induction. The formula (2) can be proven in a similar way. We first try to show that

$$\left(\frac{\partial}{\partial x} - \frac{\partial}{\partial x'} \right)^n (f(x) \circ g(x')) = \sum_{j=0}^{n} \frac{(-1)^{(n-j)} n!}{j!(n-j)!} \frac{\partial^j f(x)}{\partial x^j} \frac{\partial^{n-j} g(x')}{\partial x'^{n-j}}. \quad (3)$$

For $n = 1$, we have

$$\left(\frac{\partial}{\partial x} - \frac{\partial}{\partial x'} \right) (f(x) \circ g(x')) = \frac{\partial f(x)}{\partial x} \circ g(x') - f(x) \circ \frac{\partial g(x')}{\partial x'}$$

$$= \sum_{j=0}^{1} \frac{(-1)^{(1-j)} 1!}{j!(1-j)!} \frac{\partial^j f(x)}{\partial x^j} \frac{\partial^{1-j} g(x')}{\partial x'^{1-j}}.$$

Expression (3) obviously holds. If we assume (3) to be true for $n - 1$, then

$$\left(\frac{\partial}{\partial x} - \frac{\partial}{\partial x'} \right)^{n-1} (f(x) \circ g(x')) = \sum_{j=0}^{n-1} \frac{(-1)^{(n-1-j)} (n-1)!}{j!(n-1-j)!} \frac{\partial^j f(x)}{\partial x^j} \frac{\partial^{n-1-j} g(x')}{\partial x'^{n-1-j}}.$$

$$(4)$$

Using (4), we have

$$\left(\frac{\partial}{\partial x} - \frac{\partial}{\partial x'} \right)^n (f(x) \circ g(x')) = \left(\frac{\partial}{\partial x} - \frac{\partial}{\partial x'} \right) \left(\frac{\partial}{\partial x} - \frac{\partial}{\partial x'} \right)^{n-1} (f(x) \circ g(x'))$$

$$= \sum_{j=0}^{n-1} \frac{(-1)^{(n-1-j)} (n-1)!}{j!(n-1-j)!} \frac{\partial^{j+1} f(x)}{\partial x^{j+1}} \frac{\partial^{n-1-j} g(x')}{\partial x'^{n-1-j}}$$

$$- \sum_{j=0}^{n-1} \frac{(-1)^{(n-1-j)} (n-1)!}{j!(n-1-j)!} \frac{\partial^j f(x)}{\partial x^j} \frac{\partial^{n-j} g(x')}{\partial x'^{n-j}}$$

$$= \sum_{j=1}^{n-1} \frac{(-1)^{(n-j)} n!}{j!(n-j)!} \frac{\partial^j f(x)}{\partial x^j} \frac{\partial^{n-j} g(x')}{\partial x'^{m-j}}$$

$$+ \frac{\partial^n f(x)}{\partial x^n} + (-1)^n f(x) \frac{\partial^n g(x')}{\partial x'^n}$$

$$= \sum_{j=0}^{n} \frac{(-1)^{(n-j)} n!}{j!(n-j)!} \frac{\partial^j f(x)}{\partial x^j} \frac{\partial^{n-j} g(x')}{\partial x'^{n-j}}.$$

Thus, (4) is true for all $n = 1, 2, \ldots$. Setting $x' = x$ we have

$$D_x^n (f \circ g) = \sum_{j=0}^{n} \frac{(-1)^{(n-j)} n!}{j!(n-j)!} \frac{\partial^j f(x)}{\partial x^j} \frac{\partial^{n-j} g(x)}{\partial x^{n-j}}.$$

Problem 4.   (i) Study the *singularity structure* of the nonlinear ordinary differential equation

$$\frac{d^2u}{dt^2} + \frac{1}{2}u^3 = 0$$

where $u(t)$ is a real-valued function.
(ii) Show that the second order ordinary differential equation $d^2u/dt^2 = au + bu^3$ consider in the complex domain

$$\frac{d^2w}{dz^2} = aw + bw^3$$

passes the necessary condition of the Painlevé test. Here a, b are nonzero real constants.

Solution 4.   (i) We consider the differential equations in the complex domain

$$\frac{d^2w}{dz^2} + \frac{1}{2}w^3 = 0.$$

Inserting the ansatz (locally represented as a *Laurent expansion*)

$$w(z) = \sum_{j=0}^{\infty} a_j(z - z_0)^{j-1}$$

we find the recursion relation for the expansion coefficients a_j

$$a_j(j+1)(j-4) = -\frac{1}{2}\sum_k\sum_l a_{j-k-l}a_k a_l, \quad 0 < k+l \leq j, \quad 0 \leq k, l < j$$

where $a_0 = 2i$, $a_1 = a_2 = a_3 = 0$, $a_4 =$ arbitrary. Owing to the arbitrary pole position z_0 and coefficient a_4 we have a local representation of the general solution to this second-order differential equation. The nonlinear differential equation can be solved exactly in terms of the Jacobi elliptic functions and the movable singularities in the complex z-plane form a regular lattice of first-order poles.
(ii) We have the Laurent expansion

$$w(z) = \sum_{j=0}^{\infty} c_j(z - z_1)^{j-1}.$$

The leading terms are d^2w/dz^2 and bw^3. Only one branch appears and we have

$$c_0 = (-2/b)^{1/2}, \quad c_1 = 0, \quad c_2 = -a/(3bc_0)$$

For $m \geq 3$ we find a recursion relation for the expansion coefficients. The expansion coefficient c_4 can chosen arbitrarily since $0c\dot{c}_4 = 0$. Hence the necessary condition is satisfied.

Problem 5. Study the singularity structure of the nonlinear ordinary differential equation

$$\frac{d^2u}{dt^2} + \lambda\frac{du}{dt} + \frac{1}{2}u^3 = \epsilon f(t)$$

where $u(t)$ is a real-valued function and $f(t)$ is an analytic function of t.

Solution 5. We consider the differential equation in the complex plane, i.e.

$$\frac{d^2w}{dz^2} + \lambda\frac{dw}{dz} + \frac{1}{2}w^3 = \epsilon f(z).$$

The introduction of either the damping term or driving term leads to a breakdown of the Laurent series

$$w(z) = \sum_{j=0}^{\infty} a_j(z - z_0)^{j-1}$$

discussed in the previous problem since it is not possible to introduce an arbitrary coefficient at $j = 4$. We have to add logarithmic terms

$$w(z) = \sum_{j=0}^{\infty}\sum_{k=0}^{\infty} a_{jk}(z - z_0)^{j-1}((z - z_0)^4 \ln(z - z_0))^k$$

Computation of the recursion relation for the a_{jk} yields

$$a_{jk}((j-1)(j-2) + 4k(2j+4k-3)) + a_{j-4,k+1}(k+1)(2j+8k-3)$$
$$+a_{j-8,k+2}(k+1)(k+2) + \lambda a_{j-1,k}(j+4k-2) + \lambda a_{j-5,k+1}$$
$$= -\frac{1}{2}\sum_{p,q,r,s} a_{j-r,k-s}a_{r-p,s-q}a_{pq} + \epsilon f_{j-3}\delta_{k0}$$

where the summation is for $0 \le p \le r \le j$ and $0 \le q \le s \le k$ and

$$f_j = \frac{1}{j!}\frac{\partial^j f}{\partial z^j}.$$

The values of the first few coefficients are

$$a_{00} = 2i, \quad a_{10} = -\frac{\lambda}{3}, \quad a_{20} = -\frac{i\lambda^2}{18}, \quad a_{30} = -\frac{i\lambda^3}{27} - \frac{\epsilon f_0}{4}.$$

The coefficient a_{40} is arbitrary and therefore a_{01} is given by

$$a_{01} = \frac{4}{135}i\lambda^4 + \frac{1}{5}\epsilon(\lambda f_0 + f_1).$$

The sets of coefficients a_{0k}, $k = 0, 1, 2, \ldots$, satisfy

$$4k(k-1)a_{0k} + ka_{0k} + \frac{1}{2}a_{0k} = -\frac{1}{8}\sum_{s}\sum_{q}a_{0,k-s}a_{0,s-q}a_{0q}$$

where the summation is for $0 \leq q \leq s \leq k$. Introducing the generating function

$$\Theta(s) = \sum_{k=0}^{\infty}a_{0k}s^{k}$$

where s is some independent variable, the nonlinear differential equation for Θ is obtained

$$16s^{2}\frac{d^{2}\Theta}{ds^{2}} + 4s\frac{d\Theta}{ds} + 2\Theta + \frac{1}{2}\Theta^{3} = 0.$$

The differential equation can also be obtained by substituting

$$w(z) = \frac{1}{z - z_{0}}\Theta_{0}(s)$$

where $s = (z - z_{0})^{4}\ln(z - z_{0})$ into this second order differential equation. In the limit $z \to z_{0}$ we find that Θ_{0} satisfies this differential equation provided that there is an ordering in which $|z - z_{0}| \ll |s|$. This second order differential equation can be solved in terms of elliptic functions by making the substitution $\Theta_{0}(s) = s^{1/4}g(s^{1/4})$ which leads to the second order differential equation

$$\frac{d^{2}g(y)}{dy^{2}} + \frac{1}{2}g^{3}(y) = 0$$

where $y = s^{1/4} = z(\ln(z))^{1/4}$.

Problem 6.   (i) *Burgers' equation* is given by

$$\frac{\partial u}{\partial t} + u\frac{\partial u}{\partial x} = D\frac{\partial^{2}u}{\partial x^{2}}. \tag{1}$$

Show that Burgers' equation has the Painlevé property.
(ii) Show that one can set $u_{j} = 0$ for $j \geq 2$. Find the partial differential equation for u_{1} in this case.

Solution 6.   (i) Inserting the expansion

$$u = \phi^{n}\sum_{j=0}^{\infty}u_{j}\phi^{j}$$

into Burgers equation yields $n = -1$ and

$$(j - 2)(j + 1) \left(\frac{\partial \phi}{\partial x} \right)^2 = F_j(u_{j-1}, \dots, u_0, \partial \phi / \partial t, \partial \phi / \partial x).$$

In order that the expansion is valid F_2 has to vanish identically. We find

$$j = 0, \qquad u_0 = -2D \frac{\partial \phi}{\partial x}$$

$$j = 1, \qquad \frac{\partial \phi}{\partial t} + u_1 \frac{\partial \phi}{\partial x} = D \frac{\partial^2 \phi}{\partial x^2}$$

$$j = 2, \qquad \frac{\partial}{\partial x} \left(\frac{\partial \phi}{\partial t} + u_1 \frac{\partial \phi}{\partial x} - D \frac{\partial^2 \phi}{\partial x^2} \right) = 0.$$

The relation at $j = 2$ (so-called *compatibility condition*) is satisfied identically owing to the relation for $j = 1$. One calls $j = 2$ a resonance. Therefore the expansion is valid for arbitrary functions ϕ and u_2.

(ii) If we set the arbitrary function u_2 equal to zero

$$u_2(x, t) = 0$$

and require that the function u_1 satisfies Burgers' equation, then $u_j = 0$ for $j \geq 2$. We obtain

$$u = -2D \frac{1}{\phi} \frac{\partial \phi}{\partial x} + u_1$$

where u and u_1 satisfy Burgers' equation and

$$\frac{\partial \phi}{\partial t} + u_1 \frac{\partial \phi}{\partial x} = D \frac{\partial^2 \phi}{\partial x^2}.$$

Since $u_1 = 0$ is a solution of Burgers' equation, we find that this partial differential equation takes the form (linear diffusion equation)

$$\frac{\partial \phi}{\partial t} = D \frac{\partial^2 \phi}{\partial x^2}.$$

With $u_1 = 0$ we obtain the transformation

$$u = -2D \frac{1}{\phi} \frac{\partial \phi}{\partial x}.$$

This is the *Cole-Hopf transformation*.

Problem 7. Perform a Painlevé test for the *inviscid Burgers' equation*

$$\frac{\partial u}{\partial t} = u \frac{\partial u}{\partial x}. \tag{1}$$

Solution 7.   We find an expansion of the form

$$u = u_0 + u_1\phi^n + u_2\phi^{2n} + \cdots$$

where $n = 1/2$. For the first two expansion coefficients, i.e. u_0 and u_1 we find

$$\phi^{-1/2} : \; u_0\frac{\partial\phi}{\partial x} = \frac{\partial\phi}{\partial t}, \qquad \phi^0 : \; \frac{\partial u_0}{\partial t} + u_0\frac{\partial u_0}{\partial x} + \frac{1}{2}u_1^2\frac{\partial\phi}{\partial x} = 0.$$

If we require that u_0 satisfies the inviscid Burgers' equation, then we find from the infinite coupled system for the expansion coefficients that $u_1 = u_2 = \cdots = 0$. Then inserting the partial differential equation for $\phi^{-1/2}$ into the partial differential equation for ϕ^0 we obtain the partial differential equation

$$\left(\frac{\partial\phi}{\partial x}\right)^2\frac{\partial^2\phi}{\partial t^2} + \left(\frac{\partial\phi}{\partial t}\right)^2\frac{\partial^2\phi}{\partial x^2} - 2\frac{\partial\phi}{\partial x}\frac{\partial\phi}{\partial t}\frac{\partial^2\phi}{\partial x\partial t} = 0.$$

This equation can be linearized by applying a *Legendre transformation*.

Problem 8.   The $SO(2,1)$ invariant *nonlinear σ-model* can be written as

$$(w + \bar{w})\frac{\partial^2 w}{\partial x\partial t} - 2\frac{\partial w}{\partial x}\frac{\partial w}{\partial t} = 0 \tag{1}$$

where $\bar{w}$ denotes the complex conjugate of w. Show that the partial differential equation (1) has the Painlevé property. Introduce the real fields u and v, i.e. $w = u + iv$.

Solution 8.   Introducing the real fields u and v with $w + \bar{w} = 2u$, we find that (1) can be written as a coupled system of partial differential equations

$$u\frac{\partial^2 u}{\partial x\partial t} - \frac{\partial u}{\partial x}\frac{\partial u}{\partial t} + \frac{\partial v}{\partial x}\frac{\partial v}{\partial t} = 0 \tag{2a}$$

$$u\frac{\partial^2 v}{\partial x\partial t} - \frac{\partial u}{\partial t}\frac{\partial v}{\partial x} - \frac{\partial u}{\partial x}\frac{\partial v}{\partial t} = 0. \tag{2b}$$

Inserting the ansatz

$$u = \phi^m\sum_{j=0}^{\infty}u_j\phi^j, \qquad v = \phi^n\sum_{j=0}^{\infty}v_j\phi^j \tag{3}$$

into system (2) yields, for the first terms in the expansion $n = m = -1$ and

$$u_0^2 + v_0^2 = 0, \qquad 0 \cdot u_0 v_0\frac{\partial\phi}{\partial x}\frac{\partial\phi}{\partial t} = 0.$$

Thus u_0 or v_0 is arbitrary. Inserting the expansions (3) into (2) with the values given above we find that at $j = 1$ u_1 or v_1 can be chosen arbitrarily. For $j = 2$ we find that u_2 or v_2 can be chosen arbitrarily. Thus (1) has the Painlevé property.

Problem 9. (i) Show that the *Kadomtsev-Petviashvili equation*

$$\frac{\partial^2 u}{\partial y^2} + \frac{\partial}{\partial x}\left(\frac{\partial u}{\partial t} + u\frac{\partial u}{\partial x} + \frac{\partial^3 u}{\partial x^3}\right) = 0$$

possesses the Painlevé property.
(ii) Find an auto-Bäcklund transformation.

Solution 9. (i) The expansion about the singular manifold $\phi = 0$ is

$$u = \phi^{-2}\sum_{j=0}^{\infty} u_j\phi^j$$

with resonances at $j = -1, 4, 5, 6$. Therefore, owing to the noncharacteristic condition

$$\frac{\partial\phi}{\partial x} \neq 0$$

when $\phi = 0$ we obtain that ϕ, u_4, u_5, u_6 are arbitrary functions of x, y, t in the expansion.
(ii) The Bäcklund transformation is

$$u = u_0\phi^{-2} + u_1\phi^{-1} + u_2$$

where

$$u_0 = -12\left(\frac{\partial\phi}{\partial x}\right)^2, \qquad u_1 = 12\frac{\partial^2\phi}{\partial x^2},$$

$$u_2 + \frac{\partial\phi/\partial t}{\partial\phi/\partial x} + 4\frac{\partial^3\phi/\partial x^3}{\partial\phi/\partial x} - 3\left(\frac{\partial^2\phi/\partial x^2}{\partial\phi/\partial x}\right)^2 + \left(\frac{\partial\phi/\partial y}{\partial\phi/\partial x}\right)^2 = 0$$

and

$$\frac{\partial^2\phi}{\partial x\partial t} + \frac{\partial^4\phi}{\partial x^4} + \frac{\partial^2\phi}{\partial y^2} + \frac{\partial^2\phi}{\partial x^2}u_2 = 0.$$

Thus the *auto Bäcklund transformation* is

$$u = 12\frac{\partial^2}{\partial x^2}\ln(\phi) + u_2$$

i.e. u_2 satisfies the Kadomtsev-Petviashvili equation.

Supplementary Problems

Problem 1.   Consider the system of coupled partial differential equations (Cheng's equation)

$$\frac{\partial u}{\partial x} = -auv, \qquad \frac{\partial v}{\partial t} = b\frac{\partial u}{\partial x}.$$

The system describes the dynamics of photosensitive molecules when a light beam passes through them. Show that making the ansatz

$$u(x,t) = \frac{g(x,t)}{f(x,t)}, \qquad v(x,t) = \frac{1}{a}\frac{\partial}{\partial x}(\ln(f(x,t)))$$

and introducing the bilinear operators one can reduce the system to a linear wave equation.

Problem 2.   Let $f(x,t) = f(kx + \omega t + \phi)$. Find

$$\frac{\partial^2}{\partial x^2}\ln(1 + e^{f(x,t)}), \qquad \frac{\partial}{\partial t}\ln(1 + e^{f(x,t)}).$$

Construct a partial differential equation for f. Note that

$$\frac{\partial}{\partial x}\ln(1 + e^{f(x,t)}) = \frac{1}{1 + e^f}e^f\frac{\partial f}{\partial x}.$$

Problem 3.   Show that the second Painlevé equation

$$\frac{d^2q}{dz^2} = 2q^3 + zq + \alpha$$

can be written as Hamilton system

$$\frac{dq}{dz} = \frac{\partial H}{\partial p}, \qquad \frac{dp}{dz} = -\frac{\partial H}{\partial q}$$

with

$$H(q,p,z;\alpha) = \frac{1}{2}p^2(q^2 + z/2)p - (\alpha + 1/2)q.$$

Problem 4.   Consider the partial differential equation

$$(w + \overline{w})\frac{\partial^2 w}{\partial x \partial t} - 2\frac{\partial w}{\partial x}\frac{\partial w}{\partial t} = 0$$

where $\overline{w}$ denotes the complex conjugate. Set $w(x,t) = u(x,t) + iv(x,t)$, where u and v are real fields. Show that

$$u\frac{\partial^2 u}{\partial x \partial t} - \frac{\partial u}{\partial x}\frac{\partial u}{\partial t} + \frac{\partial v}{\partial x}\frac{\partial v}{\partial t} = 0$$

$$u\frac{\partial^2 v}{\partial x \partial t} - \frac{\partial u}{\partial t}\frac{\partial v}{\partial x} - \frac{\partial u}{\partial x}\frac{\partial v}{\partial t} = 0.$$

Show that this system of partial differential equations (considered in the complex domain) satisfies the Painlevé test with resonances -1, 0, 1, 2. Construct a Bäcklund transformation with the ansatz

$$u = u_0 \phi^{-1} + u_1, \qquad v = v_0 \phi^{-1} + v_1.$$

Construct a solution starting from the vacuum solution $u_1 = 0$, $v_1 = 0$.

Chapter 22

Lie Groups and Lie Algebras

A *Lie algebra* is defined as follows: A vector space L over a field F, with an operation $L \times L \to L$ denoted by $(x, y) \to [x, y]$ and called the commutator of x and y, is called a Lie algebra over F if the following axioms are satisfied:

(L1)   The bracket operation is bilinear.

(L2) $[x, x] = 0$ for all $x \in L$

(L3) $[x, [y, z]] + [y, [z, x]] + [z, [x, y]] = 0$ $(x, y, z \in L)$.

Axiom (L3) is called the *Jacobi identity*.

Bilinearity means that

$$[\alpha x + \beta y, z] = \alpha[x, z] + \beta[y, z], \quad [x, \alpha y + \beta z] = \alpha[x, y] + \beta[x, z]$$

where $\alpha, \beta \in F$. A Lie algebra is called real when $F = \mathbb{R}$. A Lie algebra is called complex when $F = \mathbb{C}$. If $F = \mathbb{R}$ or $F = \mathbb{C}$, then $[x, x] = 0$ and $[x, y] = -[y, x]$ are equivalent.

A Lie algebra is called abelian or commutative if $[x, y] = 0$ for all $x, y \in L$. Let L_1 and L_2 be Lie algebras and $\Phi : L_1 \to L_2$ a bijection such that for all $\alpha, \beta \in F$ and all $x, y \in L_1$

$$\Phi(\alpha x + \beta y) = \alpha\Phi(x) + \beta\Phi(y), \quad \Phi([x, y]) = [\Phi(x), \Phi(y)]$$

then Φ is called an isomorphism and the Lie algebras L_1 and L_2 are isomorphic.

Let v, w be two elements of a Lie algebra L. One defines

$$\mathrm{ad}v(w) := [v, w]. \tag{1}$$

This then provides the adjoint representation of the Lie algebra.

A *Lie group* is a group G which is also an analytic manifold such that the mapping

$$(a, b) \to ab^{-1}, \quad a, b \in G$$

of the product manifold $G \times G$ into G is analytic. Thus a Lie group is set endowed with compatible structures of a group and an analytic manifold. A subgroup H of a Lie group G is said to be a Lie subgroup if it is a submanifold of the underlying manifold of G. A group G is called a Lie transformation group of a differentiable manifold M if there is a differentiable map

$$\varphi : G \times M \to M, \quad \varphi(g, \mathbf{x}) = g\mathbf{x}$$

($\mathbf{x} \in M$) such that (i)

$$(g_1 \cdot g_2)\mathbf{x} = g_1 \cdot (g_2\mathbf{x})$$

for $\mathbf{x} \in M$ and (ii)

$$e\mathbf{x} = \mathbf{x}$$

for the identity element e of G and $\mathbf{x} \in M$.

The Lie algebra $so(n, \mathbb{R})$ consists of all $n \times n$ skew-symmetric matrices over $\mathbb{R}$ with trace equal to 0. The Lie algebra $su(n)$ consists of all $n \times n$ skew-hermitian matrices with trace equal to 0. The Lie algebra $s\ell(n, \mathbb{R})$ consists of all $n \times n$ matrices over $\mathbb{R}$ with trace equal to 0.

The compact Lie group $SO(n, \mathbb{R})$ consists of all orthogonal $n \times n$ matrices with determinant equal to 1. If $X \in so(n, \mathbb{R})$, then $\exp(X) \in SO(n, \mathbb{R})$. The compact Lie group $SU(n)$ consists of all $n \times n$ unitary matrices with determinant equal to 1. The Lie group $U(n)$ consists of all $n \times n$ unitary matrices.

If $X \in so(n, \mathbb{R})$, then $\exp(X) \in SO(n, \mathbb{R})$. If $Y \in su(n)$, then $\exp(Y) \in SU(n)$.

Note that for any $n \times n$ matrix over $\mathbb{C}$ one has

$$\exp(\mathrm{tr}(A)) \equiv \det(\exp(A)).$$

Hence if $\mathrm{tr}(A) = 0$ we have $\det(\exp(A)) = 1$. If H is a hermitian matrix, then $\exp(iH)$ is a unitary matrix.

Problem 1.   (i) Classify all real two-dimensional Lie algebras, where $F = \mathbb{R}$. Denote the basis elements by X and Y.
(ii) Find a representation of the non-abelian two-dimensional Lie algebra using differential operators.
(iii) Find a representation of the non-abelian two-dimensional Lie algebra using 2×2 matrices.

Solution 1.   (i) There exists only two non-isomorphic two-dimensional Lie algebras. Let X and Y span a two-dimensional Lie algebra. Then their commutator is either 0 or $[X, Y] = Z \neq 0$. In the first case we have a two-dimensional abelian Lie algebra. In the second case we have

$$[X, Y] = Z = \alpha X + \beta Y, \qquad \alpha, \beta \in F$$

where α and β are not both zero. Let us suppose $\alpha \neq 0$, then one has

$$[Z, \alpha^{-1}Y] = Z.$$

Hence, for all non-abelian two-dimensional Lie algebras one can choose a basis $\{ Z, U \}$ such that $[Z, U] = Z$. Consequently, all non-abelian two-dimensional Lie algebras are isomorphic.
(ii) An example is the set of differential operators

$$\left\{ \frac{d}{dx}, \, x\frac{d}{dx} \right\}$$

where $Z \to d/dx$ and $U \to xd/dx$.
(iii) Since $\text{tr}([Z, U]) = 0$ we have $\text{tr}(Z) = 0$. A representation using 2×2 matrices is

$$Z = \begin{pmatrix} 0 & 1 \\ 0 & 0 \end{pmatrix}, \qquad U = \begin{pmatrix} 0 & 0 \\ 0 & 1 \end{pmatrix}.$$

Problem 2.   Let L be the real vector space $\mathbb{R}^3$. Let $\mathbf{a}, \mathbf{b} \in L$. Define

$$\mathbf{a} \times \mathbf{b} := \begin{pmatrix} a_2 b_3 - a_3 b_2 \\ a_3 b_1 - a_1 b_3 \\ a_1 b_2 - a_2 b_1 \end{pmatrix}. \tag{1}$$

Equation (1) is the *cross* or *vector product*.
(i) Show that L is a Lie algebra.
(ii) Show that in general $\mathbf{a} \times (\mathbf{b} \times \mathbf{c}) \neq (\mathbf{a} \times \mathbf{b}) \times \mathbf{c}$.

Solution 2.   (i) Obviously, we have

$$(\mathbf{a} + \mathbf{b}) \times (\mathbf{c} + \mathbf{d}) = \mathbf{a} \times \mathbf{c} + \mathbf{a} \times \mathbf{d} + \mathbf{b} \times \mathbf{c} + \mathbf{b} \times \mathbf{d}.$$

From (1) we see that $\mathbf{a} \times \mathbf{b} = -\mathbf{b} \times \mathbf{a}$. Using definition (1) we obtain, after a lengthy calculation, that

$$\mathbf{a} \times (\mathbf{b} \times \mathbf{c}) + \mathbf{c} \times (\mathbf{a} \times \mathbf{b}) + \mathbf{b} \times (\mathbf{c} \times \mathbf{a}) = \mathbf{0}.$$

This equation is called the *Jacobi identity*.
(ii) Let

$$\mathbf{a} := \begin{pmatrix} 1 \\ 0 \\ 0 \end{pmatrix}, \qquad \mathbf{b} := \begin{pmatrix} 0 \\ 1 \\ 0 \end{pmatrix}, \qquad \mathbf{c} := \begin{pmatrix} 0 \\ 0 \\ 1 \end{pmatrix}$$

be the standard basis in $\mathbb{R}^3$. Then $\mathbf{a} \times (\mathbf{b} \times \mathbf{c}) \neq (\mathbf{a} \times \mathbf{b}) \times \mathbf{c}$.

Problem 3. Let v, w be two elements of a Lie algebra L. Then

$$\mathrm{ad}v(w) := [v, w]. \tag{1}$$

Let

$$x := \begin{pmatrix} 0 & 1 \\ 0 & 0 \end{pmatrix}, \qquad h := \begin{pmatrix} 1 & 0 \\ 0 & -1 \end{pmatrix}, \qquad y := \begin{pmatrix} 0 & 0 \\ 1 & 0 \end{pmatrix} \tag{2}$$

be an ordered basis for the semi-simple Lie algebra $s\ell(2, \mathbb{R})$.
(i) Compute the matrices $\mathrm{ad}x$, $\mathrm{ad}h$ and $\mathrm{ad}y$ relative to this basis. This is the so-called *adjoint representation*.
(ii) Calculate the eigenvalues and eigenvectors of $\mathrm{ad}x$, $\mathrm{ad}h$ and $\mathrm{ad}y$.
(iii) Do the matrices $\mathrm{ad}(x)$, $\mathrm{ad}(h)$, $\mathrm{ad}(y)$ form basis of a Lie algebra under the commutator? If so, is this Lie algebra isomorphic to $s\ell(2, \mathbb{R})$?

Solution 3. Since $[x, h] = -2x$, $[x, y] = h$, $[y, h] = 2y$ we find

$$\mathrm{ad}x(x) = 0, \qquad \mathrm{ad}x(h) = -2x, \qquad \mathrm{ad}x(y) = h$$

$$\mathrm{ad}h(x) = 2x, \qquad \mathrm{ad}h(h) = 0, \qquad \mathrm{ad}h(y) = -2y$$

$$\mathrm{ad}y(x) = -h \qquad \mathrm{ad}y(h) = 2y, \qquad \mathrm{ad}y(y) = 0.$$

From $(x, h, y)\,\mathrm{ad}(x) = (0, -2x, h)$ we find

$$\mathrm{ad}(x) = \begin{pmatrix} 0 & -2 & 0 \\ 0 & 0 & 1 \\ 0 & 0 & 0 \end{pmatrix}.$$

Analogously,

$$\mathrm{ad}h = \begin{pmatrix} 2 & 0 & 0 \\ 0 & 0 & 0 \\ 0 & 0 & -2 \end{pmatrix}, \qquad \mathrm{ad}y = \begin{pmatrix} 0 & 0 & 0 \\ -1 & 0 & 0 \\ 0 & 2 & 0 \end{pmatrix}.$$

(iii) Since

$$[\mathrm{ad}x, \mathrm{ad}h] = -2\mathrm{ad}x, \quad [\mathrm{ad}x, \mathrm{ad}y] = \mathrm{ad}h, \quad [\mathrm{ad}y, \mathrm{ad}h] = 2\mathrm{ad}y$$

we obtain that the set $\{\, \mathrm{ad}x, \ \mathrm{ad}h, \ \mathrm{ad}y \,\}$ is closed under the commutator. Thus this set forms a basis of a Lie algebra L. The vector space isomorphism $\phi : s\ell(2, \mathbb{R}) \to L$ defined on the basis elements by

$$\phi(x) \to \mathrm{ad}x \qquad \phi(h) \to \mathrm{ad}h \qquad \phi(y) \to \mathrm{ad}y$$

is a Lie algebra isomorphism. Thus the Lie algebra $s\ell(2, \mathbb{R})$ is isomorphic to L.

Problem 4.   (i) Show that the vector fields

$$\left\{ \frac{d}{dx}, \ x\frac{d}{dx}, \ x^2\frac{d}{dx} \right\} \tag{1}$$

form a Lie algebra L under the *commutator*. Recall that the commutator is given by

$$\left[f_1(x)\frac{d}{dx}, \ f_2(x)\frac{d}{dx} \right] g(x) := \left(f_1\frac{df_2}{dx} - f_2\frac{df_1}{dx} \right) \frac{dg}{dx} \tag{2}$$

where g is an arbitrary smooth function. Thus (2) can also be written as

$$\left[f_1(x)\frac{d}{dx}, \ f_2(x)\frac{d}{dx} \right] := \left(f_1\frac{df_2}{dx} - f_2\frac{df_1}{dx} \right) \frac{d}{dx}. \tag{3}$$

(ii) Determine the *centre* Z of the Lie algebra L. The centre of the Lie algebra L is defined as

$$Z(L) := \{ z \in L \ : \ [z, x] = 0 \quad \text{for all } x \in L \}.$$

(iii) Find the adjoint representation

$$\mathrm{ad}\left(\frac{d}{dx} \right), \qquad \mathrm{ad}\left(x\frac{d}{dx} \right), \qquad \mathrm{ad}\left(x^2\frac{d}{dx} \right).$$

(iv) Do the vector fields

$$\left\{ \frac{d}{dx}, \ x\frac{d}{dx}, \ x^2\frac{d}{dx}, \ x^3\frac{d}{dx} \right\}$$

form a basis of a Lie algebra? If not can the set be extended to form a Lie algebra.

Solution 4. (i) Straightforward calculation yields

$$\left[\frac{d}{dx}, x\frac{d}{dx}\right] = \frac{d}{dx}, \quad \left[\frac{d}{dx}, x^2\frac{d}{dx}\right] = 2x\frac{d}{dx}, \quad \left[x\frac{d}{dx}, x^2\frac{d}{dx}\right] = x^2\frac{d}{dx}.$$

Since the Jacobi identity is satisfied for vector fields, we find that the vector fields given by (1) form a basis of a Lie algebra.

(ii) Now an arbitrary element of L can be written as

$$f(x)\frac{d}{dx}$$

where $f(x) = a + bx + cx^2$ with $a, b, c \in \mathbb{R}$. Then we find from the condition

$$\left[f(x)\frac{d}{dx}, \frac{d}{dx}\right] = 0$$

that $df/dx = 0$. From the condition

$$\left[f(x)\frac{d}{dx}, x\frac{d}{dx}\right] = 0$$

together with condition $df/dx = 0$, we obtain $f(x) = 0$. Thus the centre Z of the Lie algebra L is given by $Z = \{0\}$.

(iii) Let $A := d/dx$, $B := xd/dx$, $C := x^2d/dx$. We introduce the order A, B, C. Then we have

$$\mathrm{ad}A(A) = 0, \qquad \mathrm{ad}A(B) = A, \qquad \mathrm{ad}A(C) = 2B.$$

Thus we find

$$\mathrm{ad}A = \begin{pmatrix} 0 & 1 & 0 \\ 0 & 0 & 2 \\ 0 & 0 & 0 \end{pmatrix}.$$

Analogously

$$\mathrm{ad}B = \begin{pmatrix} -1 & 0 & 0 \\ 0 & 0 & 0 \\ 0 & 0 & 1 \end{pmatrix}, \quad \mathrm{ad}C = \begin{pmatrix} 0 & 0 & 0 \\ -2 & 0 & 0 \\ 0 & -1 & 0 \end{pmatrix}.$$

(iv) Let $A_k := x^kd/dx$, where $k = 0, 1, \ldots$. Since

$$[A_2, A_3] = x^4\frac{d}{dx} \equiv A_4$$

we find that the set $\{A_0, A_1, A_2, A_3\}$ does not form a basis of a Lie algebra. Since $[A_j, A_k] = (k - j)A_{j+k-1}$ we find that the vector fields given by the set $S := \{A_k : k = 0, 1, 2, \ldots\}$ form an infinite-dimensional Lie algebra under the commutator.

Problem 5. Let L be any Lie algebra. If $x, y \in L$, define

$$\kappa(x, y) := \operatorname{tr}(\operatorname{ad}x \operatorname{ad}y) \tag{1}$$

where κ is called the *Killing form* and tr(.) denotes the trace.
(i) Show that

$$\kappa([x, y], z) = \kappa(x, [y, z]). \tag{2}$$

(ii) If B_1 and B_2 are subspaces of a Lie algebra L, we define $[B_1, B_2]$ to be the subspace spanned by all products $[b_1, b_2]$ with $b_1 \in B_1$ and $b_2 \in B_2$. This is the set of sums

$$\sum_j [b_{1j}, b_{2j}] \tag{3}$$

where $b_{1j} \in B_1$ and $b_{2j} \in B_2$. Let

$$L^2 := [L, L], \qquad L^3 := [L^2, L], \quad \ldots \quad , L^k := [L^{k-1}, L]. \tag{4}$$

A Lie algebra L is called *nilpotent* if $L^n = \{0\}$ for some positive integer n. Prove that if L is nilpotent, the Killing form of L is identically zero.

Solution 5. (i) For $m \times m$ matrices we have the properties for the trace

$$\operatorname{tr}(A + B) = \operatorname{tr}(A) + \operatorname{tr}(B), \qquad \operatorname{tr}(AB) = \operatorname{tr}(BA).$$

Moreover we apply $\operatorname{ad}[x, y] \equiv [\operatorname{ad}x, \operatorname{ad}y]$. Using these properties we obtain

$$
\begin{aligned}
\kappa([x, y], z) &= \operatorname{tr}(\operatorname{ad}[x, y] \operatorname{ad}z) = \operatorname{tr}([\operatorname{ad}x, \operatorname{ad}y]\operatorname{ad}z) \\
&= \operatorname{tr}(\operatorname{ad}x \operatorname{ad}y \operatorname{ad}z - \operatorname{ad}y \operatorname{ad}x \operatorname{ad}z) \\
&= \operatorname{tr}(\operatorname{ad}x \operatorname{ad}y \operatorname{ad}z) - \operatorname{tr}(\operatorname{ad}y \operatorname{ad}x \operatorname{ad}z) \\
&= \operatorname{tr}(\operatorname{ad}x \operatorname{ad}y \operatorname{ad}z) - \operatorname{tr}(\operatorname{ad}x \operatorname{ad}z \operatorname{ad}y) \\
&= \operatorname{tr}(\operatorname{ad}x [\operatorname{ad}y, \operatorname{ad}z]) = \operatorname{tr}(\operatorname{ad}x \operatorname{ad}[y, z]) = \kappa(x, [y, z]).
\end{aligned}
$$

(ii) For any $u \in L$ we have

$$(\operatorname{ad}x \operatorname{ad}y)^1(u) = [x, [y, u]] \in L^3, \quad (\operatorname{ad}x \operatorname{ad}y)^2(u) = [x, [y, [x, [y, u]]]] \in L^5.$$

In general

$$(\operatorname{ad}x \operatorname{ad}y)^k(u) \in L^{2k+1}.$$

Since L is nilpotent, say $L^n = \{0\}$ we find that for $2k + 1 \geq n$

$$(\operatorname{ad}x \operatorname{ad}y)^k(u) = 0 \quad \text{for all} \quad u \in L$$

or equivalently $(\operatorname{ad}x \operatorname{ad}y)^k = 0$ where 0 is the zero matrix. Thus the matrix $(\operatorname{ad}x \operatorname{ad}y)$ is nilpotent. If λ is an eigenvalue of $(\operatorname{ad}x \operatorname{ad}y)$ with eigenvector $\mathbf{v}$, i.e.

$$(\operatorname{ad}x \operatorname{ad}y)\mathbf{v} = \lambda \mathbf{v}$$

then $\lambda^k \mathbf{v} = (\mathrm{ad}x \, \mathrm{ad}y)^k \mathbf{v} = \mathbf{0}$. Therefore $\lambda^k = 0$ and thus $\lambda = 0$. Thus the eigenvalues $\lambda_1, \ldots, \lambda_m$ of $(\mathrm{ad}x \, \mathrm{ad}y)$ are all zero, and we find

$$\kappa(x, y) = \mathrm{tr}(\mathrm{ad}x \, \mathrm{ad}y) = \sum_{j=1}^{m} \lambda_j = 0$$

since the trace of a square matrix is the sum of the eigenvalues. This is true for any $x, y \in L$ and the result follows.

Problem 6. Let $g\ell(n, \mathbb{R})$ be the Lie algebra of all $n \times n$ matrices over $\mathbb{R}$. Suppose

$$x \in g\ell(n, \mathbb{R}) \tag{1}$$

has n distinct eigenvalues $\lambda_1, \ldots, \lambda_n$ in $\mathbb{R}$. Prove that the eigenvalues of the $n^2 \times n^2$ matrix

$$\mathrm{ad}x \tag{2}$$

are the n^2 scalars $\lambda_i - \lambda_j$, where $i, j = 1, \ldots, n$.

Solution 6. Since the eigenvalues $\lambda_1, \ldots, \lambda_n$ of x are real and distinct, the corresponding eigenvectors $\mathbf{v}_1, \ldots, \mathbf{v}_n$ are linearly independent (not necessarily orthogonal), and thus form a basis for $\mathbb{R}^n$. The existence of a basis of eigenvectors of x means that x can be diagonalized, i.e. there exists a matrix $R \in GL(n, \mathbb{R})$ such that

$$R^{-1}xR = \mathrm{diag}(\lambda_1, \ldots, \lambda_n)$$

where $GL(n, \mathbb{R})$ denotes the general linear group $(GL(n, \mathbb{R}) \subset g\ell(n, \mathbb{R}))$. Now

$$\mathrm{ad}(R^{-1}xR)(E_{ij}) \equiv [R^{-1}xR, E_{ij}] = (\lambda_i - \lambda_j)E_{ij}$$

where E_{ij} are the matrices having 1 in the (i, j) position and 0 elsewhere. Thus $\lambda_i - \lambda_j$, $i, j = 1, 2, \ldots, n$ are eigenvalues of

$$\mathrm{ad}(R^{-1}xR).$$

Since the corresponding set of eigenvectors E_{ij} $(i, j = 1, 2, \ldots, n)$ is a basis of $g\ell(n, \mathbb{R})$ we conclude that these are the only eigenvalues of $\mathrm{ad}(R^{-1}xR)$. Now if λ is an eigenvalue of $\mathrm{ad}x$ with eigenvector $\mathbf{v}$, then

$$\mathrm{ad}(R^{-1}xR)(R^{-1}\mathbf{v}R) = R^{-1}xRR^{-1}\mathbf{v}R - R^{-1}\mathbf{v}RR^{-1}xR$$
$$= R^{-1}x\mathbf{v}R - R^{-1}\mathbf{v}xR = R^{-1}(\mathrm{ad}x(\mathbf{v}))R.$$

Thus

$$\mathrm{ad}(R^{-1}xR)(R^{-1}\mathbf{v}R) = \lambda(R^{-1}\mathbf{v}R)$$

so that λ is an eigenvalue of $\text{ad}(R^{-1}xR)$. Thus we find $\lambda = \lambda_i - \lambda_j$ for some $i, j \in \{1, \ldots, n\}$. Conversely we have

$$\text{ad}(R^{-1}xR)E_{ij} = (\lambda_i - \lambda_j)E_{ij}.$$

It follows that $R^{-1}xRE_{ij} - E_{ij}R^{-1}xR = (\lambda_i - \lambda_j)E_{ij}$. Multiplying this equation from the left with R and from the right with R^{-1} yields

$$xRE_{ij}R^{-1} - RE_{ij}R^{-1}x = (\lambda_i - \lambda_j)RE_{ij}R^{-1}$$

or

$$\text{ad}x(RE_{ij}R^{-1}) = (\lambda_i - \lambda_j)(RE_{ij}R^{-1}).$$

Hence $\lambda_i - \lambda_j$ is an eigenvalue of $\text{ad}x$ for all $i, j \in \{1, \ldots, n\}$. This proves the theorem.

Problem 7. Let

$$H(\mathbf{p}, \mathbf{q}) = \frac{1}{2}(p_1^2 + p_2^2 + p_3^2) + U(\|\mathbf{q}\|) \tag{1}$$

be a Hamilton function in $\mathbb{R}^6$, where $\| \, . \, \|$ denotes the Euclidean norm, i.e.

$$\|\mathbf{q}\| = \sqrt{q_1^2 + q_2^2 + q_3^2}. \tag{2}$$

(i) Show that

$$h_1(\mathbf{p}, \mathbf{q}) = q_2 p_3 - q_3 p_2, \quad h_2(\mathbf{p}, \mathbf{q}) = q_3 p_1 - q_1 p_3, \quad h_3(\mathbf{p}, \mathbf{q}) = q_1 p_2 - q_2 p_1$$

are first integrals.
(ii) Show that the functions h_1, h_2, h_3 form a Lie algebra under the Poisson bracket. The *Poisson bracket* of two smooth functions f and g is defined as

$$\{f(\mathbf{p}, \mathbf{q}), g(\mathbf{p}, \mathbf{q})\} := \sum_{j=1}^{N} \left(\frac{\partial f}{\partial q_j} \frac{\partial g}{\partial p_j} - \frac{\partial f}{\partial p_j} \frac{\partial g}{\partial q_j} \right)$$

where $N = 3$ in the present case.
(iii) Is the Lie algebra simple? A Lie algebra is called *simple* when it contains no proper ideals.

Solution 7. (i) A function $I(\mathbf{p}, \mathbf{q})$ is called a *first integral* with respect to H if $\{I, H\} = 0$. This equation is equivalent to $dH/dt = 0$. Since

$$\{h_1, H\} = 0, \quad \{h_2, H\} = 0, \quad \{h_3, H\} = 0$$

we find that h_1, h_2 and h_3 are first integrals of H.

(ii) We obtain $\{h_1, h_2\} = h_3$, $\{h_2, h_3\} = h_1$, $\{h_3, h_1\} = h_2$. We conclude that h_1, h_2, and h_3 form a basis of a Lie algebra.

(iii) A subspace I of a Lie algebra L is called an *ideal* of L if $x \in L$, $y \in I$ together imply $[x, y] \in I$. If L has no ideals except itself and $\{0\}$, and if moreover $[L, L] \neq 0$ we call L simple. Obviously, if L is simple it implies that $Z(L) = \{0\}$, where $Z(L)$ denotes the center of the Lie algebra L. Moreover $[L, L] = L$. From the commutation relations for h_1, h_2, h_3, we find that the Lie algebra is simple.

Problem 8. Let L be a Lie algebra. If $H \in L$, then we call $A \in L$ a recursion element with respect to H with value $\mu \in \mathbb{C}$ if $A \neq 0$ and

$$[H, A] = \mu A.$$

If $\mu > 0$, then A is called a raising element with value μ and if $\mu < 0$, then A is called a lowering element with value μ. Assume that the elements of L are linear operators in a Hilbert space. Consider the eigenvalue equation

$$H\phi = \lambda\phi \tag{1}$$

and

$$[H, A] = \mu A. \tag{2}$$

Find $H(A\phi)$.

Solution 8. From (2) we obtain $HA = AH + \mu A$. Thus

$$H(A\phi) = (\mu A + AH)\phi = \mu(A\phi) + A(H\phi).$$

Using (1) we find

$$H(A\phi) = (\mu + \lambda)(A\phi).$$

This is an eigenvalue equation. Consequently, if $A\phi \neq 0$, then $A\phi$ is an eigenfunction of H with eigenvalue $\lambda + \mu$. Thus, if $\mu \neq 0$ and we have one eigenfunction ϕ for H and a recursion element A for H, then we can generate infinitely many eigenfunctions. If A and B are recursion elements with respect to H with values μ_1 and μ_2 and the product AB is well defined, then

$$[H, AB] = [H, A]B + A[H, B] = (\mu_1 + \mu_2)AB.$$

Thus AB is a recursion element with respect to H with value $\mu_1 + \mu_2$.

Problem 9. Let A, B be $m \times m$ matrices. The *Baker-Campbell-Hausdorff formula* is given by

$$\exp(A)B\exp(-A) = \sum_{n=0}^{\infty} \frac{[A, B]_n}{n!}$$

where the repeated commutator is defined by

$$[A, B]_n := [A, [A, B]_{n-1}]$$

with $[A, B]_0 = B$. Consider the nontrivial Lie algebra with two generators X and Y and the commutation relation $[X, Y] = Y$. Let

$$U = \exp(\alpha X + \beta Y), \qquad V = \exp(aX)\exp(bY).$$

Find UXU^{-1}, UYU^{-1}, VXV^{-1}, and VYV^{-1}.

Solution 9.   Since

$$[\alpha X + \beta Y, X]_n = -\frac{\beta}{\alpha}\alpha^n Y, \qquad n > 0$$

and $[\alpha X + \beta Y, Y]_n = \alpha^n Y$ we obtain

$$UXU^{-1} = X - \frac{\beta}{\alpha}(\exp(\alpha) - 1)Y, \qquad UYU^{-1} = \exp(\alpha)Y.$$

On the other hand, V transforms the generators to

$$VXV^{-1} = X - b\exp(a)Y, \qquad VYV^{-1} = \exp(a)Y.$$

Comparing the two sets of similarity transformations, one obtains

$$a = \alpha, \qquad b = \frac{\beta}{\alpha}(1 - \exp(-\alpha)).$$

Problem 10.   Let G be a Lie group with Lie algebra g. For complex z near 1 we consider the function

$$f(z) := \frac{\ln(z)}{z - 1} = \sum_{n=0}^{\infty} \frac{(-1)^n}{n+1}(z-1)^n.$$

Then for X, Y near 0 in g we have $\exp(X)\exp(Y) = \exp(C(X, Y))$, where

$$C(X, Y) = X + Y + \sum_{n=1}^{\infty} \frac{(-1)^n}{n+1} \int_0^1 \left(\sum_{\substack{k,\ell \geq 0 \\ k+\ell \geq 1}} \frac{t^k}{k!\ell!}(\mathrm{ad}X)^k(\mathrm{ad}Y)^\ell \right)^n X\,dt$$

$$= X + Y + \sum_{n=1}^{\infty} \frac{(-1)^n}{n+1} \sum_{\substack{k_1,\ldots,k_n \geq 0 \\ \ell_1,\ldots,\ell_n \geq 0 \\ k_i+\ell_i \geq 1}} \frac{(\mathrm{ad}X)^{k_1}(\mathrm{ad}Y)^{\ell_1}\cdots(\mathrm{ad}X)^{k_n}(\mathrm{ad}Y)^{\ell_n}}{(k_1 + \cdots + k_n + 1)k_1!\cdots k_n!\ell_1!\cdots \ell_n!}X$$

where $(\text{ad}X)Y := [X, Y]$. Thus

$$C(X, Y) = X + Y + \frac{1}{2}[X, Y] + \frac{1}{12}([X, [X, Y]] - [Y, [Y, X]]) + \cdots .$$

Apply this equation to the Bose operators $X = b$ and $Y = b^\dagger$.

Solution 10. The commutator of b and $b^\dagger$ is given by $[b, b^\dagger] = I$, where I is the identity operator. Since $[b, I] = 0$, $[b^\dagger, I] = 0$ we find

$$C(b, b^\dagger) = b + b^\dagger + \frac{1}{2}I.$$

Problem 11. The *Nahm equations* is a system of non-linear ordinary differential equations given by

$$\frac{dT_j}{dt} = \frac{1}{2} \sum_{k=1}^{3} \sum_{l=1}^{3} \epsilon_{jkl}[T_k, T_l](t) \tag{1}$$

for three $n \times n$ matrices T_j of complex-valued functions of the variable t. The indices j, k, l range over $1, 2, 3$. The tensor ϵ_{jkl} is the totally anti-symmetric tensor with $\epsilon_{123} = 1$. Let $\{ H_\alpha, \alpha = 1, \ldots, n-1 \}$ be generators of the Cartan-Lie subalgebra $s\ell(n, \mathbb{R})$, and let $\{ E_\alpha, E_{-\alpha} \}$ be step operators satisfying

$$[H_\alpha, E_{\pm\beta}] = \pm K_{\beta\alpha} E_{\pm\beta} \tag{2a}$$

$$[E_\alpha, E_{-\beta}] = \delta_{\alpha\beta} H_\beta \tag{2b}$$

where $\delta_{\alpha\beta}$ is the Kronecker delta. The $(n-1) \times (n-1)$ matrix $K_{\beta\alpha}$ is the *Cartan matrix* of the Lie algebra $s\ell(n, \mathbb{R})$. The *Cartan matrix* is given by

$$(K_{\beta\alpha}) = \begin{pmatrix} 2 & -1 & 0 & \cdots & 0 & 0 \\ -1 & 2 & -1 & \cdots & 0 & 0 \\ 0 & -1 & 2 & \cdots & 0 & 0 \\ \vdots & \vdots & \vdots & \ddots & \vdots & \vdots \\ 0 & 0 & 0 & \cdots & 2 & -1 \\ 0 & 0 & 0 & \cdots & -1 & 2 \end{pmatrix}.$$

Assume that

$$T_1(t) = \frac{1}{2}i \sum_{\alpha=1}^{n-1} q_\alpha(t)(E_\alpha + E_{-\alpha}), \quad T_2(t) = -\frac{1}{2} \sum_{\alpha=1}^{n-1} q_\alpha(t)(E_\alpha - E_{-\alpha})$$

$$T_3(t) = \frac{1}{2}i \sum_{\alpha=1}^{n-1} p_\alpha(t) H_\alpha$$

where q_α and p_α are smooth functions of t. Find the equations of motion for q_α and p_α.

Solution 11.   Substituting T_1, T_2 and T_3 into the Nahm equation (1) and using the commutation relations (2a) and (2b) yields

$$\frac{dp_\alpha}{dt} = q_\alpha^2, \qquad \frac{dq_\alpha}{dt} = \frac{1}{2}\sum_{\beta=1}^{n-1} p_\beta K_{\alpha\beta} q_\alpha.$$

If we set $\phi_\alpha = 2\ln(q_\alpha)$ we obtain

$$\frac{d^2\phi_\alpha}{dt^2} = \sum_{\beta=1}^{n-1} K_{\alpha\beta}\exp(\phi_\beta)$$

which are the *Toda-molecule equations*.

Problem 12.   Let A and B be $n \times n$ matrices over $\mathbb{C}$. It is known that if $[A, B] = 0_n$, then $\exp(A + B) = \exp(A)\exp(B)$. Let C and D be $n \times n$ matrices over $\mathbb{C}$. Assume that $\exp(C + D) = \exp(C)\exp(D)$. Can we conclude that $[C, D] = 0_n$?

Solution 12.   The answer is no. A counterexample is as follows. Let $\boldsymbol{\sigma} = (\sigma_1, \sigma_2, \sigma_3)$ be the Pauli spin matrices. Consider the two unit vectors $\mathbf{m}$ and $\mathbf{n}$ in $\mathbb{C}^3$

$$\mathbf{m} = (\cos(\alpha), \sin(\alpha), 0)^T, \qquad \mathbf{n} = (\cos(\beta), \sin(\beta), 0)^T.$$

Then the sum $\mathbf{p} = \mathbf{m} + \mathbf{n}$ is a unit vector if and only if $\cos(\alpha - \beta) = -1/2$. Thus $\alpha - \beta = 2\pi/3$. Moreover, for the vector product of $\mathbf{m}$ and $\mathbf{n}$ we find

$$\mathbf{m} \times \mathbf{n} = (0, 0, \cos(\alpha)\sin(\beta) - \sin(\alpha)\cos(\beta))^T$$

where $\cos(\alpha)\sin(\beta) - \sin(\alpha)\cos(\beta) = -\sin(\alpha - \beta)$. If $\alpha - \beta = 2\pi/3$ we have $-\sin(\alpha - \beta) = -\sqrt{3}/2$. Now let

$$C = 2i\pi\boldsymbol{\sigma}\cdot\mathbf{m}, \qquad D = 2i\pi\boldsymbol{\sigma}\cdot\mathbf{n}.$$

Thus $C + D = 2i\pi\boldsymbol{\sigma}\cdot\mathbf{p}$. $\exp(C)$, $\exp(D)$ and $\exp(C + D)$ now represent rotations by 4π around $\mathbf{m}$, $\mathbf{n}$ and $\mathbf{p}$, respectively. Thus

$$\exp(C) = \exp(D) = \exp(C + D) = I$$

although $[C, D] = -i4\pi^2\boldsymbol{\sigma}\cdot(\mathbf{m} \times \mathbf{n}) \neq 0_n$.

Problem 13. The semi-simple Lie algebra $s\ell(2,\mathbb{R})$ is spanned by the 2×2 matrices

$$h = \begin{pmatrix} 1 & 0 \\ 0 & -1 \end{pmatrix}, \qquad e = \begin{pmatrix} 0 & 1 \\ 0 & 0 \end{pmatrix}, \qquad f = \begin{pmatrix} 0 & 0 \\ 1 & 0 \end{pmatrix}$$

with $[e,f] = h$, $[h,e] = 2e$, $[h,f] = -2f$. Show that the bracket relations are also satisfied by the 3×3 matrices

$$\rho_2(h) = \begin{pmatrix} 2 & 0 & 0 \\ 0 & 0 & 0 \\ 0 & 0 & -2 \end{pmatrix}, \quad \rho_2(e) = \begin{pmatrix} 0 & 2 & 0 \\ 0 & 0 & 1 \\ 0 & 0 & 0 \end{pmatrix}, \quad \rho_2(f) = \begin{pmatrix} 0 & 0 & 0 \\ 1 & 0 & 0 \\ 0 & 2 & 0 \end{pmatrix}.$$

Extend the result to general n.

Solution 13. Straightforward calculation yields

$$[\rho_2(e), \rho_2(f)] = \rho_2(h), \quad [\rho_2(h), \rho_2(e)] = 2\rho_2(e), \quad [\rho_2(h), \rho_2(f)] = -2\rho_2(f).$$

For general n we have the $(n+1) \times (n+1)$ matrices

$$\rho_n(h) = \begin{pmatrix} n & 0 & \cdots & \cdots & 0 \\ 0 & n-2 & \cdots & \cdots & 0 \\ \vdots & \vdots & \ddots & \vdots & \vdots \\ 0 & 0 & \cdots & -n+2 & 0 \\ 0 & 0 & \cdots & 0 & -n \end{pmatrix}$$

$$\rho_n(e) = \begin{pmatrix} 0 & n & \cdots & \cdots & 0 \\ 0 & 0 & n-1 & \cdots & 0 \\ \vdots & \vdots & \ddots & \vdots & \vdots \\ 0 & 0 & \cdots & \cdots & 1 \\ 0 & 0 & \cdots & \cdots & 0 \end{pmatrix}, \qquad \rho_n(f) = \begin{pmatrix} 0 & 0 & \cdots & \cdots & 0 \\ 1 & 0 & \cdots & \cdots & 0 \\ \vdots & \vdots & \ddots & \vdots & \vdots \\ 0 & 0 & \cdots & 0 & 0 \\ 0 & 0 & \cdots & n & 0 \end{pmatrix}$$

as representations.

Problem 14. The Lie algebra $su(n)$ are the $n \times n$ matrices X with the conditions $X^* = -X$, $\text{tr}(X) = 0$. Hence the matrices are skew-hermitian. Find a basis for $su(3)$.

Solution 14. Let σ_1, σ_2, σ_3 be the Pauli spin matrices. Consider the eight 3×3 skew-hermitian matrices

$$S_1 = \begin{pmatrix} 0 & 0 & 0 \\ 0 & 0 & -i \\ 0 & -i & 0 \end{pmatrix}, \quad S_2 = \begin{pmatrix} 0 & 0 & 0 \\ 0 & 0 & -1 \\ 0 & 1 & 0 \end{pmatrix}, \quad S_3 = \begin{pmatrix} 0 & 0 & 0 \\ 0 & -i & 0 \\ 0 & 0 & i \end{pmatrix},$$

$$S_4 = \begin{pmatrix} -2i & 0 & 0 \\ 0 & i & 0 \\ 0 & 0 & i \end{pmatrix}, \quad S_5 = \begin{pmatrix} 0 & -1 & 0 \\ 1 & 0 & 0 \\ 0 & 0 & 0 \end{pmatrix}, \quad S_6 = \begin{pmatrix} 0 & 0 & -1 \\ 0 & 0 & 0 \\ 1 & 0 & 0 \end{pmatrix},$$

$$S_7 = \begin{pmatrix} 0 & i & 0 \\ i & 0 & 0 \\ 0 & 0 & 0 \end{pmatrix}, \quad S_8 = \begin{pmatrix} 0 & 0 & i \\ 0 & 0 & 0 \\ i & 0 & 0 \end{pmatrix}.$$

Obviously, the matrices $S_1, \ldots, S_8$ satisfy the conditions given above. From the equation

$$\sum_{j=1}^{8} c_j S_j = 0_3$$

we find that $c_1 = c_2 = \cdots = c_8 = 0$, where 0_3 is the 3×3 zero matrix. Thus the matrices are linearly independent. Therefore the matrices S_j ($j = 1, 2, \ldots, 8$) form a basis for the Lie algebra $su(3)$. Furthermore, a 9 element basis cannot exist since that would be a basis for the 3×3 matrices and would not satisfy the conditions ($\text{tr}(X) = 0$) above.

Problem 15. The Lie algebra $su(4)$ consists of all 4×4 matrices over $\mathbb{C}$ such that $A^* = -A$, $\text{tr}(A) = 0$. The first condition is that the matrices are skew-hermitian. Give a basis for the Lie algebra under the condition that the elements of the basis are orthogonal to each other with respect to the scalar product

$$\langle A, B \rangle := \text{tr}(AB^*).$$

Solution 15. Since $\text{tr}(A) = 0$ the dimension of $su(4)$ is $16 - 1 = 15$. We can built the basis on three diagonal matrices with trace 0 and 12 non-diagonal matrices. We have the three diagonal matrices

$$X_1 = \begin{pmatrix} i & 0 & 0 & 0 \\ 0 & -i & 0 & 0 \\ 0 & 0 & 0 & 0 \\ 0 & 0 & 0 & 0 \end{pmatrix}, \quad X_2 = \frac{1}{\sqrt{3}} \begin{pmatrix} i & 0 & 0 & 0 \\ 0 & i & 0 & 0 \\ 0 & 0 & -2i & 0 \\ 0 & 0 & 0 & 0 \end{pmatrix},$$

$$X_3 = \frac{1}{\sqrt{6}} \begin{pmatrix} i & 0 & 0 & 0 \\ 0 & i & 0 & 0 \\ 0 & 0 & i & 0 \\ 0 & 0 & 0 & -3i \end{pmatrix}$$

and the 12 nondiagonal matrices

$$X_4 = \begin{pmatrix} 0 & i & 0 & 0 \\ i & 0 & 0 & 0 \\ 0 & 0 & 0 & 0 \\ 0 & 0 & 0 & 0 \end{pmatrix}, \quad X_5 = \begin{pmatrix} 0 & 1 & 0 & 0 \\ -1 & 0 & 0 & 0 \\ 0 & 0 & 0 & 0 \\ 0 & 0 & 0 & 0 \end{pmatrix}, \quad X_6 = \begin{pmatrix} 0 & 0 & i & 0 \\ 0 & 0 & 0 & 0 \\ i & 0 & 0 & 0 \\ 0 & 0 & 0 & 0 \end{pmatrix},$$

$$X_7 = \begin{pmatrix} 0 & 0 & 1 & 0 \\ 0 & 0 & 0 & 0 \\ -1 & 0 & 0 & 0 \\ 0 & 0 & 0 & 0 \end{pmatrix}, \quad X_8 = \begin{pmatrix} 0 & 0 & 0 & 0 \\ 0 & 0 & i & 0 \\ 0 & i & 0 & 0 \\ 0 & 0 & 0 & 0 \end{pmatrix}, \quad X_9 = \begin{pmatrix} 0 & 0 & 0 & 0 \\ 0 & 0 & 1 & 0 \\ 0 & -1 & 0 & 0 \\ 0 & 0 & 0 & 0 \end{pmatrix},$$

$$X_{10} = \begin{pmatrix} 0 & 0 & 0 & i \\ 0 & 0 & 0 & 0 \\ 0 & 0 & 0 & 0 \\ i & 0 & 0 & 0 \end{pmatrix}, \quad X_{11} = \begin{pmatrix} 0 & 0 & 0 & 1 \\ 0 & 0 & 0 & 0 \\ 0 & 0 & 0 & 0 \\ -1 & 0 & 0 & 0 \end{pmatrix}, \quad X_{12} = \begin{pmatrix} 0 & 0 & 0 & 0 \\ 0 & 0 & 0 & i \\ 0 & 0 & 0 & 0 \\ 0 & i & 0 & 0 \end{pmatrix},$$

$$X_{13} = \begin{pmatrix} 0 & 0 & 0 & 0 \\ 0 & 0 & 0 & 1 \\ 0 & 0 & 0 & 0 \\ 0 & -1 & 0 & 0 \end{pmatrix}, \quad X_{14} = \begin{pmatrix} 0 & 0 & 0 & 0 \\ 0 & 0 & 0 & 0 \\ 0 & 0 & 0 & i \\ 0 & 0 & i & 0 \end{pmatrix}, \quad X_{15} = \begin{pmatrix} 0 & 0 & 0 & 0 \\ 0 & 0 & 0 & 0 \\ 0 & 0 & 0 & 1 \\ 0 & 0 & -1 & 0 \end{pmatrix}.$$

Problem 16. Let $X_1, X_2, \ldots, X_r$ be the basis of a Lie algebra with the commutator

$$[X_i, X_j] = \sum_{k=1}^{r} C_{ij}^k X_k$$

where the C_{ij}^k are the *structure constants*. The structure constants satisfy (third fundamental theorem)

$$C_{ij}^k = -C_{ji}^k$$

$$\sum_{m=1}^{r} \left(C_{ij}^m C_{mk}^\ell + C_{jk}^m C_{mi}^\ell + C_{ki}^m C_{mj}^\ell \right) = 0.$$

We replace the X_i's by c-number differential operators (vector fields)

$$X_i \mapsto V_i = \sum_{\ell=1}^{r} \sum_{k=1}^{r} x_k C_{i\ell}^k \frac{\partial}{\partial x_\ell}, \qquad i = 1, \ldots, r.$$

Let

$$V_j = \sum_{n=1}^{r} \sum_{m=1}^{r} x_m C_{jn}^m \frac{\partial}{\partial x_n}.$$

Show that

$$[V_i, V_j] = \sum_{k=1}^{n} C_{ij}^k V_k, \quad V_k = \sum_{n=1}^{r} \sum_{m=1}^{r} x_m C_{kn}^m \frac{\partial}{\partial x_n}.$$

Solution 16. Since $\partial x_i / \partial x_j = \delta_{ij}$ we have

$$[V_i, V_j] = \sum_{k=1}^{r} \sum_{n=1}^{r} \sum_{m=1}^{r} x_k C_{im}^k C_{jn}^m \frac{\partial}{\partial x_n} - \sum_{k=1}^{r} \sum_{m=1}^{r} \sum_{\ell=1}^{r} x_m C_{jk}^m C_{i\ell}^k \frac{\partial}{\partial x_\ell}.$$

Using the first part of the third fundamental theorem $(C_{kj}^m = -C_{jk}^m)$ and renaming the index summations we find

$$[V_i, V_j] = \sum_{n=1}^{r} \sum_{m=1}^{r} \sum_{k=1}^{r} x_m C_{nj}^k C_{ki}^m \frac{\partial}{\partial x_n} + \sum_{n=1}^{r} \sum_{m=1}^{r} \sum_{k=1}^{r} x_m C_{in}^k C_{kj}^m \frac{\partial}{\partial x_n}$$

$$= \sum_{n=1}^{r} \sum_{m=1}^{r} x_m \left(\sum_{k=1}^{r} (C_{nj}^k C_{ki}^m + C_{in}^k C_{kj}^m) \right) \frac{\partial}{\partial x_n}.$$

Using the second part of the third fundamental theorem we finally obtain

$$[V_i, V_j] = \sum_{k=1}^{r} C_{ij}^k \left(\sum_{n=1}^{r} \sum_{m=1}^{r} x_m C_{kn}^m \frac{\partial}{\partial x_n} \right) = \sum_{k=1}^{r} C_{ij}^k V_k.$$

Problem 17.   Consider the Lie group $G = O(2,1)$ and its Lie algebra $o(2,1) = \{K_1, K_2, L_3\}$, where K_1, K_2 are *Lorentz boosts* and L_3 an infinitesimal rotation. The maximal subalgebras of $o(2,1)$ are represented by $\{K_1, K_2 + L_3\}$ and $\{L_3\}$, nonmaximal subalgebras by $\{K_1\}$ and $\{K_2 + L_3\}$. The two-dimensional subalgebra corresponds to the projective group of a real line. The one-dimensional subalgebras correspond to the groups $O(2)$, $O(1,1)$ and the translations $T(1)$, respectively. Find the $o(2,1)$ infinitesimal generators.

Solution 17.   Consider the manifold $O(2,1)/T(1)$, isomorphic to the upper sheet of the cone $x_0^2 - x_1^2 - x_2^2 = 0$, $x_0 > 0$. We introduce *horospheric coordinates* (ξ, η)

$$x_0(\eta, \xi) = \eta(\xi^2 + 1)/2, \qquad x_1(\eta, \xi) = \eta(\xi^2 - 1)/2, \qquad x_2(\eta, \xi) = \xi \eta$$

where $-\infty < \xi < \infty$, $\eta > 0$. The $o(2,1)$ infinitesimal operators are

$$L_3 = \frac{1 + \xi^2}{2} \frac{\partial}{\partial \xi} - \xi \eta \frac{\partial}{\partial \eta}, \quad K_1 = \xi \frac{\partial}{\partial \xi} - \eta \frac{\partial}{\partial \eta}, \quad K_2 = \frac{1 - \xi^2}{2} \frac{\partial}{\partial \xi} + \xi \eta \frac{\partial}{\partial \eta}.$$

Problem 18.   Show that the operators

$$L_+ = \bar{z} z, \qquad L_- = -\frac{\partial}{\partial z} \frac{\partial}{\partial \bar{z}}$$

$$L_3 = -\frac{1}{2} \left(z \frac{\partial}{\partial z} + \bar{z} \frac{\partial}{\partial \bar{z}} + 1 \right), \qquad L_0 = -\frac{1}{2} \left(z \frac{\partial}{\partial z} - \bar{z} \frac{\partial}{\partial \bar{z}} + 1 \right)$$

form a basis for the Lie algebra $u(1,1)$ under the commutator.

Solution 18. We have $[L_0, L_+] = [L_0, L_-] = [L_0, L_+] = 0$ and for the nonzero commutators

$$[L_+, L_-] = -2L_3, \quad [L_+, L_3] = -L_+, \quad [L_-, L_3] = L_-.$$

The Lie algebra $u(1,1)$ can be decomposed into the direct sum $su(1,1) \oplus u(1)$.

Problem 19. Let $z \in \mathbb{C}$. Consider the vector fields

$$L_n := z^{n+1}\frac{d}{dz}, \quad n \in \mathbb{Z}.$$

Calculate the commutator $[L_m, L_n]$.

Solution 19. We obtain

$$
\begin{aligned}
[L_m, L_n] &= z^{m+1}\frac{d}{dz}z^{n+1}\frac{d}{dz} - z^{n+1}\frac{d}{dz}z^{m+1}\frac{d}{dz} \\
&= (n+1)z^{m+n+1}\frac{d}{dz} - (m+1)z^{m+n+1}\frac{d}{dz} \\
&= (n-m)z^{m+n+1}\frac{d}{dz} = (n-m)L_{m+n}.
\end{aligned}
$$

Problem 20. Let $b_1^\dagger$, $b_2^\dagger$ be Bose creation operators. The semisimple Lie algebra $su(1,1)$ is generated by

$$K_+ := b_1^\dagger b_2^\dagger, \quad K_- := b_1 b_2, \quad K_0 := \frac{1}{2}(b_1^\dagger b_1 + b_2^\dagger b_2 + I)$$

with the commutation relations

$$[K_0, K_+] = K_+, \quad [K_0, K_-] = -K_-, \quad [K_-, K_+] = 2K_0$$

where I is the identity operator. We use the ordering K_+, K_-, K_0 for the basis.
(i) Find the *adjoint representation* for the Lie algebra.
(ii) Find the *Killing form* and metric tensor (g_{ij}) for this Lie algebra.
(iii) Find the *Casimir invariant* C using

$$C = \sum_{i=1}^{3}\sum_{j=1}^{3} g^{ij}X_iX_j$$

where we set $X_1 = K_+$, $X_2 = K_-$, $X_3 = K_0$ and (g^{ij}) is the inverse matrix of (g_{ij}).

Solution 20.   (i) Using the ordering given above for the basis we have

$$(\mathrm{ad}K_+)K_+ = [K_+, K_+] = 0$$
$$(\mathrm{ad}K_+)K_- = [K_+, K_-] = -2K_0$$
$$(\mathrm{ad}K_+)K_0 = [K_+, K_0] = -K_+.$$

Thus

$$(K_+ \quad K_- \quad K_0)\begin{pmatrix} 0 & 0 & -1 \\ 0 & 0 & 0 \\ 0 & -2 & 0 \end{pmatrix} = (0 \quad -2K_0 \quad -K_+).$$

Since K_+, K_-, K_0 is a basis of the Lie algebra it follows that

$$\mathrm{ad}K_+ = \begin{pmatrix} 0 & 0 & -1 \\ 0 & 0 & 0 \\ 0 & -2 & 0 \end{pmatrix}.$$

Analogously we find

$$\mathrm{ad}K_- = \begin{pmatrix} 0 & 0 & 0 \\ 0 & 0 & 1 \\ 2 & 0 & 0 \end{pmatrix}, \qquad \mathrm{ad}K_0 = \begin{pmatrix} 1 & 0 & 0 \\ 0 & -1 & 0 \\ 0 & 0 & 0 \end{pmatrix}.$$

With $X_1 = K_+$, $X_2 = K_-$, $X_3 = K_0$ we have

$$g_{ij} = K(X_i, X_j) = \mathrm{tr}(\mathrm{ad}X_i \, \mathrm{ad}X_j)$$

where tr denotes the trace. We obtain

$$\mathrm{tr}(\mathrm{ad}K_+ \, \mathrm{ad}K_+) = 0, \quad \mathrm{tr}(\mathrm{ad}K_+ \, \mathrm{ad}K_-) = -4, \quad \mathrm{tr}(\mathrm{ad}K_+ \, \mathrm{ad}K_0) = 0,$$

$$\mathrm{tr}(\mathrm{ad}K_- \, \mathrm{ad}K_+) = 0, \quad \mathrm{tr}(\mathrm{ad}K_- \, \mathrm{ad}K_-) = 0, \quad \mathrm{tr}(\mathrm{ad}K_- \, \mathrm{ad}K_0) = 0,$$

$$\mathrm{tr}(\mathrm{ad}K_0 \, \mathrm{ad}K_+) = 0, \quad \mathrm{tr}(\mathrm{ad}K_0 \, \mathrm{ad}K_-) = 0, \quad \mathrm{tr}(\mathrm{ad}K_0 \, \mathrm{ad}K_0) = 2.$$

Thus we obtain the metric tensor

$$g = \begin{pmatrix} 0 & -4 & 0 \\ -4 & 0 & 0 \\ 0 & 0 & 2 \end{pmatrix} \Rightarrow g^{-1} = \begin{pmatrix} 0 & -1/4 & 0 \\ -1/4 & 0 & 0 \\ 0 & 0 & 1/2 \end{pmatrix}$$

where $g^{-1} = (g^{ij})$. Thus

$$C = \sum_{i=1}^{3}\sum_{j=1}^{3} g^{ij} X_i X_j = \frac{1}{2}\left(K_0^2 - \frac{1}{2}(K_+K_- + K_-K_+)\right).$$

Problem 21. Let b, $b^\dagger$ be Bose annihilation and creation operators, respectively. Let I be the identity operator. The *harmonic oscillator Lie algebra ho(1)* is spanned by $\{\ I,\ b,\ b^\dagger,\ b^\dagger b\ \}$. The Lie subalgebra spanned by $\{\ I,\ b,\ b^\dagger\ \}$ is known as the *Heisenberg-Weyl Lie algebra h(1)*.
(i) Find the commutators for the harmonic oscillator algebra.
(ii) Is the harmonic oscillator Lie algebra semisimple?

Solution 21. (i) We have $[b^\dagger b, b] = -b$, $[b^\dagger b, b^\dagger] = b^\dagger$ and

$$[b, b^\dagger] = I, \quad [b, I] = [b^\dagger, I] = [b^\dagger b, I] = 0.$$

(ii) Owing to the element I the harmonic oscillator Lie algebra is not semisimple.

Problem 22. A (global) group of $n \times n$ matrices is compact if it is a bounded, closed subset of the set of all $n \times n$ matrices. A set U of $n \times n$ matrices is bounded if there exists a constant $K > 0$ such that $|A_{ik}| \leq K$ for $1 \leq i, k \leq n$ and all $A \in U$. The set U is closed provided every Cauchy sequence in U converges to a matrix in U. A sequence of $n \times n$ matrices $\{\ A^{(p)}\ \}$ is a *Cauchy sequence* if each of the sequences of matrix elements $\{\ A_{ik}^{(p)}\ \}$, $1 \leq i, k \leq n$, is Cauchy. Show that the orthogonal group $O(3, \mathbb{R})$ is compact.

Solution 22. If $A \in O(3, \mathbb{R})$ then $A^T A = I_3$, i.e.

$$\sum_{j=1}^{3} A_{j\ell} A_{jk} = \delta_{\ell k}.$$

Setting $\ell = k$ we obtain

$$\sum_{j=1}^{3} (A_{jk})^2 = 1.$$

Thus $|A_{jk}| \leq 1$ for all j, k. Thus the matrix elements are bounded. Let $\{\ A^{(p)}\ \}$ be a Cauchy sequence in $O(3, \mathbb{R})$ with limit A. Then

$$I_3 = \lim_{p \to \infty} (A^{(p)})^T A^{(p)} = A^T A$$

so $A \in O(3, \mathbb{R})$ and the Lie group $O(3, \mathbb{R})$ is compact.

Problem 23. The Lie group $SU(1, 1)$ consists of the set of all 2×2 pseudo-unitary matrices (with determinant 1) preserving the quadratic form

$$|z_1|^2 - |z_2|^2, \qquad z_1, z_2 \in \mathbb{C}.$$

(i) Show that the 2×2 matrix

$$U(\tau, \alpha, \beta) = \begin{pmatrix} \cosh(\tau/2)e^{-i\alpha} & \sinh(\tau/2)e^{-i\beta} \\ \sinh(\tau/2)e^{i\beta} & \cosh(\tau/2)e^{i\alpha} \end{pmatrix} \qquad (1)$$

preserves the quadratic form $|z_1|^2 - |z_2|^2$ $(z_1, z_2 \in \mathbb{C})$, where $\tau, \alpha, \beta \in \mathbb{R}$.
(ii) Show that $\det(U) = 1$.
(iii) Give the inverse of U.
(iv) Show that $U_1(\tau_1, \alpha_1, \beta_1)U_2(\tau_2, \alpha_2, \beta_2)$ with $\alpha_1 + \alpha_2 = \beta_1 - \beta_2$ is again a matrix of the form (1).

Solution 23.   (i) We have

$$\begin{pmatrix} \tilde{z}_1 \\ \tilde{z}_2 \end{pmatrix} = U \begin{pmatrix} z_1 \\ z_2 \end{pmatrix} = \begin{pmatrix} z_1 \cosh(\tau/2)e^{-i\alpha} + z_2 \sinh(\tau/2)e^{-i\beta} \\ z_1 \sinh(\tau/2)e^{i\beta} + z_2 \cosh(\tau/2)e^{i\alpha} \end{pmatrix}.$$

Using the identity $\cosh^2(\tau/2) - \sinh^2(\tau/2) = 1$ we obtain

$$\begin{aligned} |\tilde{z}_1|^2 - |\tilde{z}_2|^2 &= z_1 z_1^* \cosh^2(\tau/2) + z_2 z_2^* \sinh^2(\tau/2) \\ &\quad - z_1 z_1^* \sinh^2(\tau/2) - z_2 z_2^* \cosh^2(\tau/2) \\ &= z_1 z_1^* - z_2 z_2^* = |z_1|^2 - |z_2|^2. \end{aligned}$$

(ii) Using that $\cosh^2(\tau/2) - \sinh^2(\tau/2) = 1$ we obtain $\det(U) = 1$.
(iii) The inverse of U is given by the replacements $\tau \to -\tau$, $\alpha \to -\alpha$, $\beta \to \beta$. Thus

$$U^{-1} = \begin{pmatrix} \cosh(\tau/2)e^{i\alpha} & -\sinh(\tau/2)e^{-i\beta} \\ -\sinh(\tau/2)e^{i\beta} & \cosh(\tau/2)e^{-i\alpha} \end{pmatrix}.$$

(iv) Using the identities

$$\begin{aligned} \cosh(x+y) &\equiv \cosh(x)\cosh(y) + \sinh(x)\sinh(y) \\ \sinh(x+y) &\equiv \sinh(x)\cosh(y) + \cosh(x)\sinh(y) \end{aligned}$$

and $\alpha_1 + \alpha_2 = \beta_1 - \beta_2$ we obtain

$$U_1 U_2 = \begin{pmatrix} \cosh((\tau_1 + \tau_2)/2)e^{-i(\alpha_1 + \alpha_2)} & \sinh((\tau_1 + \tau_2)/2)e^{-i(\beta_1 - \alpha_2)} \\ \sinh((\tau_1 + \tau_2)/2)e^{i(\beta_1 - \alpha_2)} & \cosh((\tau_1 + \tau_2)/2)e^{i(\alpha_1 + \alpha_2)} \end{pmatrix}.$$

Problem 24.   The group $Sp(2n, \mathbb{R})$ consists of all real $2n \times 2n$ matrices S which obey the condition $S^T J S = J$, where J is the $2n \times 2n$ skew-symmetric matrix

$$J := \begin{pmatrix} 0_n & I_n \\ -I_n & 0_n \end{pmatrix}$$

with I_n the $n \times n$ identity matrix and 0_n the $n \times n$ zero matrix. Let V be a $2n \times 2n$ real symmetric positive definite matrix. Show that there exists an $S \in Sp(2n, \mathbb{R})$ such that

$$S^T V S = D^2 > 0, \qquad D^2 = \mathrm{diag}(\kappa_1, \kappa_2, \dots, \kappa_n, \kappa_1, \kappa_2, \dots, \kappa_n).$$

Solution 24. Since $J^T = -J$, it follows that $V^{-1/2} J V^{-1/2}$ is antisymmetric. Hence there exists a $2n \times 2n$ matrix $R \in SO(2n)$ such that

$$R^T V^{-1/2} J V^{-1/2} R = \begin{pmatrix} 0_n & \Omega \\ -\Omega & 0_n \end{pmatrix}, \qquad \Omega = \text{diagonal} > 0.$$

We define a diagonal positive definite matrix

$$D = \begin{pmatrix} \Omega^{-1/2} & 0_n \\ 0_n & \Omega^{-1/2} \end{pmatrix}.$$

Then we have

$$D R^T V^{-1/2} J V^{-1/2} R D = J.$$

Now we define $S := V^{-1/2} R D$. Then $S^T J S = J$ and $S^T V S = D^2$, where D is a diagonal matrix with $d_{jj} > 0$ for $j = 1, 2, \dots, 2n$.

Problem 25. The group of complex rotations $O(n, \mathbb{C})$ is defined as the group of all $n \times n$ complex matrices O, such that $OO^T = O^T O = I_n$, where ^{T} means transpose. These transformations preserve the real scalar product

$$\mathbf{x} \cdot \mathbf{y} = \sum_{j=1}^{n} x_j y_j$$

so that $(O\mathbf{x}) \cdot O\mathbf{y} = \mathbf{x} \cdot \mathbf{y}$, where $\mathbf{x}$ and $\mathbf{y}$ are complex vectors in general, i.e. $x_j, y_j \in \mathbb{C}$.
(i) Show that the matrix ($\alpha \in \mathbb{R}$)

$$O = \begin{pmatrix} \cosh(\alpha) & i \sinh(\alpha) \\ -i \sinh(\alpha) & \cosh(\alpha) \end{pmatrix}$$

is an element of $O(n, \mathbb{C})$.
(ii) Find the partial derivatives under complex orthogonal transformations $O \in O(n, \mathbb{C})$

$$w_j(\mathbf{x}) := (O\mathbf{x})_j = \sum_{k=1}^{n} O_{jk} x_k, \qquad j = 1, \dots, n$$

i.e. $\partial / \partial w_j$ with $j = 1, \dots, n$.

Solution 25.   (i) We have

$$O^T = \begin{pmatrix} \cosh(\alpha) & -i\sinh(\alpha) \\ i\sinh(\alpha) & \cosh(\alpha) \end{pmatrix}.$$

Thus $O^T O = O O^T = I_2$.

(ii) Using the *chain rule* we have

$$\frac{\partial}{\partial w_j} = \sum_{k=1}^{n} \frac{\partial x_k}{\partial w_j} \frac{\partial}{\partial x_k} = \sum_{k=1}^{n} O_{kj}^{-1} \frac{\partial}{\partial x_k} = \sum_{k=1}^{n} O_{jk} \frac{\partial}{\partial x_k}.$$

Thus the partial derivatives transform exactly as the coordinates, since $O^{-1} = O^T$.

Problem 26.   Consider the Lie group $SL(2, \mathbb{R})$, i.e. the set of all real 2×2 matrices with determinant equal to 1. A dynamical system in $SL(2, \mathbb{R})$ can be defined by

$$M_{k+2} = M_k M_{k+1}, \qquad k = 0, 1, 2, \ldots \tag{1}$$

with the initial matrices $M_0, M_1 \in SL(2, \mathbb{R})$. Let $F_k := \text{tr}(M_k)$. Show that

$$F_{k+3} = F_{k+2} F_{k+1} - F_k \qquad k = 0, 1, 2, \ldots. \tag{2}$$

Hint. Use that property that for any 2×2 matrix A we have

$$A^2 - A\text{tr}(A) + I_2 \det(A) = 0. \tag{3}$$

Solution 26.   From (1) it follows that

$$M_{k+3} = M_{k+1} M_{k+2} = M_{k+1} M_k M_{k+1}.$$

Taking the trace of this equation and *cyclic invariance* provides

$$\text{tr}(M_{k+3}) = \text{tr}(M_{k+1} M_k M_{k+1}) = \text{tr}(M_{k+1}^2 M_k).$$

Using (3) with $\det(M_k) = 1$ we arrive at

$$\begin{aligned} \text{tr}(M_{k+3}) &= \text{tr}((M_{k+1}\text{tr}(M_{k+1}) - I_2) M_k) \\ &= \text{tr}(M_{k+1})\text{tr}(M_{k+1} M_k) - \text{tr}(M_k) \\ &= \text{tr}(M_{k+1})\text{tr}(M_{k+2}) - \text{tr}(M_k). \end{aligned}$$

Thus (2) follows.

Problem 27.   The Lie group $SO(m, n)$ consists of all real matrices S that satisfy

$$S^T g S = g$$

where $\det(S) = 1$ and $g = \operatorname{diag}(+1, +1, \ldots, +1, -1, \ldots, -1)$ with n $+1$'s and m -1's. Let V be a real symmetric positive definite matrix of dimension N. Show that for any choice of partition $N = m + n$, there exists an $S \in SO(m,n)$ such that $S^T V S = D^2 = $ diagonal (and > 0).

Solution 27. Consider the matrix $V^{-1/2} g V^{-1/2}$ constructed from the given matrix V. Since $V^{-1/2} g V^{-1/2}$ is real symmetric, there exists a rotation matrix $R \in SO(N)$ which diagonalizes $V^{-1/2} g V^{-1/2}$

$$R^T V^{-1/2} g V^{-1/2} R = \text{diagonal} \equiv \Lambda.$$

This may be viewed also as a congruence of g using $V^{-1/2} R$, and signatures are preserved under congruence. As a consequence, the diagonal matrix Λ can be expressed as the product of a positive diagonal matrix and g

$$R^T V^{-1/2} g V^{-1/2} R = D^{-2} g = D^{-1} g D^{-1}.$$

Here D is diagonal and positive definite. Taking the inverse of the matrices on both sides of this equation we find that the diagonal entries of $g D^2 = D^2 g$ are the eigenvalues of $V^{1/2} g V^{1/2}$ and that the columns of R are the eigenvectors of $V^{1/2} g V^{1/2}$. Since $V^{1/2} g V^{1/2}$, gV, and Vg are conjugate to one another, we conclude that D^2 is determined by the eigenvalues of $gV \sim Vg$. We define $S : V^{-1/2} RD$. Then S satisfies the following two equations

$$S^T g S = g, \qquad S^T V S = D^2 = \text{diagonal}.$$

The first equation tells us that $S \in SO(m,n)$ and the second tells us that V is diagonalized through congruence by S.

Problem 28. Show that the matrices

$$A(\phi, \theta, \psi) = \begin{pmatrix} e^{i\frac{\phi+\psi}{2}} \cos(\theta/2) & i e^{i\frac{\phi-\psi}{2}} \sin(\theta/2) \\ i e^{i\frac{\psi-\phi}{2}} \sin(\theta/2) & e^{-i\frac{\phi+\psi}{2}} \cos(\theta/2) \end{pmatrix}$$

form a group under matrix multiplication, where ϕ, θ, ψ are the *Euler angles*. The Euler angles are coordinates on the sphere $\mathbb{S}^3$.

Solution 28. Since

$$A_1(\phi_1, \theta_1, \psi_1) A_2(\phi_2, \theta_2, \psi_2) = A(\phi, \theta, \psi)$$

with

$$\cos(\theta) = \cos(\theta_1)\cos(\theta_2) - \sin(\theta_1)\sin(\theta_2)\cos(\phi_2 - \psi_1)$$

$$e^{i\phi} = \frac{e^{i\phi_1}}{\sin(\theta)} (\sin(\theta_1)\cos(\theta_2) + \cos(\theta_1)\sin(\theta_2)\cos(\phi_2+\psi_1) + i\sin(\theta_2)\sin(\phi_2+\psi_1))$$

$$e^{i(\phi+\psi)/2} = \frac{e^{i(\phi_1+\psi_1)/2}}{\cos(\theta/2)} \left(\cos\frac{\theta_1}{2}\cos\frac{\theta_2}{2}e^{i(\phi_2+\psi_1)/2} - \sin\frac{\theta_1}{2}\sin\frac{\theta_2}{2}e^{-i(\phi_2+\psi_2)/2}\right)$$

we find that the matrices $A(\phi, \theta, \psi)$ are closed under matrix multiplication. The neutral element is given by

$$A(\phi = 0, \theta = 0, \psi = 0) = \begin{pmatrix} 1 & 0 \\ 0 & 1 \end{pmatrix}.$$

Since $\det(A(\phi, \theta, \psi)) = \cos^2(\theta/2) + \sin^2(\theta/2) = 1$ we find that the inverse exists. The matrix A is unitary, i.e.

$$A^{-1}(\phi, \theta, \psi) = A^*(\phi, \theta, \psi) \equiv \bar{A}^T(\phi, \theta, \psi)$$

where A^* denotes the hermitian conjugate matrix of A. Since for arbitrary $n \times n$ matrices the associative law holds we have proved that the matrices $A(\phi, \theta, \psi)$ form a group under matrix multiplication.

Problem 29. The *commutator* $[A, B]$ of two elements A and B of a group G is defined by

$$[A, B] := A^{-1}B^{-1}AB.$$

Consider the Lie group $U(2)$ and $A, B \in U(2)$

$$A = \begin{pmatrix} 0 & 1 \\ 1 & 0 \end{pmatrix}, \qquad B = \frac{1}{\sqrt{2}}\begin{pmatrix} 1 & 1 \\ 1 & -1 \end{pmatrix}.$$

Calculate $[A, B]$. Is $[A, B]$ an element of the Lie group $U(2)$?

Solution 29. Since $A^{-1} = A$, $B = B^{-1}$ we have

$$[A, B] = A^{-1}B^{-1}AB = ABAB = (AB)^2 = \begin{pmatrix} 0 & -1 \\ 1 & 0 \end{pmatrix}.$$

Since $([A, B])([A, B])^* = I_2$ the commutator is an element of $U(2)$.

Problem 30. We consider the following subgroups of the Lie group $SL(2, \mathbb{R})$. Let

$$K := \left\{ \begin{pmatrix} \cos(\theta) & -\sin(\theta) \\ \sin(\theta) & \cos(\theta) \end{pmatrix} : \theta \in [0, 2\pi) \right\},$$

$$A := \left\{ \begin{pmatrix} r^{1/2} & 0 \\ 0 & r^{-1/2} \end{pmatrix} : r > 0 \right\}, \quad N := \left\{ \begin{pmatrix} 1 & t \\ 0 & 1 \end{pmatrix} : t \in \mathbb{R} \right\}.$$

It can be shown that any matrix $m \in SL(2, \mathbb{R})$ can be written in a unique way as the product $m = kan$ with $k \in K$, $a \in A$ and $n \in N$. This decomposition is called *Iwasawa decomposition* and has a natural generalization to

$SL(n, \mathbb{R})$, $n \geq 3$. The notation of the subgroups comes from the fact that K is a compact subgroup, A is an abelian subgroup and N is a nilpotent subgroup of $SL(2, \mathbb{R})$. Find the Iwasawa decomposition of the matrix

$$\begin{pmatrix} 1 & 1 \\ 1 & 2 \end{pmatrix}.$$

Solution 30. From

$$\begin{pmatrix} 1 & 1 \\ 1 & 2 \end{pmatrix} = \begin{pmatrix} \cos(\theta) & -\sin(\theta) \\ \sin(\theta) & \cos(\theta) \end{pmatrix} \begin{pmatrix} r^{1/2} & 0 \\ 0 & r^{-1/2} \end{pmatrix} \begin{pmatrix} 1 & t \\ 0 & 1 \end{pmatrix}$$

we obtain

$$\begin{pmatrix} 1 & 1 \\ 1 & 2 \end{pmatrix} = \begin{pmatrix} r^{1/2}\cos(\theta) & tr^{1/2}\cos(\theta) - r^{-1/2}\sin(\theta) \\ r^{1/2}\sin(\theta) & tr^{1/2}\sin(\theta) + r^{-1/2}\cos(\theta) \end{pmatrix}.$$

Thus we have the four conditions

$$r^{1/2}\cos(\theta) = 1, \quad r^{1/2}t\cos(\theta) - r^{-1/2}\sin(\theta) = 1$$
$$r^{1/2}\sin(\theta) = 1, \quad r^{1/2}t\sin(\theta) + r^{-1/2}\cos(\theta) = 2$$

for the three unknowns r, t, θ. We obtain the solution

$$\cos(\theta) = \sin(\theta) = \frac{1}{\sqrt{2}}, \quad r = 2, \quad t = \frac{3}{2}.$$

Problem 31. Consider the vector fields xd/dx, d/dx. The manifold is $M = \mathbb{R}$, i.e. $x \in \mathbb{R}$. Calculate

$$\exp\left(tx\frac{d}{dx}\right)\exp\left(t\frac{d}{dx}\right)x, \qquad \exp\left(t\frac{d}{dx}\right)\exp\left(tx\frac{d}{dx}\right)x.$$

Solution 31. For the first case we have

$$\exp\left(tx\frac{d}{dx}\right)\exp\left(t\frac{d}{dx}\right)x = \exp\left(tx\frac{d}{dx}\right)\left(1 + \frac{t}{1!}\frac{d}{dx} + \frac{t^2}{2!}\frac{d^2}{dx^2} + \cdots\right)x$$

$$= \exp\left(tx\frac{d}{dx}\right)(x + t)$$

$$= \exp\left(tx\frac{d}{dx}\right)x + \exp\left(tx\frac{d}{dx}\right)t$$

$$= xe^t + t.$$

Analogously for the second case we find

$$\exp\left(t\frac{d}{dx}\right)\exp\left(tx\frac{d}{dx}\right)x = e^t(x+t).$$

Problem 32. Consider the three linear operator with the commutation relation

$$[A, B] = iC, \quad [C, A] = iA, \quad [C, B] = -iB.$$

This means we have the simple Lie algebra $so(2, 1)$. Let $\epsilon \in \mathbb{R}$. Consider

$$e^{i\epsilon(A-B)} = e^{if_3(\epsilon)A}e^{if_2(\epsilon)C}e^{if_1(\epsilon)B}.$$

Applying differentiation with respect to ϵ of this expression find the autonomous system of first order differential equations. Then solve the initial value problem with $f_1(0) = f_2(0) = f_3(0) = 0$.

Solution 32. Differentiation with respect to ϵ and rearranging the exponential terms we arrive at

$$e^{i\epsilon(A-B)} = e^{if_3A}e^{if_2C}e^{if_1B}i\frac{df_3}{\epsilon}e^{f_2}\left(A - f_1C - \frac{1}{2}f_1^2B\right)$$

$$+ \frac{df_2}{d\epsilon}C + \frac{df_2}{d\epsilon}f_1B + \frac{df_1}{d\epsilon}B.$$

Separating out the coefficients of A, B, C on the two sides of the expression provides the autonomous system of first order differential equations

$$\frac{df_1}{d\epsilon} = -\frac{1}{2}(f_1^2 + 2), \quad \frac{df_2}{d\epsilon} = f_1, \quad \frac{df_3}{d\epsilon} = e^{-f_2}$$

together with the initial conditions $f_1(0) = f_2(0) = f_3(0) = 0$. The solution is

$$f_1(\epsilon) = -\sqrt{2}\tan(\epsilon/\sqrt{2}), \quad f_2(\epsilon) = \ln(\cos^2(\epsilon/\sqrt{2})), \quad f_3(\epsilon) = \sqrt{2}\tan(\epsilon/\sqrt{2}).$$

It follows that

$$e^{i\epsilon(A-B)} = e^{i\sqrt{2}\tan(\epsilon/\sqrt{2})A}e^{i\ln(\cos^2(\epsilon/\sqrt{2}))C}e^{-i\sqrt{2}\tan(\epsilon/\sqrt{2})B}.$$

Problem 33. Let $b_1^\dagger$, $b_2^\dagger$, b_1, b_2 be Bose creation and annihilation operators, respectively.
(i) The operators

$$L_+ = b_1^\dagger b_2, \quad L_- = b_2^\dagger b_1, \quad L_3 = b_1^\dagger b_1 - b_2^\dagger b_2$$

form a basis of a three-dimensional Lie algebra. The 2×2 matrices

$$S_+ = \begin{pmatrix} 0 & 1 \\ 0 & 0 \end{pmatrix}, \quad S_- = \begin{pmatrix} 0 & 0 \\ 1 & 0 \end{pmatrix}, \quad \sigma_3 = \begin{pmatrix} 1 & 0 \\ 0 & -1 \end{pmatrix}$$

form a basis of a three-dimensional Lie algebra. Show that the two Lie algebras are isomorphic.
(ii) Find the Lie algebra generated by the operators

$$\hat{A} = b_1^\dagger b_2^\dagger, \quad \hat{B} = b_1 b_2.$$

Solution 33. (i) We find $[L_+, L_-] = L_3$, $[L_+, L_3] = -2L_+$, $[L_-, L_3] = 2L_-$ and $[S_+, S_-] = \sigma_3$, $[S_+, \sigma_3] = -2S_+$, $[S_-, \sigma_3] = 2S_-$. Hence $L_+ \leftrightarrow S_+$, $L_- \leftrightarrow S_-$, $L_3 \leftrightarrow \sigma_3$.
(ii) We obtain

$$[\hat{A}, \hat{B}] = -I - b_1^\dagger b_1 - b_2^\dagger b_2 = \hat{C}.$$

$$[\hat{A}, [\hat{A}, \hat{B}]] = 2b_1^\dagger b_2^\dagger = 2\hat{A}, \quad [\hat{B}, [\hat{A}, \hat{B}]] = -2b_1 b_2 = -2\hat{B}.$$

Thus we have three-dimensional simple Lie algebra with the basis $\hat{A}$, $\hat{B}$, $\hat{C}$.

Problem 34. Let S, B be positive constants. Consider the differential operators

$$S_1 = S\cosh(\zeta) - \frac{B}{2}\sinh^2(\zeta) - \sinh(\zeta)\frac{d}{d\zeta}$$

$$S_2 = i\left(-S\sinh(\zeta) + \frac{B}{2}\sinh(\zeta)\cosh(\zeta) + \cosh(\zeta)\frac{d}{d\zeta}\right)$$

$$S_3 = \frac{B}{2}\sinh(\zeta) + \frac{d}{d\zeta}.$$

Find the commutators $[S_1, S_2]$, $[S_2, S_3]$, $[S_3, S_1]$.

Solution 34. We show the calculation for $[S_3, S_1]$. Let $: \mathbb{R} \to \mathbb{R}$ be an analytic function. Then applying the product rule

$$[S_3, S_1]f(\zeta) = S_3(S_1 f(\zeta)) - S_1(S_3 f(\zeta))$$

$$= S_3\left(S\cosh(\zeta)f(\zeta) - \frac{B}{2}\sinh^2(\zeta)f(\zeta) - \sinh(\zeta)\frac{df(\zeta)}{d\zeta}\right)$$

$$-S_1\left(\frac{B}{2}\sinh(\zeta)f(\zeta) + \frac{df(\zeta)}{d\zeta}\right)$$

$$= S\sinh(\zeta)f(\zeta) - \frac{B}{2}\sinh(\zeta)\cosh(\zeta)f(\zeta) - \cosh(\zeta)\frac{df(\zeta)}{d\zeta}$$

$$= \left(S \sinh(\zeta) - \frac{B}{2} \sinh(\zeta) \cosh(\zeta) - \cosh(\zeta) \frac{d}{d\zeta} \right) f(\zeta)$$

$$= i S_2 f(\zeta)$$

i.e. $[S_3, S_1] = i S_2$. Analogously we find $[S_1, S_2] = i S_3$, $[S_2, S_3] = i S_1$.

Problem 35.   Consider the orthogonal Lie group

$$O(n) := \{ g \in \mathbb{R}^{n \times n} \ : \ g^T g = I_n \}.$$

It is a smooth manifold of $\mathbb{R}^{n \times n}$. Consider the vector space of symmetric matrices over $\mathbb{R}$

$$\mathcal{S}_n := \{ S \in \mathbb{R}^{n \times n} \ : \ S^T = S \}.$$

We define $f : \mathbb{R}^{n \times n} \to \mathcal{S}_n$ by $f(g) = g^T g$. Show that its derivative $df(g) : \mathbb{R}^{n \times n} \to \mathcal{S}_n$ is given by $df(g)v = g^T v + v^T g$.

Solution 35.   We have

$$\begin{aligned} df(g)v &= \frac{d}{d\tau}((g + \tau v)^T (g + \tau v))_{\tau = 0} \\ &= \frac{d}{d\tau}(g^T g + \tau v^T g + \tau v^t g + \tau^2 v^T v)|_{\tau = 0} \\ &= (g^T v + v^T g + 2\tau v^T v)|_{\tau = 0} = g^T v + v^T g. \end{aligned}$$

This map is surjective for every $g \in O(n)$.

Problem 36.   Consider the vector fields in $\mathbb{R}^2$

$$V_1 = x_1^2 \frac{\partial}{\partial x_1} + x_1 x_2 \frac{\partial}{\partial x_2}, \qquad V_2 = -x_1 \frac{\partial}{\partial x_1} - \frac{x_2}{2} \frac{\partial}{\partial x_2}.$$

Show that we have a basis of a two-dimensional non-abelian Lie algebra.

Solution 36.   The commutator provides $[V_1, V_2] = V_1$.

Problem 37.   Let $c_1^\dagger$, $c_2^\dagger$, c_1, c_2 be Fermi creation and annihilation operators, respectively. Let

$$\hat{K}_1 = c_1^\dagger c_2, \qquad \hat{K}_2 = c_2^\dagger c_1$$

Find the Lie algebra generated by $\hat{K}_1$ and $\hat{K}_2$.

Solution 37.   We obtain

$$[\hat{K}_1, \hat{K}_2] = \hat{K}_3 = c_1^\dagger c_1 - c_2^\dagger c_2, \quad [\hat{K}_1, \hat{K}_3] = -2\hat{K}_1, \quad [\hat{K}_2, \hat{K}_3] = 2\hat{K}_2.$$

Hence $\hat{K}_1$, $\hat{K}_2$, $\hat{K}_3$ form a basis of a three-dimensional simple Lie algebra. Let $\alpha \in \mathbb{R}$. Find $\beta, \gamma, \delta \in \mathbb{R}$ such that (disentanglement technique)

$$e^{\alpha(\hat{K}_1 - \hat{K}_2)} = e^{\beta \hat{K}_1} e^{\gamma \hat{K}_3} e^{\delta K_2}.$$

Supplementary Problems

Problem 1. Show that both $SO(2, \mathbb{R})$ and $SO(1, 1, \mathbb{R})$ are subgroups of $SL(2, \mathbb{R})$.

Problem 2. Consider the differential operators

$$T_0 = \frac{\partial}{\partial x_0}, \quad T_1 = \frac{\partial}{\partial x_1}, \quad T_2 = \frac{\partial}{\partial x_2},$$

$$L_{01} = x_0 \frac{\partial}{\partial x_1} + x_1 \frac{\partial}{\partial x_0}, \quad L_{02} = x_0 \frac{\partial}{\partial x_2} + x_2 \frac{\partial}{\partial x_0}, \quad L_{12} = x_1 \frac{\partial}{\partial x_2} - x_2 \frac{\partial}{\partial x_1}$$

$$D = -\left(\frac{1}{2} + x_0 \frac{\partial}{\partial x_0} + x_1 \frac{\partial}{\partial x_1} + x_2 \frac{\partial}{\partial x_2} \right)$$

$$K_0 = -x_0 + (x_0^2 - x_1^2 - x_2^2 - 2x_0^2) \frac{\partial}{\partial x_0} - 2x_0 x_1 \frac{\partial}{\partial x_1} - 2x_0 x_2 \frac{\partial}{\partial x_2}$$

$$K_1 = x_1 + (x_0^2 - x_1^2 - x_2^2 + 2x_1^2) \frac{\partial}{\partial x_1} + 2x_1 x_0 \frac{\partial}{\partial x_0} + 2x_1 x_2 \frac{\partial}{\partial x_2}$$

$$K_2 = x_2 + (x_0^2 - x_1^2 - x_2^2 + 2x_2^2) \frac{\partial}{\partial x_2} + 2x_2 x_0 \frac{\partial}{\partial x_0} + 2x_2 x_1 \frac{\partial}{\partial x_1}.$$

Find the commutators and show that the differential operators form a basis of a Lie algebra.

Problem 3. Let

$$J_+ = e^{i\phi} \left(\frac{\partial}{\partial \theta} + i \cot(\theta) \frac{\partial}{\partial \phi} \right), \quad J_- = e^{-i\phi} \left(-\frac{\partial}{\partial \theta} + i \cot(\theta) \frac{\partial}{\partial \phi} \right).$$

Find the commutator $[J_+, J_-]$.

Problem 4. Let $A \in SL(2, \mathbb{C})$. Show that

$$A^2 = \mathrm{tr}(A)A - I_2, \quad A + A^{-1} = \mathrm{tr}(A)I_2, \quad A^n = U_{n-1}(x)A - U_{n-2}(x)I_2$$

where $x := \frac{1}{2}\mathrm{tr}(A)$ and $U_n(x)$ are Chebyshev's polynomial of the second kind defined by

$$U_{-1}(x) = 0, \quad U_0(x) = 1, \quad U_{n+1}(x) = 2xU_n(x) - U_{n-1}(x).$$

Apply the *Cayley-Hamilton theorem.*

Problem 5.   Consider the four dimensional Lie algebra with the basis $\{X_1, X_2, X_3, X_4\}$ and the nonzero commutation relations

$$[X_1, X_4] = 2X_1, \quad [X_2, X_3] = X_1, \quad [X_2, X_4] = X_2 + X_3, \quad [X_3, X_4] = X_3.$$

Find a representation with the vector fields

$$V_j = \sum_{k=1}^{4} \sum_{\ell=1}^{4} C^j_{k\ell} x_k \frac{\partial}{\partial x_\ell}$$

where $C^j_{k\ell}$ are the structure constants.

Problem 6.   (i) Let $c^\dagger_1$, $c^\dagger_2$, c_1, c_2 be Fermi creation and annihilation operators, respectively. Find the Lie algebra generated by the operators

$$c^\dagger_1 c^\dagger_2, \quad c_2 c_1.$$

(ii) Let $c^\dagger_1$, $c^\dagger_2$, $c^\dagger_3$, c_1, c_2, c_3 be Fermi creation and annihilation operators, respectively. Find the Lie algebra generated by

$$c^\dagger_1 c^\dagger_2 c^\dagger_3, \quad c_1 c_2 c_3.$$

Note that

$$[c^\dagger_1 c^\dagger_2 c^\dagger_3, c_1 c_2 c_3] = I - \sum_{j=1}^{3} c^\dagger_j c_j + c^\dagger_1 c_1 c^\dagger_2 c_2 + c^\dagger_1 c_1 c^\dagger_3 c_3 + c^\dagger_2 c_2 c^\dagger_3 c_3 - 2c^\dagger_1 c_1 c^\dagger_2 c_2 c^\dagger_3 c_3.$$

(iii) Consider the Fermi creation and annihilation operators $c^\dagger_\uparrow$, $c^\dagger_\downarrow$, $c_\uparrow$, $c_\downarrow$. Find the Lie algebra generated by these four operators. Note that

$$[c^\dagger_\downarrow, c_\downarrow] = 2c^\dagger_\downarrow c_\downarrow - I, \quad [c^\dagger_\uparrow, c_\uparrow] = 2c^\dagger_\uparrow c_\uparrow - I$$

$$[c^\dagger_\uparrow, c_\downarrow] = 2c^\dagger_\uparrow c_\downarrow, \quad [c^\dagger_\downarrow, c_\uparrow] = 2c^\dagger_\downarrow c_\uparrow \quad [c^\dagger_\uparrow, c^\dagger_\downarrow] = 2c^\dagger_\uparrow c^\dagger_\downarrow, \quad [c_\uparrow, c_\downarrow] = 2c_\uparrow c_\downarrow$$

and

$$[c^\dagger_\uparrow c^\dagger_\downarrow, c_\uparrow c_\downarrow] = I - c^\dagger_\uparrow c_\uparrow - c^\dagger_\downarrow c_\downarrow, \quad [c^\dagger_\uparrow c_\downarrow, c^\dagger_\downarrow c_\uparrow] = c^\dagger_\uparrow c_\uparrow - c^\dagger_\downarrow c_\downarrow$$

where I is the identity matrix.

Problem 7.   Show that the Lie algebra $s\ell(n + 1, \mathbb{R})$ can be realized by the vector fields ($j, k = 1, \ldots, n$)

$$T_j = \frac{\partial}{\partial x_j}, \quad L_{jk} = x_j \frac{\partial}{\partial x_k}, \quad C_j = x_j \sum_{k=1}^{n} x_k \frac{\partial}{\partial x_k}.$$

Problem 8. The *Chevalley bases* for the Lie algebra $so(2\ell + 1)$ is given by the $2^\ell \times 2^\ell$ matrices

$$H_\mu = \underbrace{I_2 \otimes \cdots \otimes I_2}_{\mu-1} \otimes \frac{1}{2}(\sigma_3 \otimes I_2 - I_2 \otimes \sigma_3) \otimes \underbrace{I_2 \otimes \cdots \otimes I_2}_{\ell-\mu-1}$$

$$H_\ell = \underbrace{I_2 \otimes \cdots \otimes I_2}_{\ell-1} \otimes \sigma_3$$

$$E_\mu = \underbrace{I_2 \otimes \cdots \otimes I_2}_{\mu-1} \otimes (\sigma_+ \otimes \sigma_-) \otimes \underbrace{I_2 \otimes \cdots \otimes I_2}_{\ell-\mu-1}$$

$$E_\ell = \underbrace{\sigma_3 \otimes \cdots \otimes \sigma_3}_{\ell-1} \otimes \sigma_+.$$

For $\ell = 2$ we have

$$H_1 = \frac{1}{2}(\sigma_3 \otimes I_2 - I_2 \otimes \sigma_3), \quad H_2 = I_2 \otimes \sigma_3, \quad E_1 = \sigma_+ \otimes \sigma_-, \quad E_2 = \sigma_3 \otimes \sigma_+.$$

Find the commutators and anticommutators.

Problem 9. Let $b^\dagger$, b be Bose creation and annihilation operators. Find the Lie algebra generated by $b^\dagger b$, $\sqrt{b^\dagger b}$, $b^\dagger + b$.

Problem 10. (i) Let $b_1^\dagger$, $b_2^\dagger$, b_1, b_2 be Bose creation and annihilation operators, respectively. Let

$$T_+ = b_1^\dagger b_2^\dagger, \quad T_- = b_1 b_2, \quad T_3 = \frac{1}{2}(b_1^\dagger b_1 + b_2^\dagger b_2 + I)$$

where $I = I_B \otimes I_B$ is the identity operator. Show that

$$[T_+, T_-] = -2T_3, \quad [T_3, T_+] = T_+, \quad [T_3, T_-] = -T_-.$$

(ii) Let $c_1^\dagger$, $c_2^\dagger$, c_1, c_2 be Fermi creation and annihilation operators, respectively. Let

$$T_+ = c_1^\dagger c_2^\dagger, \quad T_- = c_1 c_2, \quad T_3 = \frac{1}{2}(c_1^\dagger c_1 + c_2^\dagger c_2 + I)$$

where $I = I_F \otimes I_F$ is the identity operator. Show that

$$[T_+, T_-] = -2T_3, \quad [T_3, T_+] = T_+, \quad [T_3, T_-] = -T_-.$$

Problem 11. Let σ_1, σ_2, σ_3 be the Pauli spin matrices. Show that the fourteen 8×8 matrices

$$L_1 = \begin{pmatrix} -i\sigma_2 & 0_2 & 0_2 & 0_2 \\ 0_2 & -i\sigma_2 & 0_2 & 0_2 \\ 0_2 & 0_2 & -i\sigma_2 & 0_2 \\ 0_2 & 0_2 & 0_2 & i\sigma_2 \end{pmatrix}, \quad R_1 = \begin{pmatrix} -i\sigma_2 & 0_2 & 0_2 & 0_2 \\ 0_2 & i\sigma_2 & 0_2 & 0_2 \\ 0_2 & 0_2 & i\sigma_2 & 0_2 \\ 0_2 & 0_2 & 0_2 & -i\sigma_2 \end{pmatrix}$$

$$L_2 = \begin{pmatrix} 0_2 & -\sigma_3 & 0_2 & 0_2 \\ \sigma_3 & 0_2 & 0_2 & 0_2 \\ 0_2 & 0_2 & 0_2 & -I_2 \\ 0_2 & 0_2 & I_2 & 0_2 \end{pmatrix}, \quad R_2 = \begin{pmatrix} 0_2 & -I_2 & 0_2 & 0_2 \\ I_2 & 0_2 & 0_2 & 0_2 \\ 0_2 & 0_2 & 0_2 & I_2 \\ 0_2 & 0_2 & -I_2 & 0_2 \end{pmatrix}$$

$$L_3 = \begin{pmatrix} 0_2 & -\sigma_1 & 0_2 & 0_2 \\ \sigma_1 & 0_2 & 0_2 & 0_2 \\ 0_2 & 0_2 & 0_2 & -i\sigma_2 \\ 0_2 & 0_2 & -i\sigma_2 & 0_2 \end{pmatrix}, \quad R_3 = \begin{pmatrix} 0_2 & -i\sigma_2 & 0_2 & 0_2 \\ -i\sigma_2 & 0_2 & 0_2 & 0_2 \\ 0_2 & 0_2 & 0_2 & i\sigma_2 \\ 0_2 & 0_2 & i\sigma_2 & 0_2 \end{pmatrix}$$

$$L_4 = \begin{pmatrix} 0_2 & 0_2 & -\sigma_3 & 0_2 \\ 0_2 & 0_2 & 0_2 & I_2 \\ \sigma_3 & 0_2 & 0_2 & 0_2 \\ 0_2 & -I_2 & 0_2 & 0_2 \end{pmatrix}, \quad R_4 = \begin{pmatrix} 0_2 & 0_2 & -I_2 & 0_2 \\ 0_2 & 0_2 & 0_2 & -I_2 \\ I_2 & 0_2 & 0_2 & 0_2 \\ 0_2 & I_2 & 0_2 & 0_2 \end{pmatrix}$$

$$L_5 = \begin{pmatrix} 0_2 & 0_2 & -\sigma_1 & 0_2 \\ 0_2 & 0_2 & 0_2 & i\sigma_2 \\ \sigma_1 & 0_2 & 0_2 & 0_2 \\ 0_2 & i\sigma_2 & 0_2 & 0_2 \end{pmatrix}, \quad R_5 = \begin{pmatrix} 0_2 & 0_2 & -i\sigma_2 & 0_2 \\ 0_2 & 0_2 & 0_2 & -i\sigma_2 \\ -i\sigma_2 & 0_2 & 0_2 & 0_2 \\ 0_2 & -i\sigma_2 & 0_2 & 0_2 \end{pmatrix}$$

$$L_6 = \begin{pmatrix} 0_2 & 0_2 & 0_2 & -I_2 \\ 0_2 & 0_2 & -\sigma_3 & 0_2 \\ 0_2 & \sigma_3 & 0_2 & 0_2 \\ I_2 & 0_2 & 0_2 & I_2 \end{pmatrix}, \quad R_6 = \begin{pmatrix} 0_2 & 0_2 & 0_2 & -\sigma_3 \\ 0_2 & 0_2 & \sigma_3 & 0_2 \\ 0_2 & -\sigma_3 & 0_2 & 0_2 \\ \sigma_3 & 0_2 & 0_2 & 0_2 \end{pmatrix}$$

$$L_7 = \begin{pmatrix} 0_2 & 0_2 & 0_2 & -i\sigma_2 \\ 0_2 & 0_2 & -\sigma_1 & 0_2 \\ 0_2 & \sigma_1 & 0_2 & 0_2 \\ -i\sigma_2 & 0_2 & 0_2 & 0_2 \end{pmatrix}, \quad R_7 = \begin{pmatrix} 0_2 & 0_2 & 0_2 & -\sigma_1 \\ 0_2 & 0_2 & \sigma_1 & 0_2 \\ 0_2 & -\sigma_1 & 0_2 & 0_2 \\ \sigma_1 & 0_2 & 0_2 & 0_2 \end{pmatrix}$$

form a basis of the *exceptional Lie algebra* g_2.

Problem 12. A basis of the Lie algebra $su(1,1)$ is given by K_-, K_+, K_0 with the commutation relations

$$[K_0, K_+] = K_+, \quad [K_0, K_-] = -K_-, \quad [K_-, K_+] = 2K_0.$$

Show that the *Casimir operator* is given by

$$C = K_0^2 - \frac{1}{2}(K_+K_- + K_-K_+).$$

Show that the Lie algebra can be realized with Bose creation and annihilation operators as

$$K_0 = \frac{1}{4}(b^\dagger b + bb^\dagger) \equiv \frac{1}{2}\left(b^\dagger b + \frac{1}{2}I\right), \quad K_- = \frac{1}{2}b^2, \quad K_+ = \frac{1}{2}(b^\dagger)^2.$$

Show that from this realization one has

$$K_0|m, k\rangle = (m + k)|m, k\rangle$$

for basis states $|m, k\rangle$ $(m = 0, 1, 2, \ldots)$ and for k one obtains $k = \frac{1}{4}$ and $k = \frac{3}{4}$. The coherent states are given by

$$|\xi, k\rangle = S(z)|0, k\rangle, \quad S(z) = \exp(zK_+ - \bar{z}K_-)$$

where $z = -(\theta/2)e^{-i\phi}$, $\xi = -\tanh(\theta/2)e^{-i\phi}$ and θ and ϕ have the range $-\infty < \theta < \infty$ and $0 \le \phi \le 2\pi$. Find $\langle k, m|\xi, k\rangle$.

Problem 13. Find nonzero 2×2 matrices A, B, C over $\mathbb{C}$ such that

$$[A, B] = 0_2, \quad [C, A] = iB, \quad [C, B] = -iA.$$

Problem 14. Let $p > 0$. Given the basis $\{V_1, V_2, V_3\}$ of a Lie algebra with the commutation relation

$$[V_1, V_2] = 0, \quad [V_2, V_3] = V_1 + pV_2, \quad [V_3, V_1] = -pV_1 + V_2.$$

Show that a representation is given by the vector fields

$$V_1 = (px_1 - x_2)\frac{\partial}{\partial x_3}, \quad V_2 = (x_1 + px_2)\frac{\partial}{\partial x_3},$$

$$V_3 = -(px_1 - x_2)\frac{\partial}{\partial x_1} - (x_1 + px_2)\frac{\partial}{\partial x_2}.$$

Problem 15. Show that the differential operators

$$D_1 = -\frac{1}{2}\left(r\frac{d^2}{dr^2} + 2\frac{d}{dr} - \frac{\ell(\ell+1)}{r} + r\right), \quad D_2 = -i\left(1 + r\frac{d}{dr}\right),$$

$$D_3 = -\frac{1}{2}\left(r\frac{d^2}{dr^2} + 2\frac{d}{dr} + \frac{\ell(\ell+1)}{r} - r\right)$$

satisfy the commutation relations (Lie algebra $so(2, 1)$)

$$[D_1, D_2] = -iD_3, \quad [D_2, D_3] = iD_1, \quad [D_3, D_1] = iD_2.$$

Show that the Casimir operator is $C = D_3^2 - D_1^2 - D_2^2$.

Problem 16. Show that the vector fields

$$V = x_1\frac{\partial}{\partial x_2}, \quad W = x_2\frac{\partial}{\partial x_1}, \quad U = \frac{1}{2}\left(x_1\frac{\partial}{\partial x_1} - x_2\frac{\partial}{\partial x_2}\right)$$

form a basis of a Lie algebra under the commutator.

Problem 17.   Let $b^\dagger$, b be Bose creation and annihilation operators. Consider $b^\dagger$, b, $b^\dagger b$. Then we have the commutation relations

$$[b, b^\dagger] = I, \quad [b, b^\dagger b] = b, \quad [b^\dagger, b^\dagger b] = -b^\dagger$$

where I is the identity operator. Show that we have the following representations with differential operators

$$b \rightarrow \frac{1}{\sqrt{2}}\left(x + \frac{d}{dx}\right), \quad b^\dagger \rightarrow \frac{1}{\sqrt{2}}\left(x - \frac{d}{dx}\right)$$

$$b^\dagger b \rightarrow \frac{1}{2}\left(x^2 - \frac{d^2}{dx^2} - 1\right).$$

Problem 18.   Let $b^\dagger$, b be Bose creation and annihilation operators. Consider the operators

$$\hat{T}_1 = (b^\dagger)^2, \quad \hat{T}_2 = -b^2, \quad \hat{T}_3 = 4\left(b^\dagger b + \frac{1}{2}I\right).$$

Show that they satisfy the commutation relations

$$[\hat{T}_1, \hat{T}_2] = T_3, \quad [\hat{T}_1, \hat{T}_3] = -8T_1, \quad [\hat{T}_2, \hat{T}_3] = 8T_2.$$

Show that

$$\exp(\epsilon((b^\dagger)^2 - b^2)) =$$

$$\exp\left(-\frac{1}{2}b^2 \tanh(2\epsilon)\right) \exp\left(\left(b^\dagger b + \frac{1}{2}I\right)\ln(\cosh(2\epsilon))\right) \exp\left(\frac{1}{2}(b^\dagger)^2 \tanh(2\epsilon)\right).$$

Problem 19.   Let $N \geq 3$ and $X_1, X_2, \ldots, X_N$ be a basis of a semi-simple Lie algebra. Hence

$$[X_i, X_j] = \sum_{k=1}^{N} c_{ij}^k X_k, \quad i, j = 1, \ldots, N$$

where the *structure constants* are satisfying

$$c_{ij}^k = -c_{ji}^k, \quad \sum_{k=1}^{N}(c_{ij}^k c_{k\ell}^m + c_{j\ell}^k c_{ki}^m + c_{\ell i}^k c_{kj}^m) = 0.$$

Now we can form a metric tensor field

$$g = \sum_{i=1}^{N}\sum_{\ell=1}^{N} g_{i\ell} dx_i \otimes dx_\ell, \quad g_{i\ell} := \sum_{k=1}^{N}\sum_{j=1}^{N} c_{ij}^k c_{\ell k}^j.$$

The $N \times N$ matrix $(g_{i\ell})$ is invertible given by $(g^{\ell m})$ with $\sum_{\ell=1}^{N} g_{i\ell} g^{\ell m} = \delta_i^m$. One defines

$$X^{\ell} = \sum_{m=1}^{N} g^{\ell m} X_m$$

The second-order *Casimir operator* is given by

$$C_2 = \sum_{i=1}^{N} \sum_{\ell=1}^{N} g^{im} X_i X_m$$

A generalization of this operator is given by

$$C_p = \sum_{j_1, j_2, \ldots, j_p = 1}^{N} \sum_{i_1, i_2, \ldots, i_p = 1}^{N} c_{i_1 j_1}^{j_2} c_{i_2 j_2}^{j_3} \ldots c_{i_{p-1} j_{p-1}}^{j_p} c_{i_p j_p}^{j_1} X^{i_1} X^{i_2} \ldots X^{i_p}$$

Consider the Lie algebra $so(3)$ with the basis given by the vector fields

$$X_1 = x_3 \frac{\partial}{\partial x_2} - x_2 \frac{\partial}{\partial x_3}, \quad X_2 = x_1 \frac{\partial}{\partial x_3} - x_3 \frac{\partial}{\partial x_1}, \quad X_3 = x_2 \frac{\partial}{\partial x_1} - x_1 \frac{\partial}{\partial x_2}$$

Find the metric tensor field g and then calculate the Lie derivatives

$$L_{X_1} g, \quad L_{X_2} g, \quad L_{X_3} g.$$

Find the metric tensor field g for the exceptional Lie algebra g_2 (14 dimensional).

Problem 20. Show that the differential operators

$$T_0 = -i \frac{\partial}{\partial \phi}, \quad T_{\pm} = e^{\pm i \phi} \left(\pm \frac{\partial}{\partial x} + f(x) \left(i \frac{\partial}{\partial \phi} \mp \frac{1}{2} \right) + g(x) \right)$$

are a realization of the semisimple Lie algebra $so(2,1)$ if the smooth functions f and g satisfy the autonomous system of first order differential equations

$$\frac{df}{dx} = 1 - f^2, \quad \frac{dg}{dx} = -fg.$$

This means the differential operators have to satisfy

$$[T_+, T_-] = -2T_0, \quad [T_0, T_{\pm}] = \pm T_{\pm}.$$

Find solutions of the system of first order differential equations. First find the fixed points.

Problem 21. Show that the differential operators

$$D = \frac{\partial}{\partial w}, \quad T_+ = \frac{\partial}{\partial z_1} + \frac{1}{2} z_2 \frac{\partial}{\partial w}, \quad T_- = \frac{\partial}{\partial z_2} - \frac{1}{2} z_1 \frac{\partial}{\partial w}$$

satisfy the commutation relations $[D, T_+] = 0$, $[D, T_-] = 0$, $[T_+, T_-] = -D$.

Problem 22.   Consider the time-dependent Hamilton operator

$$\hat{H} = f_3(t)\frac{\hat{p}^2}{2m} + \frac{\omega_0}{2}f_2(t)(\hat{x}\hat{p} + \hat{p}\hat{x}) + f_1(t)\frac{m\omega_0^2}{2}\hat{x}^2$$

where f_1, f_2, f_3 are real smooth and bounded functions of t with $f_1(t_0) = f_3(t_0) = 1$, $f_2(t_0) = 0$. Introducing the Bose creation and annihilation operators

$$b^\dagger = (2m\omega_0\hbar)^{-1/2}(m\omega_0\hat{x} - i\hat{p}), \quad b = (2m\omega\hbar)^{-1/2}(m\omega x + i\hat{p})$$

show that the Hamilton operator takes the form

$$\hat{H}(t) = \hbar\omega_0\left((f_1(t) + f_3(t)\hat{K}_0 + \left(\frac{1}{2}(f_1(t) - f_2(t)) + if_2(t)\right)K_+ + \right.$$
$$\left.\left(\frac{1}{2}(f_1(t) - f_3(t)) - if_2(t)\right)K_-\right)$$

where the operators $\hat{K}_0$, $\hat{H}_+$, $\hat{K}_-$ are given by

$$\hat{K}_0 = \frac{1}{2}\left(b^\dagger b + \frac{1}{2}I\right), \quad \hat{K}_+ = \frac{1}{2}b^\dagger b^\dagger, \quad \hat{K}_- = \frac{1}{2}bb.$$

Show that $[K_0, K_+] = K_+$, $[K_0, K_-] = -K_-$, $[K_+, K_-] = -2K_0$ i.e. the operators satisfy the commutation relations of the Lie algebra $so(2, 1)$. Introduce the time-dependent Bose operators

$$b(t) = \mu(t)b + \nu(t)b^\dagger, \qquad b^\dagger(t) = \overline{\nu}(t)b + \overline{\mu}(t)b^\dagger$$

with $|\mu(t)|^2 - |\nu(t)|^2 = 1$ so that $[b(t), b^\dagger(t)] = I$ and $\mu(t_0) = 1$, $\nu(t_0) = 0$. Show that

$$T_0(t) = \frac{1}{2}\left(b^\dagger(t)b(t) + \frac{1}{2}I\right), \quad T_+(t) = \frac{1}{2}b^\dagger(t)b^\dagger(t), \quad T_-(t) = \frac{1}{2}b(t)b(t)$$

also satisfy the commutation of the Lie algebra $so(2, 1)$.

Problem 23.   Let $n > 2$ and $j, k = 1, 2, \ldots, n$. Do the vector fields

$$\frac{\partial}{\partial x_{jk}}, \quad x_{jk}\frac{\partial}{\partial x_{jk}}, \quad x_{\ell j}\frac{\partial}{\partial x_{kj}} \ (\ell \neq k), \quad x_{jk}x_{\ell m}\frac{\partial}{\partial x_{j\ell}}$$

form a Lie algebra under the commutator?

Problem 24.   Let $n \geq 2$ and $X_0, X_1, \ldots, X_{n^2-1}$ be an orthonormal basis in the Hilbert space of $n \times n$ matrices over $\mathbb{C}$ with scalar product $\langle A, B \rangle =$

$tr(AB^*)$. Let $Y \in SL(n, \mathbb{C})$, i.e. $\det(Y) = 1$. Then we can consider the map from $SL(2, \mathbb{C})$ to the vector space of $n^2 \times n^2$ matrices $\Delta = (\Delta_{\mu\nu}, (\mu, \nu = 0, 1, \ldots, n^2 - 1)$

$$\Delta_{\mu\nu} = tr(YX_\nu Y^* X_\mu), \quad \mu, \nu = 0, 1, \ldots, n^2 - 1.$$

Consider the special case $n = 2$. Then

$$\frac{1}{\sqrt{2}}\sigma_0, \quad \frac{1}{\sqrt{2}}\sigma_1, \quad \frac{1}{\sqrt{2}}\sigma_2, \quad \frac{1}{\sqrt{2}}\sigma_3$$

form an orthonormal basis with σ_1, σ_2, σ_3 be the Pauli spin matrices and $\sigma_0 = I_2$. Then

$$\Delta_{\mu\nu} = \frac{1}{2}tr(A\sigma_\nu A^* \sigma_\mu), \quad \mu, \nu = 0, 1, 2, 3.$$

Let $A_1 = \begin{pmatrix} 1 & 1 \\ 0 & 1 \end{pmatrix}$. Find Δ_1. Let $A_2 = \begin{pmatrix} 1 & 0 \\ 1 & 1 \end{pmatrix}$. Find Δ_2. Find the commutators $[A_1, A_2]$ and $[\Delta_1, \Delta_2]$.

Problem 25. Let $A \in SL(2, \mathbb{C})$ and $\sigma_0 = I_2$, σ_1, σ_2, σ_3 be the Pauli spin matrices. We define the 4×4 matrix $L = (L_{\mu\nu})$

$$L_{\mu\nu} = \frac{1}{2}tr(A\sigma_\nu A^* \sigma_\mu), \quad \mu, \nu = 0, 1, 2, 3.$$

Describe the properties of this matrix. Let $A = \begin{pmatrix} 1 & 0 \\ \epsilon & 1 \end{pmatrix}$ with $\epsilon \in \mathbb{R}$. Write a computer algebra program that finds L.

Problem 26. Consider the linear operators $\hat{E}$, $\hat{x}$, $\hat{p}$ satisfying the commutation relations

$$[\hat{E}, \hat{x}] = -i\frac{\hbar}{m}\hat{p}, \quad [\hat{E}, \hat{p}] = im\omega^2\hbar\hat{x}, \quad [\hat{x}, \hat{p}] = i\hbar\left(1 + \frac{1}{mc^2}\hat{E}\right).$$

Show that the Lie algebra can be realized by

$$\hat{E} = i\hbar\frac{\partial}{\partial t}$$

$$\hat{p} = -i\hbar\alpha\cos(\omega t)\frac{\partial}{\partial x} + i\hbar\frac{\omega x \sin(\omega t)}{\alpha c^2}\frac{\partial}{\partial t} + \frac{m\omega\sin(\omega t)}{\alpha}x$$

$$\hat{x} = \frac{\cos(\omega t)}{\alpha}x + i\hbar\frac{\alpha\sin(\omega t)}{m\omega}\frac{\partial}{\partial x} + i\hbar\frac{x\cos(\omega t)}{\alpha mc^2}\frac{\partial}{\partial t}$$

where α is the dimensionless quantity $\alpha = \sqrt{1 + \omega^2 x^2/c^2}$.

Problem 27.   Let $\hat{K}_1$, $\hat{K}_2$, $\hat{K}_3$ be three linear operators satisfying the commutation relations $[\hat{K}_1, \hat{K}_2] = \hat{K}_3$, $[\hat{K}_1, \hat{K}_3] = -2\hat{K}_1$, $[\hat{K}_2, \hat{K}_3] = 2\hat{K}_2$. Hence we have a basis of a simple three-dimensional Lie algebra. Let $\epsilon \in \mathbb{R}$. We set

$$\exp(\epsilon(K_1 - K_2)) = \exp(f_1(\epsilon)K_1) \exp(f_3(\epsilon)K_3) \exp(f_2(\epsilon)K_2).$$

Applying differentiation with respect to ϵ and comparing coefficients of K_1, K_2, K_3 we find an autonomous system of first order differential equations. Solve the system together with the initial conditions $f_1(0) = 0$, $f_2(0) = 0$, $f_3(0) = 0$. Apply this information to find $\exp(\epsilon(K_1 - K_2))$. Show that

$$e^{f_1 K_1} K_3 e^{-f_1 K_1} = K_3 - 2f_1 K_1$$

$$e^{f_3 K_3} K_2 e^{-f_3 K_3} = K_2 e^{-2f_3}$$

$$e^{f_1 K_1} K_2 e^{-f_1 K_1} = K_2 + f_1 K_3 - f_1^2 K_1.$$

Chapter 23

Differential Forms and Matrix-Valued Differential Forms

We define differential p-forms of class C^∞ on an open set Ω of $\mathbb{R}^n$ to be the expressions

$$\omega := \sum_{j_1 < j_2 < \cdots < j_p}^{n} c_{j_1 j_2 \ldots j_p}(\mathbf{x}) dx_{j_1} \wedge dx_{j_2} \wedge \cdots \wedge dx_{j_p}$$

where the functions $c_{j_1 j_2 \ldots j_p} \in C^\infty(\Omega)$ and the integers j_1, ..., j_p lie between 1 and n. Two such differential forms may be added componentwise. One defines the *Graßmann product* (also called *exterior product* or *wedge product*) of a p-form and a q-form as follows: For any permutation σ of the indices $j_1, \ldots, j_p$,

$$dx_{\sigma(j_1)} \wedge dx_{\sigma(j_2)} \wedge \cdots \wedge dx_{\sigma(j_p)} = \mathrm{sgn}(\sigma) dx_{j_1} \wedge dx_{j_2} \wedge \cdots \wedge dx_{j_p}$$

where $\mathrm{sgn}(\sigma)$ denotes the sign of the permutation σ. Let

$$\omega' = \sum_{k_1 < k_2 < \ldots < k_q}^{n} b_{k_1 k_2 \ldots k_q}(\mathbf{x}) dx_{k_1} \wedge dx_{k_2} \wedge \cdots \wedge dx_{k_q}.$$

Then

$$\omega \wedge \omega' := \sum_{\substack{j_1 < j_2 < \ldots < j_p \\ k_1 < k_2 < \ldots < k_q}} c_{j_1 j_2 \ldots j_p}(\mathbf{x}) b_{k_1 k_2 \ldots k_q}(\mathbf{x}) dx_{j_1} \wedge \cdots \wedge dx_{j_p} \wedge dx_{k_1} \wedge \cdots \wedge dx_{k_q}.$$

The *exterior derivative* d of a p-form ω is defined by

$$d\omega = \sum_{j_1 < j_2 < \cdots < j_p} dc_{j_1 j_2 \ldots j_p}(\mathbf{x}) \wedge dx_{j_1} \wedge dx_{j_2} \wedge \cdots \wedge dx_{j_p}.$$

Thus $d(d\omega) = 0$. For a smooth function f (a zero form) we have

$$df = \sum_{j=1}^{n} \frac{\partial f}{\partial x_j} dx_j.$$

Note that (associative law)

$$\alpha \wedge (\beta \wedge \gamma) = (\alpha \wedge \beta) \wedge \gamma$$

and

$$d(\alpha \wedge \beta) = (d\alpha) \wedge \beta + (-1)^{|\alpha|} \alpha \wedge (d\beta).$$

Let

$$\widetilde{\alpha} := \sum_{j=1}^{n} (a_j(x,t)dx + A_j(x,t)dt) \otimes X_j$$

be a *Lie algebra-valued differential form* where $\{X_1, X_2, \ldots, X_n\}$ forms a basis of a semi-simple Lie algebra. The exterior derivative is defined as

$$d\widetilde{\alpha} := \sum_{j=1}^{n} \left(-\frac{\partial a_j}{\partial t} + \frac{\partial A_j}{\partial x} \right) dx \wedge dt \otimes X_j.$$

The commutator is defined as

$$[\widetilde{\alpha}, \widetilde{\alpha}] := \sum_{k=1}^{n} \sum_{j=1}^{n} (a_k A_j - a_j A_k) dx \wedge dt \otimes [X_k, X_j].$$

The *covariant exterior derivative* is defined as

$$D_{\widetilde{\alpha}} \widetilde{\alpha} := d\widetilde{\alpha} + \frac{1}{2} [\widetilde{\alpha}, \widetilde{\alpha}].$$

Given a differentiable manifold M and a Lie algebra (semi-simple) valued differential form A over it. In three dimensions the Chern-Simons differential thee form is given by

$$\text{tr}(A \wedge dA + \frac{2}{3} A \wedge A \wedge A) \equiv \text{tr}(F \wedge A - \frac{1}{3} A \wedge A \wedge A)$$

where the curvature is defined as $F = dA + A \wedge A$.

Problem 1. Let $\Omega = \mathbb{R}^3$ and consider the one-form

$$\alpha := f_1(\mathbf{x})dx_1 + f_2(\mathbf{x})dx_2 + f_3(\mathbf{x})dx_3. \tag{1}$$

Calculate $d\alpha$. Let $d\alpha = 0$. Find the condition on the functions f_1, f_2 and f_3.

Solution 1. Applying $dx_j \wedge dx_k = -dx_k \wedge dx_j$ yields

$$d\alpha = \left(\frac{\partial f_2}{\partial x_1} - \frac{\partial f_1}{\partial x_2}\right) dx_1 \wedge dx_2 + \left(\frac{\partial f_3}{\partial x_2} - \frac{\partial f_2}{\partial x_3}\right) dx_2 \wedge dx_3$$
$$+ \left(\frac{\partial f_1}{\partial x_3} - \frac{\partial f_3}{\partial x_1}\right) dx_3 \wedge dx_1.$$

Thus the condition $d\alpha = 0$ leads to

$$\frac{\partial f_2}{\partial x_1} - \frac{\partial f_1}{\partial x_2} = 0, \qquad \frac{\partial f_3}{\partial x_2} - \frac{\partial f_2}{\partial x_3} = 0, \qquad \frac{\partial f_1}{\partial x_3} - \frac{\partial f_3}{\partial x_1} = 0$$

since $dx_1 \wedge dx_2$, $dx_2 \wedge dx_3$ and $dx_3 \wedge dx_1$ are linearly independent. A non trivial solution is $f_1 = x_1$, $f_2 = x_2$, $f_3 = x_3$.

Problem 2. Let $f : \mathbb{R}^n \to \mathbb{R}$ be a C^1 function with $f(\mathbf{x}) \neq 0$ for all $\mathbf{x} \in \mathbb{R}^n$ and α be a differential one-form defined on $\mathbb{R}^n$. Assume that

$$d(f\alpha) = 0. \tag{1}$$

Show that $\alpha \wedge d\alpha = 0$.

Solution 2. From (1) it follows that

$$d(f\alpha) = (df) \wedge \alpha + f d\alpha = 0.$$

Taking the exterior product with α yields $\alpha \wedge (df) \wedge \alpha + \alpha \wedge (f d\alpha) = 0$. Applying the associative law for differential forms and that $\alpha \wedge \alpha = 0$ for differential one-forms gives $\alpha \wedge (f d\alpha) = 0$. Thus $f\alpha \wedge d\alpha = 0$. Since $f(\mathbf{x}) \neq 0$ for all $\mathbf{x} \in \mathbb{R}^n$, equation $\alpha \wedge d\alpha = 0$ follows.

Problem 3. Consider a differential p-form ω defined on $\mathbb{R}^n$. If there is a $(p-1)$-form ψ defined on $\mathbb{R}^n$ such that

$$\omega = d\psi \tag{1}$$

then ω is called an *exact differential form*.
(i) Show that if ω_1 and ω_2 are exact p-differential forms then $\omega_1 \wedge \omega_2$ is also exact.

(ii) Give an example.

Solution 3.   (i) Since ω_1 and ω_2 are exact we have $\omega_1 = d\psi_1$, $\omega_2 = d\psi_2$. Since ω_1 and ω_2 are differential p-forms we find that $\omega_1 \wedge \omega_2$ is a $2p$ differential form. We have to find a $2p - 1$ differential form α defined on $\mathbb{R}^n$ such that

$$d\alpha = \omega_1 \wedge \omega_2.$$

Thus $d\alpha = d\psi_1 \wedge d\psi_2$. Since $d(\psi_1 \wedge d\psi_2) = d\psi_1 \wedge d\psi_2$, where we used that $dd\psi_2 = 0$ we find $\alpha = \psi_1 \wedge d\psi_2$.
(ii) Consider

$$\omega_1 = dx_1 \wedge dx_2, \qquad \omega_2 = dx_3 \wedge dx_4$$

defined on $\mathbb{R}^4$. Then we can choose ψ_1 and ψ_2 as $\psi_1 = x_1 dx_2$, $\psi_2 = x_3 dx_4$. Therefore

$$\psi_1 \wedge d\psi_2 = \psi_1 \wedge \omega_2 = x_1 dx_2 \wedge dx_3 \wedge dx_4 \,.$$

ψ_1 and ψ_2 are not unique, for example we could also choose

$$\psi_1 = \frac{1}{2}(x_1 dx_2 - x_2 dx_1), \qquad \psi_2 = \frac{1}{2}(x_3 dx_4 - x_4 dx_3).$$

Problem 4.   Let $e_1, \ldots, e_n$ be a moving (orthonormal) frame (orthonormal basis) in the normed space $\mathbb{R}^n$. Let ω_j^k its associated *connection forms*, i.e.

$$de_k = \sum_{j=1}^{n} \omega_k^j e_j, \quad k = 1, \ldots, n.$$

We have $\omega_j^k = -\omega_k^j$ and therefore $\omega_j^j = 0$ for $j = 1, \ldots, n$. Let α^j be the basic differential one forms associated with the moving frame by

$$d\mathbf{x} = \sum_{j=1}^{n} \alpha^j e_j$$

where $\mathbf{x} = (x_1, x_2, \ldots, x_n)$. Note that each x_j is viewed as the jth coordinate function on $\mathbb{R}^n$ defined by $x_j(\mathbf{p}) = p_j$ for each $\mathbf{p} \in \mathbb{R}^n$.
(i) Find the condition which follows from $d(d\mathbf{x}) = \mathbf{0}$.
(ii) Find the condition which follows from $d(de_k) = \mathbf{0}$ for $k = 1, \ldots, n$.

Solution 4.   (i) We have

$$\mathbf{0} = dd\mathbf{x} = \sum_{j=1}^{n} (d\alpha^j e_j - \alpha^j \wedge de_j) = \sum_{j=1}^{n} d\alpha^j e_j - \sum_{k=1}^{n} \alpha^k \wedge de_k$$

$$= \sum_{j=1}^{n} d\alpha^j e_j - \sum_{\ell=1}^{n} \alpha^\ell \wedge \left(\sum_{j=1}^{n} \omega_\ell^j e_j \right) = \sum_{j=1}^{n} \left(d\alpha^j - \sum_{\ell=1}^{n} \alpha^\ell \wedge \omega_\ell^j \right) e_j.$$

Since at each point $\mathbf{e}_1, \ldots, \mathbf{e}_n$ form an orthonormal basis we obtain

$$d\alpha^j = \sum_{\ell=1}^{n} \alpha^\ell \wedge \omega_\ell^j, \quad j = 1, \ldots, n.$$

These equations are called Cartan's first structural equations.

(ii) For each $k = 1, \ldots, n$ we have

$$\mathbf{0} = dd\mathbf{e}_k = \sum_{j=1}^{n}(d\omega_k^j \mathbf{e}_j - \omega_k^j \wedge d\mathbf{e}_j) = \sum_{j=1}^{n} d\omega_k^j \mathbf{e}_j - \sum_{\ell=1}^{n} \omega_k^\ell \wedge d\mathbf{e}_\ell$$

$$= \sum_{j=1}^{n} d\omega_k^j \mathbf{e}_j - \sum_{\ell=1}^{n} \omega_k^\ell \left(\sum_{j=1}^{n} \omega_\ell^j \wedge \mathbf{e}_j \right) = \sum_{j=1}^{n} \left(d\omega_k^j - \sum_{\ell=1}^{n} \omega_k^\ell \wedge \omega_\ell^j \right) \mathbf{e}_j.$$

Since at each point $\mathbf{e}_1, \ldots, \mathbf{e}_n$ form an orthonormal basis we obtain

$$d\omega_k^j = \sum_{\ell=1}^{n} \omega_k^\ell \wedge \omega_\ell^j, \quad j, k = 1, \ldots, n.$$

These equations are called Cartan's second structural equations.

Problem 5. (i) Consider the complex number $z = re^{i\phi}$. Calculate

$$\frac{dz \wedge d\bar{z}}{z}.$$

(ii) The differential one form $\alpha = dz/z$ with $z = x + iy$ $(x, y \in \mathbb{R})$ is defined on $\mathbb{C} \setminus \{0\}$. Find $d\alpha$. Calculate

$$\oint_C \alpha$$

where C is the unit circle around the origin in the complex plane $\mathbb{C}$.

Solution 5. (i) Note that $\bar{z} = re^{-i\phi}$. Then

$$dz = e^{i\phi} dr + ie^{i\phi} r d\phi, \qquad d\bar{z} = e^{-i\phi} dr - ie^{-i\phi} r d\phi.$$

Thus

$$dz \wedge d\bar{z} = (e^{i\phi} dr + ie^{i\phi} r d\phi) \wedge (e^{-i\phi} dr - ie^{-i\phi} r d\phi)$$
$$= -irdr \wedge d\phi + ird\phi \wedge dr = -2irdr \wedge d\phi.$$

It follows that

$$\frac{dz \wedge d\bar{z}}{z} = -2ie^{-i\phi} dr \wedge d\phi.$$

(ii) We have

$$d\alpha = d\left(\frac{1}{z}\right) \wedge dz = -\frac{1}{z^2}dz \wedge dz = 0$$

since $dz \wedge dz = 0$. From $z = x + iy$ it follows that $dz = dx + idy$. Thus the differential form α takes the form

$$\alpha = \frac{dx + idy}{x + iy} = \frac{xdx + ydy + i(xdy - ydx)}{x^2 + y^2}.$$

Introducing $z = r\exp(i\phi)$ with $r = 1$ we have $dz = ri\exp(i\phi)d\phi$. Therefore

$$\oint_C \frac{dz}{z} = i \int_0^{2\pi} d\phi = 2\pi i.$$

Problem 6.   The differential form

$$\alpha = \frac{dz_1 \wedge dz_2}{z_1 z_2}$$

where $z_j = x_j + iy_j$ $(x_j, y_j \in \mathbb{R})$ is defined on $\mathbb{C}^2 \setminus \{(z_1, z_2) : z_1 = 0 \vee z_2 = 0\}$. Find the real and imaginary part of α.

Solution 6.   Since $dz_j = dx_j + idy_j$, $j = 1, 2$ and

$$z_1 z_2 = x_1 x_2 - y_1 y_2 + i(x_1 y_2 + x_2 y_1)$$

we obtain written as a sum of real and imaginary part

$$\frac{dz_1 \wedge dz_2}{z_1 z_2} = \frac{(x_1 x_2 - y_1 y_2)\omega + (x_1 y_2 + x_2 y_1)(dx_1 \wedge dy_2 + dy_1 \wedge dx_2)}{x_1^2 x_2^2 + y_1^2 y_2^2 + x_1^2 y_2^2 + x_2^2 y_1^2}$$
$$+i\frac{(x_1 x_2 - y_1 y_2)(dx_1 \wedge dy_2 + dy_1 \wedge dx_2) - (x_1 y_2 + x_2 y_1)\omega}{x_1^2 x_2^2 + y_1^2 y_2^2 + x_1^2 y_2^2 + x_2^2 y_1^2}$$

where $\omega := dx_1 \wedge dx_2 - dy_1 \wedge dy_2$.

Problem 7.   Let

$$g := \sum_{j,k=1}^m g_{jk}(\mathbf{x})dx_j \otimes dx_k$$

be the *metric tensor field* of a Riemannian or pseudo-Riemannian real C^∞ manifold M (dim $M = m < \infty$). Let $(x_1, x_2, \ldots, x_m)$ be the local coordinate system in a local coordinate neighbourhood (U, ψ). Since $(dx_j)_p$ $(j = 1, \ldots, m)$ is a basis of T_p^* (dual space of the tangent vector space T_p)

at each point p of U, we can express an r-form ω_p $(p \in U)$ uniquely in the form

$$\omega_p = \sum_{j_1 < \cdots < j_r}^{m} a_{j_1 \cdots j_r}(p)(dx_{j_1})_p \wedge \cdots \wedge (dx_{j_r})_p.$$

The *Hodge duality operator* $\star$ is an f-linear mapping which transforms an r-form into its dual $(m-r)$-form. The $\star$-operator which is applied to an r-form defined on an arbitrary Riemannian or pseudo-Riemannian manifold M with metric tensor field g, is defined by

$$\star(dx_{j_1} \wedge dx_{j_2} \wedge \cdots \wedge dx_{j_r}) :=$$

$$\sum_{k_1,\cdots,k_m=1}^{m} g^{j_1 k_1} \cdots g^{j_r k_r} \frac{1}{(m-r)!} \cdot \frac{g}{\sqrt{|g|}} \varepsilon_{k_1 \ldots k_m} dx_{k_{r+1}} \wedge \cdots \wedge dx_{k_m}$$

where $\varepsilon_{k_1 \cdots k_m}$ is the totally antisymmetric tensor with $\varepsilon_{12 \ldots m} = +1$ and

$$g \equiv \det(g_{ij}), \qquad \sum_{j=1}^{m} g^{ij} g_{jk} = \delta_k^i$$

where δ_k^i denotes the Kronecker delta.
(i) Let $M = \mathbb{R}^2$ and $g = dx_1 \otimes dx_1 + dx_2 \otimes dx_2$. Calculate $\star dx_1$ and $\star dx_2$.
(ii) Let $M = \mathbb{R}^3$ and $g = dx_1 \otimes dx_1 + dx_2 \otimes dx_2 + dx_3 \otimes dx_3$. Calculate $\star dx_1$, $\star dx_2$, $\star dx_3$, $\star(dx_1 \wedge dx_2)$, $\star(dx_2 \wedge dx_3)$, $\star(dx_3 \wedge dx_1)$.
(iii) Let $M = \mathbb{R}^4$ and

$$g = dx_1 \otimes dx_1 + dx_2 \otimes dx_2 + dx_3 \otimes dx_3 - dx_4 \otimes dx_4.$$

This is the *Minkowski metric*. Calculate $\star dx_1$, $\star dx_2$, $\star dx_3$, $\star dx_4$ and

$$\star(dx_1 \wedge dx_2), \qquad \star(dx_2 \wedge dx_3), \qquad \star(dx_3 \wedge dx_1),$$

$$\star(dx_1 \wedge dx_4), \qquad \star(dx_2 \wedge dx_4), \qquad \star(dx_3 \wedge dx_4).$$

Solution 7. (i) Since $g_{11} = g_{22} = 1$, $g_{12} = g_{21} = 0$ we find

$$g^{11} = g^{22} = 1, \qquad g^{12} = g^{21} = 0.$$

Therefore

$$\star dx_1 = \sum_{k_1,k_2=1}^{2} g^{1k_1} \frac{1}{(2-1)!} \frac{g}{\sqrt{|g|}} \varepsilon_{k_1 k_2} dx_{k_2}.$$

We have $g = \det(g_{ij}) = 1$ and $\varepsilon_{12} = 1$, $\varepsilon_{21} = -1$, $\varepsilon_{11} = \varepsilon_{22} = 0$. It follows that

$$\star dx_1 = \sum_{j_2=1}^{2} g^{11} \varepsilon_{1j_2} dx_{j_2} = dx_2.$$

Similarly, $\star dx_2 = -dx_1$.

(ii) Since $g_{11} = g_{22} = g_{33} = 1$ and $g_{jk} = 0$ for $j \neq k$ we obtain

$$g^{11} = g^{22} = g^{33} = 1 \quad \text{and} \quad g^{jk} = 0 \quad \text{for} \quad j \neq k.$$

Therefore we find for $\star dx_1$

$$\star dx_1 = \sum_{k_1,k_2,k_3=1}^{3} g^{1k_1} \frac{1}{(3-1)!} \varepsilon_{k_1 k_2 k_3} dx_{k_2} \wedge dx_{k_3} = \frac{1}{2} \sum_{k_2,k_3=1}^{3} \varepsilon_{1k_2 k_3} dx_{k_2} \wedge dx_{k_3}.$$

Thus $\star dx_1 = dx_2 \wedge dx_3$. Analogously $\star dx_2 = dx_3 \wedge dx_1$, $\star dx_3 = dx_1 \wedge dx_2$. Furthermore

$$\star(dx_1 \wedge dx_2) = \sum_{k_1,k_2,k_3=1}^{3} g^{1k_1} g^{2k_2} \varepsilon_{k_1 k_2 k_3} dx_{k_3} = \sum_{k_3=1}^{3} \varepsilon_{12k_3} dx_{k_3} = dx_3.$$

Analogously $\star(dx_2 \wedge dx_3) = dx_1$, $\star(dx_3 \wedge dx_1) = dx_2$.

(iii) Here we have $g_{11} = g_{22} = g_{33} = 1$, $g_{44} = -1$. Therefore

$$g^{11} = g^{22} = g^{33} = 1 \qquad g^{44} = -1.$$

Moreover $g = \det(g_{jk}) = -1$. Consequently, we find

$$\star dx_1 = -dx_2 \wedge dx_3 \wedge dx_4, \qquad \star dx_2 = -dx_3 \wedge dx_1 \wedge dx_4,$$

$$\star dx_3 = -dx_1 \wedge dx_2 \wedge dx_4, \qquad \star dx_4 = -dx_1 \wedge dx_2 \wedge dx_3$$

and

$$\star(dx_1 \wedge dx_2) = -dx_3 \wedge dx_4, \qquad \star(dx_2 \wedge dx_3) = -dx_1 \wedge dx_4,$$

$$\star(dx_3 \wedge dx_1) = -dx_2 \wedge dx_4, \qquad \star(dx_1 \wedge dx_4) = dx_2 \wedge dx_3,$$

$$\star(dx_2 \wedge dx_4) = dx_3 \wedge dx_1, \qquad \star(dx_3 \wedge dx_4) = dx_1 \wedge dx_2.$$

Problem 8. In electrodynamics we have the differential two-forms

$$\beta = E_1 dx_1 \wedge dt + E_2 dx_2 \wedge dt + E_3 dx_3 \wedge dt$$
$$+ B_3 dx_1 \wedge dx_2 + B_1 dx_2 \wedge dx_3 + B_2 dx_3 \wedge dx_1$$

and

$$*\beta = \frac{E_1}{c} dx_2 \wedge dx_3 + \frac{E_2}{c} dx_3 \wedge dx_1 + \frac{E_3}{c} dx_1 \wedge dx_2$$
$$- B_3 c \, dx_3 \wedge dt - B_1 c \, dx_1 \wedge dt - B_2 c \, dx_2 \wedge dt$$

where c is a positive constant (speed of light). Here

$$\mathbf{E} = (E_1, E_2, E_3)$$

is the electric field intensity (electric field strength) with dimension

$$meter \, . \, sec^{-3} \, . \, kg \, . \, A^{-1}$$

and

$$\mathbf{B} = (B_1, B_2, B_3)$$

is the magnetic induction (magnetic flux density) with dimension

$$sec^{-2} \, . \, kg \, . \, A^{-1}.$$

(i) Calculate $d\beta$, $d(\star\beta)$ and give an interpretation of $d\beta = 0$ and $d(\star\beta) = 0$.
(ii) Calculate $\beta \wedge \beta$ and $\beta \wedge (\star\beta)$ and give an interpretation.

Solution 8. (i) We set $x_4 \equiv ct$. Since

$$dE_k = \sum_{j=1}^{4} \frac{\partial E_k}{\partial x_j} dx_j, \qquad dB_k = \sum_{j=1}^{4} \frac{\partial B_k}{\partial x_j} dx_j$$

for $k = 1, 2, 3$ and $dx_j \wedge dx_k = -dx_k \wedge dx_j$ we find

$$
\begin{aligned}
d\beta = {} & \left(\frac{\partial B_1}{\partial x_1} + \frac{\partial B_2}{\partial x_2} + \frac{\partial B_3}{\partial x_3} \right) dx_1 \wedge dx_2 \wedge dx_3 \\
& + \left(\frac{\partial B_3}{\partial t} - \frac{\partial E_1}{\partial x_2} + \frac{\partial E_2}{\partial x_1} \right) dx_1 \wedge dx_2 \wedge dt \\
& + \left(\frac{\partial B_2}{\partial t} - \frac{\partial E_3}{\partial x_1} + \frac{\partial E_1}{\partial x_3} \right) dx_3 \wedge dx_1 \wedge dt \\
& + \left(\frac{\partial B_1}{\partial t} - \frac{\partial E_2}{\partial x_3} + \frac{\partial E_3}{\partial x_2} \right) dx_2 \wedge dx_3 \wedge dt.
\end{aligned}
$$

From the condition $d\beta = 0$ we obtain

$$0 = \left(\frac{\partial B_1}{\partial x_1} + \frac{\partial B_2}{\partial x_2} + \frac{\partial B_3}{\partial x_3} \right), \qquad 0 = \left(\frac{\partial B_3}{\partial t} - \frac{\partial E_1}{\partial x_2} + \frac{\partial E_2}{\partial x_1} \right),$$

$$0 = \left(\frac{\partial B_2}{\partial t} - \frac{\partial E_3}{\partial x_1} + \frac{\partial E_1}{\partial x_3} \right), \qquad 0 = \left(\frac{\partial B_1}{\partial t} - \frac{\partial E_2}{\partial x_3} + \frac{\partial E_3}{\partial x_2} \right).$$

Calculating $d(\star\beta)$ yields

$$
\begin{aligned}
d(\star\beta) = {} & \frac{1}{c} \left(\frac{\partial E_1}{\partial x_1} + \frac{\partial E_2}{\partial x_2} + \frac{\partial E_3}{\partial x_3} \right) dx_1 \wedge dx_2 \wedge dx_3 \\
& + \left(\frac{1}{c} \frac{\partial E_3}{\partial t} - c \frac{\partial B_2}{\partial x_1} + c \frac{\partial B_1}{\partial x_2} \right) dx_1 \wedge dx_2 \wedge dt \\
& + \left(\frac{1}{c} \frac{\partial E_2}{\partial t} - c \frac{\partial B_1}{\partial x_3} + c \frac{\partial B_3}{\partial x_1} \right) dx_3 \wedge dx_1 \wedge dt \\
& + \left(\frac{1}{c} \frac{\partial E_1}{\partial t} - c \frac{\partial B_3}{\partial x_2} + c \frac{\partial B_2}{\partial x_3} \right) dx_2 \wedge dx_3 \wedge dt.
\end{aligned}
$$

From the condition $d(\star\beta) = 0$ we obtain

$$0 = \left(\frac{\partial E_1}{\partial x_1} + \frac{\partial E_2}{\partial x_2} + \frac{\partial E_3}{\partial x_3}\right), \qquad 0 = \left(\frac{1}{c}\frac{\partial E_3}{\partial t} - c\frac{\partial B_2}{\partial x_1} + c\frac{\partial B_1}{\partial x_2}\right)$$

$$0 = \left(\frac{1}{c}\frac{\partial E_2}{\partial t} - c\frac{\partial B_1}{\partial x_3} + c\frac{\partial B_3}{\partial x_1}\right), \qquad 0 = \left(\frac{1}{c}\frac{\partial E_1}{\partial t} - c\frac{\partial B_3}{\partial x_2} + c\frac{\partial B_2}{\partial x_3}\right).$$

These systems of partial differential equations are *Maxwell's equations* in free space, i.e.

$$\operatorname{div}(\mathbf{B}) = 0, \qquad -\frac{\partial \mathbf{B}}{\partial t} = \nabla \times \mathbf{E}, \qquad \operatorname{div}(\mathbf{E}) = 0, \qquad \frac{1}{c^2}\frac{\partial \mathbf{E}}{\partial t} = \nabla \times \mathbf{B}.$$

(ii) Straightforward calculation yields

$$\beta \wedge \beta = 2(B_1 E_1 + B_2 E_2 + B_3 E_3)dx_1 \wedge dx_2 \wedge dx_3 \wedge dt$$

$$\beta \wedge (\star\beta) = \left(\frac{E_1^2}{c} + \frac{E_2^2}{c} + \frac{E_3^2}{c} - B_1^2 c - B_2^2 c - B_3^2 c\right) dx_1 \wedge dx_2 \wedge dx_3 \wedge dt.$$

This equation describes the energy density of the electromagnetic field. Both equations are invariant under the Lorentz transformation. Note that $\star(\star\beta) = -\beta$.

Problem 9. Within the techniques of differential forms the *vector potential* and *scalar potential* is given by the one-form

$$\alpha = A_1 dx_1 + A_2 dx_2 + A_3 dx_3 - U dt$$

and the electromagnetic field by the two-form

$$\beta = E_1 dx_1 \wedge dt + E_2 dx_2 \wedge dt + E_3 dx_3 \wedge dt$$
$$+ B_3 dx_1 \wedge dx_2 + B_1 dx_2 \wedge dx_3 + B_2 dx_3 \wedge dx_1.$$

Find the relations which follow from $d\alpha = \beta$.

Solution 9.   We set $ct \equiv x_4$. Since $dx_j \wedge dx_k = -dx_k \wedge dx_j$ and

$$dA_k = \frac{\partial A_k}{\partial x_1}dx_1 + \frac{\partial A_k}{\partial x_2}dx_2 + \frac{\partial A_k}{\partial x_3}dx_3 + \frac{\partial A_k}{\partial t}dt$$

$$dU = \frac{\partial U}{\partial x_1}dx_1 + \frac{\partial U}{\partial x_2}dx_2 + \frac{\partial U}{\partial x_3}dx_3 + \frac{\partial U}{\partial t}dt$$

we find

$$d\alpha = \left(\frac{\partial A_2}{\partial x_1} - \frac{\partial A_1}{\partial x_2}\right) dx_1 \wedge dx_2 + \left(\frac{\partial A_3}{\partial x_2} - \frac{\partial A_2}{\partial x_3}\right) dx_2 \wedge dx_3$$

$$+ \left(\frac{\partial A_1}{\partial x_3} - \frac{\partial A_3}{\partial x_1}\right) dx_3 \wedge dx_1 + \left(-\frac{\partial A_1}{\partial t} - \frac{\partial U}{\partial x_1}\right) dx_1 \wedge dt$$

$$+ \left(-\frac{\partial A_2}{\partial t} - \frac{\partial U}{\partial x_2}\right) dx_2 \wedge dt + \left(-\frac{\partial A_3}{\partial t} - \frac{\partial U}{\partial x_3}\right) dx_3 \wedge dt.$$

Comparing the six basis elements $dx_1 \wedge dx_2$, $dx_2 \wedge dx_3$, $dx_3 \wedge dx_1$, $dx_1 \wedge dt$, $dx_2 \wedge dt$, $dx_3 \wedge dt$ we find

$$B_3 = \frac{\partial A_2}{\partial x_1} - \frac{\partial A_1}{\partial x_2}, \quad B_1 = \frac{\partial A_3}{\partial x_2} - \frac{\partial A_2}{\partial x_3}, \quad B_2 = \frac{\partial A_1}{\partial x_3} - \frac{\partial A_3}{\partial x_1}$$

$$E_1 = -\frac{\partial A_1}{\partial t} - \frac{\partial U}{\partial x_1}, \quad E_2 = -\frac{\partial A_2}{\partial t} - \frac{\partial U}{\partial x_2}, \quad E_3 = -\frac{\partial A_3}{\partial t} - \frac{\partial U}{\partial x_3}.$$

It follows that $\mathbf{B} = \nabla \times \mathbf{A}$, $\mathbf{E} = -\nabla U - \partial \mathbf{A}/\partial t$.

Problem 10. The basic quantity in electromagnetism is the differential two-form

$$\beta = E_1(\mathbf{x}, t)dx_1 \wedge dt + E_2(\mathbf{x}, t)dx_2 \wedge dt + E_3(\mathbf{x}, t)dx_3 \wedge dt$$
$$+ B_3(\mathbf{x}, t)dx_1 \wedge dx_2 + B_1(\mathbf{x}, t)dx_2 \wedge dx_3 + B_2(\mathbf{x}, t)dx_3 \wedge dx_1.$$

The system $'$ (prime) and without prime are connected by the *Lorentz transformation*

$$x_1' = \gamma(x_1 - vt), \quad t' = \gamma(t - vx_1/c^2), \quad x_2' = x_2, \quad x_3' = x_3 \quad (1)$$

where $\gamma := 1/\sqrt{1 - v^2/c^2}$ and v with $0 \le v < c$ is a constant. If β is a "physical quantity", then

$$\beta' = \beta \quad (2)$$

where

$$\beta = E_1(\mathbf{x}, t)dx_1 \wedge dt + \cdots + B_3(\mathbf{x}, t)dx_1 \wedge dx_2$$

and

$$\beta' = E_1'(\mathbf{x}'(\mathbf{x}, t), t'(\mathbf{x}, t))dx_1'(\mathbf{x}, t) \wedge dt'(\mathbf{x}, t) + \cdots$$
$$+ B_3'(\mathbf{x}'(\mathbf{x}, t), t'(\mathbf{x}, t))dx_1'(\mathbf{x}, t) \wedge dx_2'(\mathbf{x}, t).$$

Find the transformation law for $\mathbf{B}$ and $\mathbf{E}$ from the condition (2).

Solution 10. From the Lorentz transformation it follows that

$$dx_1'(\mathbf{x}, t) = \gamma(dx_1 - vdt), \quad dt'(\mathbf{x}, t) = \gamma(dt - (v/c^2)dx_1)$$
$$dx_2'(\mathbf{x}, t) = dx_2, \quad dx_3'(\mathbf{x}, t) = dx_3.$$

Consequently,

$$dx_1'(\mathbf{x}, t) \wedge dt'(\mathbf{x}, t) = \gamma^2 dx_1 \wedge dt - \gamma^2(v^2/c^2)dx_1 \wedge dt = dx_1 \wedge dt$$
$$dx_2'(\mathbf{x}, t) \wedge dt'(\mathbf{x}, t) = dx_2 \wedge \gamma(dt - (v/c^2)dx_1) = \gamma dx_2 \wedge dt + \gamma(v/c^2)dx_1 \wedge dx_2$$
$$dx_3'(\mathbf{x}, t) \wedge dt'(\mathbf{x}, t) = dx_3 \wedge \gamma(dt - (v/c^2)dx_1) = \gamma dx_3 \wedge dt - \gamma(v/c^2)dx_3 \wedge dx_1$$
$$dx_1'(\mathbf{x}, t) \wedge dx_2'(\mathbf{x}, t) = \gamma(dx_1 - vdt) \wedge dx_2 = \gamma dx_1 \wedge dx_2 + \gamma v dx_2 \wedge dt$$
$$dx_2'(\mathbf{x}, t) \wedge dx_3'(\mathbf{x}, t) = dx_2 \wedge dx_3$$
$$dx_3'(\mathbf{x}, t) \wedge dx_1'(\mathbf{x}, t) = dx_3 \wedge \gamma(dx_1 - vdt) = \gamma dx_3 \wedge dx_1 - \gamma v dx_3 \wedge dt.$$

Inserting these equations into the two-form β' yields

$$\beta = E_1'(\mathbf{x}'(\mathbf{x},t),t'(\mathbf{x},t))dx_1 \wedge dt$$
$$+E_2'(\mathbf{x}'(\mathbf{x},t),t'(\mathbf{x},t))(\gamma dx_2 \wedge dt + \gamma(v/c^2)dx_1 \wedge dx_2)$$
$$+E_3'(\mathbf{x}'(\mathbf{x},t),t'(\mathbf{x},t))(\gamma dx_3 \wedge dt - \gamma(v/c^2)dx_3 \wedge dx_1)$$
$$+B_3'(\mathbf{x}'(\mathbf{x},t),t'(\mathbf{x},t))(\gamma dx_1 \wedge dx_2 + \gamma v dx_2 \wedge dt)$$
$$+B_1'(\mathbf{x}'(\mathbf{x},t),t'(c,t))dx_2 \wedge dx_3$$
$$+B_2'(\mathbf{x}'(\mathbf{x},t),t'(\mathbf{x},t))(\gamma dx_3 \wedge dx_1 - \gamma v dx_3 \wedge dt).$$

Comparing the six basis elements of the two-forms

$$dx_1 \wedge dt, \quad dx_2 \wedge dt, \quad dx_3 \wedge dt, \quad dx_1 \wedge dx_2, \quad dx_2 \wedge dx_3, \quad dx_3 \wedge dx_1$$

we find the transformation law for $\mathbf{E}$ and $\mathbf{B}$

$$E_1(\mathbf{x},t) = E_1'(\mathbf{x}'(\mathbf{x},t),t'(\mathbf{x},t))$$
$$B_1(\mathbf{x},t) = B_1'(\mathbf{x}'(\mathbf{x},t),t'(\mathbf{x},t))$$
$$E_2(\mathbf{x},t) = \gamma E_2'(\mathbf{x}'(\mathbf{x},t),t'(\mathbf{x},t)) + \gamma v B_3'(\mathbf{x}'(\mathbf{x},t),t'(\mathbf{x},t))$$
$$E_3(\mathbf{x},t) = \gamma E_3'(\mathbf{x}'(\mathbf{x},t),t'(\mathbf{x},t)) - \gamma v B_2'(\mathbf{x}'(\mathbf{x},t),t'(\mathbf{x},t))$$
$$B_3(\mathbf{x},t) = \gamma B_3'(\mathbf{x}'(\mathbf{x},t),t'(\mathbf{x},t)) + \gamma(v/c^2)E_2'(\mathbf{x}'(\mathbf{x},t),t'(\mathbf{x},t))$$
$$B_2(\mathbf{x},t) = \gamma B_2'(\mathbf{x}'(\mathbf{x},t),t'(\mathbf{x},t)) - \gamma(v/c^2)E_3'(\mathbf{x}'(\mathbf{x},t),t'(\mathbf{x},t)).$$

Problem 11. Let the volume V of the system and the temperature T of the system be the independent variables of the given thermodynamic system. Furthermore, let the external pressure P and the internal energy U of the system be the dependent variables. All the objects under consideration are smooth. Prove the following:

Theorem. Let $\omega := dU + PdV$ be a one-form in a two-dimensional space (V,T). Assume that $\partial P/\partial T \neq 0$ and that P and U are related by the equation

$$\frac{\partial U}{\partial V} = T\frac{\partial P}{\partial T} - P \tag{1}$$

which is the so-called *thermodynamical equation of state*. Then
(i) The one-form ω is not closed.
(ii) There exists a one-form $\delta \equiv f(V,T)\omega$ such that $d\delta = 0$.
(iii) The function $f(V,T)$ is given by

$$f(V,T) = \frac{g(K(V,T))}{T} \tag{2}$$

where g is a smooth function of $K(V,T)$ and $K(V,T)$ must satisfy the partial differential equations

$$\frac{\partial K}{\partial V} - \frac{1}{T}\frac{\partial U}{\partial V} - \frac{P}{T} = 0 \tag{3}$$

$$\frac{\partial K}{\partial T} - \frac{1}{T}\frac{\partial U}{\partial T} = 0. \tag{4}$$

In particular we can choose $g(K) = 1$.

Solution 11. (i) The exterior derivative of the one-form

$$\omega = dU(V,T) + P(V,T)dV = \frac{\partial U}{\partial T}dT + \left(\frac{\partial U}{\partial V} + P\right)dV$$

yields

$$d\omega = dP \wedge dV = \frac{\partial P}{\partial T}dT \wedge dV \neq 0$$

since $\partial P/\partial T \neq 0$. $\partial P/\partial T \neq 0$ is a reasonable assumption to make.

(ii) Owing to $\omega \wedge d\omega = 0$, the *Frobenius theorem* tells us that there is a function $f(V,T)$ such that $d(f\omega) = 0$. The theorem of Frobenius can be given in this special form because there are only two independent variables. This is locally trivially true (though not necessarily globally, the Frobenius theorem itself being a local result) whether or not (1) holds, since ω is a one-form in only two variables.

(iii) The condition $d(f\omega) = 0$ yields

$$0 = d(f\omega) = (df)\wedge\omega + f d\omega = \left(\left(\frac{\partial U}{\partial V} + P\right)\frac{\partial f}{\partial T} - \frac{\partial U}{\partial T}\frac{\partial f}{\partial V} + \frac{\partial P}{\partial T}f\right)dT\wedge dV.$$

Thus we obtain a linear partial differential equation of first order

$$\left(\frac{\partial U}{\partial V} + P\right)\frac{\partial f}{\partial T} - \frac{\partial U}{\partial T}\frac{\partial f}{\partial V} = -\frac{\partial P}{\partial T}f. \tag{5}$$

Inserting (2) through (4) into (5), we find that (5) is satisfied identically. When we interpret that ω represents an "infinitesimal quantity of heat" the equation $\omega = dU + PdV$ is the *first law of thermodynamics*. If we assume that the internal energy U and pressure P (external pressure) are related by (1) (the so-called thermodynamical equation of state, where the right-hand side of the equation can be computed from the equation of state), then there is a function $f(V,T) = g(K(V,T))/T$ such that $g\omega/T$ is a closed form. $K(V,T)$ must satisfy (3) and (4). Let $g = 1$. The quantity ω/T is usually called the "infinitesimal entropy". We write $\omega = TdS$. This means that

$$dS = \frac{dU + PdV}{T}. \tag{6}$$

S is the first integral of the exterior differential equation $dU + PdV = 0$. Since

$$dS = \frac{\partial S}{\partial V}dV + \frac{\partial S}{\partial T}dT, \qquad dU = \frac{\partial U}{\partial V}dV + \frac{\partial U}{\partial T}dT$$

we can write (6) as

$$\frac{\partial S}{\partial V} - \frac{1}{T}\frac{\partial U}{\partial V} - \frac{P}{T} = 0 \tag{7}$$

and

$$\frac{\partial S}{\partial T} - \frac{1}{T}\frac{\partial U}{\partial T} = 0. \tag{8}$$

Comparing (3) and (4) with (7) and (8) we must put $K = S$. Consequently, the most general local integrating factor for ω is

$$f(V,T) = \frac{1}{T} \times \text{arbitrary smooth function of } S.$$

To prove that physically fT is actually a constant, one must invoke the zeroth law and assume that T is the temperature. The given derivation can also be considered from a converse point of view. From dS we are able to derive the thermodynamical equation of state. Since $ddS = 0$, it follows that

$$0 = d\left(\frac{dU + PdV}{T}\right).$$

Consequently,

$$0 = \frac{1}{T^2}\left(\frac{\partial U}{\partial V} + P - T\frac{\partial P}{\partial T}\right)dV \wedge dT.$$

Therefore

$$\frac{\partial U}{\partial V} + P - T\frac{\partial P}{\partial T} = 0.$$

Thus we obtain the thermodynamical equation of state.

Problem 12. Let V, T be the volume and temperature, respectively and $p(V,T)$, $E(T,V)$ be the pressure and energy depending on V, T. Consider the differential one-form

$$\alpha = dE(T,V) + p(T,V)dV.$$

From the *second law of thermodynamics* we find

$$0 = d\left(\frac{\alpha}{T}\right) = d\left(\frac{dE + pdV}{T}\right)$$

$$= \frac{1}{T}\left(\frac{1}{T}\left(\frac{\partial E}{\partial V} + p\right) - \frac{\partial p}{\partial T}\right)dV \wedge dT$$

or

$$\frac{\partial E}{\partial V} = T\frac{\partial p}{\partial T} - p. \tag{1}$$

Let $a, b, c > 0$. Consider the *van der Waals equation*

$$\left(p + \frac{a}{V^2}\right)(V - b) = cT.$$

Show that E depends on V.

Solution 12. If E is independent of V it follows that $p = T\partial T/\partial T$. However from the van der Waals equation we obtain

$$p(V - b) = cT - \frac{a}{V^2}(V - b)$$

and differentiation p with respect to T we have

$$T\frac{\partial p}{\partial T}(V - b) = cT.$$

So it follows that $T\partial p/\partial T = p + a/V^2$ and $\partial E = \partial V = a/V^2$, i.e. E depends on V. The Joule Thomson effect describes the temperature T change of a real gas (for example described by the van der Waals equation) when it is forced through a valve while keeping them insulated so that no heat is exchanged with the environment. What happens when we consider the ideal gas?

Problem 13. Let $\Omega = \mathbb{R}^3 \setminus \{(0, 0, 0)\}$. Consider

$$\alpha = \frac{\lambda}{r}(x_1 dx_2 \wedge dx_3 + x_2 dx_3 \wedge dx_1 + x_3 dx_1 \wedge dx_2)$$

where $\lambda > 0$ and $r^2 := x_1^2 + x_2^2 + x_3^2$. Consider the map f given by

$$x_1(u, v) = \sin(u)\cos(v), \quad x_2(u, v) = \sin(u)\sin(v), \quad x_3(u, v) = \cos(u)$$

where $0 \leq u < \pi$ and $0 \leq v < 2\pi$. Calculate

$$\int_{\mathbb{S}^2} f^*\alpha$$

where $\mathbb{S}^2 := \{ (x_1, x_2, x_3) : x_1^2 + x_2^2 + x_3^2 = 1 \}$.

Solution 13. Since

$$f^*(dx_1 \wedge dx_2) = d(\sin(u)\cos(v)) \wedge d(\sin(u)\sin(v)) = \sin(u)\cos(u)du \wedge dv$$
$$f^*(dx_2 \wedge dx_3) = d(\sin(u)\sin(v)) \wedge d(\cos(u)) = \sin^2(u)\cos(v)du \wedge dv$$
$$f^*(dx_3 \wedge dx_1) = d(\cos(u)) \wedge d(\sin(u)\cos(v)) = \sin^2(u)\sin(v)du \wedge dv$$

we find

$$f^*\alpha = \lambda\sin(u)du \wedge dv$$

with $r = 1$. Consequently

$$\int_{\mathbb{S}^2} f^*\alpha = \lambda \int_{u=0}^{\pi} \int_{v=0}^{2\pi} \sin(u)dudv = 4\pi\lambda.$$

Problem 14.   Let M be an oriented n-manifold and let D be a regular domain in M. Let ω be a differential form of degree $(n-1)$ of compact support. Then

$$\int_D d\omega = \int_{\partial D} i^*\omega \tag{1}$$

where the boundary ∂D of D is considered as an oriented submanifold with the orientation induced by that of M. $i : \partial D \to M$ is the canonical injection map. Equation (1) is called *Stokes theorem*.
Let $M = \mathbb{R}^3$ and

$$K := \{ \, (x_1, x_2, x_3) : \frac{x_1^2}{a^2} + \frac{x_2^2}{b^2} + \frac{x_3^2}{c^2} \le 1, \quad a, b, c > 0 \, \}$$

and

$$\omega := \frac{x_1^3}{a^2} dx_2 \wedge dx_3 + \frac{x_2^3}{b^2} dx_3 \wedge dx_1 + \frac{x_3^3}{c^2} dx_1 \wedge dx_2.$$

Calculate ∂K and $d\omega$. The orientation is x_1, x_2, x_3. Show that

$$\int_{\partial K} \omega = \int_K d\omega.$$

Solution 14.   We find that the boundary of K is given by

$$\partial K = \{ \, (x_1, x_2, x_3) : \frac{x_1^2}{a^2} + \frac{x_2^2}{b^2} + \frac{x_3^2}{c^2} = 1, \quad a, b, c > 0 \, \}.$$

The exterior derivative of ω leads to the differential three-form

$$d\omega = \frac{3x^2}{a^2} dx_1 \wedge dx_2 \wedge dx_3 + \frac{3y^2}{b^2} dx_2 \wedge dx_3 \wedge dx_1 + \frac{3z^2}{c^2} dx_3 \wedge dx_1 \wedge dx_2$$

$$= 3 \left(\frac{x_1^2}{a^2} + \frac{x_2^2}{b^2} + \frac{x_3^2}{c^2} \right) dx_1 \wedge dx_2 \wedge dx_3.$$

To calculate the right-hand side of (1) we introduce *spherical coordinates*

$$x_1(r, \phi, \theta) = ar\cos(\phi)\cos(\theta), \quad x_2(r, \phi, \theta) = br\sin(\phi)\cos(\theta),$$

$$x_3(r, \phi, \theta) = cr\sin(\theta)$$

where $0 \le \phi < 2\pi$, $-\pi/2 \le \theta < \pi/2$ and $0 \le r \le 1$. Then

$$dx_1 \wedge dx_2 \wedge dx_3 = abcr^2 \cos(\theta) dr \wedge d\phi \wedge d\theta$$

and

$$\frac{x_1^2}{a^2} + \frac{x_2^2}{b^2} + \frac{x_3^2}{c^2} = r^2(\cos^2(\phi)\cos^2(\theta) + \sin^2(\phi)\cos^2(\theta) + \sin^2(\theta)) = r^2.$$

Consequently

$$\int_K d\omega = \int_0^1 \int_{-\pi/2}^{\pi/2} \int_0^{2\pi} 3abcr^4 \cos(\theta) dr d\theta d\phi = \frac{4\pi}{5} 3abc.$$

Now we calculate the left-hand side of (1). We set

$$x(\phi, \theta) = a\cos(\phi)\cos(\theta), \quad y(\phi, \theta) = b\sin(\phi)\cos(\theta), \quad z(\phi, \theta) = c\sin(\theta)$$

where $0 \le \phi < 2\pi$ and $-\pi/2 \le \theta < \pi/2$. It follows that

$$dx_1 \wedge dx_2 = ab\cos(\theta)\sin(\theta)d\phi \wedge d\theta$$
$$dx_2 \wedge dx_3 = cb\cos(\phi)\cos^2(\theta)d\phi \wedge d\theta$$
$$dx_3 \wedge dx_1 = ac\sin(\phi)\cos^2(\theta)d\phi \wedge d\theta.$$

Therefore

$$\int_{\partial K} \omega = I_1 + I_2 + I_3$$

where

$$I_1 = \int_{-\pi/2}^{\pi/2} \int_0^{2\pi} abc\cos^4(\phi)\cos^5(\theta)d\phi d\theta = \frac{4}{5}abc\pi$$

$$I_2 = \int_{-\pi/2}^{\pi/2} \int_0^{2\pi} abc\sin^4(\phi)\cos^5(\theta)d\phi d\theta = \frac{4}{5}abc\pi$$

$$I_3 = \int_{-\pi/2}^{\pi/2} \int_0^{2\pi} abc\sin^4(\theta)\cos(\theta)d\phi d\theta = \frac{4}{5}abc\pi.$$

Thus

$$\int_{\partial K} \omega = \frac{4\pi}{5} 3abc.$$

Problem 15. *Poincaré's lemma* tells us that if ω is a p-differential form on M ($\dim M = n$) for which there exists a $(p-1)$-differential form α such that $d\alpha = \omega$, then $d\omega = 0$. The converse of Poincaré's lemma tells us that if ω is a p-differential form on an open set $U \subset M$ (which is contractible to a point) such that $d\omega = 0$, then there exists a $(p-1)$ differential form α such that $\omega = d\alpha$. The exception is $p = 0$. Then $\omega = f$ and the vanishing of df simply means f is constant.
Let

$$\omega = a(\mathbf{x})dx_{i_1} \wedge dx_{i_2} \wedge \cdots \wedge dx_{i_p}$$

and let λ be a real parameter with $\lambda \in [0,1]$. We introduce the linear operator T_λ which is defined as

$$T_\lambda \omega := \int_0^1 \lambda^{p-1} \left(x_1 \frac{\partial}{\partial x_1} + \cdots + x_n \frac{\partial}{\partial x_n} \right) \rfloor a(\lambda \mathbf{x}) dx_{i_1} \wedge dx_{i_2} \wedge \cdots \wedge dx_{i_p} d\lambda$$

where $\rfloor$ denotes the *contraction* (also called the *interior product* of a differential form and a vector field). We have

$$\frac{\partial}{\partial x_j} \rfloor dx_k = \delta_{jk}.$$

Since the operator T_λ is linear it must only be defined for a monom. Note that

$$d(T_\lambda \omega) = \omega.$$

(i) Apply the converse of Poincaré's lemma to $\omega = dx_1 \wedge dx_2$ defined on $\mathbb{R}^2$.
(ii) Apply the converse of Poincaré's lemma to $\omega = dx_1 \wedge dx_2 \wedge dx_3$ defined on $\mathbb{R}^3$.

Solution 15.   (i) We have $d\omega = 0$ and $\mathbb{R}^2$ is contractible to $0 \in \mathbb{R}^2$. From the definition for the contraction we obtain

$$T_\lambda \omega = \left(x_1 \frac{\partial}{\partial x_1} + \cdots + x_n \frac{\partial}{\partial x_n} \right) \rfloor (dx_{i_1} \wedge dx_{i_2} \wedge \cdots \wedge dx_{i_p}) \int_0^1 \lambda^{p-1} a(\lambda \mathbf{x}) d\lambda.$$

We have $n = 2$ and $p = 2$. Since

$$\left(x_1 \frac{\partial}{\partial x_1} + x_2 \frac{\partial}{\partial x_2} \right) \rfloor (dx_1 \wedge dx_2) = -x_2 dx_1 + x_1 dx_2$$

we find

$$T_\lambda \omega = (x_1 dx_2 - x_2 dx_1) \int_0^1 \lambda d\lambda = \frac{1}{2}(x_1 dx_2 - x_2 dx_1).$$

(ii) We have $d\omega = 0$ and $\mathbb{R}^3$ is contractible to $0 \in \mathbb{R}^3$. Since

$$\left(\sum_{j=1}^3 x_j \frac{\partial}{\partial x_j} \right) \rfloor (dx_1 \wedge dx_2 \wedge dx_3) = x_1 dx_2 \wedge dx_3 - x_2 dx_1 \wedge dx_3 + x_3 dx_1 \wedge dx_2$$

and

$$\int_0^1 \lambda^2 d\lambda = \frac{1}{3}$$

we find

$$T_\lambda \omega = \frac{1}{3}(x_1 dx_2 \wedge dx_3 + x_2 dx_3 \wedge dx_1 + x_3 dx_1 \wedge dx_2).$$

Problem 16.   Find the closed plane curve of a given length L which encloses a maximum area.

Solution 16. Given the curve $(x(t), y(t))$, where $t \in [t_0, t_1]$. We assume that $x(t)$ and $y(t)$ are continuously differentiable and $x(t_0) = x(t_1)$, $y(t_0) = y(t_1)$. Let A be the area enclosed and L be the given length. Then we have

$$L = \int_{t_0}^{t_1} \left(\left(\frac{dx}{dt} \right)^2 + \left(\frac{dy}{dt} \right)^2 \right)^{1/2} dt.$$

Since the exterior derivative d of $(x dy - y dx)/2$ is given by the two form

$$\frac{1}{2} d(x dy - y dx) = dx \wedge dy$$

we can apply *Stokes theorem* and find the area

$$A = \int_{t_0}^{t_1} \left(x \frac{dy}{dt} - y \frac{dx}{dt} \right) dt.$$

To find the maximum area we apply the *Lagrange multiplier method* and consider

$$H = \int_{t_0}^{t_1} \left(\left(x \frac{dy}{dt} - y \frac{dx}{dt} \right) + \lambda \left(\left(\frac{dx}{dt} \right)^2 + \left(\frac{dy}{dt} \right)^2 \right)^{1/2} \right) dt$$

where λ is the Lagrange multiplier. Thus we consider

$$\phi(x, y, \dot{x}, \dot{y}) = \frac{1}{2}(x\dot{y} - y\dot{x}) + \lambda(\dot{x}^2 + \dot{y}^2)^{1/2}$$

where ϕ satisfies the *Euler-Lagrange equations* are

$$\frac{\partial \phi}{\partial x} - \frac{d}{dt} \frac{\partial \phi}{\partial \dot{x}} = 0, \qquad \frac{\partial \phi}{\partial y} - \frac{d}{dt} \frac{\partial \phi}{\partial \dot{y}} = 0.$$

Thus, by partial differentiation

$$\frac{\partial \phi}{\partial x} = \frac{\dot{y}}{2}, \qquad \frac{\partial \phi}{\partial y} = -\frac{\dot{x}}{2},$$

$$\frac{\partial \phi}{\partial \dot{x}} = -\frac{y}{2} + \frac{\lambda \dot{x}}{(\dot{x}^2 + \dot{y}^2)^{1/2}}, \qquad \frac{\partial \phi}{\partial \dot{y}} = \frac{x}{2} + \frac{\lambda \dot{y}}{(\dot{x}^2 + \dot{y}^2)^{1/2}}.$$

It follows that

$$\frac{\dot{y}}{2} - \frac{d}{dt} \left(-\frac{y}{2} + \frac{\lambda \dot{x}}{(\dot{x}^2 + \dot{y}^2)^{1/2}} \right) = 0, \qquad -\frac{\dot{x}}{2} - \frac{d}{dt} \left(\frac{x}{2} + \frac{\lambda \dot{y}}{(\dot{x}^2 + \dot{y}^2)^{1/2}} \right) = 0.$$

If we choose a special parameter s, the arc length along the competing curves, then $(\dot{x}^2 + \dot{y}^2)^{1/2} = 1$ and $ds = (\dot{x}^2 + \dot{y}^2)^{1/2} dt$ so that Euler-Lagrange equations reduce to

$$\frac{dy}{ds} - \lambda \frac{d^2 x}{ds^2} = 0, \qquad \frac{dx}{ds} + \lambda \frac{d^2 y}{ds^2} = 0.$$

Integration of these linear system of differential equations yields

$$y - \lambda\frac{dx}{ds} = C_1, \qquad x + \lambda\frac{dy}{ds} = C_2$$

where C_1 and C_2 are constants of integration. Elimination of y yields

$$\lambda^2\frac{d^2x}{ds^2} + x = C_2$$

with the solution

$$x(s) = a\sin\left(\frac{s}{\lambda}\right) + b\cos\left(\frac{s}{\lambda}\right) + C_2$$

where a and b are constants of integration. For y we find

$$y(s) = a\cos\left(\frac{s}{\lambda}\right) - b\sin\left(\frac{s}{\lambda}\right) + C_1.$$

Thus $(x(s), y(s))$ describe a circle.

Problem 17. Consider the manifold $M = \mathbb{R}^2$ and the metric tensor field $g = dx_1 \otimes dx_1 + dx_2 \otimes dx_2$. Let

$$\omega = \omega_1(\mathbf{x})dx_1 + \omega_2(\mathbf{x})dx_2$$

be a differential one-form in M with $\omega_1, \omega_2 \in C^\infty(\mathbb{R}^2)$. Show that ω can be written as

$$\omega = d\alpha + \delta\beta + \gamma$$

where α is a $C^\infty(\mathbb{R}^2)$ function, β is a two-form given by $\beta = b(\mathbf{x})dx_1 \wedge dx_2$ $(b(\mathbf{x}) \in C^\infty(\mathbb{R}^2))$ and $\gamma = \gamma_1(\mathbf{x})dx_1 + \gamma_2(\mathbf{x})dx_2$ is a harmonic one-form, i.e. $(d\delta + \delta d)\gamma = 0$. We define

$$\delta\beta := (-1) \star d \star \beta.$$

Solution 17. We have

$$d\alpha = \frac{\partial\alpha}{\partial x_1}dx_1 + \frac{\partial\alpha}{\partial x_2}dx_2.$$

Since $\delta\beta = (-1)\star d \star \beta$ and $\star(dx_1 \wedge dx_2) = 1$, $\star dx_1 = dx_2$, $\star dx_2 = -dx_1$ we obtain

$$\delta\beta = (-1)\star d\star b(\mathbf{x})dx_1 \wedge dx_2 = (-1)\star db(\mathbf{x})(\star(dx_1 \wedge dx_2))$$
$$= (-1)\star\left(\frac{\partial b}{\partial x_1}dx_1 + \frac{\partial b}{\partial x_2}dx_2\right) = (-1)\left(\frac{\partial b}{\partial x_1}dx_2 - \frac{\partial b}{\partial x_2}dx_1\right)$$
$$= \frac{\partial b}{\partial x_2}dx_1 - \frac{\partial b}{\partial x_1}dx_2.$$

From the condition that γ is a harmonic one-form we obtain

$$(d\delta + \delta d)\gamma = d(\delta\gamma) + \delta(d\gamma)$$

$$= d(\delta\gamma_1(\mathbf{x})dx_1 + \delta\gamma_2(\mathbf{x})dx_2) + \delta\left(-\frac{\partial\gamma_1}{\partial x_2} + \frac{\partial\gamma_2}{\partial x_1}\right)dx_1 \wedge dx_2$$

$$= d((-1) \star d(\gamma_1(\mathbf{x})dx_2) - (-1) \star d(\gamma_2(\mathbf{x})dx_1)$$

$$+ \delta\left(\frac{\partial\gamma_2}{\partial x_1}dx_1 \wedge dx_2\right) - \delta\left(\frac{\partial\gamma_1}{\partial x_2}dx_1 \wedge dx_2\right)$$

$$= d\left((-1) \star \frac{\partial\gamma_1}{\partial x_1}dx_1 \wedge dx_2 - \star\frac{\partial\gamma_2}{\partial x_2}dx_1 \wedge dx_2\right)$$

$$+ (-1) \star d\left(\frac{\partial\gamma_2}{\partial x_1}\right) - (-1) \star d\left(\frac{\partial\gamma_1}{\partial x_2}\right)$$

$$= d\left(-\frac{\partial\gamma_1}{\partial x_1} - \frac{\partial\gamma_2}{\partial x_2}\right) + (-1) \star \left(\frac{\partial^2\gamma_2}{\partial x_1^2}dx_1 + \frac{\partial^2\gamma_2}{\partial x_1\partial x_2}dx_2\right)$$

$$+ \star \frac{\partial^2\gamma_1}{\partial x_1\partial x_2}dx_1 + \star\frac{\partial^2\gamma_1}{\partial x_2^2}dx_2$$

$$= -\left(\frac{\partial^2\gamma_1}{\partial x_1^2} + \frac{\partial^2\gamma_1}{\partial x_2^2}\right)dx_1 - \left(\frac{\partial^2\gamma_2}{\partial x_1^2} + \frac{\partial^2\gamma_2}{\partial x_2^2}\right)dx_2.$$

Thus from $\omega = d\alpha + \delta\beta + \gamma$ we obtain

$$\omega_1(\mathbf{x}) = \frac{\partial\alpha}{\partial x_1} + \frac{\partial b}{\partial x_2} + \gamma_1(\mathbf{x}), \qquad \omega_2(\mathbf{x}) = \frac{\partial\alpha}{\partial x_2} - \frac{\partial b}{\partial x_1} + \gamma_2(\mathbf{x})$$

and that γ is a harmonic one-form, i.e.

$$\frac{\partial^2\gamma_1}{\partial x_1^2} + \frac{\partial^2\gamma_1}{\partial x_2^2} = 0, \qquad \frac{\partial^2\gamma_2}{\partial x_1^2} + \frac{\partial^2\gamma_2}{\partial x_2^2} = 0.$$

Problem 18. Let $f : \mathbb{R}^2 \to \mathbb{R}^2$ be a smooth planar mapping with constant Jacobian determinant $J = 1$, written as

$$Q = Q(p, q), \qquad P = P(p, q).$$

For coordinates in $\mathbb{R}^2$ the (area) differential two-form is given as

$$\omega = dp \wedge dq.$$

(i) Find $f^*\omega$.
(ii) Show that $pdq - f^*(pdq) = dF$ for some smooth function $F : \mathbb{R}^2 \to \mathbb{R}$.

Solution 18. (i) Since

$$J \equiv \left(\frac{\partial P}{\partial p}\frac{\partial Q}{\partial q} - \frac{\partial P}{\partial q}\frac{\partial Q}{\partial p}\right) = 1$$

we have

$$f^*(dp \wedge dq) = (f^*dp) \wedge (f^*dq)$$
$$= \left(\frac{\partial P}{\partial q} dq + \frac{\partial P}{\partial p} dp \right) \wedge \left(\frac{\partial Q}{\partial q} dq + \frac{\partial Q}{\partial p} dp \right)$$
$$= \left(\frac{\partial P}{\partial p} \frac{\partial Q}{\partial q} - \frac{\partial P}{\partial q} \frac{\partial Q}{\partial p} \right) dp \wedge dq = dp \wedge dq.$$

(ii) Since $dp \wedge dq = d(pdq)$ and the exterior derivative commutes with the pull-back operator, i.e. $d(f^*pdq) = f^*d(pdq)$ we can write

$$d(pdq - f^*(pdq)) = 0.$$

This implies that $pdq - f^*(pdq) = dF$ for some smooth function $F : \mathbb{R}^2 \to \mathbb{R}$.

Problem 19.   (i) Calculate the covariant derivative of $\tilde{\alpha}$ and find the equation which follows from the condition

$$D_{\tilde{\alpha}} \tilde{\alpha} = 0. \tag{1}$$

(ii) Study the case where the basis of the semi-simple Lie algebra satisfies the commutation relations

$$[X_1, X_2] = 2X_2, \qquad [X_3, X_1] = 2X_3, \qquad [X_2, X_3] = X_1. \tag{2}$$

(iii) Let

$$a_1 = -\eta, \qquad a_2 = \frac{1}{2} \frac{\partial u}{\partial x}, \qquad a_3 = -\frac{1}{2} \frac{\partial u}{\partial x} \tag{3a}$$

$$A_1 = -\frac{1}{4\eta} \cos(u), \qquad A_2 = A_3 = -\frac{1}{4\eta} \sin(u) \tag{3b}$$

where η is an arbitrary constant with $\eta \neq 0$. Find the equation of motion.

Solution 19.   (i) Since

$$[X_k, X_j] = \sum_{i=1}^{n} C_{kj}^i X_i$$

it follows that

$$D_{\tilde{\alpha}} \tilde{\alpha} = \sum_{i=1}^{n} \left(\left(-\frac{\partial a_i}{\partial t} + \frac{\partial A_i}{\partial x} \right) + \frac{1}{2} \sum_{k=1}^{n} \sum_{j=1}^{n} (a_k A_j - a_j A_k) C_{kj}^i \right) dx \wedge dt \otimes X_i.$$

Since the X_j, $j = 1, \ldots, n$ form a basis of the semi-simple Lie algebra, the condition $D_{\tilde{\alpha}} \tilde{\alpha} = 0$ yields

$$\left(-\frac{\partial a_i}{\partial t} + \frac{\partial A_i}{\partial x} \right) + \frac{1}{2} \sum_{k=1}^{n} \sum_{j=1}^{n} (a_k A_j - a_j A_k) C_{kj}^i = 0$$

for $i = 1, \ldots, n$. Since $C^i_{kj} = -C^i_{jk}$, it follows that

$$\left(-\frac{\partial a_i}{\partial t} + \frac{\partial A_i}{\partial x} \right) + \sum_{k<j}^{n} (a_k A_j - a_j A_k) C^i_{kj} = 0.$$

(ii) Consider now a special case where $n = 3$ and X_1, X_2 and X_3 satisfy the commutation relations (2). We find

$$-\frac{\partial a_1}{\partial t} + \frac{\partial A_1}{\partial x} + a_2 A_3 - a_3 A_2 = 0$$

$$-\frac{\partial a_2}{\partial t} + \frac{\partial A_2}{\partial x} + 2(a_1 A_2 - a_2 A_1) = 0$$

$$-\frac{\partial a_3}{\partial t} + \frac{\partial A_3}{\partial x} - 2(a_1 A_3 - a_3 A_1) = 0.$$

A convenient choice of a basis $\{X_1, X_2, X_3\}$ is given by

$$X_1 = \begin{pmatrix} 1 & 0 \\ 0 & -1 \end{pmatrix}, \qquad X_2 = \begin{pmatrix} 0 & 1 \\ 0 & 0 \end{pmatrix}, \qquad X_3 = \begin{pmatrix} 0 & 0 \\ 1 & 0 \end{pmatrix}.$$

Consequently, the Lie algebra under consideration is $s\ell(2, \mathbb{R})$.
(iii) Inserting (3) into this system yields the sine-Gordon equation

$$\frac{\partial^2 u}{\partial x \partial t} = \sin(u).$$

Other one-dimensional soliton equations (such as the Korteweg-de Vries equation, the nonlinear Schrödinger equation, the Liouville equation) can be derived with this method.

Problem 20. Let

$$\tilde{\alpha} := \sum_{k=1}^{3} \alpha_k \otimes T_k \tag{1}$$

be a Lie algebra-valued differential one-form. Here $\otimes$ denotes the tensor product and T_k $(k = 1, 2, 3)$ are the generators given by

$$T_1 := \frac{1}{2} \begin{pmatrix} 0 & -i \\ -i & 0 \end{pmatrix}, \qquad T_2 := \frac{1}{2} \begin{pmatrix} 0 & -1 \\ 1 & 0 \end{pmatrix}, \qquad T_3 := \frac{1}{2} \begin{pmatrix} -i & 0 \\ 0 & i \end{pmatrix}.$$

$$\tag{2}$$

T_1, T_2 and T_3 form a basis of the semi-simple Lie algebra $su(2)$. The quantities α_k are differential one-forms

$$\alpha_k := \sum_{j=1}^{4} A_{kj}(\mathbf{x}) dx_j \tag{3}$$

where $\mathbf{x} = (x_1, x_2, x_3, x_4)$. The quantity $\widetilde{\alpha}$ is usually called the *connection* or *vector potential*. The actions of the Hodge duality operator $\star$ for a given metric tensor field, and the exterior derivative d, may be consistently defined by

$$\star\widetilde{\alpha} := \sum_{k=1}^{3} (\star\alpha_k) \otimes T_k \tag{4}$$

and

$$d\widetilde{\alpha} := \sum_{k=1}^{3} (d\alpha_k) \otimes T_k. \tag{5}$$

The bracket $[\,,\,]$ of Lie algebra-valued differential forms, say $\widetilde{\beta}$ and $\widetilde{\gamma}$, is defined as

$$\left[\widetilde{\beta}, \widetilde{\gamma}\right] := \sum_{k=1}^{n} \sum_{b=1}^{n} (\beta_k \wedge \gamma_b) \otimes [T_k, T_b].$$

The covariant exterior derivative of a Lie algebra-valued differential p-form $\widetilde{\gamma}$ with respect to a Lie algebra-valued one-form $\widetilde{\beta}$ is defined as

$$D_{\widetilde{\beta}}\widetilde{\gamma} := d\widetilde{\gamma} - g\left[\widetilde{\beta}, \widetilde{\gamma}\right]$$

where

$$g = \begin{cases} -1 & p \text{ even} \\ -\frac{1}{2} & p \text{ odd.} \end{cases}$$

Therefore

$$D_{\widetilde{\beta}}\widetilde{\beta} = d\widetilde{\beta} + \frac{1}{2}\left[\widetilde{\beta}, \widetilde{\beta}\right], \qquad D_{\widetilde{\beta}}(D_{\widetilde{\beta}}\widetilde{\beta}) = 0.$$

The last equation is called the *Bianchi identity*. Let $\widetilde{\alpha}$ be the Lie algebra valued-differential form given by equation (1). The *Yang-Mills equations* are given by

$$D_{\widetilde{\alpha}}(\star D_{\widetilde{\alpha}}\widetilde{\alpha}) = 0.$$

The quantity $D_{\widetilde{\alpha}}\widetilde{\alpha}$ is called the *curvature form* or *field strength tensor*. This equation is a coupled system of nonlinear partial differential equations of second order. In addition we have to impose gauge conditions. Assume the metric tensor field is given by

$$g = dx_1 \otimes dx_1 + dx_2 \otimes dx_2 + dx_3 \otimes dx_3 - dx_4 \otimes dx_4.$$

(i) Write down the Yang-Mills equation explicitly.
(ii) Impose the gauge conditions

$$A_{k4} = 0, \quad \sum_{i=1}^{3} \frac{\partial A_{ki}}{\partial x_i} = 0$$

$$\frac{\partial A_{ki}}{\partial x_j} = 0, \quad i, j, k = 1, 2, 3.$$

Solution 20. (i) The commutation relations of T_1, T_2 and T_3 are given by $[T_1, T_2] = T_3$, $[T_2, T_3] = T_1$, $[T_3, T_1] = T_2$. Since

$$D_{\widetilde{\alpha}} \widetilde{\alpha} = d\widetilde{\alpha} + \frac{1}{2} [\widetilde{\alpha}, \widetilde{\alpha}]$$

$$\frac{1}{2} [\widetilde{\alpha}, \widetilde{\alpha}] = (\alpha_1 \wedge \alpha_2) \otimes T_3 + (\alpha_2 \wedge \alpha_3) \otimes T_1 + (\alpha_3 \wedge \alpha_1) \otimes T_2$$

$$d\widetilde{\alpha} = \sum_{a=1}^{3} d\alpha_a \otimes T_a$$

and

$$\star(D_{\widetilde{\alpha}} \widetilde{\alpha}) = (\star d\alpha_1 + \star(\alpha_2 \wedge \alpha_3)) \otimes T_1 + (\star d\alpha_2 + \star(\alpha_3 \wedge \alpha_1)) \otimes T_2$$
$$+ (\star d\alpha_3 + \star(\alpha_1 \wedge \alpha_2)) \otimes T_3$$

we obtain

$$
\begin{aligned}
D_{\widetilde{\alpha}} \star (D_{\widetilde{\alpha}} \widetilde{\alpha}) = {} & d(\star d\alpha_1 + \star(\alpha_2 \wedge \alpha_3)) \otimes T_1 \\
& + d(\star d\alpha_2 + \star(\alpha_3 \wedge \alpha_1)) \otimes T_2 + d(\star d\alpha_3 + \star(\alpha_1 \wedge \alpha_2)) \otimes T_3 \\
& + (\alpha_2 \wedge (\star\alpha_3 + \star(\alpha_1 \wedge \alpha_2)) - \alpha_3 \wedge (\star d\alpha_2 + \star(\alpha_3 \wedge \alpha_1))) \otimes T_1 \\
& + (\alpha_3 \wedge (\star d\alpha_1 + \star(\alpha_2 \wedge \alpha_3)) - \alpha_1 \wedge (\star d\alpha_3 + \star(\alpha_1 \wedge \alpha_2))) \otimes T_2 \\
& + (\alpha_1 \wedge (\star d\alpha_2 + \star(\alpha_3 \wedge \alpha_1)) - \alpha_2 \wedge (\star d\alpha_1 + \star(\alpha_2 \wedge \alpha_3))) \otimes T_3.
\end{aligned}
$$

Then from the condition $D_{\widetilde{\alpha}}(\star D_{\widetilde{\alpha}} \widetilde{\alpha}) = 0$ it follows that

$$d(\star d\alpha_1 + \star(\alpha_2 \wedge \alpha_3)) + (\alpha_2 \wedge (\star d\alpha_3 + \star(\alpha_1 \wedge \alpha_2)) - \alpha_3 \wedge (\star d\alpha_2 + \star(\alpha_3 \wedge \alpha_1))) = 0$$

$$d(\star d\alpha_2 + \star(\alpha_3 \wedge \alpha_1)) + (\alpha_3 \wedge (\star d\alpha_1 + \star(\alpha_2 \wedge \alpha_3)) - \alpha_1 \wedge (\star d\alpha_3 + \star(\alpha_1 \wedge \alpha_2))) = 0$$

$$d(\star d\alpha_3 + \star(\alpha_1 \wedge \alpha_2)) + (\alpha_1 \wedge (\star d\alpha_2 + \star(\alpha_3 \wedge \alpha_1)) - \alpha_2 \wedge (\star d\alpha_1 + \star(\alpha_2 \wedge \alpha_3))) = 0.$$

Now $\star(\alpha_k \wedge \alpha_b)$ is a two-form and therefore $d \star (\alpha_k \wedge \alpha_b)$ is a three-form. We find 12 coupled partial differential equations.

(ii) Let us now impose the gauge conditions. Thus we have eliminated the space dependence of the fields. Taking into account the gauge equations we arrive at the autonomous system of ordinary differential equations of second order

$$\frac{d^2 A_{ki}}{dt^2} + \sum_{b=1}^{3} \sum_{j=1}^{3} (A_{bj} A_{bj} A_{ki} - A_{kj} A_{bj} A_{bi}) = 0$$

together with

$$\sum_{j=1}^{3}\sum_{b=1}^{3}\sum_{c=1}^{3}\varepsilon_{kbc}A_{bj}\frac{dA_{cj}}{dt}=0.$$

This system can be viewed as a Hamilton system with

$$H(A_{ki}, dA_{ki}/dt) = \frac{1}{2}\sum_{k=1}^{3}\sum_{i=1}^{3}\left(\frac{dA_{ki}}{dt}\right)^2 + \frac{1}{4}\sum_{k=1}^{3}\sum_{i=1}^{3}(A_{ki}^2)^2$$

$$-\frac{1}{4}\sum_{i=1}^{3}\sum_{j=1}^{3}\left(\sum_{k=1}^{3}A_{ki}A_{kj}\right)^2.$$

To simplify further we put $A_{ki}(t) = O_{ki}f_k(t)$ (no summation), where $O = (O_{ki})$ is a time-independent 3×3 orthogonal matrix. We put

$$\frac{df_k}{dt} \equiv p_k, \qquad f_k \equiv q_k$$

where $k = 1, 2, 3$. Then we obtain

$$\frac{d^2q_1}{dt^2}+q_1(q_2^2+q_3^2)=0, \qquad \frac{d^2q_2}{dt^2}+q_2(q_1^2+q_3^2)=0, \qquad \frac{d^2q_3}{dt^2}+q_3(q_1^2+q_2^2)=0$$

with the Hamilton function

$$H(\mathbf{p},\mathbf{q}) = \frac{1}{2}(p_1^2 + p_2^2 + p_3^2) + \frac{1}{2}\left(q_1^2q_2^2 + q_1^2q_3^2 + q_2^2q_3^2\right).$$

Problem 21. Consider the Lie group

$$G := \left\{ \begin{pmatrix} e^\alpha & \beta \\ 0 & 1 \end{pmatrix} : \alpha \in \mathbb{R},\, \beta \in \mathbb{R} \right\}. \tag{1}$$

Let

$$X := \begin{pmatrix} e^\alpha & \beta \\ 0 & 1 \end{pmatrix} \tag{2}$$

and $\Omega := X^{-1}dX$. Show that $d\Omega + \Omega \wedge \Omega = 0$.

Solution 21. From (2) we obtain the inverse matrix X^{-1} and dX

$$X^{-1} = \begin{pmatrix} e^{-\alpha} & -\beta e^{-\alpha} \\ 0 & 1 \end{pmatrix}, \qquad dX = \begin{pmatrix} e^\alpha d\alpha & d\beta \\ 0 & 0 \end{pmatrix}.$$

Thus

$$\Omega = X^{-1}dX = \begin{pmatrix} d\alpha & e^{-\alpha}d\beta \\ 0 & 0 \end{pmatrix}.$$

Therefore

$$d\Omega = \begin{pmatrix} 0 & -e^{-\alpha}d\alpha \wedge d\beta \\ 0 & 0 \end{pmatrix}, \quad \Omega \wedge \Omega = \begin{pmatrix} 0 & e^{-\alpha}d\alpha \wedge d\beta \\ 0 & 0 \end{pmatrix}$$

and the result follows.

Problem 22. Let

$$SL(2,\mathbb{R}) := \left\{ X = \begin{pmatrix} a & b \\ c & d \end{pmatrix} : ad - bc = 1 \right\} \tag{1}$$

be the semi-simple Lie group of all (2×2)-real unimodular matrices. Its right-invariant Maurer-Cartan form is

$$\omega = (dX)X^{-1} = \begin{pmatrix} \omega_{11} & \omega_{12} \\ \omega_{21} & \omega_{22} \end{pmatrix} \tag{2}$$

where $\omega_{11} + \omega_{22} = 0$.

(i) Show that ω satisfies the *structure equation* of $SL(2,\mathbb{R})$ (also called the Maurer-Cartan equation)

$$d\omega = \omega \wedge \omega. \tag{3}$$

(ii) Let U be a neighbourhood in the (x,t)-plane and consider the smooth mapping

$$f : U \to SL(2,\mathbb{R}). \tag{4}$$

The pull-backs of the Maurer-Cartan forms can be written ($\omega_{22} = -\omega_{11}$)

$$\omega_{11} = \eta(x,t)dx + A(x,t)dt,$$

$$\omega_{12} = q(x,t)dx + B(x,t)dt, \quad \omega_{21} = r(x,t)dx + C(x,t)dt$$

where the coefficients are functions of x,t. Find the differential equations for η, q, r, A, B and C.

(iii) Consider the special case that $r = +1$ and η is a parameter independent of x,t. Writing $q = u(x,t)$ find $A(x,t)$, $B(x,t)$ as functions of u and C. Let

$$C = \eta^2 - \frac{1}{2}u.$$

Find the differential equations for u.

Solution 22. (i) Since $\omega_{11} \wedge \omega_{11} = 0$, $\omega_{22} \wedge \omega_{22} = 0$, $\omega_{11} = -\omega_{22}$ we have

$$\omega \wedge \omega = \begin{pmatrix} \omega_{11} & \omega_{12} \\ \omega_{21} & \omega_{22} \end{pmatrix} \wedge \begin{pmatrix} \omega_{11} & \omega_{12} \\ \omega_{21} & \omega_{22} \end{pmatrix}$$

$$= \begin{pmatrix} \omega_{12} \wedge \omega_{21} & \omega_{11} \wedge \omega_{12} + \omega_{12} \wedge \omega_{22} \\ \omega_{21} \wedge \omega_{11} + \omega_{22} \wedge \omega_{21} & \omega_{21} \wedge \omega_{12} \end{pmatrix}$$

$$= \begin{pmatrix} \omega_{12} \wedge \omega_{21} & 2\omega_{11} \wedge \omega_{12} \\ -2\omega_{11} \wedge \omega_{21} & -\omega_{12} \wedge \omega_{21} \end{pmatrix}.$$

Or, written explicitly,

$$d\omega_{11} = \omega_{12} \wedge \omega_{21}, \qquad d\omega_{12} = 2\omega_{11} \wedge \omega_{12}, \qquad d\omega_{21} = 2\omega_{21} \wedge \omega_{11}.$$

(ii) We find

$$-\frac{\partial \eta}{\partial t} + \frac{\partial A}{\partial x} - qC + rB = 0,$$

$$-\frac{\partial q}{\partial t} + \frac{\partial B}{\partial x} - 2\eta B + 2qA = 0,$$

$$-\frac{\partial r}{\partial t} + \frac{\partial C}{\partial x} - 2rA + 2\eta C = 0.$$

(iii) From the third equation in (ii) with $r = 1$ we obtain

$$A(x,t) = \eta C(x,t) + \frac{1}{2}\frac{\partial C}{\partial x}.$$

From the first equation in (ii) we obtain

$$B(x,t) = u(x,t)C(x,t) - \eta\frac{\partial C}{\partial x} - \frac{1}{2}\frac{\partial^2 C}{\partial x^2}.$$

Inserting this expressions into the second equation in (ii) with $q = u$ yields

$$\frac{\partial u}{\partial t} = \frac{\partial u}{\partial x}C + 2u\frac{\partial C}{\partial x} + 2\eta^2\frac{\partial C}{\partial x} - \frac{1}{2}\frac{\partial^3 C}{\partial x^3}.$$

With $C = \eta^2 - u/2$ we finally obtain

$$\frac{\partial u}{\partial t} = \frac{1}{4}\frac{\partial^3 u}{\partial x^3} - \frac{3}{2}u\frac{\partial u}{\partial x}$$

which is the *Korteweg-de Vries equation*.

Problem 23.   Let $f : \mathbb{R}^4 \to \mathbb{R}$ be a smooth function $f(p_1, p_2, q_1, q_2)$. Consider the differential two form

$$\omega = dp_1 \wedge dq_1 + dp_2 \wedge dq_2.$$

We define the vector field

$$V_f = V_{q1}\frac{\partial}{\partial q_1} + V_{q2}\frac{\partial}{\partial q_2} + V_{p1}\frac{\partial}{\partial p_1} + V_{p2}\frac{\partial}{\partial p_2}$$

as $V_f \rfloor \omega + df = 0$. Find V_f and the corresponding autonomous system of first order differential equation.

Solution 23. We have

$$df = \sum_{j=1}^{2} \left(\frac{\partial f}{\partial q_j} dq_j + \frac{\partial f}{\partial p_j} dp_j \right)$$

and

$$V_f \rfloor \omega = -V_{q1} dp_1 - V_{q2} dp_2 + V_{p1} dq_1 + V_{p2} dq_2.$$

Thus from $V_f \rfloor \omega + df = 0$ it follows that

$$V_{q1} = \frac{\partial f}{\partial p_1}, \quad V_{q2} = \frac{\partial f}{\partial p_2}, \quad V_{p1} = -\frac{\partial f}{\partial q_1}, \quad V_{p2} = -\frac{\partial f}{\partial q_2}.$$

Hence we have the system of differential equations

$$\frac{dq_1}{dt} = \frac{\partial f}{\partial p_1}, \quad \frac{dq_2}{dt} = \frac{\partial f}{\partial p_2}, \quad \frac{dp_1}{dt} = -\frac{\partial f}{\partial q_1}, \quad \frac{dp_2}{dt} = -\frac{\partial f}{\partial q_2}.$$

Problem 24. Consider the Euclidean group $\mathbb{E}^2$ and

$$g(\phi, x_1, x_2) = \begin{pmatrix} \cos(\phi) & \sin(\phi) & x_1 \\ -\sin(\phi) & \cos(\phi) & x_2 \\ 0 & 0 & 1 \end{pmatrix}$$

and the differential one-forms

$$\alpha_1 = \cos(\phi) dx_1 - \sin(\phi) dx_2, \quad \alpha_2 = \sin(\phi) dx_1 + \cos(\phi) dx_2, \quad \alpha = d\phi.$$

Show that (Cartan-Maurer equations) $d\alpha_1 = -\alpha \wedge \alpha_2$, $d\alpha_2 = \alpha \wedge \alpha_1$, $d\alpha = 0$.

Solution 24. We have

$$d\alpha_1 = -\sin(\phi) d\phi \wedge dx_1 - \cos(\phi) d\phi \wedge dx_2, \quad d\alpha_2 = \cos(\phi) d\phi \wedge dx_1 - \sin(\phi) d\phi \wedge dx_2$$

and $d\alpha = 0$. With $d\phi \wedge d\phi = 0$ we have

$$\alpha \wedge \alpha_2 = \sin(\phi) d\phi \wedge dx_1 + \cos(\phi)) d\phi \wedge dx_2$$

$$\alpha \wedge \alpha_1 = \cos(\phi) d\phi \wedge dx_1 - \sin(\phi) d\phi \wedge dx_2$$

and the Cartan-Maurer equations follow.

Problem 25. Consider the three-dimensional Euclidean space $\mathbb{E}^3$ with the *vector product*

$$\mathbf{v} \times \mathbf{u} = \begin{pmatrix} v_2 u_3 - v_3 u_2 \\ v_3 u_1 - v_1 u_3 \\ v_1 u_2 - v_2 u_1 \end{pmatrix}.$$

Show that the vector product can also be written as $\star(\mathbf{v} \wedge \mathbf{u})$.

Solution 25. Let $\mathbf{e}_1$, $\mathbf{e}_2$, $\mathbf{e}_3$ be the standard basis in $\mathbb{E}^3$. Then

$$\mathbf{v} \wedge \mathbf{u} = \left(\sum_{j=1}^{3} v_j \mathbf{e}_j \right) \wedge \left(\sum_{k=1}^{3} u_k \mathbf{e}_k \right)$$

$$= (v_1 u_2 - v_2 u_1)\mathbf{e}_1 \wedge \mathbf{e}_2 + (v_2 u_3 - v_3 u_2)\mathbf{e}_2 \wedge \mathbf{e}_3 + (v_3 u_1 - v_1 u_3)\mathbf{e}_3 \wedge \mathbf{e}_1.$$

Now $\star(\mathbf{e}_1 \wedge \mathbf{e}_2) = \mathbf{e}_3$, $\star(\mathbf{e}_2 \wedge \mathbf{e}_3) = \mathbf{e}_1$, $\star(\mathbf{e}_3 \wedge \mathbf{e}_1) = \mathbf{e}_2$. Hence

$$\star(\mathbf{v} \wedge \mathbf{u}) = (v_2 u_3 - v_3 u_2)\mathbf{e}_1 + (v_3 u_1 - v_1 u_3)\mathbf{e}_2 + (v_1 u_2 - v_2 u_1)\mathbf{e}_3.$$

Supplementary Problems

Problem 1. Given the smooth differential one-forms in $\mathbb{R}^3$

$$\alpha_1 = f_1(\mathbf{x})dx_1 + f_2(\mathbf{x})dx_2 + f_3(\mathbf{x})dx_3$$
$$\alpha_2 = g_1(\mathbf{x})dx_1 + g_2(\mathbf{x})dx_2 + g_3(\mathbf{x})dx_3$$

and the vectors

$$\mathbf{f} = \begin{pmatrix} f_1 \\ f_2 \\ f_3 \end{pmatrix}, \quad \mathbf{g} = \begin{pmatrix} g_1 \\ g_2 \\ g_3 \end{pmatrix}.$$

Let $\star$ be the Hodge duality operator. Find $\alpha_1 \wedge \alpha_2$ and $\star(\alpha_1 \wedge \alpha_2)$. Find $\mathbf{f} \times \mathbf{g}$, where $\times$ denotes the vector product.

Problem 2. (i) Let α be a differential form of odd degree. Show that

$$\alpha \wedge \alpha = 0.$$

(ii) Let α_1, α_2 be differential forms of odd degree. Show that

$$\alpha_1 \wedge \alpha_2 = -\alpha_2 \wedge \alpha_1.$$

Problem 3. Let $z = x + iy$, $x, y \in \mathbb{R}$. Show that $dz \wedge d\bar{z} = -2i dx \wedge dy$.

Problem 4. A *magnetic bottle* is made up of a homogeneous magnetic (dipole) field with a superimposed octupole contribution. In cylindrical coordinates (ρ, x_3) we have

$$\mathbf{B}(\rho, x_3) = B_0 \mathbf{e}_{x_3} + B_2(-\rho x_3 \mathbf{e}_\rho + (x_3^2 - \rho^2/2)\mathbf{e}_{x_3}).$$

Find curl(**B**).

Problem 5. Let $\mathbf{v}_1$, $\mathbf{v}_2$, $\mathbf{v}_3$ be vectors in $\mathbb{R}^3$ and $\times$ be the vector product and $\cdot$ be the scalar product. Then

$$\mathbf{v}_1 \times (\mathbf{v}_2 \times \mathbf{v}_3) \equiv (\mathbf{v}_1 \cdot \mathbf{v}_3)\mathbf{v}_2 - (\mathbf{v}_1 \cdot \mathbf{v}_2)\mathbf{v}_3.$$

Express this identity using the exterior product and the Hodge duality operator for the three dimensional Euclidean space.

Problem 6. Let $n \geq 2$ and A be a nonzero $n \times n$ matrix over $\mathbb{R}$. Consider the differential one form

$$\alpha = (\, x_1 \quad \cdots \quad x_n \,) A \begin{pmatrix} dx_1 \\ \vdots \\ dx_n \end{pmatrix}.$$

Find $d\alpha$, $\alpha \wedge \alpha$ and $d\alpha - \alpha \wedge \alpha$.

Problem 7. Let $n \geq 2$. Consider the semi-simple Lie algebra $so(n)$ with dimension $n(n-1)/2$. Find the structure constants of $so(n)$ which satisfy

$$c^i_{jk} + c^i_{kj} = 0, \qquad \sum_{j=1}^{n(n-1)/2} (c^i_{jk}c^j_{rs} + c^i_{jr}c^j_{sk} + c^i_{js}c^j_{kr}) = 0.$$

Find all differential one forms α_j such that

$$d\alpha_i = \frac{1}{2} \sum_{j,k=1}^{n(n-1)/2} c^i_{jk}\alpha_j \wedge \alpha_k, \quad i = 1,\ldots,n(n-1)/2.$$

Problem 8. Let $V(x_1, x_2, x_3)$ be a smooth vector field in $\mathbb{R}^3$ and $k \in \mathbb{R}$. Find solutions of $\nabla \times V \equiv \mathrm{curl}(V) = kV$.

Problem 9. Consider the first order differential equation

$$\frac{dy}{dx} = \frac{-x + \sqrt{x^2 + y^2}}{y}$$

and the differential one-form

$$\alpha = ydy + xdx - \sqrt{x^2 + y^2}dx.$$

Find $d\alpha$. Solve the differential equation. Show that with $y = ux$ one finds

$$\frac{udu}{\sqrt{1+u^2}(1-\sqrt{1+u^2})} = \frac{dx}{x}.$$

Problem 10.   Prove the *vector identity*

$$\frac{1}{2}\nabla(\mathbf{v}^2) \equiv (\mathbf{v} \cdot \nabla)\mathbf{v} + \mathbf{v} \times (\mathrm{curl}(\mathbf{v}))$$

in $\mathbb{R}^3$. Give a version of the identity using differential forms.

Problem 11.   Let $F : \mathbb{R}^3 \to \mathbb{R}^3$. Find solution of the equation

$$\nabla \times (\nabla \times F) = k^2 F$$

where k denotes the wave number. Express the differential equation using differential forms.

Chapter 24

Metric Tensor Fields, Surfaces and Relativity

Let G be the gravitational constant and c be the speed of light in vacuum. The *Einstein equation* are given by

$$R_{\mu\nu} - \frac{1}{2}Rg_{\mu\nu} = \frac{8\pi G}{c^4}T_{\mu\nu}, \quad \mu, \nu = 0, 1, 2, 3$$

where $R_{\mu\nu}$ is the Ricci tensor, R the Ricci scalar (the tensor contraction of the Ricci tensor) and $T = (T_{\mu\nu})$ is the energy momentum tensor. In local coordinates one has

$$\sum_{\mu,\nu=0}^{3} R_{\mu\nu}dx_\mu \otimes dx_\nu - \frac{1}{2}R \sum_{\mu,\nu=0}^{3} g_{\mu\nu}dx_\mu \otimes dx_\nu = \frac{8\pi G}{c^4} \sum_{\mu,\nu=0}^{3} T_{\mu\nu}dx_\mu \otimes dx_\nu.$$

The energy-momentum tensor is given by

$$T_{\mu\nu} = \left(\rho + \frac{p}{c^2}\right)u_\mu u_\nu + pg_{\mu\nu}, \quad \mu, \nu = 0, 1, 2, 3$$

where ρ is the mass density, p the pressure and u_μ the velocity. The *Hilbert action* is given by

$$S = \frac{c^4}{16\pi} \int R\sqrt{-g}dx_0 dx_1 dx_2 dx_3$$

with $g = \det(g_{\mu\nu})$.

The metric tensor field for the Schwarzschild metric is given by

$$g = \left(1 - \frac{r_s}{r}\right) dx_0 \otimes dx_0 - \left(1 - \frac{r_s}{r}\right) dr \otimes dr - r^2 (d\theta \otimes d\theta + \sin(\theta) d\phi \otimes d\phi)$$

with $r_s = 2GM/c^2$ the Schwarzschild radius.

The Einstein-Maxwell equations with cosmological constant Λ are given by

$$R^{\mu\nu} - Rg^{\mu\nu} + \Lambda g^{\mu\nu} = \frac{8\pi G}{c^4} T^{\mu\nu}$$

with the electromagnetic stress energy given by

$$T^{\mu\nu} = -\frac{1}{\mu_0} \left(\sum_{\alpha=0}^{4} F^{\mu\alpha} F_\alpha^\nu + \frac{1}{4} g^{\mu\nu} \sum_{\alpha=0}^{3} \sum_{\beta=0}^{3} F_{\alpha\beta} F^{\alpha\beta} \right)$$

where $F^{\mu\nu}$ is the electromagnetic tensor.

Let M be a Riemannian manifold or pseudo Riemannian manifold with metric tensor field

$$g = \sum_{j,k=1}^{N} g_{jk}(\mathbf{x}) dx_j \otimes dx_k.$$

The inverse of the $N \times N$ matrix (g_{jk}) is denoted by (g^{jk}). The *Christoffel symbols* of the second kind are defined by

$$\Gamma_{mn}^a := \frac{1}{2} \sum_{b=1}^{N} g^{ab}(g_{bm,n} + g_{bn,m} - g_{mn,b})$$

where $g_{bm,n}$ denotes the partial derivative of g_{bm} with respect to the coordinates x_n, i.e. $g_{bm,n} := \partial g_{mn}/\partial x_n$. Note that

$$\Gamma_{ij}^k = \sum_{m=1}^{N} g^{km} \Gamma_{ijm}, \quad \Gamma_{ijm} = \frac{1}{2}(g_{jm,i} + g_{mi,j} - g_{ij,m})$$

where Γ_{ijm} are the Christoffel symbols of the first kind. From the Christoffel symbols we find the *curvature tensor*

$$R_{msq}^b = \Gamma_{mq,s}^b - \Gamma_{ms,q}^b + \sum_{n=1}^{N} \Gamma_{ns}^b \Gamma_{mq}^n - \sum_{n=1}^{N} \Gamma_{nq}^b \Gamma_{ms}^n.$$

The *Ricci tensor* R_{mq} is given by

$$R_{mq} = \sum_{s=1}^{N} R_{msq}^s, \quad R_q^m = \sum_{n=1}^{N} g^{mn} R_{nq}$$

and the *curvature scalar* is $R = \sum_{m=1}^{N} R_m^m$.

Problem 1. Let

$$g = dx \otimes dx + \cos(u(x,t))dx \otimes dt + \cos(u(x,t))dt \otimes dx + dt \otimes dt \qquad (1)$$

be a metric tensor field, where u is a smooth function of x and t. We set $1 \equiv x$ and $2 \equiv t$. Then $g_{11} = g_{22} = 1$ and

$$g_{12} = g_{21} = \cos(u(x,t)).$$

We write g as a 2×2 matrix and the inverse matrix g^{-1} of g as

$$g = \begin{pmatrix} g_{11} & g_{12} \\ g_{21} & g_{22} \end{pmatrix} \quad \Rightarrow \quad g^{-1} = \begin{pmatrix} g^{11} & g^{12} \\ g^{21} & g^{22} \end{pmatrix}$$

where

$$g^{11} = g^{22} = \frac{1}{\sin^2(u)}, \qquad g^{12} = g^{21} = -\frac{\cos(u)}{\sin^2(u)}.$$

(i) Calculate the *Riemann curvature scalar R*.
(ii) Find the equation which follows from the condition $R = -2$.

Solution 1. Obviously $g_{11,1} = g_{11,2} = g_{22,1} = g_{22,2} = 0$ and

$$g_{12,1} = g_{21,1} = -\frac{\partial u}{\partial x}\sin(u), \quad g_{12,2} = g_{21,2} = -\frac{\partial u}{\partial t}\sin(u).$$

Therefore the Christoffel symbols are given by

$$\Gamma^1_{11} = \frac{\partial u}{\partial x}\frac{\cos(u)}{\sin(u)}, \quad \Gamma^1_{12} = \Gamma^1_{21} = 0, \quad \Gamma^1_{22} = -\frac{\partial u}{\partial t}\frac{1}{\sin(u)}$$

$$\Gamma^2_{11} = -\frac{\partial u}{\partial x}\frac{1}{\sin(u)}, \quad \Gamma^2_{12} = \Gamma^2_{21} = 0, \quad \Gamma^2_{22} = \frac{\partial u}{\partial t}\frac{\cos(u)}{\sin(u)}.$$

Obviously $R^b_{mss} = 0$, i.e.

$$R^1_{111} = R^1_{122} = R^1_{211} = R^1_{222} = R^2_{111} = R^2_{122} = R^2_{211} = R^2_{222} = 0.$$

Moreover

$$R^1_{112} = -R^1_{121} = -\frac{\partial^2 u}{\partial x \partial t}\frac{\cos(u)}{\sin(u)}, \qquad R^1_{212} = -R^1_{221} = -\frac{\partial^2 u}{\partial x \partial t}\frac{1}{\sin(u)}$$

$$R^2_{112} = -R^2_{121} = \frac{\partial^2 u}{\partial x \partial t}\frac{1}{\sin(u)}, \qquad R^2_{212} = -R^2_{221} = \frac{\partial^2 u}{\partial x \partial t}\frac{\cos(u)}{\sin(u)}.$$

For the Ricci tensor we find

$$R_{11} = R^1_{111} + R^2_{121} = -\frac{\partial^2 u}{\partial x \partial t}\frac{1}{\sin(u)}, \qquad R_{12} = R^1_{112} + R^2_{122} = -\frac{\partial^2 u}{\partial x \partial t}\frac{\cos(u)}{\sin(u)}$$

$$R_{21} = R^1_{211} + R^2_{221} = -\frac{\partial^2 u}{\partial x \partial t} \frac{\cos(u)}{\sin(u)}, \qquad R_{22} = R^1_{212} + R^2_{222} = -\frac{\partial^2 u}{\partial x \partial t} \cdot \frac{1}{\sin(u)}.$$

From R_{nq} we obtain R^m_q via $R^m_q = g^{mn} R_{nq}$. We find

$$R^1_1 = R^2_2 = -\frac{1}{\sin(u)} \frac{\partial^2 u}{\partial x \partial t}$$

and $R^2_1 = R^1_2 = 0$. Thus for the curvature scalar we find

$$R = -\frac{2}{\sin(u)} \frac{\partial^2 u}{\partial x \partial t}.$$

If $R = -2$, then

$$\frac{\partial^2 u}{\partial x \partial t} = \sin(u).$$

This is the so-called *sine-Gordon equation*.

Problem 2.   Given the metric tensor field

$$g = g_{11}(q_1, q_2) dq_1 \otimes dq_1 + g_{22}(q_1, q_2) dq_2 \otimes dq_2. \tag{1}$$

(i) Calculate the *Riemann curvature scalar* R.
(ii) Simplify g to the special case $g_{11}(q_1, q_2) = g_{22}(q_1, q_2) = E - V(q_1, q_2)$.

Solution 2.   (i) First we have to calculate the *Christoffel symbols*. Since

$$\Gamma^1_{11} := \frac{1}{2} \sum_{k=1}^{2} g^{1k} \left(\frac{\partial g_{k1}}{\partial q_1} + \frac{\partial g_{1k}}{\partial q_1} - \frac{\partial g_{11}}{\partial q_k} \right)$$

and $g_{12} = g_{21} = 0$ we have

$$\Gamma^1_{11} = \frac{1}{2} g^{11} \left(\frac{\partial g_{11}}{\partial q_1} + \frac{\partial g_{11}}{\partial q_1} - \frac{\partial g_{11}}{\partial q_1} \right) = \frac{1}{2} g^{11} \frac{\partial g_{11}}{\partial q_1}.$$

Analogously

$$\Gamma^1_{12} = \frac{1}{2} g^{11} \frac{\partial g_{11}}{\partial q_2}, \quad \Gamma^1_{21} = \frac{1}{2} g^{11} \frac{\partial g_{11}}{\partial q_2}, \quad \Gamma^1_{22} = -\frac{1}{2} g^{11} \frac{\partial g_{22}}{\partial q_1},$$

$$\Gamma^2_{11} = -\frac{1}{2} g^{22} \frac{\partial g_{11}}{\partial q_2}, \quad \Gamma^2_{12} = \frac{1}{2} g^{22} \frac{\partial g_{22}}{\partial q_1}, \quad \Gamma^2_{21} = \frac{1}{2} g^{22} \frac{\partial g_{22}}{\partial q_1}, \quad \Gamma^2_{22} = \frac{1}{2} g^{22} \frac{\partial g_{22}}{\partial q_2}$$

where $g^{11} = 1/g_{11}$, $g^{22} = 1/g_{22}$. Next we have to calculate the curvature. We find

$$R_{12} = \sum_{h=1}^{2} \left(\frac{\partial \Gamma^h_{12}}{\partial q_h} - \frac{\partial \Gamma^h_{h1}}{\partial q_2} + \sum_{\ell=1}^{2} (\Gamma^h_{h\ell} \Gamma^\ell_{12} - \Gamma^\ell_{1h} \Gamma^h_{2\ell}) \right) = 0$$

$R_{21} = 0$ for symmetry reasons

$$R_{11} = \sum_{h=1}^{2} \left(\frac{\partial \Gamma_{11}^{h}}{\partial q_h} - \frac{\partial \Gamma_{h1}^{h}}{\partial q_1} + \sum_{\ell=1}^{2} \left(\Gamma_{h\ell}^{h} \Gamma_{11}^{\ell} - \Gamma_{1h}^{\ell} \Gamma_{1\ell}^{h} \right) \right)$$

$$= \frac{\partial \Gamma_{11}^{2}}{\partial q_2} - \frac{\partial \Gamma_{21}^{2}}{\partial q_1} + \left(\Gamma_{21}^{2} \Gamma_{11}^{1} + \Gamma_{22}^{2} \Gamma_{11}^{2} - \Gamma_{11}^{2} \Gamma_{12}^{1} - \Gamma_{12}^{2} \Gamma_{12}^{2} \right)$$

$$R_{22} = \sum_{h=1}^{2} \left(\frac{\partial \Gamma_{22}^{h}}{\partial q_h} - \frac{\partial \Gamma_{h2}^{h}}{\partial q_2} + \sum_{\ell=1}^{2} \left(\Gamma_{h\ell}^{h} \Gamma_{22}^{\ell} - \Gamma_{2h}^{\ell} \Gamma_{2\ell}^{h} \right) \right)$$

$$= \frac{\partial \Gamma_{22}^{1}}{\partial q_1} - \frac{\Gamma_{12}^{1}}{\partial q_2} + \Gamma_{11}^{1} \Gamma_{22}^{1} + \Gamma_{12}^{1} \Gamma_{22}^{2} - \Gamma_{21}^{1} \Gamma_{21}^{1} - \Gamma_{21}^{2} \Gamma_{22}^{1}.$$

Now the Riemann curvature scalar R is given by $R = g^{11} R_{11} + g^{22} R_{22}$. Consequently,

$$R = g^{11} \left(\frac{\partial \Gamma_{11}^{2}}{\partial q_2} - \frac{\partial \Gamma_{21}^{2}}{\partial q_1} \right) + g^{11} \left(\Gamma_{11}^{1} \Gamma_{21}^{2} + \Gamma_{22}^{2} \Gamma_{11}^{2} - \Gamma_{11}^{2} \Gamma_{12}^{1} - \Gamma_{12}^{2} \Gamma_{12}^{2} \right)$$

$$+ g^{22} \left(\frac{\partial \Gamma_{22}^{1}}{\partial q_1} - \frac{\partial \Gamma_{12}^{1}}{\partial q_2} \right) + g^{22} \left(\Gamma_{22}^{2} \Gamma_{12}^{1} + \Gamma_{11}^{1} \Gamma_{22}^{1} - \Gamma_{22}^{1} \Gamma_{21}^{2} - \Gamma_{21}^{1} \Gamma_{21}^{1} \right)$$

where

$$\frac{\partial \Gamma_{11}^{2}}{\partial q_2} = -\frac{1}{2} \frac{\partial}{\partial q_2} \left(g^{22} \frac{\partial g_{11}}{\partial q_2} \right) = -\frac{1}{2} \left(\frac{\partial g^{22}}{\partial q_2} \right) \left(\frac{\partial g_{11}}{\partial q_2} \right) - \frac{1}{2} g^{22} \frac{\partial^2 g_{11}}{\partial q_2^2}$$

$$\frac{\partial \Gamma_{21}^{2}}{\partial q_1} = \frac{1}{2} \frac{\partial}{\partial q_1} \left(g^{22} \frac{\partial g_{22}}{\partial q_1} \right) = \frac{1}{2} \left(\frac{\partial g^{22}}{\partial q_1} \right) \left(\frac{\partial g_{22}}{\partial q_1} \right) + \frac{1}{2} g^{22} \frac{\partial^2 g_{22}}{\partial q_1^2}$$

$$\frac{\partial \Gamma_{22}^{1}}{\partial q_1} = -\frac{1}{2} \frac{\partial}{\partial q_1} \left(g^{11} \frac{\partial g_{22}}{\partial q_1} \right) = -\frac{1}{2} \left(\frac{\partial g^{11}}{\partial q_1} \right) \left(\frac{\partial g_{22}}{\partial q_1} \right) - \frac{1}{2} g^{11} \frac{\partial^2 g_{22}}{\partial q_1^2}$$

$$\frac{\partial \Gamma_{12}^{1}}{\partial q_2} = \frac{1}{2} \frac{\partial}{\partial q_2} \left(g^{11} \frac{\partial g_{11}}{\partial q_2} \right) = \frac{1}{2} \left(\frac{\partial g^{11}}{\partial q_2} \right) \left(\frac{\partial g_{11}}{\partial q_2} \right) + \frac{1}{2} g^{11} \frac{\partial^2 g_{11}}{\partial q_2^2}.$$

Therefore

$$R = \frac{g^{11} g^{22}}{2} \left(-\frac{\partial^2 g_{11}}{\partial q_2^2} - \frac{\partial^2 g_{22}}{\partial q_1^2} - \frac{\partial^2 g_{22}}{\partial q_1^2} - \frac{\partial^2 g_{11}}{\partial q_2^2} \right)$$

$$- \frac{1}{2} \left(g^{11} \frac{\partial g^{22}}{\partial q_2} \frac{\partial g_{11}}{\partial q_2} + g^{11} \frac{\partial g^{22}}{\partial q_1} \frac{\partial g_{22}}{\partial q_1} + g^{22} \frac{\partial g^{11}}{\partial q_1} \frac{\partial g_{22}}{\partial q_1} + g^{22} \frac{\partial g^{11}}{\partial q_2} \frac{\partial g_{11}}{\partial q_2} \right).$$

(ii) If $g_{11} = g_{22} = E - V(q_1, q_2)$ we obtain $g^{11} = 1/(E-V)$, $g^{22} = 1/(E-V)$. Therefore

$$R = \frac{1}{(E-V)^3} \left(\left(\frac{\partial^2 V}{\partial q_1^2} + \frac{\partial^2 V}{\partial q_2^2} \right) (E-V) + \frac{\partial V}{\partial q_1} \frac{\partial V}{\partial q_1} + \frac{\partial V}{\partial q_2} \frac{\partial V}{\partial q_2} \right).$$

The motion of a classical mechanical system with Hamilton function

$$H(\mathbf{p}, \mathbf{q}) = \frac{1}{2} \sum_{j,k=1}^{N} \alpha^{jk} p_j p_k + V(q_1, q_2, \ldots, q_N)$$

can be represented as a *geodesic flow* on a Riemannian manifold with metric $g_{jk} = (E - V(\mathbf{q})) \alpha_{jk}$. The matrix (α^{jk}) is symmetric and the matrix (α_{kl}) is its inverse

$$\sum_{k=1}^{N} \alpha^{jk} \alpha_{kl} = \delta_l^j$$

where δ_l^j is the Kronecker symbol. In our case we have $N = 2$ and $\alpha_{jk} = \delta_{jk}$. The geodesic flow on a closed Riemannian manifold of negative curvature is a so-called *C-flow* and is therefore ergodic.

Problem 3. Let $a > b > 0$ and define $f : \mathbb{R}^2 \to \mathbb{R}^3$ by

$$f(\theta, \phi) = ((a + b\cos(\phi))\cos(\theta), (a + b\cos(\phi))\sin(\theta), b\sin(\phi))$$

where $0 \leq \phi < 2\pi$ and $0 \leq \theta < 2\pi$. The function f is a *parametrized torus* $\mathbb{T}^2$ on $\mathbb{R}^3$. Let

$$g = dx_1 \otimes dx_1 + dx_2 \otimes dx_2 + dx_3 \otimes dx_3.$$

The parametrized torus is doubly-periodic, i.e.

$$f(\theta + 2k\pi, \phi) = f(\theta, \phi), \qquad f(\theta, \phi + 2k\pi) = f(\theta, \phi), \qquad k \in \mathbb{Z}.$$

(i) Calculate $g|_{\mathbb{T}^2}$.
(ii) Calculate the Christoffel symbols Γ_{ab}^m from $g|_{\mathbb{T}^2}$.
(iii) Give the differential equations of the geodesics.

Solution 3. (i) Since

$$x_1(\phi, \theta) = (a + b\cos(\phi))\cos(\theta)$$
$$x_2(\phi, \theta) = (a + b\cos(\phi))\sin(\theta)$$
$$x_3(\phi, \theta) = b\sin(\phi)$$

and $dx_j = (\partial x_j / \partial \theta) d\theta + (\partial x_j / \partial \phi) d\phi$ $(j = 1, 2, 3)$ we find

$$dx_1 \otimes dx_1 + dx_2 \otimes dx_2 + dx_3 \otimes dx_3 = (a + b\cos(\phi))^2 d\theta \otimes d\theta + b^2 d\phi \otimes d\phi$$

where we used that $\sin^2(\theta) + \cos^2(\theta) = 1$ and $\sin^2(\phi) + \cos^2(\phi) = 1$. We identify θ with the first coordinate and ϕ with the second coordinate. Thus

$$g_{11} = (a + b\cos(\phi))^2, \quad g_{12} = 0, \quad g_{21} = 0, \quad g_{22} = b^2$$

and therefore

$$g^{11} = \frac{1}{(a + b\cos(\phi))^2}, \quad g^{12} = 0, \quad g^{21} = 0, \quad g^{22} = \frac{1}{b^2}.$$

(ii) We set $y^1 = \theta$ and $y^2 = \phi$. The *Christoffel symbols* are given by

$$\Gamma^\alpha_{\beta\gamma} = \frac{1}{2} \sum_{\delta=1}^{3} g^{\alpha\delta} \left(\frac{\partial g_{\beta\delta}}{\partial y^\gamma} + \frac{\partial g_{\gamma\delta}}{\partial y^\beta} - \frac{\partial g_{\beta\gamma}}{\partial y^\delta} \right)$$

Thus we find

$$\Gamma^1_{11} = 0, \quad \Gamma^1_{12} = -\frac{b\sin(\phi)}{a + b\cos(\phi)}, \quad \Gamma^1_{21} = -\frac{b\sin(\phi)}{a + b\cos(\phi)}, \quad \Gamma^1_{22} = 0$$

$$\Gamma^2_{11} = \frac{(a + b\cos(\phi))\sin(\phi)}{b}, \quad \Gamma^2_{12} = 0, \quad \Gamma^2_{21} = 0, \quad \Gamma^2_{22} = 0.$$

(iii) The differential equations for the *geodesics* are determined by

$$\frac{d^2 y^\alpha}{ds^2} + \Gamma^\alpha_{\beta\gamma} \frac{dy^\beta}{ds} \frac{dy^\gamma}{ds} = 0$$

where we used the summation convention (summation over β and γ, $\beta = 1, 2, 3$, $\gamma = 1, 2, 3$). Thus we find

$$\frac{d^2\theta}{ds^2} - \frac{2b\sin(\phi)}{a + b\cos(\phi)} \frac{d\theta}{ds} \frac{d\phi}{ds} = 0, \quad \frac{d^2\phi}{ds^2} + \frac{(a + b\cos(\phi))\sin(\phi)}{b} \left(\frac{d\theta}{ds} \right)^2 = 0.$$

Problem 4. The two-dimensional *de Sitter space* $\mathbb{V}$ with the topology $\mathbb{R} \times \mathbb{S}$ may be visualized as a one-sheet hyperboloid $\mathbb{H}_{r_0}$ embedded in 3-dimensional Minkowski space $\mathbb{M}$, i.e.

$$\mathbb{H}_{r_0} = \{ (y_0, y_1, y_2) \in \mathbb{M} \mid (y_2)^2 + (y_1)^2 - (y_0)^2 = r_0^2, \ r_0 > 0 \}$$

where r_0 is the parameter of the one-sheet hyperboloid $\mathbb{H}_{r_0}$. The induced metric, $g_{\mu\nu}$ ($\mu, \nu = 0, 1$), on $\mathbb{H}_{r_0}$ is the de Sitter metric.
(i) Show that we can parametrize (parameters ρ and θ) the *hyperboloid* as follows

$$y_0(\rho, \theta) = -\frac{r_0 \cos(\rho/r_0)}{\sin(\rho/r_0)}, \quad y_1(\rho, \theta) = \frac{r_0 \cos(\theta/r_0)}{\sin(\rho/r_0)}, \quad y_2(\rho, \theta) = \frac{r_0 \sin(\theta/r_0)}{\sin(\rho/r_0)}$$

where $0 < \rho < \pi r_0$ and $0 \leq \theta < 2\pi r_0$.
(ii) Using this parametrization find the metric tensor field induced on $\mathbb{H}_{r_0}$.

Solution 4. (i) Inserting y_0, y_1, y_2 into $(y_2)^2 + (y_1)^2 - (y_0)^2$ and using $\sin^2(\rho/r_0) + \cos^2(\rho/r_0) = 1$ we obtain r_0^2.

(ii) We start from the metric tensor field

$$g = dy_0 \otimes dy_0 - dy_1 \otimes dy_1 - dy_2 \otimes dy_2.$$

Since $\partial y_0/\partial \rho = \csc^2(\rho/r_0)$, $\partial y_0/\partial \theta = 0$ and

$$\frac{\partial y_1}{\partial \rho} = -\frac{\cos(\theta/r_0)\cos(\rho/r_0)}{\sin^2(\rho/r_0)}, \quad \frac{\partial y_1}{\partial \theta} = -\frac{\sin(\theta/r_0)}{\sin(\rho/r_0)}$$

$$\frac{\partial y_2}{\partial \rho} = -\frac{\sin(\theta/r_0)\cos(\rho/r_0)}{\sin^2(\rho/r_0)}, \quad \frac{\partial y_2}{\partial \theta} = \frac{\cos(\theta/r_0)}{\sin(\rho/r_0)}$$

we find

$$g = \left(\frac{1}{\sin^4(\rho/r_0)} - \frac{\cos^2(\theta/r_0)\cos^2(\rho/r_0)}{\sin^4(\rho/r_0)} - \frac{\sin^2(\theta/r_0)\cos^2(\rho/r_0)}{\sin^4(\rho/r_0)} \right) d\rho \otimes d\rho$$

$$+ \left(-\frac{\sin^2(\theta/r_0)}{\sin^2(\rho/r_0)} - \frac{\cos^2(\theta/r_0)}{\sin^2(\rho/r_0)} \right) d\theta \otimes d\theta.$$

Using $\sin^2(\theta/r_0) + \cos^2(\theta/r_0) = 1$ we arrive at

$$g = \frac{1}{\sin^2(\rho/r_0)} d\rho \otimes d\rho - \frac{1}{\sin^2(\rho/r_0)} d\theta \otimes d\theta.$$

Problem 5. The *anti-de Sitter space* is defined as the surface

$$X_1^2 + X_2^2 + X_3^2 - U_1^2 - U_2^2 = -1$$

embedded in a five-dimensional flat space with the metric tensor field

$$g = dX_1 \otimes dX_1 + dX_2 \otimes dX_2 + dX_3 \otimes dX_3 - dU_1 \otimes dU_1 - dU_2 \otimes dU_2.$$

This is a solution of Einstein's equations with the cosmological constant $\Lambda = -3$. Its intrinsic curvature is constant and negative. Find the metric tensor field in terms of the intrinsic coordinates (ρ, θ, ϕ, t) where

$$X_1(\rho, \theta, \phi, t) = \frac{2\rho}{1 - \rho^2} \sin(\theta)\cos(\phi), \quad X_2(\rho, \theta, \phi, t) = \frac{2\rho}{1 - \rho^2} \sin(\theta)\sin(\phi),$$

$$X_3(\rho, \theta, \phi, t) = \frac{2\rho}{1 - \rho^2} \cos(\theta)$$

and

$$U_1(\rho, \theta, \phi, t) = \frac{1 + \rho^2}{1 - \rho^2} \cos(t), \quad U_2(\rho, \theta, \phi, t) = \frac{1 + \rho^2}{1 - \rho^2} \sin(t)$$

where $0 \le \rho < 1$, $0 \le \phi < 2\pi$, $0 \le \theta < \pi$, $-\pi \le t < \pi$.

Solution 5. We find

$$g = -\left(\frac{1+\rho^2}{1-\rho^2}\right)^2 dt \otimes dt + \frac{4}{(1-\rho^2)^2}(d\rho \otimes d\rho + \rho^2 d\theta \otimes d\theta + \rho^2 \sin^2 \theta d\phi \otimes d\phi).$$

What are the Killing vector fields of g?

Problem 6. Consider the *Poincaré upper half-plane*

$$H_+^2 := \{ (x,y) \in \mathbb{R}^2 \; : \; y > 0 \}$$

with metric tensor field

$$g = \frac{1}{y}dx \otimes \frac{1}{y}dx + \frac{1}{y}dy \otimes \frac{1}{y}dy$$

which is conformal with the standard inner product. Find the curvature forms.

Solution 6. The orthonormal coframe and frame are

$$\sigma^1 = \frac{1}{y}dx, \quad \sigma^2 = \frac{1}{y}dy, \quad s_1 = y\frac{\partial}{\partial x}, \quad s_2 = y\frac{\partial}{\partial y}.$$

It follows that

$$d\sigma^1 = d\left(\frac{1}{y}dx\right) = \frac{1}{y^2}dx \wedge dy = \sigma^1 \wedge \sigma^2$$

and obviously $d\sigma^2 = 0$. The connection forms we compute from $d\sigma + \omega \wedge \sigma = 0$. We find

$$-d\sigma^1 = 0 + \omega_2^1 \wedge \sigma^2 = -\sigma^1 \wedge \sigma^2, \quad -d\sigma^2 = \omega_1^2 \wedge \sigma^1 + 0 = 0.$$

Therefore $\omega_2^1 = -\sigma^1 = -y^{-1}dx$. Then

$$\omega = \begin{pmatrix} 0 & -\sigma^1 \\ \sigma^1 & 0 \end{pmatrix}.$$

For the curvature forms we obtain

$$\Omega_2^1 = d\omega_2^1 + \omega_1^1 \wedge \omega_2^1 + \omega_2^1 \wedge \omega_2^2 = d(-y^{-1}dx) = -\sigma^1 \wedge \sigma^2$$

and therefore

$$\Omega = \begin{pmatrix} 0 & -\sigma^1 \wedge \sigma^2 \\ \sigma^1 \wedge \sigma^2 & 0 \end{pmatrix}.$$

Problem 7.   Consider the metric tensor field

$$g = 2H\,du_1 \otimes du_1 + du_1 \otimes du_2 + du_2 \otimes du_1 - P^2 dx_1 \otimes dx_1 - P^2 dx_2 \otimes dx_2$$

on a four dimensional manifold with coordinates (x_1, x_2, u_1, u_2). The corresponding 4×4 matrix is

$$\begin{pmatrix} -P^2 & 0 & 0 & 0 \\ 0 & -P^2 & 0 & 0 \\ 0 & 0 & 2H & 1 \\ 0 & 0 & 1 & 0 \end{pmatrix}.$$

It is assumed that $P^2 \neq 0$. The inverse of this matrix is

$$\begin{pmatrix} -1/P^2 & 0 & 0 & 0 \\ 0 & -1/P^2 & 0 & 0 \\ 0 & 0 & 0 & 1 \\ 0 & 0 & 1 & -2H \end{pmatrix}.$$

(i) Let $P = 1$, $H = 0$, $u_1 = ct - x_3$, $u_2 = \frac{1}{2}(ct + x_3)$. Show that the metric tensor field reduces to the metric tensor field of the Minkowski space.
(ii) Find the metric tensor field for

$$P = u_2(1 + \frac{1}{4}(x_1^2 + x_2^2))^{-1}, \quad H = \frac{1}{2}, \quad u_1 = ct - r, \quad u_2 = r,$$

$$x_1 = 2\cos(\phi)\cot(\theta/2), \quad x_2 = 2\sin(\phi)\cot(\theta/2).$$

Solution 7.   (i) We have $du_1 = cdt - dx_3$, $du_2 = \frac{1}{2}cdt + \frac{1}{2}dx_3$. Thus the metric tensor field takes the form

$$cdt \otimes cdt - dx_1 \otimes dx_1 - dx_2 \otimes dx_2 - dx_3 \otimes dx_3.$$

(ii) We have

$$x_1^2 + x_2^2 = 4\cot^2(\theta/2) \quad \Rightarrow \quad 1 + \frac{1}{4}(x_1^2 + x_2^2) = \frac{1}{\sin^2(\theta/2)}.$$

Hence

$$P = u_2 \sin^2(\theta/2) \quad \Rightarrow \quad P^2 = u_2^2 \sin^4(\theta/2) = r^2 \sin^4(\theta/2).$$

Now

$$du_1 \otimes du_2 + du_2 \otimes du_1 = -2dr \otimes dr + cdt \otimes dr + dr \otimes cdt$$

and

$$du_1 \otimes du_1 = cdt \otimes cdt + dr \otimes dr - dr \otimes cdt - cdt \otimes dr.$$

For dx_1 and dx_2 we obtain

$$dx_1 = -2\sin(\phi)\cot(\theta/2)d\phi - \frac{\cos(\phi)}{\sin^2(\theta/2)}d\theta,$$

$$dx_2 = 2\cos(\phi)\cot(\theta/2)d\phi - \frac{\sin(\phi)}{\sin^2(\theta/2)}d\theta.$$

It follows that

$$dx_1 \otimes dx_1 + dx_2 \otimes dx_2 = 4\cot^2(\theta/2)d\phi \otimes d\phi + \frac{1}{\sin^2(\theta/2)}d\theta \otimes d\theta.$$

Utilizing the identity $4\cos^2(\theta/2)\sin^2(\theta/2) \equiv \sin^2(\theta)$ we finally arrive at

$$cdt \otimes cdt - dr \otimes dr - r^2(d\theta \otimes d\theta + \sin^2(\theta)d\theta \otimes d\theta).$$

Problem 8. Consider the metric tensor field

$$g = -dx_0 \otimes dx_0 + dx_1 \otimes dx_1.$$

Find g expressed in *Rindler coordinates*

$$x_0(\xi, \eta) = \xi\sinh(\eta), \quad x_1(\xi, \eta) = \xi\cosh(\eta).$$

Solution 8. With

$$dx_0 = \sinh(\eta)d\xi + \xi\cosh(\eta)d\eta, \quad dx_1 = \cosh(\eta)d\xi + \xi\sinh(\eta)d\eta$$

and $\cosh^2(\eta) - \sinh^2(\eta) = 1$ we obtain $g_R = -\xi^2 d\eta \otimes d\eta + d\xi \otimes d\xi$.

Problem 9. Consider the *Robertson-Walker metric tensor field*

$$g = dx_0 \otimes dx_0 - \sum_{j=1}^{3} f(x_0)dx_j \otimes f(x_0)dx_j$$

i.e. f only depends on x_0. Find the Christoffel symbols, Ricci tensor and the curvature.

Solution 9. The corresponding 4×4 matrix is

$$g = (g_{\mu\nu}) = \begin{pmatrix} 1 & 0 & 0 & 0 \\ 0 & -f^2 & 0 & 0 \\ 0 & 0 & -f^2 & 0 \\ 0 & 0 & 0 & -f^2 \end{pmatrix}$$

with the inverse matrix

$$g^{-1} = (g^{\mu\nu}) = \begin{pmatrix} 1 & 0 & 0 & 0 \\ 0 & -1/f^2 & 0 & 0 \\ 0 & 0 & -1/f^2 & 0 \\ 0 & 0 & 0 & -1/f^2 \end{pmatrix}.$$

For the Christoffel symbols $\Gamma^a_{mn} = \Gamma^a_{nm}$ $(a, m, n = 0, 1, 2, 3)$ we have

$$\Gamma^a_{mn} = \frac{1}{2} \sum_{b=0}^{3} g^{ab}(g_{bm,n} + g_{bn,m} - g_{mn,b})$$

where $g_{bm,n}$ denotes differentiation of g_{bm} with respect to $\partial/\partial x_n$ $(n = 0, 1, 2, 3)$. The only nonzero Christoffel symbols are

$$\Gamma^0_{11} = \Gamma^0_{22} = \Gamma^0_{33} = f \frac{df}{dx_0}$$

$$\Gamma^j_{0j} = \frac{1}{f} \frac{df}{dx_0}, \quad j = 1, 2, 3$$

with $\Gamma^j_{j0} = \Gamma^j_{0j}$. The Ricci tensor follows as

$$R_{mq} = \sum_{s=0}^{3} \Gamma^s_{mq,s} - \sum_{s=0}^{3} \Gamma^s_{ms,q} + \sum_{s=0}^{3}\sum_{n=0}^{3} \Gamma^s_{ns}\Gamma^n_{mq} - \sum_{s=0}^{3}\sum_{n=0}^{3} \Gamma^s_{nq}\Gamma^n_{ms}.$$

The nonzero components are

$$R_{00} = -3\frac{1}{f}\frac{d^2 f}{dx_0^2}, \quad R_{11} = R_{22} = R_{33} = f\frac{d^2 f}{dx_0^2} + 2\left(\frac{df}{dx_0}\right)^2.$$

From $R^j_k = \sum_{m=0}^{3} g^{jm} R_{km}$ we find the nonzero components as

$$R^0_0 = -\frac{3}{f}\frac{d^2 f}{dx_0^2}, \quad R^1_1 = R^2_2 = R^3_3 = -\frac{1}{f}\frac{d^2 f}{dx_0^2} - \frac{2}{f}\left(\frac{df}{dx_0}\right)^2.$$

Hence the curvature $R = R^0_0 + R^1_1 + R^2_2 + R^3_3$ is given by

$$R = -6\left(\frac{1}{f}\frac{d^2 f}{dx_0^2} + \frac{1}{f^2}\left(\frac{df}{dx_0}\right)^2\right).$$

Apply the Cartan approach to find R starting with the four differential one-forms (tetrad)

$$\alpha_0 = dx_0, \quad \alpha_1 = f(x_0)dx_1, \quad \alpha_2 = f(x_0)dx_2, \quad \alpha_3 = f(x_0)dx_3$$

and $d\alpha_0 = 0$,

$$d\alpha_1 = \frac{\partial f}{\partial x_0}dx_0 \wedge dx_1, \quad d\alpha_2 = \frac{\partial f}{\partial x_0}dx_0 \wedge dx_2, \quad d\alpha_3 = \frac{\partial f}{\partial x_0}dx_0 \wedge dx_3.$$

Problem 10. Given a smooth surface in the Euclidean space $\mathbb{R}^3$ described by

$$\mathbf{x}(u,v) = \begin{pmatrix} x_1(u,v) \\ x_2(u,v) \\ x_3(u,v) \end{pmatrix}.$$

The *Gaussian curvature* is calculated as follows. First we calculate $E(u,v)$, $F(u,v)$, $G(u,v)$ of the first fundamental form

$$E(u,v) = \frac{\partial \mathbf{x}}{\partial u} \cdot \frac{\partial \mathbf{x}}{\partial u}, \qquad F(u,v) = \frac{\partial \mathbf{x}}{\partial u} \cdot \frac{\partial \mathbf{x}}{\partial v}, \qquad G(u,v) = \frac{\partial \mathbf{x}}{\partial v} \cdot \frac{\partial \mathbf{x}}{\partial v}$$

where $\cdot$ denotes the scalar product. Next we calculate the normal vector field

$$\mathbf{n}^+(u,v) := \frac{\frac{\partial \mathbf{x}}{\partial u} \times \frac{\partial \mathbf{x}}{\partial v}}{\left|\frac{\partial \mathbf{x}}{\partial u} \times \frac{\partial \mathbf{x}}{\partial v}\right|}$$

where $\times$ denotes the vector product. Using $\mathbf{n}^+$ we calculate $L(u,v)$, $M(u,v)$, $N(u,v)$ of the second fundamental form

$$L(u,v) = \mathbf{n}^+ \cdot \frac{\partial^2 \mathbf{x}}{\partial u^2}, \qquad M(u,v) = \mathbf{n}^+ \cdot \frac{\partial^2 \mathbf{x}}{\partial u \partial v}, \qquad N(u,v) = \mathbf{n}^+ \cdot \frac{\partial^2 \mathbf{x}}{\partial v^2}.$$

Then the Gaussian curvature $K(u,v)$ is given by

$$K := \frac{LN - M^2}{EG - F^2}.$$

The *Möbius band* embedded in $\mathbb{R}^3$ can be parametrized as

$$\mathbf{x}(u,v) = \begin{pmatrix} (2 - v\sin(u/2))\sin(u) \\ (2 - v\sin(u/2))\cos(u) \\ v\cos(u/2) \end{pmatrix}$$

where $-1/2 < v < 1/2$ and $u \in [0, 2\pi)$. Find the Gaussian curvature for the Möbius band.

Solution 10. This is a typical application for computer algebra. A SymbolicC++ program that calculates K is given by

```
// curvature.cpp
#include <iostream>
#include "symbolicc++.h"
using namespace std;

Symbolic curvature(const Symbolic &x,const Symbolic &u,
                   const Symbolic &v,const Equations &rules)
{
 Symbolic dxdu = df(x,u); Symbolic dxdv = df(x,v);
 Symbolic d2xdu2  = df(dxdu,u);
 Symbolic d2xdudv = df(dxdu,v);
 Symbolic d2xdv2  = df(dxdv,v);
 Symbolic E = (dxdu | dxdu).subst(rules);
 Symbolic F = (dxdu | dxdv).subst(rules);
 Symbolic G = (dxdv | dxdv).subst(rules);
 Symbolic n = (dxdu % dxdv).subst(rules);
 n = n/sqrt((n | n).subst(rules));
 Symbolic L = (n | d2xdu2).subst(rules);
 Symbolic M = (n | d2xdudv).subst(rules);
 Symbolic N = (n | d2xdv2).subst(rules);
 return (L*N-M*M)/(E*G-F*F);
}

int main(void)
{
 Symbolic x("x",3), u("u"), v("v");
 Equations rules = (sin(u/2)*sin(u/2)==1-cos(u/2)*cos(u/2),
                    cos(u/2)*cos(u/2)==(cos(u)+1)/2,
                    sin(u)*sin(u)==1-cos(u)*cos(u));
 x(0) = (2-v*sin(u/2))*sin(u);
 x(1) = (2-v*sin(u/2))*cos(u);
 x(2) = v*cos(u/2);
 cout << "x = " << x << endl;
 cout << "K = " << curvature(x,u,v,rules) << endl;
 return 0;
}
```

where % denotes the vector product and | denotes the scalar product. The Gaussian curvature follows as

$$K(u,v) = \frac{1}{(v^2/4 + (2 - v\sin(u/2))^2)^2}.$$

Supplementary Problems

Problem 1. Consider the metric tensor field of the unit sphere

$$g = d\theta \otimes d\theta + \sin^2(\theta)d\phi \otimes d\phi.$$

Show that under the transformation (complex stereographic coordinates)

$$\zeta(\phi, \theta) = e^{i\phi} \cot(\theta/2)$$

the metric tensor field takes the form

$$g = \frac{1}{P^2} d\zeta \otimes d\bar{\zeta}$$

where $P = \frac{1}{2}(1 + \zeta\bar{\zeta})$.

Problem 2. Let $z = x + iy$, $\bar{z} = x - iy$ with $x, y \in \mathbb{R}$. Find

$$g = dz \otimes d\bar{z} + d\bar{z} \otimes dz.$$

Problem 3. Consider the metric tensor field

$$g = dx_1 \otimes dx_1 + dx_2 \otimes dx_2 + dx_3 \otimes dx_3$$

of the Euclidean space $\mathbb{E}^3$ and the invertible chaotic map

$$f_1(x_1, x_2, x_3) = x_1 x_2 - x_3, \quad f_2(x_1, x_2, x_3) = x_1, \quad f_3(x_1, x_2, x_3) = x_2.$$

Show that the inverse map is given by

$$f_1^{-1}(x_1, x_2, x_3) = x_2, \quad f_2^{-1}(x_1, x_2, x_3) = x_3, \quad f_3^{-1}(x_1, x_2, x_3) = x_2 x_3 - x_1.$$

Show that

$$
\begin{aligned}
\mathbf{f}^*(g) = {} & (1 + x_2^2)dx_1 \otimes dx_1 + (1 + x_1^2)dx_2 \otimes dx_2 + dx_3 \otimes dx_3 \\
& + x_1 x_2(dx_1 \otimes dx_2 + dx_2 \otimes dx_1) \\
& - x_1(dx_2 \otimes dx_3 + dx_3 \otimes dx_2) - x_2(dx_1 \otimes dx_3 + dx_3 \otimes dx_1)
\end{aligned}
$$

with the corresponding matrix

$$
\begin{pmatrix}
1 + x_2^2 & x_1 x_2 & -x_2 \\
x_1 x_2 & 1 + x_1^2 & -x_1 \\
-x_2 & -x_1 & 1
\end{pmatrix}.
$$

The determinant of the matrix is $+1$.

Problem 4. Let u be smooth function of x_1, x_2. Consider the metric tensor field

$$g = \sum_{j=1}^{2} \left(1 + \left(\frac{\partial u}{\partial x_j}\right)^2\right) dx_j \otimes dx_j + \frac{\partial u}{\partial x_1}\frac{\partial u}{\partial x_2}(dx_1 \otimes dx_2 + dx_2 \otimes dx_1).$$

Find the curvature R and then set $R = -1$. Solve the resulting partial differential equation.

Problem 5.   Consider the metric tensor field

$$g = dx_0 \otimes dx_0 - dx_1 \otimes dx_1 - dx_2 \otimes dx_2 - dx_3 \otimes dx_3$$
$$+ f(x_0, x_3)dx_3 \otimes dx_0 + f(x_0, x_3)dx_0 \otimes dx_3$$

with

$$f(x_0, x_3) = x_0 \frac{dh(x_3)}{dx_3}$$

where g is a smooth function. Find the inverse of the matrix

$$g = \begin{pmatrix} 1 & 0 & 0 & f(x_0, x_3) \\ 0 & -1 & 0 & 0 \\ 0 & 0 & -1 & 0 \\ f(x_0, x_3) & 0 & 0 & -1 \end{pmatrix}.$$

Find the Christoffel symbols and the curvature.

Problem 6.   Let x_0, x_1, x_2, x_3 be the Minkowski coordinates. Show that the uniformly accelerated world line

$$x_0 = \xi \sinh(\eta), \quad x_1 = \xi \sinh(\eta), \quad x_2 = \text{const}, \quad x_3 = \text{const}$$

with ξ kept fixed is mapped under the *Wick rotation*

$$x_0 \to iX_0, \quad x_1 = X_1, \quad \eta \to i\zeta, \quad x_2 = X_2, \quad x_3 = X_3$$

to the circle

$$X_0 = \xi \sin(\zeta), \quad X_1 = \xi \cos(\zeta), \quad x_2 = \text{const}, \quad x_3 = \text{const}$$

with the metric tensor field

$$g_E = dX_0 \otimes dX_0 + dX_1 \otimes dX_1 + dX_2 \otimes dX_2 + dX_3 \otimes dX_3.$$

Problem 7.   A smooth surface $M \subset \mathbb{R}^3$ is minimal if and only if its *mean curvature* vanishes identically. The condition on the mean curvature to vanish identically can be expressed as the quasi-linear, second-order, elliptic partial differential equation

$$\left(1 + \left(\frac{\partial u}{\partial x_1}\right)^2\right)\frac{\partial^2 u}{\partial x_2^2} - 2\frac{\partial u}{\partial x_1}\frac{\partial u}{\partial x_2}\frac{\partial^2 u}{\partial x_1 \partial x_2} + \left(1 + \left(\frac{\partial u}{\partial x_2}\right)^2\right)\frac{\partial^2 u}{\partial x_1^2}.$$

Show that this partial differential equation can be written as

$$\text{div}\left(\frac{\nabla u}{\sqrt{1+|\nabla u|^2}}\right) = 0.$$

Show that the surface (*Scherk surface*) $u : (-\pi/2, \pi/2) \times (-\pi/2, \pi/2) \mapsto \mathbb{R}$

$$u(x_1, x_2) = \ln\left(\frac{\cos(x_1)}{\cos(x_2)}\right) \equiv \ln(\cos(x_1)) - \ln(\cos(x_2)).$$

Obviously $\partial^2 u/\partial x_1 \partial x_2 = 0$.

Problem 8. Consider the metric tensor field

$$g = R^2 d\chi \otimes d\chi + R^2 \sin^2(\chi)(d\theta \otimes d\theta + \sin^2(\theta)d\psi \otimes d\psi).$$

Show that the volume form Ω is given by

$$\Omega = R^3 \sin^2(\chi)\sin(\theta)d\chi \wedge d\theta \wedge d\psi.$$

Problem 9. Consider the metric tensor field

$$\tilde{g} = -f(\rho, z)(d\rho \otimes d\rho + dz \otimes dz) - \sum_{j,k=0}^{1} g_{jk}(\rho, z)dx_j \otimes dx_k$$

with the invertible 2×2 matrix $g(\rho, z) = (g_{jk}(\rho, z)$. Show that for this metric tensor field the *vacuum Einstein equation* are given by

$$\frac{\partial}{\partial\rho}\left(\rho\frac{\partial g}{\partial\rho}g^{-1}\right) + \frac{\partial}{\partial z}\left(\rho\frac{\partial g}{\partial z}g^{-1}\right) = \begin{pmatrix} 0 & 0 \\ 0 & 0 \end{pmatrix}$$

$$\frac{\partial}{\partial\rho}\ln(f) + \frac{1}{\rho} - \frac{1}{4\rho}\text{tr}(U^2 - V^2) = 0, \qquad \frac{\partial}{\partial z}\ln(f) - \frac{1}{2\rho}\text{tr}(UV) = 0$$

with the 2×2 matrices U and V given by

$$U = \rho\frac{\partial g}{\partial\rho}g^{-1}, \qquad V = \rho\frac{\partial g}{\partial z}g^{-1}.$$

Problem 10. Consider the metric tensor field

$$g = J_{00}(\rho, x_3)dx_0 \otimes dx_0 - \Omega^2(\rho, x_3)(dx_3 \otimes dx_3 + d\rho \otimes d\rho)$$
$$+ J_{11}(\rho, x_3)d\theta \otimes d\theta + J_{01}(\rho, x_3)dx_0 \otimes d\theta + J_{10}(\rho, x_3)d\theta \otimes dx_0$$

with the invertible symmetric 2×2 matrix

$$J(\rho, x_3) = \begin{pmatrix} J_{00}(\rho, x_3) & J_{01}(\rho, x_3) \\ J_{10}(\rho, x_3) & J_{11}(\rho, x_3) \end{pmatrix}$$

and $\det(J) = -\rho^2$. Show that the Einstein field equations are given by

$$\frac{\partial}{\partial \rho}\left(\rho J^{-1} \frac{\partial J}{\partial \rho}\right) + \rho \frac{\partial}{\partial x_3}\left(J^{-1} \frac{\partial J}{\partial x_3}\right) = 0$$

$$4i\frac{\partial}{\partial w}(\ln(\Omega)) = \rho \operatorname{tr}\left(\left(\frac{\partial J^{-1}}{\partial w}\right)\left(\frac{\partial J}{\partial w}\right)\right) - \frac{1}{\rho}$$

where $w = x_3 + i\rho$.

Problem 11. Let $N \geq 2$. Consider the Hilbert space $\mathbb{C}^N$ and let $\{|e_j\rangle\}$ $(j = 1, \ldots, N)$ be an orthonormal basis in $\mathbb{C}^N$. We define the (complex) numbers

$$c_j(|\psi\rangle) = \langle e_j|\psi\rangle, \quad j = 1, \ldots, N.$$

So we can construct the metric tensor field

$$g = \sum_{j=1}^{N}(d\bar{c}_j \otimes dc_j).$$

Let $N = 2$ and

$$e_1 = \frac{1}{\sqrt{2}}\begin{pmatrix} 1 \\ 1 \end{pmatrix}, \quad e_2 = \frac{1}{\sqrt{2}}\begin{pmatrix} 1 \\ -1 \end{pmatrix}, \quad |\psi\rangle = \begin{pmatrix} e^{i\phi}\cos(\theta) \\ \sin(\theta) \end{pmatrix}.$$

Show that

$$c_1(|\psi\rangle) = \langle e_1|\psi\rangle = \frac{1}{\sqrt{2}}(e^{i\phi}\cos(\theta) + \sin(\theta))$$

$$c_2(|\psi\rangle) = \langle e_2|\psi\rangle = \frac{1}{\sqrt{2}}(e^{i\phi}\cos(\theta) - \sin(\theta))$$

and

$$dc_1 = \frac{1}{\sqrt{2}}(e^{i\phi}\cos(\theta)id\phi - e^{i\phi}\sin(\theta)d\theta + \cos(\theta)d\theta)$$

$$dc_2 = \frac{1}{\sqrt{2}}(e^{i\phi}\cos(\theta)id\phi - e^{i\phi}\sin(\theta)d\theta - \cos(\theta)d\theta).$$

Find g and the Killing vector fields of g.

Chapter 25

Lie Derivative, Invariance and Killing Vector Fields

Let M be an n-dimensional C^∞ differentiable manifold with local coordinates x_j, $j = 1, \ldots, n$. Real valued C^∞ vector fields and real valued C^∞ differential forms on M can be considered. The components of the vector field V are denoted by $V_j \partial/\partial x_j$. This means

$$V := V_1(\mathbf{x}) \frac{\partial}{\partial x_1} + V_2(\mathbf{x}) \frac{\partial}{\partial x_2} + \cdots + V_n(\mathbf{x}) \frac{\partial}{\partial x_n}$$

in local coordinates. The *Lie derivative* of a differential form α with respect to V is defined by the derivative of α along the integral curve $t \mapsto \Phi_t$ of V, i.e.

$$L_V \alpha := \lim_{t \to 0} \frac{\Phi_t^* \alpha - \alpha}{t}.$$

The Lie derivative is linear and satisfies the product rule. It can be shown that this can be written as

$$L_V \alpha := d(V \rfloor \alpha) + V \rfloor (d\alpha)$$

where $d\alpha$ is the exterior derivative of the differential form α and $V \rfloor \alpha$ is the *contraction* of α by V. In local coordinates we have

$$\frac{\partial}{\partial x_j} \rfloor dx_k = \delta_{jk}$$

where δ_{jk} denotes the Kronecker delta. Furthermore, we have the *product rule*

$$V \rfloor (\alpha \wedge \beta) = (V \rfloor \alpha) \wedge \beta + (-1)^r \alpha \wedge (V \rfloor \beta)$$

where α is an r-form. The linear operators $d(.)$, $V \rfloor$ and $L_V(.)$ are coordinate-free operators.

A differential form α is called *invariant* with respect to a vector field V if

$$L_V \alpha = 0.$$

If α and β are invariant with respect to V, then $\alpha \wedge \beta$ is invariant without respect to V.

A differential form α is called *conformal invariant* with respect to a vector field V if

$$L_V \alpha = f(\mathbf{x})\alpha.$$

If α and β are invariant with respect to V, then $\alpha \wedge \beta$ is conformal invariant without respect to V.

Let M be a smooth manifold and g be a metric tensor field. Then V is called a *Killing vector field* with respect to g if

$$L_V g = 0.$$

If V and W are Killing vector fields of g, then the commutator $[V, W]$ is also a Killing vector field.

We have the following properties for the Lie derivative

$$L_V(\alpha_1 + \alpha_2) = L_V \alpha_1 + L_V \alpha_2$$
$$L_V(\alpha \wedge \beta) = (L_V \alpha) + \alpha \wedge (L_V \beta)$$
$$L_V(dx_j \otimes dx_k) = (L_V dx_j) \otimes dx_k + dx_j \otimes (L_V dx_k)$$
$$L_V(f(\mathbf{x})\alpha = (L_V f(\mathbf{x}))\alpha + f(\mathbf{x}) L_V \alpha.$$

The Lie derivative $L_V(.)$ and the exterior derivative $d(.)$ commute, i.e.

$$L_V(d\alpha) = d(L_V(\alpha)).$$

Furthermore with α be a differential one-form and V, W be vector fields

$$W \rfloor (V \rfloor d\alpha) = V(W \rfloor \alpha) - W(V \rfloor \alpha) - [V, W] \rfloor \alpha.$$

In literature we also find the notation

$$d\alpha(V, W) = V\alpha(W) - W\alpha(V) - \alpha([V, W]).$$

Problem 1. Consider the differential one form $\alpha = x_1 dx_2 - x_2 dx_1$ on $\mathbb{R}^2$. Show that α is invariant under the transformation

$$\begin{pmatrix} x_1' \\ x_2' \end{pmatrix} = \begin{pmatrix} \cos(\theta) & -\sin(\theta) \\ \sin(\theta) & \cos(\theta) \end{pmatrix} \begin{pmatrix} x_1 \\ x_2 \end{pmatrix}.$$

The matrix is an element of the compact Lie group $SO(2, \mathbb{R})$.

Solution 1. (i) We have

$$dx_1' = \cos(\theta) dx_1 - \sin(\theta) dx_2, \qquad dx_2' = \sin(\theta) dx_1 + \cos(\theta) dx_2.$$

With $\cos^2(\theta) + \sin^2(\theta) = 1$ we obtain

$$x_1' dx_2' - x_2' dx_1' = x_1 dx_2 - x_2 dx_1.$$

Let α be the $(n-1)$ differential form on $\mathbb{R}^n$ given by

$$\omega = \sum_{j=1}^{n} (-1)^{j-1} x_j dx_1 \wedge \cdots \wedge \widehat{dx_j} \wedge \cdots \wedge dx_n$$

where $\widehat{}$ indicates omission. Then ω is invariant under the orthogonal group of $\mathbb{R}^n$.

Problem 2. Let $M = \mathbb{R}^n$. Consider the volume form

$$\omega := dx_1 \wedge dx_2 \wedge \cdots \wedge dx_n.$$

Find $V \rfloor \omega$, where the vector field X is given by

$$V := V_1(\mathbf{x}) \frac{\partial}{\partial x_1} + V_2(\mathbf{x}) \frac{\partial}{\partial x_2} + \cdots + V_n(\mathbf{x}) \frac{\partial}{\partial x_n}.$$

Find $d(V \rfloor \omega)$. Give an interpretation of the result.

Solution 2. Applying the product rule we obtain

$$V \rfloor \omega = \sum_{j=1}^{n} (-1)^{j+1} V_j dx_1 \wedge \cdots \wedge \widehat{dx_j} \wedge \cdots \wedge dx_n$$

where the circumflex indicates omission. Taking the exterior derivative of this equation we obtain

$$d(V \rfloor \omega) = \left(\sum_{j=1}^{n} \frac{\partial V_j}{\partial x_j} \right) dx_1 \wedge dx_2 \wedge \cdots \wedge dx_n.$$

This is the *divergence* of the vector field V. Since $d\omega = 0$ we have

$$L_V\omega = d(V\rfloor\omega).$$

Problem 3.   Consider the autonomous nonlinear system of ordinary differential equations

$$\frac{dx_1}{dt} = x_1 - x_1x_2, \qquad \frac{dx_2}{dt} = -x_2 + x_1x_2$$

where $x_1 > 0$ and $x_2 > 0$. This equation is a *Lotka-Volterra model*. The corresponding vector field is

$$V = (x_1 - x_1x_2)\frac{\partial}{\partial x_1} + (-x_2 + x_1x_2)\frac{\partial}{\partial x_2}.$$

Let

$$\Omega = \frac{dx_1 \wedge dx_2}{x_1x_2}.$$

Calculate the Lie derivative $L_V\Omega$.

Solution 3.   Since $d\Omega = 0$ we have $L_V\Omega = d(V\rfloor\Omega)$. Thus

$$L_V\Omega = d\left[\left(\frac{x_1 - x_1x_2}{x_1x_2}\right)dx_2 - \left(\frac{-x_2 + x_1x_2}{x_1x_2}\right)dx_1\right]$$

$$= d\left(\frac{1}{x_2}dx_2 + \frac{1}{x_1}dx_1\right) = 0$$

where we have used $ddx_1 = 0$, $ddx_2 = 0$. Since $L_V\Omega = 0$ we say that Ω is *invariant* under V.

Problem 4.   (i) Let $M = \mathbb{R}^n$. Let X_1, X_2 and Y be smooth vector fields defined on M. Assume that

$$[X_1, Y] = fY, \qquad [X_2, Y] = gY \tag{1}$$

where f and g are smooth functions defined on M. Let $\Omega := dx_1 \wedge \cdots \wedge dx_n$ be the *volume form* on M. Show that

$$L_Y(X_1\rfloor X_2\rfloor Y\rfloor\Omega) = (\text{div}(Y))(X_1\rfloor X_2\rfloor Y\rfloor\Omega) \tag{2}$$

where $\text{div}(Y)$ denotes the *divergence* of the vector field Y, i.e.

$$\text{div}(Y) := \sum_{j=1}^{n}\frac{\partial Y_j}{\partial x_j}. \tag{3}$$

(ii) Let X, Y be smooth vector fields and let f and g be smooth functions. Assume that

$$L_Y g = 0, \qquad [X, Y] = fY. \tag{4}$$

Calculate $L_Y(L_X g)$. Apply the identity $L_{[X,Y]} \equiv [L_X, L_Y]$.

Solution 4. (i) Straightforward calculation shows that

$$
\begin{aligned}
L_Y(X_1 \rfloor X_2 \rfloor Y \rfloor \Omega) &= [Y, X_1] \rfloor X_2 \rfloor Y \rfloor \Omega + X_1 \rfloor L_Y(X_2 \rfloor Y \rfloor \Omega) \\
&= -fY \rfloor X_2 \rfloor Y \rfloor \Omega + X_1 \rfloor L_Y(X_2 \rfloor Y \rfloor \Omega) \\
&= X_1 \rfloor ([Y, X_2] \rfloor Y \rfloor \Omega + X_2 \rfloor L_Y(Y \rfloor \Omega)) \\
&= -X_1 \rfloor (gY) \rfloor Y \rfloor \Omega + X_1 \rfloor X_2 \rfloor L_Y(Y \rfloor \Omega) \\
&= X_1 \rfloor X_2 \rfloor (\mathrm{div}(Y)) Y \rfloor \Omega \\
&= (\mathrm{div}(Y))(X_1 \rfloor X_2 \rfloor Y \rfloor \Omega)
\end{aligned}
$$

where we have used that $Y \rfloor Y \rfloor \Omega = 0$.
(ii) Applying identity and equation (4) we find

$$L_Y(L_X g) = L_X(L_Y g) - L_{[X,Y]} g = -L_{[X,Y]} g = -L_{fY} g = -fL_Y g = 0.$$

Therefore the function $L_X g$ is invariant under the vector field Y.

Problem 5. (i) Let

$$T := \sum_{i,j=1}^{n} a_{ij}(\mathbf{x}) \frac{\partial}{\partial x_i} \otimes dx_j \tag{1}$$

be a smooth (1,1) tensor field. Let

$$V = \sum_{k=1}^{n} V_k(\mathbf{x}) \frac{\partial}{\partial x_k} \tag{2}$$

be a smooth vector field. Find $L_V T$.
(ii) Assume that $L_V T = 0$. Show that

$$L_V \sum_{j=1}^{n} a_{jj} = 0. \tag{3}$$

Recall that the Lie derivative is linear and obeys the product rule, i.e.

$$
L_V \left(a_{ij} \frac{\partial}{\partial x_i} \otimes dx_j \right) \equiv (L_V a_{ij}) \frac{\partial}{\partial x_i} \otimes dx_j + a_{ij} \left(L_V \frac{\partial}{\partial x_i} \right) \otimes dx_j
$$
$$
+ a_{ij} \frac{\partial}{\partial x_i} \otimes (L_V dx_j).
$$

Solution 5.   (i) First we calculate $L_V T$. Applying the linearity of the Lie derivative we find

$$L_V T = L_V \sum_{i,j=1}^{n} \left(a_{ij} \frac{\partial}{\partial x_i} \otimes dx_j \right) = \sum_{i,j=1}^{n} L_V \left(a_{ij} \frac{\partial}{\partial x_i} \otimes dx_j \right).$$

Applying the product rule for the Lie derivative and

$$\left[V_k \frac{\partial}{\partial x_k}, \frac{\partial}{\partial x_i} \right] = \frac{\partial V_k}{\partial x_i} \frac{\partial}{\partial x_k}$$

we find

$$L_V T = \sum_{i,j=1}^{n} \left(\sum_{k=1}^{n} \left(V_k \frac{\partial a_{ij}}{\partial x_k} - a_{kj} \frac{\partial V_i}{\partial x_k} + a_{ik} \frac{\partial V_k}{\partial x_j} \right) \right) \frac{\partial}{\partial x_i} \otimes dx_j.$$

Thus from $L_V T = 0$ we obtain

$$\sum_{k=1}^{n} \left(V_k \frac{\partial a_{ij}}{\partial x_k} - a_{kj} \frac{\partial V_i}{\partial x_k} + a_{ik} \frac{\partial V_k}{\partial x_j} \right) = 0 \qquad \text{for all} \quad i, j = 1, \dots, n.$$

(ii) For $i = j$ we obtain

$$\sum_{k=1}^{n} \left(V_k \frac{\partial a_{jj}}{\partial x_k} - a_{kj} \frac{\partial V_j}{\partial x_k} + a_{jk} \frac{\partial V_k}{\partial x_j} \right) = 0$$

for $j = 1, \dots, n$. Since

$$L_V \sum_{j=1}^{n} a_{jj} = \sum_{k=1}^{n} \sum_{j=1}^{n} V_k \frac{\partial a_{jj}}{\partial x_k}$$

we find (3).

Problem 6.   Let $M = \mathbb{R}^n$ (or any open subset of $\mathbb{R}^n$). X is a vector field on M and α an r-form on M ($r < n$). We call α a *conformal invariant r-form* of X if

$$L_X \alpha = g\alpha.$$

Here $L_X \alpha$ is the Lie derivative of α with respect to X and g an arbitrary smooth function. Prove the following

Theorem. Let $M = \mathbb{R}^n$. The volume form is given by $\omega := dx_1 \wedge \cdots \wedge dx_n$. Let X and Y be two vector fields such that $[X, Y] = fY$ and

$$\alpha := Y \rfloor \omega.$$

Then $L_X \alpha = (f + \mathrm{div}(X))\alpha$.

Solution 6. Applying the rules for the Lie derivative we find

$$
\begin{aligned}
L_X(Y \rfloor \omega) &= [X, Y] \rfloor \omega + Y \rfloor (L_X \omega) = (fY) \rfloor \omega + Y \rfloor (X \rfloor d\omega + d(X \rfloor \omega)) \\
&= f(Y \rfloor \omega) + Y \rfloor (d(X \rfloor \omega)) = f(Y \rfloor \omega) + Y \rfloor ((\mathrm{div}(X))\omega) \\
&= (f + \mathrm{div}(X))(Y \rfloor \omega).
\end{aligned}
$$

Hence we have $g = f + \mathrm{div}(X)$.

Problem 7. Let V be a smooth vector field defined on $\mathbb{R}^n$

$$
V = \sum_{i=1}^{n} V_i(\mathbf{x}) \frac{\partial}{\partial x_i}.
$$

Let T be a $(1,1)$ smooth tensor field defined on $\mathbb{R}^n$

$$
T = \sum_{i=1}^{n} \sum_{j=1}^{n} a_{ij}(\mathbf{x}) \frac{\partial}{\partial x_i} \otimes dx_j.
$$

Let $L_V T$ be the Lie derivative of T with respect to the vector field V. Show that if $L_V T = 0$ then

$$
L_V \mathrm{tr}(a(\mathbf{x})) = 0
$$

where $a(\mathbf{x})$ is the $n \times n$ matrix $(a_{ij}(\mathbf{x}))$ and tr denotes the trace.

Solution 7. The Lie derivative is linear and obeys the product rule. Thus applying the product rule we have

$$
L_V T = \sum_{i=1}^{n} \sum_{j=1}^{n} \left(L_V a_{ij} \frac{\partial}{\partial x_i} \otimes dx_j + a_{ij} \left(L_V \frac{\partial}{\partial x_i} \right) \otimes dx_j + a_{ij} \frac{\partial}{\partial x_i} \otimes L_V dx_i \right).
$$

Since

$$
L_V(a_{ij}(\mathbf{x})) = V a_{ij}(\mathbf{x}) = \sum_{k=1}^{n} V_k \frac{\partial a_{ij}}{\partial x_k}
$$

$$
L_V \frac{\partial}{\partial x_i} = \left[\sum_{k=1}^{n} V_k \frac{\partial}{\partial x_k}, \frac{\partial}{\partial x_i} \right] = -\sum_{k=1}^{n} \frac{\partial V_k}{\partial x_i} \frac{\partial}{\partial x_k}
$$

$$
L_V dx_j = \sum_{k=1}^{n} \frac{\partial V_j}{\partial x_k} dx_k
$$

we obtain for the Lie derivative

$$
L_V T = \sum_{i=1}^{n} \sum_{j=1}^{n} \sum_{k=1}^{n} \left(V_k \frac{\partial a_{ij}}{\partial x_k} \frac{\partial}{\partial x_i} \otimes dx_j - a_{ij} \frac{\partial V_k}{\partial x_i} \frac{\partial}{\partial x_k} \otimes dx_j + a_{ij} \frac{\partial V_j}{\partial x_k} \frac{\partial}{\partial x_i} \otimes dx_k \right).
$$

Using the *contraction operator* C which is f-linear and

$$C\left(\frac{\partial}{\partial x_i} \otimes dx_j\right) = \delta_{ij}$$

where δ_{ij} denotes the Kronecker delta we find $CL_V T = L_V(CT)$. Since $L_V T = 0$ by assumption we have $L_V(CT) = 0$. Therefore

$$L_V(CT) = \sum_{j=1}^{n}\sum_{k=1}^{n}\frac{\partial a_{jj}}{\partial x_k} = \sum_{k=1}^{n}\frac{\partial}{\partial x_k}\mathrm{tr}(a(\mathbf{x})) = 0\,.$$

Thus $\mathrm{tr}(a(\mathbf{x}))$ is a first integral with respect to V.

Problem 8.   Let V, W be vector fields. Let f, g be smooth functions and α be a differential form. Assume that $L_V\alpha = f\alpha$, $L_W\alpha = g\alpha$. Show that

$$L_{[V,W]}\alpha = (L_V f - L_W g)\alpha. \tag{1}$$

Solution 8.   We have

$$L_{[V,W]}\alpha = [L_V, L_W]\alpha = L_V(L_W\alpha) - L_W(L_V\alpha) = L_V(g\alpha) - L_W(f\alpha).$$

Applying the product rule we have

$$L_V(g\alpha) = (L_V g)\alpha + gL_V\alpha = (L_V g)\alpha + gf\alpha$$
$$L_W(f\alpha) = (L_W f)\alpha + fL_W\alpha = (L_W f)\alpha + fg\alpha.$$

Thus (1) follows.

Problem 9.   Consider the vector fields

$$V = x\frac{\partial}{\partial x} + y\frac{\partial}{\partial y}, \qquad W = x\frac{\partial}{\partial y} - y\frac{\partial}{\partial x}$$

defined on $\mathbb{R}^2$.
(i) Do the vector fields V, W form a basis of a Lie algebra? If so, what type of Lie algebra do we have.
(ii) Express the two vector fields in polar coordinates $x(r,\theta) = r\cos(\theta)$, $y(r,\theta) = r\sin(\theta)$.
(iii) Calculate the commutator of the two vector fields expressed in polar coordinates. Compare with the result of (i).

Solution 9.   (i) Calculating the commutator of V and W we find

$$[V, W] = 0.$$

Thus we have a basis of a commutative Lie algebra.

(ii) Applying the chain rule to $f(x(r,\theta), y(r,\theta))$ we obtain

$$\frac{\partial f}{\partial r} = \frac{\partial f}{\partial x}\frac{\partial x}{\partial r} + \frac{\partial f}{\partial y}\frac{\partial y}{\partial r} = \frac{\partial f}{\partial x}\cos(\theta) + \frac{\partial f}{\partial y}\sin(\theta)$$

$$\frac{\partial f}{\partial \theta} = \frac{\partial f}{\partial x}\frac{\partial x}{\partial \theta} + \frac{\partial f}{\partial y}\frac{\partial y}{\partial \theta} = -\frac{\partial f}{\partial x}r\sin(\theta) + \frac{\partial f}{\partial y}r\cos(\theta).$$

Thus

$$\frac{\partial f}{\partial r}r\sin(\theta) = \frac{\partial f}{\partial x}r\cos(\theta)\sin(\theta) + \frac{\partial f}{\partial y}r\sin^2\theta$$

$$\frac{\partial f}{\partial \theta}\cos(\theta) = -\frac{\partial f}{\partial x}r\sin(\theta)\cos(\theta) + \frac{\partial f}{\partial y}r\cos^2(\theta).$$

Therefore

$$\frac{\partial}{\partial y} = \sin(\theta)\frac{\partial}{\partial r} + \frac{\cos(\theta)}{r}\frac{\partial}{\partial \theta}.$$

Analogously

$$-\frac{\partial f}{\partial r}r\cos(\theta) = -\frac{\partial f}{\partial x}r\cos^2(\theta) - \frac{\partial f}{\partial y}r\sin(\theta)\cos\theta$$

$$\frac{\partial f}{\partial \theta}\sin(\theta) = -\frac{\partial f}{\partial x}r\sin^2(\theta) + \frac{\partial f}{\partial y}r\cos(\theta)\sin(\theta).$$

Therefore

$$\frac{\partial}{\partial x} = \cos(\theta)\frac{\partial}{\partial r} - \frac{\sin(\theta)}{r}\frac{\partial}{\partial \theta}.$$

Thus the vector field V takes the form $r\partial/\partial r$ and the vector field W takes the form $\partial/\partial\theta$.

(iii) For the commutator we find $[r\partial/\partial r, \partial/\partial\theta] = 0$. Thus under the transformation to polar coordinates the commutator is preserved.

Problem 10. Let V, W be two smooth vector fields defined on $\mathbb{R}^3$. We write

$$V = \sum_{j=1}^{3} V_j(\mathbf{x})\frac{\partial}{\partial x_j}, \qquad W = \sum_{j=1}^{3} W_j(\mathbf{x})\frac{\partial}{\partial x_j}.$$

Let

$$\Omega = dx_1 \wedge dx_2 \wedge dx_3$$

be the *volume form* in $\mathbb{R}^3$. Then $L_V\omega = (\operatorname{div}(V))\omega$. where $\operatorname{div}(V)$ denotes the divergence of the vector field V. Find the divergence of the vector field given by the commutator $[V, W]$. Apply it to the vector fields associated with the autonomous systems of first order differential equations

$$\frac{dx_1}{dt} = \sigma(x_2 - x_1), \qquad \frac{dx_2}{dt} = \alpha x_1 - x_2 - x_1 x_3, \qquad \frac{dx_3}{dt} = -\beta x_3 + x_1 x_2$$

and

$$\frac{dx_1}{dt} = a(x_2 - x_1), \quad \frac{dx_2}{dt} = (c - a)x_1 + cx_2 - x_1x_3, \quad \frac{dx_3}{dt} = -bx_3 + x_1x_2.$$

The first system is the *Lorenz model* and the second system is *Chen's model*.

Solution 10.   Using the properties of the Lie derivative such as linearity and the product rule we find

$$L_{[V,W]}\omega = L_V(L_W\omega) - L_W(L_V\omega)$$
$$= L_V((\mathrm{div}(W))\omega) - L_W((\mathrm{div}(V))\omega)$$
$$= (L_V(\mathrm{div}(W)) - L_W(\mathrm{div}(V)))\omega.$$

Since $\mathrm{div}(V)$ for the first system is constant and $\mathrm{div}(W)$ for the second system is also constant, we find $L_{[V,W]}\omega = 0$. Thus the vector field given by the commutator is divergenceless.

Problem 11.   Consider the smooth vector field

$$V = \sum_{j=1}^{n} V_j(\mathbf{u}) \frac{\partial}{\partial u_j}$$

defined on $\mathbb{R}^n$. Consider the smooth differential one-form

$$\alpha = \sum_{k=1}^{n} f_k(\mathbf{u}) du_k.$$

Find the Lie derivative $L_V\alpha$. What is the condition such that $L_V\alpha = 0$?

Solution 11.   Using the linearity and the product rule of the Lie derivative we have

$$L_V\alpha = \sum_{k=1}^{n} (L_V f_k du_k) = \sum_{k=1}^{n} ((L_V f_k)du_k + f_k(\mathbf{u})(L_V du_k))$$

$$= \sum_{k=1}^{n} \left(\sum_{j=1}^{n} V_j \frac{\partial f_k}{\partial u_j} \right) du_k + \sum_{k=1}^{n} f_k dV_k$$

$$= \sum_{k=1}^{n} \left(\sum_{j=1}^{n} V_j \frac{\partial f_k}{\partial u_j} \right) du_k + \sum_{k=1}^{n} f_k \sum_{j=1}^{n} \frac{\partial V_k}{\partial u_j} du_j$$

$$= \sum_{k=1}^{n} \left(\sum_{j=1}^{n} \left(V_j \frac{\partial f_k}{\partial u_j} + f_j \frac{\partial V_J}{\partial u_k} \right) \right) du_k.$$

Thus the condition is

$$\sum_{j=1}^{n}\left(V_j\frac{\partial f_k}{\partial u_j}+f_j\frac{\partial V_j}{\partial u_k}\right)=0,\qquad k=1,2,\ldots,n.$$

For $n=1$ we obtain $V\partial f/\partial u+f\partial V/\partial u=0$.

Problem 12. Let g be a metric tensor field. Assume that V and W are Killing vector fields. Show that $[V,W]$ is also a Killing vector field.

Solution 12. By assumption we have $L_V g=0$ and $L_W g=0$. Using the identity

$$L_{[V,W]}g\equiv[L_V,L_W]g\equiv L_V(L_W g)-L_W(L_V g)$$

it follows that $L_{[V,W]}g=0$.

Problem 13. Let

$$g=\frac{1}{2}\sum_{j,k=1}^{3}g_{jk}(\mathbf{x})dx_j\otimes dx_k$$

be a metric tensor field in $\mathbb{R}^3$ and let

$$V(\mathbf{x})=V_1(\mathbf{x})\frac{\partial}{\partial x_1}+V_2(\mathbf{x})\frac{\partial}{\partial x_2}+V_3(\mathbf{x})\frac{\partial}{\partial x_3}.$$

Assume that $g_{jk}=\delta_{jk}$ where δ_{jk} is the Kronecker symbol.
(i) Calculate $L_V g$.
(ii) Give an interpretation of $L_V g=0$.

Solution 13. (i) The Lie derivative is linear and satisfies the product rule

$$L_V g=L_V(dx_1\otimes dx_1)+L_V(dx_2\otimes dx_2)+L_V(dx_3\otimes dx_3)$$

$$L_V(dx_j\otimes dx_k)=(L_V dx_j)\otimes dx_k+dx_j\otimes(L_V dx_k).$$

Furthermore the Lie derivative satisfies

$$L_V dx_j=d(V\rfloor dx_j)=dV_j=\sum_{k=1}^{3}\frac{\partial V_j}{\partial x_k}dx_k.$$

Here $\rfloor$ denotes the contraction, i.e. $\frac{\partial}{\partial x_j}\rfloor dx_k=\delta_{jk}$. It follows that

$$L_V g=\frac{1}{2}\sum_{j,k=1}^{3}\left(\frac{\partial V_j}{\partial x_k}+\frac{\partial V_k}{\partial x_j}\right)dx_j\otimes dx_k.$$

Since $dx_j \otimes dx_k$ $(j, k = 1, 2, 3)$ are basic elements, the right-hand side of (11) can also be written as a 3×3 matrix, namely

$$\frac{1}{2} \begin{pmatrix} 2\dfrac{\partial V_1}{\partial x_1} & \dfrac{\partial V_1}{\partial x_2} + \dfrac{\partial V_2}{\partial x_1} & \dfrac{\partial V_1}{\partial x_3} + \dfrac{\partial V_3}{\partial x_1} \\[2mm] \dfrac{\partial V_1}{\partial x_2} + \dfrac{\partial V_2}{\partial x_1} & 2\dfrac{\partial V_2}{\partial x_2} & \dfrac{\partial V_2}{\partial x_3} + \dfrac{\partial V_3}{\partial x_2} \\[2mm] \dfrac{\partial V_1}{\partial x_3} + \dfrac{\partial V_3}{\partial x_1} & \dfrac{\partial V_2}{\partial x_3} + \dfrac{\partial V_3}{\partial x_2} & 2\dfrac{\partial V_3}{\partial x_3} \end{pmatrix}.$$

(ii) The physical meaning is as follows: Consider a deformable body B in $\mathbb{R}^3$, $B \subset \mathbb{R}^3$. A displacement with or without deformation of the body B is a diffeomorphism of $\mathbb{R}^3$ defined in a neighbourhood of B. All such diffeomorphisms form a local group generated by the so-called displacement vector field $V(\mathbf{x})$. Consequently, the strain tensor field can be considered as the Lie derivative of the metric tensor field g in $\mathbb{R}^3$ with respect to the vector field $V(\mathbf{x})$. The metric tensor field gives rise to the distance between two points, namely

$$(ds)^2 = \sum_{j,k=1}^{3} g_{jk}(\mathbf{x}) dx_j dx_k.$$

Then the strain tensor field measures the variation of the distance between two points under a displacement generated by $V(\mathbf{x})$. The equation (invariance condition)

$$L_V g = 0$$

tells us that two points of the body B do not change during the displacement. To summarize. The strain tensor is the Lie derivative of the metric tensor field with respect to the deformation (or exactly the displacement vector field).

Problem 14. Let $g = dx_1 \otimes dx_1 + dx_2 \otimes dx_2$ be the metric tensor field of the two-dimensional Euclidean space. Assume that $L_V g = 0$, where

$$V = V_1(\mathbf{x}) \frac{\partial}{\partial x_1} + V_2(\mathbf{x}) \frac{\partial}{\partial x_2}$$

is a smooth vector field defined on the two-dimensional Euclidean space. Show that $L_V(dx_1 \wedge dx_2) = 0$.

Solution 14. From $L_V g = 0$ it follows that

$$\frac{\partial V_1}{\partial x_1} = 0, \quad \frac{\partial V_2}{\partial x_2} = 0, \quad \frac{\partial V_1}{\partial x_2} + \frac{\partial V_2}{\partial x_1} = 0.$$

Now

$$L_V(dx_1 \wedge dx_2) = (L_V dx_1) \wedge dx_2 + dx_1 \wedge (L_V dx_2) = dV_1 \wedge dx_2 + dx_1 \wedge dV_2$$
$$= \frac{\partial V_1}{\partial x_1} dx_1 \wedge dx_2 + \frac{\partial V_2}{\partial x_2} dx_1 \wedge dx_2$$
$$= \left(\frac{\partial V_1}{\partial x_1} + \frac{\partial V_2}{\partial x_2} \right) dx_1 \wedge dx_2.$$

Inserting $\partial V_1/\partial x_1 = \partial V_2/\partial x_2 = 0$ into this equation yields $L_V(dx_1 \wedge dx_2) = 0$.

Problem 15. Let $\mathbb{E}^3$ be the three-dimensional Euclidean space with metric tensor field

$$g = dx_1 \otimes dx_1 + dx_2 \otimes dx_2 + dx_3 \otimes dx_3. \tag{1}$$

Let

$$M \equiv \mathbb{S}^2 = \{(x_1, x_2, x_3) : x_1^2 + x_2^2 + x_3^2 = 1\}.$$

This means the manifold M is the unit sphere.
(i) Calculate the metric tensor field $\widetilde{g}$ for M. The parametrization is given by

$$x_1(u_1, u_2) = \cos(u_1)\sin(u_2), \quad x_2(u_1, u_2) = \sin(u_1)\sin(u_2), \quad x_3(u_1, u_2) = \cos(u_2)$$

where $u_1 \in [0, 2\pi)$ and $u_2 \in [0, \pi)$.
(ii) Find the vector fields V such that $L_V \widetilde{g} = 0$.
(iii) Show that these vector fields form a Lie algebra under the commutator.

Solution 15. (i) Since

$$dx_1 = -\sin(u_1)\sin(u_2)du_1 + \cos(u_1)\cos(u_2)du_2$$
$$dx_2 = \cos(u_1)\sin(u_2)du_1 + \sin(u_1)\cos(u_2)du_2$$
$$dx_3 = -\sin(u_2)du_2$$

we obtain from (1) that

$$\widetilde{g} = \sin^2(u_2)du_1 \otimes du_1 + du_2 \otimes du_2.$$

(ii) From the condition $L_V \widetilde{g} = 0$ with

$$V = V_1(u_1, u_2)\frac{\partial}{\partial u_1} + V_2(u_1, u_2)\frac{\partial}{\partial u_2}$$

we find

$$2\left(V_2 \sin(u_2)\cos(u_2) + \frac{\partial V_1}{\partial u_1}\sin^2(u_2)\right) du_1 \otimes du_1$$

$$+ \left(\sin^2(u_2)\frac{\partial V_1}{\partial u_2} + \frac{\partial V_2}{\partial u_1}\right) du_1 \otimes du_2$$

$$+ \left(\sin^2(u_2)\frac{\partial V_1}{\partial u_2} + \frac{\partial V_2}{\partial u_1}\right) du_2 \otimes du_1 + 2\frac{\partial V_2}{\partial u_2}du_2 \otimes du_2 = 0.$$

This leads to the system of linear partial differential equations

$$V_2 \sin(u_2)\cos(u_2) + \frac{\partial V_1}{\partial u_1}\sin^2(u_2) = 0$$

$$\sin^2(u_2)\frac{\partial V_1}{\partial u_2} + \frac{\partial V_2}{\partial u_1} = 0$$

$$\frac{\partial V_2}{\partial u_2} = 0.$$

From these equations we conclude that V_2 depends only on u_1. Thus the solution of this system of partial differential equations leads to the three independent vector fields

$$V^I = \frac{\partial}{\partial u_1}, \quad V^{II} = \cos(u_1)\cot(u_2)\frac{\partial}{\partial u_1} + \sin(u_1)\frac{\partial}{\partial u_2}$$

$$V^{III} = -\sin(u_1)\cot(u_2)\frac{\partial}{\partial u_1} + \cos(u_1)\frac{\partial}{\partial u_2}.$$

(iii) We find

$$[V^I, V^{II}] = -\sin(u_1)\cot(u_2)\frac{\partial}{\partial u_1} + \cos(u_1)\frac{\partial}{\partial u_2} = V^{III}$$

$$[V^I, V^{III}] = -\cos(u_1)\cot(u_2)\frac{\partial}{\partial u_1} - \sin(u_1)\frac{\partial}{\partial u_2} = -V^{II}$$

$$[V^{II}, V^{III}] = \frac{\partial}{\partial u_1} = V^I.$$

Thus the vector fields V^I, V^{II} and V^{III} form a basis of a Lie algebra under the commutator.

Problem 16.   The nonlinear partial differential equation

$$\frac{\partial^2 u}{\partial x \partial t} = \sin(u) \tag{1}$$

is the so-called one-dimensional *sine-Gordon equation*. Show that

$$V = (4u_{xxx} + 2(u_x)^3)\frac{\partial}{\partial u} \tag{2}$$

is a Lie-Bäcklund vector field of (1).

Solution 16. Taking the derivative of (1) with respect to x we find that (1) defines the manifolds

$$u_{xt} = \sin(u)$$

$$u_{xxt} = u_x \cos(u)$$

$$u_{xxxt} = u_{xx} \cos(u) - (u_x)^2 \sin(u) \tag{3}$$

$$u_{xxxxt} = u_{xxx} \cos(u) - 3u_x u_{xx} \sin(u) - (u_x)^3 \cos(u).$$

The *prolongated vector field* $\bar{V}$ of V is given by

$$\bar{V} = V + (4u_{xxxt} + 6(u_x)^2 u_{xt}) \frac{\partial}{\partial u_t} + (4u_{xxxxt} + 12u_x u_{xx} u_{xt} + 6(u_x)^2 u_{xxt}) \frac{\partial}{\partial u_{xt}}.$$

It follows that the Lie derivative of $u_{xt} - \sin u$ with respect to $\bar{V}$ is given by

$$L_{\bar{V}}(u_{xt} - \sin(u)) = 4u_{xxxxt} + 12u_x u_{xx} u_{xt} + 6(u_x)^2 u_{xxt} - (4u_{xxx} + 2(u_x)^3) \cos(u).$$

Inserting the equations given by (3) into this equation yields (*invariance condition*)

$$L_{\bar{V}}(u_{xt} - \sin(u)) = 0.$$

Consequently, the vector field V is a Lie-Bäcklund symmetry vector field of the sine-Gordon equation.

Problem 17. Consider the *Harry-Dym equation*

$$\frac{\partial u}{\partial t} - u^3 \frac{\partial^3 u}{\partial x^3} = 0. \tag{1}$$

(i) Find the Lie symmetry vector fields.
(ii) Compute the flow for one of the Lie symmetry vector fields.

Solution 17. (i) The Lie symmetry vector field is given by

$$V = \eta_x(x, t, u) \frac{\partial}{\partial x} + \eta_t(x, t, u) \frac{\partial}{\partial t} + \phi(x, t, u) \frac{\partial}{\partial u}. \tag{2}$$

There are eight determining equations for η_x, η_t, ϕ

$$\frac{\partial \eta_t}{\partial u} = 0, \qquad \frac{\partial \eta_t}{\partial x} = 0, \qquad \frac{\partial \eta_x}{\partial u} = 0, \qquad \frac{\partial^2 \phi}{\partial u^2} = 0,$$

$$\frac{\partial^2 \phi}{\partial u \partial x} - \frac{\partial^2 \eta_x}{\partial x^2} = 0, \qquad \frac{\partial \phi}{\partial t} - u^3 \frac{\partial^3 \phi}{\partial x^3} = 0$$

$$3u^3 \frac{\partial^3 \phi}{\partial u \partial x^2} + \frac{\partial \eta_x}{\partial t} - u^3 - \frac{\partial^3 \eta_x}{\partial x^3} = 0, \qquad u \frac{\partial \eta_t}{\partial t} - 3u \frac{\partial \eta_x}{\partial x} + 3\phi = 0.$$

These determining equations can easily be solved explicitly. The general solution is

$$\eta_x = k_1 + k_3 x + k_5 x^2, \quad \eta_t = k_2 - 3k_4 t, \quad \phi = (k_3 + k_4 + 2k_5 x)u$$

where $k_1, \ldots, k_5$ are arbitrary constants. The five infinitesimal generators then are

$$G_1 = \frac{\partial}{\partial x}, \qquad G_2 = \frac{\partial}{\partial t}$$

$$G_3 = x \frac{\partial}{\partial x} + u \frac{\partial}{\partial u}, \qquad G_4 = -3t \frac{\partial}{\partial t} + u \frac{\partial}{\partial u}, \qquad G_5 = x^2 \frac{\partial}{\partial x} + 2xu \frac{\partial}{\partial u}.$$

Thus (1) is invariant under translations (G_1 and G_2) and scaling (G_3 and G_4). The flow corresponding to each of the infinitesimal generators can be obtained via simple integration.

(ii) Let us compute the flow corresponding to G_5. This requires integration of the first order system

$$\frac{d\bar{x}}{d\epsilon} = \bar{x}^2, \qquad \frac{d\bar{t}}{d\epsilon} = 0, \qquad \frac{d\bar{u}}{d\epsilon} = 2\bar{x}\bar{u}$$

together with the initial conditions $x(0) = x$, $\bar{t}(0) = t$, $\bar{u}(0) = u$, where ϵ is the parameter of the transformation group. One obtains

$$\bar{x}(\epsilon) = \frac{x}{(1 - \epsilon x)}, \qquad \bar{t}(\epsilon) = t, \qquad \bar{u}(\epsilon) = \frac{u}{(1 - \epsilon x)^2}.$$

We therefore conclude that for any solution $u = f(x,t)$ of (1), the transformed solution

$$\bar{u}(\bar{x}, \bar{t}) = (1 + \epsilon \bar{x})^2 f \left(\frac{\bar{x}}{1 + \epsilon \bar{x}}, \bar{t} \right)$$

will solve

$$\frac{\partial \bar{u}}{\partial \bar{t}} - \bar{u}^3 \frac{\partial^3 \bar{u}}{\partial \bar{x}^3} = 0.$$

Problem 18.   Consider the partial differential equation

$$\frac{\partial^2 u}{\partial x \partial t} = f(u)$$

where f is an analytic function of u. Find the Lie symmetry vector field

$$V = a(x,t,u) \frac{\partial}{\partial x} + b(x,t,u) \frac{\partial}{\partial t} + c(x,t,u) \frac{\partial}{\partial u}$$

with the corresponding vertical vector field

$$V_v = (-au_x - bu_t + c)\frac{\partial}{\partial u}.$$

Solution 18. The prolongation of the vertical vector field V_v is given by

$$\widetilde{V}_v = (-au_x - bu_t + c)\frac{\partial}{\partial u}$$

$$\left(-\frac{\partial^2 a}{\partial x \partial t}u_x - \frac{\partial^2 a}{\partial x \partial u}u_t u_x - \frac{\partial a}{\partial x}u_{xt} - \frac{\partial^2 a}{\partial u \partial t}u_x^2 - \frac{\partial^2 a}{\partial u^2}u_t u_x^2 - 2\frac{\partial a}{\partial u}u_x u_{xt}\right.$$

$$-\frac{\partial a}{\partial t}u_{xx} - \frac{\partial a}{\partial u}u_t u_{xx} - au_{xxt} - \frac{\partial^2 b}{\partial x \partial t}u_t - \frac{\partial^2 b}{\partial x \partial u}u_t^2 - \frac{\partial b}{\partial x}u_{tt}$$

$$-\frac{\partial^2 b}{\partial u \partial t}u_x u_t - \frac{\partial^2 b}{\partial u^2}u_x u_t^2 - \frac{\partial b}{\partial u}u_{xt}u_t - \frac{\partial b}{\partial u}u_x u_{tt}$$

$$\left. +\frac{\partial^2 c}{\partial x \partial t} + \frac{\partial^2 c}{\partial x \partial u}u_t + \frac{\partial^2 c}{\partial u \partial t}u_x + \frac{\partial^2 c}{\partial u^2}u_t u_x + \frac{\partial c}{\partial u}u_{xt}\right)\frac{\partial}{\partial u_{xt}}.$$

Using that

$$u_{xt} = f(u), \quad u_{xxt} = \frac{\partial f}{\partial u}u_x, \quad u_{xtt} = \frac{\partial f}{\partial u}u_t$$

we find from the condition that

$$L_{\widetilde{V}_v}(u_{xt} - f(u)) = 0$$

where $L_{\widetilde{V}_v}$ twelve conditions for the coefficients $1, u_x, u_t, u_x u_t, u_x^2, u_t^2, u_{xx}$, $u_{tt}, u_x^2 u_t, u_x u_t^2, u_x x u_t, u_x u_{tt}$. For the coefficient 1 we have

$$-c\frac{\partial f}{\partial u} - \frac{\partial a}{\partial x}f - \frac{\partial b}{\partial t} + \frac{\partial^2 c}{\partial x \partial t} + \frac{\partial c}{\partial u}f = 0.$$

From the other conditions we can conclude that the function a depends only on x and the function b depends only on t. Then we find that the function c is of the form

$$c(x, t, u) = Ku + h(x, t)$$

for some smooth function h and K is a constant. The equation for the coefficient 1 reduces to

$$-Ku\frac{\partial f}{\partial u} + h\frac{\partial f}{\partial u} - \left(\frac{da}{dx} + \frac{db}{dt} - K\right)f + \frac{\partial^2 h}{\partial x \partial t} = 0.$$

In general the functions $udf/du, df/du, f, 1$ will be linearly independent. Then it follows that $K = 0$, $h = 0$ and $da/dx + db/dt = 0$. Thus $c = 0$ and $a(x) = Ax + B$, $b(t) = -At + B$. We find the four symmetry vector fields

$$V_1 = x\frac{\partial}{\partial x} - t\frac{\partial}{\partial t}, \quad V_2 = \frac{\partial}{\partial x}, \quad V_3 = \frac{\partial}{\partial t}.$$

Problem 19.   Some quantities in physics, owing to the transformation laws have to be considered as *currents* instead of differential forms. Let M be an orientable n-dimensional differentiable manifold of class C^∞. We denote by $\Phi_k(M)$ the set of all differential forms of degree k with compact support. Let $\phi \in \Phi_k(M)$ and let α be an exterior differential form of degree $n-k$ with locally integrable coefficients. Then, as an example of a current, we have

$$T_\alpha(\phi) \equiv \alpha(\phi) := \int_M \alpha \wedge \phi.$$

Define the Lie derivative for this current.

Solution 19.   The Lie derivative of this current can be defined as

$$L_V T(\phi) = -T(L_V \phi).$$

This definition can be motivated as follows. Assume that $L_V \alpha$ exists. Since

$$(L_V \alpha) \wedge \phi \equiv L_V(\alpha \wedge \phi) - \alpha \wedge (L_V \phi)$$

we obtain

$$T_{L_V \alpha} \phi = \int_M (L_V \alpha) \wedge \phi = \int_M (L_V(\alpha \wedge \phi) - \alpha \wedge (L_V \phi))$$
$$= -\int_M \alpha \wedge (L_V \phi)$$

where we used that

$$\int_M L_V(\alpha \wedge \phi) = \int_M d(V \rfloor (\alpha \wedge \phi)) = 0$$

noting that the $(n-1)$ differential form $V \rfloor (\alpha \wedge \phi)$ has compact support.

Supplementary Problems

Problem 1.   Consider the *Poincaré upper half-plane*

$$\mathbb{H}_+^2 := \{ (x,y) \in \mathbb{R}^2 : y > 0 \}$$

with metric tensor field

$$g = \frac{1}{y} dx \otimes \frac{1}{y} dx + \frac{1}{y} dy \otimes \frac{1}{y} dy$$

which is conformal with the standard inner product. Show that the Killing vector fields are given by

$$\frac{\partial}{\partial x}, \qquad x\frac{\partial}{\partial x} + y\frac{\partial}{\partial y}, \qquad \frac{x^2 - y^2}{2} \frac{\partial}{\partial x} + xy\frac{\partial}{\partial y}.$$

Show that the vector fields form a basis of a Lie algebra under the commutator.

Problem 2.   Let V, W be smooth vector fields and γ be a smooth differential form. Show that $L_V(W \rfloor \gamma) = ([V, W]) \rfloor \gamma + W \rfloor (L_V \gamma)$.

Problem 3.   Consider the differential one forms

$$\alpha_1 = -\sin(\psi)d\theta + \cos(\psi)\sin(\theta)d\phi, \quad \alpha_2 = \cos(\psi)d\theta + \sin(\psi)\sin(\theta)d\phi,$$

$$\alpha_3 = d\psi + \cos(\theta)d\phi.$$

and the vector fields

$$V_1 = \sin(\phi)\frac{\partial}{\partial \theta} + \cot(\theta)\cos(\phi)\frac{\partial}{\partial \phi} - \frac{\cos(\phi)}{\sin(\theta)}\frac{\partial}{\partial \psi}$$

$$V_2 = \cos(\phi)\frac{\partial}{\partial \theta} - \cot(\theta)\sin(\phi)\frac{\partial}{\partial \phi} + \frac{\sin(\phi)}{\sin(\theta)}\frac{\partial}{\partial \psi}$$

$$V_3 = \frac{\partial}{\partial \phi}.$$

Find the Lie derivatives $L_{V_j}\alpha_k$ for $j, k = 1, 2, 3$.

Problem 4.   Let $H_1 : \mathbb{R}^3 \to \mathbb{R}$, $H_2 : \mathbb{R}^3 \to \mathbb{R}$ be analytic functions and let ∇H_1, ∇H_2 be the gradients of H_1, H_2, respectively. Then we can form the autonomous system of first order differential equations

$$\begin{pmatrix} du_1/d\tau \\ du_2/d\tau \\ du_3/d\tau \end{pmatrix} = (\nabla H_1) \times (\nabla H_2)$$

with the vector field

$$V = \left(\left(\frac{\partial H_1}{\partial u_2}\right)\left(\frac{\partial H_2}{\partial u_3}\right) - \left(\frac{\partial H_1}{\partial u_3}\right)\left(\frac{\partial H_2}{\partial u_2}\right) \right) \frac{\partial}{\partial u_1}$$

$$+ \left(\left(\frac{\partial H_2}{\partial u_3}\right)\left(\frac{\partial H_2}{\partial u_1}\right) - \left(\frac{\partial H_1}{\partial u_1}\right)\left(\frac{\partial H_2}{\partial u_3}\right) \right) \frac{\partial}{\partial u_2}$$

$$+ \left(\left(\frac{\partial H_1}{\partial u_1}\right)\left(\frac{\partial H_2}{\partial u_1}\right) - \left(\frac{\partial H_1}{\partial u_2}\right)\left(\frac{\partial H_2}{\partial u_1}\right) \right) \frac{\partial}{\partial u_3}.$$

Such a dynamical system is called *bi-Hamiltonian*. Show that H_1, H_2 are first integrals of the dynamical system, i.e. show that

$$L_V(H_1) = 0, \qquad L_V(H_2) = 0.$$

Apply it to $H_1(\mathbf{x}) = \frac{1}{2}(x_1^2 + x_2^2 + x_3^2)$, $H_2(\mathbf{x}) = x_1 x_2 x_3$.

Chapter 26

Spin Systems

Let S_1, S_2, S_3 be the *spin matrices* for spin

$$s = \frac{1}{2}, \ 1, \ \frac{3}{2}, \ 2, \ \frac{5}{2}, \ \ldots$$

These matrices are $(2s+1) \times (2s+1)$ hermitian matrices (S_1 and S_3 are real symmetric) with trace equal to 0 satisfying the commutation relations

$$[S_1, S_2] = iS_3, \quad [S_2, S_3] = iS_1, \quad [S_3, S_1] = iS_2.$$

The eigenvalues are $s, s-1, \ldots, -s$ for a given s. The normalized eigenvectors form an orthonormal basis in the Hilbert space $\mathbb{C}^{2s+1}$. For the spin matrix S_3 we find the standard basis in $\mathbb{C}^{2s+1}$. We have

$$S_1^2 + S_2^2 + S_3^2 = s(s+1)I_{2s+1}$$

where I_{2s+1} is the $(2s+1) \times (2s+1)$ identity matrix. Furthermore

$$\text{tr}(S_1^2) = \frac{1}{3}s(s+1)(2s+1)$$

and

$$\text{tr}(S_j S_k) = 0 \text{ for } j \neq k \text{ and } j, k = 1, 2, 3.$$

The *Pauli spin matrices* σ_1, σ_2, σ_3 are given by

$$\sigma_1 = \begin{pmatrix} 0 & 1 \\ 1 & 0 \end{pmatrix}, \quad \sigma_2 = \begin{pmatrix} 0 & -i \\ i & 0 \end{pmatrix}, \quad \sigma_3 = \begin{pmatrix} 1 & 0 \\ 0 & -1 \end{pmatrix}$$

and the spin-$\frac{1}{2}$ matrices are given by

$$S_1 = \frac{1}{2}\sigma_1, \quad S_2 = \frac{1}{2}\sigma_2, \quad S_3 = \frac{1}{2}\sigma_3.$$

The eigenvalues are S_1 are $+1/2$ and $-1/2$ with the corresponding normalized eigenvectors

$$\mathbf{v}_{1,1} = \frac{1}{\sqrt{2}}\begin{pmatrix} 1 \\ 1 \end{pmatrix}, \quad \mathbf{v}_{1,2} = \frac{1}{\sqrt{2}}\begin{pmatrix} 1 \\ -1 \end{pmatrix}.$$

The eigenvalues for S_2 are $+1/2$ and $-1/2$ with the corresponding normalized eigenvectors

$$\mathbf{v}_{2,1} = \frac{1}{\sqrt{2}}\begin{pmatrix} 1 \\ i \end{pmatrix}, \quad \mathbf{v}_{2,2} = \frac{1}{\sqrt{2}}\begin{pmatrix} 1 \\ -i \end{pmatrix}.$$

The eigenvalues for S_3 are $+1/2$ and $-1/2$ with the corresponding normalized eigenvectors

$$\mathbf{v}_{3,1} = \begin{pmatrix} 1 \\ 0 \end{pmatrix}, \quad \mathbf{v}_{3,2} = \begin{pmatrix} 0 \\ 1 \end{pmatrix}.$$

The spin-1 matrices take the form

$$S_1 = \frac{1}{\sqrt{2}}\begin{pmatrix} 0 & 1 & 0 \\ 1 & 0 & 1 \\ 0 & 1 & 0 \end{pmatrix}, \quad S_2 = \frac{1}{\sqrt{2}}\begin{pmatrix} 0 & -i & 0 \\ i & 0 & -i \\ 0 & i & 0 \end{pmatrix}, \quad S_3 = \begin{pmatrix} 1 & 0 & 0 \\ 0 & 0 & 0 \\ 0 & 0 & -1 \end{pmatrix}$$

with eigenvalues $+1, 0, -1$. The matrices S_+ and S_- are defined by

$$S_+ := S_1 + iS_2, \quad S_- := S_1 - iS_2$$

with $[S_+, S_-] = 2S_3$.

Let S_1, S_2, S_3 be the spin matrices for spin $s = 1/2, 1, 3/2, 2, \ldots$, $\mathbf{a}, \mathbf{b} \in \mathbb{R}^3$ and $\mathbf{S} = (S_1, S_2, S_3)$. Let $\times$ be the vector product and $\cdot$ the scalar product. Then

$$\mathbf{S} \cdot (\mathbf{S} \times \mathbf{a}) + (\mathbf{S} \times \mathbf{a}) \cdot \mathbf{S} = 0_{2s+1}$$

and

$$(\mathbf{S} \cdot \mathbf{a})(\mathbf{S} \cdot \mathbf{b}) - (\mathbf{S} \cdot \mathbf{b})(\mathbf{S} \cdot \mathbf{a}) = i\mathbf{S} \cdot (\mathbf{a} \times \mathbf{b})$$

with

$$\mathbf{S} \times \mathbf{a} = \begin{pmatrix} S_2 a_3 - S_3 a_2 \\ S_3 a_1 - S_1 a_3 \\ S_1 a_2 - S_2 a_1 \end{pmatrix}, \quad \mathbf{S} \cdot \mathbf{a} = a_1 S_1 + a_2 S_2 + a_3 S_3.$$

Problem 1.   (i) Find the spectrum for the four-point *Ising model*

$$\hat{H} = J \sum_{j=1}^{4} s_j s_{j+1} \tag{1}$$

with cyclic boundary condition $\sigma_5 = \sigma_1$ and $J < 0$. The quantity J is called the exchange constant. Let I_2 be the 2×2 identity matrix and σ_3 the Pauli spin matrix. Then

$$s_1 = \sigma_3 \otimes I_2 \otimes I_2 \otimes I_2, \quad s_2 = I_2 \otimes \sigma_3 \otimes I_2 \otimes I_2,$$

$$s_3 = I_2 \otimes I_2 \otimes \sigma_3 \otimes I_2, \quad s_4 = I_2 \otimes I_2 \otimes I_2 \otimes \sigma_3$$

(ii) Let $\beta = 1/(k_B T)$. Calculate the *partition function*

$$Z(\beta) := \operatorname{tr}(e^{-\beta \hat{H}})$$

and the *Helmholtz free energy*

$$F(\beta) := -\frac{1}{\beta} \ln(Z(\beta)) = -\frac{1}{\beta} \ln(\operatorname{tr}(e^{-\beta \hat{H}}))$$

for the four point Ising model.

Solution 1.   (i) We find the diagonal matrices

$$s_1 s_2 = \operatorname{diag}(1, 1, 1, 1, -1, -1, -1, -1, -1, -1, -1, -1, 1, 1, 1, 1)$$

$$s_2 s_3 = \operatorname{diag}(1, 1, -1, -1, -1, -1, 1, 1, 1, 1, -1, -1, -1, -1, 1, 1)$$

$$s_3 s_4 = \operatorname{diag}(1, -1, -1, 1, 1, -1, -1, 1, 1, -1, -1, 1, 1, -1, -1, 1)$$

$$s_4 s_1 = \operatorname{diag}(1, -1, 1, -1, 1, -1, 1, -1, -1, 1, -1, 1, -1, 1, -1, 1).$$

It follows that with $s_5 = s_1$

$$\sum_{j=1}^{4} s_j s_{j+1} = \operatorname{diag}(4, 0, 0, 0, 0, -4, 0, 0, 0, 0, -4, 0, 0, 0, 0, 4).$$

Consequently the spectrum of $\hat{H}$ is given by

$4J$ twofold degenerate,   0 12 times degenerate, $-4J$ twofold degenerate.

(ii) From the result of (i) (eigenvalues) we find that the partition function is given by

$$Z(\beta) := \operatorname{tr}(e^{-\beta \hat{H}}) = 2e^{4\beta J} + 12 + 2e^{-4\beta J}.$$

Thus the Helmholtz free energy is given by

$$F(\beta) = -\frac{1}{\beta} \ln(2e^{4\beta J} + 12 + 2e^{-4\beta J}).$$

Problem 2. Calculate the eigenvalues and eigenvectors for the two-point *Heisenberg model*

$$\hat{H} = J \sum_{j=1}^{2} \mathbf{S}_j \cdot \mathbf{S}_{j+1} \tag{1}$$

with cyclic boundary conditions, i.e. $\mathbf{S}_3 \equiv \mathbf{S}_1$ where J is the so-called exchange constant ($J > 0$ or $J < 0$).

Solution 2. It follows that

$$\hat{H} = J(\mathbf{S}_1 \cdot \mathbf{S}_2 + \mathbf{S}_2 \cdot \mathbf{S}_3) \equiv J(\mathbf{S}_1 \cdot \mathbf{S}_2 + \mathbf{S}_2 \cdot \mathbf{S}_1).$$

Therefore

$$\hat{H} = J(S_{11}S_{21} + S_{12}S_{22} + S_{13}S_{23} + S_{21}S_{11} + S_{22}S_{12} + S_{23}S_{13}).$$

Since $S_{1x} := S_x \otimes I$, $S_{2x} = I \otimes S_x$ etc., it follows that

$$\hat{H} = J[(S_1 \otimes I_2)(I_2 \otimes S_1) + (S_2 \otimes I_2)(I_2 \otimes S_2) + (S_3 \otimes I_2)(I_2 \otimes S_3) \\ + (I_2 \otimes S_1)(S_1 \otimes I_2) + (I_2 \otimes S_2)(S_2 \otimes I_2) + (I_2 \otimes S_3)(S_3 \otimes I_2)].$$

We find

$$\hat{H} = J[(S_1 \otimes S_1) + (S_2 \otimes S_2) + (S_3 \otimes S_3) + (S_1 \otimes S_1) + (S_2 \otimes S_2) + (S_3 \otimes S_3)].$$

Therefore
$$\hat{H} = 2J((S_1 \otimes S_1) + (S_2 \otimes S_2) + (S_3 \otimes S_3)).$$

Since $S_1 := \frac{1}{2}\sigma_1$, $S_2 := \frac{1}{2}\sigma_2$, $S_3 := \frac{1}{2}\sigma_3$ we obtain

$$S_1 \otimes S_1 = \frac{1}{4}\begin{pmatrix} 0 & 1 \\ 1 & 0 \end{pmatrix} \otimes \begin{pmatrix} 0 & 1 \\ 1 & 0 \end{pmatrix} = \frac{1}{4}\begin{pmatrix} 0 & 0 & 0 & 1 \\ 0 & 0 & 1 & 0 \\ 0 & 1 & 0 & 0 \\ 1 & 0 & 0 & 0 \end{pmatrix}$$

etc. Then the Hamilton operator $\hat{H}$ is given by

$$\hat{H} = \frac{J}{2}\begin{pmatrix} 1 & 0 & 0 & 0 \\ 0 & -1 & 2 & 0 \\ 0 & 2 & -1 & 0 \\ 0 & 0 & 0 & 1 \end{pmatrix}.$$

We set

$$|\uparrow\rangle := \begin{pmatrix} 1 \\ 0 \end{pmatrix} \quad \text{spin up,} \qquad |\downarrow\rangle := \begin{pmatrix} 0 \\ 1 \end{pmatrix} \quad \text{spin down.}$$

Then

$$|\uparrow\uparrow\rangle := |\uparrow\rangle \otimes |\uparrow\rangle, \quad |\uparrow\downarrow\rangle := |\uparrow\rangle \otimes |\downarrow\rangle, \quad |\downarrow\uparrow\rangle := |\downarrow\rangle \otimes |\uparrow\rangle, \quad |\downarrow\downarrow\rangle := |\downarrow\rangle \otimes |\downarrow\rangle.$$

Consequently,

$$|\uparrow\uparrow\rangle = \begin{pmatrix} 1 \\ 0 \\ 0 \\ 0 \end{pmatrix}, \quad |\uparrow\downarrow\rangle = \begin{pmatrix} 0 \\ 1 \\ 0 \\ 0 \end{pmatrix}, \quad |\downarrow\uparrow\rangle = \begin{pmatrix} 0 \\ 0 \\ 1 \\ 0 \end{pmatrix}, \quad |\downarrow\downarrow\rangle = \begin{pmatrix} 0 \\ 0 \\ 0 \\ 1 \end{pmatrix}.$$

Obviously $|\uparrow\uparrow\rangle$ and $|\downarrow\downarrow\rangle$ are eigenvectors of the Hamilton operator with eigenvalues $J/2$ and $J/2$, respectively. This means the eigenvalue $J/2$ is degenerate. The eigenvalues of the matrix

$$\frac{J}{2} \begin{pmatrix} -1 & 2 \\ 2 & -1 \end{pmatrix}$$

are given by $J/2$, $-3J/2$. The eigenvectors are linear combinations of $|\uparrow\downarrow\rangle$ and $|\downarrow\uparrow\rangle$, i.e.

$$\frac{1}{\sqrt{2}}(|\uparrow\downarrow\rangle + |\downarrow\uparrow\rangle), \qquad \frac{1}{\sqrt{2}}(|\uparrow\downarrow\rangle - |\downarrow\uparrow\rangle).$$

These eigenvectors are entangled (*Bell states*).

Problem 3. Consider the spin Hamilton operator

$$\hat{H} = a \sum_{j=1}^{4} \sigma_3(j)\sigma_3(j+1) + b \sum_{j=1}^{4} \sigma_1(j) \tag{1}$$

with cyclic boundary conditions, i.e. $\sigma_3(5) \equiv \sigma_3(1)$. Here a, b are real constants and σ_1, σ_2 and σ_3 are the Pauli matrices. Thus the underlying Hilbert space is $\mathbb{C}^{16}$.

(i) Calculate the matrix representation of $\hat{H}$. Recall that

$$\sigma_a(1) = \sigma_a \otimes I_2 \otimes I_2 \otimes I_2, \qquad \sigma_a(2) = I_2 \otimes \sigma_a \otimes I_2 \otimes I_2$$

$$\sigma_a(3) = I_2 \otimes I_2 \otimes \sigma_a \otimes I_2, \qquad \sigma_a(4) = I_2 \otimes I_2 \otimes I_2 \otimes \sigma_a.$$

(ii) Show that the Hamilton operator $\hat{H}$ admits the C_{4v} symmetry group.
(iii) Calculate the matrix representation of $\hat{H}$ for the subspace which belongs to the representation A_1.

(iv) Show that the time-evolution of any expectation value of an observable, Ω, can be expressed as the sum of a time-independent and a time-dependent term

$$\langle\Omega\rangle(t) = \sum_{n=1}^{16} |c_n(0)|^2 \langle n|\Omega|n\rangle + \sum_{\substack{m,n=1\\n\neq m}}^{16} c_m^*(0)c_n(0)e^{i(E_m-E_n)t/\hbar}\langle m|\Omega|n\rangle$$

where $c_n(0)$ are the coefficients of expansion of the initial state in terms of the energy eigenstates $|n\rangle$ of $\hat{H}$, E_n are the eigenvalues of $\hat{H}$.

Solution 3. (i) The matrix representation of the first term of the right-hand side of (1) has been calculated in problem 1. We find the diagonal matrix

$$\sum_{j=1}^{4} \sigma_3(j)\sigma_3(j+1) = \mathrm{diag}(4,0,0,0,0,-4,0,0,0,0,-4,0,0,0,0,4).$$

The second term leads to non-diagonal terms. Using (2) we find the symmetric 16×16 matrix for $\hat{H}$

$$\begin{pmatrix}
4a & b & b & 0 & b & 0 & 0 & 0 & b & 0 & 0 & 0 & 0 & 0 & 0 & 0 \\
b & 0 & 0 & b & 0 & b & 0 & 0 & 0 & b & 0 & 0 & 0 & 0 & 0 & 0 \\
b & 0 & 0 & b & 0 & 0 & b & 0 & 0 & 0 & b & 0 & 0 & 0 & 0 & 0 \\
0 & b & b & 0 & 0 & 0 & 0 & b & 0 & 0 & 0 & b & 0 & 0 & 0 & 0 \\
b & 0 & 0 & 0 & 0 & b & b & 0 & 0 & 0 & 0 & 0 & b & 0 & 0 & 0 \\
0 & b & 0 & 0 & b & -4a & 0 & b & 0 & 0 & 0 & 0 & 0 & b & 0 & 0 \\
0 & 0 & b & 0 & b & 0 & 0 & b & 0 & 0 & 0 & 0 & 0 & 0 & b & 0 \\
0 & 0 & 0 & b & 0 & b & b & 0 & 0 & 0 & 0 & 0 & 0 & 0 & 0 & b \\
b & 0 & 0 & 0 & 0 & 0 & 0 & 0 & 0 & b & b & 0 & b & 0 & 0 & 0 \\
0 & b & 0 & 0 & 0 & 0 & 0 & 0 & b & 0 & 0 & b & 0 & b & 0 & 0 \\
0 & 0 & b & 0 & 0 & 0 & 0 & 0 & b & 0 & -4a & b & 0 & 0 & b & 0 \\
0 & 0 & 0 & b & 0 & 0 & 0 & 0 & 0 & b & b & 0 & 0 & 0 & 0 & b \\
0 & 0 & 0 & 0 & b & 0 & 0 & 0 & b & 0 & 0 & 0 & 0 & b & b & 0 \\
0 & 0 & 0 & 0 & 0 & b & 0 & 0 & 0 & b & 0 & 0 & b & 0 & 0 & b \\
0 & 0 & 0 & 0 & 0 & 0 & b & 0 & 0 & 0 & b & 0 & b & 0 & 0 & b \\
0 & 0 & 0 & 0 & 0 & 0 & 0 & b & 0 & 0 & 0 & b & 0 & b & b & 4a
\end{pmatrix}.$$

(ii) To find the discrete symmetries of the Hamilton operator (1) we study the discrete coordinate transformations $\tau : \{1,2,3,4\} \to \{1,2,3,4\}$ i.e. permutations of the set $\{1,2,3,4\}$ which preserve $\hat{H}$. Any permutation preserves the second term of $\hat{H}$. The first term is preserved by cyclic permutations of 1234 or 4321. Therefore we find the following set of symmetries

$$E : (1,2,3,4) \to (1,2,3,4) \qquad C_2 : (1,2,3,4) \to (3,4,1,2)$$

$$C_4 : (1,2,3,4) \to (2,3,4,1) \qquad C_4^3 : (1,2,3,4) \to (4,1,2,3)$$

$$\sigma_v : (1,2,3,4) \to (2,1,4,3) \qquad \sigma_v' : (1,2,3,4) \to (4,3,2,1)$$

$$\sigma_d : (1,2,3,4) \to (1,4,3,2) \qquad \sigma_d' : (1,2,3,4) \to (3,2,1,4)$$

which form a group isomorphic to C_{4v}. The five classes are given by

$$\{E\} \quad \{C_2\} \quad \{C_4, C_4^3\} \quad \{\sigma_v, \sigma_v'\} \quad \{\sigma_d, \sigma_d'\}.$$

The *character table* is as follows

	$\{E\}$	$\{C_2\}$	$\{2C_4\}$	$\{2\sigma_v\}$	$\{2\sigma_d\}$
A_1	1	1	1	1	1
A_2	1	1	1	-1	-1
B_1	1	1	-1	1	-1
B_2	1	1	-1	-1	1
E	2	-2	0	0	0

(iii) From the character table it follows that the projection operator of the representation A_1 is given by

$$\Pi_1 = \frac{1}{8}(O_E + O_{C_2} + O_{C_4} + O_{C_4^3} + O_{\sigma_v} + O_{\sigma_v'} + O_{\sigma_d} + O_{\sigma_d'}).$$

We recall that $O_p f(x) := f(P^{-1}x)$. We have $C_4^{-1} = C_4^3$, $(C_4^3)^{-1} = C_4$. The other elements of the groups are their own inverse. Thus for $\mathbf{e}_1, \cdots, \mathbf{e}_4 \in \mathbb{C}^2$ we have

$$
\begin{aligned}
8\Pi_1(\mathbf{e}_1 \otimes \mathbf{e}_2 \otimes \mathbf{e}_3 \otimes \mathbf{e}_4) = {} & \mathbf{e}_1 \otimes \mathbf{e}_2 \otimes \mathbf{e}_3 \otimes \mathbf{e}_4 + \mathbf{e}_3 \otimes \mathbf{e}_4 \otimes \mathbf{e}_1 \otimes \mathbf{e}_2 \\
& + \mathbf{e}_4 \otimes \mathbf{e}_1 \otimes \mathbf{e}_2 \otimes \mathbf{e}_3 + \mathbf{e}_2 \otimes \mathbf{e}_3 \otimes \mathbf{e}_4 \otimes \mathbf{e}_1 \\
& + \mathbf{e}_2 \otimes \mathbf{e}_1 \otimes \mathbf{e}_4 \otimes \mathbf{e}_3 + \mathbf{e}_4 \otimes \mathbf{e}_3 \otimes \mathbf{e}_2 \otimes \mathbf{e}_1 \\
& + \mathbf{e}_1 \otimes \mathbf{e}_4 \otimes \mathbf{e}_3 \otimes \mathbf{e}_2 + \mathbf{e}_3 \otimes \mathbf{e}_2 \otimes \mathbf{e}_1 \otimes \mathbf{e}_4.
\end{aligned}
$$

We set

$$|\uparrow\rangle := \begin{pmatrix} 1 \\ 0 \end{pmatrix} \text{ spin up,} \qquad |\downarrow\rangle := \begin{pmatrix} 0 \\ 1 \end{pmatrix} \text{ spin down.}$$

These are the eigenfunctions of σ_3. As a basis for the total space we take all Kronecker products of the form $\mathbf{e}_1 \otimes \mathbf{e}_2 \otimes \mathbf{e}_3 \otimes \mathbf{e}_4$, where $\mathbf{e}_1, \mathbf{e}_2, \mathbf{e}_3, \mathbf{e}_4 \in \{|\uparrow\rangle, |\downarrow\rangle\}$. We define

$$|\uparrow\uparrow\uparrow\uparrow\rangle := |\uparrow\rangle \otimes |\uparrow\rangle \otimes |\uparrow\rangle \otimes |\uparrow\rangle, \qquad |\downarrow\uparrow\uparrow\uparrow\rangle := |\downarrow\rangle \otimes |\uparrow\rangle \otimes |\uparrow\rangle \otimes |\uparrow\rangle$$

and so on. From the symmetry of Π_1 it follows that we only have to consider the projections of basis elements which are independent under the symmetry operations. We find the following basis for the subspace A_1

$$\Pi_1| \uparrow\uparrow\uparrow\uparrow\rangle : |1\rangle = | \uparrow\uparrow\uparrow\uparrow\rangle$$
$$\Pi_1| \uparrow\uparrow\uparrow\downarrow\rangle : |2\rangle = \tfrac{1}{2}(| \uparrow\uparrow\uparrow\downarrow\rangle + | \uparrow\uparrow\downarrow\uparrow\rangle + | \uparrow\downarrow\uparrow\uparrow\rangle + | \downarrow\uparrow\uparrow\uparrow\rangle))$$
$$\Pi_1| \uparrow\uparrow\downarrow\downarrow\rangle : |3\rangle = \tfrac{1}{2}(| \uparrow\uparrow\downarrow\downarrow\rangle + | \uparrow\downarrow\downarrow\uparrow\rangle + | \downarrow\downarrow\uparrow\uparrow\rangle + | \downarrow\uparrow\uparrow\downarrow\rangle))$$
$$\Pi_1| \uparrow\downarrow\uparrow\downarrow\rangle : |4\rangle = \tfrac{1}{\sqrt{2}}(| \uparrow\downarrow\uparrow\downarrow\rangle + | \downarrow\uparrow\downarrow\uparrow\rangle))$$
$$\Pi_1| \uparrow\downarrow\downarrow\downarrow\rangle : |5\rangle = \tfrac{1}{2}(| \uparrow\downarrow\downarrow\downarrow\rangle + | \downarrow\downarrow\downarrow\uparrow\rangle + | \downarrow\downarrow\uparrow\downarrow\rangle + | \downarrow\uparrow\downarrow\downarrow\rangle))$$
$$\Pi_1| \downarrow\downarrow\downarrow\downarrow\rangle : |6\rangle = | \downarrow\downarrow\downarrow\downarrow\rangle.$$

Thus the subspace which belongs to A_1 is six-dimensional. Calculating $\langle j|\hat{H}|k\rangle$, where $j, k = 1, 2, \ldots, 6$ we find the 6×6 symmetric matrix

$$\begin{pmatrix} 4a & 2b & 0 & 0 & 0 & 0 \\ 2b & 0 & 2b & \sqrt{2}b & 0 & 0 \\ 0 & 2b & 0 & 0 & 2b & 0 \\ 0 & \sqrt{2}b & 0 & -4a & \sqrt{2}b & 0 \\ 0 & 0 & 2b & \sqrt{2}b & 0 & 2b \\ 0 & 0 & 0 & 0 & 2b & 4a \end{pmatrix}.$$

(iv) Let $|j\rangle$ be the eigenstates of $\hat{H}$ with eigenvalues E_j. Then we have

$$|\psi(0)\rangle = \sum_{j=1}^{16} c_j(0)|j\rangle.$$

From the *Schrödinger equation* $i\hbar\partial\psi/\partial t = \hat{H}\psi$ we obtain

$$|\psi(t)\rangle = e^{-i\hat{H}t/\hbar}|\psi(0)\rangle = \sum_{j=1}^{16} c_j(0)e^{-i\hat{H}t/\hbar}|j\rangle = \sum_{j=1}^{16} c_j(0)e^{-iE_jt/\hbar}|j\rangle$$

as $|j\rangle$ is, by assumption, an eigenstate (eigenvector) of $\hat{H}$ with eigenvalue E_j. Since the dual state of $|\psi(t)$ is given by

$$\langle\psi(t)| = \sum_{k=1}^{16} c_k^*(0)e^{iE_kt/\hbar}\langle k|$$

we obtain $\langle\Omega\rangle(t)$.

Problem 4. (i) Consider the Hamilton operator

$$\hat{H}_2 = \hbar\omega(\sigma_3 \otimes \sigma_3) + \Delta(\sigma_1 \otimes \sigma_1).$$

Find the eigenvalues and eigenvectors. Discuss whether the eigenvectors can be written as product states. Find the unitary operator $U_2(t) = \exp(-i\hat{H}_2t/\hbar)$.

(ii) Consider the Hamilton operator

$$\hat{H}_3 = \hbar\omega(\sigma_3 \otimes \sigma_3 \otimes \sigma_3) + \Delta(\sigma_1 \otimes \sigma_1 \otimes \sigma_1).$$

Find the eigenvalues and eigenvectors. Discuss whether the eigenvectors can be written as product states. Find the unitary operator $U_3(t) = \exp(-i\hat{H}_3 t/\hbar)$.

Solution 4.   (i) The Hamilton operator $\hat{H}_2$ acts in the Hilbert space $\mathcal{H} = \mathbb{C}^4$. Since $\text{tr}(\hat{H}_2) = 0$ we obtain

$$\sum_{j=1}^{2^2} E_j = 0$$

where E_j are the eigenvalues of $\hat{H}_2$. Consider the 4×4 matrices

$$\Sigma_{z,2} := \sigma_3 \otimes \sigma_3, \qquad \Sigma_{x,2} := \sigma_1 \otimes \sigma_1.$$

We have $[\Sigma_{z,2}, \Sigma_{x,2}] = 0_4$. The four eigenvalues are given by

$$E_1 = \Delta + \hbar\omega, \quad E_2 = -(\Delta + \hbar\omega), \quad E_3 = \Delta - \hbar\omega, \quad E_4 = -(\Delta - \hbar\omega)$$

with the corresponding normalized eigenvectors

$$|\Phi^+\rangle = \frac{1}{\sqrt{2}} \begin{pmatrix} 1 \\ 0 \\ 0 \\ 1 \end{pmatrix}, \quad |\Psi^-\rangle = \frac{1}{\sqrt{2}} \begin{pmatrix} 0 \\ 1 \\ -1 \\ 0 \end{pmatrix},$$

$$|\Psi^+\rangle = \frac{1}{\sqrt{2}} \begin{pmatrix} 0 \\ 1 \\ 1 \\ 0 \end{pmatrix}, \quad |\Phi^-\rangle = \frac{1}{\sqrt{2}} \begin{pmatrix} 1 \\ 0 \\ 0 \\ -1 \end{pmatrix}.$$

Note that the states do not depend on the parameters ω and Δ. These states are the *Bell states*. The Bell states are fully entangled. The measure of entanglement for bipartite states are the von Neumann entropy, concurrence and the 2-tangle. As a measure of entanglement we apply the tangle which is the squared concurrence. The *concurrence* $\mathcal{C}$ for a pure state $|\psi\rangle$ in $\mathcal{H} = \mathbb{C}^4$ is given by

$$\mathcal{C} = 2 \left| \det \begin{pmatrix} c_{00} & c_{01} \\ c_{10} & c_{11} \end{pmatrix} \right|$$

with the state $|\psi\rangle$ written in the form

$$|\psi\rangle = \sum_{j,k=0}^{1} c_{jk} |j\rangle \otimes |k\rangle$$

and $|j\rangle$ $(j = 0, 1)$ denotes the standard basis in the Hilbert space $\mathbb{C}^2$. Next we calculate $\exp(-i\hat{H}_2 t/\hbar)$. The unitary operator $U_2(t) = \exp(-i\hat{H}_2 t/\hbar)$ can easily be calculated since

$$U_2(t) = \exp(-i\hat{H}_{20} t/\hbar) \exp(-i\hat{H}_{21} t/\hbar).$$

Thus

$$U_2(t) = \exp(-i\hat{H}_2 t/\hbar) = e^{-i\omega t(\sigma_3 \otimes \sigma_3)} e^{-it\Delta(\sigma_1 \otimes \sigma_1)/\hbar}$$
$$e^{-i\omega t(\sigma_3 \otimes \sigma_3)} = I_4 \cos(\omega t) - i(\sigma_3 \otimes \sigma_3) \sin(\omega t)$$
$$e^{-i\Delta t(\sigma_1 \otimes \sigma_1)/\hbar} = I_4 \cos(t\Delta/\hbar) - i(\sigma_1 \otimes \sigma_1) \sin(t\Delta/\hbar)$$

we obtain

$$e^{-i\hat{H}_2 t/\hbar} = I_4 \cos(\omega t) \cos(t\Delta/\hbar) - i(\sigma_3 \otimes \sigma_3) \sin(\omega t) \cos(t\Delta/\hbar)$$
$$- i(\sigma_1 \otimes \sigma_1) \cos(\omega t) \sin(t\Delta/\hbar) - (\sigma_3\sigma_1) \otimes (\sigma_3\sigma_1) \sin(\omega t) \sin(t\Delta/\hbar).$$

The Hamilton operator $\hat{H}_2$ shows energy level crossing (when keeping $\hbar\omega$ fixed and varying Δ). The unitary operator $U_2(t) = \exp(-i\hat{H}_2 t/\hbar)$ can generate entangled states from unentangled states. However note that applying the unitary operator $U_2(t)$ to one of the Bell states given above cannot disentangle these states since they are eigenstates. For example, we have $U_2(t)|\Phi^+\rangle = e^{-iE_1 t/\hbar}|\Phi^+\rangle$. If we start with the unentangled state (product state)

$$|\psi\rangle = (1\,0\,0\,0)^T$$

under the evolution $U_2(t)|\psi\rangle$ depending on t, $\hbar\omega$ and Δ we can find entangled states using the concurrence as measure. For the case $\hbar\omega = \Delta$ (level crossing) the state reduces to

$$\begin{pmatrix} \cos^2(\omega t) - i\sin(\omega t)\cos(\omega t) \\ 0 \\ 0 \\ -\sin^2(\omega t) - i\cos(\omega t)\sin(\omega t) \end{pmatrix}.$$

(ii) Let $\Sigma_{3,3} = \sigma_3 \otimes \sigma_3 \otimes \sigma_3$, $\Sigma_{1,3} = \sigma_1 \otimes \sigma_1 \otimes \sigma_1$. We have for the anticommutator $[\Sigma_{3,3}, \Sigma_{1,3}]_+ = 0_8$. We find the eigenvalue $E = \sqrt{\hbar^2\omega^2 + \Delta^2}$ (four-times degenerate) with the normalized eigenvectors

$$\frac{1}{\sqrt{\Delta^2 + (E - \hbar\omega)^2}} \begin{pmatrix} \Delta \\ 0 \\ 0 \\ 0 \\ 0 \\ 0 \\ 0 \\ E - \hbar\omega \end{pmatrix}, \quad \frac{1}{\sqrt{\Delta^2 + (E - \hbar\omega)^2}} \begin{pmatrix} 0 \\ 0 \\ 0 \\ \Delta \\ E - \hbar\omega \\ 0 \\ 0 \\ 0 \end{pmatrix},$$

$$\frac{1}{\sqrt{\Delta^2 + (E + \hbar\omega)^2}} \begin{pmatrix} 0 \\ \Delta \\ 0 \\ 0 \\ 0 \\ 0 \\ E + \hbar\omega \\ 0 \end{pmatrix}, \quad \frac{1}{\sqrt{\Delta^2 + (E + \hbar\omega)^2}} \begin{pmatrix} 0 \\ 0 \\ \Delta \\ 0 \\ 0 \\ E + \hbar\omega \\ 0 \\ 0 \end{pmatrix}$$

and the eigenvalue $-E$ (four times degenerate) with the normalized eigenvectors

$$\frac{1}{\sqrt{\Delta^2 + (E + \hbar\omega)^2}} \begin{pmatrix} \Delta \\ 0 \\ 0 \\ 0 \\ 0 \\ 0 \\ 0 \\ -E - \hbar\omega \end{pmatrix}, \quad \frac{1}{\sqrt{\Delta^2 + (E + \hbar\omega)^2}} \begin{pmatrix} 0 \\ 0 \\ 0 \\ \Delta \\ -E - \hbar\omega \\ 0 \\ 0 \\ 0 \end{pmatrix},$$

$$\frac{1}{\sqrt{\Delta^2 + (E - \hbar\omega)^2}} \begin{pmatrix} 0 \\ \Delta \\ 0 \\ 0 \\ 0 \\ 0 \\ -E + \hbar\omega \\ 0 \end{pmatrix}, \quad \frac{1}{\sqrt{\Delta^2 + (E - \hbar\omega)^2}} \begin{pmatrix} 0 \\ 0 \\ \Delta \\ 0 \\ 0 \\ -E + \hbar\omega \\ 0 \\ 0 \end{pmatrix}.$$

If $\Delta = E - \hbar\omega$ then the first and second eigenstates are fully entangled. If $\Delta = E + \hbar\omega$ the third and fourth states are fully entangled. As measure we can use the 3-tangle.

Since we have the eigenvalues and eigenvectors of $\hat{H}_3$ the unitary operator $U_3(t)$ can easily be calculated. We find

$$U_3(t) = I_8 \cos(Et/\hbar) + \frac{i}{E}(\hbar\omega\sigma_3 \otimes \sigma_3 \otimes \sigma_3 + \Delta\sigma_1 \otimes \sigma_1 \otimes \sigma_1)\sin(Et/\hbar).$$

If we start with the unentangled state (product state)

$$|\psi\rangle = (1\,0\,0\,0\,0\,0\,0\,0)^T$$

under the evolution $U(t)|\psi\rangle$ depending on t, $\hbar\omega$ and Δ we can find entangled states.

Problem 5. Consider the spin-1 matrix

$$S_2 = \frac{1}{\sqrt{2}} \begin{pmatrix} 0 & -i & 0 \\ i & 0 & -i \\ 0 & i & 0 \end{pmatrix}.$$

Calculate $\exp(zS_2)$. Then substitute $z = -i\omega t$.

Solution 5. Since $S_2^3 = S_2$, $S_2^4 = S_2^2$ etc. we obtain

$$e^{zS_2} = I_3 + S_2 \left(z + \frac{z^3}{3!} + \frac{z^5}{5!} + \cdots \right) + S_2^2 \left(\frac{z^2}{2!} + \frac{z^4}{4!} + \cdots \right)$$

$$= I_3 + S_2 \sinh(z) + S_2^2(\cosh(z) - 1).$$

With $z = -i\omega t$ we obtain

$$e^{-i\omega t S_2} = I_3 - i\sin(\omega t)S_2 + (\cos(\omega t) - 1)S_2^2.$$

Since

$$S_2^2 = \frac{1}{2} \begin{pmatrix} 1 & 0 & -1 \\ 0 & 2 & 0 \\ -1 & 0 & 1 \end{pmatrix}$$

we end up with

$$e^{-i\omega t S_2} = \begin{pmatrix} 1 + (\cos(\omega t) - 1)/2 & -\sin(\omega t)/\sqrt{2} & (-\cos(\omega t) + 1)/2 \\ -\sin(\omega t) & \cos(\omega t) & -\sin(\omega t)/\sqrt{2} \\ (-\cos(\omega t) + 1)/2 & \sin(\omega t)/\sqrt{2} & 1 + (\cos(\omega t) - 1)/2 \end{pmatrix}.$$

Problem 6. Consider the spin matrices S_1, S_2, S_3 for spin s with $s = 1/2$, $s = 1$, $s = 3/2$, $s = 2$, ... They satisfy the commutation relations

$$[S_1, S_2] = iS_3, \quad [S_2, S_3] = iS_1, \quad [S_3, S_1] = iS_2.$$

So we have a basis of a simple Lie algebra. Consider the Hamilton operator

$$\hat{H} = \hbar \omega S_3$$

acting in the Hilbert space $\mathbb{C}^{2s+1}$ and ω is the frequency. Calculate the time evolution of S_1. The *Heisenberg equation of motion* is given by

$$i\hbar \frac{dS_1}{dt} = [S_1, \hat{H}](t)$$

with the initial condition $S_1(t = 0) = S_1$.

Solution 6.   Since $[S_1, S_3] = -iS_2$ we have

$$\frac{dS_1(t)}{dt} = -\omega S_2(t).$$

From $[S_2, S_3] = iS_1$ we obtain

$$\frac{dS_2(t)}{dt} = \omega S_1(t).$$

Thus we have to solve the following system of linear matrix differential equations with constant coefficients

$$\frac{dS_1(t)}{dt} = -\omega S_2(t), \quad \frac{dS_2(t)}{dt} = \omega S_1(t)$$

together with the initial conditions $S_1(t = 0) = S_1$, $S_2(t = 0) = S_2$. The solution of the initial value problem is given by

$$S_1(t) = \cos(\omega t)S_1 - \sin(\omega t)S_2, \quad S_2(t) = \cos(\omega t)S_2 + \sin(\omega t)S_1.$$

For spin-$\frac{1}{2}$ we have

$$S_1(t) = \frac{1}{2}\begin{pmatrix} 0 & \cos(\omega t) + i\sin(\omega t) \\ \cos(\omega t) - i\sin(\omega t) & 0 \end{pmatrix}$$

$$S_2(t) = \frac{1}{2}\begin{pmatrix} 0 & \sin(\omega t) - i\cos(\omega t) \\ \sin(\omega t) + i\cos(\omega t) & 0 \end{pmatrix}.$$

The solution of the Heisenberg equation of motion can also be given as

$$S_1(t) = e^{i\hat{H}t/\hbar}S_1e^{-i\hat{H}t/\hbar}, \quad S_2(t) = e^{i\hat{H}t/\hbar}S_2e^{-i\hat{H}t/\hbar}.$$

Problem 7.   Consider the Hamilton operator

$$\hat{K} = \frac{\hat{H}}{\hbar\omega} = \sigma_1 \otimes \sigma_2 \otimes I_2 + I_2 \otimes \sigma_2 \otimes \sigma_3 + \sigma_1 \otimes I_2 \otimes \sigma_3.$$

Note that $\text{tr}(\hat{K}) = 0$. Find the commutator $[\hat{K}, \sigma_1 \otimes \sigma_2 \otimes \sigma_3]$. Find the eigenvalues of $\hat{K}$. Find the commutator

$$[\hat{K}, \sigma_1 \otimes \sigma_1 \otimes I_2 + I_2 \otimes \sigma_2 \otimes \sigma_2 + \sigma_3 \otimes I_2 \otimes \sigma_3].$$

Solution 7.   Applying the Maxima program

```
/* Hamilton88.mac */
I2: matrix([1,0],[0,1]);
sig1: matrix([0,1],[1,0]);
sig2: matrix([0,-%i],[%i,0]);
sig3: matrix([1,0],[0,-1]);
T1: kronecker_product(sig1,kronecker_product(sig2,I2));
T2: kronecker_product(I2,kronecker_product(sig2,sig3));
T3: kronecker_product(sig1,kronecker_product(I2,sig3));
H: T1 + T2 + T3;
R: eigenvalues(H);
A: kronecker_product(sig1,kronecker_product(sig2,sig3));
C: H . A - A . H;
B: kronecker_product(sig1,kronecker_product(sig1,I2)) +
kronecker_product(I2,kronecker_product(sig2,sig2)) +
kronecker_product(sig3,kronecker_product(I2,sig3));
D: H . B - B . H;
```

provides the eigenvalues 3 ($2\times$) and -1 ($6\times$). For the commutators we find $[\hat{K}, \sigma_1 \otimes \sigma_2 \otimes \sigma_3] = 0_8$ and

$$[\hat{K}, \sigma_1 \otimes \sigma_1 \otimes I_2 + I_2 \otimes \sigma_2 \otimes \sigma_2 + \sigma_3 \otimes I_2 \otimes \sigma_3] \neq 0_8.$$

Supplementary Problems

Problem 1. Let $s = 1/2, 1, 3/2, \ldots$ and S_1, S_2, S_3 be the spin matrices for spin s. Let $\mathbf{a}, \mathbf{b} \in \mathbb{R}^3$, $\mathbf{S} = (S_1, S_2, S_3)$, $\times$ be the vector product and $\cdot$ the scalar product.

(i) Show that $\mathbf{S} \cdot (\mathbf{S} \times \mathbf{a}) + (\mathbf{S} \times \mathbf{a}) \cdot \mathbf{S} = 0_{2s+1}$.

(ii) Show that $(\mathbf{S} \cdot \mathbf{a})(\mathbf{S} \cdot \mathbf{b}) - (\mathbf{S} \cdot \mathbf{b})(\mathbf{S} \cdot \mathbf{a}) = i\mathbf{S} \cdot (\mathbf{a} \times \mathbf{b})$. Note that

$$\mathbf{S} \times \mathbf{a} = \begin{pmatrix} S_2 a_3 - S_3 a_2 \\ S_3 a_1 - S_1 a_3 \\ S_1 a_2 - S_2 a_1 \end{pmatrix}, \quad \mathbf{S} \cdot \mathbf{a} = a_1 S_1 + a_2 S_2 + a_3 S_3.$$

Problem 2. Let σ_1, σ_2, σ_3 be the Pauli spin matrices. Consider the 8×8 matrices $K = \sigma_1 \otimes \sigma_2 \otimes \sigma_3$, $S = \sigma_1 \otimes \sigma_1 \otimes \sigma_1$. Note that the matrices K and S are unitary and hermitian. Show that $[K, S] = 0_8$. Show that $\Pi_1 = \frac{1}{2}(I_8 + S)$, $\Pi_2 = \frac{1}{2}(I_8 - S)$ are projection matrices and $\Pi_1 \Pi_2 = 0_8$. Show that $\Pi_3 = \frac{1}{2}(I_8 - K)$, $\Pi_4 = \frac{1}{2}(I_8 + K)$ are projection matrices.

Problem 3. Consider the spin-$\frac{1}{2}$ matrices

$$S_+ = \begin{pmatrix} 0 & 1 \\ 0 & 0 \end{pmatrix}, \quad S_- = \begin{pmatrix} 0 & 0 \\ 1 & 0 \end{pmatrix} = S_+^*, \quad S_3 = \frac{1}{2}\begin{pmatrix} 1 & 0 \\ 0 & -1 \end{pmatrix}$$

with $[S_+, S_-] = 2S_3$, $[S_+, S_3] = -S_+$, $[S_-, S_3] = S_-$ and $S_+^2 = S_-^2 = 0_2$. We set (counting starts from 0)

$$S_{0,+} = S_+ \otimes I_2 \otimes I_2, \quad S_{1,+} = I_2 \otimes S_+ \otimes I_2, \quad S_{2,+} = I_2 \otimes I_2 \otimes S_+$$

$$S_{0,-} = S_- \otimes I_2 \otimes I_2, \quad S_{1,-} = I_2 \otimes S_- \otimes I_2, \quad S_{2,-} = I_2 \otimes I_2 \otimes S_-$$

$$S_{0,3} = S_3 \otimes I_2 \otimes I_2, \quad S_{1,3} = I_2 \otimes S_3 \otimes I_2, \quad S_{2,3} = I_2 \otimes I_2 \otimes S_3.$$

Consider the Hamilton operator with cyclic boundary conditions ($S_{3,+} = S_{0,+}, S_{3,-} = S_{0,-}, S_{3,3} = S_{0,3}$)

$$\hat{H} = \hbar\omega \sum_{j=0}^{2} (S_{j+1,+}S_{j-} + S_{j+1,-}S_{j,+})$$

$$= \hbar\omega(S_- \otimes S_+ \otimes I_2 + I_2 \otimes S_- \otimes S_+ s_+ \otimes I_2 \otimes s_-$$
$$+ S_+ \otimes S_- \otimes I_2 + I_2 \otimes S_+ \otimes S_- + S_- \otimes I_2 \otimes s_+).$$

(i) Find the eigenvalues and normalized eigenvectors of $\hat{H}$.
(ii) Solve the Heisenberg equations of motion

$$i\hbar \frac{dS_{j,+}}{dt} = [S_{j,+}, \hat{H}](t), \qquad i\hbar \frac{dS_{j,-}}{dt} = [S_{j,-}, \hat{H}](t).$$

Problem 4.   Let σ_1, σ_2, σ_3 be the Pauli spin matrices with the commutation relations $[\sigma_1, \sigma_2] = 2i\sigma_3$, $[\sigma_2, \sigma_3] = 2i\sigma_2$, $[\sigma_3, \sigma_1] = 2i\sigma_2$. Consider the Hamilton operator $\hat{H}$ in the Hilbert space $\mathbb{C}^4$

$$\hat{H} = \hbar\omega_1\sigma_1 \otimes \sigma_1 + \hbar\omega_2\sigma_2 \otimes \sigma_2 + \hbar\omega_3\sigma_3 \otimes \sigma_3.$$

Applying the Baker-Campbell-Hausdorff formula show that

$$e^{i\hat{H}t/\hbar}(\sigma_1 \otimes I_2)e^{-i\hat{H}t/\hbar} =$$
$$(\sigma_1 \otimes I_2)\cos(\omega_2 t)\cos(\omega_3 t) + (I_2 \otimes \sigma_1)\sin(\omega_2 t)\sin(\omega_3 t)$$
$$-(\sigma_2 \otimes \sigma_3)\cos(\omega_2 t)\sin(\omega_3 t) + (\sigma_3 \otimes \sigma_2)\sin(\omega_2 t)\cos(\omega_3 t)$$

$$e^{i\hat{H}t/\hbar}(\sigma_2 \otimes I_2)e^{-i\hat{H}t/\hbar} =$$
$$(\sigma_2 \otimes I_2)\cos(\omega_3 t)\cos(\omega_1 t) + (I_2 \otimes \sigma_2)\sin(\omega_3 t)\sin(\omega_1 t)$$
$$-(\sigma_3 \otimes \sigma_1)\cos(\omega_3 t)\sin(\omega_1 t) + (\sigma_1 \otimes \sigma_3)\sin(\omega_3 t)\cos(\omega_1 t)$$

$$e^{i\hat{H}t/\hbar}(\sigma_3 \otimes I_2)e^{-i\hat{H}t/\hbar} =$$
$$(\sigma_3 \otimes I_2)\cos(\omega_1 t)\cos(\omega_2 t) + (I_2 \otimes \sigma_3)\sin(\omega_1 t)\sin(\omega_2 t)$$
$$-(\sigma_1 \otimes \sigma_2)\cos(\omega_1 t)\sin(\omega_2 t) + (\sigma_2 \otimes \sigma_1)\sin(\omega_1 t)\cos(\omega_2 t).$$

Problem 5. Find all 8×8 hermitian matrices H such that

$$[H, \sigma_1 \otimes \sigma_2 \otimes \sigma_3] = 0_8.$$

Problem 6. Find the eigenvalues and normalized eigenvectors of the Hamilton operator (16×16 hermitian matrix)

$$\hat{H} = \hbar\omega_1(\sigma_3 \otimes \sigma_3 \otimes I_2 \otimes I_2 + I_2 \otimes I_2 \otimes \sigma_3 \otimes \sigma_3) + \hbar\omega_2(\sigma_1 \otimes \sigma_1 \otimes \sigma_1 \otimes \sigma_1).$$

Problem 7. Consider the Hamilton operator

$$\hat{H} = \hbar\omega_1(\sigma_3 \otimes I_2 + I_2 \otimes \sigma_3) + \hbar\omega_2\sigma_1 \otimes \sigma_1 + \hbar\omega_3\sigma_2 \otimes \sigma_2.$$

Find the eigenvalues and normalized eigenvectors of $\hat{H}$.

Problem 8. Let $\mu \in \mathbb{C}$, S the spin ($S = 0, 1/2, 1, 3/2, 2, \ldots$) and $|0\rangle$, $|1\rangle$, $\ldots$, $|2S\rangle$ be the standard basis in $\mathbb{C}^{2S+1}$. The *Bloch coherent states* $|\mu\rangle$ are defined by

$$|\mu\rangle = \frac{1}{(1 + |\mu|^2)^S} \sum_{p=0}^{2S} \left(\frac{(2S)!}{p!(2S - p)!} \right)^{1/2} \mu^p |p\rangle.$$

Show that

$$\langle\mu| = \frac{1}{(1 + |\mu|^2)^S} \sum_{p=0}^{2S} \left(\frac{(2S)!}{p!(2S - p)!} \right)^{1/2} (\mu^*)^p \langle p|.$$

Show that (*completeness relation*)

$$\frac{1 + 2S}{\pi} \int_{\mathbb{C}} \frac{d^2\mu}{(1 + |\mu|^2)^2} |\mu\rangle\langle\mu| = I_{2S+1}$$

where $d^2\mu = d(\Re(\mu))(d(\Im(\mu)))$. Show that the scalar product of two Bloch coherent states is given as

$$\langle\nu|\mu\rangle = \frac{(1 + \nu^*\mu)^{2S}}{(1 + |\nu|^2)^S(1 + |\mu|^2)^S}.$$

Let $S = 1/2$. Show that

$$|\mu\rangle = \frac{1}{(1 + |\mu|^2)^{1/2}} (|0\rangle + \mu|1\rangle).$$

Problem 9.   (i) Let S_1, S_2, S_3 be the spin-$\frac{1}{2}$ matrices, $\mathbf{n}$ be a normalized vector in $\mathbb{R}^3$ and $\mathbf{n} \cdot \mathbf{S} = n_1 S_1 + n_2 S_2 + n_3 S_3$. Show that $\exp(i\theta\mathbf{n} \cdot \mathbf{S}) = I_2 \cos(\theta/2)2i(\mathbf{n} \cdot \mathbf{S})\sin(\theta/2)$.
(ii) Let S_1, S_2, S_3 be the spin-1 matrices with $S_3 = \mathrm{diag}(1,0,-1)$, $\mathbf{n}$ be a normalized vector in $\mathbb{R}^3$ and $\mathbf{n} \cdot \mathbf{S} = n_1 S_1 + n_2 S_2 + n_3 S_3$. Show that

$$\exp(i\theta\mathbf{n} \cdot \mathbf{S}) = I_3 + i(\mathbf{n} \cdot \mathbf{S})\sin(\theta) + (\mathbf{n} \cdot \mathbf{S})^2(\cos(\theta) - 1).$$

Problem 10.   Let $s = 1/2, 1, 3/2, 2, \ldots$ be the spin and S_1, S_2, S_3 be the spin matrices for spin s. S_1, S_2, S_3 satisfy the commutation relation

$$[S_1, S_2] = iS_3, \quad [S_2, S_3] = iS_1, \quad [S_3, S_1] = iS_2.$$

Let $\mathbf{n}$ be a normalized vector in $\mathbb{R}^3$ and $\mathbf{n} \cdot \mathbf{S} = n_1 S_1 + n_2 S_2 + n_3 S_3$. Show that

$$(\mathbf{n} \cdot \mathbf{S})^{2s+1} = -\sum_{m=0}^{2s} t(2 + 2s, 1 + m)(\mathbf{n} \cdot \mathbf{S})^m$$

where the expansion coefficients $t(n, k)$ are central factorial numbers. This means that all powers of $(\mathbf{n} \cdot \mathbf{S})$ higher than $2s$ can be reduced to linear combinations of lower powers.

Problem 11.   Let $z \in \mathbb{C}$. Consider the spin matrices S_1, S_2, S_3 for spin $s = \frac{1}{2}, 1, \frac{3}{2}, 2, \ldots$. Show that

$$e^{zS_1} S_2 e^{-zS_1} = \cosh(z)S_2 + i\sinh(z)S_3$$
$$e^{zS_2} S_3 e^{-zS_2} = \cosh(z)S_3 + i\sinh(z)S_1$$
$$e^{zS_3} S_1 e^{-zS_3} = \cosh(z)S_1 + i\sinh(z)S_2.$$

Chapter 27

Bose and Fermi Systems

Elementary particles can be classified in two categories according to their value of the spin s. Half-integer spin particles (such as the electron and positron) obey Fermi-Dirac statistics and are called Fermions whereas integer-spin particles (such as the photon and graviton) obey Bose-Einstein statistics and are known as Bosons.

For *Fermi operators* we have to take into account the *Pauli principle*. Let $c_j^\dagger$ be Fermi creation operators and let c_j be Fermi annihilation operators, where $j = 1, \ldots, N$, with j denoting the quantum numbers (spin, wave vector, angular momentum, lattice site, etc.). Then we have

$$[c_i^\dagger, c_j]_+ \equiv c_i^\dagger c_j + c_j c_i^\dagger = \delta_{ij} I$$

$$[c_i, c_j]_+ = [c_i^\dagger, c_j^\dagger]_+ = 0$$

where δ_{ij} denotes the Kronecker delta and I is the identity operator. A consequence of these equations is

$$c_j^\dagger c_j^\dagger = 0$$

which describes the Pauli principle, i.e. two particles cannot be in the same state. The states are given by

$$|n_1, n_2, \ldots, n_N\rangle := (c_1^\dagger)^{n_1} (c_2^\dagger)^{n_2} \cdots (c_N^\dagger)^{n_N} |0, \ldots, 0, \ldots, 0\rangle$$

where, because of the Pauli principle, n_1, n_2, $\ldots$, $n_N \in \{0, 1\}$. We obtain

$$c_i |n_1, \ldots, n_{i-1}, 0, n_{i+1}, \ldots, n_N\rangle = 0$$

and
$$c_i^\dagger |n_1, \ldots, n_{i-1}, 1, n_{i+1}, \ldots, n_N\rangle = 0.$$

In the following we write $|\mathbf{0}\rangle \equiv |0, \ldots, 0\rangle \equiv |0\rangle \otimes \cdots \otimes |0\rangle$.

Consider a family of linear operators b_j, $b_j^\dagger$, $j = 1, 2, \ldots, m$ on an inner product space V, satisfying the commutation relations (Heisenberg algebra)

$$[b_j, b_k] = [b_j^\dagger, b_k^\dagger] = 0, \qquad [b_j, b_k^\dagger] = \delta_{jk} I$$

where I is the identity operator. The operator $b_j^\dagger$ is called a Bose creation operator with mode j and the operator b_j is called a Bose annihilation operator with mode j. The inner product space must be infinite-dimensional for these equations to hold. For, if A and B are $n \times n$ matrices such that

$$[A, B] = \lambda I$$

then $\text{tr}([A, B]) = 0$ implies $\lambda = 0$. Let

$$|\mathbf{0}\rangle = |00 \ldots 0\rangle$$

be the *vacuum state*, i.e.

$$b_j |\mathbf{0}\rangle = 0|00 \ldots 0\rangle$$

$\langle \mathbf{0}|\mathbf{0}\rangle = 1$, $j = 1, 2, \ldots, m$. Now a *number state* (also called Fock state) is given by

$$(b_{j_1}^\dagger)^{n_1} (b_{j_2}^\dagger)^{n_2} \cdots (b_{j_k}^\dagger)^{n_k} |\mathbf{0}\rangle$$

where $j_1, j_2, \ldots, j_k \in \{1, 2, \ldots, m\}$ and $n_1, n_2, \ldots, n_k \in \{0, 1, 2, \ldots\}$. For b_j we also use the notation

$$I \otimes I \otimes \cdots \otimes I \otimes b \otimes I \otimes \cdots \otimes I$$

where b is in the jth position. A property of bosonic coherent states $|\beta\rangle$ ($\beta \in \mathbb{C}$) is that they can closely mimic the behaviour of classical states in the classical counterpart of a quantum system.

Coherent states $|\beta\rangle$ are defined by

$$|\beta\rangle = e^{-|\beta|^2/2} e^{\beta b^\dagger} |0\rangle \equiv e^{-|\beta|^2/2} \sum_{n=0}^{\infty} \frac{\beta^n}{\sqrt{n!}} |n\rangle$$

where $\beta \in \mathbb{C}$ and $b|\beta\rangle = \beta|\beta\rangle$. *Squeezed states* are given by

$$|\zeta\rangle = S(\zeta)|0\rangle, \quad S(\zeta) = \exp\left(-\frac{\zeta}{2}(b^\dagger)^2 + \frac{\zeta^*}{2} b^2\right)$$

where $\zeta \in \mathbb{C}$ and $S(\zeta)$ is the squeezing operator.

Problem 1. (i) Let $N = 1$ and $\hat{n} = c^\dagger c$. Find the matrix representation of the operators $c^\dagger$, c, $\hat{n}$ and the states $|0\rangle$, $c^\dagger|0\rangle$, where $c|0\rangle = 0|0\rangle$.
(ii) Find the matrix representations for arbitrary N.

Solution 1. (i) A basis is given by $\{\, c^\dagger|0\rangle,\ |0\rangle \,\}$. Then the dual basis is $\{\, \langle 0|c,\ \langle 0| \,\}$. We obtain

$$c^\dagger \to \begin{pmatrix} \langle 0|cc^\dagger c^\dagger|0\rangle & \langle 0|cc^\dagger|0\rangle \\ \langle 0|c^\dagger c^\dagger|0\rangle & \langle 0|c^\dagger|0\rangle \end{pmatrix}.$$

Notice that $c|0\rangle = 0 \iff \langle 0|c^\dagger = 0$. Therefore

$$c^\dagger \to \begin{pmatrix} 0 & 1 \\ 0 & 0 \end{pmatrix} = \frac{1}{2}\sigma_+ = \frac{1}{2}(\sigma_1 + i\sigma_2).$$

In an analogous manner we find

$$c \to \begin{pmatrix} 0 & 0 \\ 1 & 0 \end{pmatrix} = \frac{1}{2}\sigma_- \equiv \frac{1}{2}(\sigma_1 - i\sigma_2)$$

and

$$c^\dagger c = \hat{n} \to \begin{pmatrix} 1 & 0 \\ 0 & 0 \end{pmatrix}.$$

The eigenvalues of $\hat{n}$ are $\{0, 1\}$. Next we give the matrix representations of $|0\rangle$ and $c^\dagger|0\rangle$. We find

$$|0\rangle \to \begin{pmatrix} \langle 0|c|0\rangle \\ \langle 0|0\rangle \end{pmatrix} = \begin{pmatrix} 0 \\ 1 \end{pmatrix}, \qquad c^\dagger|0\rangle \to \begin{pmatrix} \langle 0|cc^\dagger|0\rangle \\ \langle 0|c^\dagger|0\rangle \end{pmatrix} = \begin{pmatrix} 1 \\ 0 \end{pmatrix}.$$

As an example we have

$$c|0\rangle = 0 \iff \begin{pmatrix} 0 & 0 \\ 1 & 0 \end{pmatrix}\begin{pmatrix} 0 \\ 1 \end{pmatrix} = \begin{pmatrix} 0 \\ 0 \end{pmatrix}, \qquad c^\dagger|0\rangle \iff \begin{pmatrix} 0 & 1 \\ 0 & 0 \end{pmatrix}\begin{pmatrix} 0 \\ 1 \end{pmatrix} = \begin{pmatrix} 1 \\ 0 \end{pmatrix}.$$

(ii) For arbitrary N we have

$$c_k^\dagger \to \overbrace{\sigma_3 \otimes \cdots \otimes \sigma_3 \otimes \left(\frac{1}{2}\sigma_+\right) \otimes I_2 \otimes \cdots \otimes I_2}^{N\times}$$

where $\sigma_+/2$ is at the k-th place and

$$c_k \to \overbrace{\sigma_3 \otimes \cdots \otimes \sigma_3 \otimes \left(\frac{1}{2}\sigma_-\right) \otimes I_2 \otimes \cdots \otimes I_2}^{N\times}$$

where $\sigma_-/2$ is at the k-th place. Recall that

$$[c_k^\dagger, c_q]_+ = \delta_{kq} I, \qquad [c_k^\dagger, c_q^\dagger]_+ = [c_k, c_q]_+ = 0.$$

The commutation relations are satisfied by the faithful representations. For $N = 1$ the state $|0\rangle$ is given by

$$|0\rangle \to \begin{pmatrix} 0 \\ 1 \end{pmatrix}.$$

For arbitrary N the state $|0\rangle \equiv |0, \ldots, 0\rangle$ is given by

$$|0\rangle = \underbrace{\begin{pmatrix} 0 \\ 1 \end{pmatrix} \otimes \begin{pmatrix} 0 \\ 1 \end{pmatrix} \otimes \begin{pmatrix} 0 \\ 1 \end{pmatrix} \otimes \cdots \otimes \begin{pmatrix} 0 \\ 1 \end{pmatrix}}_{N\times}$$

where $|0\rangle$ is called the *vacuum state*. The operator $\hat{n}_k$ takes the form

$$\hat{n}_k := c_k^\dagger c_k \to I_2 \otimes I_2 \otimes \cdots I_2 \otimes \left(\frac{1}{4}\sigma_+\sigma_-\right) \otimes I_2 \otimes \cdots \otimes I_2.$$

Problem 2.   (i) Let $c^\dagger$ be a Fermi creation operator and c a Fermi annihilation operator. Let $\epsilon \in \mathbb{R}$. Show that

$$\exp(\epsilon c^\dagger c)c^\dagger \exp(-\epsilon c^\dagger c) = \exp(\epsilon)c^\dagger, \qquad \exp(\epsilon c^\dagger c)c\exp(-\epsilon c^\dagger c) = \exp(-\epsilon)c.$$

(ii) Let $\hat{n} := c^\dagger c$. Show that $\exp(-\epsilon\hat{n}) \equiv I + \hat{n}(e^{-\epsilon} - 1)$, where I is the identity operator.
(iii) Let

$$\hat{n}_\uparrow \hat{n}_\downarrow := c_\uparrow^\dagger c_\uparrow c_\downarrow^\dagger c_\downarrow.$$

Show that $\exp(-\epsilon\hat{n}_\downarrow\hat{n}_\uparrow) \equiv I + \hat{n}_\uparrow\hat{n}_\downarrow(e^{-\epsilon} - 1)$.

Solution 2.   (i) We set

$$f(\epsilon) := \exp(\epsilon c^\dagger c)c^\dagger \exp(-\epsilon c^\dagger c)$$

and therefore $f(0) = c^\dagger$. Differentiating both sides with respect to ϵ yields

$$\frac{df}{d\epsilon} = \exp(\epsilon c^\dagger c)c^\dagger cc^\dagger \exp(-\epsilon c^\dagger c) - \exp(\epsilon c^\dagger c)c^\dagger c^\dagger c\exp(-\epsilon c^\dagger c).$$

Since $c^\dagger c^\dagger = 0$ and $cc^\dagger = I - c^\dagger c$ we obtain

$$\frac{df}{d\epsilon} = \exp(\epsilon c^\dagger c)c^\dagger \exp(-\epsilon c^\dagger c) = f$$

or $df/d\epsilon = f$. The solution of this linear differential equation with the initial condition $f(0) = c^\dagger$ gives $f(\epsilon) = c^\dagger e^\epsilon$. Analogously, we can prove the second identity.

(ii) We find the identity $\hat{n}^2 \equiv c^\dagger cc^\dagger c \equiv c^\dagger (I - c^\dagger c)c \equiv c^\dagger c \equiv \hat{n}$ since $cc = 0$. Using this identity and

$$\exp(-\epsilon\hat{n}) := \sum_{k=0}^{\infty} \frac{(-\epsilon)^k \hat{n}^k}{k!}$$

we find the identity.

(iii) Applying the identity $(\hat{n}_\uparrow \hat{n}_\downarrow)^2 \equiv \hat{n}_\uparrow \hat{n}_\downarrow \hat{n}_\uparrow \hat{n}_\downarrow \equiv \hat{n}_\uparrow \hat{n}_\uparrow \hat{n}_\downarrow \hat{n}_\downarrow \equiv \hat{n}_\uparrow \hat{n}_\downarrow$ we find

$$\exp(-\epsilon\hat{n}_\uparrow \hat{n}_\downarrow) := \sum_{k=0}^{\infty} \frac{(-\epsilon)^k (\hat{n}_\uparrow \hat{n}_\downarrow)^k}{k!}.$$

Problem 3. Let $\gamma \in \mathbb{C}$ and $c^\dagger$, c be Fermi creation and annihilation operators.

(i) Calculate $(\gamma c^\dagger - \gamma^* c)^2$.

(ii) Find the operator $\exp(\gamma c^\dagger - \gamma^* c)$.

(iii) Find the state $\exp(\gamma c^\dagger - \gamma^* c)|0\rangle$.

Solution 3. (i) Since $c^\dagger c^\dagger = 0$ and $cc = 0$ we have

$$(\gamma c^\dagger - \gamma^* c)(\gamma c^\dagger - \gamma^* c) = \gamma^2 c^\dagger c^\dagger - \gamma\gamma^* cc^\dagger - \gamma\gamma^* c^\dagger c + \gamma^* \gamma^* cc$$
$$= -\gamma\gamma^* (c^\dagger c + cc^\dagger) = -\gamma\gamma^* I.$$

(ii) Using the result from (i) we find

$$\exp(\gamma c^\dagger - \gamma^* c) = I \cos(\sqrt{\gamma\gamma^*}) + \frac{\gamma c^\dagger - \gamma^* c}{\sqrt{\gamma\gamma^*}} \sin(\sqrt{\gamma\gamma^*}).$$

Note that if $\gamma\gamma^* = 0$, then $\exp(\gamma c^\dagger - \gamma^* c) = I$.

(iii) Using the result from (ii) and $c|0\rangle = 0$ we find

$$\exp(\gamma c^\dagger - \gamma^* c)|0\rangle = \cos(\sqrt{\gamma\gamma^*})|0\rangle + \frac{\gamma}{\sqrt{\gamma\gamma^*}} \sin(\sqrt{\gamma\gamma^*})c^\dagger|0\rangle.$$

Problem 4. Fermi creation and annihilation operators $c_j^\dagger$, c_j obey the anticommutation relations $[c_j, c_k]_+ = 0$, $[c_j^\dagger, c_k]_+ = \delta_{jk} I$, where I is the identity operator and $j, k = 0, 1, \ldots, N - 1$. A basis for the Hilbert space is given by

$$\prod_{j=0}^{N-1} (c_j^\dagger)^{r_j}|0\rangle, \qquad r_j = 0, 1$$

and the vacuum state $|\mathbf{0}\rangle$ is defined by $c_j|\mathbf{0}\rangle = 0|\mathbf{0}\rangle$, $j = 0, 1, \ldots, N-1$. Consider the operators

$$\rho_j = \frac{1-\nu_j}{2} c_j^\dagger c_j + \frac{1+\nu_j}{2} c_j c_j^\dagger, \qquad \nu_j \in [-1, 1].$$

Show that the ρ_j's are density matrices. Use the matrix representation for $c_j^\dagger$ and c_j

$$c_j^\dagger = \sigma_3 \otimes \sigma_3 \otimes \cdots \otimes \sigma_3 \otimes \left(\frac{1}{2}\sigma_+\right) \otimes I_2 \otimes I_2 \otimes \cdots \otimes I_2$$

$$c_j = \sigma_3 \otimes \sigma_3 \otimes \cdots \otimes \sigma_3 \otimes \left(\frac{1}{2}\sigma_-\right) \otimes I_2 \otimes I_2 \otimes \cdots \otimes I_2$$

where σ_+ and σ_- are at the j-th position ($j = 0, 1, \ldots, N-1$) and

$$\sigma_+ := \sigma_1 + i\sigma_2 = \begin{pmatrix} 0 & 2 \\ 0 & 0 \end{pmatrix}, \qquad \sigma_- := \sigma_1 - i\sigma_2 = \begin{pmatrix} 0 & 0 \\ 2 & 0 \end{pmatrix}.$$

The vacuum state is $|\mathbf{0}\rangle = |0\rangle \otimes |0\rangle \otimes \cdots \otimes |0\rangle$ with $|0\rangle = \begin{pmatrix} 0 \\ 1 \end{pmatrix}$.

Solution 4.   Since $\sigma_3^2 = I_2$ we have

$$c_j^\dagger c_j = I_2 \otimes I_2 \otimes \cdots \otimes I_2 \otimes \begin{pmatrix} 1 & 0 \\ 0 & 0 \end{pmatrix} \otimes I_2 \otimes \cdots \otimes I_2$$

and

$$c_j c_j^\dagger = I_2 \otimes I_2 \otimes \cdots \otimes I_2 \otimes \begin{pmatrix} 0 & 0 \\ 0 & 1 \end{pmatrix} \otimes I_2 \otimes \cdots \otimes I_2.$$

Thus $c_j^\dagger c_j$ and $c_j c_j^\dagger$ are diagonal matrices and the eigenvalues of $c_j^\dagger c_j$ and $c_j c_j^\dagger$ are 1 and 0. Since $\nu_j \in [-1, 1]$ it follows that $(1-\nu_j)/2$ and $(1+\nu_j)/2$ cannot be negative. Thus the eigenvalues of ρ_j cannot be negative. Furthermore $\text{tr}(\rho_j) = 1$.

Problem 5.   Consider the Fermi creation and annihilation operators $c_j^\dagger$, $c_\ell^\dagger$, c_k, c_m, where $j, \ell, k, m = 1, 2, \ldots, N$. Find the commutator

$$[c_j^\dagger c_k, c_\ell^\dagger c_m].$$

Does the set $\{c_j^\dagger c_k\}$ ($j, k = 1, 2, \ldots, N$) forms a Lie algebra under the commutator?

Solution 5.   We have $[c_j^\dagger c_k, c_\ell^\dagger c_m] = -\delta_{mj} c_\ell^\dagger c_k + \delta_{\ell k} c_j^\dagger c_m$. Thus the set $\{c_j^\dagger c_k\}$ ($j, k = 1, 2, \ldots, N$) forms a Lie algebra under the commutator.

Problem 6. Let $c_j^\dagger$, c_j ($j = 1, 2, 3$) be Fermi creation and annihilation operators. Consider the Hamilton operator

$$\hat{H} = t(c_1^\dagger c_2 + c_2^\dagger c_1 + c_2^\dagger c_3 + c_3^\dagger c_2 + c_1^\dagger c_3 + c_3^\dagger c_1) + k_1 c_1^\dagger c_1 + k_2 c_2^\dagger c_2 + k_3 c_3^\dagger c_3$$

and the *number operator* $\hat{N} = c_1^\dagger c_1 + c_2^\dagger c_2 + c_3^\dagger c_3$.
(i) Calculate the commutator $[\hat{H}, \hat{N}]$ and give an interpretation of the result.
(ii) Given a basis with two Fermi particles $c_1^\dagger c_2^\dagger |0\rangle$, $c_1^\dagger c_3^\dagger |0\rangle$, $c_2^\dagger c_3^\dagger |0\rangle$. Find the matrix representation of $\hat{H}$ and $\hat{N}$.
(iii) Given a basis with one Fermi particle $c_1^\dagger |0\rangle$, $c_2^\dagger |0\rangle$, $c_3^\dagger |0\rangle$. Find the matrix representation of $\hat{H}$.

Solution 6. (i) We have

$$[\hat{H}, c_1^\dagger c_1] = t(c_2^\dagger c_1 - c_1^\dagger c_2 + c_3^\dagger c_1 - c_1^\dagger c_3)$$
$$[\hat{H}, c_2^\dagger c_2] = t(c_3^\dagger c_2 - c_2^\dagger c_3 + c_1^\dagger c_2 - c_2^\dagger c_1)$$
$$[\hat{H}, c_3^\dagger c_3] = t(c_1^\dagger c_3 - c_3^\dagger c_1 + c_2^\dagger c_3 - c_3^\dagger c_2).$$

It follows that $[\hat{H}, \hat{N}] = 0$. From this result we find $\hat{N}$ is a constant of motion, i.e. the total number of Fermi particles remains constant in the sense that if $|n\rangle$ is an eigenstate of the number operator $\hat{N}$ with eigenvalue n at time 0, then

$$|n\rangle(t) = e^{-i\hat{H}t/\hbar}|n\rangle$$

remains an eigenstate of $\hat{N}$ with eigenvalue n for all times.
(ii) We have

$$\hat{H}c_1^\dagger c_2^\dagger |0\rangle = t(c_1^\dagger c_3^\dagger |0\rangle - c_2^\dagger c_3^\dagger |0\rangle) + (k_1 + k_2)c_1^\dagger c_2^\dagger |0\rangle$$
$$\hat{H}c_1^\dagger c_3^\dagger |0\rangle = t(c_2^\dagger c_3^\dagger |0\rangle + c_1^\dagger c_2^\dagger |0\rangle) + (k_1 + k_3)c_1^\dagger c_3^\dagger |0\rangle$$
$$\hat{H}c_2^\dagger c_3^\dagger |0\rangle = t(c_1^\dagger c_3^\dagger |0\rangle - c_1^\dagger c_2^\dagger |0\rangle) + (k_2 + k_3)c_2^\dagger c_3^\dagger |0\rangle.$$

With the dual basis $\langle 0|c_2 c_1$, $\langle 0|c_3 c_1$, $\langle 0|c_3 c_2$ the matrix representation of $\hat{H}$ is

$$\hat{H} = \begin{pmatrix} k_1 + k_2 & t & -t \\ t & k_1 + k_3 & t \\ -t & t & k_2 + k_3 \end{pmatrix}.$$

Obviously owing to the result of (i) the matrix representation of $\hat{N}$ is

$$\hat{N} = \begin{pmatrix} 2 & 0 & 0 \\ 0 & 2 & 0 \\ 0 & 0 & 2 \end{pmatrix}.$$

(iii) Since

$$\hat{H}c_1^\dagger|0\rangle = t(c_2^\dagger|0\rangle + c_3^\dagger|0\rangle) + k_1 c_1^\dagger|0\rangle$$
$$\hat{H}c_2^\dagger|0\rangle = t(c_1^\dagger|0\rangle + c_3^\dagger|0\rangle) + k_2 c_2^\dagger|0\rangle$$
$$\hat{H}c_3^\dagger|0\rangle = t(c_1^\dagger|0\rangle + c_2^\dagger|0\rangle) + k_3 c_3^\dagger|0\rangle$$

we obtain the matrix representation

$$\hat{H} = \begin{pmatrix} k_1 & t & t \\ t & k_2 & t \\ t & t & k_3 \end{pmatrix}.$$

Problem 7. Consider the Fermi operator

$$\left\{ c_{k\sigma}^\dagger, \ c_{k\sigma}, \ k = 1, 2, \cdots, N; \ \sigma \in \{\uparrow, \downarrow\} \right\}$$

where k denotes the wave vector (or lattice site) and σ denotes the spin. Find the matrix representation of $c_{k\uparrow}^\dagger$ and $c_{k\downarrow}^\dagger$. We have the commutation relation $[c_{k\sigma}^\dagger, c_{q\sigma'}]_+ = \delta_{kq}\delta_{\sigma\sigma'} I$, $[c_{k\sigma}, c_{q\sigma'}]_+ = [c_{k\sigma}^\dagger, c_{q\sigma'}^\dagger]_+ = 0$.

Solution 7. The matrix representation is given by

$$c_{k\uparrow}^\dagger \to \overbrace{\sigma_3 \otimes \sigma_3 \otimes \cdots \otimes \sigma_3 \otimes \left(\frac{1}{2}\sigma_+\right) \otimes I_2 \otimes \cdots \otimes I_2}^{2N\times}$$

where $\sigma_+/2$ is at the k-th place and

$$c_{k\downarrow}^\dagger \to \overbrace{\sigma_3 \otimes \sigma_3 \otimes \cdots \otimes \sigma_3 \otimes \left(\frac{1}{2}\sigma_+\right) \otimes I_2 \otimes \cdots \otimes I_2}^{2N\times}$$

where $\sigma_+/2$ is at the $(k + N)$-th place.

Problem 8. Let $c_{j\uparrow}^\dagger, c_{j\uparrow}, c_{j\downarrow}^\dagger, c_{j\downarrow}$ be Fermi operators, i.e.

$$[c_{j\sigma}^\dagger, c_{k\sigma'}]_+ = \delta_{jk}\delta_{\sigma,\sigma'} I, \quad [c_{j\sigma}^\dagger, c_{k\sigma'}^\dagger]_+ = 0, \quad [c_{j\sigma}, c_{k\sigma'}]_+ = 0 \qquad (1)$$

for all $j, k = 1, 2, \ldots, N$. We define the linear operators

$$\xi_{j\uparrow} := i(c_{j\uparrow} - c_{j\uparrow}^\dagger), \qquad \xi_{j\downarrow} := i(c_{j\downarrow} - c_{j\downarrow}^\dagger)(I - 2c_{j\uparrow}^\dagger c_{j\uparrow}) \qquad (2a)$$

$$\eta_{j\uparrow} := c_{j\uparrow} + c_{j\uparrow}^\dagger, \qquad \eta_{j\downarrow} := (c_{j\downarrow} + c_{j\downarrow}^\dagger)(I - 2c_{j\uparrow}^\dagger c_{j\uparrow}). \qquad (2b)$$

(i) Find the properties of these operators.
(ii) Find the inverse transformation.

Solution 8. (i) By straightforward calculations using the anticommutation relations we find that the operators $\xi_{j\sigma}$ are unitary and self-adjoint and therefore involutions

$$\xi_{j\sigma} = \xi_{j\sigma}^\dagger = \xi_{j\sigma}^{-1}, \qquad \xi_{j\sigma}^2 = I$$

for all $j = 1, 2, \ldots, N$. The linear operators $\eta_{j\sigma}$ also satisfy

$$\eta_{j\sigma} = \eta_{j\sigma}^\dagger = \eta_{j\sigma}^{-1}, \qquad \eta_{j\sigma}^2 = I$$

for all $j = 1, 2, \ldots, N$.
(ii) The inverse transformation between $c_{j\sigma}^\dagger, c_{j\sigma}$ and $\xi_{j\sigma}, \eta_{j\sigma}$ is given by

$$c_{j\uparrow} = \frac{1}{2}(\eta_{j\uparrow} - i\xi_{j\uparrow}), \qquad c_{j\uparrow}^\dagger = \frac{1}{2}(\eta_{j\uparrow} + i\xi_{j\uparrow}),$$

$$c_{j\downarrow} = \frac{1}{2}(i\eta_{j\downarrow} + \xi_{j\downarrow})\eta_{j\uparrow}\xi_{j\uparrow}, \qquad c_{j\downarrow}^\dagger = \frac{1}{2}(i\eta_{j\downarrow} - \xi_{j\downarrow})\eta_{j\uparrow}\xi_{j\uparrow}$$

where we have used that $i\eta_{j\uparrow}\xi_{j\uparrow} = I - 2c_{j\uparrow}^\dagger c_{j\uparrow}$.

Problem 9. Let $c_{j\sigma}^\dagger$, $c_{j\sigma}$ ($j = 1, \ldots, N$) be Fermi creation and annihilation operators, where $\sigma \in \{\uparrow, \downarrow\}$. We define $\hat{n}_{j\uparrow} := c_{j\uparrow}^\dagger c_{j\uparrow}$, $\hat{n}_{j\downarrow} := c_{j\downarrow}^\dagger c_{j\downarrow}$.
(i) Show that the operators

$$S_{j1} := \frac{1}{2}(c_{j\uparrow}^\dagger c_{j\downarrow} + c_{j\downarrow}^\dagger c_{j\uparrow}), \quad S_{j2} := \frac{1}{2i}(c_{j\uparrow}^\dagger c_{j\downarrow} - c_{j\downarrow}^\dagger c_{j\uparrow}), \quad S_{j3} := \frac{1}{2}(\hat{n}_{j\uparrow} - \hat{n}_{j\downarrow})$$

form a basis of a Lie algebra under the commutator.
(ii) Show that the operators

$$R_{j1} := \frac{1}{2}(c_{j\uparrow}^\dagger c_{j\downarrow}^\dagger + c_{j\downarrow} c_{j\uparrow}), \quad R_{j2} := \frac{1}{2i}(c_{j\uparrow}^\dagger c_{j\downarrow}^\dagger - c_{j\downarrow} c_{j\uparrow}), \quad R_{j3} := \frac{1}{2}(\hat{n}_{j\uparrow} + \hat{n}_{j\downarrow} - I)$$

form a Lie algebra under the commutator.
(iii) Prove the identity

$$\hat{n}_{j\uparrow}\hat{n}_{j\downarrow} \equiv \frac{1}{4}(1 - \alpha_j) + R_{j3} + \frac{1}{3}(\alpha_j - 1)\mathbf{S}_j^2 + \frac{1}{3}(\alpha_j + 1)\mathbf{R}_j^2$$

where $\mathbf{S}_j^2 := S_{j1}^2 + S_{j2}^2 + S_{j3}^2$, $\mathbf{R}_j^2 := R_{j1}^2 + R_{j2}^2 + R_{j3}^2$.

Solution 9. (i) Since

$$[c_{j\sigma}^\dagger, c_{j\sigma'}]_+ = \delta_{\sigma\sigma'}I, \quad [c_{j\sigma}^\dagger, c_{j\sigma'}^\dagger]_+ = 0, \quad [c_{j\sigma}, c_{j\sigma'}]_+ = 0$$

we obtain the commutators $[S_{j1}, S_{j2}] = iS_{j3}$, $[S_{j2}, S_{j3}] = iS_{j1}$, $[S_{j3}, S_{j1}] = iS_{j2}$. Thus the operators S_{j1}, S_{j2}, S_{j3} form a basis of a Lie algebra.
(ii) Analogously $[R_{j1}, R_{j2}] = iR_{j3}$, $[R_{j2}, R_{j3}] = iR_{j1}$, $[R_{j3}, R_{j1}] = iR_{j2}$. Thus the operators R_{j1}, R_{j2}, R_{j3} form a Lie algebra. The two Lie algebras are isomorphic.
(iii) We find

$$\mathbf{S}_j^2 = \frac{3}{4}(\hat{n}_{j\uparrow} + \hat{n}_{j\downarrow} - 2\hat{n}_{j\uparrow}\hat{n}_{j\downarrow}), \quad \mathbf{R}_j^2 = \frac{3}{2}\hat{n}_{j\uparrow}\hat{n}_{j\downarrow} - \frac{3}{4}(\hat{n}_{j\uparrow} + \hat{n}_{j\downarrow}) + \frac{3}{4}I.$$

Thus identity (1) follows.

Problem 10. Let $n_{j\uparrow} := c_{j\uparrow}^\dagger c_{j\uparrow}$, $n_{j\downarrow} := c_{j\downarrow}^\dagger c_{j\downarrow}$. Consider the Hamilton operator (*Hubbard model*)

$$\hat{H} = t \sum_{i,j} \sum_{\sigma \in \{\uparrow\downarrow\}} c_{i\sigma}^\dagger c_{j\sigma} + U \sum_{j=1}^{N} n_{j\uparrow} n_{j\downarrow}$$

where t and U is a real constant. Consider the operators

$$\hat{N}_e = \sum_{j} \sum_{\sigma} c_{j\sigma}^\dagger c_{j\sigma}, \quad \hat{S}_z = \frac{1}{2} \sum_{j}^{N} (c_{j\uparrow}^\dagger c_{j\uparrow} - c_{j\downarrow}^\dagger c_{j\downarrow})$$

where $\hat{N}_e$ is the number operator and $\hat{S}_z$ is the total spin operator in z-direction. Calculate the commutators $[\hat{H}, \hat{N}_e]$ and $[\hat{H}, \hat{S}_z]$.

Solution 10. We find $[\hat{H}, \hat{N}_e] = 0$, $[\hat{H}, \hat{S}_z] = 0$. Thus $\hat{N}_e$ and $\hat{S}_z$ are constants of motion. Thus we can consider subspaces with fixed number of electrons and fixed number of total spin in z-direction.

Problem 11. Let $c_{k\sigma}^\dagger$ and $c_{k\sigma}$ be Fermi creation and annihilation operators, where $\sigma \in \{\uparrow, \downarrow\}$ and $j = 1, 2, \ldots, N$ and k refers to the momentum. The Fermi operator obeys the anticommutation relations

$$[c_{k\sigma}, c_{p\sigma'}^\dagger]_+ = \delta_{kp}\delta_{\sigma\sigma'}I, \quad [c_{k\sigma}, c_{k\sigma'}]_+ = 0$$

where I is the identity operator and 0 the zero operator. From the second relation it follows that $[c_{k\sigma}^\dagger, c_{p\sigma'}^\dagger]_+ = 0$. Let $n_{j\uparrow} := c_{k\uparrow}^\dagger c_{k\uparrow}$, $n_{j\downarrow} := c_{k\downarrow}^\dagger c_{k\downarrow}$. Consider the four point Hamilton operator (*Hubbard model*) in *Bloch representation*

$$\hat{H} = \sum_{k \in S} \sum_{\sigma \in \{\uparrow\downarrow\}} \epsilon(k) c_{k\sigma}^\dagger c_{k\sigma} + \frac{U}{4} \sum_{k_1, k_2, k_3, k_4 \in S} \delta(k_1 - k_2 + k_3 - k_4) c_{k_1\uparrow}^\dagger c_{k_2\uparrow} c_{k_3\downarrow}^\dagger c_{k_4\downarrow}$$

where $\epsilon(k) = 2t\cos(k)$. Here

$$k, k_1, k_2, k_3 \in S = \left\{ -\frac{\pi}{2},\ 0,\ \frac{\pi}{2},\ \pi \mod 2\pi \right\}$$

and

$$\delta(k_1 - k_2 + k_3 - k_4) = \begin{cases} 1 & \text{if} \quad k_1 - k_2 + k_3 - k_4 = 0 \mod 2\pi \\ 0 & \text{otherwise} \end{cases}.$$

Consider the three operators

$$\hat{N}_e = \sum_k \sum_\sigma c^\dagger_{k\sigma} c_{j\sigma}, \quad \hat{S}_z = \frac{1}{2} \sum_k (c^\dagger_{k\uparrow} c_{k\uparrow} - c^\dagger_{k\downarrow} c_{k\downarrow}),$$

$$\hat{P} = \sum_k k(c^\dagger_{k\uparrow} c_{k\uparrow} + c^\dagger_{k\downarrow} c_{k\downarrow})$$

where $\hat{N}_e$ is the number operator, $\hat{S}_z$ is the total spin operator in z-direction and $\hat{P}$ the total momentum operator and k runs over the set S given above.
(i) Calculate the commutators $[\hat{H}, \hat{N}_e]$, $[\hat{H}, \hat{S}_z]$, $[\hat{H}, \hat{P}]$.
(ii) Find the eigenvalues of the total momentum operator $\hat{P}$.

Solution 11. (i) We find $[\hat{H}, \hat{N}_e] = 0$, $[\hat{H}, \hat{S}_z] = 0$, $[\hat{H}, \hat{P}] = 0$. Thus $\hat{N}_e$, $\hat{S}_z$ and $\hat{P}$ are constants of motion. Thus we can consider subspaces with fixed number of electrons, fixed number of total spin in z-direction and fixed total momentum.
(ii) Obviously we find $-\pi/2,\ 0,\ \pi/2,\ \pi$.

Problem 12. Consider the four point Hubbard model described in Bloch representation

$$\hat{H} = \sum_k \sum_{\sigma \in \{\uparrow\downarrow\}} \epsilon(k) c^\dagger_{k\sigma} c_{k\sigma} + \frac{U}{4} \sum_{k_1, k_2, k_3, k_4} \delta(k_1 - k_2 + k_3 - k_4) c^\dagger_{k_1\uparrow} c_{k_2\uparrow} c^\dagger_{k_3\downarrow} c_{k_4\downarrow}$$

where $\epsilon(k) = 2t\cos(k)$. Here

$$k, k_1, k_2, k_3 \in \left\{ -\frac{\pi}{2},\ 0,\ \frac{\pi}{2},\ \pi \mod 2\pi \right\}$$

Give a basis of the Hilbert space for $N_e = 4$, $S_z = 0$ and $P = 0$.

Solution 12. Owing to $N_e = 4$, $S_z = 0$ and the Pauli principle we have the states

$$c^\dagger_{k_1\uparrow} c^\dagger_{k_2\uparrow} c^\dagger_{k_3\downarrow} c^\dagger_{k_4\downarrow} |0\rangle, \quad k_1 < k_2,\ k_3 < k_4,\ k_1 + k_2 + k_3 + k_4 = 0 \mod 2\pi \}.$$

The dimensions of the subspaces with $P = -\pi/2, 0, \pi/2, \pi$ are $8, 10, 8, 10$. For $P = 0$ we obtain the ten states

$$c^\dagger_{-\pi/2\uparrow}c^\dagger_{0\uparrow}c^\dagger_{-\pi/2\downarrow}c^\dagger_{\pi\downarrow}|0\rangle, \quad c^\dagger_{-\pi/2\uparrow}c^\dagger_{0\uparrow}c^\dagger_{0\downarrow}c^\dagger_{\pi/2\downarrow}|0\rangle$$

$$c^\dagger_{-\pi/2\uparrow}c^\dagger_{\pi/2\uparrow}c^\dagger_{-\pi/2\downarrow}c^\dagger_{\pi/2\downarrow}|0\rangle, \quad c^\dagger_{-\pi/2\uparrow}c^\dagger_{\pi\uparrow}c^\dagger_{-\pi/2\downarrow}c^\dagger_{0\downarrow}|0\rangle$$

$$c^\dagger_{-\pi/2\uparrow}c^\dagger_{\pi\uparrow}c^\dagger_{\pi/2\downarrow}c^\dagger_{\pi\downarrow}|0\rangle, \quad c^\dagger_{0\uparrow}c^\dagger_{\pi/2\uparrow}c^\dagger_{-\pi/2\downarrow}c^\dagger_{0\downarrow}|0\rangle$$

$$c^\dagger_{0\uparrow}c^\dagger_{\pi/2\uparrow}c^\dagger_{\pi/2\downarrow}c^\dagger_{\pi\downarrow}|0\rangle, \quad c^\dagger_{0\uparrow}c^\dagger_{\pi\uparrow}c^\dagger_{0\downarrow}c^\dagger_{\pi\downarrow}|0\rangle$$

$$c^\dagger_{\pi/2\uparrow}c^\dagger_{\pi\uparrow}c^\dagger_{-\pi/2\downarrow}c^\dagger_{\pi\downarrow}|0\rangle, \quad c^\dagger_{\pi/2\uparrow}c^\dagger_{\pi\uparrow}c^\dagger_{0\downarrow}c^\dagger_{\pi/2\downarrow}|0\rangle.$$

Problem 13.   Consider the Hamilton operator (two-point *Hubbard model*)

$$\hat{H} = t(c^\dagger_{1\uparrow}c_{2\uparrow} + c^\dagger_{1\downarrow}c_{2\downarrow} + c^\dagger_{2\uparrow}c_{1\uparrow} + c^\dagger_{2\downarrow}c_{1\downarrow}) + U(n_{1\uparrow}n_{1\downarrow} + n_{2\uparrow}n_{2\downarrow}) \quad (1)$$

where $n_{j\uparrow} := c^\dagger_{j\uparrow}c_{j\uparrow}$, $n_{j\downarrow} := c^\dagger_{j\downarrow}c_{j\downarrow}$. The operators $c^\dagger_{j\uparrow}, c^\dagger_{j\downarrow}, c_{j\uparrow}, c_{j\downarrow}$ are Fermi operators.
(i) Show that the Hubbard Hamilton operator (1) commutes with the total number operator $\hat{N}$ and the total spin operator $\hat{S}_z$ where

$$\hat{N} := \sum_{j=1}^{2}(c^\dagger_{j\uparrow}c_{j\uparrow} + c^\dagger_{j\downarrow}c_{j\downarrow}), \qquad \hat{S}_z := \frac{1}{2}\sum_{j=1}^{2}(c^\dagger_{j\uparrow}c_{j\uparrow} - c^\dagger_{j\downarrow}c_{j\downarrow}).$$

(ii) We consider the subspace with two particles $N = 2$ and total spin $S_z = 0$. A basis in this space is given by

$$c^\dagger_{1\uparrow}c^\dagger_{1\downarrow}|0\rangle, \quad c^\dagger_{1\uparrow}c^\dagger_{2\downarrow}|0\rangle, \quad c^\dagger_{2\uparrow}c^\dagger_{1\downarrow}|0\rangle, \quad c^\dagger_{2\uparrow}c^\dagger_{2\downarrow}|0\rangle.$$

Find the matrix representation of $\hat{H}$ for this basis.
(iii) Find the discrete symmetries of $\hat{H}$ and perform a group-theoretical reduction.

Solution 13.   (i) Using the Fermi anti-commutation relations we obtain $[\hat{H}, \hat{N}] = 0$, $[\hat{H}, \hat{S}_z] = 0$. We also have $[\hat{N}, \hat{S}_z] = 0$.
(ii) Using the Fermi anti-commutation relations and $c_{j\uparrow}|0\rangle = 0|0\rangle$, $c_{j\downarrow}|0\rangle = 0|0\rangle$ we obtain the matrix representation of $\hat{H}$ with the given basis

$$\begin{pmatrix} U & t & t & 0 \\ t & 0 & 0 & t \\ t & 0 & 0 & t \\ 0 & t & t & U \end{pmatrix}.$$

(iii) The Hamilton operator (1) admits the symmetry $1 \to 2$, $2 \to 1$, i.e. swapping the sites 1 and 2 leaves the Hamilton operator invariant. Thus we have a finite group and two elements (identity and the swapping of the sites). There are two conjugacy classes and therefore two irreducible representations. We find the two invariant subspaces

$$\left\{ \frac{1}{\sqrt{2}}(c^\dagger_{1\downarrow}c^\dagger_{1\uparrow}|0\rangle + c^\dagger_{2\downarrow}c^\dagger_{2\uparrow}|0\rangle), \quad \frac{1}{\sqrt{2}}(c^\dagger_{1\downarrow}c^\dagger_{2\uparrow}|0\rangle + c^\dagger_{2\downarrow}c^\dagger_{1\uparrow}|0\rangle) \right\}$$

$$\left\{ \frac{1}{\sqrt{2}}(c^\dagger_{1\downarrow}c^\dagger_{1\uparrow}|0\rangle - c^\dagger_{2\downarrow}c^\dagger_{2\uparrow}|0\rangle), \quad \frac{1}{\sqrt{2}}(c^\dagger_{1\downarrow}c^\dagger_{2\uparrow}|0\rangle - c^\dagger_{2\downarrow}c^\dagger_{1\uparrow}|0\rangle) \right\}.$$

Problem 14. Let $b^\dagger$, b be Bose creation and annihilation operators.
(i) In a *normal ordered product* all annihilation operators are placed to the right of all creation operators. Find the normal ordering of $((b^\dagger)^2 b^2)^2$.
(ii) Find the normal ordering of $(b^2(b^\dagger)^2)^2$.
(iii) Let $b^\dagger$, b be Bose creation and annihilation operators, respectively. Calculate the commutators $[b^2, (b^\dagger)^2]$, $[b^3, (b^\dagger)^3]$. Extend the result to $[b^n, (b^\dagger)^n]$, where $n \in \mathbb{N}$.
(iv) Let $|0\rangle$ be the vacuum state. Calculate $[b^n, (b^\dagger)^n]|0\rangle$.

Solution 14. (i) Using the commutation relation $[b, b^\dagger] = I$ we find

$$((b^\dagger)^2 b^2)^2 = (b^\dagger)^4 b^4 + 4(b^\dagger)^3 b^3 + 2(b^\dagger)^2 b^2.$$

(ii) Using the commutation relation $[b, b^\dagger] = I$ we find

$$(b^2(b^\dagger)^2)^2 = (b^\dagger)^4 b^4 + 12(b^\dagger)^3 b^3 + 38(b^\dagger)^2 b^2 + 32 b^\dagger b + 4I.$$

(iii) Using $bb^\dagger = I + b^\dagger b$ we obtain

$$[b^2, (b^\dagger)^2] = 2I + 4b^\dagger b, \qquad [b^3, (b^\dagger)^3] = 6I + 18 b^\dagger b + 9 b^\dagger b^\dagger bb.$$

To calculate $[b^n, (b^\dagger)^n]$ we can apply the formula

$$[f(b), g(b^\dagger)] = \sum_{j=1}^\infty \frac{\partial^j}{\partial b^{\dagger j}} g(b^\dagger) \frac{\partial^j}{\partial b^j} f(b)$$

where f and g are analytic functions. Thus we obtain

$$[b^n, (b^\dagger)^n] = \sum_{j=1}^n \frac{(n!)^2}{j!((n-j)!)^2} (b^\dagger)^{n-j} b^{n-j}.$$

(iv) Since $b|0\rangle = 0|0\rangle$ we obtain $[b^n, (b^\dagger)^n]|0\rangle = (n!)|0\rangle$.

Problem 15.   (i) Calculate $\text{tr}(b^\dagger b \exp(-\epsilon b^\dagger b))$, where $\epsilon \in \mathbb{R}$ ($\epsilon > 0$) and $b^\dagger b$ is the infinite-dimensional diagonal matrix $b^\dagger b = \text{diag}(0, 1, 2, \ldots)$.
(ii) Calculate $\exp(-\epsilon b^\dagger) b \exp(\epsilon b^\dagger)$.
(iii) Calculate $\exp(-\epsilon b^\dagger b) b \exp(\epsilon b^\dagger b)$.

Solution 15.   (i) Since $\exp(-\epsilon b^\dagger b) = \text{diag}(1, e^{-\epsilon}, e^{-2\epsilon}, \ldots)$ and

$$b^\dagger b \exp(-\epsilon b^\dagger b) = \text{diag}(0, e^{-\epsilon}, 2e^{-2\epsilon}, 3e^{-3\epsilon}, \ldots)$$

we obtain

$$\text{tr}(b^\dagger b \exp(-\epsilon b^\dagger b)) = \sum_{n=1}^{\infty} n e^{-\epsilon n}.$$

Now

$$f(\epsilon) := \sum_{n=1}^{\infty} e^{-\epsilon n} \equiv \frac{1}{e^\epsilon - 1} \qquad \text{(geometric series)}.$$

The derivative of f yields

$$\frac{df}{d\epsilon} = -\sum_{n=1}^{\infty} n e^{-\epsilon n} = \frac{d}{d\epsilon} \frac{1}{e^\epsilon - 1} = \frac{-e^\epsilon}{(e^\epsilon - 1)^2}.$$

Consequently,

$$\text{tr}(b^\dagger b \exp(-\epsilon b^\dagger b)) = \frac{e^\epsilon}{(e^\epsilon - 1)^2}.$$

(ii) We set

$$f(\epsilon) := \exp(-\epsilon b^\dagger) b \exp(\epsilon b^\dagger).$$

We seek the ordinary differential equation for $f(\epsilon)$, where $f(0) = b$. Taking the derivative with respect to ϵ yields

$$\frac{df}{d\epsilon} = e^{-\epsilon b^\dagger}(-b^\dagger) b e^{\epsilon b^\dagger} + e^{-\epsilon b^\dagger} b b^\dagger e^{\epsilon b^\dagger} = e^{-\epsilon b^\dagger}(-b^\dagger b + b b^\dagger) e^{\epsilon b^\dagger} = I$$

where I is the identity operator. From this linear differential equation and the initial condition $f(0) = b$ we find $f(\epsilon) = \exp(-\epsilon b^\dagger) b \exp(\epsilon b^\dagger) = b + \epsilon I$.
(iii) We set $g(\epsilon) := \exp(-\epsilon b^\dagger b) b \exp(\epsilon b^\dagger b)$. We seek the ordinary differential equation for $g(\epsilon)$, where $g(0) = b$. Taking the derivative with respect to ϵ yields

$$\frac{dg}{d\epsilon} = e^{-\epsilon b^\dagger b}(-b^\dagger b) b e^{\epsilon b^\dagger b} + e^{-\epsilon b^\dagger b} b (b^\dagger b) e^{\epsilon b^\dagger b} = e^{-\epsilon b^\dagger b}(-b^\dagger b b + b b^\dagger b) e^{\epsilon b^\dagger b}$$

$$= e^{-\epsilon b^\dagger b} b e^{\epsilon b^\dagger b} = g(\epsilon).$$

From this linear differential equation and the initial condition $g(0) = b$ we obtain $g(\epsilon) = \exp(-\epsilon b^\dagger b) b \exp(\epsilon b^\dagger b) = b \exp(\epsilon)$.

Problem 16. Let b, $b^\dagger$ be Bose annihilation and creation operators. Let $m, n \geq 1$. We have the *ordering formula*

$$b^m (b^\dagger)^n = \sum_{j=0}^{\min\{m,n\}} \binom{m}{j} \frac{n!}{(n-j)!} (b^\dagger)^{n-j} b^{m-j}.$$

(i) Apply it to $b^2 b^\dagger$ and $b(b^\dagger)^2$.

(ii) Calculate the state $b^m (b^\dagger)^n |0\rangle$.

Solution 16. (i) Since $m = 2$, $n = 1$ we have $\min\{m, n\} = 1$. Thus we find

$$b^2 b^\dagger = \sum_{j=0}^{1} \binom{2}{j} \frac{1}{(1-j)!} (b^\dagger)^{1-j} b^{2-j} = b^\dagger b^2 + 2b.$$

For $m = 1$, $n = 2$ we have $\min\{m, n\} = 1$. Thus we find

$$b(b^\dagger)^2 = \sum_{j=0}^{1} \binom{1}{j} \frac{2}{(2-j)!} (b^\dagger)^{2-j} b^{1-j} = (b^\dagger)^2 b + 2b^\dagger.$$

(ii) If $m > n$ we find 0. If $m = n$ we find the state $b^m (b^\dagger)^m |0\rangle = m! |0\rangle$. If $n > m$ we obtain

$$\frac{n!}{(n-m)!} (b^\dagger)^{n-m} |0\rangle.$$

Problem 17. Let b be a Bose annihilation operator. Solve the eigenvalue problem

$$b|\beta\rangle = \beta|\beta\rangle. \tag{1}$$

Hint. Use the *number representation* $|n\rangle$, where

$$|n\rangle := \frac{(b^\dagger)^n}{\sqrt{n!}} |0\rangle \tag{2}$$

and $n = 0, 1, 2, \ldots$. Therefore

$$b|n\rangle = \sqrt{n}|n-1\rangle, \qquad b^\dagger|n\rangle = \sqrt{n+1}|n+1\rangle. \tag{3}$$

The states $|\beta\rangle$ ($\beta \in \mathbb{C}$) are called *Bose coherent states*.

Solution 17. To solve the eigenvalue problem we apply the completeness relation of the number representation. This means we expand $|\beta\rangle$ with respect to $|n\rangle$. We obtain

$$|\beta\rangle = \sum_{n=0}^{\infty} |n\rangle\langle n|\beta\rangle \equiv \sum_{n=0}^{\infty} c_n(\beta)|n\rangle \tag{4}$$

where $c_n(\beta) := \langle n|\beta \rangle$. If we insert (4) into (1) and use (3) we obtain

$$\sum_{n=1}^{\infty} c_n(\beta)\sqrt{n}|n-1\rangle = \sum_{n=0}^{\infty} \beta c_n(\beta)|n\rangle.$$

The first sum goes from 1 to ∞ since the term $n = 0$ vanishes. We can therefore shift indices and put $n \to n + 1$. It follows that

$$\sum_{n=0}^{\infty} c_{n+1}(\beta)\sqrt{n+1}|n\rangle = \sum_{n=0}^{\infty} \beta c_n(\beta)|n\rangle.$$

If we apply the dual state $\langle m|$ and use $\langle m|n\rangle = \delta_{nm}$ we obtain the linear difference equation

$$c_{n+1}(\beta)\sqrt{n+1} = \beta c_n(\beta).$$

On inspection we see that

$$c_1 = \frac{\beta}{\sqrt{1}}c_0, \quad c_2 = \frac{\beta}{\sqrt{2}}c_1 = \frac{\beta^2}{\sqrt{2!}}c_0, \quad c_3 = \frac{\beta^3}{\sqrt{3!}}c_0, \quad \ldots, \quad c_n(\beta) = \frac{\beta^n}{\sqrt{n!}}c_0.$$

Therefore

$$|\beta\rangle = c_0 \sum_{n=0}^{\infty} \frac{\beta^n}{\sqrt{n!}}|n\rangle.$$

We normalize $|\beta\rangle$ to determine c_0. From the condition $\langle\beta|\beta\rangle = 1$ we obtain $1 = |c_0|^2 e^{|\beta|^2}$, where from (1) we have $\langle\beta|b^\dagger = \langle\beta|\bar{\beta}$. It follows that

$$|\beta\rangle = e^{-|\beta|^2/2} \sum_{n=0}^{\infty} \frac{\beta^n}{\sqrt{n!}}|n\rangle \equiv e^{-|\beta|^2/2} e^{\beta b^\dagger}|0\rangle, \quad \beta \in \mathbb{C}.$$

Problem 18. Let $|n\rangle$ with $n = 0, 1, 2, \ldots$ be the number states. Let $|\beta\rangle$, $\beta \in \mathbb{C}$ be coherent states.
(i) Calculate $\langle n|\beta\rangle$ and $\langle\beta|n\rangle \equiv \overline{\langle n|\beta\rangle}$.
(ii) Calculate the distance between the number states and the coherent states $\| |n\rangle - |\beta\rangle \|^2 = (\langle n| - \langle\beta|)(|n\rangle - |\beta\rangle)$. Discuss.

Solution 18. (i) Since

$$|\beta\rangle = \exp\left(-\frac{1}{2}|\beta|^2\right) \sum_{m=0}^{\infty} \frac{\beta^m}{\sqrt{m!}}|m\rangle$$

and $\langle n|m\rangle = \delta_{nm}$ we have

$$\langle n|\beta\rangle = \exp\left(-\frac{1}{2}|\beta|^2\right)\frac{\beta^n}{\sqrt{n!}}, \quad \overline{\langle n|\beta\rangle} = \exp\left(-\frac{1}{2}|\beta|^2\right)\frac{\bar{\beta}^n}{\sqrt{n!}}$$

since $\overline{\beta^n} = \overline{\beta}^n$.

(ii) We have

$$\| \,|n\rangle - |\beta\rangle \,\|^2 = ((\langle n| - \langle\beta|)(|n\rangle - |\beta\rangle)) = 2 - \langle n|\beta\rangle - \langle\beta|n\rangle$$
$$= 2 - \langle n|\beta\rangle - \overline{\langle n|\beta\rangle}.$$

Using (i) we obtain

$$\| \,|n\rangle - |\beta\rangle \,\|^2 = 2 - \frac{e^{-|\beta|^2/2}}{\sqrt{n!}} \left(\beta^n + \overline{\beta}^n \right).$$

If we set $\beta = re^{i\phi}$ with $r \geq 0$, $\phi \in \mathbb{R}$ and apply the identity $2\cos(n\phi) \equiv e^{in\phi} + e^{-in\phi}$ we obtain

$$\| \,|n\rangle - |\beta\rangle \,\|^2 = 2 \left(1 - \frac{e^{-r^2/2}}{\sqrt{n!}} r^n \cos(n\phi) \right).$$

Problem 19. Let $\alpha \in \mathbb{R}$ and $b^\dagger$, b be Bose creation and annihilation operators. Consider the operator

$$\hat{R} = \exp(i\alpha b^\dagger b)$$

Find $\hat{R} b \hat{R}^{-1}$, where $\hat{R}^{-1} = \exp(-i\alpha b^\dagger b)$.

Solution 19. Setting $f(\epsilon) = e^{i\epsilon b^\dagger b} b e^{-i\epsilon b^\dagger b}$ and using differentiation with respect to ϵ provides the differential equation

$$\frac{df(\epsilon)}{d\epsilon} = -if(\epsilon).$$

The solution of the initial value problem ($f(0) = b$) is given by

$$f(\epsilon) = be^{-i\epsilon}.$$

With $\epsilon = \pi$ we have $f(\pi) = -1$.

Problem 20. Let $|\beta\rangle$ be a coherent state. Let $\hat{n} := b^\dagger b$. Calculate

$$\langle\beta|\hat{n}|\beta\rangle.$$

Solution 20. Since $b|\beta\rangle = \beta|\beta\rangle$, $\langle\beta|b^\dagger = \langle\beta|\beta^*$ we obtain

$$\langle\beta|\hat{n}|\beta\rangle = \beta\beta^* = |\beta|^2.$$

Problem 21.   Let $b^\dagger$, b be Bose creation and annihilation operators, respectively.

(i) Show that $e^{\epsilon b^\dagger} b = (b - \epsilon I)e^{\epsilon b^\dagger}$.

(ii) Extend the identity to $e^{\epsilon b^\dagger} b^k$, where $k \in \mathbb{N}$.

Solution 21.   Using the *Baker-Campbell-Hausdorff formula* we have

$$e^{\epsilon b^\dagger} b e^{-\epsilon b^\dagger} = b - \epsilon I.$$

It follows that $e^{\epsilon b^\dagger} b = e^{\epsilon b^\dagger} b e^{-\epsilon b^\dagger} e^{\epsilon b^\dagger} = (b - \epsilon I)e^{\epsilon b^\dagger}$.

(ii) Using the result form (i) we obtain $e^{\epsilon b^\dagger} b^k = (b - \epsilon I)^k e^{\epsilon b^\dagger}$.

Problem 22.   (i) Let $b_j^\dagger$, b_j be Bose creation and annihilation operators, respectively and $j = 1, 2, \ldots, n$. We define

$$B(x) := \sum_{j=1}^n \frac{b_j}{1 - \epsilon_j x}, \qquad f(x) := \sum_{j=1}^n \frac{1}{1 - \epsilon_j x}.$$

Show that

$$[B(x), B^\dagger(y)] = \frac{I}{x - y}(xf(x) - yf(y)), \qquad [B(x), B(y)] = 0.$$

(ii) Let

$$N(x) := \sum_{j=1}^n \frac{b_j^\dagger b_j}{1 - \epsilon_j x}.$$

Show that

$$[N(x), B^\dagger(y)] = \frac{1}{x - y}(xB^\dagger - yB^\dagger(y))$$

$$[N(x), B(y)] = -\frac{1}{x - y}(xB(x) - yB(y)).$$

Solution 22.   (i) We have

$$B^\dagger(x) = \sum_{j=1}^n \frac{b_j^\dagger}{1 - \epsilon_j x}.$$

Thus we find

$$[B(x), B^\dagger(y)] = \sum_{j=1}^n \frac{b_j}{1 - \epsilon_j x} \sum_{k=1}^n \frac{b_k^\dagger}{1 - \epsilon_k y} - \sum_{k=1}^n \frac{b_k^\dagger}{1 - \epsilon_k y} \sum_{j=1}^n \frac{b_j}{1 - \epsilon_j x}$$

$$= \sum_{j=1}^{n} \sum_{k=1}^{n} \frac{b_j b_k^\dagger}{(1-\epsilon_j x)(1-\epsilon_k y)} - \sum_{k=1}^{n} \sum_{j=1}^{n} \frac{b_k^\dagger b_j}{(1-\epsilon_k y)(1-\epsilon_j x)}$$

$$= \sum_{j=1}^{n} \sum_{k=1}^{n} \frac{b_j b_k^\dagger - b_k^\dagger b_j}{(1-\epsilon_j x)(1-\epsilon y)} = \sum_{j=1}^{n} \sum_{k=1}^{n} \frac{\delta_{jk} I}{(1-\epsilon_j x)(1-\epsilon_k y)}$$

$$= I \sum_{j=1}^{n} \frac{1}{(1-\epsilon_j x)(1-\epsilon_j y)} = \frac{I}{x-y} \sum_{j=1}^{n} \frac{x-y}{(1-\epsilon_j x)(1-\epsilon_j y)}$$

$$= \frac{I}{x-y} \sum_{j=1}^{n} \frac{x(1-\epsilon_j y) - y(1-\epsilon_j x)}{(1-\epsilon_j x)(1-\epsilon_j y)}$$

$$= \frac{I}{x-y} \left(x \sum_{j=1}^{n} \frac{1}{1-\epsilon_j x} - y \sum_{j=1}^{n} \frac{1}{1-\epsilon_j y} \right)$$

$$= \frac{I}{x-y} (x f(x) - y f(y))$$

where we used that $[b_j, b_k^\dagger] = \delta_{jk} I$. Since $[b_j, b_k] = 0$ the second result is obvious.

(ii) Using $[b_j^\dagger b_j, b_k] = -b_j \delta_{jk}$, $[b_j^\dagger b_j, b_k^\dagger] = b_j^\dagger \delta_{jk}$ and a similar calculation as in (i) we obtain the results.

Problem 23. Let $b^\dagger$, b be Bose creation and annihilation operators. Let $z, w \in \mathbb{C}$.
(i) Calculate the commutator $[z b^\dagger - \bar{z} b, w b^\dagger - \bar{w} b]$.
(ii) Let $D(z) := e^{z b^\dagger - \bar{z} b}$ be the *displacement operator*. Show that

$$D(z)D(w) \equiv e^{z\bar{w} - \bar{z} w} D(w)D(z), \quad D(z+w) \equiv e^{-\frac{1}{2}(z\bar{w} - \bar{z} w)} D(z)D(w).$$

Solution 23. (i) We have

$$[z b^\dagger - \bar{z} b, w b^\dagger - \bar{w} b] = -[\bar{z} b, w b^\dagger] - [z b^\dagger, \bar{w} b] = -\bar{z} w [b, b^\dagger] - z\bar{w} [b^\dagger, b]$$
$$= (z\bar{w} - \bar{z} w) I.$$

(ii) Using the result from (i) and that $e^{A+B} = e^{-\frac{1}{2}[A,B]} e^A e^B$ we obtain the two identities.

Problem 24. Let $z \in \mathbb{C}$ and $D(z) := e^{z b^\dagger - \bar{z} b}$ be the *displacement operator*. Let $|n\rangle$ be the number states. Calculate the matrix elements $\langle n | D(z) | m \rangle$.

Solution 24.   The matrix elements of the displacement operator $D(z)$ are

$$n \le m \quad \langle n|D(z)|m \rangle = e^{-|z|^2/2}\sqrt{\frac{n!}{m!}}(-\bar{z})^{m-n}L_n^{(m-n)}(|z|^2)$$

$$n \ge m \quad \langle n|D(z)|m \rangle = e^{-|z|^2/2}\sqrt{\frac{m!}{n!}}z^{n-m}L_m^{(n-m)}(|z|^2)$$

where $L_n^{(\alpha)}$ are the *associated Laguerre polynomials* defined by

$$L_n^{(\alpha)}(x) = \sum_{j=0}^{n}(-1)^j \binom{n+\alpha}{n-j}\frac{x^j}{j!}.$$

In particular $L_n^{(0)} = L_n$ are the usual Laguerre polynomials.

Problem 25.   Let $|\beta\rangle$ be a coherent state. Show that

$$\frac{1}{\pi}\int_{\mathbb{C}} |\beta\rangle\langle\beta| d^2\beta = I$$

where I is the identity operator and the integration is over the entire complex plane. Set $\beta = r\exp(i\phi)$ with $0 \le r < \infty$ and $0 \le \phi < 2\pi$.

Solution 25.   We have

$$\frac{1}{\pi}\int_{\mathbb{C}} |\beta\rangle\langle\beta| d^2\beta = \frac{1}{\pi}\sum_{n=0}^{\infty}\sum_{m=0}^{\infty}\frac{|n\rangle\langle m|}{\sqrt{n!\,m!}}\int_{\mathbb{C}} e^{-|\beta|^2}\beta^{*m}\beta^n d^2\beta.$$

Using $\beta = r\exp(i\phi)$ we arrive at

$$\frac{1}{\pi}\int_{\mathbb{C}} |\beta\rangle\langle\beta| d^2\beta = \frac{1}{\pi}\sum_{n=0}^{\infty}\sum_{m=0}^{\infty}\frac{|n\rangle\langle m|}{\sqrt{n!\,m!}}\int_0^{\infty} re^{-r^2}r^{n+m}dr\int_0^{2\pi} e^{i(n-m)\phi}d\phi.$$

Since

$$\int_0^{2\pi} e^{i(n-m)\phi}d\phi = 2\pi\delta_{nm}$$

we have

$$\frac{1}{\pi}\int_{\mathbb{C}} |\beta\rangle\langle\beta| d^2\beta = \sum_{n=0}^{\infty}\frac{|n\rangle\langle n|}{n!}\int_0^{\infty} e^{-s}s^n ds$$

where we set $s = r^2$ and therefore $ds = 2rdr$. Thus

$$\frac{1}{\pi}\int_{\mathbb{C}} |\beta\rangle\langle\beta| d^2\beta = \sum_{n=0}^{\infty}|n\rangle\langle n| = I$$

where we used the completeness relation for the number states.

Problem 26. Let $|\beta\rangle$ be a coherent state and $\rho = |\beta\rangle\langle\beta|$. Calculate the *characteristic function*

$$\chi(\beta) := \mathrm{tr}(\rho e^{\beta b^\dagger - \beta^* b}) \equiv \mathrm{tr}(\rho D(\beta))$$

where $D(\beta)$ is the displacement operator.

Solution 26. We apply coherent states to calculate the trace. Then

$$\chi(\beta) = \mathrm{tr}(|\beta\rangle\langle\beta|D(\beta))$$
$$= \frac{1}{\pi} \int_C d\gamma \langle\gamma|\beta\rangle\langle\beta|D(\beta)|\gamma\rangle = \frac{1}{\pi} \int_C d\gamma \langle\gamma|\beta\rangle\langle\beta|D(\beta)D(\gamma)|0\rangle$$
$$= \frac{1}{\pi} \int_C d\gamma \langle\gamma|\beta\rangle\langle\beta|e^{\frac{1}{2}(\beta^*\gamma - \beta\gamma^*)}D(\beta+\gamma)|0\rangle$$
$$= \frac{1}{\pi} \int_C d\gamma \langle\gamma|\beta\rangle e^{\frac{1}{2}(\beta^*\gamma - \beta\gamma^*)}\langle\beta|\beta+\gamma\rangle$$
$$= \frac{1}{\pi} \int_C d\gamma e^{-\frac{1}{2}(|\beta|^2+|\gamma|^2)+\beta\gamma^*} e^{\frac{1}{2}(\beta^*\gamma - \beta\gamma^*)} e^{-\frac{1}{2}(|\beta+\gamma|^2+|\beta|^2)+(\beta+\gamma)\beta^*}$$
$$= \frac{1}{\pi} \int_C d\gamma e^{-\gamma\gamma^* - \frac{1}{2}\beta\beta^* + \beta^*\gamma} = \frac{1}{\pi}e^{-\frac{1}{2}\beta\beta^*} \int_C d\gamma e^{-\gamma\gamma^* + \beta^*\gamma}$$
$$= \frac{1}{\pi}e^{-\frac{1}{2}\beta\beta^*} \int_{r=0}^{\infty} dr e^{-r^2} \int_{\phi=0}^{2\pi} d\phi e^{\beta^* r e^{i\phi}} = 2e^{-\frac{1}{2}\beta\beta^*} \int_{r=0}^{\infty} dr e^{-r^2}$$
$$= \sqrt{\pi}e^{-\frac{1}{2}\beta\beta^*}.$$

Problem 27. The *thermal mixture of states* with the mean number of photon equal to $\bar{n}$ is represented by the density operator

$$\rho_{\bar{n}} = \sum_{n=0}^{\infty} \frac{\bar{n}^n}{(\bar{n}+1)^{n+1}}|n\rangle\langle n|$$

where $|n\rangle$ are the number states. Calculate the Husimi distribution $\langle\beta|\rho_{\bar{n}}|\beta\rangle$.

Solution 27. Since

$$|\beta\rangle = e^{-|\beta|^2/2} \sum_{k=0}^{\infty} \frac{\beta^k}{\sqrt{k!}}|k\rangle, \qquad \langle\beta| = e^{-|\beta|^2/2} \sum_{j=0}^{\infty} \frac{(\beta^*)^j}{\sqrt{j!}}\langle j|$$

we have

$$\langle\beta|\rho_{\bar{n}}|\beta\rangle = e^{-|\beta|^2} \sum_{j=0}^{\infty} \frac{(\beta^*)^j}{\sqrt{j!}}\langle j| \sum_{n=0}^{\infty} \frac{\bar{n}^n}{(\bar{n}+1)^{n+1}}|n\rangle\langle n| \sum_{k=0}^{\infty} \frac{\beta^k}{\sqrt{k!}}|k\rangle$$

$$= e^{-|\beta|^2} \sum_{j=0}^{\infty} \frac{\beta^*}{\sqrt{j!}} \langle j | \sum_{n=0}^{\infty} \frac{\overline{n}^n \beta^n}{\sqrt{n!(\overline{n}+1)^{n+1}}} | n \rangle$$

$$= e^{-|\beta|^2} \sum_{n=0}^{\infty} \frac{(\beta^*)^n \beta^n \overline{n}^n}{\sqrt{n!}\sqrt{n!}(\overline{n}+1)^{n+1}} = \frac{e^{-|\beta|^2}}{(\overline{n}+1)} \sum_{n=0}^{\infty} \frac{1}{n!} \left(\frac{\beta^* \beta \overline{n}}{\overline{n}+1} \right)^n$$

$$= \frac{e^{-|\beta|^2}}{(\overline{n}+1)} e^{\overline{n}|\beta|^2/(\overline{n}+1)} = \frac{1}{\overline{n}+1} e^{-|\beta|^2/(\overline{n}+1)}$$

where we used that $\langle n|k \rangle = \delta_{nk}$ and $\langle j|n \rangle = \delta_{jn}$.

Problem 28. Consider the *two-mode state* $|\psi\rangle = e^{r(b_1^\dagger b_2^\dagger - b_1 b_2)}|00\rangle$, where $|00\rangle \equiv |0\rangle \otimes |0\rangle$ and r is the squeezing parameter. This state can also be written as

$$|\psi\rangle = \frac{1}{\cosh(r)} \sum_{n=0}^{\infty} (\tanh(r))^n |n\rangle \otimes |n\rangle.$$

This is the *Schmidt basis* for this state. The *density operator* ρ is given by $\rho = |\psi\rangle\langle\psi|$. Calculate the *partial traces* using the number states.

Solution 28. We have

$$\rho = \frac{1}{(\cosh(r))^2} \sum_{n=0}^{\infty} (\tanh(r))^n |n\rangle \otimes |n\rangle \sum_{m=0}^{\infty} (\tanh(r))^m \langle m| \otimes \langle m|$$

$$= \frac{1}{(\cosh(r))^2} \sum_{m=0}^{\infty} \sum_{n=0}^{\infty} (\tanh(r))^n (\tanh(r))^m |n\rangle\langle m| \otimes |n\rangle\langle m|.$$

Let I be the identity operator. Using that $\langle k|n \rangle = \delta_{kn}$ and $\langle m|k \rangle = \delta_{mk}$ we find

$$\rho_1 = \sum_{k=0}^{\infty} (I \otimes \langle k|) \rho (I \otimes |k\rangle)$$

$$= \frac{1}{(\cosh(r))^2} \sum_{k=0}^{\infty} (I \otimes \langle k|) \sum_{m,n=0}^{\infty} (\tanh(r))^n (\tanh(r))^m |n\rangle\langle m| \otimes |n\rangle\langle m| (I \otimes |k\rangle)$$

$$= \frac{1}{(\cosh(r))^2} \sum_{k=0}^{\infty} \sum_{m,n=0}^{\infty} (\tanh(r))^n (\tanh(r))^m |n\rangle\langle m| \delta_{kn} \delta_{mk}$$

$$= \frac{1}{(\cosh(r))^2} \sum_{k=0}^{\infty} (\tanh(r))^{2k} |k\rangle\langle k|.$$

We obtain the same result for ρ_2.

Problem 29. Let $b^\dagger$, b be Bose creation and Bose annihilation operators. Define

$$K_0 := \frac{1}{4}(b^\dagger b + b b^\dagger), \qquad K_+ := \frac{1}{2}\left(b^\dagger\right)^2, \qquad K_- := \frac{1}{2}b^2. \qquad (1)$$

(i) Show that K_0, K_+, K_- form a basis of a Lie algebra.
(ii) Define

$$L_0 := b_1^\dagger b_1 + b_2 b_2^\dagger, \qquad L_+ := b_1^\dagger b_2^\dagger, \qquad L_- := b_1 b_2. \qquad (2)$$

Show that L_0, L_+, L_- form a basis of a Lie algebra.
(iii) Are the two Lie algebras isomorphic?
(iv) Do the operators $\{I, b, b^\dagger, b^\dagger b\}$ form a Lie algebra under the commutator? Here I denotes the identity operator.

Solution 29. (i) Using the commutation relations for Bose operators we obtain $[K_0, K_+] = K_+$, $[K_0, K_-] = -K_-$, $[K_-, K_+] = 2K_0$. Thus the set $\{K_0, K_+, K_-\}$ forms a basis of a Lie algebra.
(ii) Using the commutation relations for Bose operators we find

$$[L_0, L_+] = 2L_+, \qquad [L_0, L_-] = -2L_-, \qquad [L_-, L_+] = L_0.$$

(iii) Two Lie algebras, L_1 and L_2, are called *isomorphic* if there is a map $\phi : L_1 \to L_2$ such that $\phi([a, b]) = [\phi(a), \phi(b)]$, where $a, b \in L_1$. Let

$$\phi(K_+) = L_+ \qquad \phi(K_-) = L_- \qquad \phi(K_0) = \frac{1}{2}L_0.$$

Then we find that the Lie algebras are isomorphic.
(iv) Since $[I, b] = 0$, $[I, b^\dagger] = 0$, $[I, b^\dagger b] = 0$, $[b, b^\dagger] = I$, $[b, b^\dagger b] = b$, $[b^\dagger, b^\dagger b] = -b^\dagger$ we find that the set of operators $\{I, b, b^\dagger, b^\dagger b\}$ forms a basis of a Lie algebra.

Problem 30. Let $b_1^\dagger$, $b_2^\dagger$ be the single-particle Bose creation operators for two bosonic modes. Let $\hat{N}_1 := b_1^\dagger b_1$, $\hat{N}_2 := b_2^\dagger b_2$ be the corresponding number operators. Consider the *Bose-Hubbard dimer Hamilton operator*

$$\hat{H} = \frac{k}{8}(\hat{N}_1 - \hat{N}_2)^2 - \frac{\mu}{2}(\hat{N}_1 - \hat{N}_2) - \frac{\mathcal{E}}{2}(b_1^\dagger b_2 + b_2^\dagger b_1). \qquad (1)$$

The coupling k provides the strength of the scattering interaction between bosons, μ is the external potential and $\mathcal{E}$ is the coupling for tunneling. The change $\mathcal{E} \to -\mathcal{E}$ corresponds to the unitary transformation $b_1 \to b_1$, $b_2 \to -b_2$. The change $\mu \to -\mu$ corresponds to $b_1 \leftrightarrow b_2$.
(i) Show that the Hamilton operator $\hat{H}$ commutes with the total number operator

$$\hat{N} = \hat{N}_1 + \hat{N}_2 = b_1^\dagger b_1 + b_2^\dagger b_2.$$

Thus $\hat{N}$ is a conserved quantity.

(ii) The *Jordan-Schwinger realization* in the $su(2)$ Lie algebra is given by

$$S_+ := b_1^\dagger b_2, \qquad S_- := b_2^\dagger b_1, \qquad S_3 := \frac{1}{2}(\hat{N}_1 - \hat{N}_2)$$

which is $(N+1)$-dimensional when the constraint of fixed particle number $N = N_1 + N_2$ is imposed. Express the Hamilton operator using S_+, S_-, S_3.

(iii) The same $(N+1)$-dimensional representation of $su(2)$ is given by the mapping to differential operators

$$S_+ = Nu - u^2 \frac{d}{du}, \qquad S_- = \frac{d}{du}, \qquad S_3 = u\frac{d}{du} - \frac{N}{2}$$

acting on the $(N+1)$-dimensional vector space of polynomials with basis $\{1, u, u^2, \ldots, u^N\}$. Express the Hamilton operator in this representation. Then solving for the spectrum of the Hamilton operator (1) is equivalent for solving the eigenvalue equation $Hp(u) = Ep(u)$, where p is a polynomial of u of order N. Find the energy eigenvalues $E(k, \mu, \mathcal{E})$.

Solution 30.   (i) Since

$$[b_1^\dagger b_2, b_1^\dagger b_1] = -b_1^\dagger b_2, \qquad [b_2^\dagger b_1, b_1^\dagger b_1] = b_2^\dagger b_1,$$

$$[b_1^\dagger b_2, b_2^\dagger b_2] = b_1^\dagger b_2, \qquad [b_2^\dagger b_1, b_2^\dagger b_2] = -b_2^\dagger b_1$$

we obtain $[\hat{H}, \hat{N}] = 0$.

(ii) We have

$$\hat{H} = \frac{k}{2}(S_3)^2 - \mu S_3 - \frac{\mathcal{E}}{2}(S_+ + S_-). \tag{2}$$

(iii) We obtain the second order differential operator for the Hamilton operator

$$\hat{H} = \frac{ku^2}{2}\frac{d^2}{du^2} + \frac{1}{2}((k(1-N)-2\mu)u + \mathcal{E}(u^2-1))\frac{d}{du} + \frac{kN^2}{8} + \frac{\mu N}{2} - \frac{\mathcal{E}Nu}{2}. \tag{3}$$

Solving for the spectrum of the Hamilton operator (1) is now equivalent for solving the eigenvalue equation $\hat{H}p(u) = Ep(u)$, where $\hat{H}$ is given by (3) and p is a polynomial of u of order N. We first express p in terms of its roots $\{v_j\}$

$$p(u) = \prod_{j=1}^{N}(u - v_j).$$

Now we have

$$\frac{d}{du}p(u) = p(u)\sum_{j=1}^{N}\frac{1}{u - v_j}.$$

For the second order derivative we have

$$\frac{d^2}{du^2}p(u) = p(u)\left(\left(\sum_{j=1}^{N}\frac{1}{u-v_j}\right)^2 - \sum_{j=1}^{N}\frac{1}{(u-v_j)^2}\right).$$

Furthermore we need the identity

$$\left(\left(\sum_{j=1}^{N}\frac{1}{u-v_j}\right)^2 - \sum_{j=1}^{N}\frac{1}{(u-v_j)^2}\right) \equiv 2\sum_{\substack{i<j \\ j=2 \\ i=1}}^{N}\frac{1}{v_i-v_j}\left(\frac{1}{u-v_i}-\frac{1}{u-v_j}\right).$$

Using these equations and evaluating the eigenvalue equation $\hat{H}p(u) = Ep(u)$ at $u = v_\ell$ for each ℓ we obtain the set of *Bethe ansatz equations*

$$\frac{(k(1-N)-2\mu)v_\ell + \mathcal{E}v_\ell^2 - \mathcal{E}}{kv_\ell^2} = 2\sum_{\substack{j\neq\ell \\ j}}^{N}\frac{1}{v_j-v_\ell}, \qquad \ell = 1,\ldots,N. \qquad (4)$$

Writing the asymptotic expansion $p(u) \sim u^N - u^{N-1}\sum_{j=1}^{N}v_j$ and by considering the terms of order N in the eigenvalue equation, the energy eigenvalues are given as

$$E(k,\mu,\mathcal{E}) = \frac{kN^2}{8} - \frac{\mu N}{2} + \frac{\mathcal{E}}{2}\sum_{j=1}^{N}v_j.$$

For given N, k, μ and $\mathcal{E}$ we solve the Bethe ansatz equations for $v_1, \ldots, v_N$ and insert them into this equation to find $E(k,\mu,\mathcal{E})$.

Problem 31. Consider the difference equation of first order

$$x_{n+1} = f(x_n), \qquad n = 0,1,2,\ldots \qquad (1)$$

where $f : \mathbb{R} \to \mathbb{R}$ is an analytic function. Consider the state

$$|x,n\rangle := \exp\left(\frac{1}{2}(x_n^2 - x_0^2)\right)|x_n\rangle \qquad (2)$$

where x_n satisfies (1) and $|x_n\rangle$ is a normalized coherent state. Suppose we are given a Boson operator

$$M := \sum_{k=0}^{\infty}\frac{1}{k!}b^{\dagger k}(f(b)-b)^k \qquad (3)$$

where $b^\dagger$ and b are Bose operators. Find $M|x, n\rangle$.

Solution 31.   The *coherent states* $|z\rangle$, where $z \in \mathbb{C}$, are defined as the eigenvectors of the annihilation operator b, i.e. $b|z\rangle = z|z\rangle$. The normalized coherent states are given by

$$|z\rangle = \exp(-\frac{1}{2}|z|^2)\exp(zb^\dagger)|0\rangle. \tag{4}$$

Using (1), (2) and (4) we find that

$$|x, n+1\rangle = M|x, n\rangle. \tag{5}$$

Now let $x_n(x_0)$ designate the solution of (1) and let $|x_0, n\rangle$ be the solution of (5). Taking (2) into account we find that the following eigenvalue equation holds true

$$b|x_0, n\rangle = x_n(x_0)|x_0, n\rangle.$$

Thus the solution of (1) is equivalent to the solution of the linear abstract difference equation (5).

Problem 32.   Consider the Hamilton operator $\hat{H}$ for a single two-level atom coupled to a single mode of an electromagnetic field

$$\hat{H} = b^\dagger \otimes \sigma_- + b \otimes \sigma_+$$

where

$$\sigma_+ = \begin{pmatrix} 0 & 1 \\ 0 & 0 \end{pmatrix}, \qquad \sigma_- = \begin{pmatrix} 0 & 0 \\ 1 & 0 \end{pmatrix}.$$

Let $\theta \in \mathbb{R}$ and $U = e^{i\theta\hat{H}}$. Calculate

$$A(n, \beta) = (\langle n| \otimes I_2)U(|\beta\rangle \otimes I_2)$$

where $|\beta\rangle$ and $|n\rangle$ are *coherent states* and *number states* of the electromagnetic field, respectively. This means $b|\beta\rangle = \beta|\beta\rangle$ and $b|n\rangle = \sqrt{n}|n-1\rangle$ with $\beta \in \mathbb{C}$ and $n = 0, 1, \ldots$.

Solution 32.   Since $\sigma_+\sigma_+ = 0_2$, $\sigma_-\sigma_- = 0_2$ and

$$\sigma_+\sigma_- = \begin{pmatrix} 1 & 0 \\ 0 & 0 \end{pmatrix}, \qquad \sigma_-\sigma_+ = \begin{pmatrix} 0 & 0 \\ 0 & 1 \end{pmatrix}$$

we find

$$(b^\dagger \otimes \sigma_- + b \otimes \sigma_+)(b^\dagger \otimes \sigma_- + b \otimes \sigma_+) = (I + b^\dagger b) \otimes \sigma_+\sigma_- + b^\dagger b \otimes \sigma_-\sigma_+$$
$$= I \otimes \sigma_+\sigma_- + b^\dagger b \otimes I_2.$$

Using this result we obtain

$$A(\beta, n) = e^{-|\beta|^2} \frac{|\beta|^2}{n!} \begin{pmatrix} \cos(\theta\sqrt{n}) & \frac{i\sqrt{n}}{\beta} \sin(\theta\sqrt{n}) \\ \frac{i\beta}{\sqrt{n+1}} \sin(\theta\sqrt{n+1}) & \cos(\theta\sqrt{n+1}) \end{pmatrix}$$

where $n = 0, 1, 2, \ldots$.

Problem 33.   Consider the following form of the *Jaynes-Cummings* Hamilton operator

$$\hat{H} = \hat{H}_1 + \hat{H}_2 + \frac{1}{2}\hbar\omega(I_B \otimes I_2), \quad \hat{H}_1 = \hbar\omega(b^\dagger b \otimes I_2 + \frac{1}{2}I_B \otimes \sigma_3),$$

$$\hat{H}_2 = \hbar\kappa(b^\dagger \otimes \sigma_- + b \otimes \sigma_+) - \frac{\hbar}{2}(\omega - \omega_0)I_B \otimes \sigma_3$$

with I_B the identity operator, I_2 the 2×2 identity matrix and

$$\sigma_+ := \sigma_1 + i\sigma_2 = \begin{pmatrix} 0 & 2 \\ 0 & 0 \end{pmatrix}, \quad \sigma_- := \sigma_1 - i\sigma_2 = \begin{pmatrix} 0 & 0 \\ 2 & 0 \end{pmatrix}.$$

Calculate the commutator $[\hat{H}_1, \hat{H}_2]$.

Solution 33.   Since we have the commutators

$$[b^\dagger b \otimes I_2, b^\dagger \otimes \sigma_-] = b^\dagger \otimes \sigma_-$$
$$[b^\dagger b \otimes I_2, b \otimes \sigma_+] = -b \otimes \sigma_+$$
$$[I_B \otimes \sigma_3, b^\dagger \otimes \sigma_-] = -2b^\dagger \otimes \sigma_-$$
$$[I_B \otimes \sigma_3, b \otimes \sigma_+] = 2b \otimes \sigma_+$$
$$[I_B \otimes \sigma_3, I_B \otimes \sigma_3] = 0_B \otimes 0_2$$

we obtain that $[\hat{H}_1, \hat{H}_2] = 0_B \otimes 0_2$, i.e. $\hat{H}_1$ and $\hat{H}_2$ commute.

Problem 34.   Let $b_1, b_2, b_1^\dagger, b_2^\dagger$ be Bose annihilation and creation operators. Consider the three 2×2 matrices

$$\sigma_+ = \begin{pmatrix} 0 & 1 \\ 0 & 0 \end{pmatrix}, \quad \sigma_- = \begin{pmatrix} 0 & 0 \\ 1 & 0 \end{pmatrix}, \quad \sigma_3 = \begin{pmatrix} 1 & 0 \\ 0 & -1 \end{pmatrix}.$$

Consider now the operators in the product space

$$J_+ = b_1^\dagger b_2 \otimes I_2 + I \otimes \sigma_+, \quad J_- = b_2^\dagger b_1 \otimes I_2 + I \otimes \sigma_-$$

$$J_3 = (b_1^\dagger b_1 - b_2^\dagger b_2) \otimes I_2 + I \otimes \sigma_3.$$

Find the commutators $[J_+, J_-], [J_+, J_3], [J_-, J_3]$.

Solution 34.   Note that $[b_1^\dagger b_2, b_2^\dagger b_1] = b_1^\dagger b_1 - b_2^\dagger b_2$, $[\sigma_+, \sigma_-] = \sigma_3$. Thus

$$[J_+, J_-] = [b_1^\dagger b_2, b_2^\dagger b_1] \otimes I_2 + I \otimes [\sigma_+, \sigma_-] = (b_1^\dagger b_1 - b_2^\dagger b_2) \otimes I_2 + I \otimes \sigma_3 = J_3.$$

Analogously we find $[J_+, J_3] = -2J_+$, $[J_-, J_3] = 2J_-$.

Problem 35.   Let
$$Q := (b - \epsilon(b + b^\dagger)^2) \otimes c^\dagger \tag{1}$$

be a linear operator, where b is a Bose annihilation operator, $c^\dagger$ is a Fermi creation operator and ϵ is a real parameter.
(i) Show that $Q^2 = 0$.
(ii) We define the Hamilton operator as $\hat{H}_\epsilon := [Q, Q^\dagger]_+ \equiv QQ^\dagger + Q^\dagger Q$. Find $\hat{H}_\epsilon$. Calculate the commutators $[\hat{H}_\epsilon, Q]$ and $[\hat{H}_\epsilon, Q^\dagger Q]$.
(iii) Give a basis of the underlying Hilbert space.
(iv) Can the Hilbert space be decomposed into subspaces?
The so constructed $\hat{H}_\epsilon$ is called a *supersymmetric Hamilton operator*.

Solution 35.   (i) Since $c^\dagger c^\dagger = 0$ we find that $Q^2 = 0$.
(ii) From (1) we obtain $Q^\dagger = (b^\dagger - \epsilon(b^\dagger + b)^2) \otimes c$. Applying $[b, b^\dagger] = I_B$ and $[c, c^\dagger]_+ = I_F$ we arrive at

$$\begin{aligned}
\hat{H}_\epsilon &= [Q, Q^\dagger]_+ = QQ^\dagger + Q^\dagger Q \\
&= I_B \otimes c^\dagger c + b^\dagger b \otimes I_F - 4\epsilon(b + b^\dagger) \otimes c^\dagger c \\
&\quad + \epsilon^2(b^4 + b^{\dagger 4} + 6b^2 + 6b^{\dagger 2} + 4b^{\dagger 3}b + 4b^\dagger b^3 + 6b^{\dagger 2}b^2 + 12b^\dagger b + 3) \otimes I_F \\
&\quad - \epsilon(b^3 + b^{\dagger 3} + b + b^\dagger + 3b^{\dagger 2}b + 3b^\dagger b^2) \otimes I_F
\end{aligned}$$

where I_B is the identity operator in the Hilbert space $\mathcal{H}_B$ of the Bose operators and I_F is the identity operator in the Hilbert space $\mathcal{H}_F$ of the Fermi operators. Straightforward calculation yields for the commutators

$$[\hat{H}_\epsilon, Q] = 0, \qquad [\hat{H}_\epsilon, Q^\dagger Q] = 0.$$

Thus the three operators $\hat{H}_\epsilon$, Q, $Q^\dagger Q$ may be diagonalized simultaneously.
(iii) Let
$$|n\rangle := \frac{(b^\dagger)^n}{\sqrt{n!}} |0\rangle$$

be the number states, where $n = 0, 1, 2, \ldots$. Then a basis in the Hilbert space $\mathcal{H}_B \otimes \mathcal{H}_F$ is given by $\{ |n\rangle \otimes |0\rangle_F, |n\rangle \otimes c^\dagger |0\rangle_F \}$, where $n = 0, 1, 2, \ldots$.
(iv) Since $c^\dagger c |0\rangle_F = 0|0\rangle_F$, $c^\dagger c c^\dagger |0\rangle_F = c^\dagger |0\rangle_F$ we find that the Hilbert space $\mathcal{H}_B \otimes \mathcal{H}_F$ decomposes, under the Hamilton operator, into two subspaces with the bases

$$\{ |n\rangle \otimes |0\rangle \}, \qquad \{ |n\rangle \otimes c^\dagger |0\rangle \}$$

where $n = 0, 1, 2, \ldots$.

Problem 36. Let

$$\hat{H} = -\Delta I_B \otimes \sigma_3 + \frac{k}{2}(b^\dagger \otimes \sigma_- + b \otimes \sigma_+) + \Omega b^\dagger b \otimes I_2 \qquad (1)$$

where

$$\sigma_- := \begin{pmatrix} 0 & 0 \\ 2 & 0 \end{pmatrix}, \quad \sigma_+ := \begin{pmatrix} 0 & 2 \\ 0 & 0 \end{pmatrix}, \quad \sigma_3 := \begin{pmatrix} 1 & 0 \\ 0 & -1 \end{pmatrix}, \quad I_2 := \begin{pmatrix} 1 & 0 \\ 0 & 1 \end{pmatrix}$$

and Δ, k and Ω are constants. Let

$$\hat{U} := \exp[i\pi(b^\dagger b \otimes I_2 + \frac{1}{2}I_B \otimes \sigma_3 + \frac{1}{2}I_B \otimes I_2)]$$

$$\hat{N} := b^\dagger b \otimes I_2 + \frac{1}{2}I_B \otimes \sigma_3.$$

(i) Calculate the commutators $[\hat{H}, \hat{U}]$, $[\hat{H}, \hat{N}]$, $[\hat{U}, \hat{N}]$, $[\hat{N}, I_B \otimes \sigma_3]$.
(ii) Find the eigenvalues of $\hat{U}$. Discuss how $\hat{U}$ can be used to simplify the eigenvalue problem.
(iii) Calculate the eigenvalues of $\hat{H}$.

Solution 36. (i) We find for the commutators

$$[\hat{H}, \hat{U}] = 0, \quad [\hat{H}, \hat{N}] = 0, \quad [\hat{U}, \hat{N}] = 0, \quad [\hat{N}, I_B \otimes \sigma_3] = 0.$$

Thus the operators $\hat{H}, \hat{U}, \hat{N}, I_B \otimes \sigma_3$ form a basis of an abelian Lie algebra.
(ii) A basis in the Hilbert space $\mathcal{H}_B \otimes \mathbb{C}^2$ is given by

$$\left\{ |n\rangle \otimes \begin{pmatrix} 1 \\ 0 \end{pmatrix}, \quad |n\rangle \otimes \begin{pmatrix} 0 \\ 1 \end{pmatrix} \right\}$$

where $|n\rangle$ are the Bose number states. Since the commutators of the operators $b^\dagger b \otimes I_2$, $I_B \otimes \sigma_3$ and $I_B \otimes I_2$ vanish we can write

$$\hat{U} = \exp(i\pi b^\dagger b \otimes I_2)\exp\left(\frac{1}{2}i\pi(I_B \otimes \sigma_3)\right)\exp\left(\frac{1}{2}i\pi(I_B \otimes I_2)\right).$$

We find

$$\hat{U}|n\rangle \otimes \begin{pmatrix} 1 \\ 0 \end{pmatrix} = e^{i\pi(n+1)}|n\rangle \otimes \begin{pmatrix} 1 \\ 0 \end{pmatrix}, \quad \hat{U}|n\rangle \otimes \begin{pmatrix} 0 \\ 1 \end{pmatrix} = e^{i\pi n}|n\rangle \otimes \begin{pmatrix} 0 \\ 1 \end{pmatrix}.$$

Thus the basis elements are eigenfunctions of $\hat{U}$ and the eigenvalues of $\hat{U}$ are given by $1, -1$, because $e^{i\pi m} = 1$ for m even, $e^{i\pi m} = -1$ for m odd. The eigenvalues are infinitely degenerate. Since $\hat{U}$ and $I_B \otimes I_2$ form a finite group (isomorphic to C_2) we can decompose the Hilbert space $\mathcal{H}_B \otimes \mathbb{C}^2$ into two invariant subspaces. We obtain the two subspaces

$$S_1 = \left\{ |0\rangle \otimes \begin{pmatrix} 0 \\ 1 \end{pmatrix}, \quad |1\rangle \otimes \begin{pmatrix} 1 \\ 0 \end{pmatrix}, \quad |2\rangle \otimes \begin{pmatrix} 0 \\ 1 \end{pmatrix}, \quad |3\rangle \otimes \begin{pmatrix} 1 \\ 0 \end{pmatrix}, \dots \right\}$$

$$S_2 = \left\{ |0\rangle \otimes \begin{pmatrix} 1 \\ 0 \end{pmatrix}, \quad |1\rangle \otimes \begin{pmatrix} 0 \\ 1 \end{pmatrix}, \quad |2\rangle \otimes \begin{pmatrix} 1 \\ 0 \end{pmatrix}, \quad |3\rangle \otimes \begin{pmatrix} 0 \\ 1 \end{pmatrix}, \dots \right\}.$$

The operator $\hat{U}$ is the so-called *parity operator*.

(iii) Since the Hamilton operator $\hat{H}$ commutes with $\hat{N}$, we can write the infinite matrix representation of $\hat{H}$ as the direct sum of 2×2 matrices. This can also be seen when we apply the Hamilton operator to the basis elements. We find

$$\hat{H}|n\rangle \otimes \begin{pmatrix} 1 \\ 0 \end{pmatrix} = (-\Delta + \Omega n)|n\rangle \otimes \begin{pmatrix} 1 \\ 0 \end{pmatrix} + k\sqrt{n+1}|n+1\rangle \otimes \begin{pmatrix} 0 \\ 1 \end{pmatrix}$$

$$\hat{H}|n\rangle \otimes \begin{pmatrix} 0 \\ 1 \end{pmatrix} = (\Delta + \Omega n)|n\rangle \otimes \begin{pmatrix} 0 \\ 1 \end{pmatrix} + k\sqrt{n}|n-1\rangle \otimes \begin{pmatrix} 1 \\ 0 \end{pmatrix}.$$

Thus for the subspace S_1 we have the direct sums

$$\begin{pmatrix} -\Delta & k \\ k & \Delta + \Omega \end{pmatrix} \oplus \begin{pmatrix} -\Delta + 2\Omega & \sqrt{3}k \\ \sqrt{3}k & \Delta + 3\Omega \end{pmatrix} \oplus \cdots$$

$$\oplus \begin{pmatrix} -\Delta + 2n\Omega & \sqrt{2n+1}k \\ \sqrt{2n+1}k & \Delta + (2n+1)\Omega \end{pmatrix} \oplus \cdots$$

For the subspace S_2 we find

$$\Delta \oplus \begin{pmatrix} -\Delta + \Omega & \sqrt{2}k \\ \sqrt{2}k & \Delta + 2\Omega \end{pmatrix} \oplus \cdots \oplus \begin{pmatrix} -\Delta + (2n+1)\Omega & \sqrt{2n+2}k \\ \sqrt{2n+2}k & \Delta + (2n+2)\Omega \end{pmatrix} \oplus \cdots$$

The eigenvalues of $\hat{H}$ are the eigenvalues of these 2×2 matrices.

Problem 37.   Calculate

$$f(\epsilon) = \exp(\epsilon \sigma_3 \otimes (b - b^\dagger))(\sigma_3 \otimes (b^\dagger + b)) \exp(-\epsilon \sigma_3 \otimes (b - b^\dagger)) \qquad (1)$$

where ϵ is a real parameter, σ_3 is the Pauli matrix and b and $b^\dagger$ are Bose annihilation and creation operators, i.e. $[b, b^\dagger] = I$.

Solution 37.   We find a differential equation for f and then solve the initial value problem of the linear differential equation. From (1) we find the initial condition

$$f(0) = \sigma_3 \otimes (b^\dagger + b).$$

Differentiating (1) with respect to ϵ yields

$$\frac{df(\epsilon)}{d\epsilon} =$$

$$\exp(\epsilon\sigma_3 \otimes (b - b^\dagger))(\sigma_3 \otimes (b^\dagger - b))(\sigma_3 \otimes (b^\dagger + b))\exp(-\epsilon\sigma_3 \otimes (b - b^\dagger))$$

$$- \exp(\epsilon\sigma_3 \otimes (b - b^\dagger))(\sigma_3 \otimes (b^\dagger + b))(\sigma_3 \otimes (b - b^\dagger))\exp(-\epsilon\sigma_3 \otimes (b - b^\dagger)).$$

We have $\sigma_3^2 = I_2$, where I_2 is the 2×2 unit matrix and

$$(b - b^\dagger)(b^\dagger + b) - (b^\dagger + b)(b - b^\dagger) = 2I_b$$

where I_b is the unit operator. Thus we find the linear first order differential equation

$$\frac{df(\epsilon)}{d\epsilon} = 2I_2 \otimes I_b.$$

Integrating this differential equation and inserting the initial condition we obtain

$$f(\epsilon) = 2\epsilon I_2 \otimes I_b + \sigma_3 \otimes (b^\dagger + b).$$

Problem 38. Consider the Hamilton operator

$$\hat{H} = \hat{H}_0 + g(b^\dagger \otimes S_- + b \otimes S_+) + \gamma b^\dagger b^\dagger bb \otimes I_S + \gamma(I_B \otimes S_3)^2 \qquad (1)$$

with

$$\hat{H}_0 = \omega b^\dagger b \otimes I_S + \omega_0 I_B \otimes S_3.$$

Here g and γ are real coupling constants, and

$$S_\pm := \sum_{n=1}^N \sigma_{\pm,n}, \qquad S_3 := \frac{1}{2}\sum_{n=1}^N \sigma_{3,n}$$

are collective N-atom Dicke operators, i.e. spin operators for which total spin $S \leq N/2$. The operators satisfy the $su(2)$ Lie algebra, i.e.

$$[S_+, S_-] = 2S_3, \qquad [S_3, S_\pm] = \pm S_\pm.$$

The Bose operators $b^\dagger$ and b satisfy the commutation relation $[b, b^\dagger] = I$. The number operator $\hat{M}$ is given by $\hat{M} = I_B \otimes S_3 + b^\dagger b \otimes I_S$.
(i) Find the commutator $[\hat{H}, \hat{M}]$.
(ii) Find $\hat{K} := g^{-1}(\hat{H} + (\gamma - \omega)\hat{M} - \gamma\hat{M}^2)$.
(iii) The total spin operator of the model is given by the Casimir operator

$$\hat{S}^2 = S_+S_- + S_3(S_3 - I_S).$$

Find the commutator $[I_B \otimes \hat{S}^2, \hat{K}]$.

Solution 38.   (i) From the commutation relations for the Bose operators we find $[b^\dagger b^\dagger bb, b^\dagger b] = 0$, $[b^\dagger, b^\dagger b] = -b^\dagger$, $[b, b^\dagger b] = b$. Using these commutation relations and $[\hat{A} \otimes I_S, I_B \otimes \hat{B}] = 0$ we find that $[\hat{H}, \hat{M}] = 0$.
(ii) We find

$$\hat{K} = \Delta I_B \otimes S_3 + (b^\dagger \otimes S_- + b \otimes S_+) + cb^\dagger b \otimes S_3$$

where $\Delta := g^{-1}(\omega_0 - \omega + \gamma)$, $c := -2g^{-1}\gamma$. The last term on the right-hand side of of $\hat{K}$ causes photon number dependent changes in the atomic transitions and therefore describes a *Stark shift*. Since $[\hat{H}, \hat{M}] = 0$ we obtain $[\hat{K}, \hat{M}] = 0$.
(iii) We obtain $[I_B \otimes \hat{S}^2, \hat{K}] = 0$. Since $[\hat{M}, \hat{K}] = 0$ and $[\hat{S}^2, \hat{K}] = 0$, it is convenient to decompose the Hilbert space

$$\mathcal{H} = \mathcal{H}_B \otimes \mathbb{C}_1^2 \otimes \cdots \otimes \mathbb{C}_N^2$$

in terms of the irreducible representations of the Lie algebra $su(2)$ with spin S and excitation numbers M.

Problem 39.   Consider the *uncertainty relation*

$$(\langle\psi|\hat{A}^2|\psi\rangle - \langle\psi|\hat{A}|\psi\rangle^2)(\langle\psi|\hat{B}^2|\psi\rangle - \langle\psi|\hat{B}|\psi\rangle^2) \geq \frac{1}{4}|\langle\psi|[\hat{A}, \hat{B}]|\psi\rangle|^2$$

where $\hat{A}$ and $\hat{B}$ are observable, $[\hat{A}, \hat{B}]$ denotes the commutator and $|\psi\rangle$ is a normalized state. Let $b^\dagger$, b be Bose creation and annihilation operators and

$$\hat{A} = \frac{1}{\sqrt{2}}(ib - ib^\dagger), \qquad \hat{B} = \frac{1}{\sqrt{2}}(b + b^\dagger).$$

(i) Let $|\psi\rangle = |\beta\rangle$ be a coherent state ($\beta \in \mathbb{C}$). Find the left-hand and right-hand side of the uncertainty relation.
(ii) Let $|\psi\rangle = |n\rangle$ be a number state ($n = 0, 1, 2, \ldots$). Find the left-hand and right-hand side of the uncertainty relation.
(iii) Let $|\psi\rangle = |\zeta\rangle$ be a squeezed state ($\zeta \in \mathbb{C}$). Find the left-hand and right-hand side of the uncertainty relation.

Solution 39.   (i) For the commutator we find

$$[\hat{A}, \hat{B}] = iI$$

where I is the identity operators. Thus with $\langle\beta|\beta\rangle = 1$ we obtain for the right-hand side

$$\frac{1}{4}|\langle\beta|iI|\beta\rangle|^2 = \frac{1}{4}.$$

Now $b|\beta\rangle = \beta|\beta\rangle$, $\langle\beta|b^\dagger = \langle\beta|\bar{\beta}$. It follows that

$$\langle\psi|\hat{A}|\psi\rangle^2 = \frac{1}{2}(-\beta^2 - \bar{\beta}^2 + 2\beta\bar{\beta}), \quad \langle\psi|\hat{B}|\psi\rangle^2 = \frac{1}{2}(\beta^2 + \bar{\beta}^2 + 2\beta\bar{\beta})$$

$$\langle \psi | \hat{A}^2 | \psi \rangle = \frac{1}{2}(1 - \beta^2 - \overline{\beta}^2 + 2\beta\overline{\beta}), \quad \langle \psi | \hat{B}^2 | \psi \rangle = \frac{1}{2}(1 + \beta^2 + \overline{\beta}^2 + 2\beta\overline{\beta})$$

and

$$\langle \beta | \hat{A}^2 | \beta \rangle - \langle \beta | \hat{A} | \beta \rangle^2 = \frac{1}{2}, \quad \langle \beta | \hat{B}^2 | \beta \rangle - \langle \beta | \hat{B} | \beta \rangle^2 = \frac{1}{2}.$$

It follows that the uncertainty relation is an equality for the present case.
(ii) With $\langle n | n \rangle = 1$ for the right-hand side we have again $\frac{1}{4}$. From

$$\langle n | \hat{A} | n \rangle = 0, \qquad \langle n | \hat{B} | n \rangle = 0$$

and

$$\langle n | \hat{A}^2 | n \rangle = \frac{1}{2}(1 + 2n), \qquad \langle n | \hat{B}^2 | n \rangle = \frac{1}{2}(1 + 2n).$$

Thus we have the inequality

$$\frac{1}{4}(1 + 2n)^2 \geq \frac{1}{4}.$$

So if $n = 0$ we have an equality.
(iii) We set $\zeta = se^{i\theta}$ ($s \geq 0$). From $|\zeta\rangle = S(\zeta)|0\rangle$, where $S(\zeta)$ is the squeezing operator we obtain

$$\langle \zeta | b | \zeta \rangle = \langle 0 | S^{-1}(\zeta) b S(\zeta) | 0 \rangle = \langle 0 | (\cosh(s) b - e^{i\theta} \sinh(s) b^\dagger) | 0 \rangle = 0$$

$$\langle \zeta | b^\dagger | \zeta \rangle = \langle 0 | S^{-1}(\zeta) b^\dagger S(\zeta) | 0 \rangle = \langle 0 | (\cosh(s) b^\dagger - e^{-i\theta} \sinh(s) b) | 0 \rangle = 0.$$

Now

$$\langle \zeta | b^2 | \zeta \rangle = \langle 0 | S^{-1}(\zeta) b S(\zeta) S^{-1}(\zeta) b S(\zeta) | 0 \rangle = -e^{i\theta} \cosh(s) \sinh(s)$$

$$\langle \zeta | (b^\dagger)^2 | \zeta \rangle = \langle 0 | S^{-1}(\zeta) b^\dagger S(\zeta) S^{-1}(\zeta) b^\dagger S(\zeta) | 0 \rangle = -e^{-i\theta} \cosh(s) \sinh(s).$$

Hence

$$\langle \zeta | \hat{A}^2 | \zeta \rangle = \frac{1}{2} \cosh(2s) + \frac{1}{2} \sinh(2s) \cos(\theta)$$

and

$$\langle \zeta | \hat{B}^2 | \zeta \rangle = \frac{1}{2} \cosh(2s) - \frac{1}{2} \sinh(2s) \cos(\theta)$$

and the inequality follows

$$\frac{1}{4}(\cosh^2(2s) - \sinh^2(2s) \cos^2(\theta)) \geq \frac{1}{4}.$$

With $\cosh^2(2s) - \sinh^2(2s) = 1$ we have an equality for $\theta = 0$ and $\theta = \pi$.

Problem 40. Let e be the electric charge (*sec. A*) and B_3 be a (constant) *magnetic flux density* B_3 (also called magnetic induction with dimension

($sec^2 \cdot kg \cdot A^{-1}$). Note that $B_1 = B_2 = 0$. Consider the *momentum operator* $\hat{\mathbf{P}}$ with

$$\hat{P}_1 = -i\hbar\frac{\partial}{\partial x_1} - \frac{1}{2}ex_2B_3, \quad \hat{P}_2 = -i\hbar\frac{\partial}{\partial x_2} + \frac{1}{2}ex_1B_3, \quad \hat{P}_3 = -i\hbar\frac{\partial}{\partial x_3}.$$

Let $\hat{P}_+ = \hat{P}_1 + i\hat{P}_2$, $\hat{P}_- = \hat{P}_1 - i\hat{P}_2$. Find the commutator $[\hat{P}_-, \hat{P}_+]$.

Solution 40.   We have

$$\hat{P}_- = -i\hbar\frac{\partial}{\partial x_1} - \frac{1}{2}ex_2B_3 - \hbar\frac{\partial}{\partial x_2} - \frac{1}{2}iex_1B_3$$

$$\hat{P}_+ = -i\hbar\frac{\partial}{\partial x_1} - \frac{1}{2}ex_2B_3 + \hbar\frac{\partial}{\partial x_2} + \frac{1}{2}iex_1B_3$$

and $[\hat{P}_-, \hat{P}_+] = 2e\hbar B_3 I$, where I is the identity operator. So we can introduce the (dimensionless) *Bose operators*

$$b = \frac{1}{\sqrt{2e\hbar B_3}}\hat{P}_-, \qquad b^\dagger = \frac{1}{\sqrt{2e\hbar B_3}}\hat{P}_+$$

with $[b, b^\dagger] = I$, where I is the identity operator. The dimension of $e\hbar B_3$ is $meter^2 \cdot sec^{-2} \cdot kg^2$.

Problem 41.   The one-dimensional field-quantized *nonlinear Schrödinger model* is defined by the Hamilton operator

$$\hat{H} = \int_0^L \left(\frac{\partial\psi^\dagger}{\partial x}\frac{\partial\psi}{\partial x} + c\psi^\dagger\psi^\dagger\psi\psi\right)dx \tag{1}$$

where ψ is a nonrelativistic Bose field with canonical equal-time commutation relations

$$[\psi(x,t), \psi^\dagger(x',t)] = \delta(x - x') \tag{2a}$$

$$[\psi(x,t), \psi(x',t)] = 0 \tag{2b}$$

where $\delta(x - x')$ is the delta function and c is a real constant. The Hamilton operator (1) is in the standard form of a many body problem with the second term corresponding to a two-body delta function potential. The Hamilton operator (1) commutes with the particle number operator

$$\hat{N} := \int_0^L \psi^\dagger\psi\, dx. \tag{2c}$$

We impose cyclic boundary conditions, i.e. $\psi(0) = \psi(L)$. Let

$$|\phi(k_1, k_2)\rangle := \int_0^L \int_0^L dx_1 dx_2 e^{i(k_1x_1 + k_2x_2)}(\theta(x_1 - x_2)$$

$$+ S(k_{21})\theta(x_2 - x_1))\psi^\dagger(x_1)\psi^\dagger(x_2)|0\rangle$$

be a two-particle state, where $k_{21} := k_2 - k_1$ and

$$S(k_{21}) := \frac{k_2 - k_1 - ic}{k_2 - k_1 + ic}.$$

Moreover $\langle 0|0 \rangle = 1$ and θ denotes the step function with $\theta(0) = 1$. The right-hand side of this state is called a *Bethe ansatz*. Calculate the state

$$\hat{H}|\phi(k_1, k_2)\rangle$$

and discuss. Use $\psi|0\rangle = 0$, where $|0\rangle$ is the *vacuum state*, and integration by parts. Furthermore apply

$$\frac{\partial}{\partial x_1}\theta(x_1 - x_2) = \delta(x_1 - x_2).$$

Solution 41. The Hamilton operator can be written as $\hat{H} = \hat{H}_k + \hat{H}_I$, where $\hat{H}_k$ denotes the kinetic part and $\hat{H}_I$ denotes the interacting part. First we consider the interacting part, i.e.

$$\hat{H}_I = c\int_0^L \psi^\dagger \psi^\dagger \psi\psi dx.$$

We set

$$f(x_1, x_2) := e^{i(k_1 x_1 + k_2 x_2)}(\theta(x_1 - x_2) + S(k_{21})\theta(x_2 - x_1)).$$

It follows that $f(x_1, x_1) = e^{i(k_1 x_1 + k_2 x_1)}(\theta(0) + S(k_{21})\theta(0))$. Since $\theta(0) = 1$, the parenthesis of the right-hand side of this expression can be written as

$$1 + S(k_{21}) = 1 + \frac{k_{21} - ic}{k_{21} + ic} = \frac{2k_{21}}{k_{21} + ic}.$$

Using the commutation relations we find

$$H_I|\phi(k_1, k_2)\rangle = c\int dx dx_1 dx_2 \psi^\dagger(x)\psi^\dagger(x)\psi(x)\psi(x)f(x_1, x_2)\psi^\dagger(x_1)\psi^\dagger(x_2)|0\rangle$$

$$= c\int dx dx_1 dx_2 f(x_1, x_2)\psi^\dagger(x)\psi^\dagger(x)\psi(x)\delta(x - x_1)\psi^\dagger(x_2)|0\rangle$$

$$+ c\int dx dx_1 dx_2 f(x_1, x_2)\psi^\dagger(x)\psi^\dagger(x)\psi(x)\psi^\dagger(x_1)\psi(x)\psi^\dagger(x_2)|0\rangle$$

where the integration is from 0 to L. It follows that

$$\hat{H}_I|\phi(k_1, k_2)\rangle = c\int dx_1 dx_2 f(x_1, x_2)\psi^\dagger(x_1)\psi^\dagger(x_1)\psi(x_1)\psi^\dagger(x_2)|0\rangle$$

$$+ c \int dx dx_1 dx_2 f(x_1, x_2) \psi^\dagger(x) \psi^\dagger(x) \psi(x) \psi^\dagger(x_1) \psi(x) \psi^\dagger(x_2)|0\rangle$$

$$= c \int dx_1 dx_2 f(x_1, x_2) \psi^\dagger(x_1) \psi^\dagger(x_1) \psi(x_1) \psi^\dagger(x_2)|0\rangle$$

$$+ c \int dx dx_1 dx_2 f(x_1, x_2) \psi^\dagger(x) \psi^\dagger(x) \psi(x) \psi^\dagger(x_1) \delta(x - x_2)|0\rangle$$

$$= c \int dx_1 dx_2 f(x_1, x_2) \psi^\dagger(x_1) \psi^\dagger(x_1) \delta(x_1 - x_2)|0\rangle$$

$$+ c \int dx_1 dx_2 f(x_1, x_2) \psi^\dagger(x_2) \psi^\dagger(x_2) \psi(x_2) \psi^\dagger(x_1)|0\rangle$$

$$= 2c \int dx_1 f(x_1, x_1) \psi^\dagger(x_1) \psi^\dagger(x_1)|0\rangle$$

$$= 2c \int dx_1 e^{i(k_1 x_1 + k_2 x_1)} (\theta(0) + S(k_{21})\theta(0)) \psi^\dagger(x_1) \psi^\dagger(x_1)|0\rangle$$

$$= 2c \int dx_1 e^{i(k_1 x_1 + k_2 x_1)} \left(\frac{2(k_2 - k_1)}{k_2 - k_1 + ic} \right) \psi^\dagger(x_1) \psi^\dagger(x_1)|0\rangle$$

$$= 4c \int_0^L dx_1 e^{i(k_1 x_1 + k_2 x_1)} \frac{k_2 - k_1}{k_2 - k_1 + ic} \psi^\dagger(x_1) \psi^\dagger(x_1)|0\rangle.$$

Next we consider the kinetic part. Applying integration by parts we find

$$\hat{H}_k|\phi(k_1, k_2)\rangle = \int dx dx_1 dx_2 \frac{\partial \psi^\dagger}{\partial x} \frac{\partial \psi}{\partial x} f(x_1, x_2) \psi^\dagger(x_1) \psi^\dagger(x_2)|0\rangle.$$

Thus

$$\hat{H}_k|\phi(k_1, k_2)\rangle = - \int dx dx_1 dx_2 \frac{\partial^2 \psi^\dagger}{\partial x^2} f(x_1, x_2) \psi(x) \psi^\dagger(x_1) \psi^\dagger(x_2)|0\rangle$$

where we have used the cyclic boundary condition $\psi(0) = \psi(L)$. It follows that

$$\hat{H}_k|\phi(k_1, k_2)\rangle = - \int dx dx_1 dx_2 \frac{\partial^2 \psi^\dagger}{\partial x^2} f(x_1, x_2) \delta(x - x_1) \psi^\dagger(x_2)|0\rangle$$

$$- \int dx dx_1 dx_2 \frac{\partial^2 \psi^\dagger}{\partial x^2} f(x_1, x_2) \psi^\dagger(x_1) \psi(x) \psi^\dagger(x_2)|0\rangle$$

$$= - \int dx_1 dx_2 \frac{\partial^2 \psi^\dagger}{\partial x_1^2} f(x_1, x_2) \psi^\dagger(x_2)|0\rangle$$

$$- \int dx dx_1 dx_2 \frac{\partial^2 \psi^\dagger}{\partial x^2} f(x_1, x_2) \psi^\dagger(x_1) \delta(x - x_2)|0\rangle$$

$$= - \int dx_1 dx_2 \frac{\partial^2 \psi^\dagger}{\partial x_1^2} f(x_1, x_2) \psi^\dagger(x_2)|0\rangle$$

$$- \int dx_1 dx_2 \frac{\partial^2 \psi^\dagger}{\partial x_2^2} f(x_1, x_2) \psi^\dagger(x_1)|0\rangle$$

$$= -\int dx_1 dx_2 \frac{\partial^2 f}{\partial x_1^2} \psi^\dagger(x_1)\psi^\dagger(x_2)|0\rangle$$

$$- \int dx_1 dx_2 \frac{\partial^2 f}{\partial x_2^2} \psi^\dagger(x_2)\psi^\dagger(x_1)|0\rangle.$$

From the function f we obtain the derivatives in the sense of generalized functions

$$\frac{\partial f}{\partial x_1} = ik_1 e^{i(k_1 x_1 + k_2 x_2)} (\theta(x_1 - x_2) + S(k_{21})\theta(x_2 - x_1))$$

$$+ e^{i(k_1 x_1 + k_2 x_2)} (\delta(x_1 - x_2) - S(k_{21})\delta(x_2 - x_1))$$

$$\frac{\partial^2 f}{\partial x_1^2} = -k_1^2 e^{i(k_1 x_1 + k_2 x_2)} (\theta(x_1 - x_2) + S(k_{21})\theta(x_2 - x_1))$$

$$+ 2ik_1 e^{i(k_1 x_1 + k_2 x_2)} (\delta(x_1 - x_2) - S(k_{21})\delta(x_2 - x_1))$$
$$+ e^{i(k_1 x_1 + k_2 x_2)} (\delta'(x_1 - x_2) - S(k_{21})\delta'(x_2 - x_1))$$

$$\frac{\partial f}{\partial x_2} = ik_2 e^{i(k_1 x_1 + k_2 x_2)} (\theta(x_1 - x_2) + S(k_{21})\theta(x_2 - x_1))$$

$$+ e^{i(k_1 x_1 + k_2 x_2)} (-\delta(x_1 - x_2) + S(k_{21})\delta(x_2 - x_1))$$

$$\frac{\partial^2 f}{\partial x_2^2} = -k_2^2 e^{i(k_1 x_1 + k_2 x_2)} (\theta(x_1 - x_2) + S(k_{21})\theta(x_2 - x_1))$$

$$+ 2ik_2 e^{i(k_1 x_1 + k_2 x_2)} (-\delta(x_1 - x_2) + S(k_{21})\delta(x_2 - x_1))$$
$$+ e^{i(k_1 x_1 + k_2 x_2)} (-\delta'(x_1 - x_2) + S(k_{21})\delta'(x_2 - x_1)).$$

We arrive at

$$\hat{H}_k |\phi(k_1, k_2)\rangle = (k_1^2 + k_2^2)|\phi(k_1, k_2)\rangle$$

$$- \int dx_1 2ik_1 e^{i(k_1 x_1 + k_2 x_1)} (1 - S(k_{21}))\psi^\dagger(x_1)\psi^\dagger(x_1)|0\rangle$$

$$- \int dx_1 2ik_2 e^{i(k_1 x_1 + k_2 x_1)} (-1 + S(k_{21}))\psi^\dagger(x_1)\psi^\dagger(x_1)|0\rangle$$

$$= (k_1^2 + k_2^2)|\phi(k_1, k_2)\rangle$$

$$- \int_0^L dx_1 e^{i(k_1 x_1 + k_2 x_1)} 2i(k_1 - k_1 S(k_{21}) - k_2 + k_2 S(k_{21}))$$

$$\times \psi^\dagger(x_1)\psi^\dagger(x_1)|0\rangle.$$

Since

$$2i(k_1 - k_2 - k_1 S(k_{21}) + k_2 S(k_{21})) = 4c\frac{k_2 - k_1}{k_2 - k_1 + ic}$$

the second term of the right-hand side of this equation cancels with the term $\hat{H}_I |\phi(k_1, k_2)\rangle$. Consequently we find

$$\hat{H}|\phi(k_1, k_2)\rangle = \hat{H}_k|\phi(k_1, k_2)\rangle + \hat{H}_I|\phi(k_1, k_2)\rangle = (k_1^2 + k_2^2)|\phi(k_1, k_2)\rangle.$$

Thus the state $|\phi(k_1, k_2)\rangle$ is an eigenstate of $\hat{H}$ and the eigenvalue is given by $k_1^2 + k_2^2$. The Bethe ansatz can be extended to N particles.

Problem 42. The quantum three wave interaction equation in one-space dimension is described by the Hamilton operator

$$\hat{H} = \int dx \left(\sum_{j=1}^{3} c_j b_j^* \left(\frac{1}{i} \frac{\partial}{\partial x} \right) b_j + g(b_2^* b_1 b_3 + b_1^* b_3^* b_2) \right) \qquad (1)$$

where c_j's ($j = 1, 2, 3$) are constant velocities and g is the coupling constant. We assume that all c_j's are distinct. The three fields b_j's are bosons which satisfy the equal time commutation relations

$$[b_j(x, t), b_k^*(y, t)] = \delta_{jk} \delta(x - y) \qquad (2a)$$

$$[b_j(x, t), b_k(y, t)] = [b_j^*(x, t), b_k^*(y, t)] = 0. \qquad (2b)$$

We define the *vacuum state* by $b_j(x, t)|0\rangle = 0|0\rangle$ ($j = 1, 2, 3$).
(i) Find the equation of motion for $b_j(x, t)$.
(ii) Find the commutator of $\hat{H}$ with the operators

$$\hat{N}_a := \int dx (b_1^* b_1 + b_2^* b_2), \quad \hat{N}_b := \int dx (b_2^* b_2 + b_3^* b_3).$$

We define the states

$$\|0, 0\rangle\rangle := |0\rangle$$

$$\|M, 0\rangle\rangle := \int \cdots \int dx_1 \cdots dx_M$$
$$\times \exp(i(p_1 x_1 + \cdots + p_M x_M)) b_1^*(x_1, t) \cdots b_1^*(x_M, t)|0\rangle$$

$$\|0, N\rangle\rangle := \int \cdots \int dx_1 \cdots dx_N$$
$$\times \exp(i(q_1 x_1 + \cdots + q_N x_N)) b_3^*(x_1, t) \cdots b_3^*(x_N, t)|0\rangle$$

where $M = 1, 2, \ldots$ and $N = 1, 2, \ldots$.
(iii) Find $\hat{H} \| M, 0 \rangle\rangle$. Find $\hat{H} \| 0, N \rangle\rangle$.

Solution 42. (i) The *Heisenberg equation of motion* is given by

$$\frac{\partial}{\partial t} b_j = i[\hat{H}, b_j(x, t)], \qquad j = 1, 2, 3.$$

Using the commutation relations it follows that

$$\frac{\partial b_1}{\partial t} + c_1 \frac{\partial b_1}{\partial x} = -igb_3^* b_2, \quad \frac{\partial b_2}{\partial t} + c_2 \frac{\partial b_2}{\partial x} = -igb_1 b_3$$

$$\frac{\partial b_3}{\partial t} + c_3 \frac{\partial b_3}{\partial x} = -igb_1^* b_2.$$

(ii) Using the commutation relation we find that the operators $\hat{N}_A$ and $\hat{N}_B$ commute with $\hat{H}$.

(iii) Using

$$\left(\frac{\partial}{\partial x} b_1(x,t)\right) b_1^*(x_j,t) = \frac{\partial}{\partial x}\left(b_1(x,t)b_1^*(x_j,t)\right)$$

$$= \frac{\partial}{\partial x}\delta(x-x_j) + b_1^*(x_j,t)\frac{\partial}{\partial x}b_1(x,t)$$

and the derivative of the delta function we obtain

$$\hat{H}\|M,0\rangle\rangle = \int dx \int \cdots \int dx_1 \ldots dx_M \exp(i(p_1 x_1 + \cdots + p_M x_M))$$

$$\times \left(b_1^*(x,t)c_1 \frac{1}{i}\frac{\partial}{\partial x}b_1(x,t)\right) b_1^*(x_1,t)\cdots b_1^*(x_M,t)|0\rangle$$

$$= E\|M,0\rangle\rangle$$

where the energy E is given by $E = c_1(p_1 + \cdots + p_M)$. Thus $\|M,0\rangle\rangle$ is an eigenstate. Similarly we find that $\|0,N\rangle\rangle$ is an eigenstate with the energy E given by $E = c_3(q_1 + \cdots + q_N)$.

Supplementary Problems

Problem 1. Find the matrix representation of the Hamilton operator

$$\hat{H} = \hbar\omega_1(e^{i\phi}c^\dagger + e^{-i\phi}c) + \hbar\omega_2 c^\dagger c$$

with the basis $|0\rangle$, $c^\dagger|0\rangle$. Find the eigenvalues and eigenvectors of the matrix.

Problem 2. (i) Let $c^\dagger$, c be Fermi creation and annihilation operators, respectively. Consider $\hat{R} := c^\dagger + c$. Show that $\hat{R}^2 = I$, where I is the identity operator.

(ii) Let $\sigma \in \{\uparrow,\downarrow\}$ and $c_\sigma^\dagger$, σ be Fermi operators with spin up and spin down. Consider $\hat{S} = (c_\uparrow^\dagger + c_\uparrow)(c_\downarrow^\dagger + c_\downarrow)$. Show that $\hat{S}^2 = I$, where I is the identity operator.

Problem 3. (i) Show that $e^{i\pi c^\dagger c} = I_F - 2c^\dagger c$.

(ii) Show that $e^{i\pi b^\dagger b} = \text{diag}(1,-1,1,-1,\ldots)$.

Problem 4. Let $z \in \mathbb{C}$. Show that

$$e^{z(c_1^\dagger c_2 + c_2^\dagger c_1)} c_1 e^{-z(c_1^\dagger c_2 + c_2^\dagger c_1)} = c_1 \cosh(z) - c_2 \sinh(z)$$

$$e^{z(c_1^\dagger c_2 + c_2^\dagger c_1)} c_2 e^{-z(c_1^\dagger c_2 + c_2^\dagger c_1)} = c_2 \cosh(z) - c_1 \sinh(z).$$

Problem 5. Let $c_1^\dagger$, $c_2^\dagger$, c_1, c_2 be Fermi creation and annihilation operators, respectively. Let $\alpha \in \mathbb{R}$. Show that

$$U(\alpha) = \exp(i\alpha(c_1^\dagger c_2 + c_2^\dagger c_1))$$
$$= I + i\sin(\alpha)(c_1^\dagger c_2 + c_2^\dagger c_1) + (\cos(\alpha) - 1)(\hat{N}_1 + \hat{N}_2 - 2\hat{N}_1\hat{N}_2)$$

where $\hat{N}_1 = c_1^\dagger c_1$, $\hat{N}_2 = c_2^\dagger c_2$. First show that

$$(c_1^\dagger c_2 + c_2^\dagger c_1)^2 = \hat{N}_1 + \hat{N}_2 - 2\hat{N}_1\hat{N}_2, \quad (c_1^\dagger c_2 + c_2^\dagger c_1)^2 = c_1^\dagger c_2 + c_2^\dagger c_1,$$

$$(c_1^\dagger c_2 + c_2^\dagger c_1)^2 = \hat{N}_1 + \hat{N}_2 - 2\hat{N}_1\hat{N}_2.$$

Show that $\hat{N}_1 + \hat{N}_2 - 2\hat{N}_1\hat{N}_2$ is a projection operator.

Problem 6. Let $L \in \mathbb{N}$ and odd. Let $c_j^\dagger$, c_j be Fermi creation and annihilation operators and

$$\tilde{c}_p = \frac{e^{-i\pi/4}}{\sqrt{L}} \sum_{j=1}^{L} e^{2\pi ijp/L} c_j, \quad \tilde{c}_p^\dagger = \frac{e^{i\pi/4}}{\sqrt{L}} \sum_{j=1}^{L} e^{-2\pi ijp/L} c_j^\dagger$$

be the Fourier transform. Show that the inverse transform is given by

$$c_j = \frac{e^{i\pi/4}}{\sqrt{L}} \sum_{p=0}^{L-1} e^{-2\pi ijp/L} \tilde{c}_p, \quad c_j^\dagger = \frac{e^{-i\pi/4}}{\sqrt{L}} \sum_{p=0}^{L-1} e^{2\pi ijp/L} \tilde{c}_p^\dagger.$$

Problem 7. Let $c_\ell^\dagger$, c_ℓ ($\ell = 1, \dots, n$) be Fermi creation and annihilation operators, respectively. Are the hermitian operators

$$\Pi_+ = \frac{1}{2}\left(I + \exp\left(i\pi \sum_{\ell=1}^{n} c_\ell^\dagger c_\ell\right)\right), \quad \Pi_- = \frac{1}{2}\left(I - \exp\left(i\pi \sum_{\ell=1}^{n} c_\ell^\dagger c_\ell\right)\right)$$

projection operators.

Problem 8. Let $f, g : \mathbb{C} \to \mathbb{C}$ be analytic functions. Then

$$[f(b), f(b^\dagger)] = \sum_{j=1}^{\infty} \frac{1}{j!} \frac{d^j}{d(b^\dagger)^j} \frac{d^j}{db^j} f(b).$$

Let $\tau \in \mathbb{R}$. Use this relation to calculate the commutators

$$[b, \exp(\tau(b^\dagger)^2/2)], \quad [b^2, \exp(\tau(b^\dagger)^2/2], \quad [b^3, \exp(\tau(b^\dagger)^2/2].$$

Then show that $b \exp(\tau(b^\dagger)^2/2)|0\rangle = \tau b^\dagger \exp(\tau(b^\dagger)^2/2)|0\rangle$.

Problem 9. Let $|n\rangle$ be the number states $(n \in \mathbb{N}_0)$ and $|\beta\rangle$ $(\beta \in \mathbb{C})$ be the coherent states. Then

$$|\beta\rangle = \sum_{n=0}^{\infty} \langle n|\beta\rangle |n\rangle.$$

Show that

$$|n\rangle = \frac{1}{\pi} \int_{\mathbb{C}} |\beta\rangle \langle \beta|n\rangle d^2\beta.$$

Note that

$$\langle \beta|n\rangle = \exp\left(-\frac{1}{2}|\beta|^2\right) \frac{\overline{\beta}^n}{\sqrt{n!}}.$$

Problem 10. (i) Let

$$\hat{A} = \frac{1}{2}(b^\dagger + b), \quad \hat{B} = \frac{1}{2i}(b - b^\dagger).$$

Show that $[\hat{A}, \hat{B}] = \frac{i}{2}I$.
(ii) Let $H = \frac{1}{2}(b^\dagger b + bb^\dagger)$. Show that $[b, H] = b$.
(iii) Let $n \in \mathbb{N}$. Show that $[b^\dagger b, (b^\dagger)^n] = n(b^\dagger)^n$, $[b^\dagger b, b^n] = -nb^n$.

Problem 11. Let $b^\dagger$, b be Bose creation and annihilation operators and $\gamma \in \mathbb{R}$. Calculate

$$e^{\gamma((b^\dagger)^2 - b^2)} b e^{-\gamma((b^\dagger)^2 - b^2)}, \quad e^{\gamma((b^\dagger)^2 - b^2)} b^\dagger e^{-\gamma((b^\dagger)^2 - b^2)}$$

utilizing $e^A B e^{-A} = B + [A, B] + \frac{1}{2!}[A, [A, B]] + \frac{1}{3!}[A, [A, [A, B]]] + \cdots$

Problem 12. (i) Let $b^\dagger$, b be Bose creation and annihilation operator. Find the commutators $[b, \sin(b^\dagger b)]$, $[b^\dagger, \sin(b^\dagger b)]$.
(ii) Let $b^\dagger$, b be Bose creation and annihilation operators. Show that

$$\exp(\theta b^\dagger b) = \sum_{j=0}^{\infty} \frac{(e^\theta - 1)^j}{j!}(b^\dagger)^j b^j.$$

Problem 13.   Let b be a Bose operator and I be the identity operator. Show that we can write

$$\delta(\beta I - b)\delta(\bar{\beta}I - b^\dagger) = \frac{1}{\pi^2} \int_{\mathbb{R}} \int_{\mathbb{R}} e^{-i\mu(\beta I - b)} e^{-i\bar{\mu}(\bar{\beta} - b^\dagger)} d(\Re(\mu)) d(\Im(\mu)).$$

Problem 14.   Let $b^\dagger$, b be Bose creation and annihilation operators. Show that

$$\rho(b, b^\dagger) = (1 - e^{-\lambda})e^{-\lambda b^\dagger b}$$

is a density operator, where $\lambda \equiv \beta\hbar\omega$, $\beta = 1/(k_B T)$.

Problem 15.   Let $|n\rangle$ be a number state and $|\beta\rangle$ be coherent state. Show that

$$|\langle n|\beta\rangle|^2 = \exp(-|\beta|^2)\frac{(|\beta|^2)^n}{n!}$$

which is a Poisson distribution.

Problem 16.   Consider the displacement operator, squeezing operator and rotation operator

$$D(\beta) = e^{\beta b^\dagger - \bar{\beta}b}, \quad S(\zeta) = e^{(\zeta(b^\dagger)^2 - \bar{\zeta}b^2)/2}, \quad R(\phi) = e^{-i\phi b^\dagger b}$$

where $\beta, \zeta \in \mathbb{R}$.   Find the commutators $[D(\beta), S(\zeta)]$, $[D(\beta), R(\phi)]$, $[R(\phi), S(\zeta)]$.

Problem 17.   (i) Let $|\beta\rangle$ ($\beta \in \mathbb{C}$) be a coherent state. Show that the projection operator $|\beta\rangle\langle\beta|$ can be expressed as

$$|\beta\rangle\langle\beta| = \exp(-\bar{\beta}\beta) \exp(\beta b^\dagger)|0\rangle\langle 0| \exp(\bar{\beta}b).$$

(ii) Let $|\beta\rangle$ be a coherent state and $|\zeta\rangle$ be a squeezed state. We set $\zeta = se^{i\theta}$ ($s \geq 0$). Show that

$$\langle\beta|\zeta\rangle = \sqrt{\operatorname{sech}(s)} \exp\left(-\frac{1}{2}\bar{\beta}^2 \exp(i\theta)\tanh(s)\right) \exp(-|\beta|^2/2).$$

Problem 18.   (i) Let $b^\dagger$, b be Bose creation and annihilation operators and $z \in \mathbb{C}$. Show that $[b, e^{-zb^\dagger b}] = (e^z - 1)e^{-zb^\dagger b}b$, $[b^\dagger, e^{-zb^\dagger b}] = (e^z - 1)e^{-zb^\dagger b}b^\dagger$.
(ii) Let $f : \mathbb{C} \to \mathbb{C}$ be an analytic function. Let $b^\dagger$, b be Bose creation and annihilation operators, respectively and $\hat{N} = b^\dagger b$, i.e. $\hat{N}|n\rangle = n|n\rangle$. Show that $e^{zb}f(b^\dagger)|0\rangle = f(b^\dagger + zI)|0\rangle$. Show that $e^{-zb^\dagger b}f(b^\dagger)|0\rangle = f(b^\dagger e^{-z})|0\rangle$. Show that $f(\hat{N})|n\rangle = f(n)|n\rangle$.

Problem 19. Consider the density operator

$$\rho = \frac{\exp(-\hbar\omega b^\dagger b/k_B T)}{\text{tr}(\exp(-\hbar\omega b^\dagger b/k_B T))}.$$

Show that expressed with number states $|n\rangle$ $(n = 0, 1, 2, \ldots)$ we have

$$\rho = \sum_{n=0}^{\infty} \frac{\langle n \rangle^n}{(1 + \langle n \rangle)^{n+1}} |n\rangle\langle n|$$

where $\langle n \rangle := \text{tr}(\rho b^\dagger b) = (\exp(\hbar\omega/k_B T) - 1)^{-1}$.

Problem 20. Let $|n\rangle$, $|\beta\rangle$, $|\zeta\rangle$ be number states, coherent states and squeezed states, respectively. Is

$$\langle n|(b^\dagger b(b^\dagger b - I)|n\rangle - (\langle n|b^\dagger b|n\rangle)^2 \geq 0$$

$$\langle \beta|(b^\dagger b(b^\dagger b - I)|\beta\rangle - (\langle \beta|b^\dagger b|\beta\rangle)^2 \geq 0$$

$$\langle \zeta|(b^\dagger b(b^\dagger b - I)|\zeta\rangle - (\langle \zeta|b^\dagger b|\zeta\rangle)^2 \geq 0?$$

Problem 21. Let $\zeta \in \mathbb{C}$. The squeezing operator is defined as

$$S(\zeta) = \exp\left(\frac{1}{2}\zeta(b^\dagger)^2 - \frac{1}{2}\bar{\zeta}b^2\right).$$

We set $\zeta = se^{i\theta}$. Show that

$$S^\dagger(\zeta)bS(\zeta) = \cosh(s)b + e^{i\theta}\sinh(s)b^\dagger.$$

$$S^\dagger(\zeta)b^\dagger S(\zeta) = \cosh(s)b^\dagger + e^{-i\theta}\sinh(s)b$$

$$\langle \zeta|b^\dagger b|\zeta\rangle = \langle 0|S^\dagger(\zeta)b^\dagger S(\zeta)S^\dagger(\zeta)bS(\zeta)|0\rangle = \sinh^2(s).$$

Problem 22. Study the eigenvalue problem for the Hamilton operator

$$\hat{H} = \hbar\omega_1 b^\dagger b \otimes I + \hbar\omega_2 I \otimes b^\dagger b + \hbar\omega_3 b^\dagger b \otimes (b^\dagger + b).$$

Problem 23. Consider the operators

$$J_+ = b_1^\dagger b_2 \otimes I_F + I_B \otimes c^\dagger, \qquad J_- = b_2^\dagger b_1 \otimes I_F + I_B \otimes c,$$

$$J_3 = (b_1^\dagger b_1 - b_2^\dagger b_2) \otimes I_F + I_B \otimes (2c^\dagger c - I_F).$$

Show that $[J_+, J_-] = J_3$, $[J_+, J_3] = -2J_+$, $[J_-, J_3] = 2J_-$.

Problem 24.   Let $b^\dagger$, b be Bose creation and annihilation operators, respectively. Consider the matrix-valued operator

$$B = \begin{pmatrix} b & I \\ 0 & b \end{pmatrix} \quad \Rightarrow \quad B^\dagger = \begin{pmatrix} b^\dagger & 0 \\ I & b^\dagger \end{pmatrix}.$$

(i) Find $\hat{H} = \hbar\omega B^\dagger B$.

(ii) Let $D(z) = \exp(zb^\dagger - \bar{z}b)$ be the displacement operator and

$$D = \begin{pmatrix} D(z) & 0 \\ 0 & D(z) \end{pmatrix}.$$

Find $\tilde{H} = D\hat{H}D^\dagger$.

Problem 25.   Let $b_1^\dagger$, $b_2^\dagger$, b_1, b_2 be Bose creation and annihilation operators and $N \geq 1$ be a positive integer. Consider the operators

$$\hat{T}_1 = \frac{\sqrt{2}}{N^{3/2}}(b_1^\dagger b_1^\dagger b_2 + b_2^\dagger b_1 b_1), \quad \hat{T}_2 = \frac{\sqrt{2}i}{N^{3/2}}(b_1^\dagger b_1^\dagger b_2 - b_2^\dagger b_1 b_1),$$

$$\hat{T}_3 = \frac{2}{N}(2b_2^\dagger b_2 - b_1^\dagger b_1).$$

Show that the commutators are given by

$$[\hat{T}_1, \hat{T}_2] = \frac{i}{N}(I - \hat{T}_3)(I + 3\hat{T}_3) + \frac{4i}{N^2}I$$

$$[\hat{T}_3, \hat{T}_1] = \frac{4i}{N}\hat{T}_2, \qquad [\hat{T}_3, \hat{T}_2] = -\frac{4i}{N}\hat{T}_1$$

where I denotes the identity operator.

Problem 26.   Let $b_1^\dagger$, $b_2^\dagger$, b_1, b_2 Bose creation and annihilation operators, respectively. Show that the operators

$$\hat{X}_1 = \frac{1}{2}\left(b_2^\dagger b_1 + b_1^\dagger b_2\right), \quad \hat{X}_2 = \frac{i}{2}\left(b_2^\dagger b_1 - b_1^\dagger b_2\right), \quad \hat{X}_3 = \frac{1}{2}(b_1^\dagger b_1 - b_2^\dagger b_2)$$

form a basis of the Lie algebra $su(2)$. Show that

$$U(\alpha, \beta, \gamma, \delta, \epsilon) = \exp(i(\alpha b_1^\dagger b_1 + \beta b_2^\dagger b_2))\exp(i\epsilon\hat{X}_1)\exp(i(\gamma b_1^\dagger b_1 + \delta b_2^\dagger b_2))$$

is a unitary operator.

Problem 27.   Let $b^\dagger$, b be Bose creation and annihilation operators and $f : \mathbb{R} \to \mathbb{R}$ be an analytic function.

(i) Show that $e^{\epsilon b^\dagger b}f(b, b^\dagger)e^{-\epsilon b^\dagger b} = f(be^{-\epsilon}, b^\dagger e^{\epsilon})$.

(ii) Show that $f(b^\dagger b)|n\rangle = f(n)|n\rangle$, where $|n\rangle$ are the number states.
(iii) Let $z \in \mathbb{C}$. Show that

$$\exp(zb^\dagger)|0\rangle = \sum_{j=0}^{\infty} \frac{z^j}{\sqrt{j!}}|n\rangle.$$

Problem 28. Let $b^\dagger$, b be Bose creation and annihilation operators, respectively and $n \in \mathbb{N}$. Show that

$$\exp(\gamma_1 b^\dagger + \gamma_2 b^n) = \exp(\gamma_1 b^\dagger) \exp\left(\sum_{j=1}^{n} \gamma_2 \gamma_1^j \binom{n}{j} \frac{1}{1+j} b^{n-j}\right).$$

Problem 29. (i) Let $n \geq 2$ and A be an $n \times n$ matrix over $\mathbb{C}$. What can be said about the spectrum of $\hat{H} = \hbar\omega_1(A + A^*) + \hbar\omega_2 AA^*$?
(ii) Let $c^\dagger$, c be Fermi creation and annihilation operators, respectively. What can be said about the spectrum of $\hat{H} = \hbar\omega_1(c^\dagger + c) + \hbar\omega_2 c^\dagger c$?
(iii) Let $b^\dagger$, b be Bose creation and annihilation operators, respectively. What can be said about the spectrum of $\hat{H} = \hbar\omega_1(b^\dagger + b) + \hbar\omega_2 b^\dagger b$?

Problem 30. Let $b_1^\dagger = b^\dagger \otimes I$, $b_2^\dagger = I \otimes b^\dagger$ be Bose creation operators, I is the identity operator and

$$S_+ = S_1 + iS_2 = \begin{pmatrix} 0 & 1 \\ 0 & 0 \end{pmatrix}, \quad S_- = S_1 - iS_2 = \begin{pmatrix} 0 & 0 \\ 1 & 0 \end{pmatrix},$$

$$S_3 = \frac{1}{2}\sigma_3 = \frac{1}{2}\begin{pmatrix} 1 & 0 \\ 0 & -1 \end{pmatrix}.$$

Consider the Hamilton operator

$$\hat{H} = \hbar\omega_1 b_1^\dagger b_1 \otimes I_2 + \hbar\omega_2 b_2^\dagger b_2 \otimes I_2 + \hbar\omega_0(I \otimes I) \otimes I_2$$
$$+ \lambda b_1^\ell (b_2^\dagger)^k \otimes S_+ + \lambda(b_1^\dagger)^\ell b_2^k \otimes S_-$$

where $\ell, k = 1, 2, \ldots$. Show that

$$\hat{N} = b_1^\dagger b_1 \otimes I_1 + \ell(I \otimes I) \otimes S_3, \quad \hat{M} = b_2^\dagger b_2 \otimes I_2 - k(I \otimes I) \otimes S_3$$

are constants of motion.

Problem 31. Find the matrix representation of $b^\dagger \otimes c + b \otimes c^\dagger$ applying the basis $|n\rangle \otimes |0\rangle$, $|n\rangle \otimes c^\dagger|0\rangle$.

Find the matrix representation using the basis $|\beta\rangle \otimes |0\rangle$, $|\beta\rangle \otimes c^\dagger |0\rangle$.

Problem 32.   Find the states $e^{i\pi(b^\dagger b \otimes c^\dagger c)}(|n\rangle \otimes |0\rangle)$, $e^{i\pi(b^\dagger b \otimes c^\dagger c)}(|n\rangle \otimes |0\rangle)$.

Problem 33.   Let $b^\dagger$, b be Bose creation and annihilation operators and $c^\dagger$, c be Fermi creation and annihilation operators. Consider the operators

$$\hat{N} = b^\dagger b \otimes I_F + I_B \otimes c^\dagger c, \quad \hat{M} = b^\dagger b \otimes I_F - I_B \otimes c^\dagger c + I_B \otimes I_F$$

$$\hat{Q} = b \otimes c^\dagger, \quad \hat{Q}^\dagger = b^\dagger \otimes c.$$

Show that

$$[\hat{N}, \hat{M}] = 0_B \otimes 0_F, \quad [\hat{N}, \hat{Q}] = 0_B \otimes 0_F, \quad [\hat{N}, \hat{Q}^\dagger] = 0_B \otimes 0_F$$

and

$$[\hat{M}, \hat{Q}] = -2\hat{Q}, \quad [\hat{M}, \hat{Q}^\dagger] = 2\hat{Q}^\dagger, \quad [\hat{Q}, \hat{Q}^\dagger]_+ = \hat{N}, \quad \hat{Q}^2 = (\hat{Q}^\dagger)^2 = 0_B \otimes 0_F.$$

Problem 34.   Let $c_n^\dagger$, c_n be Fermi creation and annihilation operators at lattice site n. The one-dimensional *Aubry model* is given by the Hamilton operator

$$\hat{H}(Q) = \sum_{n \in \mathbb{Z}} (V \cos(Qn) c_n^\dagger c_n + t(c_{n+1}^\dagger c_n + c_{n-1}^\dagger c_n))$$

where Q is incommensurate to π. Show that the Hamilton operator is self-dual with respect to the Fourier transform.

Problem 35.   (i) Study the spectrum of the Hamilton operator

$$\hat{H} = \hbar\omega_1 I_B \otimes \sigma_3 + \hbar\omega_2 b^\dagger b \otimes I_2 + \hbar\omega_3 (b^\dagger + b) \otimes \sigma_1.$$

(ii) Study the spectrum of the Hamilton operator

$$\hat{H} = \hbar\omega_1 (I_B \otimes \sigma_1) + \hbar\omega_2 (b^\dagger b \otimes I_2) + \hbar\omega_3 (b^\dagger + b) \otimes \sigma_3).$$

Problem 36.   Let σ_1, σ_2, σ_3 be the Pauli spin matrices and I_2 be the 2×2 identity matrix. Let $b_1^\dagger$, $b_2^\dagger$, b_1, b_2 be Bose creation and annihilation operators, respectively with $b_1^\dagger = b^\dagger \otimes I_B$ and $b_2^\dagger = I_B \otimes b$. Consider the Hamilton operator

$$\hat{H} = \hbar\omega_1 b_1^\dagger b_1 \otimes I_2 + \hbar\omega_2 b_2^\dagger b_2 \otimes I_2$$

$$\gamma_1(b_1^\dagger + b_1) \otimes \sigma_1 + \gamma_2(b_2^\dagger + b_2) \otimes \sigma_1 + (I_B \otimes I_B) \otimes \frac{1}{2}\omega_0\sigma_3.$$

Let $\hat{N} = b_1^\dagger b_1 + b_2^\dagger b_2$. Define the unitary operator

$$\hat{S} = \exp\left(i\pi(\hat{N} \otimes I_2 + \frac{1}{2}(I_B \otimes I_B) \otimes \sigma_3 + \frac{1}{2}(I_B \otimes I_B) \otimes I_2\right)$$

$$= -\cos(\pi\hat{N}) \otimes \sigma_3.$$

(i) Show that $\hat{S} = \hat{S}^\dagger$.
(ii) Show that $[\hat{H}, \hat{S}] = (0_B \otimes 0_B) \otimes 0_2$.
(iii) Define

$$\hat{T} = \exp\left(-\frac{\pi}{2}(b_1^\dagger b_1 + b_2^\dagger b_2) \otimes (\sigma_1 - I_2)\right).$$

Show that $\hat{T} = \cos^2(\pi\hat{N}/2) \otimes I_2 + \sin^2(\pi\hat{N}/2) \otimes \sigma_1 = \hat{T}^\dagger$.
Show that $\hat{T}^\dagger \hat{S} \hat{T} = -(I_B \otimes I_B) \otimes \sigma_3$.
Show that $\hat{T}^\dagger(b_1 \otimes I_2)T = b_1 \otimes \sigma_1$, $\hat{T}^\dagger(b_2 \otimes I_2)T = b_2 \otimes \sigma_1$.
Find

$$\hat{H}\left((|n_1\rangle \otimes |n_2\rangle) \otimes \begin{pmatrix} \cos(\theta) \\ \sin(\theta) \end{pmatrix}\right), \quad \hat{H}\left((|\beta_1\rangle \otimes |\beta_2\rangle) \otimes \begin{pmatrix} \cos(\theta) \\ \sin(\theta) \end{pmatrix}\right)$$

where $|n_1\rangle$, $|n_2\rangle$ are number states and $|\beta_1\rangle$, $|\beta_2\rangle$ are coherent states.

Problem 37. The eigenvalues of the Pauli spin matrix σ_1 are given by $+1$ and -1 with the corresponding normalized eigenvectors

$$|+\rangle = \frac{1}{\sqrt{2}}\begin{pmatrix} 1 \\ 1 \end{pmatrix}, \quad |-\rangle = \frac{1}{\sqrt{2}}\begin{pmatrix} 1 \\ -1 \end{pmatrix}.$$

Show that the eigenvectors of the Hamilton operator

$$\hat{H} = \hbar\omega b^\dagger b \otimes I_2 + g(b^\dagger + b) \otimes \sigma_1$$

are given by the product of displaced number states and the eigenstates $|\pm\rangle$ of σ_1, i.e.

$$|\pm n, \pm\rangle = D(\mp x)|n\rangle \otimes |\pm\rangle$$

where $x = g/(\hbar\omega)$, $D(v) = \exp(vb^\dagger \bar{v}b)$ and $|n\rangle$ are the number states with $n = 0, 1, \ldots$.

Problem 38. Let $b^\dagger$, b be the Bose creation and annihilation operators and S_3, S_+, S_- be the spin-$\frac{1}{2}$ matrices. Consider the operators

$$\hat{N} = b^\dagger b \otimes I_2 + I_B \otimes S_3, \quad \hat{K} = (b^\dagger + b) \otimes (S_+ + S_-).$$

(i) Show that $[\hat{N}, b \otimes S_+ + b^\dagger \otimes S_-] = 0_B \otimes 0_2$.
(ii) Show that $[\hat{N}, \hat{K}] = 2(b^\dagger \otimes S_+ - b \otimes S_-)$.

Problem 39.   Consider the operators

$$Q_+ = \frac{1}{2}(b_1^\dagger \otimes c + b_2 \otimes c^\dagger), \quad Q_- = \frac{1}{2}(b_1 \otimes c^\dagger - b_2^\dagger \otimes c).$$

Find Q_+Q_-, Q_-Q_+ and the anticommutator $[Q_+, Q_-]_+$.

Problem 40.   Let $b^\dagger$, b be Bose creation and annihilation operators, respectively and $c^\dagger$, c be Fermi creation and annihilation operators. Consider the Hamilton operator

$$\hat{H} = \hbar\omega_1 b^\dagger b \otimes I_F + \hbar\omega_2 I_B \otimes c^\dagger c + \hbar\omega_3 (b^\dagger b \otimes (c^\dagger + c)$$
$$+\hbar\omega_4 (b^\dagger + b) \otimes c^\dagger c + \hbar\omega_5 (b^\dagger + b) \otimes (c^\dagger + c).$$

(i) Let $|n\rangle_B$ $(n = 0, 1, \ldots)$ be number states and $|0\rangle_F$, $c^\dagger|0\rangle_F$ the two states for the Fermi system. Find

$$({}_B\langle n| \otimes {}_F\langle 0|)\hat{H}(|n\rangle_B \otimes |0\rangle_F), \quad ({}_B\langle n| \otimes {}_F\langle 0|c)\hat{H}(|n\rangle_B \otimes c^\dagger|0\rangle_F),$$

$$({}_B\langle n| \otimes {}_F\langle 0|)\hat{H}(|n\rangle_B \otimes c^\dagger|0\rangle_F), \quad ({}_B\langle n| \otimes {}_F\langle 0|c)\hat{H}(|n\rangle_B \otimes |0\rangle_F).$$

(ii) Let $|\beta\rangle$ be a coherent state. Find

$$((\langle\beta| \otimes \langle 0|)\hat{H}(|\beta\rangle \otimes |0\rangle)), \quad ((\langle\beta| \otimes \langle 0|c)\hat{H}(|\beta\rangle \otimes c^\dagger|0\rangle)).$$

(iii) Let $|\zeta\rangle$ be a squeezed state. Find

$$((\langle\zeta| \otimes \langle 0|)\hat{H}(|\zeta\rangle \otimes |0\rangle)), \quad ((\langle\zeta| \otimes \langle 0|c)\hat{H}(|\zeta\rangle \otimes c^\dagger|0\rangle)).$$

Compare the results from (i), (ii) and (iii).

Problem 41.   Let **A** be the magnetic vector potential with dimension

$$meter \,.\, sec^{-2} \,.\, kg \,.\, A^{-1}$$

E the electric field strength (electric intensity) with dimension

$$meter \,.\, sec^{-3} \,.\, kg \,.\, A^{-1}$$

H the magnetic field strength (magnetic intensity) with dimension

$$meter^{-1} \,.\, A,$$

B the magnetic flux density (magnetic induction) with dimension

$$sec^{-2} \,.\, kg \,.\, A^{-1}$$

ϕ the *electric potential* with dimension

$$meter^2 . sec^{-3} . kg . A^{-1}$$

with

$$\mathbf{B} = \nabla \times \mathbf{A}, \qquad \mathbf{E} = -\nabla\phi - \frac{\partial A}{\partial t}$$

and $\mathbf{B} = \mu_0 \mathbf{H}$. The *Poynting vector* is given by $\mathbf{S} = \mu_0^{-1}\mathbf{E} \times \mathbf{B}$. Let $\hbar$ be the reduced Planck constant with dimension $meter^2 . sec^{-1} . kg$, ω a frequency with dimension sec^{-1}, ϵ_0 the electric permittivity with dimension $meter^{-3} . sec^4 . kg^{-1} A^2$ and V a volume with dimension $meter^3$. Then $\sqrt{\hbar/(\epsilon_r\epsilon_0\omega V)}$ has the same dimension as $\mathbf{A}$. Note that ϵ_r is dimensionless. Let μ_0 be the magnetic *permeability* with dimension $meter . sec^{-2} . kg . A^{-2}$. Then $\sqrt{\hbar\omega/(\mu_0 V)}$ has the dimension of $\mathbf{H}$. Let $n \in \mathbb{N}$ and $k_n = \omega_n\sqrt{\epsilon_r\epsilon_0\mu_0} = 2\pi n/L$ and $V = L^3$. Consider the operators

$$\hat{A}_1(x_3,t) = \sqrt{\frac{\hbar}{2\epsilon_r\epsilon_0\omega_n V}} \left(b_n e^{-i(\omega_n t - k_n x_3)} + b_n^\dagger e^{i(\omega_n t - k_n x_3)} \right)$$

$\hat{A}_2 = \hat{O}$, $\hat{A}_3 = \hat{O}$. Here b_n, $b_n^\dagger$ are Bose annihilation and creation operators with $[b_n, b_{n'}^\dagger] = \delta_{nn'} I$ and I is the identity operator. Show that

$$\hat{E}_1(x_3,t) = i\sqrt{\frac{\hbar\omega_n}{2\epsilon_r\epsilon_0 V}} \left(b_n e^{-i(\omega_n t - k_n x_3)} - b_n^\dagger e^{i(\omega_n t - k_n x_3)} \right)$$

and $\hat{E}_2 = \hat{O}$, $\hat{E}_3 = \hat{O}$. Show that

$$\hat{H}_2 = i\sqrt{\frac{\hbar\omega_n}{2\mu_0 V}} \left(b_n e^{-i(\omega_n t - k_n x_3)} - b_n^\dagger e^{i(\omega_n t - k_n x_3)} \right)$$

and $\hat{H}_1 = \hat{O}$, $\hat{H}_3 = \hat{O}$.

Chapter 28

Gauge Transformation

Gauge theories study invariance of a field theory with Lagrange function or Lagrange density under internal global symmetry transformations of a Lie symmetry group G which are extended to transformations which are extended to transformations which can be different at different space-time points. So global Lie symmetry groups are independent of the space time point. Local or gauge transformations depend on (x_0, x_1, x_2, x_3) $(x_0 = ct)$. If a set of physical laws is invariant under a global Lie symmetry transformation, the requirement of invariance under gauge Lie symmetries demands the introduction of new fields, the so-called gauge fields. The simplest case is the Lie group $U(1)$. In most cases the starting point is the Lagrange function or Lagrange density.

In electrodynamics the gauge transformation is

$$\mathbf{A} \mapsto \mathbf{A} + \nabla\chi, \quad \Phi \mapsto \Phi - \frac{1}{c}\frac{\partial\chi}{\partial t}.$$

In Yang-Mills theory the Lie algebra $su(2)$ plays a central role with the basis given by $\tau_1 = i\sigma_1$, $\tau_2 = i\sigma_2$, $\tau_3 = i\sigma_3$ (σ_1, σ_2, σ_3 are the Pauli spin matrices). Starting point is the Lie algebra valued differential one-form

$$\tilde{\alpha} = \sum_{j=1}^{3}\sum_{a=1}^{3} A_{j,a}(\mathbf{x})dx_a \otimes \tau_j$$

with

$$(1 \otimes U)\tilde{\alpha}(1 \otimes U^{-1}) = \sum_{j=1}^{3}\sum_{a=1}^{3} A_{j,a}(\mathbf{x})dx_a \otimes U\tau_j U^{-1}.$$

Problem 1. Let H be a Hamilton function and

$$\omega := \sum_{j=1}^{n} p_j dq_j - H(\mathbf{p}, \mathbf{q}, t) dt \tag{1}$$

be a *Cartan form* of a Hamilton system. Consider the transformation

$$q_j \to q_j, \quad p_j \to p_j + \frac{\partial \Omega(\mathbf{q}, t)}{\partial q_j}, \quad H(\mathbf{p}, \mathbf{q}, t) \to H(\mathbf{p}, \mathbf{q}, t) - \frac{\partial \Omega(\mathbf{q}, t)}{\partial t} \tag{2}$$

where Ω is a smooth function of $\mathbf{q}$ and t. This transformation is called a *gauge transformation*.
(i) Show that under this transformation, $d\omega \to d\omega$.
(ii) Calculate the exterior derivative of ω, i.e. $d\omega$.
(iii) Calculate

$$Z \rfloor d\omega = 0$$

where $\rfloor$ denotes the *contraction* $(\partial/\partial x_j \rfloor dx_k = \delta_{jk})$ and Z is the vector field

$$Z := \sum_{j=1}^{n} \left(V_j(\mathbf{p}, \mathbf{q}, t) \frac{\partial}{\partial p_j} + W_j(\mathbf{p}, \mathbf{q}, t) \frac{\partial}{\partial q_j} \right) + \frac{\partial}{\partial t}.$$

Solution 1. (i) From (1) and transformation (2) we obtain

$$\omega \to \sum_{j=1}^{n} \left(p_j + \frac{\partial \Omega}{\partial q_j} \right) dq_j - \left(H - \frac{\partial \Omega}{\partial t} \right) dt = \omega + \sum_{j=1}^{n} \frac{\partial \Omega}{\partial q_j} dq_j + \frac{\partial \Omega}{\partial t} dt.$$

Now

$$d \left(\sum_{j=1}^{n} \frac{\partial \Omega}{\partial q_j} dq_j + \frac{\partial \Omega}{\partial t} dt \right) = \sum_{j=1}^{n} \sum_{k=1}^{n} \frac{\partial^2 \Omega}{\partial q_j \partial q_k} dq_k \wedge dq_j + \sum_{j=1}^{n} \frac{\partial^2 \Omega}{\partial t \partial q_j} dt \wedge dq_j$$
$$+ \sum_{j=1}^{n} \frac{\partial^2 \Omega}{\partial t \partial q_j} dq_j \wedge dt.$$

Since $dq_j \wedge dq_k = -dq_k \wedge dq_j$ and $dq_j \wedge dt = -dt \wedge dq_j$ we find that the right-hand side of this equation vanishes.
(ii) The exterior derivative of ω yields

$$d\omega = \sum_{j=1}^{n} \left(dp_j \wedge dq_j - \frac{\partial H}{\partial q_j} dq_j \wedge dt - \frac{\partial H}{\partial p_j} dp_j \wedge dt \right).$$

(iii) The corresponding equations of motion of the vector field Z are given by

$$\frac{dp_j}{d\epsilon} = V_j(\mathbf{p}, \mathbf{q}, t), \quad \frac{dq_j}{d\epsilon} = W_j(\mathbf{p}, \mathbf{q}, t), \quad \frac{dt}{d\epsilon} = 1$$

where $t(\epsilon = 0) = 0$. Therefore, we find

$$Z \rfloor d\omega = \sum_{j=1}^{n} \left(V_j dq_j - W_j dp_j - W_j \frac{\partial H}{\partial q_j} dt - V_j \frac{\partial H}{\partial p_j} dt + \frac{\partial H}{\partial q_j} dq_j + \frac{\partial H}{\partial p_j} dp_j \right).$$

Thus the condition $Z \rfloor d\omega = 0$ yields the *Hamilton equations of motion*

$$V_j(\mathbf{p}, \mathbf{q}, t) = -\frac{\partial H}{\partial q_j}, \qquad W_j(\mathbf{p}, \mathbf{q}, t) = \frac{\partial H}{\partial p_j}.$$

Problem 2.   The equation of motion of a charged particle with charge q in an electromagnetic field is given by

$$m \frac{d\mathbf{v}}{dt} = q(\mathbf{E} + \mathbf{v} \times \mathbf{B}). \tag{1}$$

(i) Show that the equation of motion is invariant under

$$\mathbf{A}(\mathbf{r}, t) \to \mathbf{A}(\mathbf{r}, t) + \mathrm{grad}(\phi(\mathbf{r}, t)), \qquad U(\mathbf{r}, t) \to U(\mathbf{r}, t) - \frac{\partial \phi(\mathbf{r}, t)}{\partial t} \tag{2}$$

where ϕ is a smooth function. Transformation (2) is called a *gauge transformation*.
(ii) Show that the equation of motion (1) can be derived from the *Lagrange function*

$$L(\mathbf{r}, \mathbf{v}, t) = \frac{m\mathbf{v}^2}{2} + q(-U(\mathbf{r}, t) + \mathbf{A}(\mathbf{r}, t) \cdot \mathbf{v}) \tag{3}$$

where

$$\mathbf{A} \cdot \mathbf{v} = A_1 v_1 + A_2 v_2 + A_3 v_3, \qquad \mathbf{v}^2 = v_1^2 + v_2^2 + v_3^2.$$

Is the Lagrange function (3) invariant under the gauge transformation?

Solution 2.   (i) We write

$$\mathbf{A}' = \mathbf{A} + \mathrm{grad}(\phi), \qquad U' = U - \frac{\partial \phi}{\partial t}.$$

Since $\mathbf{B} = \mathrm{curl}(\mathbf{A})$, $\mathbf{E} = -\mathrm{grad}(U) - \partial A / \partial t$ we obtain

$$\mathbf{B}' = \mathrm{curl}(\mathbf{A}') = \mathrm{curl}(\mathbf{A} + \mathrm{grad}(\phi)) = \mathrm{curl}(\mathbf{A}) = \mathbf{B}$$

where we used $\mathrm{curl}(\mathrm{grad}(\phi)) = 0$ for any smooth ϕ. Analogously $\mathbf{E}' = \mathbf{E}$. Thus $\mathbf{E}$ and $\mathbf{B}$ are invariant under the gauge transformation as well as the equation of motion (1) which only depends on $\mathbf{E}$ and $\mathbf{B}$.
(ii) The *Euler-Lagrange equations* are given by

$$\frac{d}{dt} \frac{\partial L}{\partial v_1} - \frac{\partial L}{\partial x_1} = 0, \qquad \frac{d}{dt} \frac{\partial L}{\partial v_2} - \frac{\partial L}{\partial x_2} = 0, \qquad \frac{d}{dt} \frac{\partial L}{\partial v_3} - \frac{\partial L}{\partial x_3} = 0.$$

Since

$$\frac{\partial L}{\partial v_1} = mv_1 + qA_1$$

we obtain

$$\frac{d}{dt}\frac{\partial L}{\partial v_1} = m\frac{dv_1}{dt} + q\frac{\partial A_1}{\partial t} + q\frac{\partial A_1}{\partial x_1}v_1 + q\frac{\partial A_1}{\partial x_2}v_2 + q\frac{\partial A_1}{\partial x_3}v_3.$$

Now

$$\frac{\partial L}{\partial x_1} = -q\frac{\partial U}{\partial x_1} + qv_1\frac{\partial A_1}{\partial x_1} + qv_2\frac{\partial A_2}{\partial x_1} + qv_3\frac{\partial A_3}{\partial x_1}.$$

Therefore, from the Euler-Lagrange equations we obtain

$$m\frac{dv_1}{dt} - qE_1 - q(v_2B_3 - v_3B_2) = 0.$$

Analogously we obtain the two other components of (1). Under the gauge transformation the Lagrange function L becomes $L' = L + qd\phi/dt$, where

$$\frac{d\phi}{dt} = \frac{\partial \phi}{\partial t} + \frac{\partial \phi}{\partial x_1}v_1 + \frac{\partial \phi}{\partial x_2}v_2 + \frac{\partial \phi}{\partial x_3}v_3.$$

Problem 3. Consider the linear partial differential equations

$$\frac{\partial \mathbf{\Phi}}{\partial x} = U\mathbf{\Phi}, \qquad \frac{\partial \mathbf{\Phi}}{\partial t} = V\mathbf{\Phi} \tag{1}$$

and

$$\frac{\partial Q}{\partial x} = WQ, \qquad \frac{\partial Q}{\partial t} = ZQ \tag{2}$$

where $\mathbf{\Phi} = (\Phi_1, \Phi_2)^T$ and U, V, W, Z are smooth x, t dependent 2×2 matrices. Let

$$Q(x, t) = g(x, t)\mathbf{\Phi}(x, t) \tag{3}$$

where the 2×2 matrix g is invertible. Show that

$$U = -g^{-1}\frac{\partial g}{\partial x} + g^{-1}Wg, \qquad V = -g^{-1}\frac{\partial g}{\partial t} + g^{-1}Zg. \tag{4}$$

Solution 3. From $g^{-1}g = I_2$ we obtain

$$\frac{\partial g^{-1}}{\partial x}g + g^{-1}\frac{\partial g}{\partial x} = 0, \qquad \frac{\partial g^{-1}}{\partial t}g + g^{-1}\frac{\partial g}{\partial t} = 0.$$

From $Q = g\mathbf{\Phi}$ we obtain

$$\frac{\partial Q}{\partial x} = \frac{\partial g}{\partial x}\mathbf{\Phi} + g\frac{\partial \mathbf{\Phi}}{\partial x}, \qquad \frac{\partial Q}{\partial t} = \frac{\partial g}{\partial t}\mathbf{\Phi} + g\frac{\partial \mathbf{\Phi}}{\partial t}$$

and

$$\frac{\partial \Phi}{\partial x} = \frac{\partial g^{-1}}{\partial x}Q + g^{-1}\frac{\partial Q}{\partial x}, \qquad \frac{\partial \Phi}{\partial t} = \frac{\partial g^{-1}}{\partial t}Q + g^{-1}\frac{\partial Q}{\partial t}.$$

Inserting these equations into $\partial \Phi / \partial x = U\Phi$ and using $\Phi = g^{-1}Q$ yields

$$\frac{\partial g^{-1}}{\partial x}Q + g^{-1}WQ = U\Phi = Ug^{-1}Q.$$

Thus

$$\left(-g^{-1}\frac{\partial g}{\partial x}g^{-1} + g^{-1}W - Ug^{-1}\right)Q = 0.$$

It follows that the expression in the parenthesis of the left-hand side of this equation must vanish. Multiplying this expression with g on the right-hand side, we obtain the first equation of (4). Analogously we can prove the second one.

Problem 4.   Consider the derivative *nonlinear Schrödinger equation* in one-space dimension

$$\frac{\partial \phi}{\partial t} - i\frac{\partial^2 \phi}{\partial x^2} + \frac{\partial}{\partial x}(|\phi|^2\phi) = 0 \tag{1}$$

and another derivative nonlinear Schrödinger equation in one-space dimension

$$\frac{\partial q}{\partial t} - i\frac{\partial^2 q}{\partial x^2} + |q|^2\frac{\partial q}{\partial x} = 0 \tag{2}$$

where ϕ and q are complex-valued functions. Both partial differential equations are integrable by the inverse scattering method. Both arise as consistency conditions of a system of linear differential equations

$$\frac{\partial \Phi}{\partial x} = U(x, t, \lambda)\Phi, \qquad \frac{\partial \Phi}{\partial t} = V(x, t, \lambda)\Phi$$

$$\frac{\partial Q}{\partial x} = W(x, t, \lambda)Q, \qquad \frac{\partial Q}{\partial t} = Z(x, t, \lambda)Q$$

where λ is a complex parameter and U, V, W, Z are x, t dependent 2×2 matrices. The consistency conditions have the form

$$\frac{\partial U}{\partial t} - \frac{\partial V}{\partial x} + [U, V] = 0, \qquad \frac{\partial W}{\partial t} - \frac{\partial Z}{\partial x} + [W, Z] = 0.$$

For (1) we have

$$U = -i\lambda^2\sigma_3 + \lambda\begin{pmatrix} 0 & \phi \\ -\phi^* & 0 \end{pmatrix}$$

$$V = -(2i\lambda^4 - i|\phi|^2\lambda^2)\sigma_3 + 2\lambda^3\begin{pmatrix} 0 & \phi \\ -\phi^* & 0 \end{pmatrix} + \lambda\begin{pmatrix} 0 & i\phi_x - |\phi|^2\phi \\ i\phi_x^* + |\phi|^2\phi^* & 0 \end{pmatrix}$$

where σ_3 is the Pauli spin matrix. For (2) we have

$$W = \left(-i\lambda^2 + \frac{i}{4}|q|^2\right)\sigma_3 + \lambda\begin{pmatrix} 0 & q \\ -q^* & 0 \end{pmatrix}$$

$$Z = \left(-2i\lambda^4 - i|q|^2\lambda^2 + \frac{1}{4}(qq_x^* - q^*q_x) - \frac{i}{8}|q|^4\right)\sigma_3 + 2\lambda^3\begin{pmatrix} 0 & q \\ -q^* & 0 \end{pmatrix}$$

$$+ \lambda\begin{pmatrix} 0 & iq_x - |q|^2q/2 \\ iq_x^* + |q|^2q^*/2 & 0 \end{pmatrix}.$$

Show that (1) and (2) are gauge-equivalent.

Solution 4. We look for a transformation which converts (4a) with W into the form of (3a) with U. For this purpose the transformation which eliminates the term

$$\frac{i|q|^2}{4}\sigma_3$$

in W is introduced, i.e. $Q = g\Phi$, where

$$g = \begin{pmatrix} f & 0 \\ 0 & f^{-1} \end{pmatrix}, \qquad f(x,t) := \exp\left(\frac{i}{4}\int^x |q(s,t)|^2 ds\right).$$

Note that $g(x,t) = Q(x,t,\lambda = 0)$. Thus we obtain

$$U = -g^{-1}\frac{\partial g}{\partial x} + g^{-1}Wg, \qquad V = -g^{-1}\frac{\partial g}{\partial t} + g^{-1}Zg.$$

It follows that

$$-i\lambda^2\sigma_3 + \lambda\begin{pmatrix} 0 & \phi \\ -\phi^* & 0 \end{pmatrix} = -i\lambda^2\sigma_3 + \lambda\begin{pmatrix} 0 & qf^{-2} \\ -q^*f^2 & 0 \end{pmatrix}.$$

We find that the transformation of eigenfunction and the transformation of the field variables $\phi = qf^{-2}$ keep the inverse scattering method invariant. We call the transformation which changes an eigenvalue problem into another eigenvalue problem the *gauge transformation*. Here g is the gauge transformation.

Problem 5. (i) Consider the general linear group $GL(n, \mathbb{R})$ with the usual coordinates. Let $X \in GL(n, \mathbb{R})$. We write

$$X = (x_{ij}), \qquad dX = (dx_{ij})$$

where $\det(X) \neq 0$. Consider the $n \times n$ matrix

$$\Omega := X^{-1}dX \tag{1}$$

whose entries are differential one-forms on G. Show that

$$d\Omega + \Omega \wedge \Omega = 0 \tag{2}$$

where $\wedge$ denotes the wedge product of matrices.
(ii) Show that

$$d\Omega + \Omega \wedge \Omega = 0 \tag{3}$$

is invariant under the transformation

$$\Omega \to R^{-1}dR + R^{-1}\Omega R \tag{4}$$

where R is an $n \times n$ invertible matrix. This is called a *gauge transformation*.

Solution 5.   (i) From (1) we obtain

$$dX = X\Omega. \tag{5}$$

Taking the exterior derivative of this equation we obtain

$$0 = (dX) \wedge \Omega + X \wedge d\Omega \tag{6}$$

since $ddX = 0$. Inserting (5) into (6) we yields

$$X(\Omega \wedge \Omega + d\Omega) = 0.$$

Since X is invertible ($X \in GL(n, \mathbb{R})$), we find (3).
(ii) From $R^{-1}R = I_n$, where I_n is the $n \times n$ unit matrix, we obtain

$$(dR^{-1})R + R^{-1}dR = 0_n$$

since $dI_n = 0_n$. Thus $dR^{-1} = -R^{-1}(dR)R^{-1}$. Inserting (4) into (3) we obtain

$$d(R^{-1}dR + R^{-1}\Omega R) + (R^{-1}dR + R^{-1}\Omega R) \wedge (R^{-1}dR + R^{-1}\Omega R) = 0_n.$$

From $dR = -R^{-1}(dR)R^{-1}$, $ddR = 0$ and

$$d(R^{-1}dR) = dR^{-1} \wedge dR$$
$$d(R^{-1}\Omega R) = dR^{-1} \wedge \Omega R + R^{-1}d\Omega R - R^{-1}\Omega \wedge dR$$

we find $R^{-1}(d\Omega + \Omega \wedge \Omega)R = 0_n$. Thus (2) follows.

Problem 6.   Let

$$i\hbar \frac{\partial \psi(\mathbf{x}, t)}{\partial t} = \hat{H}_0 \psi(\mathbf{x}, t) \tag{1}$$

be the *Schrödinger equation* in three space dimensions, where

$$\hat{H}_0 = -\frac{\hbar^2}{2m}\Delta \equiv -\frac{\hbar^2}{2m}\sum_{j=1}^{3}\frac{\partial^2}{\partial x_j^2} \tag{2}$$

with $j = 1, 2, 3$, $\mathbf{x} = (x_1, x_2, x_3)$. ϵ is a real dimensionless parameter.
(i) Show that (1) is invariant under the *global gauge transformation*

$$x_j'(\mathbf{x}, t) = x_j, \qquad t'(\mathbf{x}, t) = t \tag{2a}$$

$$\psi'(\mathbf{x}'(\mathbf{x}, t), t'(\mathbf{x}, t)) = \exp(i\epsilon)\,\psi(\mathbf{x}, t). \tag{2b}$$

(ii) Let

$$i\hbar\frac{\partial\psi}{\partial t} = -\frac{\hbar^2}{2m}\sum_{j=1}^{3}\left(\frac{\partial}{\partial x_j} - i\frac{q}{\hbar}A_j\right)^2\psi + qU\psi \tag{3}$$

where $\mathbf{A}$ is the vector potential and U is the scalar potential. Show that
this partial differential equation is invariant under the transformation

$$x_j'(\mathbf{x}, t) = x_j, \qquad t'(\mathbf{x}, t) = t$$

$$\psi'(\mathbf{x}'(\mathbf{x}, t), t'(\mathbf{x}, t)) = \exp(i\epsilon(\mathbf{x}, t))\,\psi(\mathbf{x}, t)$$
$$U'(\mathbf{x}'(\mathbf{x}, t), t'(\mathbf{x}, t)) = U(\mathbf{x}, t) - \frac{\hbar}{q}\frac{\partial\epsilon}{\partial t}$$
$$A_j'(\mathbf{x}'(\mathbf{x}, t), t'(\mathbf{x}, t)) = A_j(\mathbf{x}, t) + \frac{\hbar}{q}\frac{\partial\epsilon}{\partial x_j}$$

where ϵ is a smooth function of $\mathbf{x} = (x_1, x_2, x_3)$ and t, and $j = 1, 2, 3$.

Solution 6.   (i) Since ϵ is a space and time independent parameter we
find

$$\frac{\partial^2\psi'}{\partial x_j'^2} = e^{i\epsilon}\frac{\partial\psi}{\partial x_j^2}, \qquad \frac{\partial\psi'}{\partial t'} = e^{i\epsilon}\frac{\partial\psi}{\partial t}$$

where $j = 1, 2, 3$. Thus (1) is invariant under transformation(2).
(ii) Since ϵ is a smooth function of $\mathbf{x} = (x_1, x_2, x_3)$ and t we find

$$\frac{\partial\psi'}{\partial t'} = ie^{i\epsilon}\frac{\partial\epsilon}{\partial t}\psi + e^{i\epsilon}\frac{\partial\psi}{\partial t}, \qquad \frac{\partial\psi'}{\partial x_j'} = ie^{i\epsilon}\frac{\partial\epsilon}{\partial x_j}\psi + e^{i\epsilon}\frac{\partial\psi}{\partial x_j}$$

and

$$\frac{\partial^2\psi'}{\partial x_j'^2} = -e^{i\epsilon}\left(\frac{\partial\epsilon}{\partial x_j}\right)^2\psi + ie^{i\epsilon}\frac{\partial^2\epsilon}{\partial x_j^2}\psi + 2ie^{i\epsilon}\frac{\partial\epsilon}{\partial x_j}\frac{\partial\psi}{\partial x_j} + e^{i\epsilon}\frac{\partial^2\psi}{\partial x_j^2}.$$

Therefore

$$\frac{\partial \psi}{\partial x_j} = e^{-i\epsilon} \left(\frac{\partial \psi'}{\partial x'_j} - i\frac{\partial \epsilon}{\partial x_j} \psi' \right)$$

$$\frac{\partial \psi}{\partial t} = e^{-i\epsilon} \left(\frac{\partial \psi'}{\partial t'} - i\frac{\partial \epsilon}{\partial t} \psi' \right)$$

$$\frac{\partial^2 \psi}{\partial x_j^2} = e^{-i\epsilon} \left(\frac{\partial^2 \psi'}{\partial x'^2_j} - \left(\frac{\partial \epsilon}{\partial x_j} \right)^2 \psi' - i\frac{\partial^2 \epsilon}{\partial x_j^2} \psi' - 2i\frac{\partial \epsilon}{\partial x_j}\frac{\partial \psi'}{\partial x'_j} \right).$$

Since

$$\left(\frac{\partial}{\partial x_j} - i\frac{q}{\hbar}A_j \right)^2 \psi \equiv \frac{\partial^2 \psi}{\partial x_j^2} - \frac{2iq}{\hbar}A_j\frac{\partial \psi}{\partial x_j} - \frac{iq}{\hbar}\psi\frac{\partial A_j}{\partial x_j} - \frac{q^2}{\hbar^2}A_j^2\psi$$

we obtain using

$$\frac{\partial A'_j}{\partial x'_j} = \frac{\partial A_j}{\partial x_j} + \frac{\hbar}{q}\frac{\partial^2 \epsilon}{\partial x_j^2}$$

that

$$\left(\frac{\partial}{\partial x_j} - i\frac{q}{\hbar}A_j \right)^2 \psi = e^{-i\epsilon}\left(\frac{\partial}{\partial x'_j} - i\frac{q}{\hbar}A'_j \right)^2 \psi'.$$

Moreover

$$i\hbar\frac{\partial \psi}{\partial t} - qU\psi = i\hbar e^{-i\epsilon}\left(\frac{\partial \psi'}{\partial t'} - i\frac{\partial \epsilon}{\partial t}\psi' \right) - qe^{-i\epsilon}\left(U' + \frac{\hbar}{q}\frac{\partial \epsilon}{\partial t} \right)\psi'$$

$$= e^{-i\epsilon}\left(i\hbar\frac{\partial \psi'}{\partial t'} - qU'\psi' \right).$$

Since $e^{-i\epsilon} \neq 0$, we obtain

$$i\hbar\frac{\partial \psi'}{\partial t'} = -\frac{\hbar^2}{2m}\sum_{j=1}^{3}\left(\frac{\partial}{\partial x'_j} - i\frac{q}{\hbar}A'_j \right)^2 \psi' + qU'\psi'.$$

Consequently, (3) is invariant under the gauge transformation.

Problem 7. Apply the approach described in the previous problem to the linear Dirac equation. The linear *Dirac equation* with nonvanishing rest mass m_0 is given by

$$\left(i\hbar\left(\gamma_0\frac{\partial}{\partial x_0} + \gamma_1\frac{\partial}{\partial x_1} + \gamma_2\frac{\partial}{\partial x_2} + \gamma_3\frac{\partial}{\partial x_3} \right) - m_0 c \right)\psi(\mathbf{x}) = 0$$

where $\mathbf{x} = (x_0, x_1, x_2, x_3)$, $x_0 = ct$ and the complex valued *spinor* ψ is

$$\psi(\mathbf{x}) := \begin{pmatrix} \psi_1(\mathbf{x}) \\ \psi_2(\mathbf{x}) \\ \psi_3(\mathbf{x}) \\ \psi_4(\mathbf{x}) \end{pmatrix}.$$

The *gamma matrices* are defined by

$$\gamma_0 \equiv \beta := \begin{pmatrix} 1 & 0 & 0 & 0 \\ 0 & 1 & 0 & 0 \\ 0 & 0 & -1 & 0 \\ 0 & 0 & 0 & -1 \end{pmatrix}, \quad \gamma_1 := \begin{pmatrix} 0 & 0 & 0 & 1 \\ 0 & 0 & 1 & 0 \\ 0 & -1 & 0 & 0 \\ -1 & 0 & 0 & 0 \end{pmatrix},$$

$$\gamma_2 := \begin{pmatrix} 0 & 0 & 0 & -i \\ 0 & 0 & i & 0 \\ 0 & i & 0 & 0 \\ -i & 0 & 0 & 0 \end{pmatrix}, \quad \gamma_3 := \begin{pmatrix} 0 & 0 & 1 & 0 \\ 0 & 0 & 0 & -1 \\ -1 & 0 & 0 & 0 \\ 0 & 1 & 0 & 0 \end{pmatrix}.$$

The linear Dirac equation and the *free wave equation*

$$\left(\frac{\partial^2}{\partial x_0^2} - \frac{\partial^2}{\partial x_1^2} - \frac{\partial^2}{\partial x_2^2} - \frac{\partial^2}{\partial x_3^2} \right) A_\mu = 0$$

can be derived from the Lagrange densities

$$\mathcal{L}_D = c\bar{\psi} \sum_{\mu=0}^{3} \left(i\hbar\gamma_\mu \frac{\partial}{\partial x_\mu} - m_0 c \right) \psi, \quad \mathcal{L}_E = -\frac{1}{2\mu_0} \sum_{\mu,\nu=0}^{3} \left(\frac{\partial A_\mu}{\partial x^\nu} - \frac{\partial A_\nu}{\partial x^\mu} \right) \frac{\partial A^\mu}{\partial x_\nu},$$

respectively. We used $\bar{\psi} = \psi^\dagger \gamma_0$, $A_\mu = (A_0, A_1, A_2, A_3)$, $A^\mu = (A_0, -A_1, -A_2, -A_3)$, $x_\mu = (x_0, -x_1, -x_2, -x_3)$, $x^\mu = (x_0, x_1, x_2, x_3)$.

Solution 7. From gauge theory described above we obtain the Lagrangian density

$$\mathcal{L} = -\frac{1}{2\mu_0} \sum_{\mu,\nu=0}^{3} \left(\frac{\partial A_\mu}{\partial x^\nu} - \frac{\partial A_\nu}{\partial x^\mu} \right) \frac{\partial A^\mu}{\partial x_\nu} + c\bar{\psi} \left(i\hbar \sum_{\mu=0}^{3} \gamma_\mu \frac{\partial}{\partial x_\mu} - e \sum_{\mu=0}^{3} \gamma_\mu A^\mu - m_0 c \right) \psi$$

where $\bar{\psi} = \psi^\dagger \gamma_0 \equiv (\psi_1^*, \psi_2^*, -\psi_3^*, -\psi_4^*)$ and

$$\sum_{\mu=0}^{3} \gamma_\mu A^\mu = \gamma_0 A^0 + \gamma_1 A^1 + \gamma_2 A^2 + \gamma_3 A^3 = \gamma_0 A_0 - \gamma_1 A_1 - \gamma_2 A_2 - \gamma_3 A_3.$$

From this Lagrangian density and the *Euler-Lagrange equation*

$$\frac{\partial \mathcal{L}}{\partial \phi_r} - \sum_{\mu=0}^{3} \frac{\partial}{\partial x^\mu} \frac{\partial \mathcal{L}}{\partial(\partial \phi_r / \partial x^\mu)} = 0$$

we obtain the *Maxwell-Dirac equation*

$$\left(\frac{\partial^2}{\partial x_0^2} - \frac{\partial^2}{\partial x_1^2} - \frac{\partial^2}{\partial x_2^2} - \frac{\partial^2}{\partial x_3^2}\right) A_\mu = \mu_0 ec\bar{\psi}\gamma_\mu\psi$$

$$\left(i\hbar\left(\gamma_0\frac{\partial}{\partial x_0} + \sum_{j=1}^{3}\gamma_j\frac{\partial}{\partial x_j}\right) - m_0 c\right)\psi = e\left(\sum_{j=1}^{3}\gamma_j A^j + \gamma_0 A_0\right)\psi$$

$$\frac{\partial A_0}{\partial x_0} + \frac{\partial A_1}{\partial x_1} + \frac{\partial A_2}{\partial x_2} + \frac{\partial A_3}{\partial x_3} = 0 \qquad (\textit{Lorentz gauge condition})$$

is the Dirac equation which is coupled with the vector potential A_μ where $A_0 = U/c$ and U denotes the scalar potential. Here $\mu = 0, 1, 2, 3$ and

$$\bar{\psi} = (\psi_1^*, \psi_2^*, -\psi_3^*, -\psi_4^*)$$

c is the speed of light, e the charge, μ_0 the permeability of free space, $\hbar = h/(2\pi)$ where h is Planck's constant and m_0 the particle rest mass. We set $\lambda = \hbar/(m_0 c)$. The Maxwell-Dirac equation can be written as the following coupled system of thirteen partial differential equations

$$\left(\frac{\partial^2}{\partial x_0^2} - \frac{\partial^2}{\partial x_1^2} - \frac{\partial^2}{\partial x_2^2} - \frac{\partial^2}{\partial x_3^2}\right) A_0 = (\mu_0 ec)\sum_{j=1}^{4}(u_j^2 + v_j^2)$$

$$\left(\frac{\partial^2}{\partial x_0^2} - \frac{\partial^2}{\partial x_1^2} - \frac{\partial^2}{\partial x_2^2} - \frac{\partial^2}{\partial x_3^2}\right) A_1 = (2\mu_0 ec)(u_1 u_4 + v_1 v_4 + u_2 u_3 + v_2 v_3)$$

$$\left(\frac{\partial^2}{\partial x_0^2} - \frac{\partial^2}{\partial x_1^2} - \frac{\partial^2}{\partial x_2^2} - \frac{\partial^2}{\partial x_3^2}\right) A_2 = (2\mu_0 ec)(u_1 v_4 - u_4 v_1 + u_3 v_2 - u_2 v_3)$$

$$\left(\frac{\partial^2}{\partial x_0^2} - \frac{\partial^2}{\partial x_1^2} - \frac{\partial^2}{\partial x_2^2} - \frac{\partial^2}{\partial x_3^2}\right) A_3 = (2\mu_0 ec)(u_1 u_3 + v_3 v_1 - u_2 u_4 - v_2 v_4)$$

$$\lambda\frac{\partial u_1}{\partial x_0} + \lambda\frac{\partial u_3}{\partial x_3} + \lambda\frac{\partial u_4}{\partial x_1} + \lambda\frac{\partial v_4}{\partial x_2} - v_1 = \frac{e}{m_0 c}(A_0 v_1 - A_3 v_3 - A_1 v_4 + A_2 u_4)$$

$$-\lambda\frac{\partial v_1}{\partial x_0} - \lambda\frac{\partial v_3}{\partial x_3} - \lambda\frac{\partial v_4}{\partial x_1} + \lambda\frac{\partial u_4}{\partial x_2} - u_1 = \frac{e}{m_0 c}(A_0 u_1 - A_3 u_3 - A_1 u_4 - A_2 v_4)$$

$$\lambda\frac{\partial u_2}{\partial x_0} + \lambda\frac{\partial u_3}{\partial x_1} - \lambda\frac{\partial v_3}{\partial x_2} - \lambda\frac{\partial u_4}{\partial x_3} - v_2 = \frac{e}{m_0 c}(A_0 v_2 - A_1 v_3 - A_2 u_3 + A_3 v_4)$$

$$-\lambda\frac{\partial v_2}{\partial x_0} - \lambda\frac{\partial v_3}{\partial x_1} - \lambda\frac{\partial u_3}{\partial x_2} + \lambda\frac{\partial v_4}{\partial x_3} - u_2 = \frac{e}{m_0 c}(A_0 u_2 - A_1 u_3 + A_2 v_3 + A_3 u_4)$$

$$-\lambda\frac{\partial u_1}{\partial x_3} - \lambda\frac{\partial u_2}{\partial x_1} - \lambda\frac{\partial v_2}{\partial x_2} - \lambda\frac{\partial u_3}{\partial x_0} - v_3 = \frac{e}{m_0 c}(A_3 v_1 + A_1 v_2 - A_2 u_2 - A_0 v_3)$$

$$\lambda\frac{\partial v_1}{\partial x_3} + \lambda\frac{\partial v_2}{\partial x_1} - \lambda\frac{\partial u_2}{\partial x_2} + \lambda\frac{\partial v_3}{\partial x_0} - u_3 = \frac{e}{m_0 c}(A_3 u_1 + A_1 u_2 + A_2 v_2 - A_0 u_3)$$

$$-\lambda\frac{\partial u_1}{\partial x_1} + \lambda\frac{\partial v_1}{\partial x_2} + \lambda\frac{\partial u_2}{\partial x_3} - \lambda\frac{\partial u_4}{\partial x_0} - v_4 = \frac{e}{m_0 c}(A_1 v_1 + A_2 u_1 - A_3 v_2 - A_0 v_4)$$

$$\lambda\frac{\partial v_1}{\partial x_1} + \lambda\frac{\partial u_1}{\partial x_2} - \lambda\frac{\partial v_2}{\partial x_3} + \lambda\frac{\partial v_4}{\partial x_0} - u_4 = \frac{e}{m_0 c}(A_1 u_1 - A_2 v_1 - A_3 u_2 - A_0 u_4)$$

$$\frac{\partial A_0}{\partial x_0} + \frac{\partial A_1}{\partial x_1} + \frac{\partial A_2}{\partial x_2} + \frac{\partial A_3}{\partial x_3} = 0.$$

Problem 8. In *string theory* one considers 2-dimensional surfaces in d-dimensional Minkowski space called the world sheet of strings. As a string propagates through space-time, it describes a surface in space-time which is called its world sheet. The action S is proportional to an area of the world sheet. This is the simplest generalization of the mechanics of relativistic particles with action, proportional to the length of the world line. One considers a fixed background pseudo-Riemannian space-time manifold M of dimension D, with coordinates $X = (X^\mu)$ and $\mu = 0, 1, \ldots, D - 1$. The metric is $G_{\mu\nu}$ and we take the signature $(-, (+)^{D-1})$. In Minkowski space we have $D = 4$ and $(-, +, +, +)$. The motion of a relativistic string in M is described by its generalized world line, a two-dimensional surface Σ, which is called the *world-sheet*. For a single non-interacting string the world-sheet has the form of an infinite strip. One introduces coordinates $\boldsymbol{\sigma} = (\sigma^0, \sigma^1)$ on the world-sheet. Sometimes one also writes $\sigma^0 = \tau$ and $\sigma^1 = \sigma$. The embedding of the world-sheet into space-time is given by the smooth maps

$$X : \Sigma \to M : \boldsymbol{\sigma} \to X(\boldsymbol{\sigma}).$$

The background metric induces a metric on the world-sheet

$$G_{\alpha\beta} := \sum_{\mu,\nu=0}^{D-1} \frac{\partial X^\mu}{\partial \sigma^\alpha} \frac{\partial X^\nu}{\partial \sigma^\beta} G_{\mu\nu}$$

where $\alpha, \beta = 0, 1$ are the world-sheet indices. The induced metric $G_{\alpha\beta}$ is to be distinguished from the intrinsic metric $h_{\alpha\beta}$ on Σ. An intrinsic metric is used as an auxiliary field in the Polyakov formulation of the bosonic string. This setting can be viewed from two perspectives. In the space-time perspective we interpret the system as a relativistic string moving in the space-time M. On the other hand it can be considered as a two-dimensional field theory living on the world sheet, with fields X which take values in the target-space M. This is the world-sheet perspective. Thus we can use methods of the two-dimensional field theory. The natural action for a relativistic string is its area, measured with the induced metric

$$S_{NG} = \frac{1}{2\pi\alpha'} \int_\Sigma d\sigma_0 d\sigma_1 |\det(G_{\alpha\beta})|^{1/2}.$$

This is the *Nambu-Goto action*. The prefactor $(2\pi\alpha')^{-1}$ is the energy per length or tension of the string, which is the fundamental parameter of the theory. The *Polyakov action*, is equivalent to the Nambu-Goto action. It is a standard two-dimensional field theory action. For this action one introduces an intrinsic metric on the world-sheet, $h_{\alpha\beta}(\sigma)$. The action takes the form of a *nonlinear sigma model* on the world-sheet

$$
S_P = \frac{1}{4\pi\alpha'} \int_\Sigma d\sigma^0 d\sigma^1 \sqrt{h} \sum_{\alpha,\beta=0}^{1} \sum_{\mu,\nu=0}^{D-1} h^{\alpha\beta} \frac{\partial X^\mu}{\partial\sigma^\alpha} \frac{\partial X^\nu}{\partial\sigma^\beta} G_{\mu\nu}(X)
$$

where $h := |\det(h_{\alpha\beta})|$ and $h_{\alpha\beta}$ is the inverse of $h^{\alpha\beta}$.
(i) Let

$$
G_{\mu\nu} = \eta_{\mu\nu} = \mathrm{diag}(-1, +1, \ldots, +1)
$$

be a flat Minkowski metric and let $h_{00} = -1$, $h_{11} = 1$, $h_{01} = h_{10} = 0$. Find the Polyakov action for this special case. Give the Lagrangian density and the equations of motion.
(ii) Let $D = 3$ and use the results from (i). Consider

$$
\begin{aligned}
X^0(\tau,\sigma) &= A\tau \\
X^1(\tau,\sigma) &= A\cos(2\tau)\cos(2\sigma) \\
X^2(\tau,\sigma) &= A\sin(c\tau)\cos(c\sigma)
\end{aligned}
$$

where A is a nonzero constant and c is a positive constant. Find the condition on c such that the equations of motion are satisfied. Are the boundary conditions $X^\mu(\tau,\sigma+\pi) = X^\mu(\tau,\sigma)$ satisfied? Find the condition on c such that

$$
\frac{\partial X}{\partial\tau} \cdot \frac{\partial X}{\partial\sigma} := -\frac{\partial X^0}{\partial\tau}\frac{\partial X^0}{\partial\sigma} + \frac{\partial X^1}{\partial\tau}\frac{\partial X^1}{\partial\sigma} + \frac{\partial X^2}{\partial\tau}\frac{\partial X^2}{\partial\sigma} = 0.
$$

Solution 8.   (i) We have $h = 1$ and

$$
\begin{aligned}
\sum_{\mu,\nu=0}^{D-1}\sum_{\alpha,\beta=0}^{1} h^{\alpha\beta}\frac{\partial X^\mu}{\partial\sigma^\alpha}\frac{\partial X^\nu}{\partial\sigma^\beta}\eta_{\mu\nu} &= \sum_{\mu=0,\nu=0}^{D-1}\left(h^{00}\frac{\partial X^\mu}{\partial\sigma^0}\frac{\partial X^\nu}{\partial\sigma^0}\eta_{\mu\nu} + h^{11}\frac{\partial X^\mu}{\partial\sigma^1}\frac{\partial X^\nu}{\partial\sigma^1}\eta_{\mu\nu}\right) \\
&= \sum_{\mu,\nu=0}^{D-1}\left(-\frac{\partial X^\mu}{\partial\sigma^0}\frac{\partial X^\nu}{\partial\sigma^0}\eta_{\mu\nu} + \frac{\partial X^\mu}{\partial\sigma^1}\frac{\partial X^\nu}{\partial\sigma^1}\eta_{\mu\nu}\right) \\
&= \frac{\partial X^0}{\partial\sigma^0}\frac{\partial X^0}{\partial\sigma^0} - \sum_{\mu=1}^{D-1}\frac{\partial X^\mu}{\partial\sigma^0}\frac{\partial X^\mu}{\partial\sigma^0} \\
&\quad - \frac{\partial X^0}{\partial\sigma^1}\frac{\partial X^0}{\partial\sigma^1} + \sum_{\mu=1}^{D-1}\frac{\partial X^\mu}{\partial\sigma^1}\frac{\partial X^\mu}{\partial\sigma^1}.
\end{aligned}
$$

The *Euler-Lagrange equations* are given by ($\sigma^0 = \tau, \sigma^1 = \sigma$)

$$\frac{\partial \mathcal{L}}{\partial X^\mu} - \frac{\partial}{\partial \tau}\left(\frac{\partial \mathcal{L}}{\partial(\partial X^\mu/\partial \tau)}\right) - \frac{\partial}{\partial \sigma}\left(\frac{\partial \mathcal{L}}{\partial(\partial X^\mu/\partial \sigma)}\right) = 0$$

with the *Lagrange density*

$$\mathcal{L} = -\frac{\partial X^0}{\partial \tau}\frac{\partial X^0}{\partial \tau} + \sum_{\mu=1}^{D-1}\frac{\partial X^\mu}{\partial \tau}\frac{\partial X^\mu}{\partial \tau} + \frac{\partial X^0}{\partial \sigma}\frac{\partial X^0}{\partial \sigma} - \sum_{\mu=1}^{D-1}\frac{\partial X^\mu}{\partial \sigma}\frac{\partial X^\mu}{\partial \sigma}.$$

Since $\partial \mathcal{L}/\partial X^\mu = 0$ for $\mu = 0, 1, \ldots, D-1$ we obtain a system of one-dimensional linear wave equations

$$\frac{\partial^2 X^\mu}{\partial \tau^2} - \frac{\partial^2 X^\mu}{\partial \sigma^2} = 0$$

where $\mu = 0, 1, \ldots, D-1$.

(ii) Inserting the ansatz for X^0, X^1, X^2 into these linear wave equations we find that the equations are satisfied with c arbitrary. We obtain

$$\frac{\partial X}{\partial \tau} \cdot \frac{\partial X}{\partial \sigma} := -\frac{\partial X^0}{\partial \tau}\frac{\partial X^0}{\partial \sigma} + \frac{\partial X^1}{\partial \tau}\frac{\partial X^1}{\partial \sigma} + \frac{\partial X^2}{\partial \tau}\frac{\partial X^2}{\partial \sigma}$$
$$= 4A^2 \sin(2\tau)\cos(2\tau)\sin(2\sigma)\cos(2\sigma)$$
$$- A^2 c^2 \sin(c\tau)\cos(c\tau)\sin(c\sigma)\cos(c\sigma).$$

Thus c must be 2. Since $\cos(2\pi) = 1$ the boundary conditions $X^\mu(\tau, \sigma + \pi) = X^\mu(\tau, \sigma)$ are satisfied.

Problem 9. The AdS/CFT correspondence is the equivalence between a string theory or supergravity defined on some sort of anti de Sitter space (AdS) and a conformal field theory (CFT) defined on its conformal boundary whose dimension is lower by one. Therefore it refers to the existence of dualities between theories with gravity and theories without gravity. An example of such a correspondence is the exact equivalence between type IIB string theory compactified on $AdS^5 \times \mathbb{S}^5$, and four-dimensional $N = 4$ supersymmetric Yang-Mills theory. Here AdS^5 refers to an Anti-de Sitter space in five dimensions, S^5 refers to a five-dimensional sphere

$$S^5 := \{\, (X_1, X_2, \ldots, X_6) \ : \ X_1^2 + X_2^2 + \cdots + X_6^2 = 1 \,\}.$$

Anti-de Sitter spaces are maximally symmetric solutions of the Einstein equations with a negative cosmological constant. The large symmetry group of 5d anti-de Sitter space matches with the group of conformal symmetries of the $N = 4$ super Yang-Mills theory. Integrable nonlinear dynamical systems can be derived from this correspondence. Consider the bosonic part of the classical closed string propagating in the $AdS_5 \times \mathbb{S}^5$ space time. The two metrics have the standard form in terms of the $5 + 5$ angular coordinates

$$(ds^2)_{AdS_5} = -\cosh^2(\rho)dt^2 + d\rho^2 + \sinh^2(\rho)(d\theta^2 + \sin^2(\theta)d\phi^2 + \cos^2(\theta)d\varphi^2)$$

$$(ds^2)_{\mathbb{S}^5} = d\gamma^2 + \cos^2(\gamma)d\varphi_3^2 + \sin^2(\gamma)(d\psi^2 + \cos^2(\psi)d\varphi_1^2 + \sin^2(\psi)d\varphi_2^2).$$

Consider the compact Lie group $O(6)$ and the non-compact Lie group $SO(4,2)$. The bosonic part of the sigma model action as an action S for the $O(6) \times SO(4,2)$ sigma-model is given by

$$S = \frac{\sqrt{\lambda}}{2\pi} \int d\tau d\sigma (\mathcal{L}_S + \mathcal{L}_{AdS}), \qquad \sqrt{\lambda} \equiv \frac{R^2}{\alpha'}$$

with the Lagrange densities

$$\mathcal{L}_S = -\frac{1}{2} \sum_{M=1}^{6} \sum_{\alpha=0}^{1} \partial_\alpha X_M \partial^\alpha X_M + \frac{1}{2}\Lambda \sum_{M=1}^{6} (X_M X_M - 1)$$

$$\mathcal{L}_{AdS} = -\frac{1}{2} \sum_{M=0}^{5} \sum_{N=0}^{5} \sum_{\alpha=0}^{1} \eta_{MN} \partial_\alpha Y_M \partial^\alpha Y_N + \frac{1}{2}\tilde{\Lambda} \sum_{M=0}^{5} \sum_{N=0}^{5} (\eta_{MN} Y_M Y_N + 1).$$

Here X_M, $M = 1, 2, \ldots, 6$ and Y_M, $M = 0, 1, \ldots, 5$ are the embedding coordinates of $\mathbb{R}^6$ with the Euclidean metric in $\mathcal{L}_S$ and with

$$\eta_{MN} = (-1, +1, +1, +1, +1, -1)$$

in $\mathcal{L}_{AdS}$, respectively. Λ and $\tilde{\Lambda}$ are the Lagrange multipliers. The world-sheet coordinates are σ^α, $\alpha = 0, 1$ with $(\sigma^0, \sigma^1) = (\tau, \sigma)$ and the world-sheet metric is $\mathrm{diag}(-1, 1)$. The action is to be supplemented with the usual conformal gauge constraints. The embedding coordinates are related to the angular ones as follows

$$X_1 + iX_2 = \sin(\gamma)\cos(\psi)e^{i\varphi_1}, \qquad X_3 + iX_4 = \sin(\gamma)\sin(\psi)e^{i\varphi_2}$$
$$X_5 + iX_6 = \cos(\gamma)e^{i\varphi_3}$$
$$Y_1 + iY_2 = \sinh(\rho)\sin(\theta)e^{i\phi}, \qquad Y_3 + iY_4 = \sinh(\rho)\cos(\theta)e^{i\varphi}$$
$$Y_5 + iY_0 = \cosh(\rho)e^{it}.$$

Consider the case that the string is located at the centre of AdS_5 and rotating in $\mathbb{S}^5$. This means it is embedded in AdS_5 as $Y_5 + iY_0 = e^{i\kappa\tau}$ with $Y_1 = Y_2 = Y_3 = Y_4 = 0$. The $\mathbb{S}^5$ metric has three translational isometries in φ_i which give rise to three global commuting integrals of motion (spins) J_i. When one considers period motion with three J_i non-zero, one chooses the following ansatz for X_M

$$X_1 + iX_2 = x_1(\sigma)e^{i\omega_1\tau}, \qquad X_3 + iX_4 = x_2(\sigma)e^{i\omega_2\tau}, \qquad X_5 + iX_6 = x_3(\sigma)e^{i\omega_3\tau}$$

where the real radial functions x_i are independent of time and as a consequence of the condition $X_M^2 = 1$ lie on a 2-sphere $\mathbb{S}^2$, $x_1^2 + x_2^2 + x_3^2 = 1$. Then the spins $J_1 = J_{12}$, $J_2 = J_{34}$, $J_3 = J_{56}$ forming a Cartan subalgebra of the compact Lie algebra $so(6)$ are

$$J_i = \sqrt{\lambda} \omega_i \int_0^{2\pi} \frac{d\sigma}{2\pi} x_i^2(\sigma).$$

Substitute the ansatz for X_1+iX_2, X_3+iX_4 and X_5+iX_6 into the Lagrange density $\mathcal{L}_S$ and find the Lagrange function and the corresponding equations of motion.

Solution 9. Inserting the ansatz for X_1+iX_2, X_3+iX_4, X_5+iX_6 into the Lagrange density $\mathcal{L}_S$ we find the Lagrange function of a one-dimensional nonlinear system

$$L(\mathbf{x}', \mathbf{x}) = \frac{1}{2} \sum_{i=1}^3 (x_i'^2 - \omega_i^2 x_i^2) + \frac{1}{2} \Lambda \sum_{i=1}^3 (x_i^2 - 1).$$

This Lagrange function describes an $n = 3$ harmonic oscillator constrained to remain on a unit $n-1 = 2$ sphere. It is a special case of the n-dimensional Neumann dynamical system which is know to be integrable. Solving the equation of motion for the Lagrange multiplier Λ we obtain the autonomous nonlinear system of ordinary differential equations

$$\frac{d^2 x_j}{d\sigma^2} + \omega_j^2 x_j = -x_j \sum_{k=1}^3 \left(\left(\frac{dx_k}{d\sigma} \right)^2 - \omega_k^2 x_k^2 \right), \qquad j = 1, 2, 3.$$

The canonical momenta conjugate to x_j are defined by

$$\pi_j := \frac{dx_j}{d\sigma}, \qquad \sum_{j=1}^3 \pi_j x_j = 0.$$

The n-dimensional Neumann system has the following n first integrals

$$F_j(\mathbf{x}, \boldsymbol{\pi}) = x_j^2 + \sum_{k \neq j} \frac{(x_j \pi_k - x_k \pi_j)^2}{\omega_j^2 - \omega_k^2}, \qquad \sum_{j=1}^n F_j(\mathbf{x}, \boldsymbol{\pi}) = 1$$

where $n = 3$ for the present case.

Problem 10. The *Nambu-Goto action* in a flat target space of signature $(2, 2)$ is given by

$$S_{NG} = -\frac{T}{2} \int d\tau d\sigma \sqrt{-\det(\gamma)}$$

where $(\tau = \sigma^0, \sigma = \sigma^1)$

$$\gamma_{A''B''} := \sum_{\mu=0}^{3}\sum_{\nu=0}^{3} \frac{\partial X^\mu(\tau,\sigma)}{\partial \sigma^{A''}} \frac{\partial X^\nu(\tau,\sigma)}{\partial \sigma^{B''}} \eta_{\mu\nu}$$

with $\eta = \mathrm{diag}(1,1,-1,-1)$, $A'',B'' = 0,1$ and $\mu,\nu = 0,1,2,3$. Consider the 2×2 matrix

$$X^{AA'} := \frac{1}{\sqrt{2}} \begin{pmatrix} -X^0 + X^2 & X^1 - X^3 \\ -X^1 - X^3 & -X^0 - X^2 \end{pmatrix}$$

where $A, A' = 0,1$. We define the $2 \times 2 \times 2$ hypermatrix a by

$$a_{AA'0} := \frac{\partial X^{AA'}}{\partial \tau}, \qquad a_{AA'1} := \frac{\partial X^{AA'}}{\partial \sigma}$$

where $A, A' = 0,1$. The *hyperdeterminant* $\mathrm{Det}(a)$ of a $2 \times 2 \times 2$ hypermatrix a defined by

$$\mathrm{Det}(a) := -\frac{1}{2}\epsilon^{AB}\epsilon^{A'B'}\epsilon^{CD}\epsilon^{C'D'}\epsilon^{A''D''}\epsilon^{B''C''} a_{AA'A''}a_{BB'B''}a_{CC'C''}a_{DD'D''}$$

where $\epsilon^{00} = \epsilon^{11} = 0$ and $\epsilon^{01} = -\epsilon^{10} = 1$. Find the connection between the 2×2 matrix γ and the $2 \times 2 \times 2$ hypermatrix a. Express $\det(\gamma)$ using the hypermatrix and hyperdeterminant.

Solution 10.   Obviously we have

$$\gamma_{A''B''} = \sum_{A,A',B,B'=0}^{1} \epsilon^{AB}\epsilon^{A'B'} a_{AA'A''}a_{BB'B''}.$$

We obtain $\det(\gamma) = -\mathrm{Det}(a)$. The non-compact Lie group $SL(2,\mathbb{R})$ acting on the index A and the $SL(2,\mathbb{R})$ acting on the index A' is the $O(2,2) \sim SL(2,\mathbb{R}) \times SL(2,\mathbb{R})$ space-time symmetry. The Lie group $SL(2,\mathbb{R})$ acting on the index A'' is the worldsheet symmetry

$$\begin{pmatrix} \partial X^\mu/\partial\tau \\ \partial X^\mu/\partial\sigma \end{pmatrix} \to \begin{pmatrix} c_{11} & c_{12} \\ c_{21} & c_{22} \end{pmatrix} \begin{pmatrix} \partial X^\mu/\partial\tau \\ \partial X^\mu/\partial\sigma \end{pmatrix}$$

where for the Lie group $SL(2,\mathbb{R})$ the c_{jk} are real constants with $c_{11}c_{22} - c_{12}c_{21} = 1$.

Supplementary Problems

Problem 1.   Let $\{X_1, \ldots, X_n\}$ be a basis of a semi-simple Lie algebra. Let $\tilde{\alpha} = \sum_{j=1}^{n} \alpha_j \otimes X_j$ be a Lie algebra valued differential one form and let β be a Lie algebra valued p-form $\widetilde{\beta} = \sum_{k=1}^{n} \beta_k \otimes X_k$. The *covariant exterior derivative* of a Lie algebra valued p-form β with respect to a Lie algebra valued differential form $\widetilde{\alpha}$ is defined as

$$D_{\widetilde{\alpha}}\widetilde{\beta} : d\widetilde{\beta} - g[\widetilde{\alpha}, \widetilde{\beta}]$$

where $g = -1$ if p is even and $g = -1/2$ if p is odd and

$$d\widetilde{\alpha} = \sum_{j=1}^{n}(d\alpha_j) \otimes X_j, \quad [\widetilde{\alpha}, \widetilde{\beta}] = \sum_{j=1}^{n}\sum_{k=1}^{n}(\alpha_j \wedge \beta_k) \otimes [X_j, X_k].$$

So for $\widetilde{\beta} = \widetilde{\alpha}$ we have $D_{\widetilde{\alpha}}\widetilde{\alpha} = d\widetilde{\alpha} + \frac{1}{2}[\widetilde{\alpha}, \widetilde{\alpha}]$ and $D_{\widetilde{\alpha}}(D_{\widetilde{\alpha}}) = 0$ (Bianchi identity). The Yang-Mills equation is given by

$$D_{\widetilde{\alpha}}(\star(D_{\widetilde{\alpha}}\widetilde{\alpha})) = 0$$

where $\star$ is the Hodge duality operator. In physics $\widetilde{\alpha}$ is referred to as vector potential and $D_{\widetilde{\alpha}}\widetilde{\alpha}$ the field strength tensor. Assume the Lie algebra L are represented by $r \times r$ matrices. L is associated by the Lie group G, where G is a Lie subgroup of $GL(r, \mathbb{R})$. Let $U \in C^{\infty}(M, G)$. We define (gauge transformation)

$$\widetilde{\alpha} \cdot U := U^{-1}\alpha U + U^{-1}dU.$$

Show that if

$$D_{\widetilde{\alpha}}(\star D_{\widetilde{\alpha}}\widetilde{\alpha}) = 0$$

then

$$D_{\widetilde{\alpha} \cdot U}(\star D_{\widetilde{\alpha} \cdot U}(\widetilde{\alpha} \cdot U)) = 0.$$

Note that

$$(1 \otimes U)(\widetilde{\alpha})(1 \otimes U^{-1}) = \sum_{j=1}^{n}(\alpha_j \otimes (UX_jU^{-1})).$$

We use the short cut notation $U\widetilde{\alpha}U^{-1}$ instead of $1 \otimes U)\widetilde{\alpha}(1 \otimes U^{-1})$. Utilize that

$$dI_r = d(UU^{-1})(dU)U^{-1} + U(dU^{-1}) = 0_n$$

i.e. $dU^{-1} = -U^{-1}(dU)U^{-1}$.

Problem 2. Consider the Hubbard model on a simple cubic lattice

$$\hat{H} = t\sum_{n,j}(c_{n+j,\uparrow}^{\dagger}c_{n,\uparrow} + c_{n+j,\downarrow}^{\dagger}c_{n,\downarrow}) + V\sum_{n}c_{n,\uparrow}^{\dagger}c_{n,\uparrow}c_{n,\downarrow}^{\dagger}c_{n,\downarrow}$$

where $c_{n,\uparrow}^{\dagger}$, $c_{n,\downarrow}^{\dagger}$ are Fermi creation operators with spin up and down, respectively, at lattice site n and $c_{n,\uparrow}$, $c_{n,\downarrow}$ are Fermi annihilation operators with spin up and down, respectively at lattice side n. The vector j points to the nearest neighbours. To introduce a vector potential A (dimension $meter \cdot sec^{-2} \cdot kg \cdot Ampere^{-1}$) we denote by $A(n)_j$ be the vector potential related to the edge connecting lattice site n with nearest lattice site $n + j$. We can approximate as

$$A(n)_j = \int_n^{n=j} A \cdot dr.$$

A gauge transformation $A \mapsto A + \mathrm{grad}(\phi)$ provides

$$A(n)_j \mapsto A(n)_j + \phi(n + j) - \phi(n).$$

With the gauge transformation

$$c_{n,\uparrow}^\dagger \mapsto \exp(i\phi(n))c_{n,\uparrow}^\dagger, \quad c_{n,\downarrow}^\dagger \mapsto \exp(i\phi(n))c_{n,\downarrow}^\dagger$$

we find that the interacting term $V \sum_n c_{n,\uparrow}^\dagger c_{n,\uparrow} c_{n,\downarrow}^\dagger c_{n,\downarrow}$ stays invariant, whereas for the kinetic term we find a factor of the form $\exp(i(\phi(n+j)_-\phi(n))$. Hence the Hamilton operator takes the form

$$\hat{H} = t \sum_{n,j} \exp(ie_0\hbar A(n)_j)(c_{n+j,\uparrow}^\dagger c_{n,\uparrow} + c_{n+j,\downarrow}c_{n,\downarrow} + V \sum_n c_{n,\uparrow}^\dagger c_{n,\uparrow}c_{n,\downarrow}^\dagger c_{n,\downarrow}$$

The magnetic field induces currents $I(n)_j$ flowing from lattice site n to lattice site $n + j$. Show that with the electric charge

$$q(n) = e_0(c_{n,\uparrow}^\dagger c_{n,\uparrow} + c_{n,\downarrow}^\dagger c_{n,\downarrow})$$

and (Heisenberg equation of motion)

$$\frac{dq(n}{dt} = \frac{i}{\hbar}[\hat{H}, q(n)] = -\sum_j I(n)_j$$

one finds

$$I(n)_j = ie_0 t\hbar^{-1}(\exp(ie_0\hbar^{-1}A(n)_j)(c_{n+j,\uparrow}^\dagger c_{n,\uparrow}^\dagger + c_{n+j,\downarrow}^\dagger c_{n,\downarrow})$$
$$- \exp(-ie_0\hbar^{-1}A(n)_j)(c_{n,\uparrow}^\dagger c_{n+j,\uparrow} + c_{n,\downarrow}^\dagger c_{n+j,\downarrow})).$$

Problem 3. Let $\hbar$ be the reduced Planck constant and ϵ_0 be the dielectric constant. The *vector potential* operator in the *Coulomb gauge* is in the box-normalized $V = L^3$ form given by

$$\hat{A}(x) = (\hbar/(2\epsilon_0 L^3))^{1/2} \sum_k \sum_{\lambda=1}^2 (\omega_k)^{-1/2}(b_{k,\lambda} \exp(ik \cdot x) + b_{k,\lambda}^\dagger \exp(-ik \cdot x))e_\lambda(k)$$

where $\omega_{\mathbf{k}} = c|\mathbf{k}|$, $\mathbf{k} = 2\pi\mathbf{n}/L$, $n_j \in \mathbb{Z}$ and $j = 1, 2, 3$. Here $b_{\mathbf{k},\lambda}$ and $b^{\dagger}_{\mathbf{k},\lambda}$ are Bose creation and annihilation operators with the commutator

$$[b_{\mathbf{k},\lambda}, b^{\dagger}_{\mathbf{k}',\lambda'}] = \delta_{\mathbf{k}\mathbf{k}'}\delta_{\lambda\lambda'}I$$

and $\mathbf{e}_{\lambda}(\mathbf{k})$ ($\lambda = 1, 2$) are the two polarization vectors orthogonal to the wave vector $\mathbf{k}$ and to each other. Find $\hat{\mathbf{B}} = \mathrm{curl}(\hat{\mathbf{A}})$.

Bibliography

Barbeau, E. J., *Polynomials*, Springer-Verlag, New York (1989)

Becker K., Becker M. and Schwarz J. H., *String Theory and M-Theory: A Modern Introduction*, Cambridge University Press (2007)

Berger, M., Pansu, P., Berry, J. P. and Saint-Raymond, X., *Problems in Geometry*, Springer, New York (1984)

Bernstein, D. S., *Matrix Mathematics: Theory, Facts, and Formulas with Application to Linear Systems Theory*, Princeton University Press (2005)

Constantinescu F. and Magyari E., *Problems in Quantum Mechanics*, Pergamon Press, Oxford (1971)

Cronin, J. A., Greenberg, D. F. and Telegdi, V. L., *Graduate Problems in Physics*, Addison Wesley, Reading (1967)

Davis H. T., *Introduction to Nonlinear Differential and Integral Equations*, Dover Publication, New York (1962)

de Souza, P. N. and Silva J.-N., *Berkeley Problems in Mathematics*, Springer, New York (1998)

Flügge S., *Practical Quantum Mechanics*, Springer, Berlin (1974)

Gelbaum, B., *Problems in Analysis*, Springer, New York (1982).

Hall B. C., *Lie Groups, Lie Algebras and Representations*, Springer, New York (2004)

Hardy G. H. and Wright E. M., *An Introduction to the Theory of Numbers*, Clarendon, Oxford (1980)

Hardy, Y., Kiat Shi Tan and Steeb W.-H., *Computer Algebra with SymbolicC++*, World Scientific, Singapore (2008)

Helgason S., *Differential Geometry, Lie Groups and Symmetric Spaces*, American Mathematical Society (2001)

Hirota R., *The Direct Method in Soliton Theory*, Cambridge University Press (2004)

Howson A. G., *A Handbook of terms used in algebra and analysis*, Cambridge University Press (1972)

Ince E. L., *Ordinary Differential Equations*, Dover Publication, New York (1956)

Knapp A. W., *Lie Groups: Beyond an Introduction*, 2nd edition, Birkhäuser, Boston 2002

Knoop, K., *Problem Book in the Theory of Functions*, Volume I, Dover, New York (1952)

Kriele M., *Spacetime: Foundations of General Relativity and Differential Geometry*, Springer (2001)

Krzyz, J. G., *Problems in Complex Variables Theory*, Elsevier, New York (1971)

Larson, L. C., *Problem Solving Through Problems*, Springer, New York (1983)

Lemons D. S., *Dimensional Analysis*, Cambridge University Press, Cambridge (2017)

Louisell W. H., *Quantum Statistical Properties of Radiation*, Wiley, New York (1973)

Manteanu L. and Donescu S., *Introduction to Soliton Theory: Applications to Mechanics*, Springer (2004)

Misner C. W., Thorne K. S. and Wheeler J. A., *Gravitation*, Freeman, San Franciso (1973)

Morse P. H. and Feshbach H., *Methods of Theoretical Physics*, McGraw-Hill, New York (1953)

Moloney J. V. and Newell A. C., *Nonlinear Optics*, CRC Press (2018)

Olver P. J., *Applications of Lie Groups to Differential Equations*, Springer, New York (1986)

Polyanin A. D. and Zaitsev V. F., *Handbook of Nonlinear Partial Differential Equations*, Chapman and Hall/CRC (2003)

Procesi C., *Lie Groups: An Approach through Invariants and Representations*, Springer (2009)

Rassias J. M., *Counter Examples in Differential Equations and Related Topics*, World Scientific, Singapore (1991)

Rogers C. and Shadwick W. F., *Bäcklund Transformations and Their Applications*, Academic Press, New York 1982

Sakurai J. J., *Modern Quantum Mechanics*, revised edition, Addison Wesley (1993)

Spiegel M. R., *Advanced Calculus*, Schaum's Outline Series, McGraw Hill, New York (1974)

Spiegel M. R., *Finite Differences and Difference Equations*, Schaum's Outline Series, McGraw Hill, New York (1971)

Spiegel M. R., *Complex Variables*, Schaum's Outline Series, McGraw Hill, New York (1971)

Spiegel, M. R., *Complex Variables*, Schaum's Outline Series, McGraw Hill, New York (1971)

Steeb W.-H., *Continuous Symmetries, Lie Algebras Differential Equations and Computer Algebra*, 2nd edition, World Scientific, Singapore (2007)

Steeb W.-H., *The Nonlinear Workbook*, 6th edition, World Scientific, Singapore (2018)

Steeb W.-H. and Hardy Y., *Matrix Calculus and Kronecker Product with Applications and C++ Programs*, second edition, World Scientific, Singapore (2011)

Steeb W.-H. and Hardy Y., *Problems and Solutions in Introductory and Advanced Matrix Calculus*, second edition, World Scientific, Singapore (2015)

Steeb W.-H. and Hardy Y., *Bose, Spin and Fermi Systems: Problems and Solutions*, World Scientific, Singapore (2015)

Steeb W.-H. and Hardy Y., *Problems and Solutions in Quantum Computing and Quantum Information*, World Scientific, fourth edition, Singapore (2018)

Steele J. M., *The Cauchy-Schwarz Master Class*, Cambridge University Press, Cambridge (2004)

Tai-Pei Cheng and Ling-Fong Li, *Gauge Theory of Elementary Particle Physics: Problems and Solutions*, Oxford University Press, USA (2000)

Thorpe, J. A., *Elementary Topics in Differential Geometry*, Springer, New York (1979)

Titchmarsh E. C., *The Theory of Functions*, Oxford University Press, Oxford (1939)

Tomescu I., *Problems in Combinatorics and Graph Theory*, Wiley, New York (1985)

Wells D. R., *Lagrange Dynamics*, Schaum's Outline Series, McGraw-Hill, New York (1967)

Zeidler E., *Quantum Field Theory II: Quantum Electrodynamics: A Bridge between Mathematicians and Physicists*, Springer, Heidelberg (2008)

Index